PERSPECTIVES ON ENERGY

PERSPECTIVES ON ENERGY

ISSUES, IDEAS, AND ENVIRONMENTAL DILEMMAS

THIRD EDITION

Edited by
LON C. RUEDISILI
UNIVERSITY OF TOLEDO

MORRIS W. FIREBAUGH
UNIVERSITY OF WISCONSIN–PARKSIDE

New York • Oxford • OXFORD UNIVERSITY PRESS • 1982

Copyright © 1975, 1978, 1982 by Oxford University Press, Inc.

Library of Congress Cataloging in Publication Data
Main entry under title:
Perspectives on energy.
   1. Power resources.   2. Power (Mechanics)   3. Energy policy.
I. Ruedisili, Lon C.   II. Firebaugh, Morris W.
TJ163.2.P47   1982   333.79   82-12558
ISBN 0-19-503289-6
ISBN 0-19-503038-9 (pbk.)

Printing (last digit): 9 8 7 6 5 4 3 2 1

Printed in the United States of America

To my wife, Susan, and to Stephen and Robert and future generations whose environment and welfare will depend upon the responsible energy decisions of our society   L.C.R.

To Joyce and our children, Steve and Susan, in the hope that wise energy management policies today will assure productive and rewarding lives tomorrow   M.W.F.

Foreword

The year 1973 represented a turning point in our lives: we never again shall be in a position to take our supply of energy for granted. The energy problem took a long time in coming—even though general recognition came suddenly—and it will take a long time in going. We face difficult choices in the years to come, choices that will cut to the fabric of our way of life.

To make these choices wisely, and take the actions that will be necessary to implement them, we need not only a sense of determination, but also a coherent and realistic energy policy. Prerequisite to this is a public understanding of our dilemma and conflicting choices. In this book of readings, Professors Ruedisili and Firebaugh have assembled a spectrum of authoritative information and opinion on our energy problem. The balance of articles is sufficiently broad to stimulate and educate the reader, college student, or other interested citizen on this vital issue.

The articles make it clear that our only hope is to work on both elements of the energy problem—supply and demand. We must increase the supply and decrease the demand or at least reduce the rate of increase of the demand. At the same time, because it is all part of the same problem, we must not compromise the efforts that are under way to improve both our natural and our manmade environments and to protect the public health and safety. Our conflicting choices include those between the public goals of abundant energy and environmental quality, and these conflicts will have to be reconciled as best they can with regard to the total social welfare.

The energy technologies that can be expected to play a major role in the American energy economy in the coming decades include the use of fossil fuels and their improvement and interconversion, nuclear fission, nuclear fusion, geothermal energy, and solar energy. These sources, and others covered in this book's articles, should be developed. It is not a question of whether geothermal energy can produce electricity, or whether solar energy is feasible, but rather how much can we expect from these sources, when and under what conditions can we expect their utilization, and, importantly, what will it cost us to ensure their development in an expenditious, timely, and safe manner.

We are in trouble because of inadequate national planning in the past, and this includes insufficient research on and development of new energy sources. The development of new technologies, where the systems are very complex, requires much basic research before they can be utilized and institutionalized. The cost of basic research is small compared with the value of the technologies it spawns, and we should not neglect the lessons of the past in this regard, even as we search for immediate solutions to our energy problems.

We are accustomed to the idea of unlimited growth from our classical theories of economics, in which supply always rises to meet demand. But unlike capital, which is a construct of man, the natural resources of our earth are finite. In fact, we are going to be faced next with a series of resource crises—metals crises (copper, aluminum, chromium, nickel, tin, manganese, and so on), a food crisis, a water crisis—if we do not plan better for the future than we have planned in the past for our future energy and material needs.

In trying to solve our energy problems, we must not neglect the questions of fragile values in our society. We need to understand more fully the role of our institutions in the resolution of specific value conflicts in order to understand how the patterns of these decisions, if there are patterns, themselves create new values and new institutions that perpetuate those values. Such questions as these should be addressed: In crises, do fragile values that are without institutional embodiment get lost, even though they may be deeply held by a substantial number of citizens? With what kind of analysis do we approach decisions involving competition and compromise between the satisfaction of present and future welfare? How is our concern for the future, itself a fragile value, concretely manifested?

The articles in this book should help in the understanding of and deliberations on all these issues.

University of California   Glenn T. Seaborg
Berkeley, California

Preface

It has been a decade since the initial "energy crisis" awakened the world to its dependence on this critical commodity. During this time, as literally hundreds of studies on all aspects of energy have appeared, our response to the problem of energy has evolved from initial panic and searches for scapegoats to a much clearer picture and deeper understanding of the role of energy in society. What emerges is the image of energy as the "life blood" of a society whose scientific and technical outlines are coming into focus but one whose policy goals remain blurred and contradictory as a result of the divergent social forces at work. However, energy as a social phenomenon reflects cultural, economic, and political forces determined by differing values and world perspectives. The actual pattern of energy production and consumption is the result of these social forces working within the boundary conditions set by technical and scientific limitations. In this book we attempt to provide both a generally accepted survey of the status of energy technology and a broad cross section of opinion and interpretation on important issues of energy policy.

We have attempted to present a balanced and representative analysis in areas of controversy, but we readily admit that this particular selection of articles reflects our own interpretation of the most significant energy resources and policy issues. We have attempted to present current analyses in an area of rapid change and new information in both scientific and policy areas. Thus, we have presented the material in the authors' words in order to preserve the tone and content of the more subjective interpretations. Several of the articles have been prepared specifically for this book in order to present a more complete picture of the range of interpretation on energy policy.

The objectives and structure of the third edition are the same as those of the first two. However, as serious students of energy are well aware, the energy situation is a very dynamic, ephemeral thing. Within a decade we have seen gasoline pump lines and oil gluts come and go. Energy supply and demand is a function of a very complex system of interacting social, economic, and political forces on a worldwide scale. We have therefore expanded the international emphasis in the third edition and extended the discussion of economics and resource availability. We have expanded the solar energy section to include passive solar design and the prospect for photovoltaic electrical production and have included, in the nuclear section, a detailed analysis of the Three Mile Island accident and its significance. Finally, since much of the debate on energy policy issues arises from philosophical differences summarized by the "hard" versus "soft" technology conflict, we examine this issue from several perspectives in this edition.

In the final analysis, public policy in the critical area of energy management must be determined by enlightened public opinion as it is worked out through the political process. We believe that the present generation of college students is capable of the critical "sifting and winnowing" required to reach intelligent and informed opinions on energy policies. We hope this book will assist them in the process.

At this point we want to acknowledge the fine cooperation of many people that made this book possible. First we wish to thank all the authors for permission to include their articles. We appreciate the efforts of Professor Glenn T. Seaborg of the University of California–Berkeley in preparing an appropriate and timely foreword. We thank Professor John S. Steinhart of the University of Wisconsin–Madison for his contributions to the text and for several valuable and critical discussions on the selection of articles. We appreciate the feedback of our students at the University of Wisconsin–Parkside and the University of Toledo in evaluating previous editions and for preliminary input on selected articles in this edition. In addition to the University of Wisconsin–Parkside and the University of Toledo, the Institute for Energy Analysis of Oak Ridge, Tennessee, provided valuable library support for which we are grateful. Finally, we appreciate the continued assistance and encouragement of Ms. Joyce Berry of Oxford University Press.

University of Toledo   L.C.R.
University of Wisconsin–Parkside   M.W.F.
August 1982

Contents

I

BACKGROUND
AND LIMITATIONS

Zepher I, a semi-submersible drilling rig, being towed into the North Sea to seek out sea-bottom oil. (Skyfotos and Texaco, Inc.)

This rotating-tower wind turbine was designed by Charles Schachle and erected by Bendix in San Gorgonio Pass near Palm Springs, California, for Southern California Edison. It is designed for the strong winds there and is nominally rated at 3 MW. Its blades sweep an area 174 feet in diameter.

Human muscles still provide much indispensable power throughout the developing world. Here peasants from Hebei Province in the North China Plain are building a canal to control the flood-prone Hai River; each winter about 100 million Chinese are engaged in similar mass construction activities. (China Pictorial)

INTRODUCTION

A sound understanding of the boundary conditions applicable to energy issues is essential to the study of energy and its role in human affairs. Such limitations include the obvious physical constraints of thermodynamic laws, geological resource data, and energy flow patterns with their associated physical facilities. Less obvious, but equally important, constraints involve environmental costs, health effects, and economic factors that are intrinsic to various energy alternatives. Finally, we must be aware of social values and their expression in the political and legal systems. These values in combination with the inertia of existing corporate and institutional infrastructures are the most important determinants of our energy system. Any discussion of current energy issues must be informed of both the physical and institutional constraints on energy patterns in society.

As a background for the interpretation of specific energy issues, we must understand the effects of these boundary conditions not only as independent constraints but also as interacting forces. For example, a change in environmental law affecting source A may cause a reduced thermodynamic efficiency, leading to a reduced economic advantage and resulting in a switch to source B, causing increased dependence on foreign suppliers, which may increase international tensions in times of crisis, and so on. The networks of these cause-effect relationships are very complex, and only recently have computer programs been developed to study the dynamics of such systems. As we consider the environmental, health, economic, political, and social aspects of energy, it is important to remain aware of these relationships.

The contributions of energy in heating, lighting, and motive power for industrial society are obvious. There is an even more fundamental role for energy which stems from basic physical law. This law states that unless an outside agent intervenes, a given system tends to increase its entropy (e.g., disorder or randomness). For example, water in the mountains runs downhill; high- and low-temperature objects in contact tend to equalize their temperatures; and high-quality energy, such as electricity, always ends up as low-quality energy, heat. This tendency, interpreted as eventually resulting in the "heat death of the universe" on a large scale, is apparent on a smaller scale as thermal pollution of streams and air, exhaustion of high-grade (organized) ores, and consumption of fossil fuels.

In this thermodynamic analogy, many of our environmental problems stem from the increase in disorder and randomness. The outside agent that can restore order on a local basis is energy. Solar energy can lift water back to the mountains. Energy enables us to reorganize such disordered collections as solid wastes into burnable fuels and primary metals. With appropriate expenditures of energy, automobile exhausts and sulfur dioxide from coal-burning power plants may be "organized" out of the environment. So although energy production and consumption are the cause of many environmental problems, energy is also an important element of the solution.

To begin our examination of the boundary conditions on energy set by ecological systems, C. S. Holling discusses, in *Myths of Ecology and Energy,* the myths of "Nature Benign" and "Nature Perverse," which are interpreted as logical human attempts to cope with the uncertainties inherent in complex systems. He presents interesting analogies between such myths and the stability of geometric systems and stresses the value of developing a resilient global energy system.

In *Energy and the Environment,* John M. Fowler presents an excellent historical summary of the development of various energy sources and the flow of energy through the social system. He discusses basic thermodynamic relations and outlines the environmental costs of various forms of energy. Several future energy options are introduced, and these are examined in greater detail later in this volume.

Albert A. Bartlett emphasizes the absolute constraints posed by the finite nature of traditional energy resources in *Forgotten Fundamentals of the Energy Crisis.* The undisputed arithmetic of a steady growth rate (i.e., exponential growth) is graphically illustrated and applied to fossil-fuel consumption. The interaction between the price mechanism and the estimate of these resources is considered in detail in later chapters. Professor Bartlett concludes this chapter with an intriguing suggestion on how we can make such finite resources last forever.

In *Economic Analysis of Energy Policy Issues,* Henry Steele analyzes energy policy issues from a classical free market economist's perspective. The concept of maximizing the overall economic welfare by the free market mechanism of balancing the supply and demand functions is described and applied to policy issues including conservation, environmental protection, national security, price controls, and regulation. The article begins and concludes with a provocative and highly critical analysis of OPEC pricing policy as the major unresolved energy policy issue. This critical issue is reexamined from different perspectives in the next section, on fossil fuels.

John and Carol Steinhart review the enormously complex food system and its energy implications in *Energy Use in the U.S. Food System*. The possible failure of the energy-dependent food chain is one of the gravest concerns of many who have studied recent climatic changes. Although high-energy agriculture may be necessary to carry out North America's "bread basket" mission, the Steinharts raise serious questions about the wisdom of exporting this food system to energy-poor nations.

Specific environmental stresses which expanded national energy production will place on the western states are analyzed by M. D. Devine, S. C. Ballard, and I. L. White in *Conflicts and Constraints: Energy from Western United States*. The key role of water in limiting new energy resource development is highlighted. In addition to technical improvements in resource extraction, the authors call for flexible and equitable institutional responses to anticipated problems resulting from rapid expansion of energy production facilities. Examples are offered of successful experiences with a task force approach to resolve siting issues.

In *Energy Flows in the Developing World*, V. Smil forcefully reminds us that our perspective on energy issues depends strongly on our social environment. Fully 75 per cent of the world's population lives in developing nations in peasant societies based primarily on solar energy economies. For these societies, the energy crisis is more a shortage of wood, straw, and charcoal than of gasoline or electricity. This analysis establishes the extent to which human, animal, and phytomass (plant matter) energy drives the economy of developing nations such as the People's Republic of China. The author suggests that incremental improvements of these "soft technologies" may be the preferable path for developing nations, rather than development of the excessive dependence on fossil fuels demonstrated by the highly industrialized societies.

Myths of Ecology and Energy   1

C. S. HOLLING

The key issue for the design of an energy policy is how to cope with the uncertain, the unexpected, the unknown. The price and availability of imported oil are uncertain in precisely the terms that international relations are. The social and environmental consequences of emphasizing nuclear power, coal, solar or geothermal power, or a mix are uncertain. Even the ultimate objective is uncertain. Community wisdom has held that an increasing of the energy consumption per capita will increase the standard of living. And yet at least one analysis suggests[1] the reverse has been true over the recent past. And in his espousal of small and diverse solutions, Schumacher[2] has raised the heresy again, suggesting that quality of living is perhaps not improved by unending growth in gross national product.

Originally, I intended to explore specific environmental constraints of alternative energy proposals in this article. But the central constraint is their very uncertainty. That, therefore, will be my theme—how can we deal with the unknown and uncertain?

After exploring aspects of uncertainty, I will treat the subject under three topics: the response of systems to disturbance, the role of hierarchies of function, and the consequence of aggregations of scale. Throughout I will draw examples largely from ecological systems—in part to reap any benefits from analogy and in part to help define possible environmental consequences of energy developments. From that, I point a direction that suggests neither that small is beautiful nor that big is necessary, but something of both.

THE PROBLEM OF UNCERTAINTY

Man has always lived in a sea of the unknown and yet has prospered. His customary method of dealing with the unknown has been

Dr. C. S. Holling is a professor at the Institute of Animal Resource Ecology at the University of British Columbia, Vancouver, British Columbia, Canada V6T 1W5. He has taught at the University of California—Berkeley, the University of Washington, and the University of Idaho and headed the systems ecology group at the International Institute for Applied Systems Analysis, Laxenburg, Austria. He is editor of International Journal of Environmental Sciences and Journal of Theoretical Population Biology.

From Future Strategies for Energy Development—A Question of Scale, ORAU-130, pp. 36–49. Reprinted by permission of the author and Oak Ridge Associated Universities. Copyright © 1977 by Oak Ridge Associated Universities.

trial and error. Existing information is used to set up a trial. Any errors provide additional information to modify subsequent efforts. Such "failures" create the experience and information upon which new knowledge feeds. Both prehistoric man's exploration of fire and the scientist's development of hypotheses and experiments are in this tradition. The success of this time-honored method, however, depends on some minimum conditions. The experiment should not, ideally, destroy the experimenter—or at least someone must be left to learn from it. Nor should the experiment cause irreversible changes in its environment. The experimenter should be able to start again, having been humbled and enlightened by a "failure."

We are now ignoring those minimum conditions. Our trials are capable of producing errors larger and more costly than society can afford. This leads to the dilemma of "hypotheticality" posed by Haefele,[3] who argues that the design of energy policies is locked in a world of hypothesis because we dare not conduct the trials necessary to test and refine our understanding.

One of the reactions to this dilemma is, paradoxically, the generation of a plethora of alternative hypotheses that are, in principle, untestable. They appear in a variety of guises—as the hidden or explicit assumptions behind scenarios, alternative forecasts, exploratory calculations, or "world" models. Such exercises can be salutary by forcing questions that otherwise would not be raised, but they can be just as pernicious by tempting us to use the exercise to make decisions.

Consider, for example, even such a noble task as the prediction of environmental constraints. Haefele and Sassin[4] have made a preliminary and yet effective projection of the constraints to an energy policy that emphasizes a shift from oil to coal. Accumulation of carbon dioxide in the atmosphere suggests that global climate would be affected so much that new technologies would be essential within 60 years. New technologies, however, are not created overnight. Other analyses suggest a lead time of the same duration. Hence the pressure to design and implement a new technology now. But once such a decision was made, the magnitude of the effort would foreclose the options of withdrawing from it if other, unexpected problems arose. And they would arise.

The problem, therefore, is not only one of inadequate knowledge of environmental constraints or of social consequences. There is a more fundamental issue, suggested by the theme set by Schumacher[2]—small is beautiful versus big is necessary. Both have the flavor of myths. Myths are a way in which mankind captures some essence of experience or wisdom in a simple and elegant form. Myths are necessary to guide us in our actions, and they protect us from that larger reality of the frightening unknown. Myths may be as all-embracing as "there is a God" or as narrow as "the free market model will correct all inequities."

But myths are only a partial representation of reality. Through their acceptance, man becomes bold enough to accumulate new knowledge that eventually exposes their incomplete nature. That introduces confusion; old myths are dusted off, and new ones developed. In many ways, the present emergence of a multitude of hypotheses is a process of evolving new myths. But let us call them myths and not pretend they are recipes for action.

Moreover, it might be useful to structure consciously alternative myths and trace their logical consequences in relation to dealing with the unexpected. The way systems respond to unexpected events is determined by their stability properties. Hence the myths might usefully be myths of stability.

STABILITY OF SYSTEMS

Initially, it is less important how variables respond to disturbances than whether a system is stable at all and under what conditions. These are questions of structural stability, and there is little interest in kinds of equilibria (fixed points, limit cycles, nodes, etc.) and even less in the details of a trajectory as it evolves after a disturbance. If we don't strain the analogy too far, these structural properties of stability can be imagined as forming a topographic surface representing some aspect of the time-independent forces of a system.

One of the simplest and most common images would be a surface with a valley shaped like a bowl, within which a ball moved, partially as a consequence of its own acceleration and direction and partially as a consequence of the forces exerted by the bowl and by gravity. If the bowl were infinitely large, or events beyond its rim meaningless, then this would be an example of global stability. That is, no matter how far from the bottom the ball was displaced, eventually it would return and come to rest there again.

Now this myth of global stability has been pervasive. It underlies much of the presumptions of economics, with its focus on equilibrium and near-equilibrium conditions. After all, the dominant property seems to be the final condition of rest or equilibrium, and any other behavior is transient and seems trivial. Moreover, equilibrium mathematics is easy!

It is also a comforting myth, for it represents a benign and infinitely forgiving Nature. Trials and mistakes of any scale can be made in this world, and the system will recover once the disturbance is removed. Since there are no penalties to size, only benefits of scale, this myth of Nature Benign comfortably and logically leads to big is necessary.

An opposing myth is that of instability, where the imagined surface is now dominated by a smoothly convex hill rather than a bowl. The top of the hill represents an unstable equilibrium, for if the ball is only slightly displaced from this point, it will roll away. Ultimately, complete instability of this kind leads to all variables becoming zero—to extinction. Since we know that systems persist, this myth of instability would seem to be impossible and uninteresting. Yet it has a rich tradition in philosophy and science.

Persistence is conceivable if we introduce

true spatial dimensions. Imagine a mosaic of spatial elements, or sites, each occupied by such an unstable system, each out of phase with the other and contributing excess individuals by movement to neighboring sites. The result could be as much as was actually observed in a classic experiment by Huffaker et al.[5] They examined the interaction between populations of a plant-eating mite and a predatory mite. In the relatively small enclosures used, when there was unimpeded movement throughout the experimental universe (a homogeneous world, therefore), the system was unstable and the populations became extinct. When barriers were introduced to impede dispersal between parts of the universe, small-scale heterogeneity was introduced and the interaction persisted. Thus populations in one small area that suffered extinction were reestablished by invasion from other populations that happened to be at the peak of their numbers.

This myth of ephemeral Nature leads to a concentration on issues of spatial heterogeneity, diversity over space, and fine-scaled, local autonomy. Nature Ephemeral is in harmony with the persistent traditions of anarchism and of small is beautiful. It is in complete opposition to big is necessary, since any effort to increase size, to homogenize, will lead to collapse as the underlying instability of Nature becomes obvious.

Both these myths of stability and instability used to be espoused by two separate schools within ecology. One school showed that many population processes, such as predation, parasitism, and competition, have important regulatory properties. Such processes have a density-dependent, or negative feedback, influence such that if numbers become excessively high, there will be a proportionate increase in mortality to damp the upswing. The reverse will occur if numbers become too low. That, essentially, is the myth of the single equilibrium and global stability. The opponents of this school saw evidence in Nature of what Huffaker demonstrated in the laboratory. That is, populations are ephemeral, chance events profoundly influence numbers, and invasion–reinvasion is the rule.

The familiar Lotka–Volterra equations demonstrate a stability condition precisely midway between these two myths. These equations are coupled differential equations representing the interactions between two populations—a predator and prey, for example, or two competitors—in a highly simplified form. They result in neutral stability as if the topography was represented as a flat plane. It is the stability of a frictionless pendulum. But the relationships represented in the Lotka–Volterra equations are not only simple, they are simplistic. Adding any kind of realistic negative feedback (or density dependence) results in the myth of Nature Benign and global stability. Adding any kind of realistic positive feedback, or explicit lags, results in the myth of Nature Ephemeral and instability.

Yet both kinds of addition have been demonstrated to exist.[6] The component structure of key population processes leads to classes of behavior that can be formally defined[7] and that have mixed stabilizing and destabilizing properties. When all of these are included, a different myth emerges.

The dominant feature of this myth is the appearance of more than one equilibrium state.[6] Each equilibrium state, or attractor, is separated from its neighbors so that two or more basins of attraction, or domains of stability, are formed. As long as variables remain within one basin of attraction, they will tend to the same attractor. If, however, variables happen to be close to the boundaries of these basins, then an incremental disturbance could shift the variables into another basin, thereby causing radically altered behavior. Or, returning to the theme of trial-and-error learning, incremental trials could seem not to be causing significant error and yet could accumulate until one more trial "flips" the system into a different stability region. Consider eutrophication, where progressive phosphate loading of aquatic systems can be accommodated only up to a point. Then algal blooms are triggered, and subsequent rotting of the plant material can lead to anaerobic conditions.

Rather than the image of hill or bowl, a better analogy for this myth would be a mesa with a depression at its top. As long as the ball is in the depression, the system appears qualitatively stable. If the ball is tipped over the edge of the mesa, it will move to a different position, one that could well represent extinction. So, if we perceived reality as being the myth of Nature Benign, this would seem to represent a rather perverse Nature. On the other hand, if we perceived reality as being the myth of Nature Ephemeral, it would seem to represent a quite tolerant Nature—more stable than we thought.

The burden of evidence suggests that the multiequilibria world of Nature Perverse/Tolerant is common—and not only for ecological systems. Like all good myths, it can be usefully captured by some simple differential equations, as long as some effort is made to represent relationships simply and not simplistically. Such models have been proposed for ecological systems,[8] institutional systems,[9] and societal systems.[10] Even these simple structures exhibit rich topologies with the dominant feature of multiple stability regions.

Moreover, many empirical studies of specific systems suggest a multiequilibria structure and sharp behavioral shifts between equilibria. A preliminary survey[6,11] shows numerous examples in the ecological, water resource, engineering, and anthropological literature. We know enough about the functional form of ecological processes at least to know that their qualitative properties can produce multiple equilibria. This makes topological approaches in ecology so important,[12] since they emphasize equilibria manifolds and conditions that trigger sharp shifts of behavior. Simple examples are the catastrophe manifolds, familiar now even to readers of *Newsweek*.

The myth of Nature Perverse/Tolerant can

accommodate either small is beautiful or big is necessary. The small-is-beautiful theme operates much as before, with perhaps a more formal sense of optimal scale. It is still contrary to big is necessary. In a spatial world with multiequilibria systems, a distinctive spatial scale can be identified. Steele,[13] for example, in a pioneering analysis of spatial attributes of plankton in the ocean, shows that there is a distinctive size to a patch of plankton that results from a balance between the kind of stability described by this third myth and diffusion forces that tend to dissipate the patch. This primary scale is measured in kilometers, and he points out the essential constraints this places on the magnitude of intrusion of pollutants or of thermal discharge from power plants. Terrestrial systems show the same scale definitions—insect defoliators, grazing ungulates, plant colonizers. Again, the minimum viable "patch" size, or scale, is set by the balance between magnitude and extent of dispersal and the endogenous stabilizing and destabilizing forces.

But note that ecological and environmental scales can be very small (the soil bacteria's world) or very large (global climatic systems). Are the big ones ugly and the small ones beautiful? There is an inconsistency, and we will turn to that later when we consider the role of hierarchies in buffering the unknown.

For now, the principal point is that this multiple equilibria myth can also be accommodated by big is necessary; one only has to be a little more cautious. Thus, if boundaries exist separating "desirable" from "undesirable" stability regions (that is, if separatrices exist), then the task is to carefully control the variables to keep them well away from the dangerous separatrix. To do this effectively, big might well be necessary as the only way to gain sufficient knowledge of the separatrix, to monitor the distance to it, and to institute control procedures to maximize that distance. Haefele and Buerk[10] suggested this in their search for a resilient energy policy that can absorb unexpected events. They start with a brilliant articulation of a simple socie-

tal model that clearly has two equilibria— one representing high population and low energy per capita and one the reverse. That representation has great value as a myth—as one partial representation of one view of reality that serves as a paradigm but not as a blueprint for action. But they go on to use it as such by treating this powerful myth as a model of reality. To use it in this way (even with the careful and responsible qualifications they emphasize) is to destroy its value as myth, as hypotheses with untestable consequences.

The goal of maximizing the distance from an undesirable stability region is exactly in the highly responsible tradition of engineering for safety, for example, of nuclear safeguards, of environmental and health standards. It demands and presumes knowledge. It works beautifully if the system is simple and known—say, the design of a bolt for an aircraft. Then the stress limits can be clearly defined, these limits can be treated as if they are static, and the bolt can be crafted so that normal or even abnormal stresses can be absorbed. The goal is to minimize the probability of failure. And in that, the approach has succeeded. The probability of failure of nuclear plants is extremely small. But in parallel with that achievement is a high cost of failure—the very issue that now makes trial-and-error methods of dealing with the unknown so dangerous. Far from being resilient solutions, they seem to be the opposite, when applied to large systems that are only partially known.

For this reason, I did not address myself to the issue of environmental constraints to energy developments. To be able to identify environmental constraints presumes sufficient knowledge, presumes the possibility of testing by hypotheses, presumes a myth of Nature Tolerant and Static. It reinforces the trap of hypotheticality. It emphasizes a fail-safe design at the price of a safe-fail one.

One final myth is therefore needed to make a Compleat Mythology. The three myths discussed so far have been described in three steps of increasing reality and comprehensiveness and implicitly assumed constancy and determinacy. That is, the topography of hills, valleys, and mesas was fixed and immutable, and the ball followed a predetermined path to some constant condition. But ecological—and, for that matter, economic, institutional, and social—systems are neither static nor completely determined. Variability and change are the rule and provide the next step toward reality.

Stochastic events dominate some ecosystems. Fire, rather than being a disaster, is the source of maintenance of some grassland ecosystems. Shifting patterns of drought determine the structure of some savannah systems in Africa. The internal variables themselves can move, through endogenous mechanisms, from one equilibrium to another. Such seems to be the lesson from our recent studies of forest insect pests. There, periodic outbreaks can be triggered by stochastic events, by spatial events, or by the growth of the forest itself. Populations then increase explosively from a lower to a higher equilibrium. While the higher one is stable for the insects, it is unstable for the forest. Consequently, large swings and movements between stability regions contribute to forest renewal and the maintenance of diversity.

Hence for natural ecosystems, residence in one stability region, far from separatrices, is not necessarily the rule. Locally, a system may even move to an extinction region, and stochastic or deterministic events will reinstate it. Hence the ball on our surface is moving continuously, and the stability boundaries are continually being tested.

But the topography itself is not static. In the mathematical representations, changing the values of parameters can change the stability landscape. Hence some regions in parameter space can be identified that indeed represent global stability, others that represent instability, and still others that show various multiequilibria configurations. Now, parameters in equations are presumed to be constant only as a convenience and first approximation. They can better be viewed as variables that normally change so slowly that,

within limits, they can be presumed to be constant or, at most, stochastic variables.

The values of these parameters at any moment define the form of the topographic landscape. They assume these values in nature as a consequence of long- and short-term adaptation to variation—both genotypic and phenotypic. In essence, natural selection produces a balanced set of parameters whose value is, in part, the consequence of the historic variation of the system. Changing the pattern of variability can thus change this balance. Moreover, our mathematical models show that if key parameters are arbitrarily moved, often there is little change in the topography until a certain point, when the topography suddenly shifts: stability regions implode, new regions of stability and instability appear. There are, in short, separatrices in parameter space as well as in state-variable space.

This dynamic pattern of the variables of the system and of its basic stability structure lies at the heart of coping with the unknown. However much we may be sure of the stability landscape of a physical system, rarely will we know the societal or ecological stability landscape in great detail. Policies often attempt to reduce variability within these partially understood systems, either as a goal in itself or as an effort to meet standards of safety, health, or environment. That changed variability in turn may itself shift the balance of natural, cultural, or psychological selection so that stability regions will contract. Paradoxically, success in maximizing the distance from a dangerous stability boundary may cause collapse because the boundary may evolve to meet the variables. That is, if surprise, change, and the unexpected are reduced, systems of organisms, or people, and of institutions can "forget" the existence of limits until it is too late.

This final myth is the myth of Nature Resilient. Originally, resilience was defined[6] as the property that allowed a system to absorb change and still persist. But if that is followed with the myth of Nature Tolerant and Static in mind, resilience becomes, as Haefele has translated it, *Schlagabsorptionsfae-higkeit,* or "strike absorption capability." Using that as a goal leads to avoidance of separatrices, followed, in partially known systems, by their possible collapse. A better definition, therefore, is that resilience is a property that allows a system to absorb *and utilize* (or even benefit from) change. Thus, it can lead to radically different kinds of approaches—for example, the dynamic management of environmental standards [14] or of environmental impact assessment. [15]

Such a myth explicitly recognizes the unknown and the ability to survive and benefit from "failures." It certainly would allow for a world that can accommodate trial-and-error learning. Since it is a myth of a natural, evolved system, we might profit from looking deeper into the organization of such systems, as Simon [16] has done in an important essay on hierarchical organization.

HIERARCHIES OF FUNCTION VERSUS AGGREGATIONS OF SCALE

Hierarchy, as defined in theories of complex systems, is not the linear rank structure of some human organizations. Rather, it can be visualized as the roots of a tree that branch into distinct levels of function. Each level consists of a number of components with different functions whose integration produces the higher level. Hierarchies seem to be universal within a large class of evolved systems—physical, chemical, biological, or social—because they have properties that facilitate the evolution and survival of complexity. If that is true, then it touches precisely the issue of energy, society, and environment.

The familiar biological hierarchy runs from organic compounds to macromolecules to cells to tissues and organs to whole organisms. This hierarchy can be extended to include at one end a microhierarchy of molecules, atoms, and so on, and at the other an ecological hierarchy. The individual organisms form populations. These populations are organized into distinct functions—primary energy acquisition (the plants, or primary producers), herbivory, predation, and decomposition. These populations are connected by energy transfer, mass transfer, and competition; together they form ecosystems. We could, as some have done, continue the hierarchy to communities, biogeocoenoses, and biomes. But beyond the ecosystem, the concept becomes fuzzy and is, at the moment, more an exercise in classification than understanding.

Simon [16] shows that certain attributes are found in all hierarchies. Each level has a distinctive time scale and, we might add, spatial scale. The higher the level in the hierarchy, the slower the time scale and the larger the spatial scale. Hence, to return to an earlier point, the parameters of one level are, in reality, slow variables of the next level.

Each level exists as the consequence of interactions between components of the lower level. That interaction results in a buffering of the extremes and an easing of the demand for adaptation at the lower level. Thus components need not be infinitely adaptable.

Communication between components and levels is simple compared with the interactions within components. As long as that simple communication is maintained, the details of operations within the components can shift, change, and adapt without threatening the whole. In ecological systems, for example, there is a great diversity of populations performing any one function, such as primary production. Part of that diversity is due to heterogeneity—solar energy, while diffuse, is not homogeneous. Hence "sun-loving" trees and shade-tolerant plants live together. Part of that diversity is due to uncertainty. Any particular function represents a role that at different times can be performed by different actors (species) that happen to be those available and best suited for the moment.

All these attributes make for flexibility and resilience. Additionally, the simple mathematics of hierarchical structures allow rapid evolution and the absorption and utilization of unexpected events. Simon's example of the two Swiss watchmakers demonstrates this.

One watchmaker assembles his watch as a sequence of subassemblies—a hierarchical approach. The other has no such approach, but builds from the basic elements. Each watchmaker is frequently interrupted by phone calls, and each interruption causes an assembly to fall apart if it has not yet reached a stable configuration. If the interruptions are frequent enough, the second watchmaker, having always to start from scratch, might never succeed in making a watch. The first watchmaker, however, having a number of organized and stable levels of assembly, is less sensitive to interruption. The probability of surprise (of failure) is the same for each. The cost of surprise (of failure) is very different.

Let me now draw some temporary conclusions from this treatment of stability and hierarchies. First, in terms of function, the value attached to big versus small is meaningless. Higher levels in a hierarchy are slower and larger in scale than are lower levels. We could even say that embedding in a larger system buffers the operation of smaller components. Thus, the reason why there is a flea on the frog on the bump on the log in the hole at the bottom of the sea is because there is a sea, a set of holes, a number of logs, and other objects, etc., etc. Equally, we could reverse the argument. The higher, larger level exists as a consequence of the interactions between the smaller components. In short, small is necessary for big to be.

Second, missing levels in a hierarchy will force bigger steps. And bigger steps will take a longer time. An example might be the one mentioned in my introduction—that is, the 60-year lead time to develop technologies that avoid excessive accumulation of carbon dioxide in the atmosphere. Bigger steps also presume the greatest knowledge and require the greatest investment. Hence, once initiated, they are more likely to persist even in the face of obvious inadequacy. Finally, bigger steps will produce a larger cost if failure does occur. To avoid that, the logical effort will be to minimize the probability of the unexpected, of surprise, or of failure.

For example, another solution to the watchmaker's problem of interruption might be for a number of watchmakers to join together, pool their resources, occupy a large building, and hire a secretary to handle the phone calls. This would control the disturbance and interruptions, and both watch-building strategies would succeed. Without the interruptions, there is not that much to gain by maintaining very many steps in a hierarchy of subassemblies. The hierarchically organized watchmaker could well begin a process of aggregation of scale, reducing the number of steps and having larger subassemblies at each step. That might increase efficiency in use of time and produce economies of scale, but it is totally dependent on complete and invariant control of disturbance. If the secretary was sick for one day, production would halt.

This is the same conclusion I reached in the discussion of the stability myths, and the circle of argument is now closed. The engineering-for-safety, fail-safe approach, while appropriate for totally known systems, is not so for uncertain, heterogeneous, and partially known ones. Whither nuclear safeguards, then, and the proliferation of nuclear power plants? The best safeguard system conceived by the mind of man can be deceived by another mind. That line of argument leads to the proposals for nuclear parks, floating energy islands that serve whole hemispheres of the globe via pipelines carrying hydrogen. It is not likely that we need treat these proposals as anything more than bad science fiction. But they do suggest an alternative that might even be of the same scale. Is it not possible, by formally restructuring a hierarchical organization to energy production, supply, and use, to have a global energy system? Then trial-and-error learning would become possible again, and the fears of the unknown could, temporarily, hide behind the myth of Nature Resilent.

REFERENCES

1. K. E. F. Watt, Y. L. Hunter, J. E. Flory, P. J. Hunter, and N. J. Mosman in Part I, *Middle and Long-Term Energy Policies and Alternatives*. pp.

303–362. Hearings before the Subcommittee on Energy and Power of the Committee on Interstate and Foreign Commerce, 94th Congress. March 25–26, 1976. Serial No. 94063. Washington, D.C.: U.S. Government Printing Office, 1976.

2. E. F. Schumacher, *Small Is Beautiful: A Study of Economics as if People Mattered*. New York: Harper & Row, 1973.

3. W. Haefele, "Hypotheticality and the New Challenges: The Pathfinder Role of Nuclear Energy." *Minerva* **10**, 303 (1974).

4. W. Haefele, and W. Sassin, *Energy Strategies*. IIASA Research Report RR-76-8. Laxenburg, Austria: International Institute for Applied Systems Analysis, 1976.

5. C. D. Huffaker, K. P. Shea, and S. S. Herman. "Experimental Studies in Predation. Complex Dispersion and Levels of Food in an Acarine Predator-Prey Interaction." *Hilgardia* **34**, 305 (1963).

6. C. S. Holling, "Resilience and Stability of Ecological Systems." *Annual Review of Ecology and Systematics* **4**, 1 (1973).

7. C. S. Holling, and S. Buckingham, "A Behavioral Model of Predator-Prey Functional Responses." *Behavioral Science* **3**, 183 (1976).

8. A. D. Bazykin in V. A. Ratner, ed., *Problems in Mathematical Genetics*. Novosibirsk: U.S.S.R. Academy of Sciences, 1974. (Available in English as *Structure and Dynamic Stability of Model Predator-Prey Systems*. Vancouver: Institute of

Resource Ecology Research Report R-3-R, University of British Columbia, 1975.)

9. C. S. Holling, C. C. Huang, and I. Vertinsky in R. Trappl, ed., *Progress in Cybernetics and Systems Research* (Austrian Institute for Cybernetics, Vienna). Washington, D.C.: Hemisphere, 1975.

10. W. Haefele, and R. Buerk, "An Attempt of Long-Range Macroeconomic Modeling in View of Structural and Technological Change." IIASA Research Memorandum RM-76-32. Laxenburg, Austria: International Institute for Applied Systems Analysis, 1976.

11. C. S. Holling in E. Jantsch and C. H. Waddington, ed., *Evolution and Consciousness: Human Systems in Transition,* pp. 73–92. Reading, Mass.: Addison-Wesley, 1976.

12. D. D. Jones in G. S. Innis, ed., *Simulation in Systems Ecology*. Simulation Council, Logan, Utah.

13. J. H. Steele, "Spatial Heterogeneity and Population Stability." *Nature* **248,** 83 (1974).

14. M. B. Fiering and C. S. Holling, "Management and Standards for Perturbed Ecosystems." *Agro-Ecosystems* **1,** 301 (1974).

15. R. Hilborn, C. S. Holling, and C. J. Walters in J. J. Reisa, ed., *Biological Analysis of Environmental Impacts*. Council on Environmental Quality, Washington, D.C.

16. H. A. Simon in Howard H. Pattee, ed., *Hierarchy Theory,* pp. 1–28. New York: George Braziller, 1973.

Energy and the Environment 2

JOHN M. FOWLER

ENERGY: WHERE IT COMES FROM AND WHERE IT GOES

The world runs on energy, both literally and figuratively. It spins on its axis and travels in its orbit about the sun; winds blow, waves crash on the beaches, volcanoes and earthquakes rock its surface. Without energy it would be a dead world. Energy was needed to catalyze the beginning of life; energy is needed to sustain it.

For most of life, animal and plant, energy means food, and most of life turns to the sun as ultimate source. The linked-life patterns—the ecosystems—that have been established between plants and animals are very complex; the paths of energy wind and twist and double back, but ultimately they all begin at that star that holds us in our endless circle.

When man crossed the threshold of consciousness that separates him from animals, his uses of energy began to diversify. He, too, needed food, but poorly furred as he was, he also needed warmth. With the discovery of fire he was able to warm himself. He also found that fire could make his food digestible and thus increase its efficiency as an energy source. After a while he began to use fire to make the implements through which he slowly started to dominate nature.

Man's use of energy grew very slowly. In the beginning he required only 2000 Calories or so[1] per day for food; the convenience of warmth added a few thousand more Calories from easily obtainable wood. The first big

Dr. John M. Fowler is Director of Special Projects and Manager of Publications of the National Science Teachers Association, Washington, D.C. 20009. He directs the NSTA's Project for an Energy-Enriched Curriculum, which is providing energy-education materials for teachers and students throughout the country. Earlier, he was the Director of the Commission on College Physics (1965–1972) and in 1969 was awarded the Millikan Medal by the American Association of Physics for contributions to the teaching of physics. During a tenure at Washington University, St. Louis, he helped start the Committee on Environmental Information and the magazine *Environment*. He is currently on the Advisory Board of *Environment* and is a director of the Scientist's Institute for Public Information. His publications include *Fallout: A Study of Superbombs, Strontium 90, Survival*, Basic Books (1960), *Energy and the Environment*, McGraw-Hill (1975), and the NSTA *Energy-Environment Source Book* (1980). He has lectured and written extensively on energy topics.

This article was prepared by the author in April 1981. An earlier version was published in *The Science Teacher* **39** (9), 10 (1972).

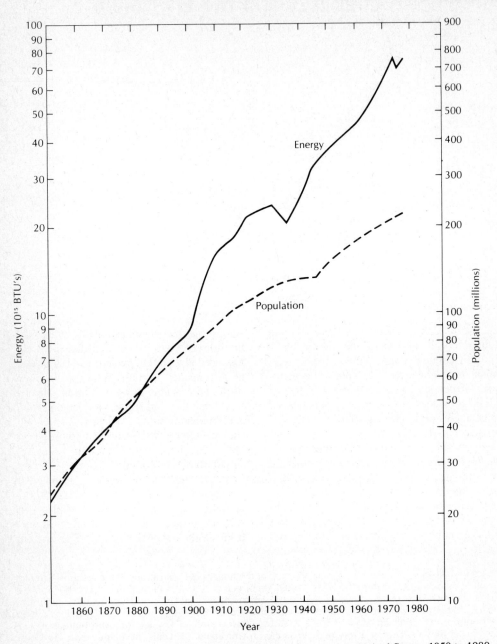

Figure 2-1. Comparison of energy consumption and population growth, United States, 1850 to 1980.

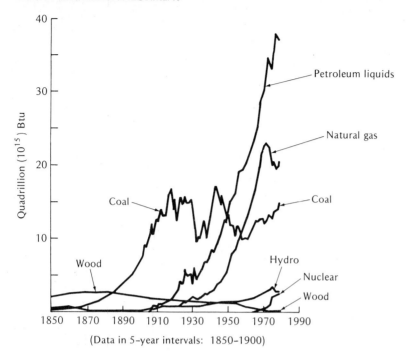

Figure 2-2. Contribution of various sources to U.S. primary energy consumption, 1850–1979. (data in 5-year intervals: 1850–1900) (From Henry R. Linden and Joseph D. Parent, *Perspectives on U.S. and World Energy Problems*. Chicago: Gas Research Institute, April 1980)

jump in energy use came about 4000 or 5000 years B.C., when man domesticated several animals and—at the cost of a little food, much of which the animal gathered—was able to triple the amount of energy at his service.

The waterwheel was introduced in the first century B.C. and again multiplied the amount of energy available to man. Its introduction was perhaps even more significant because it was an inanimate source of energy. For a long time the waterwheel was the most important source of energy for nascent industry. It was not until the twelfth century A.D., more or less, that the analogy between flowing air and flowing water led to the use of windmills.

The beginning of the modern era of industry coincided with the development of the steam engine. Since that time the world's use of energy, which until then had been very nearly proportional to the number of people,

has grown in the industrial countries more rapidly than the population has increased. This historical growth pattern for the U.S. is shown in Figure 2-1.

The second historical trend, which, together with the increasing per capita use, has brought us to our present state, is the constant change in the energy mix. Wood was the dominant fuel in the 1850s but had lost all but 20 per cent of the market to coal by 1900. Coal in turn lost out after 75 years to oil and natural gas which now account for 75 per cent of the energy used, but they in turn will be (and *must* be, as we shall see) replaced by other sources (Figure 2-2).

Energy as a Commodity

In the early stages of man's history, energy was food, something to be found and con-

Table 2–1.   Energy Costs for Some Foods and Fuels (Retail, Washington, D.C., April 1977).

	Price per Unit (dollars)	Energy (Cal/kg)	Cost per Million Calories (dollars)
Bread	0.69/lb	2660	571.00
Butter	2.19/lb	7950	606.20
Sugar	0.48/lb	4100	258.50
Sirloin steak	2.59/lb	1840	3097.80
Scotch	15.00/half gal	2580	3840.00
Coal	140.00/ton	7040	21.30
Fuel oil	1.32/gal	10,660	43.10
Natural gas	3.89/MBtu	12,450	16.30
Gasoline	1.42/gal	12,280	44.70
Electricity	0.05/kWh	—	58.10

sumed. But as life became more complex, and early barter systems were followed by a money-based economy, energy had to be bought. At first it was purchased indirectly as food or fuel. With the introduction of electricity, energy could be run directly into the house or the factory.

Energy is a commodity; it can be measured, bought, and sold. But its price depends on the form in which it is purchased— as food or fuel or electricity. Table 2-1 is an "energy shopping list." It is clear that we pay for the good taste of energy in the form of steak.

Where It Goes

The energy crisis is not a crisis caused by the "using up" or the disappearance of energy. The first law of thermodynamics assures us of that. Energy is conserved, at least in the closed system of the universe. The crisis must then be found in the pathways of energy conversion.

We use energy, in its kinetic form, as mechanical energy, heat, or radiant energy. The form in which it is stored is potential energy. We know from physics that the potential energy of a system is increased by ΔE when we operate against a force over a distance ΔX, i.e.,

$$\Delta E = \vec{F} \cdot \Delta \vec{X}$$

In the infinite variety of the universe we have discovered, so far, only three types of forces: gravitational, electrical, and nuclear (there seem to be two nuclear forces corresponding to the weak and the strong nuclear interaction). It follows, therefore, that there are three primary sources of energy: gravitational, electrical (chemical), and nuclear. In the phenomena of the universe, these are the most important, and the weak one, the gravitational force, and the strong one, the nuclear force, give us the most visible effects.

At earth's scale we choose other primary sources of energy. Solar energy, produced by the thermonuclear processes in the sun, is the most important of these. It gives us the kinetic energy of water power and wind power, warms us, is stored as chemical energy in growing things, and is preserved in the fossil fuels.

We store the gravitational energy of lifted water in reservoirs. We also make use of some of the mechanical energy stored in the earth-moon system. Through the intermediary of gravitational force, this energy causes the oceans' tides.

The chemical energy of fossil fuels is at present far and away the most important of the primary sources of energy, but in addition to tidal energy there are two other non-solar sources, geothermal energy from the earth's heated interior (originally heated by gravitational contraction and kept warm by radioactivity) and the new entrant, nuclear energy.

Excepting solar energy, the sources are of little use to us in their primary form; they must be converted to the intermediate forms and often converted again to the end uses.

The major source of energy in this country is the chemical energy of the fossil fuels. From them we get 93 per cent of the energy we use.[2] They are fuels; their chemical energy is released by burning. Thus, the major energy conversion pathway is from primary chemical energy to intermediate thermal energy. In fact, most of the conversion pathways go through the thermal intermediate form.

We shall look later in detail at the distribution of energy among the end uses of energy. We know in advance, however, that the major end uses are thermal (space heating, for example) and mechanical. Mechanical energy is also a major intermediate form and is converted to that most important intermediate form, electrical energy. The convenience of electrical energy shows up in its ready conversion to all the important end uses.

Conversion Efficiency

The most important conversion pathway is thus chemical→thermal→mechanical, and here we enter into the domain of the second law of thermodynamics. It is this "thermal bottleneck" through which most of our energy flows that contributes mightily to the various facets of the energy crisis. We burn to convert, and this causes pollution. We are doomed to low efficiencies by the second law and waste much of this heat energy. Let us consider the efficiency problem first.

Efficiency, the ratio of output work to input energy, varies greatly from conversion to conversion. Generally speaking, we can convert back and forth from electrical energy

to other forms of energy with high efficiency, but when we convert other forms of energy to heat and then try to convert heat to mechanical energy we enter the one-way street of the second law.

The efficiency of a "heat engine" (a device for converting heat energy to mechanical energy) is governed by the equation:

$$\text{Eff.} = \left(1 - \frac{T_{out}}{T_{in}} \right) 100$$

where T_{in} and T_{out} are the temperatures of intake and exhaust (in absolute degrees). This equation sets an upper limit of efficiency (it is for a "perfect" Carnot cycle). Since we are forbidden $T_{out} = 0$ K or $T_{in} = \infty$, we are doomed to an intermediate range of modest efficiencies. For example, most modern power plants use steam at 1000°F (811 K) and exhaust at about 212°F (373 K), with a resulting upper limit of efficiency of 54 per cent. The actual efficiency is closer to 40 per cent. Nuclear reactors presently operate at a T_{in} of about 600°F (623 K) and T_{out} of 212°F (373 K) for an upper efficiency limit of 40 per cent. They actually operate at about 32 per cent. In an automobile the input temperature of 5400°F (3255 K) and output of 2100°F (1433 K) would allow an efficiency of 56 per cent. The actual efficiency of an automobile engine is about 25 per cent.

So far we have talked about the efficiency of the major conversion process, heat to mechanical work. What is more important to an understanding of the entire energy picture, however, is the system efficiency; for example, the overall efficiency with which we use the energy stored in the underground petroleum deposit to move us down a road in an automobile. Table 2-2 shows the system efficiency for transportation by automobile and the use of electric power for illumination. One can see that over all there are large leaks in the system and that most of the energy is lost along the way.

"Lost" does not, of course, describe precisely what happens to energy. We know what happens: it is converted to heat. The

Table 2–2.   Energy System Efficiency.

Step	Efficiency of Step (per cent)	Cummulative Efficiency (per cent)
Automobile		
Production of crude oil	96	96
Refining of gasoline	87	84
Transportation of gasoline	97	81
Thermal to mechanical-engine	25–30	20–24
Mechanical efficiency-transmission (includes auxiliary systems)	50–60	10–15
Rolling efficiency	60	6–9
Electric Lighting (from Coal-Fired Generation)		
Production of coal	96	96
Transportation of coal	97	93
Generation of electricity	33	31
Transmission of electricity	85	26
Lighting, incandescent (fluorescent)	5 (20)	1.3 (5.2)

inexorable second law describes the one-way street of entropy. All energy conversion processes are irreversible; even in the highly efficient electrical generator some of the mechanical work goes into unwanted heat. The conversion of other forms of energy to heat is a highly efficient process—ultimately 100 per cent. It is a downhill run. But the reverse is all uphill; heat energy can never be completely converted to mechanical work. The potential energy available to us, whether it be chemical, nuclear, or gravitational, is slowly being converted to the random motion of molecules. We cannot reverse this process, we can only slow it down. This fact has resulted in the increasing use of another measure of efficiency—the so-called second law efficiency, which is, instead of the ratio of desired output to input energy, the smallest amount of work or energy that could have been used to achieve the desired output, divided by the input energy.

This second law efficiency shows us where improvements can be made. For example, we use an oil furnace, producing air at 600°F, to heat a room to about 70°F. The first law efficiency of such a furnace may be 75 per cent. The same result could be accomplished with much less expenditure of energy by the use of a heat pump. The second law efficiency of the oil furnace is about 7 per cent.

Patterns of Consumption

Ever since former President Johnson turned off the lights in the White House there has been a small (too small) but growing effort to save energy. From 1950 to 1970 energy consumption in this country averaged a growth rate of 3.5 to 4 per cent per year. For the last decade, however, the average rate of growth was only about 1.5 per cent and from 1978 through 1980 energy consumption did not grow at all. The 1980 domestic total of 76.1 quadrillion Btu is less than the 1978 total of 78.2 quadrillion Btu.

A gross flow chart of energy in our economy is shown in Figure 2-3. One sees the thermal bottleneck. Heat is the desired end product from about one half of our energy. We do use that amount of energy efficiently. Of the one half that goes to provide mechanical work, however, large amounts are lost in the production of electrical energy and energy for transportation. The net result is that over all our system is about 50 per cent efficient.

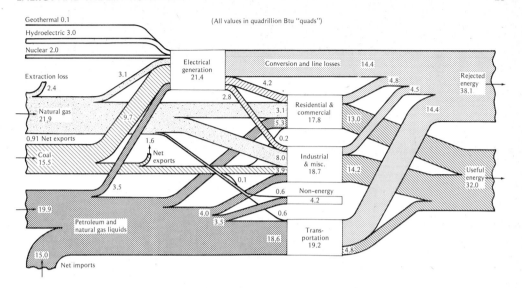

Figure 2-3. The flow of energy in the United States (all values in quadrillion Btu, "quads").

ENVIRONMENTAL EFFECTS OF ENERGY USE

Patterns of Consumption

The intimate connections between energy, our way of life, and the natural environment are seen in many places. The most important places are, of course, in the production of energy—in the mines and wells, refineries and generating plants—and at the place of consumption. Figure 2-3 gave a crude picture of consumption; we need to look at it in more detail.

Figure 2-4 gives both a crude breakdown by category and the details of the 74.7 quadrillion Btu of energy. One sees that industry and transportation use the lion's share. The importance of space heating and water heating also is clear.

These data are for 1973, the most recent year for which we have a complete study. The latest data for the large categories show the residential and commercial sectors combining to account for 34.9 per cent, the industrial sector 39.7 per cent, and the transportation sector 25.4 per cent.

Electrical Energy—The People's Choice

What isn't clear in this presentation is the special case of electrical energy. It is there, contributing heavily to all categories except transportation, and it shares with transportation most of the blame for energy's role in environmental degradation.

The growth rate of electrical energy consumption, shown in Figure 2-5, has been the highest of all the various forms of energy. In discussing growth, a most useful concept is "doubling time." The energy curve of Figure 2-1 shows several different periods of growth and therefore several different doubling times. In the late 1800s the doubling time was about 30 years; by the early 1900s this had been cut in half, to about 16 years. The doubling time during the growth period from 1950 to 1960 was 25 years, and for the period 1960 to 1970 it dropped to 18 years.

Electrical energy can be said to have arrived commercially with the startup of the Pearl Street Station by Thomas Edison in 1882. (The energy curve in Figure 2-1 breaks away from the "people" curve by about 1885.) The doubling time for per capita elec-

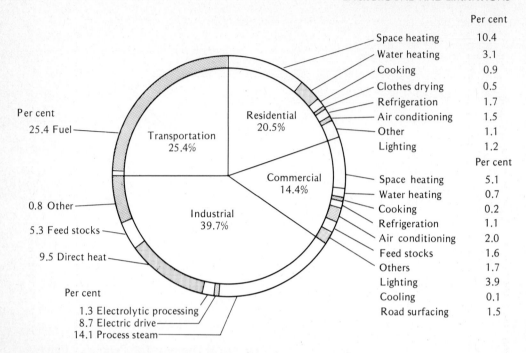

Figure 2-4. United States energy consumption, 1973 (total 74.7 × 10^{15} Btu).

trical energy consumption of Figure 2-5 was only 7½ years during the start-up period of 1910 to 1920, and about 14 years in the 1950s. It decreased to 10 years in the last decade. Electrical energy shared in the decline of 1974 and 1975. It dropped by 0.9 per cent in 1974 and increased by only 0.6 per cent in 1975. In 1976 and 1977 growth resumed by 5.6 and 6.1 per cent, respectively—still less than the 6.7 per cent growth rate of 1973. Since 1977 electric growth rate has been low and erratic: 2.6 per cent in 1978, 4.1 per cent in 1979.

The reasons for the rapid increase in demand for electrical energy are several. Electricity is in many ways the most convenient of the forms of energy. It can be transported by wire to the point of consumption and then turned into mechanical work, heat, radiant energy, and so on.

It cannot be very effectively stored, and this has also contributed to its increasing use.

Generating facilities have to be designed for peak use. In the late 1950s and early 1960s this peak came in the winter, when nights were longer and more lighting and heating were needed. It was economically sound to heavily promote off-peak use, such as summer use of air conditioners. This promotion was so effective that the summer is now the time of peak use, and the sales effort seems to be going into selling all-electric heating for off-peak winter use.

The rate structure, the so-called declining block, in which each additional block of electrical energy used during a specified period is less expensive, has also contributed to demand growth.

Promotion, of course, made sense when each new generating plant was more efficient than the last one and the cost of electrical energy continued to drop. After 30 years or so of declining prices, the cost of a kilowatt-hour of electricity began to rise in 1971 and

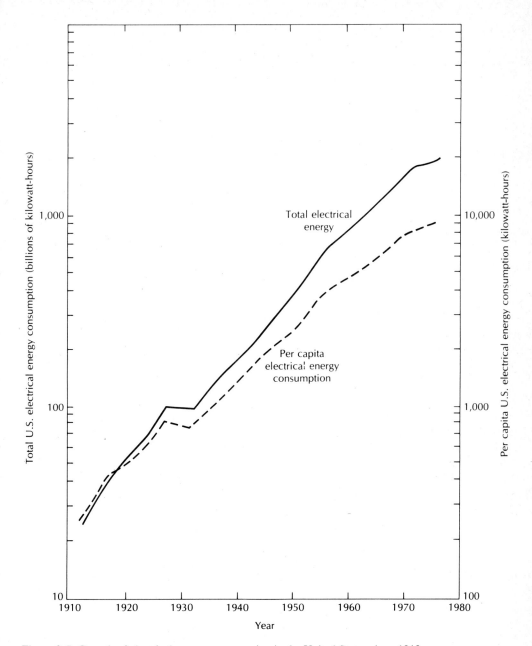

Figure 2.5. Growth of electrical energy consumption in the United States since 1910.

has increased each year since then. Even though the rising cost is expected to continue into the future, the electrification of the country is being accelerated and most of the new energy sources, which are discussed later, are associated with generation of electricity.

Environmental Effects: Air Pollution

Almost all airborne pollution—the familar smog—is due to energy consumption of one form or another. We show, in Table 2-3, a breakdown of the various categories of polluters. One sees that the generation of electric power is the major source of sulfur oxides, whereas the automobile leads for three other pollutants.

It is of course not possible to determine the importance of these pollutants from their gross weight because they have very different effects. Some, like carbon monoxide, affect health in even minute concentrations; others, like the particulates, largely add to cleaning bills. This article is not the place for a detailed discussion of the effects of air pollution.[3] We shall simply summarize the costs, which come from its effects on health and damage to crops and exposed materials and property values, by quoting the Second Annual Report of the Council on Environmental Quality, August 1971:

The annual toll of air pollution on health, vegetation and property values has been estimated by EPA at more than 16 billion dollars annually—over $80 for each person in the United States.

Our dependence on the automobile for transportation presents us with a complex mix of problems; in addition to polluting the air, the automobile uses one quarter of our energy total in a very inefficient way, leads to the covering of more and more of our countryside with concrete, contributes to many aspects of the problems of our cities, and takes a high toll in human life. The discussion of these problems and suggestions for solutions are fascinating and important, but cannot be undertaken here.

The generation of electric power at present depends predominantly on the burning of the fossil fuels. Sulfur oxides come from the sulfur impurities in these fuels. The burning of these fuels also converts large amounts of carbon to carbon dioxide. This familiar gas is not a pollutant in the ordinary sense, but its steady increase in the atmosphere is a cause for concern. Carbon dioxide is largely transparent to the incoming short-wave solar radiation, but reflects the longer-wave radiation by which the earth's heat is radiated outward, producing the so-called greenhouse effect. At present about six billion tons of carbon dioxide are being added to the earth's atmosphere per year, increasing its carbon dioxide content by 0.5 per cent per year. By the year 2000 the increase could be as much as 25 per cent. Our understanding of the atmosphere is not sufficient to predict the eventual changes in climate that might be produced by this increase and by a related increase in water vapor and dust, but small changes in the average temperature could

Table 2–3.   Estimated Pollutant Emissions by Source, 1977 (millions of metric tons).

Source	Particles		Sulfur Oxides		Nitrogen Oxides		Volatile Hydrocarbons		Carbon Monoxide	
Transportation (autos, trucks)	1.1	9%	0.8	3%	9.2	40%	11.5	41%	85.7	83%
Combustion (power, heating)	4.8	39%	22.4	82%	13.0	56%	1.5	5%	1.2	1%
Industrial processes	5.4	43%	4.2	15%	0.7	4%	10.1	36%	8.3	8%
Solid waste (incinerators)	0.4	3%	—	—	0.1	—	0.7	2%	2.6	3%
Miscellaneous (fires, solvents)	0.7	6%	—	—	0.1	—	4.5	16%	4.9	5%
Total	12.4	—	27.4	—	23.1	—	28.3	—	102.7	—

have catastrophic effects. For a detailed discussion of the carbon dioxide problem, see Article 13.

Nuclear Reactors—Clean Power?

In 1980, we obtained about 11 per cent of our electrical energy from nuclear power plants. At the end of that year some 74 nuclear plants, with a capacity for generating 52,953 megawatts of electricity, were in operation. We are thus well into the nuclear age.

The light-water reactors (LWR) now in use obtain energy by fission (splitting) of the rare isotope of the heavy metal uranium, ^{235}U. The by-products of this fission are stopped in the fuel rods, heating them, and this heat is transferred by some type of heat exchanger to a conventional steam-powered electric generator.

The fission products are radioactive, dangerously so. They have many different half-lives, but the whole mess averages a half-life of perhaps 100 to 150 years. The switch to nuclear energy for the generation of electricity will be accompanied by a growing problem of disposal for this radioactive waste. The 16 metric tons of radioactive fission products from reactors accumulated in 1970 grew to 2300 metric tons of spent fuel in 1978 and could grow to at least 71,000 metric tons by the year 2000.

Nuclear reactors are carefully designed to securely contain these products, which are collected and stored for safety. But the storage problem itself is a far from negligible one, with no generally agreed-upon solution in sight.

So far, the radioactivity associated with nuclear reactors seems to have been handled safely. Any exposure to the general population from this source is in the range of present exposure from past nuclear testing. It is in all likelihood causing damage, but so are all the other forms of power generation. What really must concern us when we consider substituting uranium fission for the burning of fuel is the possibility of accident.

When discussing accidents, we are not talking about a real nuclear explosion in which a "critical mass" of fissionable material accumulates and reacts. The low enrichment densities preclude that. But since the reactor core is a witches' caldron of radioactive waste products, any accident that opens that up and spreads it over the countryside will be catastrophic. The accident that designers fear is cooling system failure. If the cooling water somehow did not reach the fuel rods, in only a matter of minutes they would begin to melt, leaving the reactor core an uncontrollable blob of melting metal, heated internally so that it would continue to melt. The resulting steam pressure explosions could then spray the radioactive products into the environment. It is this small but troublesome chance of accident that keeps reactors away from the cities where their products, electricity and heat, are needed.

What we have said so far is specific to the LWR using ^{235}U. It has long been planned by the nuclear power industry to eventually switch to plutonium (^{239}Pu) for part of the fuel. Plutonium-239 is a man-made radioactive element; it is "bred" from the more common ^{238}U, which makes up most of the fuel in the LWR. It is also the fuel that would be provided by the so-called breeder reactors. The use of ^{239}Pu as a fuel will add much energy to our nuclear fuel stockpile, but it will also add some considerable complications. Unlike the mixtures of ^{238}U and ^{235}U of the LWR, ^{239}Pu-containing fuel is practically the raw material for nuclear bombs. The steps to separate out the ^{239}Pu are not very difficult. Entering the "Plutonium Age" will thus be dangerous. Nuclear policy and the effect of the accident at Three Mile Island are discussed in the nuclear section of this book.

Heat as a Pollutant

As we stressed earlier, energy conversion is largely a one-way street: all work eventually produces heat. The "heat engines," because of their inefficiency, however, are particularly troublesome. A steam power plant, which is only 30 per cent efficient, dumps two units of heat energy for every one it converts to electricity. As our appetite for electricity grows in its apparently unbounded way, so also does the problem of heat discharged to the environment.

If we express our consumption of electricity in terms of energy released per square foot of U.S. land area, we obtain, for 1970, 0.017 watt per square foot. At the doubling time of 10 years for electric power consumption, in 100 years we would have gone through 10 doubling periods, and the energy release would be 17 watts per square foot—almost the same as the 18 or 19 watts per square foot of incoming solar energy (averaged over 24 hours). Long before we reach such a level, something will have to be changed.

Energy use, particularly electric power use, cannot be allowed to continue to grow as it has. There are other data that reinforce this conclusion. Electricity means power plants and transmission lines; doubling consumption means doubling these. There are now about 300,000 miles of high-voltage transmission lines occupying 4 million acres of countryside in the United States. By 1990, this is projected to be 500,000 miles of lines occuping 7 million acres.

All this serves to make the point that exponential growth cannot continue. But we could have learned that from nature. Exponential growth is unnatural; it occurs only temporarily when there is an uncoupling from the constraints of supply and of control. For instance, it can be demonstrated by the growth of a bacterial population with plenty of food, but it will eventually be turned over either by exhaustion of the food supply or by control from environmental processes that resist the growth of the population. We have examined some of the areas of environmental damage that may bring us to resist continued growth of energy production and consumption. What about our energy sources; are they likely to be the controlling factor?

RESOURCES AND NEW SOURCES

Before we ask for a statement of the lavish deposits nature has made to our energy ac-

count, we must abandon our parochial view and briefly look at energy consumption as the world problem it is.

Energy and the Gross National Product

It can and will be argued that man can live happily and productively at rather low levels of energy consumption. The fact remains, however, that today per capita energy consumption is an indicator of national wealth and influence—of the relative state of civilization as we have defined it. That this is so is seen most clearly by plotting that talisman of success, the (per capita) gross national products (GNP) against the (per capita) energy consumption, shown in Figure 2-6. There appears a rough proportionality between per capita GNP and per capita energy consumption, with the United States at the top and countries like Uruguay and Mexico near the bottom. To the left of the "band of proportionality" lie the countries that manage a relatively large GNP with a relatively small energy expenditure.

Countries of this type are worthy of study, and many of our present plans for industrial and other conservation strategies are suggested by practices in, for instance, West Germany and Sweden.

The United States, of course, consumes much more than its share of world energy. The United States, with 6 per cent of the world's population, uses 27 per cent of the world's energy. If we look at comparative rates of growth, we see that the U.S. per capita energy consumption is much larger (by a factor of about 30) than that of India, for instance, and is growing more rapidly. The world figure is about six times smaller but is growing a bit more rapidly than is the U.S. figure.

Even if the United States were to stabilize at the present per capita figure of 250 kilowatt-hours per day, it would take about 120 years for the world per capita average to equal it and hundreds of years for India, at its present rate of growth, to catch up. If we were to set the 1975 U.S. level of consumption as a world target for the year 2000, and

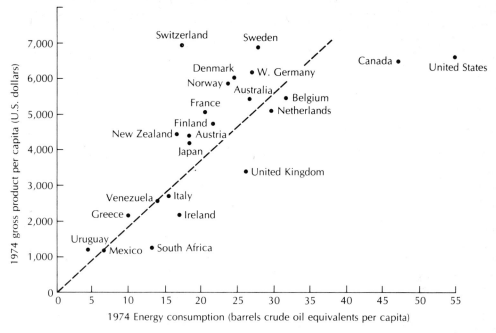

Figure 2-6. Per capita consumption of energy and gross national product for some countries of the world. (From Earl Cook, "The Flow of Energy in an Industrial Society," *Scientific American* **224**, 142; September 1971)

if we take into account the population growth by then, the world would then be consuming ten times as much energy as it does now. And this brings us to energy resources.

How Long Will They Last?

As someone said, "Prophecy is very difficult, especially when it deals with the future." Predicting the lifetime of energy resources is doubly difficult. Energy-use curves must be projected and then unknown resource potentials guessed at. It is difficult to hope for much accuracy in either of these procedures.

The estimation of resources is based on a general knowledge of the kind of geological conditions associated with the resource and on a detailed knowledge of the distribution and extent of a resource within a favorable geological area. Coal is the easiest to work with, for it seems almost always to be found

where it is predicted to be. Oil and natural gas, on the other hand, are erratic in distribution within favorable areas and are found only by exploration. In addition to coal, oil, and natural gas, there are two other sources of organic carbon compounds that are potential fuel sources, the so-called tar sands and oil shale. In the tar sands, which so far have been found in appreciable amounts only in Canada, a heavy petroleum compound (bitumen) binds the sands together. Canadian refineries are currently producing oil products from this material. Oil shale is shale rock containing considerable amounts of a solid organic carbon compound (kerogen). Oil can be extracted by heating the rocks and commercial demonstration of several processes is under way.

In Figure 2-7 we show the estimates of world fossil fuel resources of various types and the global distribution of these resources. The unit used to measure these re-

ENERGY CONTENT OF WORLD RESOURCES

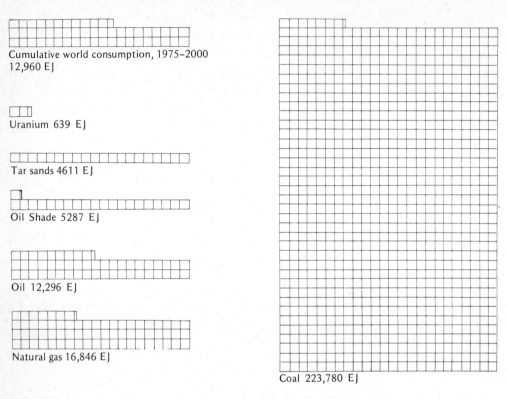

Cumulative world consumption, 1975–2000
12,960 EJ

Uranium 639 EJ

Tar sands 4611 EJ

Oil Shade 5287 EJ

Oil 12,296 EJ

Natural gas 16,846 EJ

Coal 223,780 EJ

☐ = 250 exajoules

Figure 2-7. Energy content of world resources. The 639 EJ for uranium is very low because it includes only the "reserves" and "probable" resources of uranium. In proportion, U.S. resources in these categories total 1.7 million metric tons, or 249 EJ. The value given for oil shale represents identified resources; the world figure is surely a considerable underestimate. (From U.S. Geological Survey Professional Paper 820, 1973)

sources is the exajoule (1 EJ = 10^{18} joules). As a crude reference, it would take about 2 EJs of energy to boil Lake Michigan. Perhaps more useful is the fact that U.S. total energy consumption in 1980 was about 81 EJ (76 quads), whereas world consumption was 300 EJ (280 quads).

One sees from these data that most of the remaining fossil-fuel resources, for the United States and for the world, are in the form of coal.

Presenting data on resources does not by itself answer the question "How long will they last?" To answer that question one has to look at the rate at which the resources are being used. A simplified but very graphic way of displaying this has been adopted by M. King Hubbert of the U.S. Geological Survey. Since supplies of fossil fuels are finite, the curve tracing their production rate will be pulse-like, that is, it will rise exponentially in the beginning, turn over when the resources come into short supply, and then decay exponentially as the resources become

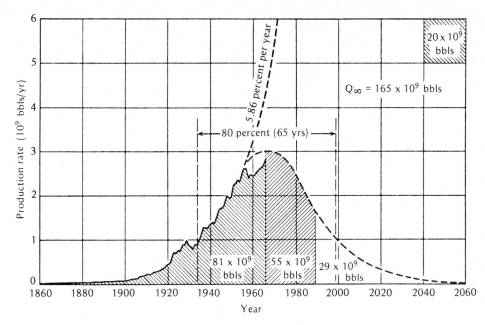

Figure 2-8. Use rate for U.S. oil resources. (From M. K. Hubbert, *Resources and Man,* p. 183. San Francisco: W. H. Freeman.)

Figure 2-9. Use rate for U.S. coal resources. (From M. K. Hubbert, *Resources and Man,* p. 200. San Francisco: W. H. Freeman.)

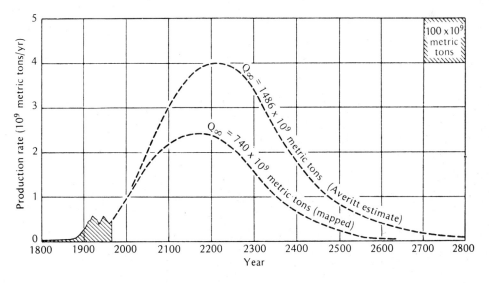

harder and harder to find. Such data for U.S. oil and U.S. coal are presented in Figures 2-8 and 2-9.

The curve for U.S. oil is of particular interest. It shows that the actual rate of production from 1880 to the present does have roughly the predicted shape and that, in fact, we are already past the peak of production. Although the estimate of total resources, 165 billion barrels, may be a bit pessimistic (current estimate range between 112 and 189 billion barrels), it is the shape of the curve that has the real lesson in it. We are on the down side of that curve and already feeling the pinch of impending shortages.

Figure 2-9 for coal is consistent with the data of Figure 2-7. We have a large amount of coal and are just started up the slope to the peak in production. The time scale is much larger, the time to produce 80 per cent of U.S. coal is on the order of 300 to 400 years.

In combination, Figures 2-7, 2-8, and 2-9 give a representative view of the state of U.S. and world fossil-fuel resources. They give a qualified answer to "How long will they last?" The answer: "Not very long" if we are talking about crude oil and natural gas; "Long enough for us to find other sources" if we are talking about coal.

The answer might also have been "Long enough for us to wreck our environment." Without much improvement in the protection we give our environment, about two more doubling periods of energy consumption may be sufficient. Each doubling not only reduces the resources but, for instance, doubles the generating capacity (more plants), doubles (almost) the number of transmission lines, doubles coal-mining activity, doubles (unless rigid controls are implemented) the air and water emissions, and so on.

Energy for the Future

Those of us who write about energy are no longer "prophets crying in the wilderness." The oil embargo, the cold winter of 1976–1977, and the gasoline shortage of 1979–1980 have finally made believers of politi-

cians and citizens alike. It is now clear that we are in one of those rare times of transition that mark human progress; that we are voyaging from familiar shores toward an unknown future home. What kind of energy future can we expect?

One thing, at least, is certain. We must kick the fossil-fuel habit. If we are again to reach a time of energy abundance, we must turn to other primary sources. In our hazy crystal ball we can see only three candidates, fission energy based largely on conversion of ^{238}U to ^{239}Pu, fusion reactions with hydrogen, and solar energy. In addition, there are other, smaller, continuous energy sources—compared in Table 2-4—which can help.

Solar Energy

A look at Table 2-4 allows us to place the continuous energy sources in perspective. One, direct solar energy, is sufficient to promise future abundance. It is also clear that we must find a new method of conversion. The three existing "natural converters"—

Table 2–4. Comparison of Continuous Energy Sources—United States.

	kw/acre	kw/hectare
Solar energy at ground level[a]	720	1800
All the winds (at surface)[b]	43	110
U.S. tides[c]	0.02	0.05
U.S. geothermal[d]	0.02	0.05
All photosynthesis[e]	0.4	1.0
U.S. hydroelectric power	0.2	0.4
U.S. consumption, 1977	1.1	2.8

[a] This is an annual average over day and night. The U.S. range is from 600 kw/acre in Vermont to 1040 kw/acre in Arizona and New Mexico.
[b] Includes Aleutians, Eastern seaboard. *Wind Machines*, NSF, 1975.
[c] Maximum for practical sites includes Bay of Fundy and Cook Inlet in Alaska. *The Energy Source Book*, Center for Compliance Information, December 1977.
[d] Maximum of geothermal sources recoverable with present technology. *Western Energy Resources and the Environment: Geothermal Energy*, EPA-600/9-77-010, May 1977.
[e] C. C. Burwell, "Solar Biomass Energy: An Overview of U.S. Potential," *Science* **199**, 1041 (1978).
(From John M. Fowler, *Energy-Environment Source Book*. Washington, D.C.: NSTA, 1980).

photosynthesis, water power, and wind power—even at maximum utilization, would just barely meet our present needs.

Research into methods of utilizing solar energy has greatly increased during the past five years as federal financial support increased from about $1 million in 1971 to about $570 million in 1980. This research is seeking answers in several directions. Bioconversion (use of photosynthesis) is being studied, and, even if we do not decide to deliberately grow energy crops, we are turning to organic waste (paper, crop residues, wood waste, manure, etc.) for energy.

The solar heating and cooling industry is experiencing a growth spurt as solar becomes a more important primary source of heat. The technology exists and the market is growing, despite competition from the fossil fuels.

To convert solar energy to electricity economically will take considerably more research and development. Several methods are under investigation: photovoltaic devices (solar cells), large areas of solar collectors producing steam, mirrors focused on a boiler, and a solar sea plant, which works off of the thermal gradient between the surface layer and the deep layers of the ocean. The focusing mirrors ("power tower") are being given top priority, and a pilot 10-megawatt plant in California went into operation in 1982.

Nuclear Energy

The only other potentially large source of primary energy is nuclear energy. There are two possibilities here. We pointed out earlier that the ^{235}U needed by the conventional LWR is in relatively short supply. If, however, the so-called breeder reactor were to be developed it could create a fissionable fuel, ^{239}Pu, out of the relatively abundant ^{238}U. If we were to go into this "plutonium economy" and build not only breeder reactors but large fuel reprocessing plants to remove the ^{239}Pu from the reactor fuel, we could conceivably provide for our energy needs from

known uranium supplies for 1000 years or so.

We mentioned earlier some of the hazards of plutonium use that make this route the least attractive of the three routes to energy abundance that are open to us.

The third potential source of a large amount of energy also involves the nucleus. In fission, energy is released when a large nucleus is split in two. Energy is also released when two small nuclei are "fused" together. This is the fusion process we have been trying to master for 30 years or so.

The fuel for fusion is the heavy isotope of hydrogen, deuterium (^{2}H). It is present in unimaginable abundance in the ocean; there is enough fusion fuel there to last for hundreds of thousands of years. We seem to be on the threshold of proving that the fusion process is scientifically feasible. The 1980s should see this breakthrough. Beyond that, however, are the unknown problems of engineering and of demonstration of commercial feasibility.

Fusion may be a power source for the next century. It will offer the advantages of unlimited fuel and, compared with fission, much less radioactivity and essentially no explosion hazard. It is an attractive dream—but it may be only a dream.

CONCLUSION

Mankind is in the midst of an energy crisis. It has several dimensions. On a short time scale, we are faced with a serious shortage of natural gas and with the necessity of importing more and more oil.

On an intermediate time scale, 20 to 40 years, we are faced with the necessity of finding a substitute for the petroleum products that now dominate the energy mix.

Finally, on a larger scale of a few hundred years, we are faced with the exhaustion of fossil fuels.

What can we do? Some of the answers are obvious. We can plan more carefully and project further into the future. We can try to reduce energy expenditures and increase research and development. As concerned citizens we must demand this. We must also demand a role in the decision-making process and decide, for instance, whether we will rely most on the breeder reactor, with its dangers, or on coal and coal products (gases and liquids).

The science teachers of this country have a vital role to play in the next few years. The decisions just mentioned must be based on information, and it is the science teacher who must make sure that developing citizens have access to the necessary information without bias and distortion. It is my hope that this summary, brief and sketchy as it is, can serve as a part of the foundation of that necessary effort.

NOTES

1. We here deal with Calories (kilocalories), the amount of heat energy needed to raise the temperature of 1 kilogram of water 1 Celsius degree (1 Cal = 3.96 Btu).
2. See Jackalie Blue and Judith Arehart, "A Pocket Reference of Energy Facts and Figures," Energy Division, Oak Ridge, National Laboratory, Oak Ridge, Tenn., 1980.
3. See for instance, J. M. Fowler, *Energy and the Environment,* Chap. 7. New York: McGraw-Hill, 1975.

SUGGESTED READINGS

Electricity: Today's Technologies, Tomorrow's Alternatives. Palo Alto, Calif.: Electric Power Research Institute, 1981.

Energy: Facing Up to the Problem, Getting Down to Solutions. National Geographic Special Report. Washington, D.C.: National Geographic Society, 1981.

Energy in Transition 1985–2010: Final Report of the Committee on Nuclear and Alternative Energy Systems (CONAES). National Research Council, National Academy of Sciences. San Francisco: W. H. Freeman, 1980.

Morris W. Firebaugh, "Perspectives on Energy and the Environment," *The Physics Teacher* **15,** 78 (1977).

John M. Fowler, *Energy and the Environment.* New York: McGraw-Hill, 1975.

John M. Fowler, *Energy-Environment Source Book.* Washington, D.C.: National Science Teachers Association, 1980.

Henry W. Kendall and Steven J. Nadis, eds., *Energy Strategies: Toward a Solar Future.* Union of Concerned Scientists. Cambridge, Mass.: Ballinger, 1980.

Henry R. Linden and Joseph D. Parent, *Perspectives on U.S. and World Energy Problems.* Chicago: Gas Research Institute, April 1980.

Amory Lovins, *Soft Energy Paths: Toward a Durable Peace.* Cambridge, Mass.: Ballinger, 1977.

Sam H. Schurr et al., *Energy in America's Future: The Choices Before Us.* Baltimore: Johns Hopkins University Press, 1979.

Robert Stobaugh and Daniel Yergin, eds., *Energy Future: Report of the Energy Project at the Harvard Business School.* New York: Random House, 1979.

Forgotten Fundamentals
of the Energy Crisis   3

ALBERT A. BARTLETT

INTRODUCTION

The world energy system involves a dozen or more types of fossil and nuclear fuels, thousands of uses of these fuels, hundreds of governmental jurisdictions and entities, thousands of businesses and related enterprises, and dozens of variations of political and economic systems that are all woven into a complex and irregular fabric that affects the lives of the earth's approximately 4.6 billion people as of 1981. From this point of view, the energy situation may appear to be so intricate and complex as to defy understanding. If we are to understand the energy picture, we must first focus on the fundamentals, for only when we understand the fundamentals can we begin to understand the complexities which often seem to be so overwhelming. The goal of this article is to establish the validity of the following two propositions:

1. The *energy shortages* are the result of long periods of growth in the rates of consumption of finite reserves of fuels.

2. The *energy crisis* is the result of our failure to understand the arithmetic and the consequences of steady growth.

ARITHMETIC
Background

When a quantity such as the rate of consumption of a resources is growing a fixed percentage per year, the growth is said to be *exponential*. The important property of the growth is that the time required for the growing quantity to increase its size by a fixed fraction is constant. For example, a growth of 5 per cent (a fixed fraction) per year (a constant time interval) is exponential. If it takes a fixed length of time to grow 5 per cent, it follows that it will take a longer fixed time to grow 100 per cent (double in size). This time is called the doubling time, T_2, and it is related to P, the per cent growth per unit time, by a very simple relation that should be a central part of the educational repertoire of every person:

Original article by Dr. Albert A. Bartlett. Dr. Bartlett is Professor of Physics at the University of Colorado, Boulder, Colorado 80309, and has published extensively on energy problems and the nature of exponential growth. He was President of the American Association of Physics Teachers in 1978 and received the organization's Robert A. Millikan Distinguished Service Award in 1981. This article is based upon articles of the same title which appeared in *Journal of Geological Education* 28, 4 (1980), *American Journal of Physics* 46, 876 (1978), and the *Proceedings* of the Third Annual UMR-MEC Conference on Energy, held at the University of Missouri-Rolla in 1976.

$$T_2 = 70/P \qquad (1)$$

As an example, if a quantity is growing 10 per cent per year, it will double every $70/10 = 7$ years. In two doubling times (14 years), it will double twice (quadruple) in size. In three doubling times, its size will increase eightfold ($2^3 = 8$); in four doubling times (28 years), it will increase 16-fold ($2^4 = 16$), etc. It is natural then to talk of steady growth in terms of powers of 2.

The Power of Powers of 2

Legend has it that the game of chess was invented by a mathematician who worked for an ancient king. As a reward for the invention, the mathematician asked for the amount of wheat that would be determined by the following process. He asked the king to place one grain of wheat on the first square of the chessboard, two grains on the second square, and continue this way, putting on each square twice the number of grains on the preceding square. The filling of the chessboard is shown in Table 3-1. We see that on the last square there will be 2^{63} grains and the total number of grains on the board will then be one grain less than 2^{64} grains.

How much wheat is 2^{64} grains? Simple arithmetic shows that it is approximately 500 times the 1976 annual worldwide harvest of wheat! This amount is probably larger than

all the wheat that has been harvested by humans in the history of the earth! How did we get to this enormous number? It is simple; we started with one grain of wheat and we let it grow until it had doubled a mere 63 times!

The example of the chessboard (Table 3-1) shows us another important aspect of steady growth: the increase in any doubling is approximately equal to the sum of all the preceding growth! Note that when eight grains are placed on the fourth square, the number on this square (eight) is greater than the total of seven grains already on the board. The 32 grains placed on the sixth square are more than the total of 31 grains already on the board. Covering any square requires one grain more than the total number of grains already on the board.

On April 18, 1977, President Carter told the American people,

And in each of these decades [the 1950s and 1960s], more oil was consumed than in all of man's previous history combined.

We can now see that this astounding observation is based on the simple arithmetic of long-term steady growth at a rate whose doubling time is one decade. The annual growth rate that has this doubling time is $P = 70/10 = 7$ per cent, which is the growth rate of world petroleum consumption from 1890 to 1970.

In 1975, an advertisement told the American people that the demand for electrical power was expected to continue its historical growth so that "our need for electricity will actually double in the next 10–12 years." From this statement one can draw two important conclusions:

1. The expected rate of growth of demand for electricity would be between 5.8 and 7 per cent per year.
2. The expected consumption of electrical energy in those next 10–12 years would be approximately equal to the total of all of the electrical energy that had been consumed in the entire previous history of the electrical industry in the United States.

Table 3–1.   Filling the Squares on a Chessboard.

Square	No. of Grains on This Square	Cumulative No. of Grains
1	1	1
2	2	3
3	4	7
4	8	15
5	16	31
6	32	63
7	64	127
. . .	. . .	. . .
64	2^{63}	$2^{64} - 1$

Combining these two, we see that a modest rate of growth of consumption that many people would find ''acceptable'' can have an enormous consequence. *Many people find it difficult to believe that when a rate of consumption has been growing steadily at a mere 7 per cent per year the consumption in one decade exceeds the total of all the previous consumption!*

It is very useful to remember that a steady growth rate of n per cent per year for a period of 69.3 years (which is $100 \times \ln2$ and is also approximately one human lifetime) will produce an overall increase by a factor of 2^n. We can illustrate this with an example. The city of Boulder, Colorado, today has one overloaded sewage treatment plant. If Boulder has an annual population growth rate of 5 per cent, then in 70 years it would need $2^5 = 32$ overloaded sewage treatment plants!

Populations tend to grow in this steady (exponential) manner. The world population in 1975 reached 4 billion, and it was growing approximately 1.9 per cent per year. Your reaction to 1.9 per cent per year might be to say, ''That's so small, nothing bad could ever happen at 1.9 per cent per year.'' But when you do the arithmetic you find that if this small rate of growth continued unchanged in the future,

1. The world population would increase by one billion people in just under 12 years.
2. The world population would double in 36.5 years.
3. The world population would reach a density of one person per square meter on the dry land surface of the earth in 550 years.
4. The mass of people would equal the mass of the earth in 1620 years.

This arithmetic permits us to draw a very profound conclusion. Zero population growth (ZPG) is going to happen! We can debate whether we like ZPG or do not like it, but it is going to happen whether we debate it or not, whether we like it or not. Today's very high birthrate is going to drop and today's very low death rate is going to rise until they have the same numerical value, and this will certainly happen in much less than 550 years.

When one is making a graphic representation of steady growth, it is common to use a semilogarithmic plot. In such a graph, equal spaces on the vertical axis each represent increases of a factor of 10. The convenience of a semilogarithmic graph is that *steady growth is represented by a straight line* sloping upward to the right.

This exponential arithmetic has a profound effect on many aspects of our lives: in compound interest, inflation, cost of living, etc. It has even been shown that the number of kilometers of highway in the United States tends to grow exponentially!

A financier who admitted to being unable to name the seven wonders of the world said that he did, however, know the eighth wonder. It is *compound interest,* i.e., exponential arithmetic.

The reader can begin to see that our long history of national and world population growth and of growth in our per capita rates of consumption of resources lies at the heart of our world and national energy problems.

Steady Growth in a Finite Environment

The fossil-fuel and mineral resources of the earth are finite. It is therefore important that we look next at the arithmetic of steady growth in a finite environment.

Bacteria grow by division so that one bacterium becomes two, the two divide to become four, the four divide to give eight, etc. Suppose we had a hypothetical strain of bacterium that doubled in number this way every minute. Suppose that we put one of these bacteria in an empty bottle at 11:00 A.M. and that we observed that the bottle was full at 12:00 noon. Here is a simple case of steady growth with a doubling time of one minute in the finite environment of one bottle. We wish to ask three questions.

Question No. 1: At what time was the bottle half full?

Answer: At 11:59 A.M. (with steady growth, the bacteria double in number every minute!).

Question No. 2: If you were an average bacterium in the bottle, at what time would you first realize that you were running out of space?

Answer: There is no unique answer, and so let us ask, "At 11:55 A.M., when the bottle is only 3 per cent filled and is 97 per cent open space (just yearning for development!), how many of you would realize that there was a problem?" Some years ago someone wrote into a Boulder, Colorado, newspaper to say that there was no problem with population growth in Boulder because there was then 15 times as much open space as had already been developed. When one thinks of the bacteria in the bottle, one can see that the time in Boulder Valley is four minutes before noon! This is illustrated in Table 3-2.

To lead up to the third question, suppose that at 11:58 some farsighted bacteria realize that they are running out of space and so they launch a great and costly search for new bottles. They search offshore on the outer continental shelf, they search in the overthrust belt and in the Arctic, and they discover three new empty bottles. Great sighs of relief come from all of the worried bacteria because this new discovery is three times the number of bottles that had hitherto been known. The discovery quadruples the total space resource known to the bacteria. Surely now the bacteria can become self-sufficient in space. Surely this will satisfy the needs of their bacterial "Project Independence"!

Question No. 3: How long can the steady growth continue as a result of this magnificent discovery?

Table 3–2.   The Last Minutes in the Bottle.

11:54 A.M.	1/64 full (1.5%)	63/64 empty
11:55	1/32 full (3%)	31/32 empty
11:56	1/16 full (6%)	15/16 empty
11:57	1/8 full (12%)	7/8 empty
11:58	1/4 full (25%)	3/4 empty
11:59	1/2 full (50%)	1/2 empty
12:00 noon	Full (100%)	

Answer: Only two more minutes (two more doubling times)! This is illustrated in Table 3-3.

Table 3–3.   The Effect of the Discovery of Three New Bottles.

11:58 A.M.	Bottle 1 is one-quarter full
11:59	Bottle 1 is half full
12:00 noon	Bottle 1 is full
12:01 P.M.	Bottles 1 and 2 are both full
12:02	Bottles 1, 2, 3, and 4 are all full

James R. Schlesinger, secretary of energy in President Carter's cabinet, noted that, in the energy crisis, "we have a classic case of exponential growth agaiñt a finite source." This statement seemed to demonstrate a profound understanding of the energy situation, but later statements by the secretary contradicted this appearance of understanding:

The entire program [President Carter's] is based upon the premise that the economy must continue to expand. . . . We can have nothing to do with that kind of attitude which is antigrowth.

Our next task is to calculate how long a finite resource would last if one had steady growth in the rate of consumption until the last bit of the resource was used up. From the appendix at the end of this article, we find that the exponential expiration time (abbreviated T_e) *is:*

$$T_e = \frac{1}{k}\ln\left(\frac{kR}{r_0} + 1\right) \qquad (2)$$

where k = fractional growth rate per year = $0.01P$, r_0 = current rate of consumption, and R = size of known resource. This is illustrated in Figure 3-1. As we shall see later, it is not realistic to imagine that the rate of consumption of a nonrenewable resource could grow steadily until the last bit of the resource was used, as is modeled in Figure 3-1. The reader may then ask, "In that case, why bother calculating T_e?" The answer is seen in our society, where leaders in business, industry, politics, and government, at the local and national levels, place the highest priority on maintaining unending growth of our gross national product (GNP) and of all of the min-

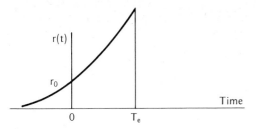

Figure 3-1. Curve of steady growth.

eral and energy resources that go into the GNP. This blind devotion to endless growth is demonstrated by the remark that a recession is a period during which the annual growth rate of the GNP falls below 4 per cent. Since steady, unending growth is a central goal of contemporary American society, we should understand the consequences as they are indicated in Figure 3-1 and in the expression for T_e.

If we now have reserves of R metric tons of a resource and are using the resource today at a rate of r_0 metric tons per year, then *at present rates of consumption* (i.e., zero growth) the resource would last R/r_0 years. If the rate of consumption is growing steadily, as in Figure 3-1, the resource will last fewer than R/r_0 years, but *this obvious fact is frequently overlooked*. For instance, we read of the great effort being made to achieve *rapid growth* of U.S. coal production and we are then assured that *we have enough coal to last over 500 years at present rates of consumption*. It is seriously misleading for an article to concentrate on the importance of rapid growth of U.S. coal production and to reassure the readers not to worry "because we have enough coal to last 500 years at present rates of consumption."

The very important relationship between lifetimes of a resource at present rates of consumption (zero growth) and several rates of steady growth are shown in Table 3-4. Each vertical column represents a given amount of resource for which R/r_0 is the number at the top of the column. This number (opposite zero growth) is the number of years the quantity of the resource would last "at current rates of use." The numbers be-

Table 3–4. Exponential Expiration Times (T_e) for Various Steady Annual Rates of Growth.

Annual Growth Rate Zero	Lifetimes (years)						
	10	30	100	300	1000	3000	10,000
1%	9.5	26	69	139	240	343	462
2%	9.1	24	55	97	152	206	265
3%	8.7	21	46	77	115	150	190
4%	8.4	20	40	64	93	120	150
5%	8.1	18	36	56	79	100	124
6%	7.8	17	32	49	69	87	107
7%	7.6	16	30	44	61	77	94
8%	7.3	15	28	40	55	69	84
9%	7.1	15	26	37	50	62	76
10%	6.9	14	24	34	46	57	69

low are the lifetimes for the annual growth rates indicated at the left. For instance, an amount of a resource that would last 1000 years "at current rates of use" would last 115 years at an annual growth rate of 3 per cent. An amount of a resource that would last 206 years at an annual growth rate of 2 per cent would last 69 years if the growth rate were 8 per cent per year.

Summary: Arithmetic of Steady Growth

We can summarize the important characteristics of steady growth as follows:

The growing quantity increases in size by a fixed fraction per unit time, i.e., a fixed per cent per year.

It follows that the growing quantity will take different fixed lengths of time to grow, for example, by 50, 100, 200, 1000 per cent. The time to grow (increase) by 100 per cent is called the doubling time.

The doubling time in years is calculated by dividing the number 70 by the per cent growth per year.

In a relatively few doubling times, the size of the growing quantity will increase enormously.

The increase during any doubling time is approximately equal to the sum of all proceeding growth.

When steady growth takes place in a finite environment, as in the case of steadily growing rates of consumption of a finite resource,

The end comes frighteningly fast.

At one minute before 12 noon (one doubling time before the end), the amount of the remaining resource is equal to all that had been used in all of previous history.

Enormous discoveries of new quantities of the resource can extend the period of steady growth only by very short times.

An understanding of these points leads us to a very powerful conclusion in regard to populations and the use of resources: *steady growth cannot continue indefinitely*.

The reader will recognize that this conclusion is contrary to the goals and expectations of business, industrial, and political leaders. Here lies the heart of the energy crisis.

APPLICATION TO FUEL RESOURCES

Introduction

Through careful analytical studies, Dr. M. King Hubbert has estimated the "ultimate total production" of several fossil-fuel resources. This gives the total size of all of the resource that we will be able to recover.

Decades ago people would tabulate all of the known reserves of a fossil fuel and divide the sum by the present rate of consumption to tell us (50 years ago) that we would run out of oil, for example, in 18 years. We did not run out because new discoveries enlarged the known reserves. People today say that the "experts" 50 years ago have been proved wrong, and from this they would have us believe that today's estimates will also be proved wrong if we simply unleash free enterprise so that more discoveries are made. There is an enormous difference between the estimates of 50 years ago and the estimates that are available today. The "reserves" of 50 years ago were all of the resource that had been discovered. In contrast, Hubbert's "ultimate total production" is the total amount of the resource that can be recovered, including all that has been recovered in the past *plus all that will ever be recovered in the future*. Subtract the production to date from the ultimate total production and you have the size of the remaining resource including all future discoveries that can be expected! The "ultimate total production" will not be enlarged by the amount of new discoveries, but it can be adjusted, or "fine-tuned," by repeated calculations based on new exploration and drilling data. The ultimate total production cannot be changed by the removal of bureaucratic or other impediments to exploration and production. It cannot be changed by the offer of incentives to producers, and, perhaps most frustrating of all, it cannot be changed by political debate.

Let us look now at the numbers for a few fossil fuels.

U.S. Crude Oil

Table 3-5 gives data showing the production and reserves of U.S. crude oil based on Hubbert's work. From these data, we can calculate that 50.8 per cent of ultimate total production had been produced by 1972. Thus, the "U.S. petroleum time" is about 59 seconds before noon. Roughly, half of the re-

Table 3–5.   Data on U.S. Crude Oil.

	Units (10^9 metric tons)	Units (10^9 barrels)
Ultimate total production (Hubbert[1])	25.7	190
Produced to 1972	13.0	96.6
Remaining after 1972	12.6	93.4
Annual production rate 1970	0.445	3.29

Table 3–6. Exponential Expiration Times (T_e) of Various Estimates of U.S. Oil Reserve for Different Rates of Growth of Annual Production.

Growth Rate (%)	T_{e1} (years)	T_{e2} (years)	T_{e3} (years)
Zero	28.5	31.7	63.4
1	25.1	27.5	49.1
2	22.6	24.5	40.9
3	20.6	22.3	35.5
4	19.0	20.5	31.6
5	17.7	19.0	28.6
6	16.6	17.8	26.2
7	15.7	16.7	24.2
8	14.9	15.8	22.5
9	14.1	15.0	21.1
10	13.5	14.3	19.9

coverable oil in the U.S. has already been used up.

The vast oil shale deposits of Colorado and Wyoming represent an enormous resource. Hubbert[1] reports that the oil recoverable under 1965 techniques is 11×10^9 metric tons (80×10^9 barrels). He also notes that there are other higher estimates. What happens to T_e if we add estimates of Alaskan crude and oil shale to the ultimate total production of the lower 48 states? This is illustrated in Table 3-6. This table is prepared by using Equation 2 with $r_0 = 0.445 \times 10^9$ metric tons per year. *Note that this is domestic production, which provides only about half of our domestic consumption.* Column 1 is the percent annual growth of annual production. Column 2 is the lifetime (T_{e1}) of the resource, which is calculated using $R = (25.7 - 13.0) \times 10^9 = 12.7 \times 10^9$ metric tons as the estimated oil ultimately recoverable in the lower 48 states. The lifetime T_{e2} is calculated

using $R = (12.7 + 1.4) \times 10^9 = 14.1 \times 10^9$ metric tons to include an estimate (10 billion barrels) of Alaskan oil, and T_{e3} is the lifetime using $R = (12.7 + 1.4 + 14.1) \times 10^9$ metric tons, which arbitrarily assumes that oil from U.S. oil shale is 14.1×10^9 metric tons, a value chosen for purposes of illustration only because it is equal to the total of the estimated remaining oil in the continental United States and Alaska.

World Crude Oil

Figure 3-2 by Hubbert shows the semilogarithmic graph of world oil production. The line is approximately straight for 70 years, indicating steady growth with an average annual growth rate of 7 per cent. Table 3-7 gives Hubbert's 1973 data[1] for world crude oil. These data show that 13.4 per cent of the ultimate total production was produced to 1972. Roughly one eighth of the world's petroleum has been used. The time (in 1973) was 2 minutes 54 seconds before noon. Table 3-8 shows the T_e values for world crude oil, world crude oil plus an estimate of world shale oil, and world crude plus hypothetical shale reserves assumed to be four times as large as the known shale reserve. This table gives yet another demonstration of the fact that enormous new sources make little difference in T_e when one has steady growth in consumption. This table is prepared by using Equation 2 with $r_0 = 2.26 \times 10^9$ metric tons/year. Column 1 is the per cent annual growth of the rate of production. The T_{e1} is the lifetime of the resource using $R = 229 \times 10^9$ metric tons as the estimate of the remaining oil; T_{e2} is the lifetime calculated using $R = (229 + 25.7) \times 10^9$ metric tons to repre-

Table 3–7. World Crude Oil.

	Units (10^9 metric tons)	Units (10^9 barrels)
Ultimate total production (Hubbert[1])	264	1592
Produced to 1972	35.3	261
Remaining after 1972	229	1691
Annual production rate 1970	2.26	16.7

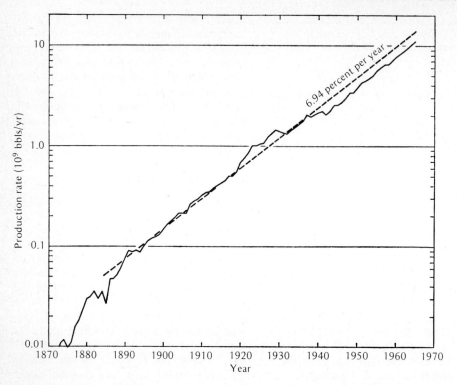

Figure 3-2. World production of crude oil (semilogarithmic scale). [From M. K. Hubbert, *Scientific American* **224** (3), 60 (1971)]

sent crude oil plus oil shale; T_{e3} is the lifetime calculated using $R = (229 + 4 \times 25.7) \times 10^9$ metric tons, which arbitrarily assumes that the amount of shale oil is four times the amount known now. This arbitrary assumption is made solely for the purpose of demonstrating the consequence of the large increase in the resource.

Figure 3-3 is an interesting graphic representation of the effects of a steady growth of 7 per cent per year, as was the case from 1890 to 1970 for world oil production. Suppose the area of the small rectangle in the upper left represents all the oil we used on this earth before 1950. Then with roughly 7 per cent annual growth, the oil we used in the decade of the 1950s is represented by the rectangle to the right, which is equal to all we used in all of previous history. The amount we used in the 1960s is represented by the rectangle labeled 1960s, which again

is equal to the sum of all of the consumption in all of previous history. Our quote from President Carter refers to this. We demonstrated this concept with the chessboard.

Table 3–8.　Life Expectancy of Various Estimates of World Oil Reserves for Different Rates of Growth of Annual Production.

Growth Rate (%)	T_{e1} (years)	T_{e2} (years)	T_{e3} (years)
Zero	101	113	147
1	70.0	75.5	90.4
2	55.4	59.0	68.5
3	46.5	49.2	56.3
4	40.5	42.7	48.2
5	36.1	37.8	42.4
6	32.6	34.2	38.1
7	29.9	31.2	34.6
8	27.6	28.8	31.8
9	25.7	26.8	29.5
10	24.1	25.1	27.5

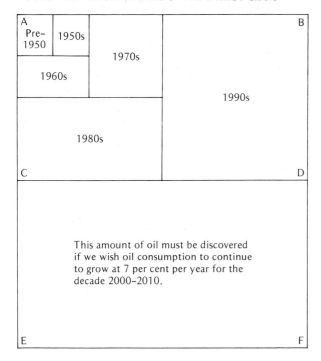

Figure 3-3. The area of each rectangle represents the quantity of petroleum consumed in the labeled decade. This makes it clear, as was seen in the example of the chessboard (Table 3-1), that if the doubling time is one decade, the consumption in any decade is equal to the total of all previous consumption. The area of the rectangle ABDC represents the known world petroleum resource.

If the historic growth rate of about 7 per cent per year continued, our consumption in the 1970s, 1980s, and 1990s would be as shown. The rectangle ABCD represents the ultimate total production. This is in accord with Table 3-8, which shows a T_e of about 30 years from 1972 for an annual growth rate of 7 per cent. The rectangle CDFE represents the quantity of oil that we must discover if we would have the annual growth rate of 7 per cent continue for just one extra decade!

U.S. Coal

Table 3-9 shows data on U.S. coal, and Table 3-10 shows the lifetime of these reserves for various rates of steady growth. The figures are based on a 1974 production rate (r_o) of 0.6×10^9 metric tons per year, and they tell how long U.S. coal will last if coal pro-

duction grows at various rates from this rate. In parentheses beside each figure is the T_e, which shows how long coal could "satisfy present U.S. energy needs." Since coal supplies only about one fifth of present U.S. energy needs, the figures in parentheses are based on an r_o of $5 \times 0.6 \times 10^9$ metric tons per year. We should note from Table 3-10 that the "Reserve Base" is the conservative and well-documented estimate of U.S coal reserves. The quantities added in the columns labeled "Total Identified Resources" and "Total Including Hypothetical" are to some degree speculative and uncertain, and these added quantities will be difficult and costly to use. It is unlikely that they could all be made available for use.

We can see from Table 3-10 that:

If we want to have steady growth of U.S. coal production for 100 years (from 1975),

Table 3–9.   U.S. Coal Data. Resources *Remaining* as of January 1, 1974.

	10^9 Metric Tons
1) Identified Resources	
A) Reserve base	379
B) Additional	1167
C) Total identified resources	1546
2) Hypothetical Resources	
A) 0–914 meters overburden	1651
B) 914–1829 meters overburden	346
C) Total hypothetical resources	1997
3) Total Remaining Resources	
(including hypothetical resources)	3543
4) Produced Through January 1, 1974	45
5) Annual Production Rate 1974	0.6
6) Annual Rate of Export of Coal 1974	0.06
From these data we can calculate:	
7) A) Based on the reserve base of 1A per cent of ultimate	
total production	10.6
B) U.S. "coal time" is 3 minutes 14 seconds before noon	
8) Based on the total identified resources of 1C	
A) Per cent of ultimate total production produced to 1974	2.82
B) U.S. "coal time" is 5 minutes 8 seconds before noon	
9) Based on the total remaining resources	
(including hypothetical resources)	
of 3	1.25
A) U.S. "coal time" is 6 minutes 19 seconds before noon	

Items 1, 2, and 3 are taken from Averitt.[2]
Item 4 is author's estimate based on Hubbert's graph (reproduced here as Figure 3-4).

the growth rate should not exceed 3 per cent per year.

If we want U.S. coal production to grow steadily for two more American centennials, the growth rate should not exceed 1 per cent per year.

Coal could supply total present U.S. energy needs (i.e., zero growth in coal production) for 126 years.

U.S. coal production grew steadily by 6.69 per cent per year from the close of the American Civil War to about 1910. If we had returned to this growth rate in 1975, U.S. coal would last 56 years.

President Gerald Ford set U.S. coal production goals (2.5×10^9 metric tons/year by 1985) that called for an annual growth rate of 11 per cent. If this could continue until

the reserve base was used up, U.S. coal would last for 39 years (from 1975).

The Carter administration set coal production goals (1×10^9 metric tons/year by 1985) that called for an annual growth rate of production of 5 per cent per year. If this could continue, the reserve base would last 70 years.

Figure 3-4 shows the history of U.S. coal production. Note the steady growth in coal production of 6.69 per cent per year from 1870 to 1910. There then followed 60 years of approximately zero growth in U.S. coal production. If we had continued the historic annual growth rate of 6.69 per cent in the years after 1910,

1. The reserve base would have been used up by the year 1966.

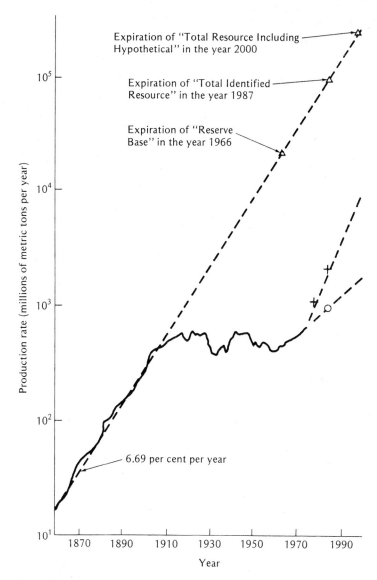

Figure 3-4. History of U.S. coal production. This figure is redrawn from Hubbert.[1] From the close of the American Civil War to about the year 1910, coal production grew steadily at a rate that averaged 6.69 per cent per year. Production then leveled off (with wide fluctuations) for 60 years and has now started to climb again. The crosses in the steep dashed curve at the right show the coal production goals set by the administration of President Gerald Ford while the circle in the lower dashed line represents the production goals of the Carter administration. The long dotted straight line shows what would have happened if the growth of consumption had not stopped or slowed in 1910. The three triangles mark the expiration times (T_e) of Averitt's[2] three estimates of the size of U.S. coal resources. *Coal is an energy option for the United States in 1980 only because coal production had zero growth for 60 years following 1910!*

Table 3–10.   Exponential Expiration Time of Three Different Estimates of U.S. Coal (from Table 3–9).

Annual Growth (%)	Reserve Base (379×10^9 metric tons)	Total Identified Resources (1546×10^9 metric tons)	Total Including Hypothetical (3543×10^9 metric tons)
Zero	632 (126)	2577 (515)	5905 (1181)
1	199 (82)	329 (182)	410 (255)
2	131 (63)	198 (121)	239 (160)
3	100 (52)	145 (93)	173 (120)
4	82 (45)	116 (77)	137 (97)
5	70 (40)	97 (66)	114 (82)
6	61 (36)	84 (58)	98 (71)
6.69	56 (34)	77 (53)	89 (66)
7	54 (33)	74 (52)	86 (63)
8	49 (30)	67 (47)	77 (57)
9	45 (28)	61 (43)	70 (52)
10	42 (26)	56 (40)	64 (48)
11	39 (25)	51 (37)	59 (44)
12	36 (23)	48 (35)	55 (41)
13	34 (22)	45 (32)	51 (39)
14	32 (21)	42 (31)	48 (37)
15	30 (20)	40 (29)	45 (35)

Note: The right-hand column, based on hypothetical estimates of U.S. coal, should be used only for purposes of illustration and should not be used as the basis for planning.

2. The total identified resource would be used up by the year 1987.

These facts make it clear that *the only reason coal is an energy option in the United States in the 1980s is that we enjoyed 60 years of zero growth of coal production in this country* (1910–1970).

MISREPRESENTATIONS

The most confusing thing about the energy situation is the widespread misrepresentations that are offered by many who speak and write on the subject.

In a three-hour national television special (CBS) on energy (August 31, 1977), a reporter said:

Most experts believe that oil shale deposits like those near Rifle, Colorado, could provide a 100-year supply.

This optimistically imprecise statement does not tell us whether the oil from U.S. shale could supply our U.S. oil imports, domestic production or consumption, or some fraction of them "for more than 100 years." Nor does it tell us what rate of growth of production of oil from shale is used in making the estimate. The comparison of this statement with the results of the simple arithmetic of Table 3-6 will introduce the reader to the disturbing divergence between the facts and the authoritative statements that are so common. An ad circulated nationally by a major oil company proclaimed, in *bold type,*

Vast oil potential in the U.S. untapped and unavailable.

In fine print, the ad said,

The U.S. may still have as much unproduced and undiscovered oil as has been produced in America's entire history.

The ad was undoubtedly intended to be optimistic. (We probably have as much remaining oil as all we have used in all of history!) It is sobering to realize that this means that we have already used half of the recoverable petroleum that was ever in the ground in the United States. The U.S. petroleum time is one minute before noon!

Prominent people repeat the statement "We have as much oil remaining as all we have used in all of history . . ." without showing any recognition of its full meaning.

The "energy" crisis has always been more political than real. . . . The U.S. Geological Survey says there is in the United States as much oil waiting to be discovered as has been discovered until now.[3]

Coal is the largest fossil-fuel reserve in the United States. Let us look in detail at several misrepresentations of the facts about U.S. coal.

In the mid 1970s, an ad from a major electric power company assured us not to worry too much about the Arab oil embargo, saying that

We're sitting on half of the world's known supply of coal, enough for over 500 years.

Where did this "500 years" statement come from? It may have had its origin in a report from the Committee on Interior and Insular Affairs of the U.S. Senate,

At current levels of output and recovery, these [American] coal reserves can be expected to last more than 500 years.

Here is one of the most dangerous statements in the literature. It is dangerous because it is true. But it is not the truth that makes it dangerous. The danger lies in the fact that people take the sentence apart. They just tell us that "U.S. coal will last 500 years" and they forget the caveat with which the sentence started: "At current levels of output and recovery. . . ."

The caveat means "zero growth of U.S. coal production." Table 3-10 shows that the reserve base is indeed adequate to last "over 500 years—at current levels of production." It is irresponsible to cite this figure as a realistic estimate of the lifetime of U.S coal when recent presidents of the United States have advocated growth rates of U.S coal

ranging from 5 to 11 per cent per year. For example, *Newsweek* magazine, in a cover story on energy, said that "at current rates of consumption we have enough coal to last 666 years." The article explained in detail the progress being made in achieving rapid growth in U.S. coal production without telling the readers that if we have rapid growth in production, U.S. coal will not last anywhere near 666 years.

Time magazine reported that "to achieve some form of energy self-sufficiency, the U.S. must mine all of the coal it can." We can paraphrase this: the more rapidly we consume our resources, the more self-sufficient we will be. David Brower, former president of the Sierra Club, has referred to this as the policy of strength through exhaustion. A booklet by the Energy Fuels Corporation[4] tells us:

As reported by *Forbes* magazine, the United States holds 437 billion tons of known [coal] reserves. That is equivalent to 1.8 trillion barrels of oil in British thermal units, or *enough energy to keep 100 million large electric generating plants giving for the next 800 years or so.* [emphasis added]

This is an accurate quotation from *Forbes*, the business magazine.[5] Long division is all that is needed to show that 437×10^9 tons of coal would supply our 1976 production of 0.6×10^9 tons per year for only 728 years, and we have fewer than 500 large electric generating plants in the United States today.

The *Journal of Chemical Education* said,

Our proved coal reserves are enormous (at least 120 billion recoverable tons); these could satisfy present U.S. energy needs for nearly a thousand years.[6]

Comparing their statement with the facts, we note:

1. The quantity of coal they cite is only about one third of the reserve base.
2. If the correct figure for the reserve base had been used, it could supply "present U.S. energy needs" for only about 126 years.

The *Journal* published a "correction" in which they state that reserves of U.S. coal are estimated to be 1.58×10^{15} tons.[7] This is very nearly 1000 times larger than the total identified resources quoted in Table 3-9.[8]

Time magazine (April 17, 1978, p. 74) said:

Beneath the pitheads of Appalachia and the Ohio Valley and under the sprawling strip mines of the west lie coal seams rich enough to supply the country's power needs for centuries, *no matter how much energy consumption may grow*. [emphasis added]

This is but a small sample of the misrepresentations that characterize much of what is written about the energy situation. These examples lead us to recognize the validity of the following law:

Do not believe any prediction of the life expectancy of a nonrenewable resource until you have confirmed the prediction by repeating the calculation.

As a corollary let us note that:

The more optimistic the prediction, the greater the probability that it is based on faulty arithmetic or on no arithmetic at all.

RECENT DEVELOPMENTS IN U.S. COAL

There has been an alarming trend in recent developments with U.S. coal. In his speech accepting the Democratic nomination, President Carter said that the United States should increase its coal exports "so we can become the Saudi Arabia of coal." A news story (Denver *Post*, January 20, 1981) reported that:

A government task force predicts a rosy future for U.S. coal exports, saying this country will be supplying 38 per cent of the world demand by the end of the century.

The capacity to load another 23 million tons of coal annually is already being built at the nation's ports and there are plans for facilities for another 160 million tons. . . .

The task force of 60 officials from the Energy Department and various other federal agencies was appointed six months ago by [President] Carter to recommend government actions to spur coal exports.

National Geographic devoted an entire issue to energy (February 1981). A photograph (pp. 19a, 19b) carried this caption:

The Persian Gulf of Coal? Virginia's Hampton Roads may rival Mideast oil ports as an energy depot by the end of this century. But the United States must first spend hundreds of millions of dollars improving port facilities to realize its potential as a coal exporter. On this moonlit night . . . some 50 foreign freighters lie at anchor off Cape Henry, their holds hungry for U.S. coal. . . .

The United States seems eager to spend enormous sums of money in order to expedite our export of coal. Throughout much of the first half of this century, the United States was a major exporter of petroleum and crude oil. We now wish we had some of that oil. It appears that we are now well along toward repeating the process for U.S. coal. Coal is the last large reserve of fossil fuel in the United States. The facts about the reserves of coal are given in Tables 3-9 and 3-10. The massive export program should be reevaluated in the light of our own long-term need for energy.

AN ABSOLUTE LIMIT TO GROWTH

If the earth were a spherical tank of oil, the oil would last 4×10^{11} years at the 1970 rate of consumption and only 342 years at the 7.04 per cent annual growth rate of world petroleum production that was maintained from 1890 to 1970.

GROWTH PROBLEMS IN OTHER AREAS

The daily news brings a continuous spectrum of stories of growing populations, of food shortages and famine, of inflation and rising costs, of shortages of firewood and fossil fuels, of shortages of strategic minerals, of shortages of water and even of clean air to breathe. These problems are all brought about by the growth of populations and of our per capita demands for more things. Often these shortages are related to one another. The world situation in regard to population, energy, farming, and food was put into perspective by David Pimental:

As a result of overpopulation and resource limitations, the world is fast losing its capacity to feed itself. More alarming is the fact that while the world population doubled its numbers in about 30 years, the world doubled its energy consumption within the past decade. Moreover, the use of energy in food production has been increasing faster than its use in many other sectors of the economy.[9]

It has been noted that modern agriculture is the use of land to convert petroleum into food. It has also been noted that the greatest shortcoming of the human race is our inability to understand the exponential function (i.e., the arithmetic of growth).

It is clear that steady growth as pictured in Figure 3-1 cannot continue. Hubbert has shown that production graphs rise and fall in a symmetrical, bell-shaped curve. The next step in understanding the energy problem is to study Hubbert's paper, which is discussed in Article 8 in this volume. In studying this paper, realize that when the time for a particular resource is one minute before noon we are at the peak of the production curve for that resource and that the rate of production can only go down in the future.

WHAT CAN WE DO? A SUGGESTED SOLUTION

In the words of Winston Churchill, ''Sometimes it is necessary to do what is required.'' We must educate all of our people to an understanding of the arithmetic and consequences of growth, especially in terms of populations and the earth's finite resources. We must remember the words of Aldous Huxley: ''Facts do not cease to exist because they are ignored.''

An Exponential Solution

When a quantity such as the rate $r(t)$ of consumption of a resource grows a fixed per cent per year, the growth is exponential:

$$r(t) = r_0 e^{kt} = r_0 2^{t/T_2} \qquad (3)$$

where r_0 is the current rate of consumption at $t = 0$, e is the base of natural logarithms, k is the fractional growth per year, and t is the time in years. The growing quantity will increase to twice its initial size in the doubling time T_2, where

$$T_2 = (\ln 2)/k \approx 70/P \qquad (4)$$

and where P, the per cent growth per year, is $100k$. The total consumption of a resource between the present $(t = 0)$ and a future time T is

$$C = \int_0^T r(t)\, dt \qquad (5)$$

The consumption in a steady period of growth is

$$C = r_0 \int_0^T e^{kt}\, dt$$
$$= (r_0/k)(e^{kT} - 1) \qquad (6)$$

If the known size of the resource is R, then we can determine the exponential expiration time (T_e) by finding the time T_e at which the total consumption C is equal to R:

$$R = (r_0/k)(e^{kT_e} - 1) \qquad (7)$$

We may solve this for the exponential expiration time:

$$T_e = (1/k)\ln[(kR/r_0) + 1] \qquad (8)$$

This equation is valid for all positive values of k and for those negative values of k for which the argument of the logarithm is positive.

The exponential function provides an interesting ''solution'' to the energy and resource problems. If we make the rate of production of a resource follow the curve

$$r(t) = r_o e^{-(r_o/R)t} \qquad (9)$$

then the total production from now to infinity is exactly equal to the size R of the resource. *We can use the resource and have it last forever!* This is the ultimate self-sufficiency. For the large estimate of U.S. coal, the production rate would have to *decrease* approximately 3 per cent per century in order to assure that U.S. coal would last forever. The difficulty our nation would face with this declining rate of production can be expressed in the question "Could we live one year from now with coal production at the rate it was a day and a half ago?" This program can be applied to any resource. If r_o/R for world petroleum is $1/101$, then the decline in petroleum consumption would have to be about 1 per cent per year to make petroleum last forever.

The greatest act of responsibility we could perform for our descendants for all time would be to put our consumption of coal and other resources on this declining curve. Not only is it proper to save some resources for those who will follow us, but it may make the difference in national self-sufficiency and, ultimately, of national survival.

NOTES AND REFERENCES

1. M. K. Hubbert, U.S. Energy Resources, A Review as of 1972. Background paper prepared at the request of Senator Henry M. Jackson, Chairman, Senate Committee on Interior and Insular Affairs. A National Fuels and Energy Policy Study. Serial 93-40 (92-75) Part 1, Washington, D.C.: U.S. Government Printing Office. This document is an invaluable source of data on consumption rates and trends in consumption for both the United States and the world. In it, Hubbert also sets forth the simple calculus of his methods of analysis. He does not confine his attention solely to exponential growth and predicts that the rate of rise and subsequent fall

of consumption of a resource will follow a symmetrical curve that looks like the normal error curve. Several figures in this article are redrawn from Hubbert's paper.

2. P. Averitt, "Coal Resources of the United States," January 1, 1974. U.S. Geological Survey Bulletin 1412, 1975.

3. W. Armstrong, *Rocky Mountain Journal* **30** (49), p. 1 of Sec. 2 (1979).

4. "Energy and Economic Independence," p. 3. Denver: Energy Fuels Corporation, 1976.

5. *Forbes,* December 15, 1975, p. 28.

6. Editorial Staff, "Energy Review," *Journal of Chemical Education* **55** (4), 263 (1978). Chem I Supplement for Secondary School Chemistry.

7. *Journal of Chemical Education* **56** (3), 188 (1979). See also A. A. Bartlett, *Journal of Chemical Education* **59** (6), 501 (1981).

8. The *Journal* attributed this large figure to the U.S. Bureau of Mines Information Circular 8531, *Strippable Reserves of Bituminous Coal and Lignite in the U.S.* The author is unable to find the quoted figure in this publication.

9. David Pimentel, "Fossil Energy and Food Supplies," *Bulletin of the American Physical Society* **22** (1), 20 (1977).

SUGGESTED READINGS

Albert A. Bartlett, "The Exponential Function" series in *The Physics Teacher.* (a) **14,** 393 (1976); (b) **14,** 485 (1976); (c) **15,** 37 (1977); (d) **15,** 98 (1977); (e) **15,** 225 (1977); (f) **16,** 23 (1978); (g) **16,** 92 (1978); (h) **16,** 158 (1978).

A. A. Bartlett, *Bulletin of the American Physical Society* **2** (21), 42 (1976).

D. Brower, *Not Man Apart,* Vol. 6 (20). San Francisco: Friends of the Earth, 1976.

M. K. Hubbert, in *Committee on Resources and Man,* National Academy of Sciences–National Research Council. San Francisco: W. H. Freeman, 1969.

M. K. Hubbert, *Canadian Mining and Metallurgical Bulletin* **66** (735), 37 (1973).

D. H. Meadows, D. L. Meadows, J. Randers, and W. W. Behrens, *The Limits to Growth,* New York: New American Library, 205 p., 1972.

U.S. Bureau of Mines, "Strippable Reserves of Bituminous Coal and Lignite in the United States." U.S. Bureau of Mines Information Circular 8531, 1971.

Economic Analysis of Energy Policy Issues 4

HENRY STEELE

INTRODUCTION

In October 1973 the Arab members of the
Organization of Petroleum Exporting Countries (OPEC) imposed an embargo on shipments of oil to countries they regarded as their
enemies. They reduced shipments of oil to
other countries by 25 per cent and indicated
a further reduction of 5 per cent per month
until Arab political objectives were achieved.
The embargo thus reduced supplies and
stimulated panic buying, which increased
demand and resulted in frantic bidding for
scarce supplies of exported oil. As a result
of high price bids during the embargo period, OPEC raised the posted price of its reference crude oil (Saudi Arabian light oil) from
$3.01 per barrel in September 1973 to $11.65
as of January 1, 1974. Later price hikes by
OPEC raised prices to about $14 per barrel
as of late 1978, but supply interruptions during the ensuing Iranian revolution created an
environment of opportunistic pricing that resulted in a further doubling of prices during
1979, to an average of close to $30 per bar-
rel. The Iran–Iraq conflict of 1980 created

additional supply uncertainties which fos-
tered further price increases so that average
OPEC prices, after peaking at $40 per barrel
in early 1981, have recently stood in the $30–
35 per barrel range.

Such enormous price hikes confronted the
oil-importing nations with unprecedented
economic challenges that many of them were
unable to cope with. The economically
weaker oil importers, primarily in the Third
World, were bankrupted by the first price
hike, although the willingness of certain
lenders, including some OPEC members, to
accumulate their bad debts in lieu of cash
payments for oil has tended to conceal the
fact of their bankruptcy. Economic growth
virtually ceased in those parts of the world
where such growth was crucially dependent
upon continued low energy prices. Growth
elsewhere in the world has been seriously reduced since cheap energy has been one of the
most important factors responsible for economic growth since the beginning of the industrial revolution. With energy no longer
cheap, there is no assurance of continued
economic growth. The initial quadrupling of

Dr. Henry Steele is Professor of Economics at the University of Houston,, Houston, Texas 77004.

This paper was specifically prepared for *Perspectives on Energy* and is based on the analysis and policy
recommendations contained in *Energy Economics and Policy* (New York: Academic Press, 1980), which
was written by Professor Steele and his colleague Dr. James M. Griffin.

the price of oil raised the market value of every source of energy that could be substituted for crude oil and its products and therefore cut the economic value of every energy-intensive item of fuel-using equipment in the world, whether a capital good or a durable consumer good.

OPEC price hikes have become a tremendous engine of world price inflation with the consequence that the stability of the world economy and the world monetary system is being severely strained. As long as OPEC members will accept paper money for oil exports, and as long as the continued existence of the governments of oil-importing countries is dependent upon large volumes of oil imports, the money will be printed or otherwise brought into being. This fuels inflationary price spirals from the money supply side. But as long as investors can foresee continuing price inflation, they will hold back on investments (particularly long-term investments) for fear that future price controls or low demand will not permit them to recover their investments after adjustment has been made for the impact of inflation on the lowering of the purchasing power of future income streams. Hence the actions of OPEC members raise the money supply and, by discouraging investment, reduce the supply of goods and services. This contributes to inflation in the most direct way by causing more paper money to be used in bidding for the purchase of relatively fewer actual goods and services.

It is not an exaggeration to state that OPEC members have, perhaps in spite of themselves, engaged in de facto economic warfare against the oil importers because since 1974 they may have priced oil above the monopoly price that would bring them maximum profits in the long run. While the logic of monopoly is to maximize the benefits to the seller, this cannot be done by destabilization of the market to such an extent that demand is undermined. But the logic of warfare is to inflict more injury than you sustain, sacrificing if need be some economic advantages for the sake of furthering extra-economic goals. With oil politically identified as the ''life blood'' of the OPEC members, exporters cannot limit themselves to charging oil prices consistent with equating supply and demand relationships at competitive levels. Thus the problem the oil importers face is quite grim: if the cartel cannot be overthrown, the importers must either find ways to reduce their dependence upon it or submit to economic suffocation at its hands.

What forces might overthrow OPEC? It is very unlikely that the cartel can be destabilized at this late date by joint importer actions, although something might be done along these lines with adequate cooperation. In the past, however, very strong cartels have been overcome only by forces with very strong economic repercussions, notably wars and world depression—and OPEC is the strongest cartel the world has ever seen.

Another disastrous impact of OPEC is its contribution toward increasing world political and military tensions. The most important OPEC members are in the Persian Gulf region, one of the most politically unstable regions in the world. As long as militarily weak regimes were not particularly wealthy, the costs of undermining them as seen by radical groups might not be worth the gains of conquest. But now that their oil-exporting revenues have enriched OPEC members' treasuries without corespondingly increasing their defense postures, the scales may be tipping toward military adventurism. Furthermore, some OPEC members have diverted some of their added oil revenues into financing terrorist activities, which adds dangerously to preexisting tensions.

DIFFERING PROFESSIONAL PERSPECTIVES ON ENERGY POLICY

It is therefore obvious that the problems that confront us in the sphere of energy policy have political and military as well as economic dimensions. Some problems are primarily economic, such as determination of the effect of different price levels on the consumption of energy. Some problems are primarily political, such as estimates of the range of impacts of a particular energy policy on the distribution of income of various income groups in the society. Some problems are primarily military, such as assessment of the impact of the national security posture of a certain degree of dependence upon oil imports from insecure sources. The virtue of economic analysis, however, is that its methods can be employed to estimate the costs and benefits of any proposed program and to identify (within the limits of data availability) the point at which costs begin to exceed benefits, such that further persistence in the program would involve net economic losses.

Economists naturally see energy issues primarily in terms of economic forces and regard their solutions as dependent upon the supplying of proper economic incentives and disincentives such as those provided by a freely working market price system. Politicians no less naturally view energy matters primarily in political terms, favoring solutions involving direct governmental controls, with considerations of economic costs being somewhat secondary. Military spokesmen are concerned with the security of energy supplies and, not surprisingly, are inclined to think in terms of military strategies involving the achievement of credible defense and deterrence postures and their employment in the prevention of such contingencies as excessive use of oil-embargo powers by exporters. Other participants in the energy policy debates—technologists, scientists, environmentalists, ecologists, business people, conservationists, attorneys, and others—tend to think in terms of their own professional specialties providing the best basis for solutions of energy policy problems. To the economist, however, the fundamental distinction between the program preferences of different participants appears to be between market-financed versus budget-financed activities, between resource allocation via an impersonal price mechanism versus resource allocation via direct controls. From the standpoint of economic efficiency, the former

approach is generally superior, but the latter approach has advantages in some contexts.

The structural conditions required to permit perfect competition in a given market include not only the absence of external constraints of a political nature but also the presence of conditions conducive to competitive supply and demand. Competitive supply requires the presence of a large number of small suppliers, none of which is dominant over any of the others; a homogeneous product; the absence of effective collusions between sellers; and good market information regarding technology, costs, and prices. Competitive demand requires a large number of small independent buyers, absence of collusion, and the provision of buyers with good information about the market. Under circumstances where these conditions prevail, the operation of markets will be efficient. The market price will equate the incremental private benefit to buyers with the incremental private cost to sellers, such that no more is produced than buyers are willing to purchase at the existing market price and no more is demanded than sellers are willing to produce at that price. This is the implication of supply being equated with demand at the equilibrium price in the market.

Demand is defined as a functional relationship between price charged and quantity purchased. A demand function is basically a schedule of maximum demand price that buyers will pay to purchase a given total quantity of some commodity. Buyers are arrayed from the most eager purchasers (those with very strong preferences for the good in question, and with high enough incomes to make those preferences effective through purchases at high prices) down through the least involved potential purchasers (those who have less-intense preferences for the good and/or who lacks the income to finance purchases except at low prices). This arraying is cumulative, so that as price steadily falls, the quantity demanded steadily increases and buyers who were already willing to make some purchases at higher prices now purchase even more at lower prices, while buyers who were not in the market at higher prices now make purchases at a new lower price.

A supply function is basically a schedule of the minimum supply price that producers require in order to incur the costs of production and produce a given total quantity of some commodity. Suppliers are arrayed from the lowest-cost producers (those who are very efficient at producing or who own resources that are very well adapted to producing the good in question) up through the highest-cost producers (those who are relatively inefficient in organizing production or who possess resources not well adapted to producing the particular good). This array is also cumulative, so that as price steadily increases, the quantity supplied steadily rises as suppliers who were already able to produce a certain output at lower prices now produce even more at higher prices, while producers who were not efficient enough to enter the market at lower prices now produce at least a small output at a new higher price.

Thus rising prices encourage productivity but discourage consumption, and an equilibrium is achieved where the market price is just high enough to equate the maximum demand price of the least-eager actual purchaser with the minimum supply price of the highest-cost actual producer. Since all buyers pay the same price for the good in a competitive market, but since the maximum demand price of all the buyers except the least-eager buyer, who is right at the margin, is above the actual price paid, each supramarginal buyer obtains what is called a "consumer surplus" from being able to buy at a price below his maximum demand price. Similarly, since all sellers receive the same price for selling the good, while the minimum supply price of all sellers except the highest-cost seller, who is right at the margin, is below the actual price paid, each supramarginal seller obtains what is called a "producer surplus" from being able to sell his output at a price above his minimum supply price, which can be equated with his minimum full-production cost. "Producer

surplus" can be understood more easily than "consumer surplus" since it is essentially the same thing as the producer's profit, over and above all his production costs. The chief advantage of fully competitive markets is that they maximize the sum of producer and consumer surplus, or what in this simple context is called "economic welfare." Thus, subject to many qualifications, we can say that perfect competition maximizes economic welfare. (It should be noted that it does not maximize producer surplus and consumer surplus individually. In general, as price falls, consumer surplus increases while producer surplus falls. Pure monopoly maximizes producer surplus, whereas the maximization of consumer surplus would usually require special subsidies to sellers.)

We can identify for any given output rate, a maximum demand price, which would be paid by buyers if only that particular amount were to be made available on the market, and a minimum supply price, which would have to be paid if that particular amount was all that the market was to produce. Thus for any given output rate we select, we can speak of the maximum demand price as measuring the marginal private benefit from consuming that output rate, while the minimum supply price measures the marginal private cost of producing that output rate. The equilibrium output is where the two functions intersect, so marginal private benefit equals marginal private cost at the equilibrium price.

Economists would ordinarily be critical of political devices that substitute the often arbitrary outcomes of political decision-making processes for the impersonal and informed collective judgment of all the individuals participating in the market. If political controls decree too low a price, output will be inadequate since at a relatively low price more will be demanded than will be supplied. Since more cannot be consumed than is produced, there will be excess demand at the regulated price. At a relatively low output level, marginal private benefit will exceed marginal private cost, such that buyers will be willing to pay more than the controlled price for the product. There will be a tendency for a so-called black market to develop, where those who buy at the controlled lower price can resell to others at the higher market price consistent with short supplies. In order to prevent this from happening, the authorities have to work out some sort of system of quantity rationing to substitute for the price-rationing function of the market even though quantity-rationing systems are more costly and inefficient. If it is desired to restrict the output of some commodity, such as a pollution-producing energy source, it is better to increase its price by placing a tax upon it such that price-induced conservation will result from the operations of the market as buyers react to the higher price inclusive of the tax by reducing their purchases.

Conversely, if policy makers wish to increase the output level of some good, such as a new and nonpolluting source of energy, it is impossible to simply decree a high output since at high production levels marginal private cost may exceed marginal private benefit, and consumers will not buy all that the suppliers can produce. What is required in this instance is the payment of subsidies to producers, which will allow them to reduce the price of the desired output rate to the level at which price equals marginal private benefit. The unit subsidy required is the amount that makes up the difference between the higher production cost and the lower demand price. Thus we see that by using taxes or subsidies, markets can bring about the desired degree of resource reallocation through the impact on prices.

It must be recognized, however, that the situation now becomes more complicated. If taxes are used, only a part of the proceeds from the sale of the goods goes to the suppliers, while the rest goes to the taxing authorities. The disposition of the tax receipts must also be considered. These receipts can be used in various ways, such as financing of programs to abate pollution or develop new energy resources, to compensate some income groups for the impact of the tax on their purchasing power, or in other ways that may either help or hamper the achievement of the goals of the basic program. In the case of a subsidy, only part of the receipts to producers come from consumers; the rest comes from the groups paying the subsidy. If taxpayers as a group pay the subsidy, then the consumers of the product are being subsidized by taxpayers. If the product consumed is widely purchased in proportion to buyer income, and if the subsidy is financed by income tax, then there may be little net subsidy to buyers; if not, certain buyer groups may be highly subsidized and questions of equity arise. It is the task of politicians to resolve conflicts involving social perceptions of equity, although economists can suggest the best measures by which the distribution of income might be compensatorily altered through the various devices available to the government.

ECONOMIC ANALYSIS OF ENERGY POLICY PROBLEM AREAS

Let us now turn to specific aspects of the economic analysis of the major energy policy problem areas. Each area will be discussed individually. First the economic goals will be described and compared with political and other goals sought within the confines of each policy area. Second, the economic measures that might be used to achieve the goals described earlier will be analyzed and compared with other measures that might be used.

Conservation Policy

We begin with conservation policy, since it is here that the conflict between price and nonprice policy methods is sharpest and clearest. Let us begin by assuming no elements of market failure to be present. If an economist were asked to draw up an optimal conservation policy for nonrenewable energy resources, he would begin by requiring that the purchase and sale of the resources take place to a perfectly competitive market and would then be very likely to advocate simply that there be no regulation of the prices at which these energy resources were bought and sold in this competitive market. His argument would be that, under competition, existing buyers will bid for resources in accordance with their marginal private benefit realized from consumption, while sellers will hold out for prices that cover not only marginal private production costs but also so-called user costs of selling now rather than waiting and selling in a future period when market conditions may be more advantageous. It is the seller, therefore, who provides the effective link between present and future supply. He must make forecasts of present versus future supply of the resource, of demand for the resource, of marginal costs of production, of interest rates, and of other relevant variables. If supply is currently scarce but he foresees new discoveries in the future, he will be inclined to sell more oil now while the market is inclined to place a high scarcity premium upon it. By selling more now, he reduces current prices through increasing current supply and thus tolerates lower prices today since he anticipates possibly still lower prices tomorrow.

Conversely, if he regards supplies as being shorter tomorrow than today, he will sell less today, hoping for a higher price tomorrow. But selling less today will raise today's price. If less is sold today, more will be available tomorrow, lowering tomorrow's price. Hence well-informed competitive speculation tends to reduce price fluctuations over time. If uninformed but pessimistic authorities require that less oil be sold today and more saved for tomorrow because they fear a greater shortage tomorrow, today's prices will rise as a result of reduced availability of marketable supplies. If, however, new supplies are unexpectedly developed tomorrow, then tomorrow's prices will be still lower because, by the artificial curtailment of today's supplies and raising of today's prices, tomorrow's supplies will be artificially augmented and tomorrow's price arbitrarily depressed.

We can further elaborate upon this analysis by considering the impact of accurate forecasts for other of the key variables. If sellers predict higher demand for oil today and lower demand tomorrow, as a result, example, of the development of substitute energy sources tomorrow, then they will tend to sell relatively more oil today and save less for tomorrow, depressing today's prices with added supplies in the expectation that tomorrow's price will be low anyhow because of declining oil demand. Conversely, if they expect higher demand tomorrow, they will sell less oil today and raise today's price, selling more oil tomorrow and preventing tomorrow's price from rising as much as it would be if more oil had been sold today. Thus the operation of a competitive market tends to "save resources for the future" to the extent that higher demand is foreseen in the future, other things being equal.

Objections to this approach are made by those who submit that depletable resources should be regarded as the common property of the present and all future generations jointly. Obviously, however, only the current generation is involved in present market transactions, which tend—through bringing about current consumption—to reduce the total remaining for the future. This is certainly true, but this observation in itself gives no clue as to a more desirable policy. A simple-minded approach might be to estimate the total number of future generations of human life the earth can support, estimate the total recoverable reserves in the resource base, divide the latter by the former, and establish a quota for the consumption of each generation. If desired, the population in each future generation could be estimated and a constant per-capita consumption quota for every individual in each generation could be assigned. A moment's reflection should raise the suspicion that any reasonable forecast of the probable number of future consumers would reduce the per-capita entitlement to any finite depletable resource to a virtually negligible amount. Thus it could be argued that the cost of the nonuse to a limited number of generations in the immediate future might outweigh the minute benefits to a nearly infinite number of future generations. This is particularly true if the next few generations succeed in using today's depletable resources as an input in the devising of a technology that will produce tomorrow's products from undepletable resources.

From a broader viewpoint, future generations are beneficiaries, not victims, of the current generation to the extent that we pass on to them a more flexible technology, a more comprehensive educational system, and a more productive capital stock, even though we reduce the stock of certain depletable resources. Although many among the large number of individual resource owners may lack adequate knowledge on which to forecast the future, the mistakes of some may be such as to offset the mistakes of others. If decisions on natural resource use policy are centralized in a fallible agency, we may save resources for future generations which will have no use for them as a result of the advance of technology.

The true role of public policy in the conservation area is to correct for any market failures that may exist and then provide reliable market information to guide the decisions of individual resource owners and the buyers of their resources and products. Chief among the externalities that must be dealt with is monopoly and oligopoly power. Any monopolies or cartels that prevent the competitive buying or selling of any energy resource or of its products should be done away with and competition established. If it is determined that the provision of information regarding future trends can be done more readily in the public than in the private sector, then the provision of such information has positive externalities and market participants might be subjected to a small tax to support public collection and dissemination of data useful in establishing expectations regarding future reserves and discoveries, future market conditions, future interest rates, future technology and production costs, and so on.

It is by no means certain, however, that public sector activities could invariably provide information superior to that available to individual market participants, although they might be able to disseminate it more widely. (Such dissemination, however, would not be an advantage if the information proved to be incorrect.) Public investment in studies of the total resource base might be quite valuable, possibly less valuable but still helpful in projecting future technology, of some value in projecting interest rates, but perhaps of little value in detailed forecasting of future market conditions since these are matters about which individual market participants as a group are likely to have better information. Thus from the economist's standpoint the decisions made by a large group of competitive buyers and sellers who have been provided with reliable market data projections would provide the best basis for maximization of the present discounted value of depletable energy resources and thus for achievement of conservation in the economic sense.

When we turn to the current debate on conservation policy, we realize that we are unfortunately in an entirely different world. While domestic energy resources in the United States appear to be held sufficiently widely that monopoly power in their private use is not particularly likely, we find that until early 1981 owners of "old oil" under price controls were compelled to sell their reserves at very low prices relative to the world market prices set by OPEC. This underpricing of these reserves led to an excess supply of low-cost oil on the U.S. market, hence to an excess production of refined products from low-cost domestic petroleum and to an overconsumption of these products by buyers. Since price controls on refined products were also acting to keep prices to buyers lower than the true cost of energy under a world regime controlled by OPEC crude oil pricing, there was an additional bias in the regulatory program toward keeping prices low, and thus, by encouraging excessive consumption, the controls were working against price-induced conservation. We consistently find the same phenomena wherever primarily political controls have determined energy prices in the oil-importing countries in the post-embargo era. Since politicians are under pressure from their constituents to minimize energy prices, and since these politicians have no control over OPEC prices, they tend to underprice all the sources of domestic energy over which they do have some control. But this leads to lower domestic production and higher domestic consumption and simply increases the nation's demand for OPEC oil and thus fosters OPEC'S ability to further raise its prices.

Economists would prefer that conservation be achieved by prices being allowed to increase to market-clearing levels. If for some reason it is desired to raise prices above competitive levels, which would be super-induced conservation, fuel taxes would be an efficient method. If it is desired to increase the energy efficiency of fuel-using equipment, various types of taxes, individually or in combination, could be effective in different ways. A tax on automobiles as such, for example, would decrease the number of cars on the road to some extent, but would in itself not be conducive to greater fuel efficiency or to reduced utilization per car. Per-car utilization might very well increase with a reduction in the number of automobiles. A tax on automobiles that varies with fuel efficiency would increase the average fuel efficiency of cars and might lower the total stock of automobiles but would have no clear impact on utilization.

Let us inquire into the various ways which taxes could reduce the consumption of gasoline by automobiles. Consumption can be reduced by (1) a reduction of the number of cars in existence, (2) a reduction in the number of miles driven per car, and (3) an increasing of the fuel efficiency of automobiles so that more miles per gallon are achieved. A direct fuel tax that increased the cost of gasoline per gallon would, by tending to make all driving more expensive, have some impact toward achieving all three of the goals listed above. A tax on the fuel efficiency of cars that was heaviest for low-mileage cars and amounted to a subsidy for the highest-mileage cars would primarily increase fuel efficiency, might or might not have an impact on the total number of cars sold, and would tend to increase the share of total mileage driven in fuel-efficient cars. While the direct fuel tax would probably have the greatest conservation impact, a mix of gasoline taxes and fuel-efficiency taxes would probably have about the same effects and would stand a better chance politically of being put into effect. People naturally oppose high gasoline taxes, whereas there is more toleration for fuel-efficiency taxes, where the burden of the tax can be avoided by buying a higher-mileage car.

Nonprice methods of conservation would include direct quantity rationing with its drawbacks in the form of higher administrative costs, curtailment of higher-valued as well as lower-valued uses of the rationed fuel, and high policing costs if a black market in rationing coupons is to be avoided. It has the advantage that it conceals the impact on income distribution of resource scarcity. If prices were increased, it would be obvious that a greater unit burden would be placed on lower-income individuals. If quantities are restricted but rationing tickets are distributed in the same proportion to high- and low-income individuals, there is still a burden, but its impact in different income groups is not visible.

Finally, there is the conservation method favored by those who regard conservation as a moral duty, regardless of prices or scarcity: public exhortation. Prices are not raised, but appeals are made to the public to consume less in the interests of the greater welfare of the total community. This method tends to be most effective when the nature of the emergency is obvious to everyone, when it is widely regarded as important to cope with it, and when the duration of the emergency is relatively brief. Voluntary reductions in electrical consumption of about 5 per cent were registered during the Arab oil embargo of 1973–74, but after the embargo was lifted, consumption quickly returned to normal levels despite the continuance of public exhortation efforts. As experience with conservation and rationing during the Second World War has shown, voluntary compliance with nonprice conservation methods tends to falter during a prolonged emergency. The economist views public exhortation to make conservation efforts as, in effect, asking people to imagine a price higher than the existing price and then reduce their consumption in accordance with the imaginary rather than the real price. Different people have different imaginations, and as soon as widespread evidence exists that others are failing voluntarily to deny themselves, those more inclined to comply will cease doing so.

Environmental Policy Issues and Control Measures

In theory, the economics of pollution abatement is simple. Just estimate the costs and benefits of pollution abatement and abate pollution up to the point where the marginal social benefit of abatement equals the marginal social cost of abatement. The problems arise when you try to measure costs and benefits, when you propose some particular measure for internalizing the externalities of pollution costs, and when you try to obtain political support for economically efficient means of abatement. Finally, it must be realized that, given the uncertainties associated with estimates of costs and benefits and of the resulting doubtful effectiveness of achieving certain goals with precision, the use of political means may be superior to economic means in some instances.

Let us begin, however, by assuming that problems of measuring costs and benefits have been solved and it has been determined that the marginal social benefit from abatement should be equated with the marginal social cost of abatement at a level of $1 per ton of pollution abated, where abatement is at the rate of 50,000 tons per year. Assume that initially there are no abatement efforts and total pollution is 100,000 tons per year. With no abatement, the air is very dirty and the marginal social benefit from abating the

first ton of pollutants would be $2. But as the air becomes increasingly clean, the need for further abatement is less urgent, so that abating one half of the pollution, or 50,000 tons per year, will reduce marginal social benefits to $1 per ton, whereas the marginal social benefit from abating the last ton of pollutants is essentially zero. With the costs of pollution control, it is just the opposite. We assume that the first ton of abatement can be accomplished at practically zero cost, but from there on costs increase, so that the cost of abating the 50,000th ton is $1 and abating the last ton cost $2.

One way of efficiently abating the pollution to the economically optimal degree would be to impose a $1 tax per ton on the amount of pollutants emitted. Polluters would then be motivated to abate all pollution emissions for which the cost of abatement was less than the tax charged for failure to abate. They would thus increase their abatement efforts from zero abatement before the tax to 50,000 tons per year with the tax. They would still emit 50,000 tons of pollutants after abatement, but it would cost them more to abate this remaining pollution than it would to pay the $1 per ton tax. But it would not be socially optimal to require the abatement of more than 50,000 tons per year since beyond that point the costs of abatement are greater than the social benefits. In our example, the abatement of the first 50,000 tons costs the polluters $25,000 and social benefits in reduced pollution damages are $75,000. But to require the abatement of the remaining 50,000 tons would impose costs of $75,000 and the social benefits would be only $25,000. With the tax, however, revenues of $50,000 are raised, and this can be used to compensate the victims of the remaining pollution damage. Since in our example these damages are only $25,000, the pollution tax "overcompensates" for the damages resulting from the remaining pollution. Assuming the polluter to be engaged in a perfectly competitive industry where there are no excess profits to absorb the costs of pollution abatement and of the pollution taxes paid, the price of the product will have to be increased to cover these pollution control costs. This is, however, an efficient way of internalizing the pollution externalities in the market price of the product. Without the tax, the price of the product would be too low, such that its buyers would be subsidized, as a group, by the group that suffers from pollution resulting from the sale and use of the product. With the price increase, not only can pollution victims be compensated for remaining pollution, but the higher price will in itself cut consumption to some extent and thus further reduce the remaining unabated pollution.

Alternatively, the polluters could be induced to clean up by giving them a $1 per ton subsidy for every ton of pollutants abated. Under these circumstances they would elect to cut pollution from 100,000 tons to 50,000 tons by abating it to the extent of 50,000 tons since they would make a profit on the abatement subsidy by abating up to the point where the cost of abatement was equal to the unit subsidy. Here, too, the abatement of the first 50,000 tons of pollutants would result in a social benefit of $75,000 at an abatement cost of $25,000—but this would be canceled by the $50,000 subsidy paid for abatement, which gave producers an extra profit of $25,000 since abatement efforts costing only $25,000 were sufficient to qualify for a $50,000 subsidy. The social cost of the unabated pollution is $25,000, as in the earlier case. Here the industry is "overcompensated" for undertaking pollution control efforts, and there are no tax receipts that might be used to compensate the victims of the remaining pollution. Unless competition among polluters is so strong that they must pass on the net subsidy of $25,000 to consumers in the form of lower prices, it would appear that taxes are superior to subsidies, even though both methods achieve the same measure of pollution abatement.

Finally, let us assume that the authorities wish to use direct controls rather than taxes or subsidies. They can simply decree that pollution emissions must be reduced by 50,000 tons per year. Under these circumstances, polluters spend $25,000 to install equipment abating this amount of pollution, and social benefits total $75,000. But now there is no tax revenue from taxes on the remaining emissions of pollutants, and so no funds are available to compensate pollution victims for the remaining $25,000 worth of social costs. Some people object to a policy of placing taxes on pollutants emitted by characterizing it as "selling the right to pollute" for the unabated remainder. But a policy of limiting abatement to some given level can be with equal justice characterized as "giving away the right to pollute" for the uncontrolled remainder of pollution.

The above example, of course, assumed perfect knowledge about the costs and benefits of abatement, so that the optimal amount of abatement is known and it can be achieved by either an appropriate tax rate, an appropriate subsidy rate, or an appropriate mandated level of abatement. Where adequate knowledge is lacking, the amount of abatement induced by a tax or subsidy is uncertain, whereas that achieved by a well-enforced standard will be much more nearly certain. While the cost of administering mandated standards are generally higher than the cost of administering taxes or subsidies, in the case of potentially very harmful pollutants such as high-level nuclear wastes, the greater certainty of limiting pollutant outputs to desired levels under mandated standards renders this method superior to taxes or subsidies on public welfare grounds.

Energy and National Security Policy

During the period 1945 to 1970, imported oil prices declined and imports became much cheaper than domestic production in the United States, leading to an increasing degree of dependence upon foreign oil. Dependence on imports increased from zero in 1945 to about 37 per cent of total oil consumption by the time of the Arab embargo of 1973 and has since risen to close to 50 per cent by 1981. The increase since 1973 has come about not because foreign oil is cheaper than domestic oil but because domestic supplies are simply inadequate to satisfy domestic demand at the controlled prices that prevailed prior to 1981. In a perfectly competitive market where there is no danger of interruption in supplies of imported oil, there would be little reason for restricting oil imports in terms of national security implications. But where imports come from politically unstable regions and are subject to the threat of interruption either by embargoes or by the outbreak of war, the supply risks associated with oil imports are much greater than those associated with domestic production. In order to limit oil imports to the moderate volumes consistent with military security and economic stabilization, the costs of varying degrees of supply interruptions should be computed, risks assigned for each degree of supply interruption from each import source, and additional charges should be levied upon the importation of insecure oil, presumably in the form of a variable import tariff. High tariffs on oil imported from areas like Iran would reduce our vulnerability to economic and military damage in the event of politically imposed embargos. Lower tariffs could be charged for oil from more secure sources like Venezuela, and zero tariffs from import sources like Canada, where risks of punitively motivated supply interruptions are negligible.

The problems of estimating the national security posture costs of import interruptions are probably more difficult than those encountered in any other energy policy problem area. At one extreme, a nuclear war need not be lost because of even severe cutbacks in fuel availability during an embargo since little fuel is needed to produce a thermonuclear holocaust. On the other hand, embargos and threats of embargos may so reduce our ability to conduct a rational foreign policy or fight more limited and fuel-intensive wars that the danger of a nuclear war is increased. But the purely economic impact of an embargo is subject to somewhat closer estimation. It has been estimated that, during the embargo of 1973–74, the GNP fell by about $100 for every barrel of oil we were unable to import. Hence, to a first approximation which tends to be conservative, the costs of a long and severe embargo that would reduce national income, output, and employment would be quite large. If we estimated the damage to national security and to macroeconomic stability resulting from a supply interruption at perhaps $200 per barrel, we would then have to estimate the likelihood that oil from any particular source would be cut off during an unforeseen emergency. Let us assume that the likelihood that oil imported from Iran will be cut off at any particular time is 5 per cent; in that case a tariff of 5 per cent of $200, or $10 per barrel, should be added to the cost of Iranian imports. If we assume that the risk of interruption in supplies from a safer supplier like Mexico is only 1 per cent, then a $2-per-barrel risk-based tariff should be applied to Mexican imports. Such a structure of tariffs would lead us to reduce our reliance on the most insecure sources and increase our imports from more secure sources, as well as stimulate greater supplies from the domestic petroleum industry.

There are, however, real difficulties with such a proposal. For one thing, oil is a homogeneous commodity, and as long as imports are available it does not matter to the buyer where the oil comes from. It would be easy for oil suppliers to misrepresent the source of their imports, and policing imports to prevent this situation would be very expensive. The most that could be hoped for would probably be a two-tariff system which distinguished only in a rough way between low-risk oil and higher-risk oil. Furthermore, it is not certain that even a well-informed competitive market could achieve a risk-minimizing mix of reliance upon various foreign sources which would bring risks to an acceptably low level. In part, this is because the impact of a given tariff rate on the volumes imported from any given country will be hard to predict. Thus, on national security grounds, it might be better to rely upon a combination of several programs. Specifically, tariffs could be supplemented by some reliance upon import quotas, emergency storage capacity, reserve standby producing capacity, and international sharing agreements.

Emergency storage capacity could provide a margin of reserve supplies to supplement domestic production during a period of import interruption from one or more supply sources. The economic problem is the determination of the marginal social costs and social benefits of increasing volumes of oil held in storage. How much is stored would depend in part upon the amounts being imported from the riskiest suppliers; if an embargo of up to six months is predicted with a 10 per cent likelihood from all Arab suppliers, then emergency storage of volumes equal to six months' imports from all Arab sources would provide good protection, and storage of a smaller amount, such as three months' supply, would still add considerably to oil supply security in view of the relatively low likelihood of an effective embargo by all Arab exporters. Costs of storage are relatively high and the best natural storage locations (salt domes) are often located distant from refineries and markets.

Reserve standby producing capacity would provide additional insurance against import interruptions, although at relatively high cost. Owners of existing oil fields could be paid not to produce, saving this productive capacity for an emergency. These shut-in fields would have to be kept in producing condition, and payments for nonproduction would have to be high enough to compensate producers for the deferment of revenues from production and for maintenance of the pumping and pipeline facilities required to transport the oil to refineries. Only fields with high daily producing rates and low drilling costs per well, and fairly close to refineries would be reasonably economical. For the few fields possessing such characteristics, like the Elk Hills field in California and the Yates field in Texas, standby producing capacity might be maintained as a way of supple-

menting other measures to guard against import interruptions.

International sharing agreements may be of some assistance in preventing selective embargos aimed against individual countries, as when the Arabs singled out the United States and the Netherlands in 1973. Membership in the International Energy Agency (IEA), formed in response to the Arab embargo, obligates countries to share oil supplies jointly with any members suffering from an embargo. If this agreement proves workable in practice, then exporters cannot hope to embargo individual "enemies" without embargoing all IEA members. It is possible, however, that individual members might withdraw from the IEA if this prevented them from suffering an embargo.

Price Controls and Energy Policy Problems

From the economic point of view, price controls have created more energy policy problems than they have solved. As stated earlier, price controls on the wellhead prices of interstate natural gas, dating back to 1954, represent what virtually all economists now recognize to be an instance of "regulatory failure." The market structure was competitive prior to regulation and could not be improved upon by controls, since competitive markets achieve maximum efficiency in the allocation of resources. The demand for regulation was largely political and regional, with the predictable result that regulated prices were set at levels that were too low to equate long-run supply and demand. Production was reduced while consumption was stimulated, and since incentives were lacking to replace depleted reserves, supplies available to consumers in the interstate market eventually declined. New discoveries were diverted toward the intrastate market, where prices were more nearly competitive. The higher prices raised production and moderated consumption so that intrastate customers obtained supplies at market-clearing prices while interstate customers paid less per unit but could not consume all that they wanted at that price. In 1978 price control was extended to the intrastate market, and this created the same sort of distortions that had long plagued the interstate market. At the same time, a plan to phase out gas price controls over a number of years was devised. But the regulation of natural gas wellhead prices has been a total failure economically, from beginning to end.

While the federal government was reducing gas prices below competitive levels, however, a combination of state and national policies kept U.S. crude oil prices above the levels that would have prevailed in a free-trade world. State conservation commission regulation in the major crude-oil-producing states after 1935 limited production rates relative to demand so as to stabilize U.S. prices, while the federal import control program during 1959–73 limited import levels and thus made the price-stabilizing tasks of the state commissions easier to fulfill without the severe curtailment of domestic output. An early rationale for the interaction of these programs was that in order to keep the United States from becoming too dependent upon insecure foreign oil, imports should be limited to no more than about 12 per cent of domestic consumption (in markets east of the Pacific Coast states) and domestic production should be stimulated by keeping U.S. prices sufficiently above world prices. In practice, U.S. prices were stabilized but not increased to the levels required to maintain 88 per cent domestic self-sufficiency. After U.S. oil production peaked in 1970 while U.S. demand kept increasing, imports increased and it was no longer possible to impose quota limits after 1972.

After 1969, U.S. crude oil prices began increasing slowly, but in 1971 President Nixon's wage and price control program froze oil as well as other prices. These controls were lifted in early 1973 but were reimposed for oil in August 1973, two months before the embargo, with a price ceiling of about $4.35 per barrel. During the embargo, when the cost of imported oil rose above $12 per barrel, the price ceiling for "old" oil was raised to a maximum of $5.35 while the price of domestic "new" oil was allowed to increase to an average of about $9 per barrel. A two-price system for a homogeneous commodity is always economically anomalous as far as the consumer is concerned. The government's concern with holding down prices reflected two types of policy considerations. First, there was the desire to prevent domestic oil companies from realizing so-called windfall profits from the selling of existing oil reserves at the new, higher OPEC price level. Second, it was felt that allowing the prices of crude oil and of refined petroleum products to rise to the high levels required to equate supply and demand during the emergency period of the embargo would mean very high prices and a considerable transfer of income from consumers to producers, a change that would be politically unpopular. It was furthermore believed that the period of high prices would be relatively brief, lasting no longer than the embargo.

If these expectations had been borne out by later developments, a policy of holding down prices by controls would have been justified. It would be economically and politically disruptive to have energy prices double or triple during the embargo emergency and then fall back to pre-embargo levels within a few months. But after the embargo was lifted and OPEC prices remained at their record high levels, policy makers should have changed their plans.

As stated earlier, if a country is dependent upon imported oil and if OPEC sets the world price for oil exports, then the cost to that country of using oil is the OPEC price, whether the oil being used is domestic or imported. The underpricing of domestic oil and refined products simply reduces domestic production while it increases domestic consumption and results in greater import levels. Thus the remedy would be to junk the price controls and price oil and energy at their true cost to the economy—the cost of imported oil. So-called windfall profits could be controlled by requiring that the oil companies invest them in new domestic energy supply projects or else have them taxed away at very high rates. But instead of abandoning price controls in early 1974, an elaborate system of multiple pricing of various categories of oil was developed over the years; phased-in price decontrol was not provided for until 1978 and was not fully implemented until 1981. In the meantime, persistent low prices for some resources frustrated conservation efforts, reduced supply-promoting activities in the domestic industry, and resulted in the importation of hundreds of millions of barrels of otherwise unneeded OPEC oil. The windfall profits tax of 1978 was in many ways inferior to a simple high tax on oil company profits in excess of a certain level which were not reinvested in domestic energy activities of a supply-increasing nature.

The chief reason politicians prefer price controls to permitting market-clearing prices to prevail is the perceived impact of higher prices on lower-income individuals. And economists must concede that under circumstances like the OPEC revolution, where oil prices suddenly rise from competitive to monopoly levels, there are real problems facing politicians who advocate free markets. Much of the opposition to higher prices on grounds of harm to the poor is probably hypocritical; the problems of genuine poverty which afflict less than 10 per cent of the population are probably not of equally crucial concern to the other 90 per cent. Voters, politicians, and lobbyists simply hope to defeat forces conducive to price increases by arguing about damage to the poor. But this cynically realistic observation does not negate the fact that large price increases for energy sources can have substantially different impacts on the budgets of different income groups. The politician's usual preference is for some non-price means of coping with a problem so that the real costs of what may be just temporizing with the problem are hard to determine. The economic position is that two sets of policies can be employed simultaneously. First, price-related policies can be introduced to solve the resource alloca-tion aspects of a problem efficiently. Second, compensatory economic measures can be introduced to mitigate the impact of the pricing solution on lower-income people and thus solve the problem of income distribution effects equitably.

Policies on the Optimal Structure and Regulation of Energy Industries

When OPEC imposed its first great price hike in January 1974, the profits of the major oil companies increased considerably. Much, but not all, of this increase was due to an unrepeatable cause: the taking of inventory profits on the large volumes of oil that had been purchased at the old prices and were in inventory or in transit at the time of the price hike, which meant that large volumes of oil purchased at pre-embargo prices could be sold at prices reflecting the 1974 price hike. There was already a strong suspicion that the major oil companies were in collusion with OPEC over the price hike, and the fact of high reported oil earnings in 1974 seemed, in the minds of many, to confirm this suspicion. Accordingly, efforts to control the companies by punitive regulation were introduced into Congress in the form of a number of bills that sought four major types of structural "reforms": (1) outright nationalization of oil and other energy industries, (2) public utility regulation, (3) vertical divestiture of the major oil companies, and (4) horizontal divestiture of companies having investments in two or more different energy industries.

No such energy industry reorganization laws have yet been passed. In part this is because in the years since 1974 it has become increasingly clear that the OPEC price hike worked against the interests of the oil companies in many ways. While OPEC per-barrel revenues increased from about $1 in 1970 to over $40 in 1981, oil company earnings on the production of foreign oil declined from a range of 30 to 40 cents per barrel just before the embargo to 15 to 25 cents after the price hike, as OPEC governments nationalized the properties of their former concessionaires.

Thus, it has become more apparent that monopoly power in the world crude oil market belongs to OPEC, not to the major companies. More important, however, advocates of specific reforms have been unable to convince members of Congress of the economic virtues of reform programs.

Government ownership of energy industries was never taken too seriously since there are horrible examples of the inefficiency of governmental management of businesses, such as the post office and railway passenger service, which critics can use to cast doubt upon the probable cost-effectiveness of governmentally owned and operated energy industries. The public utility regulation of energy industries also found little support since it was reasonably clear that the necessary market characteristics required for successful public utility regulation (monopoly power over a specific market area in which volumes sold are relatively insensitive to significant increases in price) did not exist. Since the electric utility industry typically has such characteristics, commission regulation has been tolerably satisfactory in the past for the setting of electric power rates in individual markets. But few if any of the many other markets in the energy sector have the necessary characteristics. The attempt to impose price regulation of the public utility variety on the natural gas industry showed how mistaken further attempts along this line would be.

Vertical divestiture proposals contemplated the separation of the successive functional phases of major vertically integrated petroleum companies into entirely separate crude oil production companies, pipeline companies, refining companies, and marketing companies. It was felt by the backers of these proposals that vertical integration was a main source of the alleged monopoly power of the major oil companies, and that once they were broken up the severed segments would compete with each other and greater competition would result. Critics pointed out that, relative to other major industries, oil companies do not have significant monopoly power, as evidenced by relatively low concentration of ownership of reserves and control over sales on the part of the top four or eight firms, and that there are substantial economies of vertical integration, such as savings in storage facilities and refinery design, which would be sacrificed in the event of vertical disintegration. Furthermore, disintegrated companies might be at an added disadvantage in dealing with OPEC. It thus appeared that vertical divestiture would occasion substantial economic costs and might not achieve any significant benefits.

Horizontal divestiture proposals contemplated the prevention of any company from owning productive facilities in more than one energy industry. Thus oil companies would not be allowed to acquire coal mines or uranium properties. While economists can readily point to economies of vertical integration, those of horizontal integration are not so prominent and would seem to consist chiefly in the applicability of the research and development capabilities of oil companies to the problems faced by mining companies and, in particular, by companies in the synthetic fuels sector, where synthetic oil or gas might be made by processing coal, oil shale, or tar sands. On the other hand, the chief drawback of horizontal integration would be the use of alleged existing oil company monopoly power in petroleum to extend this monopoly power to coal, uranium, and new sources of energy, such as synthetic fuels. But if the evidence is that oil companies lack monopoly power in their own market, there is no real danger that such power will be extended to other energy industries.

Some critics urge the application of both horizontal and vertical divestiture to all energy industries. For economists who are convinced that the number one problem of domestic energy policy is reduction of our reliance upon the OPEC cartel, the best policy toward the structure and regulation of domestic energy industries would be one that retains existing elements of competition and tries to extend them, where necessary, while focusing on an efficient expansion of output in all parts of the energy sector, from new industries as well as existing industries. Such a policy on the structural side would rule out nationalization, utility regulation, and divestiture, and might lean toward toleration or even encouragement of "energy conglomerates," or horizontally integrated energy companies, operating in both old and new energy industries, since the most expeditious way to create new production capability in synthetic fuels industries may be to permit the entry of existing petroleum companies into this market. Wherever anticompetitive problems exist, however, as in the instance of the recent attempt of certain companies to create a cartel for uranium ore, they should be vigorously attacked by such weapons as antitrust suits.

Economic Policy on New Energy Sources

As stated above, it is urgent that the United States develop new sources of domestic energy in order to reduce its dependence upon OPEC. Accordingly, reasonable efforts should be undertaken to facilitate the development of economically promising new energy resources by a variety of means, ranging from the removal of arbitrary government-imposed obstacles to entrance into new energy industries or expansions of old industries to the partial or complete subsidy of research and development for new technologies whenever the economic benefits of such programs show a reasonable probability of exceeding the economic costs. This is not to say that vast amounts should be spent in public subsidy of new approaches—only that, for a given total budgetary appropriation for such research, there should be a serious attempt to compare costs and benefits for proposed programs and to fund those which seem the most promising.

The heading "new energy sources" would cover not only new fuels like those coming from coal and oil shale but also new methods of finding and developing petroleum and other "conventional" energy resources. Per-

fected tertiary recovery methods would make available large reserves of very heavy crude oil previously regarded as uneconomical and would increase recovery percentages from existing oil fields. The discovery and development of geopressurized methane fields along the Texas and Louisiana Gulf coasts might increase gas reserves by enormous amounts. Better formation fracturing techniques might make available large new reserves of natural gas currently trapped in tight Devonian shale strata. There are thus a number of ways in which the application of research and development efforts to increasing energy production from known petroleum provinces might yield large returns.

If we turn to genuinely new sources of energy, synthetic fluid fuels from coal and oil shale would appear to offer the greatest immediate economic promise since these fuels will be produced, transported, and processed in fluid form and can thus use existing facilities (with some adaptations) designed to handle gaseous and liquid petroleum and petroleum products. Synthetic fluid fuels also have the advantage that pollutants such as sulfur can be removed during the processing of solid materials into fluid form, rather than burning the fuels (such as high-sulfur coal) in solid form and cleaning up the combustion fumes at great expense. Shale oil is technologically closer to the commercial application stage than coal liquefaction but has greater environmental pollution hazards.

While the chief emphasis of this policy area is on increasing the supply of domestic energy sources, advances in technology which increase the ability of the economy to derive more usable energy from the same resources can, to some extent, substitute for the discovery of new sources. Thus, research should be pursued along other lines in addition to those outlined above. Ways could be devised to produce existing energy reserves more efficiently and recover a greater fraction of their contained energy potential. Energy resources could be processed more efficiently by generators and industrial and other intermediate users. In this respect, the more widespread use of combined cycles in electric power generation might achieve real economies. Efforts should be made to increase the efficiency of energy transmission and distribution systems. Finally, studies should be made regarding ways to increase the efficiency of energy use by the final consumer.

Economic analysis of energy policy for increasing supply by these means would stress two points. First, there are many barriers of a regulatory nature to developing new domestic energy industries, particularly from coal and oil shale. Wherever these barriers impose social costs that exceed probable benefits, they should be reduced or relaxed. Second, not every possible new source of energy should be subsidized by grants of public research and development funds. While the total size of such subsidies, direct or indirect, should be of considerable magnitude, this is essentially a matter of budgetary allocation and the amount should be decided by political processes. Once the amount is determined, however, proposals should be ranked in accordance with their expected social value adjusted for risks of failure. Certainty of success cannot be a criterion since it is the essence of true research that failures will outnumber successes. If there is no risk attached to a particular project, then there is little if any need for public subsidy since private investors should be able to raise funds for such projects in private capital markets.

One very important policy issue is to what extent the development of synthetic fluid fuels industries should be publicly subsidized. It is important that such subsidy be efficient, and it should not be available to only one or two firms so that competition in the new synthetics industry is less than vigorous at the outset. One promising possibility would be to have government guarantee a minimum price for the output of first-generation synthetics plants. If the market price was at or above the guaranteed price, no subsidy would be paid; if the market price was below the price guarantee, the government would make up the difference. The government might ask for minimum guaranteed price bids from

companies interested in entering the synthetics industry.

It is to be expected that political inputs into the process of evaluating energy research and development proposals may often be counterproductive. Politicians naturally like to finance research that will bear fruit during their own terms of office. Since most research is risky and long-term, politicians logically fear that they will be criticized for "wasting" money on risky research that does not succeed. If long-term projects do succeed, then the chances are good that the politicians who supported the research will by that time be out of office and their successors will claim the credit. While economists can sympathize with their feelings, it is desirable that the purely political input into risky research and development project evaluation be minimized.

Policy Problems in Coping with the OPEC Cartel

Although the policy problems encountered in coping with the monopoly power of the OPEC cartel are the most important of all, they are discussed after all the others since the prospects for effective action by the United States seem smaller here than in any other area. The basic problem is that a strong cartel can be opposed most efficiently by a strong countercartel, but virtually perfect coordination of the policies of all of the oil-importing countries would be required and such cooperation among a large number of sovereign governments with diverse views is simply out of the question. The most appropriate economic response to an exporter group embargo of shipments of a vitally needed commodity is for the importers to impose a counterembargo on the exporters, depriving them of all goods they import from any of their oil customers. Since a naval blockade would be necessary to support such a general embargo, however, the problem ceases to be simply economic in nature (i.e., coping with an act of economic warfare) and passes over into the military area. It is only too likely that the tensions involved in such a confrontation with a militarily much inferior adversary would lead to importer attempts to occupy the oil fields and exporter attempts to sabotage them. The risk of a great power confrontation and of a nuclear war would be very grave, since the Soviet Union will certainly not stand by and see the oil importers occupy the Persian Gulf oil fields and thus preempt the Soviets from taking them over themselves at some later time when they really need them. In the Persian Gulf, all the elements of military logistics strongly favor the Soviet forces.

Milder responses might be sufficient to handle milder provocations. If instead of an embargo there had only been a huge price hike, a countercartel might be drawn up that would unanimously refuse to pay more than a minimum competitive price for oil. Furthermore, as long as the price hike was in effect, the members of the countercartel would not sell anything to the oil exporters except at prices multiplied by the same factor present in the oil price hike. Here, too, the degree of cooperation required would be impossible to achieve, and the danger of military confrontations too high to accept.

We are therefore left with the hope that if the other policy problem areas are dealt with effectively, our reliance upon OPEC will steadily decline. As stressed in the earlier discussion, we should raise domestic energy prices to, or even above, the OPEC level, achieving price-based conservation and the reduction of environmental problems. Then, through higher prices and a program of stimulating research and development in new energy sources, we can increase domestic energy output. Higher prices will raise our energy production and decrease our consumption, cutting into our demand for OPEC exports and reducing OPEC's market power. Measures to promote national security in the face of continuing oil imports will reduce our vulnerability to foreign manipulation in the form of short-run embargos or opportunistic price hikes during periods of supply interruption. As a final measure, we might impose high tariffs on OPEC oil imports and raise the tariffs $2 per barrel for every $1 per barrel price hike by OPEC, thus penalizing OPEC by having its market contract strongly each time it raises prices. OPEC's greatest strength is that it is hard for importers to hurt the cartel without hurting themselves at the same time. Once we realize that this is the strategy of economic warfare rather than free trade, we may achieve the resolve not only to impair the market power of the cartel but also to help ourselves through the adoption of energy policies that will restore efficient conditions in our own energy sector.

Energy Use in the U.S. Food System   5

JOHN S. STEINHART AND CAROL E. STEINHART

In a modern industrial society, only a tiny fraction of the population is in frequent contact with the soil and an even smaller fraction of the population raises food on the soil. The proportion of the population engaged in farming halved between 1920 and 1950 and then halved again by 1962. Now it has almost halved again, and more than half of these remaining farmers hold other jobs off the farm.[1] At the same time the number of work animals has declined from a peak of more than 22×10^6 in 1920 to a very small number at present.[2] By comparison with earlier times, fewer farmers are producing more agricultural products and the value of food in terms of the total goods and services of society now amounts to a smaller fraction of the economy than it once did.

Energy inputs to farming have increased enormously during the past 50 years,[3] and the apparent decrease in farm labor is offset in part by the growth of support industries for the farmer. With these changes on the farm have come a variety of other changes in the U.S. food system, many of which are now deeply embedded in the fabric of daily life. In the past 50 years, canned, frozen, and other processed foods have become the principal items of our diet. At present, the food-processing industry is the fourth largest energy consumer of the Standard Industrial Classification groupings.[4] The extent of transportation engaged in the food system has grown apace, and the proliferation of appliances in both numbers and complexity still continues in homes, institutions, and stores. Hardly any food is eaten as it comes from the fields. Even farmers purchase most of their food from markets in town.

Present energy supply problems make this

Dr. John S. Steinhart is a professor of geology and geophysics and a professor at the Institute for Environmental Studies, University of Wisconsin-Madison, Madison, Wisconsin 53706. He also is the associate director of the University of Wisconsin-Madison Sea Grant Program.

Dr. Carol E. Steinhart, formerly a biologist with the National Institutes of Health, is now a science writer and editor.

growth of energy use in the food system worth investigating. It is our purpose in this article to do so. But there are larger matters at stake. Georgescu-Roegen notes that "the evidence now before us—of a world which can produce automobiles, television sets, etc., at a greater speed than the increase in population, but is simultaneously menaced by mass starvation—is disturbing".[5] In the search for a solution to the world's food problems, the common attempt to transplant a small piece of a highly industrialized food system to the hungry nations of the world is plausible enough, but so far the outcome is unclear. Perhaps an examination of the energy flow in the U.S. food system as it has developed can provide some insights that are not available from the usual economic measures.

MEASURES OF FOOD SYSTEMS

Agricultural systems are most often described in economic terms. A wealth of statistics is collected in the United States and in most other technically advanced countries indicating production amounts, shipments, income, labor, expenses, and dollar flow in the agricultural sector of the economy. But, when we wish to know something about the food we actually eat, the statistics of farms are only a tiny fraction of the story.

Energy flow is another measure available to gauge societies and nations. It would have made no sense to measure societies in terms of energy flow in the eighteenth century, when economics began. As recently as 1940, four fifths of the world's population were still on farms and in small villages, most of them engaged in subsistence farming.

Only after some nations shifted large portions of the population to manufacturing, specialized tasks, and mechanized food production, and shifted the prime sources of energy to move society to fuels that were transportable and usable for a wide variety of alternative activities, could energy flow be used as a measure of a society's activities. Today it is only in one fifth of the world

where these conditions are far advanced. Yet we can now make comparisons of energy flows even with primitive societies. For even if the primitives, or the euphemistically named "underdeveloped" countries, cannot shift freely among their energy expenditures, we *can* measure them, and they constitute a different and potentially useful comparison with the now traditional economic measures.

What we would like to know is this: How does our present food supply system compare, in energy measures, with those of other societies and with our own past? Perhaps then we can estimate the value of energy flow measures as an adjunct to, but different from, economic measures.

ENERGY IN THE U.S. FOOD SYSTEM

A typical breakfast includes orange juice from Florida, by way of the Minute Maid factory, bacon from a midwestern meat packer, cereal from Nebraska and General Mills, eggs and milk from not *too* far away, and coffee from Colombia. All of these things are available at the local supermarket (several miles each way in a 300-horsepower automobile), stored in a refrigerator-freezer, and cooked on an instant-on stove.

The present food system in the United States is complex, and the attempt to analyze it in terms of energy use will introduce complexities and questions far more perplexing than the same analysis carried out on simpler societies. Such an analysis is worthwhile, however, if only to find out where we stand. We have a food system, and most people get enough to eat from it. If, in addition, one considers the food supply problems present and future in societies where a smaller fraction of the people get enough to eat, then our experience with an industrialized food system is even more important. There is simply no gainsaying that many nations of the world are presently attempting to acquire industrialized food systems of their own.

Food in the United States is expensive by world standards. In 1970 the average annual per capita expenditure for food was about

$600.[3] This amount is larger than the per capita gross domestic product of more than 30 nations of the world which contain the majority of the world's people and a vast majority of those who are underfed. Even if we consider the diet of a poor resident of India, the annual cost of his food at United States prices would be about $200—more than twice his annual income.[3] It is crucial to know whether a piece of our industrialized food system can be exported to help poor nations, or whether they must become as industrialized as the United States to operate an industrialized food system.

Our analysis of energy use in the food system begins with an omission. We shall neglect that crucial input of energy provided by the sun to the plants upon which the entire food supply depends. Photosynthesis has an efficiency of about 1 per cent; thus, the maximum solar radiation captured by plants is about 5×10^3 kilocalories per square meter per year.[3]

Seven categories of energy use on the farm are considered here. The amounts of energy used are shown in Table 5-1. The values given for farm machinery and tractors are for the manufacture of new units only and do not include parts and maintenance for units that already exist. The amounts shown for direct fuel use and electricity consumption are a bit too high because they include some residential uses of the farmer and his family. On the other hand, some uses in these categories are not reported in the summaries used to obtain the values for direct fuel and electricity usage. These and similar problems are discussed in the references. Note the relatively high energy cost associated with irrigation. In the United States less than 5 per cent of the crop land is irrigated.[1] In some countries where the "green revolution" is being attempted, the new high-yield varieties of plants require irrigation where native crops did not. If that were the case in the United States, irrigation would be the largest single use of energy on the farm.

Little food makes its way directly from field and farm to the table. The vast complex of processing, packaging, and transport has been grouped together in a second major subdivision of the food system. The seven categories of the processing industry are listed in Table 5-1. Energy use for the transport of food should be charged to the farm in part, but we have not done so here because the calculation of the energy values is easiest (and we believe most accurate) if they are taken for the whole system.

After the processing of food there is further energy expenditure. Transportation enters the picture again, and some fraction of the energy used for transportation should be assigned here. But there are also the distributors, wholesalers, and retailers, whose freezers, refrigerators, and very establishments are an integral part of the food system. There are also the restaurants, schools, universities, prisons, and a host of other institutions engaged in the procurement, preparation, storage, and supply of food. We have chosen to examine only three categories: the energy required for refrigeration, cooking, and for the manufacture of the heating and refrigeration equipment (Table 5-1). We have made no attempt to include the energy used in trips to the store or restaurant. Garbage disposal has also been omitted, although it is a persistent and growing feature of our food system; 12 per cent of the nation's trucks are engaged in the activity of waste disposal,[1] of which a substantial part is related to food. If there is any lingering doubt that these activities—both the ones included and the ones left out—are an essential feature of our present food system, one need only ask what would happen if everyone should attempt to get on without a refrigerator or freezer or stove. Certainly the food system would change.

Table 5-1 and the related references summarize the numerical values for energy use in the U.S. food system from 1940 to 1970. As for many activities in the past few decades, the story is one of continuing increase. The totals are displayed in Figure 5-1 along with the energy value of the food consumed by the public. The food values

Table 5–1. Energy Use in the U.S. Food System.[a]

Component	1940	1947	1950	1954	1958	1960	1964	1968	1970	References
On Farm										
Fuel (direct use)	70.0	136.0	158.0	172.8	179.0	188.0	213.9	226.0	232.0	(13–15)
Electricity	0.7	32.0	32.9	40.0	44.0	46.1	50.0	57.3	63.8	(14, 16)
Fertilizer	12.4	19.5	24.0	30.6	32.2	41.0	60.0	87.0	94.0	(14, 17)
Agricultural steel	1.6	2.0	2.7	2.5	2.0	1.7	2.5	2.4	2.0	(14, 18)
Farm machinery	9.0	34.7	30.0	29.5	50.2	52.0	60.0	75.0	80.0	(14, 19)
Tractors	12.8	25.0	30.8	23.6	16.4	11.8	20.0	20.5	19.3	(20)
Irrigation	18.0	22.8	25.0	29.6	32.5	33.3	34.1	34.8	35.0	(21)
Subtotal	124.5	272.0	303.4	328.6	356.3	373.9	440.5	503.0	526.1	
Processing Industry										
Food processing industry	147.0	177.5	192.0	211.5	212.6	224.0	249.0	295.0	308.0	(13, 14, 22)
Food processing machinery	0.7	5.7	5.0	4.9	4.9	5.0	6.0	6.0	6.0	(23)
Paper packaging	8.5	14.8	17.0	20.0	26.0	28.0	31.0	35.7	38.0	(24)
Glass containers	14.0	25.7	26.0	27.0	30.2	31.0	34.0	41.9	47.0	(25)
Steel cans and aluminum	38.0	55.8	62.0	73.7	85.4	86.0	91.0	112.2	122.0	(26)
Transport (fuel)	49.6	86.1	102.0	122.3	140.2	153.3	184.0	226.6	246.9	(27)
Trucks and trailers (manufacture)	28.0	42.0	49.5	47.0	43.0	44.2	61.0	70.2	74.0	(28)
Subtotal	285.8	407.6	453.5	506.4	542.3	571.5	656.0	787.6	841.9	
Commercial and Home										
Commercial refrigeration and cooking	121.0	141.0	150.0	161.0	176.0	186.2	209.0	241.0	263.0	(13, 29)
Refrigeration machinery (home and commercial)	10.0	24.0	25.0	27.5	29.4	32.0	40.0	56.0	61.0	(14, 30)
Home refrigeration and cooking	144.2	184.0	202.3	228.0	257.0	276.6	345.0	433.9	480.0	(13, 29)
Subtotal	275.2	349.0	377.3	416.5	462.4	494.8	594.0	730.9	804.0	
Grand total	685.5	1028.6	1134.2	1251.5	1361.0	1440.2	1690.5	2021.5	2172.0	

[a] All values are multiplied by 10^{12} kilocalories.

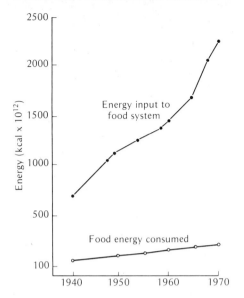

Figure 5-1. Energy use in the food system, 1940 through 1970, compared to the caloric content of food consumed.

were obtained by multiplying the daily caloric intake by the population. The differences in caloric intake per capita over this 30-year period are small,[1] and the curve is primarily an indication of the increase in population in this period.

OMISSIONS AND DUPLICATIONS FOR FOOD SYSTEM ENERGY VALUES

Several omissions, duplications, and overlaps have been mentioned. We shall now examine the values in Table 5-1 for completeness and try to obtain a crude estimate of their numerical accuracy.

The direct fuel and electricity usage on the farm may be overstated by some amounts used in the farmer's household, which, by our approach, would not all be chargeable to the food system. But about 10 per cent of the total acreage farmed is held by corporate farms for which the electrical and direct fuel use is not included in our data. Other estimates of these two categories are much higher (see Table 5-1)[15,16]

No allowance has been made for food exported, which has the effect of overstating the energy used in our own food system. For the years prior to 1960 the United States was at times a net importer of food, at times an exporter, and at times there was a near balance in this activity. But during this period the net flow of trade was never more than a few per cent of the total farm output. Since 1960 net exports have increased to about 20 percent of the gross farm product.[1,3] The items making up the vast majority of the exports have been rough grains, flour, and other plant products with very little processing. Imports include more processed food than exports and represent energy expenditure outside the United States. Thus, the overestimate of energy input to the food system might be 5 per cent with an upper limit of 15 per cent.

The items omitted are more numerous. Fuel losses from the well head or mine shaft to end use total 10 to 12 per cent.[6] This would represent a flat addition of 10 per cent or more to the totals, but we have not included this item because it is not customarily charged to end uses.

We have computed transport energy for trucks only. Considerable food is transported by train and ship, but these items were omitted because the energy use is small relative to the consumption of truck fuel. Small amounts of food are shipped by air, and, although air shipment is energy intensive, the amount of energy consumed appears small. We have traced support materials until they could no longer be assigned to the food system. Some transportation energy consumption is not charged in the transport of these support materials. These omissions are numerous and hard to estimate, but they would not be likely to increase the totals by more than 1 or 2 per cent.

A more serious understatement of energy usage occurs with respect to vehicle usage (other than freight transport) on farm business, on food-related business in industry and commercial establishments, and in the supporting industries. A special attempt to estimate this category of energy usage for 1968 suggests that it amounts to about 5 per cent of the energy totals for the food system. This estimate would be subject to an uncertainty of nearly 100 per cent. We must be satisfied to suggest that 1 to 10 per cent should be added to the totals on this account.

Waste disposal is related to the food system, at least in part. We have chosen not to charge this energy to the food system, but, if one half of the waste disposal activity is taken as food related, about 2 per cent must be added to the food system energy totals.

We have not included energy for parts and maintenance of machinery, vehicles, buildings, and the like, or lumber for farm, industry, or packaging uses. These miscellaneous activities would not constitute a large addition in any case. We have also excluded construction. Building and replacement of farm structures, food industry structures, and commercial establishments are all directly part of the food system. Construction of roads is in some measure related to the food system, since nearly half of all trucks transport food and agricultural items (see Table 5-1).[27] Even home construction could be charged in part to the food system since space, appliances, and plumbing are, in part, a consequence of the food system. If 10 per cent of housing, 10 per cent of institutional construction (for institutions with food service), and 10 per cent of highway construction are included, about 10 per cent of the total construction was food related in 1970. Assuming that the total energy consumption divides in the same way that the gross national product does (which overstates energy use in construction), the addition to the total in Table 5-1 would be about 10 per cent, or 200 $\times 10^{12}$ kilocalories. This is a crude and highly simplified calculation, but it does provide an estimate of the amounts of energy involved.

The energy used to generate the highly specialized seed and animal stock has been excluded because there is no easy way to estimate it. Pimentel et al.[3] estimate that 1800 kilocalories are required to produce 1 pound (450 grams) of hybrid corn seed. But in addition to this amount, some energy use should be included for all the schools of agriculture, agricultural experiment stations, the far-flung network of county agricultural agents [one local agent said he traveled over 50,000 automobile miles (80,000 kilometers) per year in his car], the U.S. Department of Agriculture, and the wide-ranging agricultural research program that enables man to stay ahead of the new pest and disease threats to our highly specialized food crops. These are extensive activities, but we cannot see how they could add more than a few per cent to the totals in Table 5-1.

Finally, we have no attempt to include the amount of private automobile usage involved in the delivery system from retailer to home, or other food-related uses of private autos. Rice[7] reports 4.25×10^{15} kilocalories for the energy cost of autos in 1970, and shopping constitutes 15.2 per cent of all automobile usage.[8] If only half of the shopping is food-related, 320×10^{12} kilocalories of energy use is at stake here. Between 8 and 15 per cent should be added to the totals of Table 5-1, depending on just how one wishes to apportion this item.

It is hard to take an approach that might calculate smaller totals, but, depending upon point of view, the totals could be much larger. If we accumulate the larger estimates from the above paragraphs as well as the reductions, the total could be enlarged by 30 to 35 per cent, especially for recent years. As it is, the values for energy use in the food system from Table 5-1 account for 12.8 per cent of the total U.S. energy use in 1970.

PERFORMANCE OF AN INDUSTRIALIZED FOOD SYSTEM

The difficulty with history as a guide for the future or even the present lies not so much in the fact that conditions change—we are continually reminded of that fact—but in the fact that history is only one experiment of the many that might have occurred. The U.S. food system developed as it did for a variety

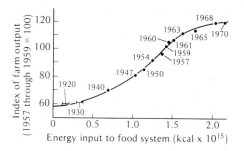

Figure 5-2. Farm output as a function of energy input to the U.S. food system, 1920 through 1970.

of reasons, many of them not understood. We would do well to examine some of the dimensions of this development before attempting to theorize about how it might have been different, or how parts of this food system can be transplanted elsewhere.

ENERGY AND FOOD PRODUCTION

Figure 5-2 displays features of our food system not easily seen from economic data. The curve shown has no theoretical basis but is suggested by the data as a smoothed recounting of our own history of increasing food production. It is, however, similar to most growth curves and suggests that, to the extent that the increasing energy subsidies to the food system have increased food production, we are near the end of an era. Like the logistic growth curve, there is an exponential phase which lasted from 1920 or earlier until 1950 or 1955. Since then, the increments in production have been smaller despite the continuing growth in energy use. It is likely that further increases in food production from increasing energy inputs will be harder and harder to come by. Of course, a major change in the food system could change things, but the argument advanced by the technological optimist is that we can always get more if we have enough energy, and that no other major changes are required. Our own history—the only one we have to examine—does not support that view.

ENERGY AND LABOR IN THE FOOD SYSTEM

One farmer now feeds 50 people, and the common expectations is that the labor input to farming will continue to decrease in the future. Behind this expectation is the assumption that the continued application of technology—and energy—to farming will substitute for labor. Figure 5-3 shows this historic decline in labor as a function of the energy supplied to the food system, again the familiar S-shaped curve. What it implies is that increasing the energy input to the food system is unlikely to bring further reduction in farm labor unless some other major change is made.

The food system that has grown in this period has provided much employment that did not exist 20, 30, or 40 years ago. Perhaps even the idea of a reduction of labor input is a myth when the food system is viewed as a whole instead of from the point of view of the farm worker only. When discussing inputs to the farm, Pimentel et al.[3] cite an estimate of two farm support workers for each person actually on the farm. To this must be added employment in food-processing industries, in food wholesaling and retailing, as well as in a variety of manufacturing en-

Figure 5-3. Labor use on farms as a function of energy use in the food system.

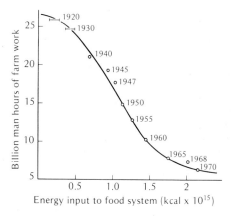

terprises that support the food system. Yesterday's farmer is today's canner, tractor
mechanic, and fast-food carhop. The process
of change has been painful to many ordinary
people. The rural poor, who could not quite
compete in the growing industrialization of
farming, migrated to the cities. Eventually
they found other employment, but one must
ask if the change was worthwhile. The answer to that question cannot be provided by
energy analysis anymore than by economic
data, because it raises fundamental questions
about how individuals would prefer to spend
their lives. But if there is a stark choice between long hours as a farmer or shorter hours
on the assembly line of a meat-packing plant,
it seems clear that the choice would not be
universally in favor of the meat-packing
plant. Thomas Jefferson dreamed of a nation
of independent small farmers. It was a good
dream, but society did not develop in that
way. Nor can we turn back the clock to recover his dream. But, in planning and preparing for our future, we had better look
honestly at our collective history, and then
each of us should closely examine his dreams.

THE ENERGY SUBSIDY TO THE FOOD SYSTEM

The data in Figure 5-1 can be combined to
show the energy subsidy provided to the food
system for the recent past. We take as a measure of the food supplied the caloric content
of the food actually consumed. This is not
the only measure of the food supplied, as the
condition of many protein-poor peoples of the
world clearly shows. Nevertheless, the comparison between caloric input and output is a
convenient way to compare our present situation with the past, and to compare our food
system with others. Figure 5-4 shows the
history of the U.S. food system in terms of
the number of calories of energy supplied to
produce 1 calorie of food for actual consumption. It is interesting and possibly
threatening to note that there is no real suggestion that this curve is leveling off. We appear to be increasing the energy input even

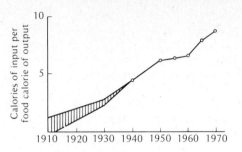

Figure 5-4. Energy subsidy to the food system
needed to obtain 1 food calorie.

more. Fragmentary data for 1972 suggest that
the increase continued unabated. A graph like
Figure 5-4 could approach zero. A natural
ecosystem has no fuel input at all, and those
primitive people who live by hunting and
gathering have only the energy of their own
work to count as input.

SOME ECONOMIC FEATURES OF THE U.S. FOOD SYSTEM

The markets for farm commodities in the
United States come closer than most to the
economist's ideal of a "free market." There
are many small sellers and many buyers, and
thus no individual is able to affect the price
by his own actions in the market place. But
government intervention can drastically alter
any free market, and government intervention in the prices of agricultural products (and
hence of food) has been a prominent feature
of the U.S. food system for at least 30 years.
Between 1940 and 1970, total farm income
ranged from $4.5 to $16.5 billion, and the
national income originating in agriculture
(which includes indirect income from agriculture) has ranged from $14.5 to $22.5 billion.[1] Meanwhile, government subsidy programs, primarily farm price supports and soil
bank payments, grew from $1.5 billion in
1940 to $6.2 billion in 1970. In 1972 these
subsidy programs grew to $7.3 billion, de-
spite foreign demand of agricultural products. Viewed in a slightly different way, direct government subsidies have accounted for

30 to 40 per cent of the farm income and 15 to 30 per cent of the national income attributable to agriculture for the years since 1955. This point emphasizes once again the striking gap between the economic description of society and the economic models used to account for that society's behavior.

This excursion into farm price supports and economics is related to energy questions in this way: first, as far as we know, government intervention in the food system is a feature of all highly industrialized countries (and, despite the intervention, farm incomes still tend to lag behind national averages) and, second, reduction of the energy subsidy to agriculture (even if we could manage it) might decrease the farmer's income. One reason for this state of affairs is that the demand for food quantity has definite limits, and the only way to increase farm income is then to increase the unit price of agricultural products. Consumer boycotts and protests in the early 1970s suggest that there is considerable resistance to this outcome.

Government intervention in the functioning of the market in agriculture products has accompanied the rise in the use of energy in agriculture and the food supply system, and we have nothing but theoretical suppositions to suggest that any of the present system can be deleted.

SOME ENERGY IMPLICATIONS FOR THE WORLD FOOD SUPPLY

The food supply system of the United States is complex and interwoven into a highly industrialized economy. We have tried to analyze this system on account of its implications for future energy use. But the world is short of food. A few years ago it was widely predicted that the world would suffer widespread famine in the 1970s. The adoption of new high-yield varieties of rice, wheat, and other grains has caused some experts to predict that the threat of these expected famines can now be averted, perhaps indefinitely. Yet, despite increases in grain production in some areas, the world still seems to be headed toward famine. The adoption of these new varieties of grain—dubbed hopefully the "green revolution"—is an attempt to export a part of the energy-intensive food system of the highly industrialized countries to nonindustrialized countries. It is an experiment because, although the whole food system is not being transplanted to new areas, a small part of it is. The green revolution requires a great deal of energy. Many of the new varieties of grain require irrigation where traditional crops did not, and almost all the new crops require extensive fertilization.

Meanwhile, the agricultural surpluses of the 1950s have largely disappeared. Grain shortages in China and Russia have attracted attention because they have brought foreign trade across ideological barriers. There are other countries that would probably import considerable grain, if they could afford it. But only four countries may be expected to have any substantial excess agricultural production in the next decade. These are Canada, New Zealand, Australia, and the United States. None of these is in a position to give grain away, because each of them needs the foreign trade to avert ruinous balance of payments deficits. Can we then export energy-intensive agricultural methods instead?

ENERGY-INTENSIVE AGRICULTURE ABROAD

It is quite clear that the U.S. food system cannot be exported intact at present. For example, India has a population of 550×10^6 persons. To feed the people of India at the U.S. level of about 3000 food Calories per day (instead of their present 2000) would require more energy than India now uses for all purposes. To feed the entire world with a U.S. type of food system, almost 80 per cent of the world's annual energy expenditure would be required just for the food system.

The recourse most often suggested to remedy this difficulty is to export methods of increasing crop yield and hope for the best. We must repeat as plainly as possible that this is an experiment. We know that our food system works (albeit with some difficulties and warnings for the future). But we cannot know what will happen if we take a piece of that system and transplant it to a poor country, without our industrial base of supply, transport system, processing industry, appliances for home storage and preparation, and, most important, a level of industrialization that permits higher costs for food.

Fertilizers, herbicides, pesticides, and in many cases machinery and irrigation are needed for success with the green revolution. Where is this energy to come from? Many of the nations with the most serious food problems are those nations with scant supplies of fossil fuels. In the industrialized nations, solutions to the energy supply problems are being sought in nuclear energy. This technology-intensive solution, even if successful in advanced countries, poses additional problems for underdeveloped nations. To create the bases of industry and technologically sophisticated people within their own countries will be beyond the capability of many of them. Here again, these countries face the prospect of depending upon the goodwill and policies of industrialized nations. Since the alternative could be famine, their choices are not pleasant and their irritation at their benefactors—ourselves among them—could grow to threatening proportions. It would be comfortable to rely on our own good intentions, but our good intentions have often been unresponsive to the needs of others. The matter cannot be glossed over lightly. World peace may depend upon the outcome.

CHOICES FOR THE FUTURE

The total amount of energy used on U.S. farms for the production of corn is now near 10^3 kilocalories per square meter per year,[3] and this is more or less typical of intensive agriculture in the United States. With this application of energy we have achieved yields of 2×10^3 kilocalories per square meter per year of usable grain—bringing us to almost half of the photosynthetic limit of produc-

tion. Further applications of energy are likely to yield little or no increase in this level of productivity. In any case, no amount of research is likely to improve the efficiency of the photosynthetic process itself. There is a further limitation on the improvement of yield. Faith in technology and research has at times blinded us to the basic limitations of the plant and animal material with which we work. We have been able to emphasize desirable features already present in the gene pool and to suppress others that we find undesirable. At times the cost of the increased yield has been the loss of desirable characteristics—hardiness, resistance to disease and adverse weather, and the like. The farther we get from characteristics of the original plant and animal strains, the more care and energy are required. Choices need to be made in the directions of plant breeding. And the limits of the plants and animals we use must be kept in mind. We have not been able to alter the photosynthetic process or to change the gestation period of animals. In order to amplify or change an existing characteristic, we shall probably have to sacrifice something in the overall performance of the plant or animal. If the change requires more energy, we would end with a solution that is too expensive for the people who need it most. These problems are intensified by the degree to which energy becomes more expensive in the world market.

WHERE TO LOOK FOR FOOD NEXT?

Our examination in the foregoing pages of the U.S. food system, the limitations on the manipulation of ecosystems and their components, and the risks of the green revolution as a solution to the world food supply problem suggests a bleak prospect for the future. This complex of problems should not be underestimated, but there are possible ways of avoiding disaster and of mitigating the severest difficulties. These suggestions are not very dramatic and may be difficult to accept.

Figure 5-5 shows the ratio of the energy

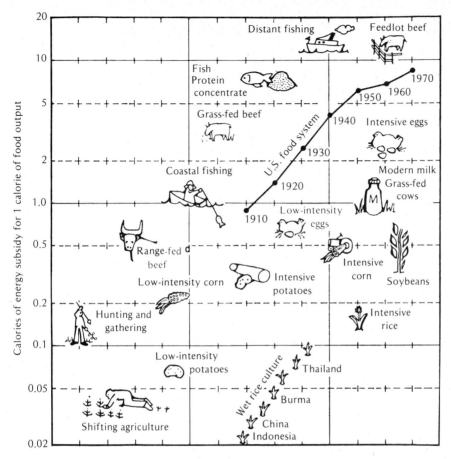

Figure 5-5. Energy subsidies for various food crops. The energy history of the U.S. food system is shown for comparison. (Source of data: ref. 31)

subsidy to the energy output for a number of widely used foods in a variety of times and cultures. For comparison, the overall pattern for the U.S. food system is shown, but the comparison is only approximate because, for most of the specific crops, the energy input ends at the farm. As has been pointed out, it is a long way from the farm to the table in industrialized societies. Several things are immediately apparent and coincide with expectations. High-protein foods such as milk, eggs, and especially meat have a far poorer energy return than plant foods. Because protein is essential for human diets and the amino

acid balance necessary for good nutrition is not found in most of the cereal grains, we cannot take the step of abandoning meat sources altogether. Figure 5-5 does show how unlikely it is that increased fishing or fish protein concentrate will solve the world's food problems. Even if we leave aside the question of whether the fish are available—a point on which expert opinions differ somewhat—it would be hard to imagine, with rising energy prices, that fish protein concentrate will be anything more than a by-product of the fishing industry, because it requires more than twice the energy of production of grass-fed beef or eggs.[9] Distant fishing is still less likely to solve food problems. On the other hand, coastal fishing is relatively low in energy cost. Unfortunately, without the benefit of scholarly analysis, fishermen and housewives have long known this, and coastal fisheries are threatened with overfishing as well as pollution.

The position of soybeans in Figure 5-5 may be crucial. Soybeans possess the best amino acid balance and protein content of any widely grown crop. This has long been known to the Japanese, who have made soybeans a staple of their diet. Are there other plants, possibly better suited for local climates, that have adequate proportions of amino acids in their proteins? There are about 80,000 edible species of plants, of which only about 50 are actively cultivated on a large scale (and 90 per cent of the world's crops come from only 12 species). We may yet be able to find species that can contribute to the world's food supply.

The message of Figure 5-5 is simple. In "primitive" cultures, 5 to 50 food Calories were obtained for each calorie of energy invested. Some highly civilized cultures have done as well and occasionally better. In sharp contrast, industrialized food systems require 5 to 10 calories of fuel to obtain 1 food Calorie. We must pay attention to this difference—especially if energy costs increase. If some of the energy subsidy for food production could be supplied by on-site, renewable sources—primarily sun and wind—we might be able to continue an energy-intensive food system. Otherwise, the choices appear to be either less energy-intensive food production or famine for many areas of the world.

ENERGY REDUCTION IN AGRICULTURE

It is possible to reduce the energy required for agriculture and the food system. A series of thoughtful proposals by Pimentel and his associates[3] deserves wide attention. Many of these proposals would help ameliorate environmental problems, and any reductions in energy use would provide a direct reduction in the pollutants due to fuel consumption as well as more time to solve our energy supply problems.

First, we should make more use of natural manures. The United States has a pollution problem from runoff from animal feed lots, even with the application of large amounts of manufactured fertilizer to fields. More than 10^6 kilocalories per acre (4×10^5 kilocalories per hectare) could be saved by substituting manure for manufactured fertilizer[3] (and, as a side benefit, the soil's condition would be improved). Extensive expansion in the use of natural manure will require decentralization of feed-lot operations so that manure is generated closer to the point of application. Decentralization might increase feed-lot costs, but, as energy prices rise, feed-lot operations will rapidly become more expensive in any case. Although the use of manures can help reduce energy use, there is far too little to replace all commercial fertilizers at present.[10] Crop rotation is less widely practiced than it was even 20 years ago. Increased use of crop rotation or interplanting of winter cover crops of legumes (which fix nitrogen as a green manure) would save 1.5×10^6 kilocalories per acre by comparison with the use of commercial fertilizer.

Second, weed and pest control could be accomplished at a much smaller cost in energy. A 10 per cent saving in energy in weed control could be obtained by the use of the rotary hoe twice in cultivation instead of herbicide application (again with pollution abatement as a side benefit). Biologic pest control—that is, the use of sterile males, introduced predators, and the like—requires only a tiny fraction of the energy of pesticide manufacture and application. A change to a policy of "treat when and where necessary" pesticide application would bring a 35 to 50 per cent reduction in pesticide use. Hand application of pesticides requires more labor than machine or aircraft application, but the energy for application is reduced from 18,000 to 300 kilocalories per acre.[3] Changed cosmetic standards, which in no way affect the taste or the edibility of foodstuffs, could also bring about a substantial reduction in pesticide use.

Third, plant breeders might pay more attention to hardiness, disease and pest resistance, reduced moisture content (to end the wasteful use of natural gas in drying crops), reduced water requirements, and increased protein content, even if it should mean some reduction in overall yield. In the long run, plants not now widely cultivated might receive some serious attention and breeding efforts. It seems unlikely that the crops that have been most useful in temperate climates will be the most suitable ones for the tropics where a large portion of the undernourished peoples of the world now live.

A dramatic suggestion, to abandon chemical farming altogether, has been made by Chapman.[11] His analysis shows that, were chemical farming to be ended, there would be much reduced yields per acre, so that most land in the soil bank would need to be put back into farming. Nevertheless, output would fall only 5 per cent, and prices for farm products would increase 16 per cent. Most dramatically, farm income would rise 25 per cent, and nearly all subsidy programs would end. A similar set of propositions treated with linear programming techniques at Iowa State University resulted in an essentially similar set of conclusions.[12]

The direct use of solar energy farms, a return to wind power (modern windmills are now in use in Australia), and the production of methane from manure are all possibilities. These methods require some engineering to become economically attractive, but it should be emphasized that these technologies are now better understood than the technology of breeder reactors. If energy prices rise, these methods of energy generation would be attractive alternatives, even at their present costs of implementation.

ENERGY REDUCTION IN THE U.S. FOOD SYSTEM

Beyond the farm, but still far from the table, more energy savings could be introduced.

The most effective way to reduce the large energy requirements of food processing would be a change in eating habits toward less highly processed foods. The current aversion of young people to spongy, additive-laden white bread, hydrogenated peanut butter, and some other processed foods could presage such a change if it is more than just a fad. Technological changes could reduce energy consumption, but the adoption of lower-energy methods would be hastened most by an increase in energy prices, which would make it more profitable to reduce fuel use.

Packaging has long since passed the stage of simply holding a convenient amount of food together and providing it with some minimal protection. Legislative controls may be needed to reduce the manufacturer's competition in the amount and expense of packaging. In any case, recycling of metal containers and wider use of returnable bottles could reduce this large item of energy use.

The trend toward the use of trucks in food transport, to the virtual exclusion of trains, should be reversed. By reducing the direct and indirect subsidies to trucks we might go a long way toward enabling trains to compete.

Finally, we may have to ask whether the ever-larger frostless refrigerators are needed, and whether the host of kitchen appliances really means less work or only the same amount of work to a different standard.

Store delivery routes, even by truck, would require only a fraction of the energy used by autos for food shopping. Rapid transit, giving some attention to the problems with shoppers with parcels, would be even more energy-efficient. If we insist on a high-energy food system, we should consider starting with coal, oil, garbage—or any other source of hydrocarbons—and producing in factories bacteria, fungi, and yeasts. These products could then be flavored and colored appropriately for cultural tastes. Such a system would be more efficient in the use of energy, would solve waste problems, and would permit much or all of the agricultural land to be returned to its natural state.

ENERGY, PRICES, AND HUNGER

If energy prices rise, as they have already begun to do, the rise in the price of food in societies with industrialized agriculture can be expected to be even larger than the energy price increases. Slesser, in examining the case for England, suggests that a quadrupling of energy prices in the next 40 years would bring about a sixfold increase in food prices.[9] Even small increases in energy costs may make it profitable to increase labor input to food production. Such a reversal of a 50-year trend toward energy-intensive agriculture would present environmental benefits as a bonus.

We have tried to show how analysis of the energy flow in the food system illustrates features of the food system that are not easily deduced from the usual economic analysis. Despite some suggestions for lower-intensity food supply and some frankly speculative suggestions, it would be hard to conclude on a note of optimism. The world drawdown in grain stocks which began in the mid-1960s continues, and some food shortages are likely all through the 1970s and early 1980s. Even if population control measures begin to limit world population, the rising tide of hungry people will be with us for some time.

Food is basically a net product of an ecosystem, however simplified. Food production starts with a natural material, however modified later. Injections of energy (and even brains) will carry us only so far. If the population cannot adjust its wants to the world in which it lives, there is little hope of solving the food problem for mankind. In that case the food shortage will solve our population problem.[32]

NOTES AND REFERENCES

1. *Statistical Abstract of the United States.* Washington, D.C.: Government Printing Office, various annual editions.
2. *Historical Statistics of the United States.* Washington, D.C.: Government Printing Office, 1960.
3. D. Pimentel, L. E. Hurd, A. C. Bellotti, M. J. Forster, I. N. Oka, O. D. Scholes, and R. J. Whitman, *Science* **182,** 443 (1973).
4. A description of the system may be found in *Patterns of Energy Consumption in the United States* (report prepared for the Office of Science and Technology, Executive Office of the President, by Stanford Research Institute, Stanford, Calif., January 1972), Appendix C. The three groupings larger than food processing are primary metals, chemicals, and petroleum refining.
5. N. Georgescu-Roegen, *The Entropy Law and the Economic Process*, p. 301. Cambridge, Mass.: Harvard University Press, 1971.
6. *Patterns of Energy Consumption in the United States* (report prepared for the Office of Science and Technology, Executive Office of the President, by Stanford Research Institute, Stanford, Calif., January 1972).
7. R. A. Rice, *Technol. Rev.* **75** (1), 32 (1972).
8. Federal Highway Administration, Nationwide Personal Transportation Study Report No. 1, p. 151 (1971) [as reported in Energy Research and Development, hearings before the Congressional Committee on Science and Astronautics, May 1972].
9. M. Slesser, *Ecologist* **3** (6), 216 (1973).
10. J. F. Gerber, personal communications (we are indebted to Dr. Gerber for pointing out that manures, even if used fully, will not provide all the needed agricultural fertilizers).
11. D. Chapman, *Environment* (*St. Louis*) **15** (2), 12 (1973).
12. L. U. Mayer and S. H. Hargrove, *CAED Rep. No. 38* (1972), as quoted in Slesser (9).
13. We have converted all figures for the use of electricity to fuel input values, using the average efficiency values for power plants given by C. M. Summers, *Sci. Am.* **224** (3), 148 (1971). Self-generated electricity was converted to fuel inputs at an efficiency of 25 per cent after 1945 and 20 per cent before that year.
14. Purchased material in this analysis was converted to energy of manufacture according to the following values derived from the literature or calculated. In doubtful cases we have made what we believe to be conservative estimates: steel (including fabricated and castings), 1.7×10^7 kcal/ton (1.9×10^4 kcal/kg); aluminum (including castings and forgings), 6.0×10^7 kcal/ton; copper and brass (alloys, millings, castings, and forgings), 1.7×10^6 kcal/ton; paper, 5.5×10^6 kcal/ton; plastics, 1.25×10^6 kcal/ton; coal, 6.6×10^6 kcal/ton; oil and gasoline, 1.5×10^6 kcal/barrel (9.5×10^3 kcal/liter); natural gas, 0.26×10^3 kcal/cubic foot (9.2×10^3 kcal/m^3); petroleum wax, 2.2×10^6 kcal/ton; gasoline and diesel engines, 3.4×10^6 kcal/engine; electric motors over 1 horsepower, 45×10^3 kcal/motor; ammonia, 2.7×10^7 kcal/ton; ammonia compounds, 2.2×10^6 kcal/ton; sulfuric acid and sulfur, 3×10^6 kcal/ton; sodium carbonate, 4×10^6 kcal/ton; and other inorganic chemicals, 2.2×10^6 kcal/ton.
15. Direct fuel use on farms: Expenditures for petroleum and other fuels consumed on farms were obtained from *Statistical Abstracts* (1) and the *Census of Agriculture* (Bureau of the Census, Government Printing Office, Washington, D.C., various recent

editions) data. A special survey of fuel use on farms in the 1964 *Census of Agriculture* was used for that year and to determine the mix of fuel products used. By comparing expenditures for fuel in 1964 with actual fuel use, we were able to calculate the apparent unit price for this fuel mix. Using actual retail prices and price indexes from *Statistical Abstracts* and the ratio of the actual prices paid to the retail prices in 1964, we derived the fuel quantities used in other years. Changes in the fuel mix used (primarily the recent trend toward more diesel tractors) may understate the energy in this category slightly in the years since 1964 and overstate it slightly in years before 1964. S. H. Schurr and B. C. Netschert (*Energy in the American Economy, 1850–1975*, p. 774. Baltimore: Johns Hopkins Press, 1960), for example, using different methods, estimate a figure 10 per cent less for 1955 than that given here. On the other hand, some retail fuel purchases appear to be omitted from all these data for all years. M. J. Perelman [*Environment* (*St. Louis*) **14** (8), 10 (1972)], from different data, calculates 270×10^{12} kcal of energy usage for tractors alone.

16. Electricity use on farms: Data on monthly usage on farms were obtained from the Report of the Administrator, Rural Electrification Administration (U.S. Department of Agriculture, Government Printing Office, Washington, D.C., various annual editions). Totals were calculated from the annual farm usage multiplied by the number of farms multiplied by the fraction electrified. Some nonagricultural uses are included, which may overstate the totals slightly for the years before 1955. Nevertheless, the totals are on the conservative side. A survey of on-farm electricity usage published by the Holt Investment Corporation, New York, May 18, 1973, reports values for per farm usage 30 to 40 per cent higher than those used here, suggesting that the totals may be much too small. The discrepancy is probably the result of the fact that the largest farm users are included in the business and commercial categories (and excluded from the U.S. Department of Agriculture tabulations used).

17. Fertilizer: Direct fuel use by fertilizer manufacturers was added to the energy required for the manufacture of raw materials purchased as inputs for fertilizer manufacture. There is allowance for the following: ammonia and related compounds, phosphatic compounds, phosphoric acid, muriate of potash, sulfuric acid, and sulfur. We made no allowance for other inputs (of which phosphate rock, potash, and "fillers" are the largest), packaging, or capital equipment. Source, *Census of Manufactures*. Washington, D.C.: Government Printing Office, various recent editions.

18. Agricultural steel: Source, *Statistical Abstracts* for various years (1). Converted to energy values according to note 14.

19. Farm machinery (except tractors): Source, *Census of Manufacturers*. Totals include direct energy use and the energy used in the manufacture of steel, aluminum, copper, brass, alloys, and engines converted according to note 14.

20. Tractors: numbers of new tractors were derived from *Statistical Abstracts* and the *Census of Agriculture* data. Direct data on energy and materials use for farm tractor manufacture was collected in the *Census of Manufactures* data for 1954 and 1947 (in later years these data were merged with other data). For 1954 and 1947, energy consumption was calculated in the same way as for farm machinery. For more recent years, a figure of 2.65×10^6 kcal per tractor horsepower calculated as the energy of manufacture from 1954 data (the 1954 energy of tractor manufacture, 23.6×10^{12} kcal, divided by sales of 315,000 units divided by 28.7 average tractor horsepower in 1954). This figure was used to calculate energy use in tractor manufacture in more recent years to take some account of the continuing increase in tractor size and power. It probably slightly understates the energy in tractor manufacture in more recent years.

21. Irrigation energy: Values are derived from the acres irrigated from *Statistical Abstracts* for various years; converted to energy use at 10^6 kcal per acre irrigated. This is an intermediate value of two cited by Pimentel et al. (3).

22. Food processing industry: Source, *Census of Manufactures;* direct fuel inputs only. No account taken for raw materials other than agricultural products, except for those items (packaging and processing machinery) accounted for in separate categories.

23. Food processing machinery: Source, *Census of Manufactures* for various years. Items included are the same as for farm machinery (see note 13).

24. Paper packaging: Source, *Census of Manufactures* for various years. In addition to direct energy use by the industry, energy values were calculated for purchased paper, plastics, and petroleum wax, according to note 14. Proportions of paper products having direct food usage were obtained from *Containers and Packaging*, Washington, D.C.; U.S. Department of Commerce, various recent editions. [The values given include only proportional values from Standard Industrial Classifications 2651 (half), 2653 (half), 2654 (all).]

25. Glass containers: Source, *Census of Manufactures* for various years. Direct energy use and sodium carbonate (converted according to note 14) were the only inputs considered. Proportions of containers assignable to food are from *Containers and Packaging*. Understatement of totals may be more than 20 per cent in this category.

26. Steel and aluminum cans: Source, *Census of Manufactures* for various years. Direct energy use and energy used in the manufacture of steel and aluminum inputs were included. The proportion of cans used for food has been nearly constant at 82 per cent of total production (*Containers and Packaging*).

27. Transportation fuel usage: Trucks only are included in the totals given. After trucks used solely for personal transport (all of which are small trucks) are

subtracted, 45 per cent of all remaining trucks and 38 per cent of trucks larger than pickup and panel trucks were engaged in hauling food or agricultural products, or both, in 1967. These proportions were assumed to hold for earlier years as well. Comparison with ICC analyses of class I motor carrier cargos suggests that this is a reasonable assumption. The total fuel usage for trucks was apportioned according to these values. Direct calculations from average mileage per truck and average number of miles per gallon of gasoline produces agreement to within ±10 per cent for 1967, 1963, and 1955. There is some possible duplication with the direct fuel use on farms, but it cannot be more than 20 per cent considering on-farm truck inventories. On the other hand, inclusion of transport by rail, water, air, and energy involved in the transport of fertilizer, machinery, packaging, and other inputs of transportation energy could raise these figures by 30 to 40 per cent if ICC commodity proportions apply to all transportation. Sources: *Census of Transportation* (Washington, D.C.: Government Printing Office, 1963, 1967); *Statistical Abstracts* (1); *Freight Commodity Statistics of Class I Motor Carriers* (Interstate Commerce Commission, Government Printing Office, Washington, D.C., various annual editions).

28. Trucks and trailers: Using truck sales numbers and the proportions of trucks engaged in food and agriculture obtained in note 27, we calculated the energy values at 75×10^6 kcal per trucks for manufacturing and delivery energy [A. B. Makhijani and A. J. Lichtenberg, *Univ. Calif. Berkeley Mem. No. ERL-M310* (revised) (1971)]. The results were checked against the *Census of Manufacturers* data for 1967, 1963, 1958, and 1939 by proportioning motor vehicles categories between automobiles and trucks. These checks suggest that our estimates are too small by a small amount. Trailer manufacture was estimated by the proportional dollar value to truck sales (7 per cent). Since a larger fraction of aluminum is used in trailers than in trucks, these energy amounts are also probably a little conservative. Automobiles and trucks used for personal transport in the food system are omitted. Totals here are probably significant, but we know of no way to estimate them at present. Sources: *Statistical Abstracts, Census of Manufactures,* and *Census of Transportation* for various years.

29. Commercial and home refrigeration and cooking: Data from 1960 through 1968 (1970 extrapolated) from *Patterns of Energy Consumption in the United States* (6). For earlier years sales and inventory in-use data for stoves and refrigerators were compiled by fuel and converted to energy from average annual use figures from the Edison Electric Institute (*Statistical Year Book.* New York: Edison Electric Institute, various annual editions) and American Gas Association values (*Gas Facts and Yearbook.* Arlington, Va.: American Gas Association, various annual editions) for various years.

30. Refrigeration machinery: Source, *Census of Manufactures.* Direct energy use was included and also energy involved in the manufacture of steel, aluminum, copper, and brass. A few items produced under this SIC category for some years perhaps should be excluded for years prior to 1958, but other inputs, notably electric motors, compressors, and other purchased materials, should be included.

31. There are many studies of energy budgets in primitive societies. See, for example, H. T. Odum, *Environment, Power, and Society* (New York: Wiley, Interscience, 1970) and R. A. Rappaport, *Sci. Am.* **224** (3), 104 (1971). The remaining values of energy subsidies in Fig. 5-5 were calculated from data presented by Slesser (9), Table 1.

32. Some of this research was supported by the U.S. Geological Survey, Department of the Interior, under grant No. 14-08–0001-G-63. Contribution 18 of the Marine Studies Center, University of Wisconsin-Madison. Since this article was completed, the analysis of energy use in the food system by E. Hirst has come to our attention ("Energy Use for Food in the United States," *ONRL—NSF-EP-57,* October 1973). Using different methods, he assigns 12 per cent of total energy use to the food system for 1963. This compares with our result of about 13 per cent in 1964.

SUGGESTED READINGS

David Pimental and Marcia Pimental, *Food, Energy and Society.* New York: Wiley, 1979.

David Pimental and Elinor C. Terhune, "Energy and Food," *Annual Review of Energy* **2,** 171 (1977).

Conflicts and Constraints: Energy from Western United States   6

M. D. DEVINE, S. C. BALLARD, AND I. L. WHITE

The United States is under considerable pressure, both at home and abroad, to reduce its imports of foreign oil. To do so will require the limiting of energy consumption growth rates and increased production from its own domestic energy reserves. The eight western states of the United States (see Figure 6-1) contain a large fraction of undeveloped U.S. energy reserves. The significance of these states' reserves can be judged from Table 6-1. The technologies being considered for their exploitation are listed in Table 6-2. The list includes technologies currently being used commercially as well as newer energy conversion systems not yet commercially feasible, at least in the United States.[1]

CHARACTERISTICS OF THE REGION

Energy resource development in the West is, and will continue to be, strongly influenced by a number of regional factors, including the availability of natural resources other than energy, population patterns, sectors of economic activity, and attitudes towards industrial development. Among the most important of these regional characteristics are:

Water scarcity. Most surface waters are already allocated, often for irrigated agriculture. In the Upper Missouri region the surface waters are poorly located for on-site energy resource development. In some areas ground water is being used more quickly than the recharge rate. Precipitation levels range from less than 10 inches per year in the southwestern desert area to 10 to 20 inches in the northern Great Plains region.

Federal and Indian land ownership. More than half the land in Arizona, Utah, and Wyoming is under federal government and Indian ownership (see Table 6-3). Data on energy resource ownership are difficult to obtain, but it appears that the federal government owns half the coal, geothermal, and uranium resources, and about 80 per cent of

Dr. Michael Devine and Dr. Steven Ballard are, respectively, Director and Assistant Director of the Science and Public Policy Program, University of Oklahoma, Norman, Oklahoma 73019.

Dr. Irvin L. White is Executive Director of the New York State Energy Research and Development Authority, Albany, New York 12223.

From *Energy Policy* **8** (3), 229 (1980). Reprinted by permission of the authors and *Energy Policy*. Copyright © 1980 by the IPC Business Press Ltd., New York.

Figure 6-1. The eight western states: Montana, North Dakota, South Dakota, Wyoming, Utah, Colorado, Arizona, and New Mexico.

the oil shale reserves in the eight-state area. The Council of Energy Resources Tribes (CERT)[2] estimates that its member tribes own over 40 per cent of the total U.S. uranium reserves and roughly 30 per cent of all strippable western coal.

Table 6–1. Proven Reserves of Six Energy Resources in the Eight-State Study Area.[a]

Resources	Reserves in the Eight Western States (10^{15} Btu)	Proportion of Total U.S. Reserves (%)
Coal	3430.0	35.9
Crude oil	12.2	6.4
Natural gas	19.9	7.8
Oil shale (high grade)	464.0	100.0[b]
Uranium	246.6	90.6
Geothermal[c]	650.0	22.1

[a] See Figure 6–1.
[b] Roughly.
[c] This figure includes both reserves and resources in identified hydrothermal convection systems which are not currently economically competitive.
From: Irvin L. White et al., *Energy from the West: Policy Analysis Report,* Washington, D.C.: Environmental Protection Agency, 1979.

Table 6–2.   Development Options.

Coal
Surface and underground mining
Direct export by unit train and slurry pipeline
Electric power generation
Gasification
Liquefaction
Transport by pipeline
High voltage transmission

Oil Shale
Underground mining
Surface retorting
Modified *in situ* retorting
Transport by pipeline

Uranium
Surface, underground, and solutional mining
Milling
Transport by road

Oil and Natural Gas
Conventional drilling and production
Enhanced oil recovery
Transport by pipeline

Geothermal
Hot water and hot rock
Electric power generation
High-voltage transmission

Table 6–3. Federal and Indian Lands in the Eight Western States.

State	Total State Area (thousands of acres)	Federal (%)	Indian (%)	Federal and Indian (%)
Arizona	72,688	43	29	72
Utah	52,697	66	4	70
Wyoming	62,343	48	3	51
New Mexico	77,766	34	8	42
Colorado	66,486	36	1	37
Montana	93,271	30	6	36
South Dakota	48,882	7	12	19
North Dakota	44,452	5	5	10
Total	518,585	35	9	44

Sparse population. Except for five large metropolitan areas (Albuquerque, Denver, Phoenix, Salt Lake City, and Tucson), the region is sparsely populated. In contrast to the U.S. average population density of 60 persons per square mile, only three of the eight states have population densities greater than ten persons per square mile, and Wyoming has fewer than four persons per square mile.

Scenic beauty. The eight-state area includes some of the most scenic and highly valued recreation areas in the United States. These include parks, forests, and recreation, historic, and wilderness areas as well as a wide range of plant and animal communities.

Economic activities. Agriculture, mining, tourism, and government agencies are the major sectors of economic activity. Manufacturing employs more than 10 per cent of the labor force only in Arizona, Colorado, and Utah.

Opposition to outside intervention. There is a general opposition in the region to outside intervention, whether by the federal government or by environmentalists. Opposition ranges from an unwillingness to be treated as an energy-exporting colony for the rest of the nation to resentment that development is restricted in many areas because of environmental legislation and the activities of environmental interest groups.

THE APPROACH

To examine the effects of greatly increased energy resource development, scenarios were constructed both for specific sites and for the region as a whole. The six site-specific scenarios combined different local conditions (e.g., topography, meteorology, water availability, current population, and community service and facilities) with alternative energy development technologies. For example, a scenario for Farmington, New Mexico, considered surface coal mines, a coal-fired plant, a Lurgi high-Btu coal gasification plant, a Synthane high-Btu coal gasification plant, a coal liquefaction plant, a uranium mining and milling facility, and the necessary energy transport systems (high-voltage transmission lines and pipelines).

Two regional scenarios (nominal-demand and low-demand) were developed using Stanford Research Institute's interfuel competition model.[3] The nominal- and low-demand scenarios assume a national energy supply of 155.1×10^{15} Btu and 124.0×10^{15} Btu, respectively, in the year 2000. The contribution from the eight western states in 2000 is 60.70×10^{15} Btu and 42.56×10^{15} Btu for the nominal- and low-demand cases, respectively. A breakdown of this contribution into various fuel types is shown in Table 6-4. The low-demand scenario is closer to current projections of U.S. energy supply for 2000

Table 6–4. Low-Demand and Nominal-Demand Scenarios for Energy Production in the Eight-State Region.

Energy Type	Nominal-Demand Case (10^{15} Btu)				Low-Demand Case (10^{15} Btu)		
	1975	1980	1990	2000	1980	1990	2000
Coal production	1.57	5.04	12.60	29.90	4.28	9.80	22.10
Direct use	1.50	3.73	8.30	14.80	3.13	6.38	11.10
Unit train	1.50	3.14	5.33	7.07	2.71	4.13	5.28
Slurry pipeline	—	0.52	2.97	7.75	0.42	2.26	5.79
Gasification	—	—	0.23	3.83	—	0.15	2.30
Liquefaction	—	—	0.001	0.31	—	0.001	0.38
Electrical generation	0.03	0.48	1.08	1.43	0.41	·0.84	1.08
Oil shale	—	0.001	0.95	4.76	0.001	0.19	0.95
Uranium fuel	1.95	4.77	12.70	23.80	4.15	9.46	17.10
Gas (methane)	1.71	1.97	2.08	1.06	1.99	1.89	1.19
Domestic crude oil	1.43	1.67	1.32	1.03	1.74	1.34	1.20
Geothermal	—	—	0.09	0.15	—	0.01	0.02
Total	6.66	13.45	29.74	60.70	12.16	22.69	42.56

than the nominal-demand scenario. However, the higher figure for supply from the western states may still be reasonable, given the probability that a national synthetic fuels program will be implemented. Such a program will almost certainly result in increased coal and oil shale development in the eight-state western area.

The scenarios were constructed to reflect the range of energy resources, technologies, and site conditions which could occur in the West. They were constructed to help answer "what if" questions rather than to be actual projections of future development. Analysis of the range of potential effects of energy resource development is based on these scenarios. Four categories of impact are analysed here: water availability and quality, air quality, growth management and housing, and land use.

WATER AVAILABILITY AND QUALITY

Water availability and water quality impacts are among the most critical associated with expanded energy development. This is because of both physical and sociopolitical factors. If large-scale energy resource develop-

ment does take place, water supplies are likely to be inadequate to meet the increased demands of all users (including the energy industries) over the next 25 years, and this will intensify existing political conflicts over water resource management.

Supply and Demand

Water shortages are likely to be particularly severe in the Colorado River Basin. When requirements for water estimated on the basis of the nominal-demand scenario are added to nonenergy requirements for the year 2000, the total may exceed minimum availability estimates by as much as one million acre-feet/year (see Figure 6-2). Even using optimistic water availability assumptions and low water requirement assumptions, energy resource development will consume a large fraction of remaining unappropriated surface waters.

While available supplies are larger in the Upper Missouri River Basin than in the Upper Colorado, serious conflicts have already developed in that basin over water use. One manifestation of this conflict is the four-year moratorium (1975–78) on new water allo-

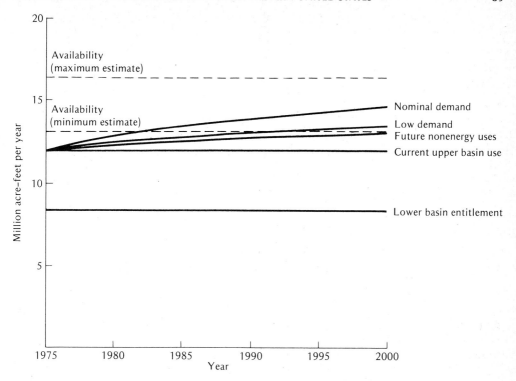

Figure 6-2. Projections of water requirements and availability for the Colorado River area in 2000. (From Irvin L. White et al., *Energy From the West: Policy Analysis Report,* p. 103.)

cations from the Yellowstone River and a recently enacted Wyoming law which effectively bans the use of surface waters in coal slurry pipelines.

The major water use conflict is between energy resource development and agriculture. Agriculture is by far the largest water user in the west. In 1974, for example, agriculture accounted for 58 per cent of total water depletion in the Upper Colorado River Basin. On the other hand, municipal, industrial, and energy resource development together accounted for only 5.4 per cent. As water is diverted to energy resource development, irrigated agriculture will be restricted or, in some areas, eliminated. Since this could change the social and economic structure of the region, there will almost certainly be increased political conflicts between the federal and state governments, between regions of the United States, and between the individual western states.

Water Regulations

Water availability issues are complicated by the complex regulatory system governing water use. This system includes international treaties, interstate compacts, state appropriation systems, court decisions, and federal reclamation and pricing policies. One of the most important water rights conflicts in the West concerns federal and Indian reserved water rights. Reserved rights recognize that when the United States establishes a federal reservation, such as a national park or Indian reservation, a sufficient quantity of unappropriated water is reserved to accomplish the

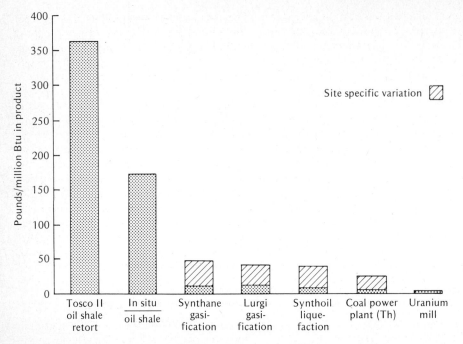

Figure 6-3. Quantities of solid and liquid effluents from various energy technologies (dry basis).

purpose for which the land was reserved. However, in most cases, the amount of such reserved water rights has not been quantified, and this leaves current and future users uncertain as to how much, if any, water might be available to fulfil their rights.

Environmental Damage

Another major issue raised by increased water use for energy development is the environmental damage resulting from reduced stream flows. Threatened interests include recreation, in-stream habitat and lands, and scenic and aesthetic values. The importance of these issues is illustrated by the fact that recently the Montana Fish and Game Commission requested an appropriation of 7.2 million acre-feet/year from unappropriated water of the Yellowstone River Basin. This request for water to protect in-stream values is larger than the amount of water currently estimated to be available from the river.

Of course, environmental damage can also occur more directly as a result of lowered water quality resulting from the exploitation of energy resources. This can occur in several ways. Strip mining and *in situ* oil shale development can contaminate aquifers which feed surface streams, and mining run-off can pollute streams directly. Also, although the Federal Water Pollution Control Act limits the direct discharge of industrial pollutants into surface water, energy facilities can still pose a threat to both surface and ground waters through the current methods of disposing of liquid and solid effluents. Figure 6-3 indicates the relative quantities of these residuals per unit energy produced. Spent shale from above-ground oil shale retorting facilities poses the largest problem in terms of magnitude; a 100,000-barrel/day facility will generate approximately 13.8 million tons of spent shale annually. Run-off and leaching through spent shale disposal piles could be serious water quality problems; at present there is great uncertainty about the extent to which these wastes can be stabilized and re-vegetated. At coal conversion facilities, wastes will be discharged to holding ponds, and over a 25-year lifetime, 13 to 100 million tons of wastes can be accumulated, depending on the type of facility. These wastes can contain heavy metals and organic compounds which may threaten water quality if there is seepage from holding ponds, flooding, or retaining wall failures. The amount of waste generated by a uranium mill is small compared with the other types of plant included in Figure 6-3, and therefore the regional environmental problems associated with waste disposal are fewer than with coal and oil shale. Nevertheless, because of the character of these wastes, they do represent a significant environmental risk. For example, in July 1979 a tailings pond dam failure at a uranium production site in New Mexico resulted in 100 million gallons of radioactive tailings waste water being released.[4]

Salinity, a major concern in the western region, will also be affected by the increased water demands of energy resource development. When water is consumed by energy facilities (or any other users), salinity levels are generally concentrated downstream. Not only is the United States obliged by treaty to limit the salinity of the Colorado River flowing into Mexico, but also the Environmental Protection Agency (EPA) has required that the states act together to maintain salinity in the Lower Colorado River Basin at or below 1972 levels.

Reducing the Conflict

The conflicts over water use and quality reflect both the physical scarcity of the resource and the rapidly increasing demands for its use. The traditional technological solution to western water availability—to divert, transfer, and store water to make it more generally usable—appears to be much less viable today than in the past. This is, in part, because few acceptable locations for these projects remain in the Colorado River Basin.[5]

However, many other problems exist with this alternative, including questions about economic efficiency and environmental costs, which are likely to limit its application in both the Colorado and Upper Missouri river basins.

Several other options exist to reduce the conflicts described above. These include several technological approaches, such as weather modification, conservation of water by altering process designs or by using some degree of dry cooling in energy conversion facilities, and using low-quality saline water for energy conversion facilities. Legal and institutional policy options include changes in state appropriation systems, water quality control plans, and development of regional water management institutions.

Water Conserving Technology

One of the most promising of these alternatives for dealing with water availability issues is to promote the use of water conserving technologies in the various energy conversion facilities. Table 6-5 lists the water savings and costs for utilizing various degrees of dry cooling instead of wet cooling. As the data indicate, minimum wet cooling systems can reduce the water consumption of synthetic fuel plants by 25 to 40 per cent with very little economic penalty. For power stations, intermediate wet cooling can be used to reduce water consumption by about two thirds with a cost increase of around 3 to 7 per cent at typical electricity prices. Minimum wet cooling is generally not economically practical for power plants unless water gets very expensive—more than $1900 per acre-foot.

Another approach available to policy makers is to promote water conservation in agriculture (through such means as sprinkler systems or trickle irrigation) with the water savings being diverted for energy uses. While this is an attractive approach technically, in many cases there are several obstacles to its implementation. Most important, these more efficient irrigation systems are expensive

Table 6–5.   Two Water Conservation Technologies for Energy Facilities.

Cooling Technology	Facility			
	Electric Power Plant (3000 MWe)	Lurgi High-Btu Gas Plant (250 million ft³/day)	Synthane High-Btu Gas Plant (250 million ft³/day)	Synthoil Liquefaction Plant (100,000 bbl/day)
Intermediate Wet Cooling				
Water savings				
Acre-feet per year	6700–20,300	1500–1700	1800–2000	1700–2100
Percentage	67–77	21–32	23–24	16–19
Increased cost	0.11–0.18¢/kWh	0.15–1.03¢/10⁶ Btu	0.18–1.22¢/10⁶ Btu	0.14–1.00¢/10⁶ Btu
Minimum Wet Cooling				
Water savings				
Acre-feet per year	n.a.	1900–2100	2150–2400	2200–2600
Percentage	n.a.	27–42	27–29	21–25
Increased cost	n.a.	1.18–1.32¢/10⁶ Btu	1.33–1.47¢/10⁶ Btu	1.06–1.27¢/10⁶ Btu

Note: Based on water costs of 20¢ per 1000 gallons. Ranges reflect site-specific differences within the eight-state region.
From: Irvin L. White et al., *Energy from the West: Policy Analysis Report*, and Water Purification Associates, *Wet/Dry Cooling and Cooling Tower Blow-Down Disposal in Synthetic Fuel and Steam-Electric Power Plants*. Washington, D.C.: Environmental Protection Agency, 1979.

(estimates range from $64 to $148 total annualized cost per acre per year) and beyond the financial capabilities of most farmers without some form of government assistance. Also, there are several institutional and legal barriers, including the water appropriation systems, which may not allow a farmer to put the water saved to some other beneficial use, since a downstream user may depend on the return flow.

AIR QUALITY

Along with water availability and water quality, air quality issues have a pervasive influence on energy development in the western states of the United States. In most areas, the air is very clean and visibility distances are great. The desire to protect the currently pristine conditions means that air quality will continue to affect which resources will be developed as well as where, when, and at what development and product costs. However, of the six resources considered in this article, four (uranium, geothermal power, crude oil, and natural gas) have relatively low emission levels, and thus their development is not likely to be affected in any major way by air quality regulations.

Best Available Control Technology

It is interesting to note that the 1970s boom in demand for western coal was largely due to federally established sulphur dioxide (SO_2) emission limits for new power plants of 1.2 pounds per million Btu of fuel burned. Many users (especially electric power companies in the central region of the nation) found they could meet this standard most economically by transporting low-sulphur western coals (which required no sulphur controls to meet the standard) over long distances rather than by using higher-sulphur local coals along with flue gas desulphurization.

However, this regulation was changed by the 1977 Clean Air Act Amendments.[6] Basically the new regulation requires the use of the "best available control technology" (BACT) regardless of the original sulphur content of the coal. This change was intended not only to reduce sulphur emission levels but also to remove some competitive advantages western coals were enjoying relative to central and eastern coals. Analysis of the economic trade-offs indicates that, in fact, BACT requirements could substantially reduce the demands for western coals. For example, one study indicates that by 1990, under the previous standard, Northern Great Plains coal production would grow to an annual production rate of 388 million tons, whereas with the BACT standard it will reach only about 200 million tons.[7] However, these projections are highly uncertain because of such factors as the questionable reliability of supply from unionized eastern mines (most western coal mining operations are non-unionized and less labor intensive) and the possibility that increased pollution control costs will cause a substitution of nuclear power for coal.

In addition to limiting export of western coal to other regions, the BACT requirement will tend to reduce the air degradation caused by energy facilities located in the western region. One study indicates that by 1995, SO_2 emissions from power plants will be approximately half what they would be without the new BACT standard.[8] However, it should be pointed out that even without the BACT requirement, emission levels will be relatively small in the area as a result of federally established prevention of significant deterioration (PSD) standards and state standards which, for the eight-state area, are often more stringent than the federal ones. To illustrate, Table 6-6 shows the SO_2 control requirements that would be necessary to meet all applicable standards (excluding BACT) for large power plants located at different sites in the region. As shown, even without BACT, high degrees of SO_2 control will be required (ranging from 78 to 96 per cent) for power plants of this size.

Prevention of Significant Deterioration

Another major issue concerns the effects of federally established PSD regulations on energy development. These regulations require all areas to be placed in one of three classes of air quality. Class I areas allow the smallest increases in air pollution and Class III the largest. The eight western states contain a large number of mandatory Class I areas, including national parks, wilderness areas, and other areas of particular natural beauty. The regulations were established to protect such areas. The issue is that energy development

Table 6–6. Sulphur Removal Efficiencies Required for Coal-Fired Power Plants To Meet All Federal and State Sulphur Dioxide Standards Except Federal BACT.

State, Town, and Installed Capacity	Sulphur Removal Required (%)	Governing Regulation
Colorado		
Rifle, 1000 MW and mine	96.2	State Category II Ambient
Montana		
Colstrip, 3000 MW and mine	80[a]	Federal 24-hour Class II PSD[b]
North Dakota		
Beulah, 3000 MW and mine	85	Federal 24-hour Class II PSD[b]
New Mexico		
Farmington, 3000 MW and mine	78	Federal 24-hour Class II PSD[b]
Utah		
Escalante, 3000 MW and mine	94	Federal 24-hour Class II PSD[b]
Wyoming		
Gillette, 3000 MW and mine	82	State NSPS[c]

[a] Maximum control capability is determined by the State New Source Performance Standards. Technical and economic feasibility is determined by the Air Quality Bureau.
[b] PSD denotes Prevention of Significant Deterioration standard.
[c] NSPS denotes New Source Performance Standard.

Table 6–7.   Distance from Class I Areas
Required To Meet *Prevention of Significant
Deterioration* Regulations for Sulphur Dioxide,
80 Per Cent Sulphur Dioxide Control.

Station	Size (MWe)	Required Distance from Class I Areas (miles)
Escalante	3000	>25.0[a]
Kaiparowits	3000	>25.0[a]
Rifle	1000	13.6
Farmington	3000	58.0
Beulah	3000	75.0
Gillette	3000	44.0
Colstrip	3000	75.0

[a]Class I buffer zone size cannot be established because of the
complex terrain in the Kaiparowits/Escalante area.

may be severely constrained by the strictness
and the complexity of the regulations. Before a plant can be given a construction permit it must demonstrate, using an accepted
pollution dispersion model, that the facility
will not cause the allowable PSD increments
for any surrounding areas to be exceeded. The
allowable increments for Class I areas effectively establish a "buffer zone" between the
area and an energy facility, the size of which
depends on meteorological conditions, facility size and type, and other factors. To illustrate, Table 6-7 lists separation distances that
would be required for large coal-fired power
plants under conditions existing at the specific sites studied. These PSD regulations
have already delayed for at least 18 months
(and possibly blocked) the construction of two
new coal-fired power plant units at Colstrip,
Montana, because they would have caused
the Class I increments to be exceeded on the
nearby Northern Cheyenne Indian Reservation. In early 1977, the Northern Cheyenne
elected to have their reservation designated
Class I, but a legal battle ensued over the
issue of whether the two new units required
PSD permits since they were already under
construction.

Visibility Standards

The effect of large-scale energy development
in the western United States on visibility is

another major concern. Although visibility
standards for Class I PSD areas are to be established, they have not yet been promulgated.[9] While scientific knowledge concerning the relationship between air emissions and
visibility reduction is incomplete, it does appear that visibility considerations will be quite
important. Table 6-8 shows, for the six site-
specific scenarios, estimates for the worst-
case short-term visibility reductions due to
sulphates, based on an assumed SO_2 to sulphates conversion rate of 1 per cent per
hour.[10] As the data show, these reductions
can be quite high (80–90 per cent) in many
cases. These worst-case conditions will occur only a few times each year, but nevertheless, given the many scenic areas in the region, the potential for conflict is great.

Resolving the Conflicts

While there are other important air quality
issues, this brief description illustrates the
conflict between protecting and enhancing the
West's air on the one hand and increasing
domestic energy production on the other. It
also highlights the uncertainty of scientific
knowledge about air quality impacts and the
complexity of the rapidly changing regulatory system. Because of this uncertainty and
complexity, it is difficult if not impossible to
suggest a single policy approach or alternative for best dealing with these conflicts.
Several technological options do exist—
ranging from alternative definitions of BACT
and the encouraging of less polluting technologies (such as synthetic fuel plants or
geothermal power plants) to the construction
of smaller conversion facilities. However, the
success of these options will depend on
whether or not imposed procedural mechanisms for energy facility siting can be established.

In this regard, one option with demonstrated success is to establish siting task forces
which represent various parties at interest (for
example, state and federal agencies, environmental groups, farmers and ranchers, and
energy companies). The general purpose of
such task forces would be to create an overtly

Table 6–8. Worst-Case Short-Term Effects on Visibility.

Type of Facility[a]	Site	Background Visibility (miles)	Worst-Case Short-Term Visibility (miles)	Visibility Reduction (%)
Coal-fired power plant	Kaiparowits	70	8.6	87.7
	Rifle	60	43.6	27.3
Coal-fired power plant and mine	Gillette	70	9.6	86.3
	Beulah	60	4.8	92.0
Synthane gasification and mine	Gillette	70	59.5	15.0
	Colstrip	60	48.0	18.5
TOSCO II oil shale	Rifle	60	44.4	26.0
Lurgi and Synthane gasification, coal-fired power plants, and mines	Farmington	60	9.3	84.5
	Colstrip	60	8.1	86.5
Lurgi and Synthane gasification, Synthoil liquefaction, coal-fired power plant, and mines	Farmington	60	8.2	86.3
Lurgi gasification and mine (two plants)	Beulah	60	37.4	37.7

[a] Facilities modeled are 3000-MWe coal-fired power plants (except for Rifle, where a 1000-MWe plant was modeled), 250-million-ft³/day Lurgi gasification plant, 250-million-ft³/day plant for Synthane gasification, 100,000-bbl/day Synthoil liquefaction plant, and a 50,000-bbl/day TOSCO II oil shale plant with the associated mine. The power plants were modeled with 99 per cent removal of particulates and 80 per cent removal of SO_2.
Source: Irvin L. White et al., *Energy from the West: Impact Analysis Report.*

political process for reaching an accommodation among conflicting interests. The specific purpose would be to reach a consensus· on future sites for energy facilities and thus avoid the expensive and time-consuming delays often encountered under present siting procedures. Such an approach has been used in Utah, resulting in an apparently successful agreement to avoid siting power plants in Utah's scenic southeast quadrant, with eight sites in the central part of the state designated for future development.

This procedural approach to resolution of air quality conflicts obviously does not reduce air pollution. However, it has several advantages. First, it has the potential to accommodate competing interests in a more systematic fashion than is currently typical. That is, in the current system, interests are typically accommodated through ad hoc, fragmented, and time-consuming court proceedings. Second, the economic costs of the task forces are apparently small, unless extensive environmental studies are undertaken.[11] Third, implementation of the alternative should be straightforward; it could be convened at the request of a governor, requiring no new laws or institutions.

GROWTH MANAGEMENT AND HOUSING

Energy resource development in the West can produce serious growth management and housing problems for small, isolated communities. These communities often experience rapid and large population fluctuations (i.e., boom and bust cycles) when energy development takes place nearby. The greatest effects result from labor intensive facilities, such as coal gasification and electric power plants. Population increases from these

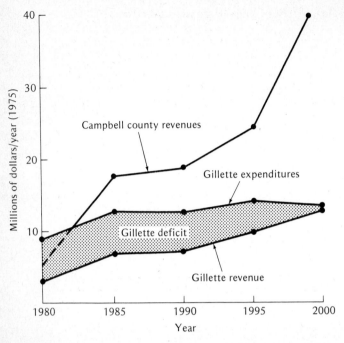

Figure 6-4. Projected revenue-expenditure imbalances for Campbell County and Gillette, Wyoming, in 2000.

developments will require rapid increases in expenditures for new public facilities and social services, housing, and higher-quality mobile home sites.

Perhaps the most serious problem in this area is the imbalance between revenues received from energy development and public expenditure needs across local government units. Counties and school districts can generally anticipate rapid and large tax revenues because energy facilities are usually located within their boundaries. However, municipal authorities will often experience serious problems arising from increased growth because they will face immediate demands for services and facilities but will realize new revenues only after energy facilities begin operation. This problem is illustrated in Figure 6-4, showing projected tax revenues versus needs for the Gillette, Wyoming, scenario.

Boom and Bust

Table 6-9 shows projected direct population increases attributable to energy facilities. In addition to indicating peak population increases, these data suggest that communities near coal gasification plants and electric power plants will experience serious boom and bust cycles, since these facilities have the largest ratio of construction to operation employment. This means that these communities will not only have a very difficult time in meeting immediate growth needs, but could also have large excesses of services and facilities after the construction stage.

These problems are exacerbated by several legal and institutional problems. Among the most serious of these is the inadequacy of public and private sector cooperation. The tendency is to deal only with the employees of an individual company, rather than also

Table 6–9. Population Increases Caused by Building and Operation of Energy Facilities.

| Type of Facility | Peak Population [a] | | Ratio of Construction to Operation Population |
	Construction	Operation	
Coal gasification (250 million ft³/day)	14,040	2360	6.0
Power plant (3000 MWe)	7620	1760	4.3
Oil shale retort (100,000 bbl/day)	8040	2600	3.1
Natural gas production (250 million ft³/day)	5100	3160	1.6
Coal liquefaction (100,000 bbl/day)	15,660	12,240	1.3
Crude oil production (100,000 bbl/day)	11,760	12,200	1.0
Uranium milling (1200 tons/day)	270	440	0.6

[a] Based on population/employment multiplier of 3.0 for the construction phase and 4.0 for the operation phase.

considering the overall effects on permanent residents and service workers. Planning and growth management are made more difficult by a shortage of information from the private sector about the timing, magnitude, and location of development, and by resistance to planning in many western towns.

Public sector assistance to local areas is also inadequate. Few federal programs are directed towards the problems of rapid growth resulting from energy resource development. Those that are, usually give funds to county authorities rather than towns. Although several western states have established housing finance agencies, they have had little impact on housing markets.

Housing

The risks of new-home construction in energy resource development areas are apparently too great for lenders, developers, and energy firms. Relatively small western housing markets cannot compete with major urban lenders and are often unable to finance construction and mortgages. The resultant low level of building has had a dramatic effect on housing prices, causing further housing shortages. The 1977 average price of new homes in areas of the western states affected by energy resource development was about 30 per cent higher than the national average for nonmetropolitan areas, and apartment units (in towns where they were available) were nearly double the national median. Energy developers have tried to alleviate housing shortages for their employees by supporting mobile home parks, single- and multi-family developments, and developing company towns such as Colstrip, Montana. However, most newcomers still live in mobile home parks, which are usually crowded and unmaintained and contribute relatively little to local taxes.

Changing the Tax Structure

Several alternatives are available to policy makers to deal with these problems. These include tax structure changes, assistance for rapid-growth areas, reductions in the number of workers living on-site, and improving the incentives for housing construction. Changes in the state and local tax structure appear to be the most direct and effective means of dealing with the "front-end problems" of affected communities, i.e., providing revenue during the construction phase in order for needed facilities to be built in time to meet increased population demands. Since the direct beneficiaries of energy development would share the costs, this alternative provides some equity. Also, tax structure changes can be easily adjusted to specific cases and combined with other alternatives.

The key variable for these taxes is how they are distributed among local governments. If significant portions of severance tax revenues, for example, were allocated to local governments, jurisdictional imbalances between revenues and expenditures could be overcome. But the trend in the West is for the states to retain most of the severance tax revenues for state purposes. Further, methods for distributing severance tax revenues to local governments can be inflexible. For example, although Miles City, which is about 70 miles from coal projects at Colstrip, Montana, has apparently experienced increased population growth as a result of energy extraction activities, it has had trouble paying for a census to document the growth and, as a result, getting assistance from the Montana Coal Board to deal with the effects. In contrast, towns closest to the projects had little trouble getting assistance.

Assistance Programs

Thus, assistance will also probably be required to provide necessary finances to communities, particularly for water and sewage facilities. Impact assistance programs appear to be easier to administer in the critical near term than state and local-level tax changes, but they would be costly and would not necessarily change the fundamental imbalance between town and state governments.

The cost of constructing houses for those employed on energy resource development projects can range from $5000 per house for site development to more than $30,000 per house for construction (1975 dollars). Since many energy industries are unwilling to assume the risk of holding unsold homes if development plans are delayed or cancelled, a more feasible alternative would be to improve the quality of mobile home parks—for example, by providing amenities typical of residential areas. If this kind of industry investment was made a provision of the siting permit for energy facilities, it would help to solve the housing problems created by such facilities.

LAND USE

Although the land used for mine workings, access roads, support and conversion facilities, waste disposal, housing, and recreation will affect only a very small percentage of land in the region, such land use will intensify conflicts between the energy industries, farmers, ranchers, environmentalists, recreationists, etc. Dealing with these land use and reclamation issues will challenge planning, monitoring, and enforcement capabilities at all levels of government.

As shown in Figure 6-5, the most serious land disturbances will result from surface oil shale retorts and surface coal mines. Oil shale retorts require at least twice as much land as any other conversion facility. About 70 per cent of this land is for spent shale disposal. Because this disposal will occur in scenic Rocky Mountain areas of western Colorado, eastern Utah, and southwestern Wyoming and because of the associated water quality problems, land use appears to be a significant barrier to oil shale development.

Figure 6-5 also illustrates the effect of surface mining, which disrupts large areas and produces more visible land degradation than underground mining. Although the 1977 Surface Mine and Reclamation Act[12] requires all strip-mined lands to be reclaimed, long-term reclamation success is uncertain. The potential for reclaiming mined lands depends on important locational features, in-

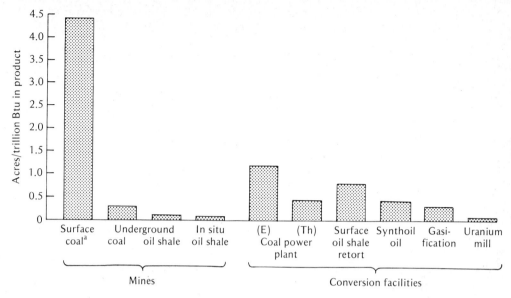

Figure 6-5. The land requirements of various energy technologies.

cluding climate (especially annual rainfall and its seasonal distribution), soil composition, topography of the land, and existing biological communities. For example, the arid climate and poor topsoil characteristics of the southwestern region provide for an ecosystem that is more difficult to restore than in the Northern Great Plains. Siting choices affect land requirements for coal mining because of the heating value (Btu content) of the resource and its seam thickness. As shown in Table 6-10, the number of acres disturbed at Beulah, North Dakota, and at Farmington, New Mexico, could be more than twice that disturbed at Gillette, Wyoming, or Colstrip, Montana, for the same level of energy production.

Using land to meet the needs of the expanded population resulting from an energy facility produces more significant ecological

Table 6–10.   Land Use by Mines in Six Locations.[a]

Type of Mine and Site	Typical Mine Size (acres over 30 years)	Acres Used per 10^{12} Btu Output
Surface coal		
Navajo/Farmington	4000–27,820	0.5–3.2
Gillette	4030	0.5
Colstrip	9680	1.2
Beulah	24,210	3.0
Underground coal		
Kaiparowits	1700	0.2
Underground oil shale		
Rifle	4200	0.5

[a] Average includes all surface lands disturbed, including disturbance caused by disposal of wastes such as spent oil shale.

impacts than land used directly by the energy facility. Public lands (national parks, forests, recreation and wilderness areas), particularly those in their natural state, are likely to experience the greatest changes primarily as a result of increased use due to population growth.

Thus, energy resource development will compete with other economically productive land uses, such as grazing and row crops, and with preservation, conservation, and leisure time uses. Conflicts are likely to increase between a variety of constituents and between government agencies as to how increased demands on land use should be accommodated.

Reclamation and Redevelopment

We have evaluated two ways of dealing with the effects of surface mined lands: reclamation to predevelopment use and redevelopment to recreation areas. Redevelopment of surface mined land appears more likely to succeed than does reclamation, primarily because of the uncertainty of reclamation success in arid and semi-arid areas. In general, redevelopment can put lands back to productive uses three to four years earlier than reclamation, is much more adaptable to particular sites, and allows local officials much more discretion in interpreting local needs.

The economic costs of reclamation and redevelopment will depend on local conditions, particularly climate, contour of the land, quantity of spoils, distance spoils and topsoil must be moved, and ultimate use to which the land will be put. Given this variation, the costs of these two alternatives appear to be about the same. Reclamation is estimated to cost about $1000 to $7000 per acre; redevelopment will cost $1000 to $6000 per acre.

Overall, there does not appear to be a "best choice" between reclamation and redevelopment. There are areas where reclamation will probably be impractical, if not impossible. The redevelopment option is limited by the demand for recreation and the suitability

of the site. However, given the high degree of uncertainty about the success of reclamation in arid and semi-arid lands, present federal policies seem to be an oversimplified response to a complex problem. In particular, the fact that the populations associated with energy facilities need recreational areas challenges the appropriateness of a policy emphasizing only reclamation.

CONCLUSIONS

The western region of the United States is full of conflicts, some of which are unavoidable: it is, for example, rich in natural resources but poor in one of the most essential resources, water; it is rich in energy resources but development of these will conflict directly with the desire to protect the relatively pristine air and natural beauty of much of the area; and there is wide disagreement among residents (withing both the Indian and the non-Indian populations) about the best way to develop its energy resources.[13]

Given these conflicts and the uncertainty and complexity of the U.S. energy policy-making system, conclusions regarding future levels of development in the region are hazardous. Nevertheless, it is clear that increased levels of energy production will occur in the region, and despite many uncertainties, much is already known about the problems and trade-offs associated with various policies for dealing with the problems and issues. The challenge for policy makers will be to develop and implement improved mechanisms for enhancing the economic benefits which increased energy development will bring while minimizing the adverse environmental and socioeconomic impacts.

We have identified only a few of the policy options in this article. Several technological options appear promising, particularly in protecting air and water resources. For example, water-saving cooling techniques can reduce the risk of water shortages and stringent sulphur controls can significantly reduce potential air quality problems.

However, the resolution of conflicts associated with large-scale energy development in the West will also require a variety of legal and institutional changes. This is particularly true regarding growth management problems, which necessitate state tax structure changes, federal impact assistance, and improved relationships between the public and private sectors.

Resolving conflicts over air and water resources will be particularly difficult because of the complexity of the regulatory system and the uncertainty associated with predicting impacts and, in the case of water, accurately describing the quantity and quality of the resource. Thus, in addition to technical approaches, policy makers will need to develop better research programs and new institutional mechanisms to accommodate the diverse interests and values at stake *before* large-scale impacts are felt. One apparently successful institutional mechanism is the siting task force. For water resource management, new institutional arrangements which provide basinwide or regionwide management may be required to address the interrelationships between water quality and water availability.

Recommendations for new and more comprehensive institutional mechanisms do not suggest that uniform approaches to these conflicts are appropriate. In fact, differences between the western states, variations in impacts from different energy development technologies, and the uncertainty associated with these impacts suggest that flexible policy responses will be needed. This is perhaps best exemplified by land use problems, which are likely to require a diversified approach rather than a strict requirement for reclamation to predevelopment use.

Energy resource development can provide substantial benefits at the local, state, and national levels. However, much of this development could be blocked or delayed by several economic and environmental issues. While many of these conflicts are manageable, a mix of technical, legal, and institutional responses will be required.

NOTES AND REFERENCES

1. For example, coal liquefaction and gasification, oil shale retorting, and the use of liquid geothermal resources to produce electricity.
2. CERT (Council of Energy Resources Tribes) is a nonprofit corporation founded in 1974. It represents the interests of 25 Indian tribes in the western United States.
3. E. Cazalet et al., *A Western Regional Energy Development Study: Economics,* Project 4000. Stanford, Calif.: Stanford Research Institute, November 1976.
4. The wastes referred to here are only those associated with uranium milling and do not include those from the entire fuel cycle. H. Kurtz, *Denver Post,* September 2, 1979, p. 61.
5. National Water Commission, *Water Policies for the Future,* Final Report to the President and the Congress of the United States, June 1973, pp. 319–33.
6. 95th U.S. Congress, *Clean Air Act Amendments,* PL 95-95, August 7, 1977, Section 109.
7. G. Krohm, C. Dux, and J. Van Kuiken, *Effects on Regional Coal Markets of the Best Available Control Technology Policy for Sulfur Emissions,* ANL/AA-16, National Coal Utilization Assessment, Argonne National Laboratory, December 1977.
8. Teknekron, Inc., *Review of New Source Performance Standards for Coal-Fired Utility Boilers, Volume 1: Emissions and Non-Air Quality Environmental Impacts,* TK 77-2192, March 1978.
9. 95th U.S. Congress, *op. cit.,* Ref. 6.
10. I. L. White et al., *Energy from the West: Impact Analysis Report.*
11. Utah Energy Office staff, personal communication, June 1978.
12. 95th U.S. Congress, *Surface Mining Control and Reclamation Act,* PL 95-87, August 3, 1977.
13. For those interested in more details, three major project reports are available: I. L. White et al., *Energy from the West: Impact Analysis Report, Energy from the West: Energy Resource Development Systems Report* and *Energy from the West: Policy Analysis Report.* These reports are available from the National Technical Information Service, Springfield, Va., as well as from the Environmental Protection Agency.

SUGGESTED READINGS

J. Russell Boulding, "Savior or Demon?", *National Parks and Conservation Magazine* **53** (6), 15 (1979).

Alvin M. Josephy Jr., "Agony of the Northern Plains" in L. C. Ruedisili and M. W. Firebaugh, eds., *Perspectives on Energy,* 2nd edn., p. 205. New York: Oxford University Press, 1978.

Energy Flows in the Developing World 7

VACLAV SMIL

Analytical studies of energy flows through national economies are of relatively recent origin. It was only in 1971 that a Stanford Research Institute report, outlining the movements of fossil fuels and electricity, provided a deeper look at the nature of energy consumption in the United States (Stanford Research Institute 1971). Since 1973, under the impetus of rising energy prices, anticipated fuel shortages, and economic disruptions, such studies have become more frequent and detailed, tracing national and regional flows either by sector (e.g., households, steel production) or by final use (direct heat, electricity for lighting, process steam) and more recently offering also revealing international comparisons (e.g., Basile 1976, Schipper and Lichtenberg 1976). These analyses are essential for understanding the dynamics of energy systems, without which rational management and choice of conservation strategies are not possible.

The approaches used to study energy flows in advanced economies, however, have only limited application for the world's developing nations, which make up some 75 per cent of mankind. Most of the population in these countries is rural and is either completely separated from or just marginally involved in the flows of modern commercial fuels and electricity, the only energies regularly accounted for by the national and international statistics.

To understand the actual situation in the developing nations, and to evaluate the appropriateness of alternative modernization paths, Third-World energetics—that is, the study of energy sources, flows, and transformations—must be put into the framework of the ecosystems of these countries. In contrast to the overwhelmingly fossil-fueled societies of North America, Europe, the USSR, Japan, and Australia, the peasants of Asia, Latin America, and Africa depend largely on solar energy, converted by photosynthesis,

Dr. Vaclav Smil is professor of geography at the University of Manitoba, Winnipeg, Manitoba R3T 2N2, Canada. His research interests focus on China's energy, economy, food, and environment, on energy in the Third World, on energy analysis of major grain crops, and the global potential of renewable energy resources. He is author of *China's Energy* (1976) and editor of *Energy in the Developing World* (1980).

From *American Scientist* 67 (5), 522 (1979). Reprinted by permission of the author and Sigma Xi, The Scientific Research Society, Incorporated. Copyright © 1979 by Sigma Xi.

to produce not only their food and feed for their animals but also fuel and raw materials.

Consequently, meaningful energy-flow analyses in developing nations must deal, above all, with plants, people, and animals, rather than with refined oil products and power generation. They should start with accounts of insolation and primary productivity, encompass food production and human and animal energetics, and trace the uses of harvested phytomass (plant matter) and organic wastes. In practice, such involved interdisciplinary exercises are hampered by the lack of quantitative information as well as of a fair understanding of numerous national particularities. I shall first discuss the general considerations and some estimating procedures necessary to overcome the limitations of the data, I shall then review the principal energy flows in populous developing nations, and, finally, I shall introduce a detailed case study.

INSOLATION AND PRIMARY PRODUCTION

All renewable energies are simply different displays of converted solar radiation, and thus it is logical, although not necessary, to start tracing a nation's energy flows with total insolation. Average annual clear-sky radiation, a function of geographical latitude, is readily determined from standard tables (Budyko 1974). Actual duration and intensity of sunshine is, of course, affected by cloudiness and atmospheric turbidity. As a result, regional and local variations may be extremely large, and only a reasonably dense observation network can assure a good national estimate of solar radiation received at the ground.

Even the United States, however, has long-term radiation records for only about fifty stations, and thus it is often necessary to estimate insolation on the basis of more readily obtainable data regarding cloud cover or sunshine duration. Various techniques to perform such conversions are available and give surprisingly good results for averages

over fairly long periods (Gates 1962, Reifsnyder and Lull 1965).

Nations along the Tropic of Cancer in North Africa and in the Middle East from Mauritania to Pakistan have by far the highest average insolation: the global peak in southern Egypt and northern Sudan is in excess of 220 kcal/cm² annually. (This high-intensity insolation is equivalent to approximately 2500 kWh/m², which is the amount of thermal energy available from about 1.5 barrels of crude oil; the author has chosen to use kilocalories to facilitate the relation of the figures given in the paper to food energy.) In contrast, the climatically deprived tropical regions, whose potential crop productivity is much lower than that of the temperate zones because of cloudiness and short daylight (Chang 1968), have annual insolation below 150 kcal/cm², with large areas in the western Amazon Basin, Gulf of Guinea, and Indonesia receiving below 120 kcal/cm² (Table 7-1).

Much of the radiation is reflected or absorbed and reradiated by barren subtropical or high-altitude deserts, which constitute a large part of many developing regions, such as China, Iran, North Africa, and the Sahel. Only a small fraction—globally in net terms no more than 0.3 per cent of the annual full spectrum of radiation reaching the earth's continents—is converted into new phytomass by land plants, primarily forest vegetation. While recent advances in the study of primary productivity of principal ecosystems still do not allow detailed, reliable accounts of net energy fixation for individual developing nations, they are good enough to provide estimates of at least the right order of magnitude (Eckardt 1968, Reichle 1970, Duvigneaud 1971, Bazilevich et al. 1971, Reichle et al. 1975, Lieth and Whittaker 1975, Olson et al. 1978).

The inability to calculate precise figures for energy fixation is due not only to the scarcity of actual primary-productivity field measurements and to the differences arising from theoretical modeling based on environmental variables (temperature, precipitation,

Table 7–1.   Magnitude and Range of Insolation and Net Primary Productivity.

Energy Flow	Range (kcal/m²/year)		
	Low	Medium	High
Insolation[a]	$1.0–1.4 \times 10^6$ Indonesia, Cameroon, Peru	$1.4–1.8 \times 10^6$ India, Thailand, Tanzania	$1.8–2.2 \times 10^6$ Egypt, Pakistan, Haiti
Forest productivity[b]	$1.0–1.3 \times 10^3$ Boreal forest (China), woodland (Kenya), chaparral (Mexico)	$4.0–6.0 \times 10^3$ Mangroves (Burma), temperate deciduous forest (Mexico), broadleaf forest (China)	$7.0–9.0 \times 10^3$ Tropical seasonal forest (Brazil), tropical rain forest (Brazil, Zaire, Indonesia)
Grassland productivity[b]	$0.5–1.5 \times 10^3$ Stressed subtropical and montane formations (China, Sahel, Andes)	$1.5–2.5 \times 10^3$ Temperate grassland (China, Turkey, Argentina)	$2.5–3.5 \times 10^3$ Tropical grassland (Brazil, Sudan, India)
Cereal productivity[c]	150–300 Upper Volta, Nigeria, Zaire	450–600 Pakistan, Brazil, Mexico	750–900 Vietnam, China, Indonesia

[a] From Budyko 1974.
[b] According to Duvigneaud 1971, Lieth and Whittaker 1975, Reichle et al. 1975.
[c] Calculated from unmilled grain data in FAO 1978. Inclusion of by-product and root mass would about double these values.

radiation) but also to a seemingly simple, yet intractable, obstacle: lack of any reliable knowledge about the areal extent of individual ecosystems. We know that most of the world's terrestrial phytomass energy flows through the tropical forests of Latin America, Africa, and Asia, but the best highest and lowest estimates of forest and woodland in these regions differ by some 6 million km² (Food and Agriculture Organization 1966 and 1971, Sommer 1976). Until these discrepancies are resolved we shall have to work with the best available figures for area and productivity and accept an uncomfortable margin of error in primary-production estimates for most natural ecosystems. Yet such estimates are essential to appraise both the fuel wood and timber harvest potential of forests and the ecologically sound densities of large animal herds that can be supported by grasslands.

Net energy conversion in agricultural crops can be established more reliably by collecting detailed production statistics, available either from national sources or from the Food and Agriculture Organization production yearbooks, and enlarging these totals by appropriate amounts of by-products and roots and expressing the harvested and the residual mass in energy equivalents. Estimates of by-product (straw, stems, vines, leaves) and root production are more open to error than the fairly reliable values for edible harvest of principal crops, especially cereals, which provide the bulk of food energy throughout the developing world (see Table 7-1).

FOOD AND HUMAN ENERGETICS

Energy contained in food, even including the numerous and growing direct and indirect fuel and electricity subsidies going into the modern nutritional system, accounts for only a fraction of the total energy flow through the developed economies (see, for example, Article 5, this volume, Hirst 1974). In contrast, many developing nations use an overwhelming share of their aggregate energy flows in farming and food preparation. Consequently, tracing the average food consumption, nutritional requirements, and useful energy expenditure of the population is among the key tasks in studying the energetics of the developing world.

To establish the mean energy consumption levels of a nation, it is necessary to construct a food balance sheet (Schulte et al. 1973). This complex exercise starts with production figures for all important edible crops, with total counts of domestic animals and total fish catches. After these figures have been corrected for trade in raw or processed foodstuffs, domestic utilization of crops is subdivided into food and nonfood uses (seed, feed, industrial manufacture), storage and transportation wastes are subtracted, and appropriate extraction (milling, processing) rates are used to convert the raw products into table foods. Typical carcass weights and production rates are used to derive meat, milk, and egg output. Finally, all mass values must be multiplied by proper energy equivalents.

For some developing countries, detailed food balance sheets are available in national statistical sources, and the FAO has regularly published its own itemized calculations for most Third-World nations (FAO 1978). The results are usually presented in terms of daily per capita consumption of total energy, proteins, glycids, and lipids and subdivided according to origin (plant or animal foods). Daily energy intakes in developing countries in the mid-1970s ranged from 1700 to 1800 kcal in Upper Volta, Mali, and Chad, through intermediate values of 2100 to 2300 kcal in Nigeria, China, and Pakistan, to high average inputs in excess of 2500 kcal in Brazil, Mexico, Egypt, and Turkey.

Inevitably, average per capita consumption figures mask significant socioeconomic, regional, and seasonal disparities and are also insufficient to evaluate the status of a nation's food supply, which can be appreciated only by comparing the available nutrition with average food-energy needs. Food-energy requirements are determined by age, body size

and composition, and physical activity, as well as by climate and other ecological factors (Joint FAO/WHO 1973).

Colder climates undoubtedly demand more energy than warmer ones, but it would be inappropriate to relate physiological parameters only to the average annual ambient temperature, because the thermal environment cannot be defined precisely by any single climatic measure. Consequently, there is no readily quantifiable basis for correcting the energy requirements according to the climate, and this consideration, which might cause variations no greater than 5 per cent for each 10 C° away from the mean reference temperature of 10 °C, may be disregarded in tracing the energy needs of a developing nation.

Body size and composition affect both resting metabolism and activity expenditures, but detailed anthropometric data at the national level are rarely available. Information on body composition (relative mass of muscle and fat) is scarce even for the developed nations, although there are quite a few studies of body weight, which should be consulted before making any calculations because national differences in typical body weights are very significant. Mean masses of fifteen-year-old boys range from 36–38 kg in Vietnam and Bangladesh to 44–46 kg in Egypt and Nigeria (the value for the United States is nearly 59 kg); average mass of healthy adult males in India is over 7 kg less than that in Turkey and 15 kg below the mean for the United States (Meredith 1969, Chapanis 1975). Consequently, use of Western or international reference weights is inappropriate (see Table 7-2).

Knowledge of age and sex distributions of the population is necessary in order to use the typical body-weight figures for a nationwide calculation of energy needs. For countries with continuous demographic investigations (India, Brazil, Egypt), the most reliable distributions can be obtained from recent censuses, and satisfactory approximations can be derived from older data for those developing nations that had at least one

Table 7–2.   Body Mass and Energy Requirements of Adults in
Various Nations.

Country	Adult Body Masses (kg)		Daily Energy Requirements with Moderate Exertion (kcal)[a]	
	Male	Female	Male	Female
United States[b]	70	58	3220	2320
International[c]	65	55	2990	2200
China[d]	60	50	2760	2000
India[e]	55	45	2530	1800
Philippines[f]	53	45	2440	1800

[a] Calculated according to Joint FAO/WHO 1973.
[b] Food and Nutrition Board 1968.
[c] Joint FAO/WHO 1973.
[d] Whyte 1972.
[e] Patwardhan 1960.
[f] Pascual et al. 1976.

modern census in the recent past. For Ethiopia, Vietnam, or Nepal, it appears to be anybody's guess.

Although there is no precise categorization of human activities according to their energy requirements, and although variations in caloric cost of a particular activity performed by different individuals (or even by the same person at different times) may be quite large, rough classification of tasks is possible. Most daily tasks around a house, on a farm, and in a factory fall into the light to moderate categories, but in traditional farming, still the dominant activity throughout the developing world, there are many heavy tasks: mowing with a scythe, binding and stacking sheaves, plowing, hoeing, clearing brush and scrub, and ridging, digging, and cleaning irrigation canals.

In preparing nationwide estimates of energy requirements, more realistic calculations can thus be made by adjusting the needs of the part of the population that does not participate in moderate activity by applying typical correction multipliers—0.9 for groups with light activity, 1.17 for very active, and 1.34 for exceptionally active groups (Joint FAO/WHO 1973). If there are more precise occupational data available, it is possible to

go a step further: using indirect calorimetry, thousands of measurements have been made over the decades for hundreds of work and leisure activities, and these values, preferably corrected for specific circumstances, can be used to make some very detailed calculations (Durnin and Passmore 1967).

After establishing how much nutrition is necessary to supply adequate energy, the key question remaining is the amount of actual work accomplished by human exertion. Approximate yet satisfactory and revealing calculations of useful annual energy expenditures can be made in three different ways that result in very similar totals. The first approach is to multiply reasonable estimates of typical work rate by the aggregate number of work hours. During sustained work, the average force a person can exert is equal to about one tenth of his body weight at a speed of one meter per second (Hopfen 1969)—that is, an average power of 45–60 newton-meters per second (about equal to 50 watts). The total number of hours worked annually may range widely (and seasonal fluctuations may be very large), but it is usually around 2400. Annual useful output per person derived in this way is 91,000–121,000 kcal.

The second method is to treat a person as

an energy convertor, with a predetermined physiological performance, and to calculate useful work as a proportion of total food consumption. Assuming a typical all-day energetic efficiency of 15 per cent for eight hours of work (Brody 1964), this procedure yields annual useful energy totals of 81,000–124,000 kcal per person.

In the third approach, used by the World Health Organization, energy available for activity is calculated as the residual after subtracting maintenance energy cost (about 1.5 times the basal metabolic rate for a given body mass) from average food intake; this method results in 70,000–110,000 kcal for economically active adults.

DOMESTIC ANIMALS

Domestic animals, especially large ruminant livestock such as cattle, water buffalo, and camels, are an extremely valuable renewable resource that fulfills several essential functions within the predominantly solar energetics of the developing countries. Ruminants consume mostly phytomass unsuitable for human digestion (grasses, crop by-products, and processing residues), supplemented with small quantities of grain; they provide much useful motive power for field work, irrigation, food processing, and transportation; they recycle valuable nutrients in often copious wastes; and they are a critical source of exceptionally good protein as well as of fiber, leather, and other by-products.

The relative importance of these functions differs widely in various developing nations. Where dung is used as fuel, in India, Bangladesh, and Turkey, for example, its equivalent heat content greatly surpasses the energy value of labor or milk. In pastoral societies, like those in the Sahel and East Africa, the food (milk, meat, and also blood) is the most important contribution. In China and Vietnam, where farming is heavily dependent on fermented manure, the return of nitrogen, phosphorus, and potassium to the soil is of primary importance.

Approximate calculations of useful work performed by draft animals require information on head counts, the proportion that is actually working, typical body weights, work speeds, and number of hours worked per year. Some developing nations have regular censuses of domestic animals, and the FAO provides annual estimates for all animals and all countries, but to avoid serious miscalculation any assumptions regarding the other variables should be based on knowledge of specific conditions. The proportion of working animals is about two thirds of the total count in China, but among India's more than 250 million cattle, there are only some 75 million working bullocks (Reddy and Prasad 1977, Smil 1978). Body weights vary widely even within the same species: water buffalo, the principal source of animate energy in monsoon Asia, can weigh from 250 to 700 kg (Cockrill 1974, Epstein 1969).

It is necessary to know the weight of the animal to estimate its average performance. In sustained effort, cattle, buffalo, and asses can exert a force equal to about 10 per cent of their body weight, and this ratio is as much as 15 per cent for horses and mules; but the draft power also depends on the animal's height. Work speeds in field work range from 0.5 m/sec for some water buffaloes to just over 1 m/sec for good horses (Hopfen 1969, Cockrill 1974). Owing to these differences, average power may range from 0.25 horsepower for weak Ethiopian cattle or donkeys to over 0.75 horsepower for good North China horses. Annual work hours may fall anywhere within a wide span of from two to eight hours per day for 100 to 300 days.

Most of the working animals in developing countries subsist almost exclusively on forage (grass, hay, crop residues), with very little supplemental feeding. There are, of course, notable differences in availability, quality, and kinds of feed. In densely populated regions, grazing is limited to canal banks or roadsides, no special fodder plants are cultivated, and virtually all feed comes from crop by-products such as cereal straws, cane leaves, and potato vines or from processing residues such as grain brans or pressed oil cakes (Odend'hal 1972). Where pasture and range forage is available, its quality, expressed by protein content and total digestible nutrients, may be excellent during periods of rapid growth but very poor in other seasons (Byerly 1977). In calculating the energy requirements of draft animals in the Third World, as well as those of pigs and poultry, specific feeding customs and body weights generally lower than those in developed countries must be kept in mind. Estimates based on feeding standards of developed nations would be very misleading.

TRADITIONAL FUELS

In any traditional society motive power comes from human and animal muscles, supplemented in places by water and wind. Thermal energy for households (cooking, heating, ironing, preparation of feed) and for local manufacturing (ore smelting, blacksmithing, ceramics, brick and glass making, food processing) is supplied by burning a variety of phytomass. Firewood is still the most important traditional fuel in global terms, though it is becoming increasingly less accessible. The annual worldwide consumption of firewood can be only roughly estimated, as most of it is collected by the users themselves, and only a small part of the total harvest enters commercial exchange. As a result, estimates of fuel wood use in the developing world in the mid-1970s range from around 1.3 billion to about 2.5 billion m^3 (Arnold 1978, Openshaw 1978).

Only a handful of developing countries—Brazil and South Korea, for example—have fairly reliable continuous data on fuel wood and charcoal use. In a few other cases, including Thailand, Sudan, Gambia, Tanzania, and Upper Volta, special consumption surveys have been carried out in some locations and for limited periods of time (Digernes 1977, Ernst 1978, Openshaw 1978). The Forestry Division of the FAO prepares annual fuel wood and charcoal consumption estimates for all developing nations, but, as local surveys have shown, these figures un-

derrate the actual use of forest phytomass in most of the poor countries. In fact, the very concept of fuel wood consumption as usually understood in forestry studies is inappropriately restrictive. Whereas in commercial forestry it is the merchantable bole (stem wood) that counts (Young 1975), fuel collected in developing countries contains a large proportion of branch wood, twigs, bark, stumps, and nonwoody tissue such as fresh twigs, leaves, and even grasses.

Consequently, any consumption estimates that assume annual harvesting of a certain share of the forest growth expressed only in the terms of stem wood, rather than as the total above-ground ecosystem productivity, must be unduly conservative. The best available statistics of consumption show that people in rural areas who depend on forest energy consume daily 2.5–4.5 kg of forest fuels per capita for household use and that in the cities and in many industrial applications charcoal remains the preferred fuel. This continuing dependence on forest phytomass is increasingly accompanied by dangerous deforestation and erosion, longer trips and greater effort to gather the fuel, and prohibitive prices in many rural and urban markets. It has also spurred a wave of a long-overdue research in the ways this essential resource is used and depleted and how it can be properly managed and conserved.

In deforested regions any remaining phytomass is sought for fuel: branches and bark are torn off the remaining trees; shrubbery, leaves, reeds, and grasses are collected. By far the most important plant fuel in such areas is furnished by crop residues—in particular, cereal straws and, to a much lesser extent, tuber and legume vines; in many regions sugar cane stalks and leaves are widely used. No systematic data are available on the use of crop residues in any developing nation, but rough calculation of potential by-product output is easy, given the knowledge of a nation's crop harvest and of the appropriate residue ratios—for example, unmilled grain and legume harvest should be multiplied by 1.5, tuber crops and sugar cane output by

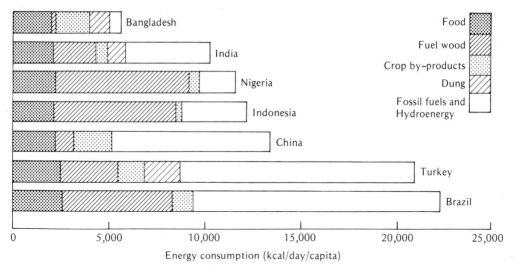

Figure 7-1. The consumption of food and renewable fuels is contrasted with the use of modern energies in seven developing countries, ranging from impoverished Bangladesh to relatively rich Brazil. The bar for the United States could not be shown on the same scale: it would measure almost one meter, with some 3200 kcal/day/capita for food, about 500 for fuel wood, and over 210,000 for fossil fuels and primary (hydro and nuclear) electricity.

0.2. More reliable calculations require a fairly detailed knowledge of specifics: because the residue yield is determined by a host of factors, such as plant varieties, maturation periods, soil characteristics, fertilization, irrigation, and weather conditions, application of approximate generalized multipliers may often result in major errors.

A very large portion of the potential crop-residue harvest is not, however, available for combustion. All principal by-products—rice, wheat, and millet straws; corn, cotton, and cane stalks and leaves; potato, vegetable, and legume vines—are widely used as animal feed and bedding, in composting, mulching, thatching, and packing, as well as for the manufacture of many household articles (Tanaka 1973). Preferably, a part of the residues should not be harvested at all but rather left on the field to recycle the nutrients, activate microbial soil processes, improve the tilth, and reduce soil erosion and water losses (Mannering and Meyer 1963). I know of only one nation—Taiwan—that has detailed and

reliable figures on all uses to which its main crop residue, rice straw, is put. Partial accounts are available for a few countries (South Korea, Philippines), but in most cases it is possible to offer only rough estimates based on the knowledge of farming and household customs.

SOLAR ENERGETICS

Table 7-3 illustrates the relative importance of energy produced from biomass (phytomass and organic wastes) in the developing world by comparing a set of average daily per capita flows of food and traditional fuels with the mean aggregate use of fossil fuels and hydroelectricity (see also Fig. 7-1). Because all these values are national averages they blur the distinction between urban and rural consumption. For Bangladesh, China, India, Indonesia, and Nigeria, with their overwhelmingly (more than 75 per cent) rural populations, the figures for traditional fuels come closest to the modal use; for Turkey, where just over 50 per cent of the people live in villages, and for Brazil, where the majority of the population is urban, the figures undervalue the actual rural consumption of traditional fuels while exaggerating the use of modern energies in the countryside. Unfortunately, the available information does not allow reliable partitioning between rural and urban consumption. To avoid the appearance of unwarranted accuracy, all values in Table 7-3 are rounded.

With these caveats, a classification of energetics in developing countries can be offered. In the first group are the poorest countries, where almost no modern energy sources are used in the countryside—Bangladesh, Ethiopia, the Sahel region. In these areas the daily consumption of biomass fuels is between 3000 and 9000 kcal per capita, just 1.5–4.5 times the average daily food energy intake and only enough to cook two simple meals and, perhaps, to warm a room for a few hours. These are still overwhelmingly solar economies, in which energy affluence is determined by the number and quality of meals, possession of domestic animals, and the degree of access to vanishing fuel wood

and to the seasonally available crop by-products. India, Pakistan, and Thailand represent nations in the second category, in which modern forms of energy are just about equivalent to the use of traditional fuels; however, most of the villagers are still heavily dependent on phytomass for fuel and on animate energy for motive power.

China and Brazil are one more step removed: combined flow of food and phytomass fuels is smaller than the consumption of modern energy sources. Although significant segments of the rural population in these nations continue to rely on phytomass for household use, cities run on fossil fuel and refined oil products are increasingly powering field, irrigation, and processing machinery in the villages. Nations in the fourth category, exemplified by Taiwan (see Table 7-3), have accomplished the transition to fossil fuels. Traditional fuels are still burned in some rural areas and animate power has not been completely displaced by machinery, but modern energies dominate the economy, some two thirds of the population are already urban, and the next important energy transition is already under way, one in which

Table 7-3. Traditional and Modern Energy Consumption in Some Developing Nations.

| Country | Traditional Energies (kcal/day/capita) | | | | | Modern Energies [b] (kcal/day/capita) | Total Nonfood Energy Consumption Filled by Traditional Energies (%) |
	Food [a]	Fuel Wood	Crop By-Products	Dung	Total		
Bangladesh [c]	1900	200	1900	1100	3200	500	86
Nigeria	2100	7000	600	—	7600	1700	82
Indonesia	2000	6500	200	—	6700	3400	66
India [d]	2000	2400	500	1000	3900	4200	48
Brazil [e]	2500	5800	1100	—	6900	12,800	35
Turkey	2800	2600	1300	2000	5900	12,100	33
China [f]	2100	1100	2000	—	3100	8300	27
Taiwan [g]	2700	100	200	—	300	29,200	1

[a] All values except for China and Taiwan from FAO 1978.
[b] All values except for China and Taiwan from United Nations 1977.
[c] Calculated from data in Tyers 1978.
[d] Calculated from data in Henderson 1975.
[e] Calculated from data in Muthoo 1978.
[f] All values for China are from Smil 1978.
[g] Calculated from the official statistics of the Republic of China and estimates of the Joint Commission on Rural Reconstruction.
All other values are the author's estimates based on a variety of fragmentary data and appraisals.

Table 7–4. Classification of Developing Nations According to Their Primary Energy Base

Proportion of Total Energy Supplied by Biomass	Number of Countries and Territories	Total Population (millions)	Population as a Percentage of the Third World	Most Populous Nations in This Category
More than two thirds	61	690	22	Indonesia, Bangladesh, Nigeria, Burma
Two fifths to three fifths	17	930	30	India, Pakistan, Philippines, Thailand
One sixth to two fifths	17	1250	40	China, Brazil, Colombia, Algeria
Less than one tenth	32	270	8	Mexico, South Korea, Iran, Argentina

Population figures are mid-1978 estimates from the Population Reference Bureau's list of 163 countries and territories, except for China, where the estimates of the Foreign Demographic Analysis Division, Bureau of Census, were used.

an increasing share of the fossil fuel is converted to electricity.

An approximate division of developing nations into these four categories shows that the primary energy supply in some 60 countries and territories with nearly 700 million people (42 of these countries, with some 310 million inhabitants, are in Africa) still comes predominantly from biomass and that only about 8 per cent of the Third-World population lives in the nations that have shifted to fossil-fueled energetics (Table 7-4). I do not think it will ever be possible to account with satisfactory precision for the global flow of biomass energies, but, on the basis of the best current evidence, I would suggest that the total annual use of firewood in the developing world amounts to just over 7×10^{15} kcal, and that the crop by-products and dung burned as fuel contribute, respectively, nearly 1.5×10^{15} and 0.4×10^{15} kcal. In aggregate, the three kinds of biomass fuel contain energy equivalent to about 1.3×10^9 metric tons of hard coal (or approximately 0.9×10^9 metric tons of crude oil), an amount equal to the developing world's annual use of fossil fuels.

This comparison, perhaps better than any other, indicates the magnitude of the task involved in the transition from traditional to modern energetics—and suggests the undesirability of a total displacement of biomass fuels and animate energy sources. China

provides a unique illustration of why total displacement is undesirable.

CHINA: A CASE STUDY

China, the planet's most populous developing nation, is currently engaged in an ambitious modernization effort, in an unprecedented shift from traditional to modern high-technology energetics. In the countryside, where four fifths of China's one billion people live, the modernization effort is more than a decade old, but energy-flow analysis (Fig. 7-2) shows the continuing domination of biomass and animate power (Smil 1978, 1979). In the mid-1970s, crop by-products and fuel wood still contributed more than six times the total of fossil fuels and hydroelectricity flowing directly into China's farming and rural households, and the difference is still more than double when the fuel and power consumed by numerous local small-scale industries are added.

Combined capacity of mechanical pumps and tractors in 1974 was still about 20 per cent smaller than the aggregate power of China's rural labor force and draft animals, and it was only in 1975–76 that domestically produced and imported synthetic nitrogen fertilizers surpassed the nitrogen applied to fields in fermented animal and human wastes. Even when urban and industrial uses are taken

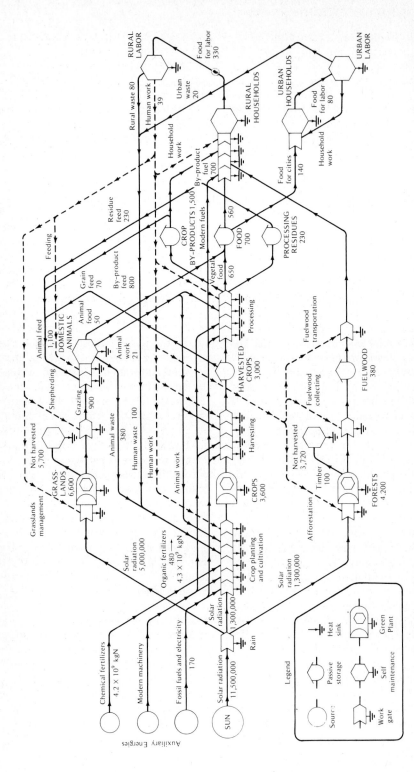

Figure 7-2. Major energy flows are shown for rural China in 1974. Extensive recycling of organic wastes and a maximum use of crop byproducts and processing residues have made Chinese agriculture perhaps the most efficient traditional farming system. Now the use of auxiliary energies such as fertilizers, machinery, and fossil fuels are raising the output further. The chart uses the symbols of H. T. Odum's energy-circuit language (1971). All values except for nitrogenous fertilizers are in 10^{12} kcal.

into account, solar energy recently transformed by green plants still predominates in China. Although the country is now the world's third largest coal producer and ranks, respectively, eighth and fifth in the extraction of crude oil and natural gas, the total flow of fossil fuels in the mid-1970s was still at least 30 per cent lower than the phytomass energy consumed as feed, food, fuel, and raw material.

Rural energy-flow analysis makes clear the very high energy cost of the planned modernization and provides a useful quantitative foundation for the assessment of alternative development strategies. Perhaps the key Chinese economic modernization aim is complete farm mechanization before the end of the 1980s. This means that almost all the current 60×10^{12} kcal of useful human and animal energy spent in agriculture must be replaced by machinery. Gross efficiencies of small- and medium-sized pumps, tractors, and motors suitable for the Chinese environment are, on the average, not higher than 15 per cent. To supply 60×10^{12} kcal of useful energy, this machinery would thus have to consume some 400×10^{12} kcal annually, an equivalent of about 40 million metric tons of refined oil products.

To replace all nitrogen currently applied in animal and human wastes with synthetic ammonia and urea would require total feedstock and process energy equivalent to more than 7 million metric tons of crude oil annually. About the same amount of energy would have to be invested to produce the related field and processing machinery, trucks, spare parts, storage sheds, and repair shops and to supply adequate amounts of phosphorus, potassium, and pesticides. Replacement of animate power and organic fertilizers in Chinese farming would thus require no less than the equivalent of about 50–55 million metric tons of oil annually. And should the Chinese seek to eliminate almost completely the use of fuel wood and crop by-products in the countryside, the equivalent of another 50 million metric tons of oil would be needed, even if the modern fuels could be burned in stoves

and boilers with twice the combustion efficiency of the current ones.

The cost of replacing animate power and biomass fuels in rural China is clearly enormous—the equivalent of at least 100 million metric tons of oil annually. And more energy subsidies would be needed to increase crop yields, modernize inefficient irrigation, build roads and countless rural industrial enterprises, boost fossil-fuel output and power generation, and raise living standards. A prudent conclusion from China's rural energy-flow analysis would be that the country, in its quest for modernization, should not discard all the renewable energy sources, as there are many advantages in maintaining and improving certain facets of solar energetics.

Not a few of the aspects of traditional energetics deserve to disappear: peasants pumping water by treading dragonwheels, pulling incredibly overloaded carts, or chopping down newly planted trees for firewood; oxen or water buffaloes threshing grain by trampling; primitive, inefficient stoves that need almost continuous feeding with straw. But many others should be preserved and perfected. Chinese draft animals do not put a damaging strain on the country's ecosystem, and many of them should be kept to contribute useful and flexible work, excellent fertilizer, and, if better managed, valuable protein. The complete use of logging residues for fuel should be encouraged in wooded areas, and massive afforestation and establishment of fuel wood plots of fast-maturing species should be pursued in all suitable locations. Anaerobic fermentation of animal and human wastes should be expanded to produce good fertilizer and clean biogas.

At a time when the most advanced industrial powers are searching for ways to reduce their dependence on fossil fuels by turning to renewable energy sources, it would be ironic if most of the poor nations pursued developmental strategies that tied them to nonreplaceable resources. Detailed studies of energy use in developing countries are becoming more frequent (see, for example,

Palmedo et al. 1978, Smil and Knowland, 1980), but most of the task lies ahead, and proper long-range choices that might balance traditional solar and modern fossil-fueled energetics can be made only after a thorough analysis of the existing and the potential energy flows.

REFERENCES

Arnold, J.E.M., 1978. "Wood Energy and Rural Communities," in *Eighth World Forestry Congress*. Jakarta.

Basile, P. S., ed., 1976. *Energy Demand Studies: Major Consuming Countries*. Cambridge, Mass.: MIT Press.

Bazilevich, N. I., L. Y. Rodin, and N. N. Rozov, 1971. "Geographical Aspects of Biological Productivity," *Soviet Geography* 12:293.

Brody, S., 1964. *Bioenergetics and Growth*. Hafner.

Budyko, M. I., 1974. *Climate and Life*. New York: Academic Press.

Byerly, T. C., 1977. "Ruminant Livestock R & D," *Science* 195:450.

Chang, J. H., 1968. *Climate and Agriculture*. Hawthorne, N.Y.: Aldine.

Chapanis, A., ed., 1975. *Ethnic Variables in Human Factors Engineering*. Baltimore: John Hopkins University Press.

Cockrill, W. R., ed., 1974. *The Husbandry and Health of the Domestic Buffalo*. FAO.

Digernes, T. H., 1977. "Wood for Fuel: Energy Crisis Implying Desertification. The Case of Bara, the Sudan." 1977 dissertation, University of Bergen, Norway.

Durnin, J.V.G.A., and R. Passmore. 1967. *Energy, Work and Leisure*. Heinemann Educational Books.

Duvigneaud, P., ed., 1971. *Productivity of Forest Ecosystems*. UNESCO.

Eckardt, F. E., ed., 1968. *Functioning of Terrestrial Ecosystems at the Primary Production Level*. UNESCO.

Epstein, H., 1969. *Domestic Animals of China*. Commonwealth Agricultural Bureaux.

Ernst, E., 1978. "Fuel Consumption Among Rural Families in Upper Volta, West Africa," in *Eighth World Forestry Congress*. Jakarta.

Food and Agriculture Organization, 1966. *World Forest Inventory*. FAO.

Food and Agriculture Organization, 1971. "Supply of Wood Materials for Housing," *Unasylva* 25(2–4):28.

Food and Agriculture Organization, 1978. *Production Yearbook*. FAO.

Food and Nutrition Board, 1968. *Recommended Dietary Allowances*. Washington, D.C.: National Academy of Sciences.

Gates, D. M., 1962. *Energy Exchange in the Biosphere*. New York: Harper and Row.

Henderson, P. D., 1975. *India: The Energy Sector*. World Bank.

Hirst, E., 1974. "Food-Related Energy Requirements," *Science* 184:134.

Hopfen, H. J., 1969. *Farm Implements for Arid and Tropical Regions*. FAO.

Joint FAO/WHO Ad Hoc Expert Committee, 1973. *Energy and Protein Requirements*. World Health Organization.

Lieth, H., and R. H. Whittaker, eds., 1975. *Primary Productivity of the Biosphere*. New York: Springer.

Mannering, J. V., and L. D. Meyer, 1963. "The Effects of Various Rates of Surface Mulch on Infiltration and Erosion," *Soil Sci. Soc. Am. Proc.* 27:84.

Meredith, H. V., 1969. *Body Size of Contemporary Youth in Different Parts of the World*. Society for Research in Child Development.

Muthoo, M. K., 1978. "Forest Energy and the Brazilian Socio-Economy, with Special Reference to Fuelwood," in *Eighth World Forestry Congress*. Jakarta.

Odend'hal, S., 1972. "Energetics of Indian Cattle in Their Environment," *Human Ecology* 1:3.

Odum, H. T., 1971. *Environment, Power, and Society*. New York: Wiley-Interscience.

Olson, J. S., H. A. Pfuderer, and Y. H. Chan, 1978. *Changes in the Global Carbon Cycle and the Biosphere*. Oak Ridge, Tenn.: Oak Ridge National Laboratory.

Openshaw, K., 1978. "Woodfuel: A Time for Reassessment," *Natural Resources Forum* 3:35.

Palmedo, P. F., R. Nathans, E. Beardsworth, and S. Hale Jr., 1978. *Energy Needs, Uses and Resources in Developing Countries*. Brookhaven, N.Y.: Brookhaven National Laboratory.

Pascual, C. R., et al., 1976. "Nutritional Requirements: Dietary Allowances and Requirements for Calories and Nutrients," in K. Rao, ed., *Food Consumption and Planning*. New York: Pergamon.

Patwardhan, V. N., 1960. *Dietary Allowances for Indians*. Indian Council of Medical Research.

Reddy, A.K.N., and K. K. Prasad, 1977. "Technological Alternatives and the Indian Energy Crisis," *Economic and Political Weekly*, Special Number 1465–1502.

Reichle, D. E., ed., 1970. *Analysis of Temperate Forest Ecosystems*. New York: Springer.

Reichle, D. E., S. F. Franklin, and D. W. Goodall, eds., 1975. *Productivity of World Ecosystems*. Washington, D.C.: National Academy of Sciences.

Reifsnyder, W. E., and H. W. Lull, 1965. *Radiant Energy in Relation to Forests*. Washington, D.C.: Government Printing Office.

Schipper, L., and A. J. Lichtenberg, 1976. "Efficient

Energy Use and Wellbeing: The Swedish Example," *Science* **194**:1001.

Schulte, W., et al., 1973. "Food Balance Sheets and World Food Supplies," *Nutrition Newsletter* **11**(2):15.

Smil, V., 1978. "China's Energetics: A System Analysis," in *Chinese Economy Post-Mao*. Washington, D.C.: Joint Economic Committee, United States Congress.

Smil, V., 1979. "Energy Flows in Rural China," *Human Ecology* **7**:119.

Smil, V., and W. E. Knowland, eds., 1980. *Energy in the Developing World*. New York: Oxford University Press.

Sommer, A., 1976. "Attempt of an Assessment of the World's Tropical Forests," *Unasylva* **28**(112–113):5.

Stanford Research Institute, 1971. *Patterns of Energy Consumption in the United States*. Stanford, Calif.: Stanford Research Institute.

Steinhart, J. S., and C. E. Steinhart, 1974. "Energy Use in the U.S. Food System," *Science* **184**:307.

Tanaka, A., 1973. "Methods of Handling the Rice Straw in Various Countries," *International Rice Committee Newsletter* **2**(2):1.

Tyers, R., "Optimal Resource Allocation in Transitional Agriculture. Case Studies in Bangladesh." 1978 dissertation, Harvard University.

United Nations, 1977. *World Energy Supplies 1971–1975*. UNO.

Whyte, R. O., 1972. *Rural Nutrition in China*. New York: Oxford University Press.

Young, H. E., 1975. "The Enormous Potential of the Forests," *Journal of Forestry* **73**(2):99.

SUGGESTED READINGS

Richard J. Barnet, "The World's Resources," *The New Yorker,* March 17, March 31, and April 7, 1980.

Wolfgang Sassin, "Energy," *Scientific American* **243**(3), 119 (1980).

Vaclav Smil, *China's Energy*. New York: Praeger, 1976.

FOSSIL-FUEL ENERGY SOURCES

World's first commercial well. This 1861 photograph taken near Titusville, Pennsylvania, shows, in top hat and frock coat, Edwin L. Drake, the man who conceived the idea of drilling for oil. On August 27, 1859, he proved his theory with the primitive rig pictured in the background. With Drake is Peter Wilson, a Titusville druggist who encouraged him in the venture. (American Petroleum Institute Historical photographs)

Supertanker unloading at a refinery. (Photo Courtesy of Texaco, Inc.)

A stripping shovel operator's view of the Peabody Coal Company surface pit, Sinclair Mine, Kentucky. The dipper of this big shovel holds 125 cubic yards of earth and rock. The loading shovel in the background scoops coal into 135-ton trucks. (National Coal Association)

TVA's Bull Run Steam plant, Oak Ridge, Tennessee. This large coal-fired unit has a capacity of 950 megawatts (electric), burning over 5 tons of coal a minute. (Tennessee Valley Authority)

Oil shale. Retorts near Grand Valley, Colorado, operated by a consortium of oil companies. (American Petroleum Institute)

An LMG bucketwheel excavator extracting the Athabasca Oil Sands, For McMurray, Alberta, Canada. This $4.6 million excavator can dig 54,000 tons of sand daily, yielding 50,000 to 55,000 barrels of oil. (Photo courtesy Great Canadian Oil Sands, Ltd.)

INTRODUCTION

The period 1900 to 2000 has been accurately called the Chemical Century. Despite all the discussion, pro and con, of fission, fusion, and solar energy, fossil fuels still dominate the energy supply picture with over 90 per cent of the market. Because of the enormous social inertia discussed in Section I, all energy scenarios foresee this dominance as continuing to the end of the twentieth century. So it is essential that we understand these energy sources and the role they play in society.

The inevitable dilemmas surrounding fossil fuels involve the finite nature of the resources and the environmental impact of their use. From a long-term perspective, these dilemmas will be resolved. When we have consumed all the economically accessible oil and gas, their production, transportation, and consumption will no longer threaten the environment. Since we are, in fact, over the peak in U.S. production of oil and gas and possibly in consumption as well, we conclude that, at most, we can do only as much additional environmental damage as we've already done. As environmental legislation enacted in the 1970s is implemented, the environmental insult per unit of fossil-fuel consumption should substantially decrease.

Diseases such as black lung, the ravages of strip mining, and oil pollution of our beaches are obvious social and environmental costs imposed by society's demand for energy. Less apparent are such elusive phenomena as "acid rain" resulting from coal-burning plants. Sulfur and nitrogen oxides, which were previously partially neutralized by the alkaline particulate matter from "dirty stacks," are even more serious problems after such particulate pollution is eliminated. Tall stacks now disperse sulfur dioxide over a larger area, turning what was once a local problem into a national one. Rain then reacts with the SO_2 to produce sulfuric acid; there is evidence that this leaches out soil nutrients, destroys coniferous forest over wide areas, and has killed fish in lakes in New England, Canada, and Sweden.

125

Even less well understood and potentially more serious are possible climatic changes resulting from unavoidable carbon dioxide production in the combustion of fossil fuels. Some scientists feel the "greenhouse effect" may melt the polar ice caps, flooding most of the major coastal cities of the world; others claim human activity may cause a new ice age with its attendant human misery and mass starvation. Although the direction of this manmade climactic change is unclear, there is little doubt as to its possibility. Carbon dioxide production is intrinsic to fossil-fuel combustion—no technological fix exists.

It has frequently been observed that the United States does not have an energy problem—it has an oil problem. That is, the source on which we place greatest reliance in terms of energy work load is also the least abundant domestically. This has led to our present heavy dependence on foreign oil with the associated problems already discussed in Section I. Because of the importance of oil in our economy, we devote two articles to it, the first examining the facts of its role in society and the second a penetrating examination of its future availability. Then we turn to three other fossil fuels proposed to bridge the gap between depletable and renewable energy economies. Finally, we consider the CO_2 question.

The largest single source of energy in the United States is petroleum with about 43 per cent of the market. In *Facts About Oil* by the American Petroleum Institute, a wealth of information on the history, production, transportation, and uses of petroleum is presented. The great versatility, low cost, and ease of using this resource, primarily as a fuel, have resulted in its assumption of a central role in industrial society. The critical consideration is that this valuable resource on which we are most dependent is the least abundant domestic fossil fuel. As with many chemical addictions, life appears beautiful while under the influence, but the withdrawal symptoms have proved painful.

In *Future Availability of Oil for the United States,* John Steinhart and Barbara J. McKellar examine more closely the most troublesome characteristic of oil—its availability. The work of M. King Hubbert and his famous logistic curves are presented, and the importance of the underlying assumptions on total resource estimates is emphasized. Not even the price mechanism can change the limits imposed by geology, since "if high prices bring a surge of exploratory drilling we shall simply find more sooner and less later." Some international implications of oil trade are discussed and the role and purposes of OPEC members interpreted in a far more sympathetic light than in the economic analysis of Steele (Article 4).

Dr. Carroll L. Wilson summarizes the findings of the World Coal Study (WOCOL) in his article, *Coal—Bridge to the Future.* In recognizing both the reduced availability of oil and the finite nature of the coal resource, he proposes coal as an acceptable intermediate-term fuel to carry us from the present into an energy future of renewable and essentially unlimited resources. Moderate growth of world economies is possible through a tripling of coal production, and without such expansion the outlook is seen as bleak. The article examines in detail the nature of coal resources, environmental standards for its use, and the possibilities for coal conversion to liquid and gaseous fuels. Throughout the discussion, the importance of developing the "infrastructure" necessary for getting coal from mine to end user is emphasized. Conventional power stations take six to seven years to build and new technologies (e.g., coal gas) require 30 years to develop. For coal to serve the role of bridge to the future envisioned here, early decisions are required.

Although natural gas is already our second largest source of energy with about 27 per cent of the market, Henry R. Linden presents the case for expanding the role of natural gas in his article, *Gas Supply: The Future Looks Bright.* By carefully distinguishing between resources and reserves, he shows how the shortage of natural gas ($0.20–0.30/million cubic feet) was eliminated when the price was allowed to rise to $3–6/million cubic feet. The author discusses the importance of unconventional resources, technological innovation, and international trade in expanding gas supplies and concludes that natural gas could be substituted for much of the 4.7 million barrels of oil per day currently being burned by stationary heat energy plants. This would free up these liquid fuels for the transport fuel market and reduce our dependence on imported oil. The author concludes that a free market economy, an even-handed federal research and development program, and enlightened foreign trade policies toward Canada and Mexico will help keep the "gas option" open as a feasible alternative to reduced oil supply.

The third major source of fossil fuel suggested to replace conventional petroleum is *Synfuels: Oil Shale and Tar Sands.* John Ward Smith presents a detailed characterization of these two resources, estimates of the enormous size of each resource, and describes the difficulties of production. While we often view these resources as "exotic fuels of the future," the author indicates that small shale oil industries actually predated the original Drake well of 1859. The Athabasca tar sands now produce over 100,000 barrels of oil per day, but oil shale remains in the pilot-plant stage since "the estimated price of a barrel of shale oil newly produced has scooted along just above world petroleum prices for more than 30 years." After examining environmental and other sociopolitical costs, the author concludes that these synfuels, while not solving our energy problems forever, can help us buy time enough to find a solution.

We conclude the discussion of fossil fuels with a detailed discussion by Gregg Marland and Ralph Rotty of the perplexing problem, *Atmospheric Carbon Dioxide: What To Do?* Other environmental hazards (e.g., acid rain, black lung disease, oil spills) can be minimized or their risks reduced to acceptable limits by the spending of money on the appropriate technological fixes. The chemistry of the process dictates that for every ton of coal burned, between two and four tons of CO_2 are produced and no known fixes exist. This is one of the sources in the measured increase of 12 to 15 per cent in atmospheric CO_2 during this century. The authors discuss the nature of models used to calculate the inventory of CO_2 in the atmosphere, the more complex models predicting possible climatic changes caused by this increase, and the great uncertainties in each. What emerges is a picture of a threat so poorly understood that the authors recommend against any quickly contrived ban of fossil-fuel burning since the costs of such a policy are well understood and considerable while the benefits are uncertain at best.

The ecological balance of nature is at the same time both delicate and sturdy. Man in his ignorance or greed may abuse his natural habitat, causing enormous short-term damage, but nature is resilient and will generally recuperate in the long run. Abuses in fossil-fuel production and consumption, the most severe environmental disruptions, will certainly disappear—the question is simply when and how. Will it be sooner, through responsible human decisions, or later, through the total depletion of available resources or zonal climactic changes? As one wit has observed, "Nature always bats last."

Fossil fuels are still the mainstay of our energy economy and will certainly remain so for the midterm future. As such they serve as benchmarks for benefit and environmental cost analysis; we frequently measure alternative energy sources against them in the following sections.

Facts About Oil   8

AMERICAN PETROLEUM INSTITUTE

OIL HISTORY

Since prehistoric days, man has sought to convert earth's resources to his own use—to bring him warmth, cook his food, and ease his workload. When he first began to use petroleum remains a secret of the ages. However, seepages of crude oil and natural gas are thought to have furnished fuels for the "sacred" fires of primitive peoples, who ascribed divine powers to the flames.

Both the Babylonian and Judaic versions of the Great Flood record that pitch—a natural form of asphaltic petroleum—was used to caulk the vessels that saved Utnapishtim and Noah. Pitch was also used to grease the axles of the Pharoahs' chariots in ancient Egypt. And the Greeks record the destruction of a Scythian fleet when oil was poured on the water and set afire.

The ancient Chinese used both oil and gas. In fact, they actually drilled for oil as early as the third century B.C. Using bamboo tubes and bronze bits, they successfully reached depths of over 3000 feet. This was long before Western civilization began to search underground for oil.

The Early Years

The oil industry in the United States had its beginning at Titusville, Pennsylvania, on August 27, 1859. On that date, the first well drilled specifically for the purpose of finding oil was successful. Oil, however, was not new to the North American continent.

White men exploring the area of New York and Pennsylvania found Indians using crude oil for medicinal purposes. The Indians obtained the oil by skimming it from the springs and streams, where it appeared as a thick scum. George Washington described a spring on his property, which was probably seeping natural gas, as "of so inflammable a nature as to burn freely as spirits."

In Washington's lifetime, and even in the first half of the nineteenth century, Americans had only tallow candles and whale oil for light. Whaling became a major industry as New England ships combed the seas. But the hunting was too good, for over the years the supply of whales became scarcer.

The Stage Is Set

At the same time, other forces were working to provide a new source of illumination. In

This article is an abbreviated version of the latest edition of *Facts About Oil* prepared by The American Petroleum Institute, 2101 L Street, N.W., Washington, D.C. 20037. Reprinted by permission of the American Petroleum Institute.

America and in Europe during the 1850s, some illuminating oils were being distilled from coal and from petroleum skimmed off ponds and streams. Although the amounts of oil that could be gathered in this way were extremely limited, the stage was set for the industry's birth. Americans needed and wanted a plentiful source of oil for light. And far-sighted men were beginning to see the possibilities of petroleum to fill this need.

The Drake Well

Armed with a favorable report on petroleum as an illuminant, which was written by Yale University Professor Benjamin Silliman, a group of businessmen led by James M. Townsend formed the Pennsylvania Rock Oil Company, which later became the Seneca Oil Company. Their objective was to recover petroleum in large quantities by drilling for it just as men drilled for salt.

Townsend and his associates hired Edwin L. Drake, a retired railroad conductor, to undertake drilling operations on the banks of Oil Creek at Titusville, Pennsylvania, near an old oil spring. By the time Drake and his men had struck rock below the 30-foot level, their drilling rig—an old steam engine and an iron bit attached by a rope to a wooden windlass—had won the title "Drake's Folly." And when the rock slowed drilling to 3 feet a day, even Townsend became discouraged.

Drake and his men persisted, however, and late one summer afternoon in 1859, after the bit had been withdrawn from the well at 69½ feet, a dark green liquid rose to a few feet below the top. The drillers were jubilant. They had struck oil.

The events leading to the first shout of "We've struck oil!" were of vast significance, because they proved to the Western world that it was sound and practical to drill for oil.

The Oil Boom

The immediate results of Drake's success were dramatic. An oil boom, which took on many aspects of California's gold rush a decade before, began. The Pennsylvania hills echoed to the sound of thousands of drill bits pounding their way into the earth. Towns mushroomed.

As the trickle of oil became a stream, facilities for storage and transportation became a pressing problem. Coopers worked overtime making barrels for crude oil. Thousands of teamsters churned roads into quagmires as they lurched and splashed toward the nearest railroad. Rafts laden with barrels of oil were floated down Oil Creek.

In 1861, the first refinery went into operation in the oil region. Like other early refineries, it produced kerosine* and little else. There were some local markets for greases and lubricating oils, but refiners concentrated on producing a kerosine that was a virtually odorless and smokeless illuminant.

To supply refineries at a distance, railroads built spurs into the producing region. At first, the barrels of oil were transported in flatcars. Then, in 1865, a railroad tank car was developed especially for carrying crude oil. A primitive forerunner of modern tank cars, it was a flatcar on which two vertical wooden tanks had been built.

That same year, the first oil pipeline was laid. It brought crude oil five miles from Pithole City to the Oil Creek Railroad, impressing producers with its efficiency. And in 1879, the first major pipeline was completed. It extended 110 miles across the Allegheny Mountains to Williamsport, Pennsylvania, and was regarded as the engineering marvel of the age.

The Coming of the Automobile

The practical development of the internal combustion engine gave the petroleum industry its greatest impetus. Inventors had experimented with "horseless carriages" for many years. But it was not until 1892 that the first American gasoline-powered auto-

*Spelling preferred by certain technical authorities and used by the American Petroleum Institute to conform to the spelling of gasoline.

mobile was built by Frank and Charles Duryea. Although in 1900 many people still thought that steam or electricity would power the automobile of the future, it turned out that gasoline proved to be the most practical fuel for automotive transportation because, when it burns, it delivers direct power in the form of a high amount of heat energy in relation to its weight and leaves almost no ash.

The oil industry had a growing outlet for what was previously a useless by-product. Until the development of the automobile, gasoline had been a waste product in the distillation of crude oil. Gasoline-filling stations began to appear; several cities claim to have had the first.

Meanwhile, the search for new oil fields continued. Oil was found in West Virginia (1860), Colorado (1862), Texas (1866), California (1875), Ohio and Illinois (1880s), Oklahoma (1905), Louisiana (1906), and Kansas (1916).

The Need for More and Better Fuels

As the number of automobiles increased and the demand for more and better motor fuels grew, it became necessary to find more oil and to improve refining processes. Only so much gasoline, however, could be obtained from crude oil by straight distillation. It became apparent that even the tremendously increased production of crude oil would not be enough to meet the rising demand for motor fuels.

A young chemist, Dr. William M. Burton, stepped into the breach with a significant new application of chemistry and engineering. He developed the thermal cracking process that was patented in 1913. By applying intense heat and pressure to the heavier fuels, Dr. Burton found that their larger molecules broke up into smaller molecules of lighter fuels, including gasoline.

Cracking units were soon put into commercial operation. Within a few decades, improved cracking processes more than doubled the yield of gasoline from crude oil. Equally important, cracking produces a

product far superior in antiknock characteristics—its ability to burn properly in an engine—than the "straight-run" gasoline produced from the primary distillation process. This permitted the development of higher compression ratio engines with greatly improved efficiency.

Later Developments

In 1937, the catalytic cracker was introduced, increasing substantially the quality and yield of gasoline from each barrel of crude oil.

By cracking, adding tetraethyl lead, and making other improvements in the refining process, the antiknock characteristics of gasoline were increased.

At the same time, automotive engineers were improving engines to use the higher-quality fuels. Thus, fuel and automobile engine improvements went hand in hand. And hundreds of other new and improved products began to appear during this period as a result of this research.

The farmer also took to oil. New highways, surfaced with petroleum asphalt, got him "out of the mud" and gave him a ready means of delivering his produce to market.

More and more rapidly, mechanized farm equipment replaced the horse and mule. As it did, agricultural productivity took giant strides forward.

During World War II, the petroleum industry proved its ability to meet sudden demands for specialized products. Huge quantities of oil were produced and converted to motor and high-octane aviation gasoline, lubricants, butadiene (for synthetic rubber), toluene (for TNT), medicinal oils, and an expanding number of other critically needed products.

Despite the tremendous wartime needs, peace brought even greater demands for petroleum products. Americans, no longer faced with wartime rationing of gasoline, took to the roads in droves. Motor vehicle registrations increased by one half during the five years following the war. The number of domestic oil burners doubled, and farm tractors increased by almost 50 per cent. Consumption of all petroleum products rose to record levels, jumping from 1.8 billion barrels (a barrel equals 42 gallons) in 1946 to 2.4 billion barrels in 1950 and 6.4 billion barrels in 1976.

Natural gas also came into its own after World War II. Transmission pipelines were constructed, linking remote producing areas to large population centers, and natural gas established itself as a major fuel for industry and home heating and as a feedstock in making glass and other products.

So tremendous was the increase in oil and natural gas consumption that in 1946, for the first time, petroleum supplied more of our nation's energy needs than coal. By 1950, when the Korean War broke out, oil and natural gas were supplying 57.8 per cent of our energy requirements. And by the 1973–74 embargo, oil and gas accounted for three quarters of the energy we used. Today, even with the cut off of Iranian oil supplies and the uncertainties of other foreign sources, we import nearly half of the oil that we use.

Functions of the Industry

Today's oil industry is engaged not only in finding oil and gas and getting them out of the ground, but also in transporting oil, making it into useful products, and marketing and delivering these products to other industries and to private consumers.

The *production* branch of the industry is concerned with the science and mechanics of exploring for new oil and gas fields, drilling wells, and bringing oil and gas to the surface.

Once crude oil is above ground, the *transportation* branch must deliver it to refineries in large volume often over long distances. From the refinery, oil products must then be delivered to bulk depots, to distributors, and finally to consumers. Pipelines, ocean-going tankers, canal and river barges, railroad tank cars, and highway tank trucks are all part of the petroleum transportation network. Trans-

portation and distribution of gas is generally independently handled by the gas industry.

The oil industry's *manufacturing* branch concentrates on refining and petrochemical processing. At the refinery, crude oil is separated by a complex series of processing methods into gasolines, fuel oils, lubricants, and many other products. Petrochemical plants process the gaseous by-products of refining into many useful chemicals.

The *marketing* end of the oil business sells and distributes oil products to consumers, who range from industrial users to homeowners. Sales departments of oil companies, independent jobbers or wholesalers, service station operators, and fuel oil dealers are all engaged in marketing activities.

Performing these diverse functions are several thousand companies, ranging from small one-man businesses, such as the consulting geologist or the single-truck fuel oil dealer, to large integrated firms that operate throughout the world.

Integrated firms are engaged in all aspects of the industry's operations, from exploration to marketing. There are over 50 integrated U.S. oil companies.

However, a far larger number of companies are "semi-integrated"—engaged in one, or sometimes two or three, of the industry's operating functions. Many companies work exclusively in the production of crude oil and gas, or transport oil exclusively. There are separate refining companies and various types of marketers. A semi-integrated company may be engaged in production and transportation, refining and marketing, or production, transportation, and refining.

Oil Installations

In the more than 120 years since the birth of the U.S. oil industry, oil companies have had to develop and expand their facilities to supply the vast amounts of energy required by our highly industrialized society.

As a consequence, the oil industry is no longer the localized business it was in the early years. Oil and/or natural gas are now produced in 33 states. The leading oil-producing states are Texas, Louisiana, Alaska, California, Oklahoma, Wyoming, New Mexico, and Kansas. Some states that produce oil do not produce natural gas. Nevada, for example, produces oil but has no commercial natural gas production. Maryland, on the other hand, produces natural gas only.

In 1859, the year of Colonel Drake's pioneer well, total United States production of crude oil was about 2000 barrels. In 1978, 3.2 billion barrels of crude oil and condensate were produced in this country from more than 516,000 wells.

The United States was for many years the leading oil-producing country in the world. It is now third behind the Soviet Union and Saudi Arabia in production. Our 1978 production represented 14.3 per cent of the estimated 22 billion barrels produced throughout the world.

Of the world's other major oil-producing nations, the Soviet Union ranked first in 1978 with an output of 4.1 billion barrels. Iran with 1.9 billion barrels and Iraq with 1.0 billion barrels ranked fourth and fifth.

Processing our nation's output of crude oil is the task of 290 refineries that currently operate in 42 states and Puerto Rico. In the aggregate, these refineries have a capacity to process more than 17 million barrels of crude oil every day. This capability represents 21.7 per cent of the total world refining capacity.

The petroleum transportation network in the United States is vast and complex. Today, more than 227,000 miles of pipeline carry crude oil and crude oil products. The pipeline network, along with thousands of petroleum tank cars, tank trucks, ocean tankers, and barges, serves to carry petroleum and its products quickly and economically from producer to consumer.

Petroleum marketing is also a large-scale operation. The wholesale distribution of petroleum products is handled, for the most part, by about 15,000 jobbers. These companies operate many of the more than 25,000 bulk plants and terminals in this country.

Their job is to fill the bulk orders for petroleum products from service stations, commercial consumers, public utilities, transportation companies, factories, and rural farm accounts.

At the retail level, there are approximately 13,700 fuel oil and liquefied petroleum gas establishments and about 158,000 gasoline service stations. More than 80 per cent of these service stations are owned or operated by local businesspeople.

Supply and Demand

The United States is one of the world's greatest producers and refiners of crude oil. It is also the greatest consumer of petroleum products. In 1978, it consumed about 6.9 billion barrels of petroleum, or approximately 29.6 per cent of the total world demand.

Total U.S. production in 1978 was 3.75 billion barrels—3.18 billion barrels of crude oil and lease condensates and 572 million barrels of natural gas liquids. The balance was supplied by imports of crude oil and oil products.

Major energy consumers in the United States can be divided into four major sectors—household and commercial, industrial, transportation, and electrical generation. It is estimated that in 1978 the household and commercial sector consumed 16.9 per cent of the U.S. demand for petroleum products. The industrial market consumed 20.2 per cent and the transportation sector used 52.6 per cent. Generation of electric power used the remaining 10.3 per cent.

In 1981, oil accounted for 43 per cent of all U.S. energy needs and natural gas accounted for 27 per cent. Coal, on the other hand, contributed 22 per cent and water power, 4 per cent. Nuclear power is, at the present time, a minor factor in the total energy picture—contributing only 4 per cent.

Supplying petroleum energy requires not only specialized facilities but also vast amounts of capital. Here are some examples: The new Trans-Alaska pipeline is 48 inches in diameter and 789 miles long and cost about $7.7 billion. A 100,000-barrel refinery can cost about $500 million. A complete drilling rig in Alaska may represent an investment of over $2.8 million. And an offshore installation can cost over $100 million.

Comparable high capital costs are faced throughout the petroleum industry. During the past decade, about 80 per cent of all oil company expenditures have come from cash earnings. In 1977 alone, industry earmarked an estimated $61.6 billion for worldwide capital expenditures to maintain, replace, and expand production, refining, transportation, and marketing facilities. This is an increase of over 395 per cent of what they invested ten years earlier.

Petroleum's Employees

Petroleum companies and dealer-operated service stations employ approximately 1.6 million people. Thus, the industry provides a source of personal income for one out of every 60 persons in the U.S. civilian labor force. Wages paid by the petroleum industry are among the highest paid by industry in general.

SEARCHING FOR OIL

Exploring for oil and natural gas begins with geology and geophysics. It is based on theories of the origin of oil and on its behavior underground.

The petroleum we use today has been in the earth for millions of years. Since it was formed long before men appeared on our planet, we can only theorize about its origins.

In Drake's time men had strange ideas about oil. Some thought it ran in a vast underground river from Pennsylvania to Texas. A few even protested that oil drillers were removing the oil that kept the earth's axle greased.

Today, the "organic" theory of petroleum's origin is the one most widely accepted. This theory holds that crude oil and natural gas are organic minerals formed by the decay and alteration of the remains of prehistoric marine animals and plants.

Oil is most often found in sediments laid down in ancient seas. These seas once covered much of the continental land area we now know. According to the organic theory, oil began with the remains of countless tiny creatures and plants that lived in the sea or washed down into it with mulch and silt from streams. This residue settled to the bottom of the ancient seas and accumulated there, layer by layer.

As the old layers were buried deeper and deeper, they were compressed by the tremendous weight of the sediments above them. Under this pressure, heat was generated. This heat, along with chemical and bacterial activity and radioactivity, gradually reformed the organic matter into the compounds of hydrogen and carbon we know as petroleum.

Through geologic ages, the sands, marine forms, and silts that settled to the ocean floor were transformed into sedimentary rocks. These loose particles, cemented into solid masses, are now called sandstone, limestone, and shale.

Some of these sedimentary rock layers were too dense to permit easy penetration by gases or liquids. Others were more loosely compacted and porous enough to let the newly formed oil and gas migrate slowly through them.

Since gas and oil are lighter than water, they tended to rise upward through the water that filled the porous spaces when the sedimentary formations were laid down.

Where a formation of dense, nonporous rock lay above a porous layer, the migration stopped, and there oil and gas collected. Lightweight gas filled the small voids present in the upper parts of the porous rock layer. Oil settled beneath the gas, and heavier water remained in the lower areas.

Originally, the deposits laid down on the sea floors were nearly horizontal. But through the centuries, movements and strains in the earth's crust folded or disrupted them. Seas shifted and land areas formed. Mountain

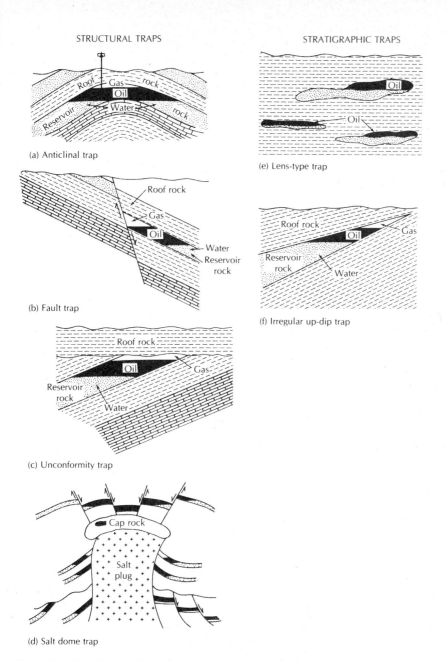

STRUCTURAL TRAPS

(a) Anticlinal trap

(b) Fault trap

(c) Unconformity trap

(d) Salt dome trap

STRATIGRAPHIC TRAPS

(e) Lens-type trap

(f) Irregular up-dip trap

Figure 8-1. Idealized sections through various types of petroleum traps: a, b, c, and d are structural traps; e and f are stratigraphic traps. Gas (white) overlies oil (black) that floats on water (stippled), which saturates the reservoir rock and is sealed by roof rock (shale or clay). Gas and oil fill only the void (pore) spaces of the rock.

ranges appeared and ancient sea bottoms became dry land.

These upheavals often caused shifts or breaks in the rock layers to create traps that collected migrating gas and oil.

The most common types of oil and gas traps are called structural traps (e.g., unconformities, faults, anticlines, salt domes) and stratigraphic traps (e.g., lens-type, irregular up-dip) (Figure 8-1).

An *anticline* is an arch-like upward fold in the earth's strata.

A *fault* is a fracture in the earth's crust that vertically shifts a nonporous layer next to a porous one, cutting off the porous layer.

Oil and natural gas are frequently found adjacent to, or on top of, an underground *salt dome*. This type of structure results from the upward thrust of a great mass of salt far below the earth's surface. Sometimes the salt enters the overlying strata. When a salt dome rises through a layer of oil-bearing sedimentary rock, oil may be trapped above the salt dome or in structures similar to faults along its flanks.

An *unconformity* is a break in the geologic sequence and is marked by a surface of erosion, or a period of nondeposition, separating two groups of strata.

A *stratigraphic trap* is one in which there are lateral changes formed as a result of rock types rather than structural deformation.

Exploring for Oil and Gas

The men who searched for oil during the first decades after Drake's well knew little about underground rock formations and potential oil traps. But as more and more oil was discovered, oil geology became a more precise science.

Geologists learned to take careful notes on underground structures and formations where oil was found. And they learned to categorize and compare them. In time, new scientific accomplishments helped geologists develop various instruments for mapping underground rock formations more accurately.

Today, oil geologists use many tools when they set out to find a new area for exploration. First, however, they must be granted access to the land they want to explore. Most often this is federal land where permission to explore is constrained by strict laws and regulations imposed for environmental reasons. In the cases where permission is granted to explore, they start out by doing preliminary work in the office and library, checking records and data. Since most of the oil fields developed to date are located in the world's great sedimentary basin regions, they begin with maps of these basins, which are fairly well documented for most of the world.

One of these regions is the intercontinental basin of the Western Hemisphere, including the U.S. Gulf Coast, Mexico, and the Caribbean area.

Recent discoveries of crude oil and natural gas in the Overthrust Belt—a highly complex geologic structure in the Rocky Mountain area—suggest that it might prove to be an extremely important petroleum province.

Another is in the Middle East area, including the oil fields of southern Russia, Iraq, Iran, and the countries of the Arabian peninsula.

Still another is in the countries of northern and western Africa.

A new region, just discovered in 1968, is the North Sea Field. Scientists predict it contains from 10 to 40 billion barrels of oil and 26 trillion cubic feet of natural gas.

New oil discoveries in Indonesia, both offshore and onshore, are occurring with increasing frequency.

Alaska, both offshore and onshore, is also thought to be a significant area of future petroleum discovery. The U.S. Geological Survey has estimated that the offshore basins of Alaska may contain two thirds of the undiscovered, recoverable petroleum resources of America's Outer Continental Shelf. The Prudhoe Bay Field, discovered in 1968 on Alaska's North Slope, has estimated proved recoverable reserves of 9.6 billion barrels of crude oil and 26 trillion cubic feet of natural gas.

Geophysical surveys suggest that sedimentary formations underlying the Georges Bank, Baltimore Canyon, and Blake Plateau troughs in the U.S. Atlantic Outer Continental Shelf may also contain oil and gas reserves.

When the geologists have selected a particular area, they check local geological offices for maps showing rock formations. They look at the data obtained from wells already drilled in the area for clues about the underlying strata. They are looking primarily for records of sandstone and limestone, for this is where oil and gas are most likely to be found.

When they have pinpointed an area that appears favorable, the geophysicist enters the picture. His tools, the magnetometer, gravimeter, and seismograph, provide a more detailed picture of subsurface formations.

The *magnetometer* measures variations in the earth's magnetic field. Sedimentary rocks generally have low magnetic properties compared to other rocks, particularly basement rocks, which are much denser and contain a higher concentration of iron and other magnetic materials. Thus, the magnetometer gives geophysicists a clue to their location. The "flying magnetometer" is trailed on a cable from a surveying plane. It records the varying magnetic intensities of the earth as the plane flies overhead.

The *gravimeter* measures minute differences in the pull of gravity at the earth's surface. Because large masses of dense rock increase the pull of gravity, gravimeter readings at a number of surface points provide a key to the underlying structures.

The *seismograph,* originally developed to record earthquakes, is now also widely used as an oil-prospecting tool. An explosive is set off in a shallow, small-diameter hole. The waves from the explosion travel downward, where they strike successive rock formations, and are reflected back to the surface. As these waves return, geophones (detectors) pick up and record the impulses. By correlating wave intensity and the time intervals required for them to travel down and

back, the geophysicist learns the general characteristics of the underground structure.

Newer developments in seismic survey on land include the use of vibratory, or percussion, devices which do away with the drilling of holes and the use of explosives. Vibratory devices are used to provide more accurate information and minimize environmental damage. In water, they send out pulses in the form of electrical discharges or contained explosions of propane gas or air so that there will be no harm to marine animal life, fewer hazards to personnel, and lower operating costs.

Photography has also become an invaluable tool, helping geologists find more telltale signs that indicate possible oil deposits. The use of photography has expanded in recent years to include satellite photos that provide large-scale pictures of land forms that may be rich in oil and gas.

The oil explorer has still another tool. Often he drills a well, known as a stratigraphic test, near a likely site to get a core of the rock strata. Geologists can then examine these core samples for traces of oil and gas. And paleontologists can study the core for fossils that might indicate the ages of various strata. This information often provides important clues on the possible location of reserves of crude oil or natural gas.

But even if all the preliminary reports are favorable, they simply indicate the possibility of finding oil. There is no sure way to know whether the oil is there until a drill has actually found it. And the odds are seven to one against the possibility that the driller of an exploratory well in a new field will find any oil or natural gas, and 68 to one that the petroleum will be found in commercial quantity.

The next step is to secure permission to drill on the property. Under most agreements, both the surface and mineral rights to the property are leased for a stipulated period. The land owner is usually given an initial "bonus" and an annual rental fee for the term of the lease, or until production is begun. Once production commences he receives a percentage royalty on any oil or gas that may be produced.

Oil companies spent approximately 1.3 billion in 1978 in the United States for geological and geophysical exploration and lease rentals. In addition, payments to individual landowners and federal and state governments in the form of royalties, including Outer Continental Shelf lease sale bonuses, can amount to several billion dollars a year.

There is little chance of environmental pollution from exploratory operations prior to actual drilling into the earth for oil and gas. But numerous safety devices and techniques are used to minimize the risk of pollution-causing incidents during the production, refining, and marketing stages.

DRILLING FOR OIL AND GAS

Before drilling can start, a great deal of work must be done. In remote or rough and swampy terrain, site preparation alone is a major engineering job. The land must be surveyed, cleared, and graded.

Often, roads must be built to move in derricks and other drilling equipment, fuel, and supplies. Roadbuilders must follow environmental requirements in locating these roads. For example, they must respect game trails in woodland areas.

A reliable source must be provided for operating the on-site equipment, and provisions must be made for a constant supply of water, for both the workers and the drilling operation. If the site is far from a town, a camp may have to be built to house the workers. And sometimes, temporary fences must be built to keep out inquisitive animals.

Finally, the drilling rig is brought to the site. Two basic types of drilling equipment, a rotary tool and a cable tool, are used today.

In both cases, machinery is set up and a derrick is erected for handling the tools and pipe that go down into the well. Most rigs now use a portable hinged, or "jackknife," derrick, or mast, that can be raised and lowered intact rather than steel girder "standard" derricks that must be erected and dismantled, piece by piece. Standard derricks, however, are still used for some very deep wells and in other special situations.

Cable Tool Drilling

Of the two drilling methods, the cable tool is much older; it is seldom used today. Five centuries before Christ, the Chinese invented a cable drilling method essentially similar to early American techniques.

A cable tool rig is made up of machinery and gear that raise and drop a "string" of tools, consisting of a bit and stem on the end of a cable. The heavy bit pounds its way into the earth, pulverizing soil and rock. At intervals, the string of tools is removed, the hole is flushed, and the resulting "slurry" of drilling cuttings is removed. Periodically the hole is lined with steel casing to prevent caving in and to protect fresh ground water encountered during drilling.

The cable tool rig is used today primarily to drill shallow water wells. Most petroleum wells are now drilled by the rotary method.

Rotary Drilling

In rotary drilling, a bit is attached to the lower end of a string of pipe, called "drill pipe." Bits range in size from less than 4 inches to more than 22 inches in diameter. They are made of very hard steel and may cost thousands of dollars. Some bits have two or three rotating cones covered with sharp teeth for grinding through rock. Others have industrial diamonds embedded in them.

The bit and pipe pass vertically through a turntable on the derrick floor. As the pipe is turned and lowered into the earth, the bit bores a hole deeper and deeper. As the hole deepens, the drilling crew adds new lengths of drill pipe.

When a bit becomes dull, the entire length of drill pipe must be removed, disconnected in stands of two or three joints each, and stacked in the derrick. After a new bit is attached, the drilling crew reconnects the stands of pipe one at a time and runs the drillstring into the hole again. This operation demands skill, speed, and precision and must be repeated many times in drilling a deep well.

Extreme care must be taken during the entire drilling operation to avoid having the string of drilling tools part and drop to the bottom of the hole. If this happens, it could mean the loss of the well. At the very best, it becomes a costly "fishing" operation to retrieve whatever has been dropped.

Drillers must also take great care not to let the drilling tools get stuck in the hole. To lessen the chances of this happening, and for other reasons, rotary drilling follows a nonstop schedule seven days a week.

A mixture of water, clay, and chemical additives called drilling mud is pumped under pressure down through the drill pipe during rotary drilling. When it reaches the bottom, it is forced out through openings in the bit, and returns to the surface outside the drill pipe. This constantly circulating fluid cools and cleans the bit and transports cuttings from the well. It also cakes the sides of the hole, preventing cave-ins and, by its weight, controls the pressure of any gas, oil, or water that may be encountered by the drill bit.

Setting casing is an important job in drilling wells. Casing is run in the hole to shut off water-permeated sands and high-pressure gas zones, prevent cave-ins, and protect fresh water strata. The casing consists of a number of lengths of heavy steel pipe joined together, usually by threaded couplings.

After the casing has been lowered to the desired point, it must be securely sealed to the walls of the hole. This is done by pumping a cement slurry through the string of casing and forcing it out through the bottom, so that it rises to fill the space between the outside of the casing and the walls of the hole.

Blowout Prevention

The casing and drilling muds normally keep the flow from the well under control, but other safeguards are provided. After the surface casing is placed in the well, further drilling is protected by one or more large valves, called blowout preventers, which are attached to the top of the casing in what is known as a stack.

These valves can be closed to seal off the well bore so that the gases and liquids under pressure in the hole can be controlled. The blowout preventers in a stack vary in size, design, and number. There are usually from four to seven on a deep well.

One commonly used control is the annular (bag-type) blowout preventer. It uses an inflatable bag to close the area between the casing and the drill pipe in an emergency. Hydraulic fluid is forced into the bag, expanding it and sealing off any flow from the well outside the pipe.

Another frequently used control is the ram type. Hydraulic pressure activates piston-like rams, causing them to close in the well. Rams are designed either to close around drill pipe (pipe rams) or to completely shut in the well (blind rams). These rams can be reopened, tested, and set again by reversing the fluid pressure.

While a well is being drilled, instruments measure and record critical aspects of the drilling operation, what is occurring within the well, and the behavior of the drilling machinery. Any unexpected change indicating the threat of a blowout triggers an automatic alarm. Crews immediately close the blowout preventer, adjust the weight of the drilling mud, or take other steps to control the flow of fluids.

Drilling Facts

Most of the wells drilled in the early days were relatively shallow. Drake's well, which was 69½ feet deep, took about 12 to 15 drilling days.

Today's wells average nearly 5000 feet. And many go beyond 15,000 feet. The world's deepest well, drilled in Oklahoma, was 31,441 feet. Drilling time now commonly ranges from a few days to a few months, with some wells taking longer than a year to drill.

Today, drilling an oil well involves large

quantities of equipment and supplies as well as substantial capital investment. As one example, the cost of drilling a single well may range from an average of $185,420 for an onshore well to over $1.7 million for an offshore well. Ten years ago, costs ranged from $55,846 for an onshore well to $449,928 for an offshore well.

PRODUCING OIL

Between the time the bit drills into a formation containing oil and/or natural gas and the time the well begins to produce, several operations, called "completing the well," take place. During this period of completion and testing, the pressure of the drilling fluid and the use of special surface equipment restrict and control the flow from the well.

First, the drill pipe and bit are removed. Cementing operations then set the final string of casing. Next, a special instrument, containing sockets holding either shaped explosive charges or bullets similar to those used in a gun, is lowered into the well. This perforating instrument fires charges or bullets by electric impulse through the casing into the producing formation to open passages through which the oil and natural gas can flow into the well bore. Tubing is installed inside the casing and the oil and natural gas flow through it to the surface.

To control the flow, a set of valves and control equipment is put in place at the top of the well. In oil country, this mechanism is called a "Christmas tree" because of its many branchlike fittings. It controls the flow of oil and natural gas from the moment the well starts producing.

Other safety devices in the production system may include automatic and manually operated valves, surface and subsurface alarms, and monitoring and recording equipment.

Sometimes in drilling a well, more than one commercially productive formation is found. In such cases, a separate tubing string is run inside the casing for each productive formation. Production from the separate formations is directed through the proper tubing strings and is isolated from the others by packers that seal the annular space between the tubing strings and casing. These are known as multiple completion wells.

When the oil and natural gas reach the surface, they are separated. The gas is sent to a gas processing plant; once the water and sediment are removed from the oil, it is transported to a refinery.

Wells that are primarily natural gas producers are of two types. If they produce gas containing dissolved liquids, they are called wet gas wells. These natural gas liquids are taken out before the gas is delivered to a pipeline company for transmission. Other wells produce only "dry" gas, that is, gas composed of hydrocarbons that cannot easily be liquefied. Since natural gas moves easily through porous and permeable rock, it can be produced simply and efficiently under its own pressure.

Life of an Oil Well

The life of a producing well begins with the first barrel of oil brought to the surface. It ends when the well is abandoned as uneconomical because the cost of producing the oil is greater than the price received for it. The life of a well varies greatly from field to field. A small pool may be in production for only a few years. Others may produce for 75 years or more.

The recovery of oil is basically a displacement process. Oil alone does not have the ability to expel itself from the reservoir. It must be moved from the rock formation to the well bore by a displacing agent. Fortunately, oil has two natural displacement agents that usually occur with it—gas and water.

The varying pressures and drives of the natural displacement agents provide a general basis for the different phases of a well's producing life. These phases are commonly called the flush, artificial lift, and stripper periods of production.

Flush production refers to the phase during which the rate of flow is governed by natural pressure within the reservoir. It is usually the first stage in a well's life, although not always, and occurs when the drill taps an oil-bearing formation that has enough natural pressure to enable the petroleum to flow by itself. With variations, three types of "drives" can generate this force.

(1) *Dissolved Gas Drive.* In virtually all oil accumulations, gas is found dissolved in the oil. Under certain conditions of pressure, all the gas is dissolved in the oil. When the formation is penetrated, the gas can expand and drive the mixture to the surface. In principle, this condition is similar to the action of gas dissolved in soda pop; it expands when the bottle is opened.

(2) *Gas Cap Drive.* In many reservoirs the pressure is such that a considerable cap of gas is trapped above the oil in addition to the gas dissolved in the oil. When the rock is penetrated, this gas (both dissolved gas and cap gas) expands and exerts enough pressure on the oil to move it toward the only escape hatch available—the well bore leading to the surface.

(3) *Water Drive.* In many oil reservoirs, there is water under hydrostatic pressure beneath the oil. When a drill penetrates the reservoir, the resulting release of pressure drives the oil to the well bore and, in some cases, upward to the surface. As the natural water pressure in the reservoir is reduced by oil production, water from the surrounding porous rock tends to flow into the reduced pressure zone, displacing the oil and driving it toward the well bore.

Artificial lift methods are applied when the initial pressure of a flush well expends itself. At this point, the well is usually put on pump. There are several varieties of pumps. One kind of surface pump connects the power source by means of a rocker arm assembly to a rod extending into the well, which is attached to a valved plunger. Submersible electric pumps can be lowered into the well. The gas lift method uses pressurized gas to raise the fluids up the well bore. Many wells never flow naturally and must be pumped from the start. Others drop in flow rate shortly after production begins, and go on artificial lift fairly early in their lives.

Stripper, or marginal, production exists when a well reaches the point when it can be produced only intermittently. Stripper wells are usually older wells that produce only a few barrels of oil a day, but are kept in production because their output is marginally above the costs of operation. Many are pumped intermittently to allow the oil to accumulate in the well bore.

Today there are more than 374,000 stripper wells in this country. Their slow but sure production gives us 12.3 per cent of our total domestic production. In 1978, stripper wells added over 391 million barrels of crude oil to the nation's energy supply. During the ten-year period from 1967 to 1978, oil produced from stripper wells totaled almost 4.6 billion barrels. Thus, while production per stripper well is slow, they play a significant part in meeting our energy demands.

Abandoning a Well

Under current practice, when an oil or gas reservoir is depleted, the site is cleaned up and the well abandoned. The hole is plugged with cement to protect all underground strata, prevent any flow or leakage at the surface, and protect water zones. Salvageable equipment is removed, pits used in the operations are filled in, and the site is regraded. Where practical, the ground is replanted with grass or other kinds of vegetation.

TRANSPORTING OIL AND GAS

To make products from petroleum, tremendous quantities of crude oil—more than 10 million barrels daily—are moved from producing fields in the United States to refineries. A large part of this volume must be transported over long distances. The output of refineries, in turn, must be delivered wherever petroleum products are used. Every branch of the petroleum industry depends upon economical and efficient transportation facilities.

Crude oil and petroleum products are moved by five major means of transportation: pipelines, tankers, barges, highway tank trucks, and railroad tank cars.

. Each method was devised and developed to meet the unique problems of petroleum transportation. No other industry is required to move such enormous volumes of liquids and gases over such long distances.

In its movement from the oil field to the refinery, and on to the consumer, a petroleum product may use any combination of these methods of transportation. Today, not counting gathering lines, domestic crude oil and products movement is about 47 per cent by pipeline, 29 per cent by tank truck, 22 per cent by water, and 1 per cent by railroad.

Oil and Gas Move by Pipeline

Buried underground, pipelines move great volumes of crude oil, products, and natural gas across the nation so silently and efficiently that few people are aware of their existence. They are the largest single movers of oil, carrying 986 million tons annually. Pipelines rank third among all types of domestic freight carriers in tonnage handled.

Most crude oil travels at least part of its journey to a refinery by pipeline. It is carried from producing wells to field storage tanks by flow lines. Gathering lines connect these tanks to trunk pipelines. Trunk lines are the long-distance arteries that usually end at a refinery or at waterside terminals. From the refinery, oil products begin their journey to consuming centers, also by pipeline.

There is now a vast network of more than 227,000 miles of crude oil and products pipelines serving 50 states and the District of Columbia. Today about 25 per cent of all intercity freight (including coal and food) moves by pipeline.

With the tremendous expansion of natural gas use after World War II, the mileage of our natural gas pipelines system had grown to a total of 1,013,000 miles by the end of 1978.

Modern pipelines vary in diameter from small, 2-inch flow, or gathering, lines to large, 48-inch trunk lines. Until World War II, very few pipelines were larger than 12 inches in diameter. With heavy wartime demands, pipeliners came up with a technological innovation, the "big inch" pipe.

As a result, pipelines have become one of the most economical means of transporting petroleum. It now costs little more than one cent to ship a gallon of oil by large-diameter pipeline from Texas to New York. Sending a postcard that distance costs nine times as much.

Pipelines are constructed to meet safety standards set by industry specifications and government regulations. These standards include pipewall thickness, welds, testing procedures, and pipe placement. Valves are installed along the line as needed to lessen spill damage in the event of an accident and to isolate the pipeline from storage and pump facilities.

To lay a pipeline, a ditch is dug deep enough to ensure adequate cover over the pipe for safe operation. Then various lengths of pipe are joined by welding. They are scrubbed clean, and corrosion-resistant coverings are applied around the outside. The line is laid in the ditch by hoists mounted on tractors called "side-booms" and the ditch backfilled to provide firm support for the pipe. Once the trench is filled with earth, the right of way is cleaned up and regraded. The pipeline may be further protected by the application of a low external voltage that keeps the pipe relatively free of corrosion almost indefinitely.

Although most pipelines are made of steel, there has been a marked increase in the use of plastic pipe in recent years. This is primarily used for small-diameter gathering lines, saltwater disposal lines, and water injection lines in oil and gas fields.

Pipelines go almost anywhere, through swamps and forests, across desert wastes, and even over mountain ranges. Sometimes the ditch must be dug through solid rock or, in some cases, the line may lie on the surface. Frequently, pipeliners must contend with other obstacles, such as rivers, lakes, and other large bodies of water. They may have to build spans to carry pipelines over streams or highways. Or, they may have to weight the pipe so that it will lie in a trench beneath the bed of a lake, a river, or an ocean.

Building a pipeline is a costly venture. For example, the Colonial Pipeline laid between the Texas Gulf Coast and New York cost over $370 million. Construction was begun in 1971 on an additional 461 miles of 36-inch loop line and 128 miles of 10-inch spur lines, plus pump stations and modifications of existing stations, for an estimated cost of $113.8 million. The nearly 800-mile crude oil pipeline across Alaska cost about $7.7 billion. The Northern Tier pipeline will cost an estimated $1.23 billion. Its 1491 miles will run from Port Angeles, Washington, to Clear Brook, Minnesota.

Oil and Gas Transport by Water

Many pipelines end at water terminals, where crude oil is transferred to tankers or barges for the second phase of its long journey to refineries. And refined products from refineries or coastal storage centers are often sent on their way to consuming centers by tanker or barge.

The tanker, a special type of merchant vessel, has a hull divided into compartments for carrying liquid cargoes. The largest modern tankers are more than 1300 feet in overall length, with capacities of 1.75 to 3.50 million barrels of crude oil or petroleum products. Their cruising speeds average 15 knots.

At the end of 1979, the U.S. flag tanker fleet consisted of about 300 vessels. Most of the American tanker fleet operates in coastal and intercoastal trade. The most frequent run is between the Gulf Coast producing areas and the Atlantic seaboard, because eastern refineries depend primarily on tankers for their supplies. Most of the crude oil and heavy fuel oil supplying East Coast refineries is imported by tanker, primarily from points in South America and the Middle East.

On a worldwide basis, water-borne traffic is also a vital factor. The world tankship fleet of approximately 5000 tankers, flying the flags of many countries, currently carries petroleum from producing nations to those which must import part or all of their oil requirements. Our country must import nearly 50 per cent of its petroleum needs, and nearly all of that oil comes by tanker.

On any given day, about 750 million barrels of crude oil and products, 31.5 billion gallons, are in transit in tankers on the high seas. A large, modern-day tanker costs about $100 million. Specially designed tankers, capable of transporting liquefied natural gas (LNG) at subfreezing temperatures, are now in operation. When the temperature of the gas is lowered to $-260\,°F$, it occupies 1/600th the volume it would at normal temperatures. This permits the shipment of greater quantities of gas in these tankers.

Oil Moves by Tank Truck and Railroad

Tank trucks are a major means of transporting oil products from bulk plants to consumers. Direct deliveries from the refiner to large bulk consumers are also frequently made by tank truck. At one point or another, between the refinery and its final destination, probably every oil product is carried by tank truck.

Today there are approximately 158,000 tank trucks of all kinds in operation. Some are built for long-distance hauls, others for local deliveries.

Trucks roll from refineries, terminals, and bulk plants to farms, factories, service stations, and homes. They supply more than 13 million households with more than 19 billion gallons of heating oil a year. They supplied 171,000 service stations and other outlets across the nation with almost 112 billion gallons of gasoline in 1979.

Most over-the-road tank trucks operate on diesel fuel, which is more economical than gasoline for long hauls. Modern tank trucks are lighter and stronger than their forerunners, because their compartments are now made of aluminum alloys, stainless steel, and reinforced fiber glass. They are also larger.

Today, tank trucks usually carry loads of 5000 to 6000 gallons.

Railroad tank cars carry a smaller volume of oil than other modes of petroleum transportation but are nevertheless important. There are some 165,000 privately owned railroad tank cars in operation, excluding those owned by the railroads themselves. About 50,000 of those carry petroleum products. These tank cars range in carrying capacity from 4000 to 60,000 gallons. Some are specifically designed to transport particular products.

Each Method Has Its Advantages

When large volumes of crude oil or products must be moved overland, pipelines are usually the most economical means of transportation, despite the large initial investment they require.

Over long distances that are nearly equal by land or sea, large ocean tankers are usually more economical than pipelines. If volume is relatively small and distance great, barge transportation offers advantages when points of origin and destination are near inland water routes.

If pipelines or water routes are not available, shipment by tank truck and railroad tank car, which are the most expensive modes of transport, may be necessary.

With distance, volume, and many other factors in mind, oil transportation experts must select not only the most efficient means of transportation for a particular shipment but also the most economical. In a business as highly competitive as oil, each company realizes the importance of saving even a fraction of a cent on every gallon it ships.

REFINING AND PETROCHEMICALS

There were 311 refineries owned by more than 140 different companies operating in the United States on January 1, 1979. Refineries range in size from small plants capable of processing only 190 barrels of crude oil daily to modern complex giants with daily crude capacities of more than 500,000 barrels. At the beginning of 1979, the total crude oil capacity of all operating U.S. refineries amounted to more than 17.5 million barrels daily.

Many factors influence the choice of a refinery location. Some refineries are built close to oil fields; others are constructed near large consuming areas. Frequently a controlling factor is easy access to water transportation.

To understand some of the processes of modern refining, it is necessary to know something about the nature of crude oil.

Crude oil, as it is delivered to the refinery, is a mixture of thousands of different hydrocarbons—compounds of hydrogen and carbon. The mixture varies widely from one oil field to another. Not all hydrocarbons are present in every crude oil.

The assortment of hydrocarbons in a crude oil and the proportions in which they are mixed determine its particular character and type. Crude oils are generally classified into three basic types. *Paraffin-base* crude oils contain a high degree of paraffin wax and little or no asphalt. Besides wax, they yield large amounts of high-grade lubrication oil. *Asphalt-base* crude oils contain large proportions of asphaltic matter, and *mixed-base* crude oils contain quantities of both paraffin wax and asphalt.

This is why crude oils do not always look alike. Some are almost colorless, and some are pitch black. Others may be amber, brown, or green. They may flow like water or creep like molasses. Some crude oils, containing over 1 per cent of sulfur and other mineral impurities, are called sour crudes. Crude oils having a sulfur content below 1 per cent are called sweet crudes.

Crude oil's basic unit is a molecule of one carbon atom linked with four hydrogen atoms. This is the molecule of methane, or marsh gas. Millions of variations of carbon-hydrogen pairing are possible, and millions of different hydrocarbon compounds can be formed.

This is where the petroleum chemist steps in. His job is to rearrange and juggle the number of atoms to make new combinations that can open up a whole new world of products.

The refining process performs on a large scale what the chemist has already done in the laboratory. By a series of processes, crude oil is separated into various hydrocarbon groups. These are then combined, broken up, or rearranged, and perhaps other ingredients are added.

Among the several refining operations known today, the three major processes are *separation, conversion,* and *treatment.*

Separation

The most common separation processes are solvent extraction, adsorption, crystallization, and the most important, fractional distillation.

In *solvent extraction,* different components of a mixture are separated from one another by a liquid solvent that dissolves certain compounds but not others.

Adsorption is quite similar to solvent extraction. The important difference is that a solid rather than a liquid solvent is used. The solid substance must be porous in order to adsorb, or hold, the undesired petroleum components on its surface.

In *crystallization,* cooling the mixture causes some of its compounds to solidify, or crystallize, and separate out of the liquid.

Fractional distillation is the fundamental process of refining. In distillation, volatility characteristics of hydrocarbons are particularly important. Volatility refers to the ease with which a liquid vaporizes. It depends on the boiling points of the various hydrocarbon compounds of crude oil. Hydrocarbon boiling points range from below minus 250 °F up to several hundred degrees above zero. Some crudes contain a significant amount of material boiling above 1300 °F. Because different hydrocarbons compounds have different boiling points, they condense at different temperatures.

In modern refineries, crude oil is pumped through rows of steel tubes inside a furnace, where it is heated to as much as 725 °F depending on the type of crude. The resulting mixture of hot vapors and liquid passes into the bottom of a closed, vertical tower, sometimes as high as 100 feet. This is a fractionating, or "bubble," tower.

As the vapors rise, they cool and condense at various levels in the tower. The liquid "residue" is drawn off at the bottom of the tower to be used as asphalt or heavy fuel. Higher up on the column, lubricating oil is drawn off at a lower temperature. Next come fuel oils, including gas oil, light heating oil, and light diesel fuel at still lower temperatures. Kerosine condenses still higher in the column, and gasoline condenses at the top. Those gases that do not condense are carried from the top of the column.

The condensed liquids are caught by a number of horizontal trays, placed one above the other, inside the tower. Each tray is designed to hold a few inches of liquid and the rising vapors bubble up through the liquids. At each condensation level, the separated fractions are drawn off by pipes running from the sides of the tower. The fractions obtained by this distillation process are known as "straight-run" products.

Conversion

At the beginning of the twentieth century, the market for petroleum products began to change radically. Automobiles became popular, and the demand for gasoline increased. But only a relatively small amount of gasoline can be distilled from the average crude oil. So refiners had to find some way of producing more gasoline from each barrel of crude processed. This problem was overcome by the development of conversion processes that enabled refiners to produce gasoline from groups of hydrocarbons that are not normally in the gasoline range.

These basic conversion processes are *thermal cracking, catalytic cracking, and polymerization.*

Thermal Cracking

Two American chemists, William M. Burton and Robert E. Humphreys, introduced the thermal cracking process in 1913. With this process, the less volatile heating oil fractions are subjected to higher temperatures under increasing pressure. The heat puts a strain on the bonds holding the larger, complex molecules together and causes them to break up into smaller ones, including those in the gasoline range.

With this discovery, refiners also found that cracking not only increased gasoline *quantity* per barrel of crude oil processed, but also produced a substantial improvement in its *quality*. The product obtained by thermal cracking was found to be far superior in antiknock characteristics than the gasoline obtained by straight distillation.

Catalytic Cracking

Thermal cracking awakened refiners to what could be done by altering the petroleum molecule. This led to extensive probing into the physical and chemical properties of hydrocarbons.

Catalytic cracking was the next great advance. It was brought to the United States in 1937 by Eugene Houdry of France.

Catalytic cracking does essentially the same thing as thermal cracking, but by a different method. A catalyst is a substance that causes or accelerates chemical changes without itself undergoing change. Unfinished heating oils are exposed to a fine granular catalyst, and the result is the breakup of these heavier hydrocarbons into light fractions, including gasoline. The catalyst enables this breakup to be accomplished under only moderate pressure.

Catalytic cracking produces a gasoline with an even higher octane (antiknock) rating. This became extremely important in meeting the special needs of World War II. Today there are many versions of this process, in which catalysts in the form of beads, powders, or pellets are used. Such catalysts range from aluminum and platinum to acids and processed clay.

Altering Molecular Structure

Refining processes that alter the molecular structure of hydrocarbons include *alkylation, polymerization, catalytic reforming, isomerization,* and *hydrocracking*.

Alkylation and polymerization combine light hydrocarbon gases to form liquids in the gasoline range having high performance characteristics as engine fuels.

Catalytic reforming converts low-quality naphthas to high-quality gasoline blending stocks, primarily aromatics. Isomerization accomplishes the same ends by changing straight-chain molecules to branched-chain molecules having higher antiknock characteristics.

Hydrocracking is essentially catalytic cracking in the presence of hydrogen to give products of higher quality and lower sulfur content.

As a result of these advances, refinery efficiency has increased substantially. In 1920, the gasoline yield from each 42-gallon barrel of crude oil refined was only 11 gallons. Today, refiners obtain an average of 19 gallons from each barrel of crude. Far higher yields could be obtained if market demand called for more gasoline in relation to fuel oil and other petroleum products.

If no improvements had been made in gasoline processing techniques since 1920, refiners would have to run over 3.5 billion barrels of additional crude oil per year to meet our gasoline requirements today.

PETROLEUM PRODUCTS AND THEIR USES

At the turn of the century, it was relatively simple to pinpoint the major markets of the petroleum industry. Grease was the major lubricant. Kerosine was used for illumination. Coal, the major energy source, was used for heating. Today, petroleum supplies the power for our transportation network and our factories and farms; the heat for our homes and large buildings; the lubricants to keep the wheels of industry turning; the material with which much of our highway, road, and street network is paved; and a host of products derived from petrochemicals.

Product Quality Assessment

Approximately 3000 products are currently produced wholly or in part from petroleum, in addition to 3000 or so petrochemicals. The companies that manufacture these products and the customers who purchase them want to have some general idea of their properties, quality, and performance. In the early days of the industry, each customer and each major supplier had his own methods of assessing performance of quality, and disagreement was widespread over these various proprietary methods. In more recent years, the excellent programs of such organizations as the American Society for Testing and Materials, Society of Automotive Engineers, National Lubricating Grease Institute, Acid Processors Association, Coordinating Research Council, Inc., Department of Defense, National Bureau of Standards, Bureau of Mines, and others have led to the development of technology that permits the writing of tests. Once these standardized tests have been agreed upon, it is possible to come up with specifications. The lead taken by U.S. institutions in this area of petroleum testing is being followed by the rest of the world through the work of such international organizations as the International Organization for Standardization (ISO).

Petroleum Fuels

Among the products derived from petroleum are the fuels that now supply three quarters of all the energy consumed in the United States. Petroleum energy provides the power for supersonic jet aircraft, the fuel for small space heaters, and a multitude of uses in between.

Gasoline

Motor and aviation gasolines are blends of straight-run gasoline (obtained by primary distillation), natural gasoline (one of the liquids processed out of natural gas), cracked gasoline, reformed gasoline, polymerized gasoline, and alkylate.

To these gasoline blends, refiners add a wide variety of chemicals, called additives, to further improve the quality of the fuel.

Antiknock compounds are one such gasoline additive. Their purpose is to reduce or eliminate the "knock," or "ping," that occurs when the fuel is not being properly burned in the engine. The measure of gasoline's resistance to engine knock is its octane number. This is determined by comparing a gasoline with fuels of known composition and knock characteristics under specified conditions. Two different types of tests conducted in a single-cylinder laboratory engine yield antiknock measures called Research or Motor Octane Number. When the test is run in a test car, it is called Road Octane Number.

With the advent of octane posting requirements by the government, a new octane number was selected to be displayed on all dispensing pumps. This new octane number is derived by adding the Research and Motor Octane numbers and then dividing by two. The formula is commonly expressed as $(R + M)/2$; R represents the Research Octane Number and M represents the Motor Octane Number.

Because of emission control devices and regulations on motor vehicles, refiners are now producing gasolines that contain little or no lead-based antiknock additives. Instead, they use hydrocarbon compounds as antiknock additives.

Over the years, the octane numbers of gasoline have increased as automotive engineers have developed cars with higher compression ratios. In 1935, regular-grade gasoline had a Research Octane Number of 72. By 1976, it had risen to 93.3. Premium gasolines rose during the same period from a Research Octane of 78 to almost 100.

Gasoline is the petroleum industry's principal product. Some 112 billion gallons were consumed by motor vehicles in this country during 1979. Of this total, about 75 per cent was used in automobiles alone. Trucks, buses, and motorcycles accounted for the rest. Passenger cars consume, on the average, about 688 gallons of gasoline yearly. The average truck uses 1270 gallons of motor fuel a year, and the average commercial bus, an estimated 2041 gallons.

Aviation Gasoline

In the early days of aviation, airplanes were fueled with the same kind of gasoline burned in automobiles. It was not until World War I that serious research was begun on aviation gasoline that was distinctly different from the fuel consumed in automobiles.

After the war, research continued on ways to develop increased power without enlarging the size of the airplane engine. Researchers realized that the key was a standardized fuel of high antiknock characteristics. Refiners went to work and developed an aviation gasoline with an octane number of about 87. In 1934, an aviation fuel of 200 octane was developed. The engine's size remained the same, but its power output was considerably increased.

Commercial aviation began to grow significantly during the early 1930s, but it was somewhat curtailed as World War II approached. The war generated an immediate military demand for large quantities of 100-octane aviation gasoline. The prewar peak annual production of all aviation fuel had been 14.7 million barrels. By 1945, 100-octane aviation fuel production alone had reached a level of more than 124 million barrels.

Commercial aviation came into its own as Americans took to the air after the war. Mail and all kinds of freight were carried by air. Private aviation surged upward, as more and more business organizations bought and maintained their own airplanes. Small-plane flying became popular, and aircraft became

important to farmers for seeding, dusting, and spraying their crops and defoliating and fertilizing.

During the 1950s, commercial airlines began to switch over to the faster jets. Even though the private use of piston-engine aircraft continued to increase steadily, the demand for aviation gasoline began to slacken. There is still a substantial market for it, however—over 594 million gallons in 1978.

Jet Fuels

When research first began on jet fuels, commercially available kerosine was used because of its relatively low volatility—an important jet fuel requirement. Today's commercial jet airliners still depend on a highly refined kerosine for most of their fuel supply.

Military jets require a somewhat more complex fuel to withstand the severe conditions of supersonic flight. To meet the military's jet fuel requirements, scientists have developed carefully compounded blends of kerosine and gasoline.

Jet aircraft consume enormous amounts of fuel. Some of the jumbo-sized jets now in operation consume an average of 3335 gallons of fuel during each hour of flight. Their fuel tanks can hold up to 47,210 gallons of fuel.

The extent of the changeover from pistons to jets by the military and by commercial airlines is evidenced by the rapid rise in jet fuel consumption during the last 16 years. In 1959, U.S. jet fuel consumption totaled 4.4 billion gallons. By 1978, it had risen to over 13.2 billion gallons.

Kerosine

In the early days of the petroleum industry, kerosine was the refiners' principal product. Once used primarily as an illuminant, kerosine is now used mostly for cooking, space heaters, farm equipment, and jet fuel.

But this product also has many other uses. It is an ingredient in insecticides, paints, polishes, and cleaning and degreasing compounds. On farms, kerosine not only powers tractors and other equipment but also provides the fuel for heaters used in the curing of tobacco.

Total consumption of kerosine in 1978, exclusive of its use as jet fuel, amounted to more than 64 million barrels.

Diesel Fuels

The diesel engine is fundamentally different from the gasoline engine. A diesel's ignition is caused by the heat of compressed air in a cylinder, whereas the spark plug performs that function in a gasoline engine.

Early diesel engines were massive, built to withstand tremendous heat and pressure. At first, they served primarily as stationary power sources in factories and ships, and they used almost any oil as fuel.

Large diesel engines are still important stationary power sources, but through the years diesels have gradually been adapted for other uses. Today, they provide economical power for heavy road equipment, such as trucks, buses, and tractors. Railroads have turned to diesel engines for locomotives and now represent a large market for diesel fuels. Advances in metallurgy and rapid-fire fuel delivery made possible the development of the diesel automobile.

Just as modern diesel engines have become specialized, the diesel fuels that run them are highly refined for their specialized uses. Modern diesel fuels are manufactured in several grades, ranging from heavy oils to light kerosine-type oils.

Fuel Oils

Today, fuel oils are designed to meet the needs of residential and commerical heating, manufacturing processes, industrial steam and electrical generation, and marine engine and many other uses.

Fuel oils are generally classified as distillates or residuals.

Distillates are the lighter oils, some of which are used in space heating, water heating, and cooking. But the major market for distillates is in the automatic central heating of homes and smaller apartment houses and buildings.

One of the big advances in home heating has been the development of the Degree Day—the system worked out by oil companies to determine just when each customer's oil tank needs filling. This system practically eliminates the possibility of an oil-user finding himself without oil on a cold morning.

Total distillate consumption (including diesel fuel) in 1978 amounted to 1.3 billion barrels.

Residuals are the heavier, high-viscosity fuel oils which usually need to be heated before they can be pumped and handled conveniently. Gas and electric utilities are residual oil's major market. Industry is the second largest consumer of residual oil, using it to fire open-hearth furnaces, steam boilers, and kilns. Large apartment and commercial buildings make up the third largest residual oil market.

The total market of residual oil during 1978 amounted to over 1.1 billion barrels.

Petroleum Coke

Petroleum coke is almost pure carbon. As such, it has many useful properties. It burns with little or no ash, conducts electricity, is highly resistant to chemical action, does not melt, and is an excellent abrasive.

Coke is invaluable in the manufacture of electrodes for electric furnaces and in electrochemical processes. Carbon, or graphite, often made from petroleum coke, is used in flashlight and radio batteries. Sandpaper and knife-sharpening whetstones are made by the fusing of sand and coke.

But petroleum coke is primarily a fuel, valuable in refining aluminum, nickel, special steels, and chemicals. The fact that it is almost pure carbon reduces the chance of contamination in the metals or chemicals being refined.

Liquefied Petroleum Gas (LPG)

Familiar to many as "bottled gas," LPG consists primarily of propanes and butanes, highly volatile gases that are extracted from refinery and natural gases. LPG has a unique double characteristic. Under moderate pressure, LPG becomes liquid and can be easily transported by pipelines, railroad tank cars, or trucks. Released from its storage tank, LPG reverts to vapor form, burning with high heat value and a clean flame.

Part of the growing demand for LPG is as fuel for internal combustion engines, primarily those used in buses, tractors, forklift trucks, and in-plant equipment. A principal use for the fuel is in the manufacture of petrochemicals.

LPG is also used extensively in the home— for stoves, refrigerators, water heaters, space heaters, and furnaces. At present, cooking is the predominant household use for LPG, with clothes drying and air conditioning gaining in importance.

On farms LPG heats incubators and brooders, sterilizes milk utensils, dehydrates fruit and vegetables, prevents frost damage, and controls weeds by flame weeding. In industry, it is used in metal cutting and welding.

Total consumption of liquefied petroleum gases and ethane (excluding those amounts used in the production of gasoline) reached the 25-billion-gallon level in 1978.

Lubricants and Greases

Lubricating Oils

All equipment with moving parts require lubrication. Finished lubricating oils range from the clear, thin oil that is placed by a hypodermic needle on tiny compass bearings to the thick, dark oil that is poured into the massive gears of giant machines.

As the needs of industry have become more complex, lubricating oils have become more specialized. American companies manufacture hundreds of different oils to exacting specifications. Industrial lubricating oils must maintain their friction-fighting properties without thickening or coagulating for thousands of hours between shutdowns.

Lubricating oils for automotive engines are designed to show a relatively small change in viscosity with temperature, neither thinning out unduly in searing heat nor thickening excessively in below-zero winter weather. They are designed to prevent formation of excessive carbon, corrosive acids, or sticky deposits. Chemical detergents enable these oils to hold in suspension harmful matter that would otherwise accumulate on engine parts.

Greases

The essential ingredients of grease are lubricating oil base stocks, soaps that act as thickening agents, additives to improve performance and stability, and fillers to increase wearability. Technologists have developed greases in hundreds of consistencies to meet innumerable performance needs. Some greases must resist the pounding heat of steel rolling mills or the subzero temperatures encountered by high-altitude aircraft. Others must withstand acids or water or stand up to the grinding friction of railroad roller bearings.

Waxes

The two types of wax derived from petroleum, paraffin and microcystalline, are extracted from lubricating oil fractions by chilling, filtering, and solvent washing.

Paraffin wax is a colorless, more or less clear, crystalline substance, without odor or taste, and is slightly greasy to the touch. Microcrystalline waxes do not crystallize like paraffin because they are composed of finer particles.

Wax is used for the most part in packaging. It waterproofs and vapor-proofs such articles as milk containers and wrappers for bread, cereals, and frozen foods. Wax is also used in the casting of intricate parts and components of machinery, jewelry, and dentures.

Asphalt

For centuries, men obtained asphalt from natural deposits, or "lakes," where it remained as residue after the air and sun had evaporated the lighter petroleum fractions from it. Solid or semi-solid at normal temperatures, asphalt liquefies when it is heated. It is a powerful binding agent, a sticky adhesive, and a highly waterproof, durable material.

Today, asphalt is an important petroleum product, extracted as a refining residue or by solvent precipitation from residual fractions. By careful selection of crude oils, controlled air oxidation, and blending, modern asphalt is given several added properties such as inertness to most chemicals and fumes, weather and shock resistance, toughness, and flexibility over a wide temperature range.

There are a multitude of uses for asphalt. It is a major road-paving material. It also surfaces sidewalks, airport runways, and parking lots. It goes into such products as floor coverings, roofing materials, protective coatings for pipelines, and underbody coatings for automobiles.

Petroleum's Total Market

All told, a level of 6.9 billion barrels of petroleum products were consumed in the United States during 1978—a 40 per cent increase over demand 10 years ago. This works out to be an average of 1342 gallons per person—or more than 103 gallons per week for an average family of four. Much of this, of course, would be in the form of petroleum products used by industry and transportation to provide the goods and services that such a family requires.

The major portion of petroleum's total market is accounted for by gasoline—over 2.7 billion barrels in 1978. Petroleum's second largest market is the fuel oils—the distillates and residuals—which were consumed at a rate of over 2.4 billion barrels. Jet fuel demand totaled 313 million barrels, and kerosine, once the industry's prime product, accounted for 64 million barrels.

The remaining 1.4 billion barrels were accounted for by LPG, asphalt, lubricating oils and greases, and many other products.

ENERGY AND THE ENVIRONMENT

The petroleum industry is well aware of the damage that petroleum and its products could do to the environment. That is why the industry has for many years conducted extensive research on the problems of atmospheric wastes and industrial water pollution.

The ever-growing need for energy in this country, with a rising population, more automobiles, and the universal desire to raise living standards, signals the need for petroleum companies to expand operations. At the same time, environmental quality is a major concern.

The cost to the oil industry during the ten-year period ending in 1978 for pollution abatement was almost $15.5 billion. During 1978 alone, it is estimated that oil companies' capital investment to combat pollution was over $2.4 billion.

While exploring for and producing oil, the industry is careful that its operations do not contaminate the environment. Seismic crews in many situations replace explosives with air guns to produce acoustical waves during exploration. Oil well blowouts, almost a trademark of early days in the industry, almost never happen today, thanks to advances in petroleum technology.

No longer are natural gases burned off or wells drilled side by side. Instead, wells are carefully spaced far apart and production is controlled to conserve reservoir energy for maximum recovery before pumps are used.

The industry has also improved oil field equipment design, including oil-water separators and remote emergency equipment, as well as the cement plugging of abandoned wells, all in the interest of conservation and environmental concerns.

Ocean-going tankers bring to our shores nearly one half of the oil used in the United States. Pipelines, railroad tank cars, barges, and trucks move tremendous quantities across the country every day, rarely disturbing the environment. The people who carry out each transportation phase are prepared to prevent and clean up oil spills quickly and efficiently.

Oil pollution at sea results from discharges of bilge and waste, spills from tanker accidents, and discharges of dirty water from tanker washings. Over four fifths of the world's tankers no longer discharge dirty water at sea, but store it and either off-load it at the next port for processing in refinery oil-water separators or make it part of the next crude oil cargo without any of the oil being spilled into the sea. And the percentage of tankers which do this is expected to rise constantly in the 1980s to almost 100 per cent.

New techniques have also been developed for use at sea and on beaches for the cleanup of possible oil spills. The American Petroleum Institute was instrumental in the founding of a school at the Texas Engineering Extension Service where oil company personnel learn the skills of oil-spill cleanup. In addition, API has sponsored eight years of research on the behavior and effects of oil on the marine environment. This research shows that the effects of oil on the marine environment are far less than some have predicted and that such effects are reversible.

Three fourths of American crude oil and almost one third of our refinery products move around the country in pipelines, with minor amounts of leakage—about six-thousandths of 1 per cent of the volume moved annually. Safety devices that help prevent pollution include ultrasonic leak detectors, corrosion preventives, automatic shutdown techniques that go into operation in case of a leak, and regular pipeline inspections. Computerized control of the pipelines helps make possible the safe movement of batches through the lines. Pipelines themselves, of course, have to comply with very strict government safety regulations.

Furthermore, when a new pipeline is laid, construction crews work to restore the right of way to its original form, with almost no perceptible change in the environment.

Although refineries in years past con-

tributed to the pollution of air and water, tremendous improvements have taken place. Noxious emissions of fumes and odors have been eliminated to a great extent. A refinery uses great quantities of water for cooling and processing, usually from a river, and that water is now purified in multiple-stage effluent treatment to remove oil and other wastes. The water is then returned to rivers in the same state of purity and at about the same temperature at which it was removed. And thanks to technological advances in many cases, river water is returned in a greater state of purity. Also, air is replacing water as a coolant in some systems, with a resultant decrease in thermal pollution.

Refineries are now better neighbors, reducing the light and noise of refining with shielded, smokeless flares and equipment silencers. In all, refiners have spent billions of dollars to stop air and water pollution, and additional millions have been spent in fuel and process research.

Automobile Emissions

A decrease in automobile emissions has taken place with improvements in automobile systems. According to EPA estimates, there have been reductions of 83 per cent in carbon monoxide emissions, and emissions of nitrogen oxides have declined 43 per cent. The automobile industry's estimates run even higher, and figure a reduction of over 90 per cent in hydrocarbon emissions, an 84 per cent reduction in carbon monoxide emissions, and a 63 per cent reduction in emissions of nitrogen oxides.

Since 1974 lead-free gasoline has been made by refiners throughout the industry and is available at service stations in every state. In addition, lead emissions from leaded grades of gasoline are going down and will go down even further because of the lead phasedown in gasoline ordered by EPA.

ENERGY CONSERVATION

The petroleum industry strongly supports the wise and efficient use of energy. Since energy conservation has the effect of increasing domestic energy supplies—much like a new oil recovery process or opening a new coal mine—it can be a key element in reducing U.S. oil imports.

Today, Americans are using energy more carefully than ever before. During the last six years, as the cost of energy has gone up, Americans have decided to buy less. Before the 1973–74 oil embargo, most forecasters agreed that energy consumption in the United States would grow at a rate of about 3.5 per cent each year. In contrast, forecasts made in 1979 project the growth rate at about 2 per cent or less, reflecting the conservation effect of higher prices in general and oil prices in particular. In fact, consumption of oil in the United States was virtually stable in 1979.

With more fuel efficient new cars, with consumers turning more and more to smaller vehicles, with businesses introducing more efficient equipment, with homeowners installing more insulation, evidence is piling up to demonstrate the potential savings of conservation.

Residential

Since home heating accounts for 50 per cent of all residential energy use, significant energy savings can be achieved by cutting back on home energy use.

Energy can be saved through better insulation, weather stripping, caulking, and storm windows.

More efficient heating systems such as heat pumps can be installed in new homes and used to replace older conventional electrical systems.

And improved house designs and layouts with features such as double-wall construction and windows with southern exposures can also help cut residential energy use.

Other energy-saving features include internal venting of clothes dryers for winter heat, in-home utility meters that allow residents to monitor their use of energy and its costs on a daily basis, and flue dampers that close when the furnace is off. Such devices can help reduce total household energy consumption in the years ahead with little or no effect on individual lifestyles.

Automotive

Energy used in transportation—almost entirely from oil—has accounted for about one fourth of the nation's energy consumption in recent years. Of that amount, private passenger cars account for about 50 per cent.

Significant energy savings are expected by the year 2000 as a result of government-imposed fleet fuel standards and higher fuel prices despite an increase in the number of cars and people.

Other initiatives being considered to cut the amount of energy used in personal transportation include increased public transportation systems, more commuter car and van pools, as well as more controversial programs such as gasoline rationing, service station closings, day-of-week driving restrictions, and higher excise taxes on fuel.

Industrial

Substantial energy savings could be achieved in certain industrial processes, as well. Energy for industrial process steam, for example, accounts for about 14 per cent of total U.S. energy consumption. Fuel consumption in the production of this steam could be reduced by as much as 30 per cent by employing cogeneration—the combined production of electrical or mechanical energy and heat.

Other Energy Savings

Since 1972, business and industry have become much more energy efficient than ever before. Production has increased nearly 21 per cent, while energy use has increased only 7 per cent. Oil companies have demonstrated their commitment to conservation. Back in 1974, U.S. petroleum refiners pledged to join in a federally sponsored volunteer energy conservation program. Their goal was to try to cut energy use in refineries by 15 per cent by 1980. By the end of 1977, the refiners had already surpassed the 15 per cent mark, which in the meantime had been raised to a new 20 per cent goal. By the end of 1979, refiners had achieved a reduction of 19.6 per cent. They achieved these savings through better training and better housekeeping features and through capital investment in equipment such as improved instrumentation and heat exchangers.

Clearly, energy conservation has become a part of the national effort to reduce oil imports. Forecasts of U.S. energy consumption have been dropping steadily over the past four or five years. The estimates for 1990 are around 42 to 46 million barrels a day of oil equivalent—10 million barrels a day below what had earlier been projected. It should be noted, however, that even with continued conservation we shall still need 5 to 10 million barrels of oil more in 1990 than today. In other words, we can't "save our way out." However, energy conservation can make a big contribution toward U.S. energy self-sufficiency. How much more will depend on voluntary savings by Americans and their willingness to let government impose policies that force them to conserve.

ENERGY AND THE FUTURE

Over the past century there have been many gloomy predictions about this country's energy potential.

In the 1880s and 1890s a government bureau said there was little or no chance of finding oil in California, Kansas, or Texas. Today, they are all leading oil-producing states.

In 1939 a government official said U.S. oil supplies would last only 13 years.

In 1949 a federal official announced that the end of U.S. oil supplies was almost in sight.

If oilmen had allowed themselves to be discouraged by some of the "expert opinions" of the past, they would have stopped searching for oil and gas in this country many years ago. Instead, drillers keep on working and hoping. As a result, today 33 of our 50

states produce either oil or natural gas, or both.

More recent oil and gas reserve estimates are more optimistic. Studies by government agencies, scientific organizations, universities, and oil companies have concluded that this country has the potential to produce large amounts of oil and gas for many years to come.

Dr. Charles D. Masters, chief of the Office of Energy Resources in the U.S. Geological Survey, reviewed several such studies in a speech in January 1979. He said that although the estimates vary in detail, they indicate that at least for the next 30 to 50 years, resource *potential* will not be a limiting factor in sustaining U.S. oil and gas production.

Government and industry studies indicate that the United States could produce about 150 billion barrels of crude oil and natural gas liquids in the future—more than this country has produced in the past 120 years.

The same studies indicate future natural gas production of about 800 trillion cubic feet—40 per cent more gas than the nation has ever produced and the equivalent of 142 billion barrels of oil.

These amounts would be more than 40 times the current annual production of oil and gas.

The limiting factor will be the rate at which this country can find those supplies and convert them from potential to proved reserves. How much of this oil and gas will be found and produced, and how soon, will depend on many factors, including pricing, incentives for investment, technological progress, environmental restrictions, access to promising new areas to explore, and the overall political and economic climate.

These factors will also affect other energy sources—many of which we have in abundant quantities.

Alternate Energy Sources

Total recoverable reserves of coal in the United States are at least 250 billion tons (the equivalent of about 1000 billion barrels of oil)—more than three times greater than our recoverable resources of oil and gas.

This country also has vast reserves of other fuels that could contribute increasingly to the nation's energy supply in the 1980s and beyond:

Recoverable reserves of uranium "concentrate" are about 700,000 tons—enough to sustain the nuclear power industry well into the twenty-first century.

Resources of relatively high-grade shale oil will begin to make an appreciable contribution by 1990. Shale oil could contribute 800 billion barrels initially and as much as 600 billion barrels eventually—twice as much as our oil and gas resources.

In addition to these fuels, "renewable" sources, such as solar and geothermal energy and biomass, will be used increasingly in the future.

Beyond these sources of energy, the United States has another tremendous resource—conservation—that is already reducing our dependence on imported oil.

The solutions to America's energy questions need not depend on miracles or great self-sacrifice. They require only that we approach our energy problems with common sense and use effectively the vast resources that are available to us.

SUGGESTED READINGS

Michel Grenon, Global Energy Resources, *Annual Review of Energy* **2,** 67 (1977).

Michel Grenon, On Fossil Fuel Reserves and Resources: International Institute for Applied Systems Analysis, RM 78–35, June 1978.

Michel Grenon, World Oil Resources Assessment and Potential for the 21st Century: International Institute for Applied Systems Analysis, WP 80–6, January 1980.

Earl T. Hayes, Energy Resources Available to the United States, 1985 to 2000, *Science* **203,** (4377), 233 (1979).

M. King Hubbert, Energy Resources, A Scientific and Cultural Dilemma, *Bulletin of the Association of Engineering Geologist* **13,** (2) 81 (1976).

M. King Hubbert, World Oil and Natural Gas Reserves and Resources, Part IV, B., in U.S. Energy Demand and Supply, 1975–1980: Congressional Research Service of the Library of Congress, 1977.

M. King Hubbert, Measurement of Energy Resources, *Journal of Dynamic Systems, Measurement, and Control* (American Society of Mechanical Engineers) **101** (March), 16 (1979).

Richard A. Kerr, How Much Oil? It Depends on Whom You Ask, *Science* **212** (4493), 427 (1981).

Richard Nehring, *The Discovery of Significant Oil and Gas Fields in the United States.* Santa Monica, Calif.: The Rand Corporation, 1981.

D. H. Root and E. D. Attanasi, World Petroleum Resource Estimates and Production Forecasts: Implications for Government Policy, *Natural Resources Forum* **4** (2), 181 (1980).

Future Availability of Oil for the United States   9

JOHN S. STEINHART AND BARBARA J. MCKELLAR

INTRODUCTION

Because the United States derives more than 45 per cent of its primary energy from oil, substitutions for oil require extensive changes in our present energy supply and use systems and will not come quickly. In particular, many of these changes cannot be made in the next 20 years. We intend to focus on oil availability in the next 20 years, and in that time span we conclude that availability will be pressing no matter what measures are taken to seek substitution. The policy goal of imports reduction ensures that outcome in the short run, and we conclude that imports may continue at some level beyond the end of the century.

Future availability of oil depends on four things: production from known reserves, discoveries and extensions to reserves, improved recovery proportions from oil originally in place, and, for an importing nation like the United States, amounts available for import from the world petroleum market. In the short run, the amounts demanded and supplied are determined by price, technological capabilities, and government policies and, in the longer term, controlled by the geological facts of what is available and what is possible. Of course, other sources of energy can substitute for oil over various time periods, and "synthetic" oil can be produced from sources such as coal and oil shale.

Reserves are certain. This term is used to describe oil that has been discovered and the amount assessed, and that can be produced under current economic and technological conditions. But future production will also be from oil fields that are not even, as yet, discovered. In addition, technological improvements at every step in the process could enlarge recovery from known fields and enhance the yield of useful products from existing production. The approach to this problem has been to estimate the amounts of ultimately producible oil. This, of course, involves estimation of discoveries yet to be made as well as assumptions about achieva-

Dr. John S. Steinhart is a professor of geology and geophysics and a professor at the Institute of Environmental Studies, University of Wisconsin, Madison, Wisconsin 53706. He also holds a teaching appointment in the Department of Political Science at UW-Madison.

Barbara J. McKellar is a graduate student in geology and geophysics at the University of Wisconsin-Madison. She holds degrees in biology and energy policy analysis and was formerly a member of the Ford Foundation Energy Policy Project.

ble recovery rates and future technology. Some new techniques have been brought to bear on these questions in recent years, and we have more of a record against which to test past predictions.

Rates of domestic production were, until the early 1970s, dictated by production quotas, prices, and estimated demand. In the future, however, production will also be determined by what is possible. Knowledge that the oil is there is not enough. Recovery of oil from the porous rocks that constitute the oil reservoir can be done only at rates that are physically limited. Thermodynamics of the flow of fluids through porous media suggests that the maximum ultimate recovery can be obtained if the oil is produced infinitely slowly. Since that is unrealistic, the industry has devised a measure of recovery rates (called the Maximally Efficient Rate, or MER) that is a compromise between the physical limitations and the economics of slowing production. Thus, when the amounts that may ultimately be produced are being considered, it should be recalled that at any time only a small portion of known amounts can be produced quickly and that the remaining stream of oil will come at steadily declining rates into the future. The pessimistic side of this result is that we cannot merely produce the remaining oil whenever we wish to and contemplate an abrupt end to the oil industry on a Thursday afternoon, but the optimistic side is that, whatever may happen, there will be steady, but declining, oil production for a long time even from existing reserves.

Finally, the world oil market has been transformed in the past decade. We shall explore these changes and attempt to estimate their impact on oil availability on the world market in the next 20 years.

HISTORICAL REVIEW OF RESOURCE ESTIMATES

Methodology

Estimates of ultimately producible oil extend as far back as the first production of domestic oil in 1859.[1] So do predictions of running out of oil in the near future. In 1909, David T. Day, of the U.S. Geological Survey (USGS), predicted that by 1943 most of the total maximum oil available from fields would be "used up" given the then present-day technology and rates of production. Needless to say, time proved Day's prediction incorrect.

Day's predictions were based solely on the areas then producing oil. In the early 1900s, this was limited to the northeastern parts of the United States, extending to Indiana and Illinois; a few fields were producing oil west of the Mississippi River. The large oil fields in Texas, one of the United States's largest producing areas today, as well as oil fields in Oklahoma, Colorado, and Alaska, had not been discovered.

Since Day's time many estimates of ultimately available oil have been made, including more geographical area and using a variety of methods. At the risk of oversimplification, these methods fit into three basic categories:

1. Projections based on geologic analogy, in which similar geology is taken to indicate similar potential for oil production.
2. Extrapolation methods based on historical data.
3. More complex models combining historical data with geologic data.

Geologic analogy is one method of estimating oil potential in unexplored areas. Simply stated, this method compares the geology of a producing area with that of a geologically similar but unexplored area. Many nonquantifiable geological observations enter this comparison. Until a well is drilled, more quantitative information is simply not available. Unlike coal, which is often exposed at the surface, oil is located beneath the surface.

Many USGS resource analysts using this method have assumed that all the remaining unexplored areas are the same as the explored areas with respect to potential producible oil (Zapp 1962, Duncan and McKelvey 1963, Theobald et al. 1972). This raises an important issue. If all the geologically favorable areas yet to be explored have a potential for producing oil equal to that of those areas already explored, this implies that companies exploring for oil in the past have been unable to select the most likely oil-bearing areas because similar amounts of oil *have not* been found. Additionally, many areas that have been assessed as favorable have not yielded commercial amounts of oil.

Methods based upon historical data include consideration of past experience in annual production or discovery rates. Future discovery and production rates, and ultimately producible oil, are estimated by the fitting of a curve to the available data and extension of the fitted curve into the future. The application of this method requires a time series of data in order to determine a trend. Available data are frequently of different quality. While production data are factual within a ± 1–2 per cent error margin, distortions may occur in reserve estimates due, for example, to tax laws.

Linear extrapolation is a common technique that has been used in estimates of future oil availability. As recently as 10 years ago, most oil company estimates predicted that future U.S. annual oil production would continue as it had in the past, implying a straight line into the future.

In recent years other methods have been used to fit historical data to a curve, providing a base for extrapolation into the future. One example is the logistic growth curve that has been used to explain a wide variety of growth patterns seen in the natural world.

Geologist Donald Foster Hewett was the first person to discuss metals in a cyclical fashion resembling the logistic growth curve. In 1929 Hewett presented a paper entitled "Cycles in Metal Production" to the American Institute of Metallurgical Engineers. He studied historical records of largely depleted European mines, which he presented in graphs of metal production (Figure 9-1). The one criterion of the many Hewett developed that is applicable to fossil fuels is that exploitation of fossil fuels starts at zero, rises continuously to a peak with increasing time, and then declines (Hubbert 1973).

Figures 9-2 shows two kinds of logistic growth curves which represent an important mathematical treatment of Hewett's criterion. Figure 9-2A is sigmoid, or logistic, curve and represents the theoretical cumulative production of a finite resource. The term Q_∞ is used to describe the ultimately producible amount that this curve approaches. Figure 9-2B shows time incremental rates of production (usually annual). The area beneath the incremental production curve is equal to Q_∞.[2]

In 1956, M. King Hubbert, then with Shell Oil, presented a paper to the Southwestern Section of the American Petroleum Institute (API) at San Antonio, Texas, entitled "Nuclear Energy and Fossil Fuels." In this paper

Figure 9-1.

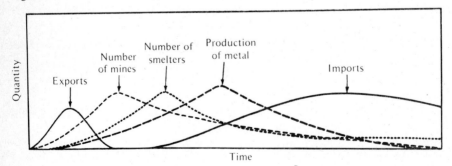

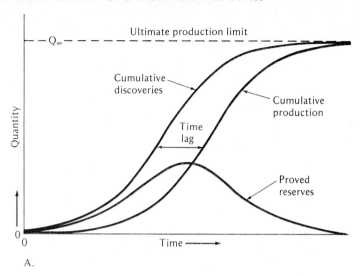

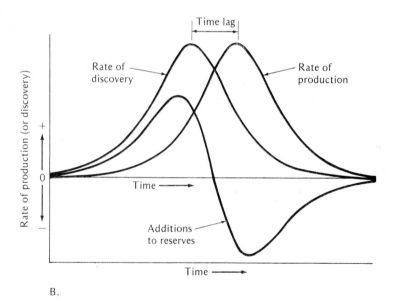

Figure 9-2. Logistic growth curves.

Hubbert applied Hewett's criterion to the fossil fuels. Hubbert determined Q_∞ from a consensus of estimates from colleagues in the petroleum field. Taking into account the range for Q_∞, and using the logistic growth curve as his model, Hubbert predicted that a peak in domestic annual oil production would occur about 1970.

In order to reduce the subjectivity of his earlier work, Hubbert (1962) did a second analysis using available production and discovery data from the API. The data showed that annual discoveries had peaked in about 1957 and that production was likely to peak at the end of the 1960s, and that ultimately producible reserves would be 170–175 billion barrels. Hubbert's prediction of a peak year for domestic annual oil production has proved correct.

The logistic curves provide a mechanism that allows something to be said about how much of a resource may ultimately be produced. Production of a finite resource such as crude oil begins, expands, peaks, and ultimately declines. These curves capture this inevitable sequence.

More complex models combine historical and geologic information used in the two previous methods as well as assumptions based on economic theory. This often requires computers for handling the large information base that is used in the analysis. These models are likely to consider more variables in a given analysis than either geologic analogy or projection of historical data. A complex model projecting ultimately producible oil may include information on basin evaluation studies with geologic models, statistical models, and/or economic models (for example, Miller et al. 1975). More complex models require much more time and energy but produce findings similar to earlier works done in the 1950s and 1960s which used a simpler analysis. This may indicate that the disagreement in results does not result from differences in method.

Estimates of Ultimately Producible Oil

Figure 9-3 compares ultimately producible oil in the United States estimated by various resource analysts with the year the forecast was made. No attempt has been made to reflect differences in areas covered, assumptions, or methods used. The estimates are simply represented here as they were in the original forecast. Note that the majority of projections from 1950 to the present ranged from 170 to 240 billion barrels. Hubbert (USGS), Weeks (Standard Oil), and Pratt (geologic consultant) performed more than one analysis.

During 1960–75, the USGS figures of ultimately producible oil were among the highest. In USGS Circular 650, the authors state that the USGS figures were among the largest published because ". . . our estimates include the largest proportion of favorable ground for exploration" (Theobald et al. 1972). This is a puzzling statement because most authors of forecasts say, or imply, that all potential producing areas have been included in their projections.

Cumulative production plus proven reserves over time is shown in Figure 9-4. These are two data sets that resource analysts accept in their estimates. Beyond these two points of general agreement, opinions begin to diverge concerning future discoveries and the resultant ultimately producible oil. Ultimately producible oil includes both discovered and undiscovered resources that are expected to be produced in the future. A satisfactory classification system for oil reserves and resources has not been devised as yet.

Vincent E. McKelvey, former director of the USGS, derived a three-by-four matrix system for resource classification (Figure 9-4). The degree of geologic certainty increases from right to left, and the feasibility of economic recovery increases from bottom to top. The McKelvey matrix has been criticized for providing no information for determining the magnitude of resources within given categories, except the reserves categories (Hubbert 1974). A further shortcoming in the McKelvey matrix is the lack of quantitative rules for the limits of "submarginal" economic viability and for speculative undiscovered resources. Most forecasters using the McKelvey matrix have been associated with the U.S. Geological Survey. More commonly, resource estimates are broken down in the following manner: cumulative production, identified reserves, and a single number for undiscovered and/or dis-

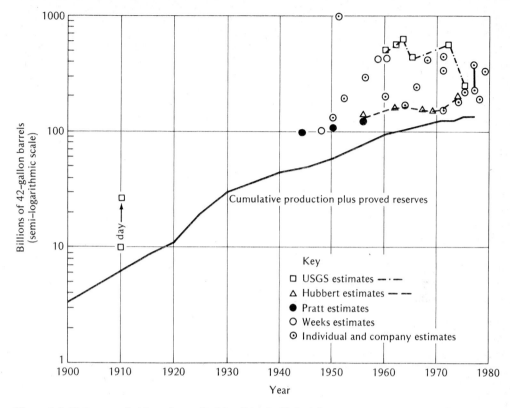

Figure 9-3. Estimates of ultimately producible oil in the United States.

covered resources that the analysts expect will be produced in the future.

Assumptions

Important components of any forecast are the assumptions used in the analysis. Assumptions are often chosen to serve the purpose of the forecast, but they are always the foundation of the analysis. It is therefore disturbing when oil forecasts are published with little mention or discussion of the assumptions. Changes in the assumptions usually lead to changes in a forecast's outcome. For example, an assumption made by the USGS that the same amounts of oil production will be found in all geologically similar areas led to overestimates of ultimately producible oil (Zapp 1962).

Estimates of ultimately producible oil depend upon several key assumptions. Explicit or implicit assumptions are needed for future discoveries and production, technology, recovery factors, and the price of crude oil. These are not necessarily all the assumptions that may be needed, but to varying degrees they will determine ultimate oil availability, as well as the oil available in the next 20 years. This poses a problem for those who are trying to make energy policy decisions using the results of forecasts which may vary widely. One is then dependent upon an "expert's" opinion. Hubbert (1974) has discussed the validity of assumptions used by various estimators.

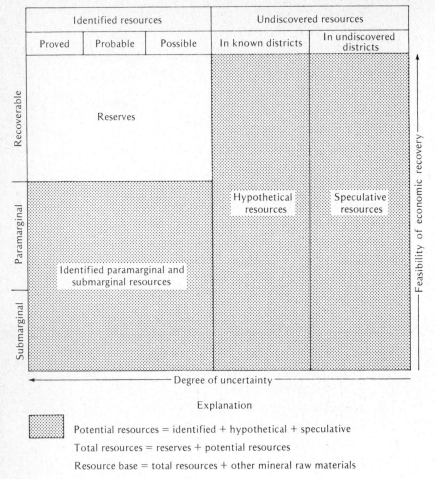

Figure 9-4. Diagram showing categories of a resource. The area of each box is proportional to the amount of the resource so classified.

A more complete examination of the available data will show that many "optimistic" assumptions about discovery and production technology, for example, may not have been warranted. This becomes particularly evident when discovery and production trends are evaluated.

DISCOVERY AND PRODUCTION TRENDS

From 1859 to 1979, approximately 460 billion barrels of crude oil were found in the United States. Of this amount, 125 billion barrels have been produced and 27.1 billion barrels are proven reserves as of December 31, 1979. The amount of remaining oil in known reservoirs is approximately 308 billion barrels.

Yardsticks for analyzing discovery and production trends include the amount of oil found per exploratory foot drilled and the size of oil fields discovered. Both of these indicators have been declining exponentially; no reversal of these trends is seen in the near future, even with decontrol of prices.

Method

Crude oil is found in sedimentary basins. The rock formations containing crude oil are called reservoir rocks. An oil pool is an accumulation of crude oil within the reservoir rock. The term *oil pool* is somewhat misleading. The oil is contained within the porous rock formation, frequently sands, and not in a pool like water in a pond.

Because oil is only rarely visible on the surface, as coal is, wells must be drilled to see if oil exists in a potential oil-bearing area. Exploratory activity ends with the drilling of wells in currently nonproducing underground formations. Despite the growing sophistication of geologic and geophysical methods of seeking potential oil-producing areas, it is only with the drill that oil is proved to be present. Further drilling is usually required to prove that commercial amounts are present.

Development continues with the drilling of additional wells in areas surrounding previously discovered reservoirs. The information gathered from development work helps to define the limits of a newly discovered reservoir and to determine the productive capacity of that reservoir. This is the upper physical bound on the capacity to produce petroleum from the reservoir during any particular time period. The additional oil found through development drilling that is producible is reported as reserve extensions.

When an oil field is discovered, estimates initially are made of the oil recoverable from the field. Subsequent examination of the field both in areal extent and at greater depths usually expands the first estimates as the limits of the field and deeper pay zones are examined. Revisions in reserve estimates are usually made annually in the years that follow discovery. Figure 9-5 shows the growth of the reserves as a multiple of the original estimates for several fields found at different times and places (Hubbert 1974). The dots show the average for the four fields. These four fields yield approximately four to six times the amount of reserves originally estimated. Methods of estimating oil reserves are continually improving, and newer fields are unlikely to have more than five times as much

Figure 9-5. Extensions to reserves. (From Hubbert 1974)

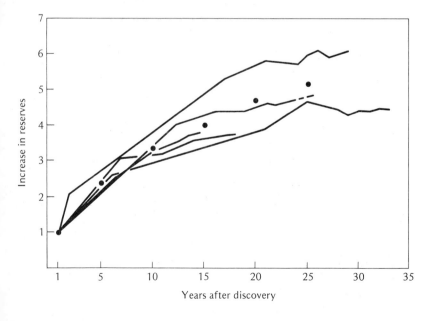

Years after discovery

recoverable oil as the original estimates.

Discovery precedes production by approximately 10 to 12 years in the United States. The discovery curve is an early warning of impending production changes and an independent check on ultimate production calculated from annual production alone. The most recent discovery data may underestimate what was discovered unless a way can be found to correct for this bias. Figure 9-5 provides the empirical observations to make such a correction based on past experience.

Oil Discovered per Unit Effort

Exploratory activity has been extensive in the lower 48 states and continues to be active. Cumulative footage drilled exceeds two billion feet. The year 1979 was the biggest drilling year in the United States since 1957. One estimate places total wells drilled in 1979 at 51,263, and total footage drilled is estimated at 143.9 million feet (McCaslin 1981). Drilling in 1980 showed even more activity. Estimated totals are 64,628 wells with 294.3 $\times$ 10^6 cumulative footage.

Representatives of oil companies have said frequently that if the price of oil increased, so, too, would the exploration for oil. The first evidence of this was in 1974 following the 1973 Arab oil embargo, when prices rose. An increase in exploration activity was seen in 1974. The total footage drilled and the total number of wells drilled increased 10.2 per cent and 19.2 per cent, respectively, from the previous year. The success ratio, which is the ratio of new fields found to total new field wildcats was 14.2 per cent, the highest in the history of the oil industry. Yet, with all this activity, the *amount* of oil found was the *lowest* in many years. It would appear that this trend is continuing today. The amount of new oil reserves discovered in new reservoirs in 1979 was 239 million barrels.

The information and knowledge for locating oil has grown over the years. In the early years of oil exploration, wells were drilled in areas of surface shows of oil. Around the turn of the century, geologic theory began to be used as well as geologic knowledge gained through experience. In the 1920s and the early 1930s, the tools of geophysics, especially the study of seismic waves reflected and refracted by subsurface geological formations, began to be used for locating oil.

Figure 9-6 shows the annual proven discoveries for the United States over time credited to the year the field was discovered. During the early years of exploration, the increased knowledge and experience appear to have led to an increase in the annual discovery of oil. However, from the late 1950s to the present, this trend has been downward. There has been no decrease in knowledge and experience during the past 20 to 30 years. Assuming the knowledge and experience enabled oil companies to look in the most promising areas first, the amount of oil discovered declines as the area is more completely explored. Geographical expansion in the exploration for crude oil has been extensive.

As mentioned earlier, USGS forecasts were among the highest available from 1960 to 1975. The basis for these inflated estimates began with a USGS study by A. D. Zapp (1962). Zapp estimated that, as of 1961, approximately one billion cumulative exploratory footage had been drilled out of a potential five billion feet, leaving four billion feet that still could be drilled in search of oil. This would imply that a good deal of oil was still to be found; all the oil companies exploring for crude oil had to do was drill. Zapp made the assumptions that unexplored areas geologically similar to explored areas contained proportional amounts of oil and that the amount of oil found per exploratory foot drilled was a constant (approximately 118 barrels per foot). Zapp applied his constant finding rate to the four billion feet that remained to be drilled, and this application gave his result of 590 billion barrels of ultimately producible oil.

Hubbert (1967) showed that the amount of oil discovered can be measured directly as a function of effort (cumulative footage drilled), which eliminates price fluctuations

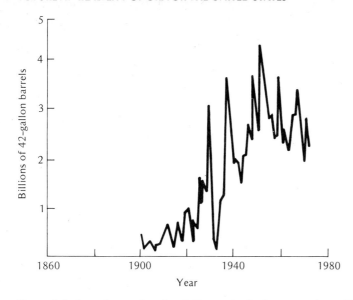

Figure 9-6. Annual proved crude-oil discoveries in the conterminous United States.

and time. The data show that, although increased effort has been expended looking for oil, less oil is now being found per foot drilled. This is a declining and *not* a constant finding rate. Figure 9-7 compares Zapp's hypothesis with Hubbert's hypothesis. Hubbert's estimate for ultimately producible oil was 165 billion barrels. This estimate is similar to his earlier works which used the logistic growth model on production data to estimate ultimately producible oil.

Figure 9-8 shows cumulative discoveries versus cumulative exploratory footage. Note once again the similarity to the logistic curve.

Figure 9-7. Comparison of U.S. crude-oil discoveries according to the Zapp hypothesis with actual discoveries. The difference between the areas beneath the two curves represents an overestimate of about 425 billion barrels. (From Hubbert 1969)

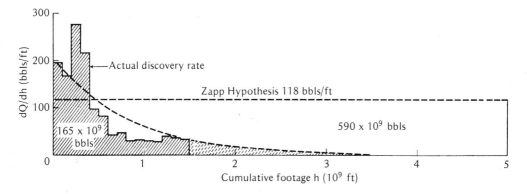

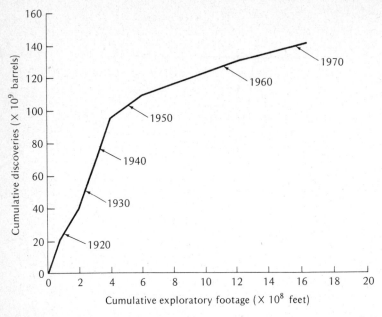

Figure 9-8. Cumulative crude-oil discoveries in the United States vs. cumulative exploratory footage.

If the Zapp hypothesis was correct, the curve should be a straight line of positive slope. Instead the data seem to indicate a plateau is being reached in oil found even though cumulative exploratory footage drilled keeps increasing (Hubbert 1974).

The decline in the rate of finding oil was confirmed by Menard and Sharman (1975) with an interesting technique. A random exploration program in which successive random choices of drilling locations were selected from the entire area of sedimentary rocks in the United States was prepared for a computer. As these random choices coincided with locations underlain by known oil fields, a simultated history of discovery from random exploration was produced. Many such simultated histories can be generated quickly and compared with the actual history of success in oil exploration. Before 1900, the random model found oil more successfully than actual exploration (which was often little more than random), partly because historical exploration was concentrated in the

eastern United States. As surface geological techniques became more prominent after the turn of the century and as subsurface geophysical techniques were added to the tools of actual exploration after World War I, the industry was more successful than with purely random drilling, although only marginally more successful for the very largest fields (greater than 100 million barrels). Finding rates were notably better than random for medium-size fields. Projections of these results confirm the bleak prospects for future finds and amounts of oil given by Hubbert. It should be kept in mind that these results are per unit effort, and if high prices bring a surge of exploratory drilling we shall simply find more sooner and less later.

The random exploration simulation can also be used to test the credibility of estimates of oil remaining to be discovered. Menard listed one such widely circulated estimate that 25 to 28 very large undiscovered oil fields remain in North America (Moody et al. 1970). The result was that random drilling would have been nearly 50 times as successful by 1955 than the industry actually was. Thus, belief in numerous undiscovered large oil fields requires the conclusion that scientific exploration not only is ineffective but also biases companies to drill in the *wrong place* most of the time.

There is an air of optimism in the exploration industry concerning the potential contribution of new discoveries to domestic crude oil supplies. Recent decontrol of crude oil prices has served to heighten this enthusiasm. For example, a considerable amount of exploration is already under way in the Overthrust Belt. Implications are that more than 500 million barrels of oil resources have already been discovered. Less than this amount will be recoverable, but even 500 million barrels is only a 30-day supply at present consumption levels.

It is crucial to keep in mind that increased prices may spur exploration activity, but there is no guarantee that increased amounts of oil will be discovered proportional to the effort or costs expended. Hubbert showed that in the lower 48, the amount of oil found per foot drilled has declined considerably since the beginning of oil exploration, irrespective of the price of oil.

Large Versus Small Reservoirs

A further discovery trend is the declining field size. Production from a single field varies from less than one million barrels to more than one billion barrels. Giant fields have reserves greater than 100 million barrels, and super-giant fields have reserves greater than 500 million barrels of oil. Fields found with reserves smaller than one million barrels are not currently commercial. This may change in the future as the price of oil continues to rise.

Menard (1976) showed that the average size of fields discovered has decreased with time. The majority of fields currently being discovered have less than one million barrels in reserves. Figure 9-9 is a graph of oil discovery by four categories of field sizes. The data suggest that mainly smaller fields will be discovered in the future. Smaller fields are always more expensive to find and produce, and, as exploration ventures into harsher environments, it will cost still more to find those smaller fields.

More significant, overall, is the number of giant fields to be discovered and the impact on production likely in the near future. Figure 9-10 shows a profile of the number of giant fields in the United States discovered between 1870 and 1968. The bulk of giant fields was discovered in the 1930s. Costs are proportionally higher per barrel of production for a small field than for a large field, so fewer giant fields mean increased costs. Secondary and enhanced recovery costs per barrel produced are also higher for small fields than for large.

Figure 9-11 shows cumulative oil production in the United States. Compare the graph with the logistic growth curves (Figures 9-2A and 9-2B). Cumulative production looks very similar to Figure 9-2A; production appears to have passed the exponential phase of the logistic growth curve. Annual production appears to have peaked in 1970 with an offset in 1976, when the trans-Alaskan pipeline opened, followed by renewed decline. After a time, annual production does appear to follow annual discoveries (Figure 9-6). Hubbert (1962) used this time lag to predict the peak year in annual oil production. Neither increasing exploration activities nor increased demand has reversed the downward trend, first in discoveries then in production.

Just as the giant oil fields play a major role in the U.S. discovery profile, they also dominate U.S. recent production. In 1979, approximately 197 giant oil fields were responsible for 63 per cent of total domestic oil production (*Oil and Gas Journal* 1981). The five largest fields provided 26 per cent of 1980 production, and just one field, Prudoe Bay, accounted for 19 per cent. Figure 9-12 shows the annual production for all domestic crude oil fields from 1935 to the present, as well as the contribution from giant fields. The path of giant field production follows closely

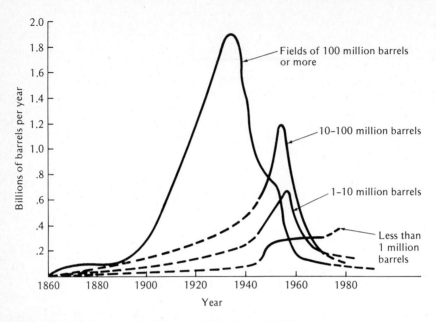

Figure 9-9. Oil discovery by field size. (From Menard 1976)

Figure 9-10. Giant oil field discoveries in the United States. (From Halbouty 1970)

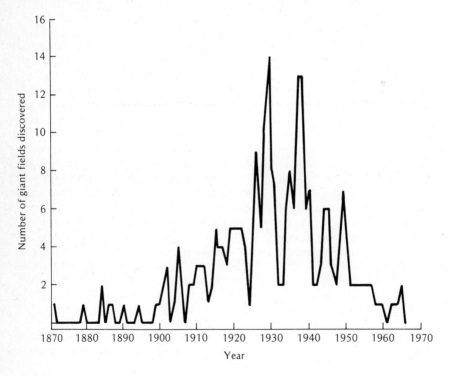

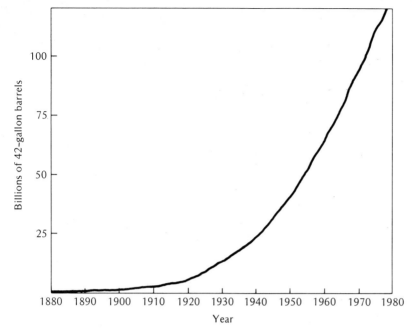

Figure 9-11. Cumulative crude-oil production in the United States, 1880 to 1980. (From *Oil and Gas Journal,* forecast review, various years)

total domestic annual production. Although the bulk of the giant fields was discovered in the 1930s, or earlier, their role in current production is still dominant.

Enhanced Oil Recovery Trends

Primary and secondary oil production are two stages routinely used commercially. During the primary phase of production, oil leaves the ground under naturally existing pressure in the field. When the field pressure drops, additional methods are needed to get the oil out. Secondary recovery, which is currently used in more than 50 per cent of domestic fields, involves the introduction of water or natural gas into wells surrounding the producing well(s) in order to drive out more oil. Frequently, secondary flooding is required to minimize subsidence. Water flooding is a more common practice today largely because

of the rising cost of natural gas. Primary and secondary production phases recover an average 32 to 33 per cent of the original oil in place, although recovery proportions for individual fields vary widely depending upon host rocks.

Enhanced oil recovery (EOR) methods require the injection of fluids into a reservoir to produce additional oil not recovered in the primary or secondary production phase. *Enhanced oil recovery* replaces the older term *tertiary recovery,* since additional oil can be recovered by applying EOR techniques either with or following the primary and secondary phases of production. Most enhanced oil recovery techniques are currently noncommercial, as much testing and field application are still needed.

The optimism for advanced production technology is frequently expressed in forecasters' assumptions for future recovery rates.

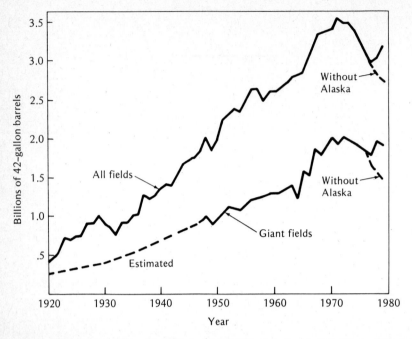

Figure 9-12. Annual U.S. oil production (all fields and giant fields). (From Bureau of Census, Department of Commerce, *The U.S. Fact Book,* various years, and *Oil and Gas Journal,* forecast/review, various years)

The high expectations for additions to oil reserves from improved recovery factors extends back to the early 1950s—a Presidential commission noted enhanced oil recovery to be just around the corner and estimated 75 per cent recovery of the original oil in place by 1975 (U.S. President's Materials Policy Commission 1952). American Petroleum Institute (1975) shows that recovery efficiency has increased, but only from 25 per cent in 1945 to 32–33 per cent today.

Figure 9-13 shows recovery factors for all fields credited to the year the field was discovered. Fields discovered in 1930 have an average recovery factor of 56 per cent, whereas those fields discovered in 1949 have an average recovery factor of 19 per cent. On a year-by-year basis, recovery factors appear to be more variable than the overall average figure. This is influenced by the geologic characteristics of individual fields. Many of

the reservoirs being discovered now are deeper and smaller. A deeper reservoir is likely to have reduced permeability, which is a measure of a fluid's ability to flow as a result of available interconnected pore space. Low permeability leads to low recovery rates.

Enhanced oil recovery production requirements for a field with a small amount of reserves are almost the same as those for a field with a large amount of reserves. Consider one small field (10 million barrels of proven reserves). The number of injection wells required is approximately the circumference of the area times the spacing between the wells. More wells are needed for ten fields of 10 million barrels each than for one field of 100 million barrels, perhaps as many as five times the number of wells. *Each* small field, furthermore, requires as much testing and planning as one large field.

It is not difficult to understand why many

analysts predict the great potential of advanced production technologies. As of December 31, 1979, the estimated original oil in place was 460 billion barrels. Of this amount, approximately 147 billion barrels were estimated to be ultimately recoverable under current economic technical conditions (excluding any contribution from EOR); this is approximately 32 per cent of the original oil in place. Lewin and Associates (1978), the National Petroleum Council (1976), and the Office of Technology Assessment (1978) prepared estimates of the potential contribution of enhanced oil recovery to crude oil reserves (Table 9-1). All three studies concluded that enhanced oil recovery is going to

be far more expensive than conventional production. Also, technological problems and varying geologic parameters from field to field add uncertainty to any prediction of EOR's contribution in the near future.

There are currently 226 EOR projects in the United States. Thermal, miscible, and chemical drives are the three major groups of EOR methods. In 1979, these technologies produced 385,070 barrels per day, or 4.5 per cent of total domestic oil production.

Thermal enhanced recovery raises the temperature of the oil remaining in the reservoir. This reduces the viscosity of the oil so that it flows more easily when fluids are injected. Steam is one such fluid and the most

Figure 9-13. U.S. oil recovery efficiency (credited to year of discovery). (From API, *Reserves of Crude Oil, Natural Gas Liquids, and Natural Gas in the United States and Canada,* various years)

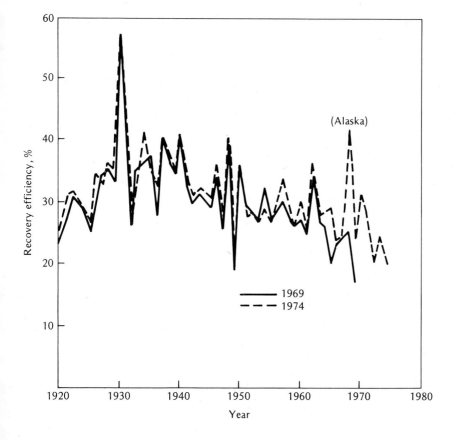

Table 9–1.   Three Estimates of Enhanced Oil Recovery Potential.

	Billions of Barrels of Oil[a]			
	Lewin[b]	OTA[b]	NPC[b]	Range
Light Oil				
CO$_2$ Miscible Recovery	13	12–21	7	7–21
Chemical Recovery				
Micellar-Polymer	5	7–12	6	
Polymer	N/A	0.3–0.4	0.5	
Caustic	N/A	N/A	0.4	5–12
				12–33
Heavy Oil				
Thermal Recovery				
Steam	17	4–6	5	
In Situ	2	1.6–1.9	2	1.6–17
Total Light and Heavy Oil				14–50

[a] All figures are based on oil prices of $20 to 22 per barrel of oil in 1976 constant dollars.
[b] Sources listed at end of article.

widely used of all the EOR technologies. In 1979, steam drive accounted for 77 per cent of total EOR production.

In an evaluation of thermal processes, Prats (1978) of Shell Development Company maintains that there are no real technological reasons thermal processes should not be considered for oil production. Two problems deterring potential use of thermal processes might be cost of residual fuel oil for combustion and possible formation damage from steam condensate. Prats maintains that only societal, governmental, and economic problems remain. These problems may not be easy to overcome.

Miscibility refers to the ability of two fluids to mix. When oil comes in contact with a fluid with which it is miscible, a liquid of lower viscosity is formed and this liquid flows more readily than the original crude left in the ground. Some potential miscible fluids for crude oil are carbon dioxide (CO_2), alcohol, and other hydrocarbons. Carbon dioxide is receiving the most attention, yielding 19 per cent of the total 1979 EOR production.

Not enough data have been generated to fully evaluate the potential for increased recovery. Experimental data have shown recovery factors of 90 per cent of the remaining oil in place (Stalkup 1978). However, there have been only a few field trials in both sandstone and carbonate formations. Of all the fluids, testing has shown that carbon dioxide is definitely effective in displacing and recovering residual oil, but less so following water-flooding operations.

Part of the uncertainty surrounding the potential for carbon dioxide is that the phase behavior in the reservoir is far more complicated than that for the steam-flooding operation. The carbon dioxide miscible phases can be multi-equilibria with as many as four parts, such as two liquids, a gas, and a solid. This may lead to problems because a heavy liquid or solid precipitate can result, which would lead to less oil recovery and a lower potential for fluids injected later (Stalkup 1978).

Of the possible solvents mentioned earlier, petroleum hydrocarbons have received the least attention. Petroleum hydrocarbons are often more valuable as a chemical feedstock than as a solvent for oil production. Industry sees carbon dioxide sources more readily available. Much attention has focused on natural sources of carbon dioxide.

A pipeline is being planned that will link natural sources of carbon dioxide in Colorado to production centers in Texas (Stalkup 1978, Renfro 1979). Currently, power plant stack gases are a diffuse source of carbon dioxide. A 250-megawatt gas-fired plant produces 90 million cubic feet per day of carbon dioxide, whereas the same plant fired by coal produces 250 million cubic feet per day of carbon dioxide. Coal gasification plants have an even higher potential as sources of carbon dioxide. This source is not currently available.

Chemical drives are the third major group of EOR technologies. Primarily three chemical processes are being considered within the chemical group (surfactants, detergents, and alkaline flooding). All three processes use some form of chemical to enhance displacement of oil from a reservoir. Although chemical drives represent only 4 per cent of total EOR daily production, interest in chemical methods is growing, with 42 reported projects in 1980, an increase from 19 in 1971 (Matheny 1980). There have been few major tests conducted using this process, and so its economic and technological feasibility is uncertain. Also, not all the process parameters are completely understood (OTA 1978).

There is one point concerning EOR that is a complicating factor in its development. The geology of different fields is variable, and each one must be evaluated so that the correct technology is chosen. The costs of this planning and evaluation are likely to limit EOR use on small fields. As a result of changes within a reservoir, the use of one EOR method may preclude the use of another technology at a later date.

Environmental considerations and/or energy requirements may prohibit the development of some EOR projects.

EOR processes contained within a reservoir pose no environmental problems. Environmental problems could arise from transit spills, on-site spills, well-system failure, reservoir migration, and general operations. Regulations designed to minimize environmental damage directly affect EOR projects on federal lands. Two important problems for EOR development are air quality and water availability. The greatest impact on air quality will be in those areas already in violation of current federal air quality standards. For example, many steam generators for EOR are being used in Kern County, California, which already suffers from air quality problems. Air quality laws will serve to delay and/or add large sums of money to the expense of EOR projects.

A second environmental concern is water: pollution, availability, and possible conflicts over use. The chemical processes could pose a major environmental problem should a spill occur and water be contaminated as a result, in particular ground water sources. Although reservoir leaks to date have been rare with secondary recovery methods, concern is warranted when chemical drive EOR processes are being considered. The large increase in costs of injection fluids may serve as a stimulus for extra care during the operation. However, Brown (1979), writing about chemical waste disposal, paints a grim picture of past and present chemical operations. Lack of concern for human health and safety is evident. EOR projects are likely to occur in the southern and southwestern parts of the United States, where water is scarce and likely to become scarcer by the end of the century. Although most oil processes do not use fresh water, some chemicals used in the chemical drive methods currently work best with fresh water, requiring an estimated 6 gallons of fresh water per barrel of oil (OTA 1978). EOR processes would be competing for water not only with agriculture but also with other energy development projects in the western and southern states.

Energy requirements for oil production are not frequently discussed in the literature. EOR may increase crude oil reserves, but it will also increase energy costs. More specifically, the production of 30 million barrels of oil from the Wilmington field by steam drive could require as much as 9 million barrels per year of residual fuel oil in the generators (OTA 1978). This is comparable to 30 per cent of the available energy in 30 million barrels of crude oil.

INTERNATIONAL IMPLICATIONS
Who Owns Subsurface Resources?

There are many ways in which the United States is unusual among the nations of the modern world. One difference seldom noted is in ownership of subsurface resources. In almost all of the other nations of the world, by law and custom the government owns subsurface resources, even if surface land ownership is private. This is expected in centrally planned, socialist economies, but it is also true in much of the free world. This means that development, production, and trade policies are those of governments rather than those of private firms or individuals. Some serious failure of expectations seem to have resulted from this difference. The tradition of private enterprise and capitalism in the United States and the aspirations for international free trade have combined with the models of neoclassical economics to provide a basis for policy that has little to do with the actual state of resource trade in the world.

The question is: In what ways will resource trade differ when the parties to the trade are governments instead of competitive firms? Of course, there never has been a simple free trade among a large number of purely competitive, profit-maximizing firms. For a long time, large, multinational firms have conducted much of the international mineral trade. Although occasionally conspiring as cartels, most such trade is best described as oligopolistic or, as the economist Chamberlain (1950) labeled it, monopolistic competition. As recently as the early 1970s, the largest oil companies controlled 70 per cent of internationally traded oil production (Steinhart and Steinhart 1974), and national oil companies controlled only about 15 per cent. Governments have influenced international mineral trade for an even longer time through taxes and tarriffs, quotas, and trade agreements. The history of trade modifica-

tion and control by private firms has been described from several different points of view (Penrose 1968, Adelman 1972, Blair 1976, Solberg 1976). Intervention in mineral trade by nations is discussed in some of the references just cited and in Bohi and Russell (1978) and Krasner (1977).

Although these histories and analyses of the past are important to an understanding of the changing international oil trade, the perceptions and myths of the history may be equally important to an understanding of present policies. We shall return to this subject later, but first a detour to a more contentious matter.

The Oil-Exporting Countries

Is OPEC a cartel? Some political analysts say so, and their counterparts in the oil-producing countries hotly deny the label. Some economists use the cartel label, seeking to bring to bear the theoretical expectations that economic theory offers for cartels, while other economists trained in the same range of institutions argue that it is more like a trade association. In all this acrimonious discussion, there is an unmistakable overtone of judgment, of which side has the right and justice with them. There is, too, a dangerous tendency to confuse the word with the reality. Cartels, monopolies, and trade associations are usually thought of as associations of firms. Whatever descriptor is attached to OPEC ought not to disguise the basic difference between an association of firms and an association of sovereign national governments.

Some of the disarray of OPEC over pricing policy in the meetings of 1979 and 1980 seems understandable. What is astonishing is that OPEC is able to stay together at all. The political variety of governments included among its members covers the spectrum of conservative hereditary rulers, military governments of both the left and the right, an occasionally shaky but durable elective democracy, and fundamentalist religious states.

While the popular press at times uses

OPEC and *Arabs* as synonyms, there is a centuries-long history of distrust and conflict between non-Arab Iran and the adjacent Arab states. If the overarching culture of Islam is invoked as the common glue of OPEC, we have seen in recent years the differences between different kinds of Islamic believers. Beyond the Middle East, the founder of OPEC is staunchly Catholic Venezuela. Catholic Ecuador is also a member. Nigeria and Indonesia also are not primarily Islamic nations. Neither religious nor ethnic solidarity does much to explain why OPEC stays together.

Governments as Market Traders

After the assumption of control of oil production in the early 1970s, the national governments have set production policies more and more often. At first glance, the wide variety in type and stated aim among these governments would seem to confound any attempt at outlining common purpose beyond the simple attainment of maximum financial returns. Needs for revenue differ widely between small producers with large populations and ambitious development plans and large producers with sparse populations and sharply limited capacities to spend gigantic sums of money in fruitful development projects.

Saudi Arabia, Kuwait, Qatar, the United Arab Emirates, and Libya are able to control production amounts in pursuit of objectives other than maximization of return without serious damage to their financial position or development plans. Together these nations account for more than 50 per cent of all OPEC production (in 1979). Other OPEC members suffer at least some financial pain from reduced production, but that has not always deterred them from doing so.

Some of the common purpose among OPEC members is to be found in shared views of the real and imagined injustices and betrayals of the past. Common to all OPEC countries is their keen recollection of history and experience of foreign domination or as actual colonies. Many Middle East OPEC leaders are of an age that causes them to single out the promises—as they understood them—of freedom and self-determination by the Allied governments in exchange for their cooperation during World War II. Yet after World War II, colonies and less explicit forms of foreign domination continued. Most of the postwar political control of the Middle East was mapped in Paris and London. Oil production and marketing control was determined by the large oil companies with the active or tacit assistance of the governments of Europe and the United States. This was not new for the Middle East. Agreements about concessions, production, and marketing date from 1929 or earlier, and had long been disliked by nationally minded Middle East leaders. Blair (1976) has discussed this history in considerable detail.

Many other issues are raised by the leaders of oil-producing nations as illustrative of a history of victimization by Europe and the United States. For example, North Africa resented being the site of some of World War II. One Arab official sought comprehension by asking what we in the United States would feel like if China and the U.S.S.R. went to war—in Colorado. Persian Gulf nations frequently bring up the British and U.S. intervention in the Mossadegh regime in Iran in the 1950s and the related response to Iran's attempt to nationalize its own oil production. Venezuelans are more likely to give their account of the heavy-handed policies of the United States in Latin America in much of the past. Conflict over Israel and Palestine often acts as a focal point for these and other perceptions of past wrongs. Experienced observers on the scene have remarked on the extent to which these feelings are misunderstood or ignored (Nutting 1964, Akins 1978).

Government policies of the oil-consuming countries have also been a direct source of irritation. The oil import quota system of the United States from 1956 to 1972 (Bohi and Russell 1978) is often given as the main reason for Venezuela's leading role in the founding of OPEC. Criticism of the concerted policies of OPEC are answered not only with the history of control of the multinationals but also with a discussion of the International Energy Agency (IEA). By producers, the IEA is seen as a customers' "cartel" designed to destroy OPEC, and former Secretary of State Kissinger's announced objective for the IEQ *was* to "neutralize" OPEC.

It is not our intention to comment on the validity, accuracy, or merit of these old and new grievances. It is important to know that they are strongly believed by the oil-producing governments who now control oil production for international trade and who are moving steadily into all phases of the oil industry.

Objectives and Policies of the Oil Producers

Past and present grievances, justified or not, do not provide policy. The oil-producing nations have made no secret of their policy aims. Indeed, there are policy statements at so many levels that Western observers often hear whatever they wish to hear and interpret what they do not hear. At the most general level, the objectives of the oil producers are linked directly to the large aims of the New International Economic Order (NIEO). These objectives on behalf of the entire Third World have been discussed many times (Myrdal 1971, Tinbergen 1976, Ul Haq 1976), and the failure to achieve these objectives despite several years of North-South talks has also been lamented (Barraclough 1978, Tomeh 1980). As they relate to resource trade, the most important of the NIEO goals call for (1) stabilization of world raw materials markets, (2) indexation of raw materials prices to inflation in the industrialized countries, (3) better terms of trade in raw materials markets, (4) technology transfer, including the most modern technology and access to the training to use it, and (5) less restrictive entry into the resource-based industries— the "downstream" activities—for resources originating in the developing countries.

While the record attests to OPEC's vocal support of NIEO goals, including the substantial portion of their oil earnings provided to poor nations as aid, the oil-producing nations will certainly pursue these ends with special vigor on their own behalf. The secretary general of OPEC was much more explicit when suggesting that the price of oil must advance toward the cost of alternative energy sources. He concluded that an orderly advance requires "(1) an index reflecting the impact of inflation on international trade, both merchandise and services, (2) an automatic rate adjustment factor based on the main currencies used in world trade, and (3) oil prices should rise in real terms" (Ortiz 1980). To these economic aims must be added the desire for better access to downstream oil technology. The oil producers' view is that

transnational oil companies kept their activities dependent on centers of technology abroad and hence left an inadequately developed technological base for the succeeding national oil companies. . . . This situation weakens the position of the national oil companies via-à-vis oil and technology transnationals in world markets. The technological sources of productivity of the industry, as developed by the petroleum transnationals and others, are not generally available at competitive prices for the oil states. (Al-Wattari 1980)

The technology referred to by Al-Wattari extends beyond merely that of production to refining, petrochemical production, and the whole range of related downstream activities.

To economic and developmental goals must be added the political aims of the governments that own and control oil production. As we have seen, more than half of OPEC's production comes from nations that have some latitude in controlling production, and these nations could use those controls as a political force in international affairs. Will they again, as they did in 1973? Some voices from oil-producing countries are very explicit:

Finally, a landmark has been our actual utilization of oil as a strategic commodity of great political potency as well as developmental power. Now the

West, finding itself at a disadvantage, labels this "blackmail" and complains about this "mixture" of economics with politics. The linkage has always existed—it is a lesson we learned from the West. (Sayigh 1975)

After that remark, Sayigh goes on to discuss the trade manipulations undertaken by the West against Cuba, Rhodesia, the U.S.S.R., and China. The list of political aims is often led by the search for resolution to the Israeli-Palestinian conflict, but many other issues are also possible determinants of oil negotiations: for example, the United States is seen as favoring Egypt in the struggle for leadership in the Arab world, and the attempts to fill the U.S. strategic oil reserve is seen as a hostile act that attempts to deprive the oil-producing nations of their principal influence in international affairs.

One very important outcome of these aims and objectives is that, to have any hope of achievement, negotiations about oil production and prices must be carried out government to government instead of according to the past pattern of negotiations for concessions between nations and oil companies. Besides the aversion and distrust of the major oil companies based on past performance, the considerable mismatch in such government-to-company negotiations has often been to the disadvantage of the concessionary government, as outlined by Smith (1977). Companies are simply not in a position to discuss or deliver agreements on the issues of importance to OPEC. Evidence of the growing exclusion of major oil companies is easy to find. At the first Arab Energy Conference in 1979 the national oil companies of Europe were full participants, but not a single representative of the private oil companies was to be seen. This reduced role for the oil companies extends well beyond OPEC, as may be seen by the increasingly assertive role of governments in England and Norway for North Sea oil production. The 1980 increased role of the national oil company in Canada means more government ownership and policy making, whether the outcome of the dispute puts more of the power in the provincial governments or the central government. Although Mexico ejected foreign oil companies more than 40 years ago, it is also seeking government-to-government agreements for exports. This was illustrated by Mexico's reaction to the announcement by the short-lived conservative government of Canada that PetroCanada would be returned to the private sector. After Mexico suggested that the long negotiations for Mexican oil exports to Canada would have to begin again from zero, little more was heard of returning PetroCanada to the private sector.

Thus, the future is likely to move the world further from free market expectations rather than closer to them. The OPEC nations' objectives are clear enough and include political goals as well as protection from inflation and better access to development technology and downstream oil activities. As we have already seen (Figure 9-1), the movement of downstream industries to the source of raw materials supply is nothing new in the mineral industries and would very likely happen anyway—soon or late. The rapid shift from concessions and control by the multinational oil companies in the 1970s to national oil companies and government control and international agreements will be met most easily by Japan among the consumer nations, but the growing role of the national oil companies in Europe and Canada suggests that they, too, are adapting. Only the United States seems still to hope for some reversal in this trend. It seems unlikely.

Price and Political Stability in the 1980s and 1990s

The modern history of oil prices has two eras. From 1945 to 1973 oil prices were almost constant in current dollars. This meant a slow decline in real price as modest rates of inflation pushed up the cost of other goods and services. By the late 1950s, international oil prices were sufficiently low in comparison to domestic prices that oil import quotas were put in place (justified on national security grounds) in order to allow domestic producers to maintain a price that was about twice that of internationally traded oil. Although there was conflict between the multinational oil companies and producer governments about prices and royalties in this era, the low prices now seem like quaint nostalgia. This long period led some students of the oil industry to conclude that the real price of oil would continue to decline for the rest of the century (Adelman 1972). Small price increases in 1971 were followed by the abrupt increases of 1973 ushering in the new era of control by the producer governments. The modest price increases between 1973 and 1979 were always less than the inflation in prices for commodities in world trade. Some adjustments were made in production amounts from OPEC countries so that surpluses did not accumulate on world markets, and when shortages threatened, some OPEC producers, especially Saudi Arabia, increased production to meet world demand. They seem to have been motivated in these increases partly by a desire to assert their leading role in price determination by OPEC and partly by a concern that shortage and the possible wild price fluctuations would bring disastrous problems to the world financial system in which they also have a stake.

Western analysts, especially economists, have generally related oil prices to the costs of production and subsequent operations to produce a finished product. Yet economists and analysts in oil-producing countries scarcely mention production costs when discussing appropriate price, even though they have been professionally trained in the same Western universities. Jamali (1978) relates appropriate price to opportunity cost. He adds in terms to account for the expected higher future value of the oil, the foregone returns from downstream processing of crude oil, and a number of other terms that produce a much higher price expectation. Others have mentioned the marginal replacement cost of liquid fuels from coal or oil shale (Ortiz 1980) as an appropriate price goal. Al-Janabi (1979), chief economist of OPEC, begins by

noting that the price of crude oil is not the issue since no one uses crude oil. One must begin with the price paid for the finished products by the actual consumers of those products. He notes that in much of Europe the final price of gasoline contains taxes in excess of 50 per cent of the price. In the United States, the taxes are in the 10 to 15 per cent range (states vary). By comparison, the oil producers get only about one quarter of the final price. Starting from this point of view, the oil producers conclude that the price must be much too low if consuming governments can still skim substantial government revenues from oil products. The purpose here is not to review these and many other disputes about the ''right'' price, but to suggest that there are plausible points of view that lead to a justification not only of present prices but also of further price increases.

More cynical observers conclude that all these theoretical arguments are merely justifications for customers seeking lower prices and sellers seeking the highest possible price. It is indeed a matter of U.S. law that oil from public lands must be sold to the highest bidder. One result was that oil from the Elk Hills Petroleum Reserve was sold at the highest price paid for any oil in the United States in 1979—substantially above OPEC prices. However these motivations may play a role, the policy is clear: OPEC intends that prices will increase, and, as they enlarge their share of the downstream activities, product prices will do the same. North Sea producers, Mexico, Canada, and other producers for export have generally set prices above OPEC and can be expected to continue to do so.

But prices cannot be raised easily when there is a surplus to market. As has been indicated, there is some flexibility in production cutbacks by the leading OPEC producers and ample evidence of willingness to reduce production—at least to reduce by some significant amounts. Thus far, however, the episodes of rapid price escalation in 1973 and 1979–80 have been facilitated by war and political conflict (1973) or revolution and war (1979–80).

There have been substantial political changes in oil-producing countries in the years since World War II. In the Western press, this is often labeled political instability. It is hard to know what terms to use, since Italy has experienced an average of one new government every year since World War II. To examine the potential for oil production disruption due to unexpected government changes, we have examined the record since independence or World War II, whichever was later. We counted revolutions, coups, unexpected deaths of authoritarian leaders, and similar disorderly changes of government. Preliminary results indicate that such potential disruptions occurred on the average of every six or seven years for the average producing country. Some countries have experienced more, and some less, and not every such event would reduce oil production. In the era of oil company control, many such events did not. But now governments are firmly in control of policy, and the 1980s are likely to include a number of such unpredictable events. Since the political conflicts that have provoked reductions in production have not markedly diminished, some unexpected supply interruption is all but certain. These supply disruptions are also likely to be accompanied by further price increases.

U.S. PETROLEUM AVAILABILITY

The United States produced approximately 3 billion barrels of crude oil in 1979 and consumed more than 5 billion. With reserves of 27.8 billion barrels (as of December 31, 1979), this production rate is unlikely to continue unless additional crude oil is added to reserves through new discoveries or enhanced oil recovery methods. Imported oil will be required until the end of the century, but unless demand is reduced still more imported oil will be needed.

The forecasts presented earlier are concentrated between 170 and 225 billion barrels of ultimately producible oil.[3] Of this amount, approximately 125 billion barrels

have been produced. The logistic growth curve combined with estimates of ultimately producible oil aids in conceptualizing what may be required to meet projected production rates. Estimates of future annual oil production from the 1960s through the late 1970s forecasted increases in future oil production. A close look at the recent estimates, which are lower, shows that the amount of oil needed to meet these lower estimates is substantial.

Figure 9-14 is a projection of U.S. annual oil production based upon the best fit to U.S. production, which in turn gives a Q_∞ equal to 192 billion barrels. This curve shows that additions to reserves of two billion barrels in 1980 would be needed simply to produce two billion barrels in 1990. In 1990, additions to reserves of 1.4 billion barrels would be necessary for production in the year 2000.[4] Additions to reserves have been below two billion barrels per year for seven of the past ten years.

Figure 9-15 is an Atlantic Richfield Company (*Oil and Gas Journal* 1979) projection of future U.S. annual oil production. The production profile is broken down into EOR, future discoveries, North Slope, and appreciation of known reserves. In their projection, estimated daily production rates of approximately 8.2 million barrels are maintained through 1995, followed by an *increase* through 2005. For annual production to remain near 8 million barrels per day (3 billion barrels per year), North Slope oil production is estimated to be approximately 1.2 million barrels per day through the year 2000. As of January 1, 1981, 17 per cent of the known reserves in Alaska had been produced. Production from existing Alaskan reserves start to decline after 1985. Unless considerable new oil is discovered in that area, Atlantic Richfield's prediction seems too optimistic.

From 1971 to 1978, new field discoveries *totaled* 1.16 billion barrels (API 1979). New field discoveries plus revisions to prior reserve estimates, extensions of fields, and new reservoir discoveries in old fields averaged

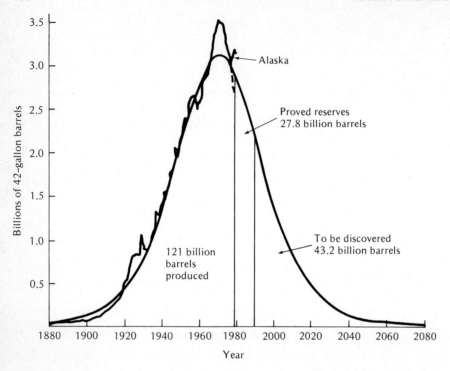

Figure 9-14. Annual crude petroleum production in the United States, 1880–2080 (projected). (From J. Steinhart et. al. 1980)

1.65 billion barrels per year for the same period. This amount added to reserves in 1980 would be short of meeting 1990 production in Figure 9-14. Any optimism about finding many new giant fields should be tempered by Menard's (1976) work. If many giant fields remain to be discovered, this would imply that exploration efforts to date have been less successful than a random process.

EOR is an alternative way of increasing production. From a technological standpoint there are many possibilities. The technologies for various processes are understood to varying degrees, and most application of these processes has taken place in the laboratory with small-scale field testing. To date, there has been only limited large-scale application.

A late ARCO estimate suggests that by

1990, EOR will contribute 540,000 barrels per day in addition to 1980 daily production (ARCO 1980). Most of the expected increase is to come from thermal EOR; carbon dioxide methods will contribute an additional 200,000 barrels per day in 1990. While the estimate for thermal seems reasonable, the CO_2 estimate may be on the high side unless certain technological problems are worked out and sufficient quantities of CO_2 become available. This information shows that continued development in advanced production technologies plus new discoveries will barely serve to meet a declining production profile in 1990 and 2000. Unless there is a reduction in demand for crude oil, the United States is likely to put more pressure on the world oil markets.

Figure 9-16 shows production profiles of

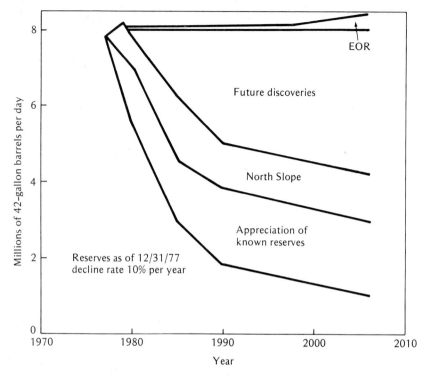

Figure 9-15. Outlook for U.S. crude oil production (Atlantic Richfield). (From *Oil and Gas Journal,* November 12, 1979)

two giant oil fields. The production starts from zero. However, the production paths for the fields vary. Following a peak in production for each field, the back slopes of the curves are not symmetrical. The shape of the back slope of the curve is governed by physical conditions in the field, including the appropriate timing of secondary flooding and EOR, and not by what production may be desired. An increase in the price of oil may indicate a higher pumping rate. Thermodynamics, however, says that if most of the oil available in a reservoir is to be recovered, then the pumping rate should be infinitesimally low. Based upon this and engineering considerations, maximally efficient recovery (MER) rates are determined for each new field. Pumping rates higher than the MER may preclude ever getting out the full oil po-

tential of a field. High pumping rates were seen in the 1920s, when the only concern was to get as much oil out as quickly as possible. Many of these fields are still pumping today, but in the long run less will come from those fields because a lower rate was not used from the beginning.

For the U.S. annual oil production curve to be skewed to the right, extending the production profile, discovery and production slopes would have to become less steep to the right of the peak than they now are. This means increased annual discoveries and production beyond what is currently being discovered. The declining size of oil fields discovered and the declining amount of oil found per foot drilled seem to indicate that new discoveries will serve to prevent a too precipitous decline in future oil production, but do

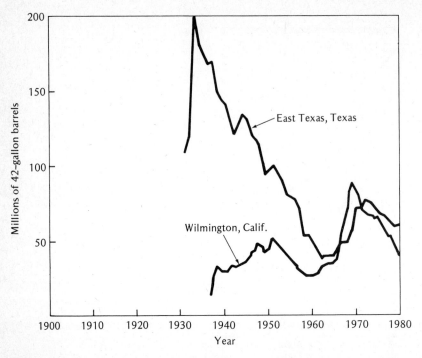

Figure 9-16. Annual oil production of two giant U.S. oil fields: East Texas, Texas, and Wilmington, California, 1900–1980. (From *Oil and Gas Journal,* forecast review, various years)

not guarantee a leveling off of oil production.

If technology is to have a role in changing the back slope of the curve, the rate of technological development will have to be *higher* than it has been in the past. From what is known of the time to develop technology and make it commercial, help will not come soon. Higher prices may provide the stimulus to increase technological research and development, but it is too early to tell. The end result is still the same—production will eventually show a decline, similar to the down side of the Hubbert curve.

The forecasts of domestic oil production and especially of oil available for the international market in the next 20 years are quite varied. We examined 20 forecasts for domestic oil production in 1990 and found a range of U.S. production estimates from six to more than 11 million barrels per day in 1990 (DOE 1979, API 1980, DOE 1980, OTA 1980). These estimates were prepared by industry, government, universities, and other independent groups. There is as much disagreement between various sophisticated and complex computer-based models as between forecasts based on simpler models and reasoning. Pessimistic forecasts come from oil companies, government, and independent sources, and so do optimistic ones. It is hard to escape the conclusion that the forecasters' assumptions determine the outcome since all have the same data before them.

In Table 9-2, we present our estimates for domestic production through the year 2000. These estimates were derived from the Hubbert curve shown in Figure 9-14 for the conterminous United States as a base. In light of recent discoveries this choice may be slightly optimistic. To that base was added production from the North Slope of Alaska (south-

Table 9–2.   Oil Availability in the 1980s and 1990s.

	1980	1985	1990	1995	2000	Source
	Projected Production (million bbl/day)					
United States	10.7	10.5	9.5	8.4	7.3	See text
OPEC	29.5	31.0	31.0	31.0	31.0	OTA 1980
						DOE 1980
OECD (except U.S.)	4.7	5.6	5.8	5.7	5.6	DOE 1980
Mexico	1.8	3.5	4.0	4.0	4.0	DOE 1980
Other LDCs	3.5	5.3	6.5	7.5	8.0	DOE 1980
	Projected Consumption (million bbl/day)					
United States	17.0	15.5	15.7	15.3	14.0	DOE 1980
OECD (except U.S.)	22.2	20.4	21.3	22.4	23.5	DOE 1980
OPEC	2.5	3.9	6.3	10.4	16.7	OPEC Bulletin
						Jan. 1981
LDCs (includes Mexico)	10.0	8.7	11	14	16	DOE 1980
	Projected Trade Balance[a] *(million bbl/day)*					
United States	−6.3	−5.0	−6.2	−6.9	−6.7	
OECD (less U.S.)	−17.5	−14.8	−15.5	−16.7	−17.9	
OPEC	27	27.1	24.7	20.6	14.3	
LDCs (includes Mexico)	−4.7	0.1	−0.5	−2.5	−4.0	
Centrally planned economies	1.0	0.0	−1.0	−1.7	−2.4	Chase Manhattan
						Bank 1980
Trade Balance	−0.5	+7.2	+1.5	−7.2	−16.7	

[a] Minus sign indicates imports.

ern Alaskan production is already declining). Again being cautiously optimistic about new discoveries, we conclude that production from this region will not begin to decline until 1990, even though Prudoe Bay must begin to decline several years earlier. For enhanced recovery, we have used estimates from Atlantic Richfield Company which we regard as cautiously optimistic (and larger than those shown in Figure 9-15). To these amounts we have added our estimates of natural gas liquid (NGL) production. Reflecting the widely held belief that price increases will prevent decline in natural gas production in the 1980s and even slow decline in the 1990s, we have chosen larger NGL amounts than the Hubbert curve would suggest (see Article 11 by H. R. Linden). The results are shown in Table 9-2. As can be seen, the decline of total liquid hydrocarbon production is slow in the 1980s and accelerates in the 1990s. In our opinion, the actual production is more likely

to be lower than these totals rather than higher.

Estimates for world oil production and consumption are far less certain. We have chosen midrange estimates from the recent literature to illustrate when problems may arise. These projections assume no wars or other political disruption of supply and moderate price increases. Both assumptions are likely to be false. We do not think surpluses will occur in the mid-1980s, either because of political events or because large exporters cut back on production. By about 1990 shortages seem likely unless consumption amounts are much reduced by conservation or fuel shifting. Since both are technically possible, these developments cannot be ruled out.

The projected oil consumption rates in developing countries reflect a substantial reduction in the growth rates of the past two decades. This means development will also be slowed. Besides postponing the fulfillment of the aspirations of the three quarters of the world's population that are poor, the slowing of development may also mean increased political turbulence.

Neither can we rule out the discovery of some rare super-giant fields such as Campeche Bank in Mexico, Prudhoe Bay in Alaska, or Ghawar in Saudi Arabia. Such extremely large fields (10 billion barrels or more) are very rare. The likelihood of finding such a field in the well-explored United States, however, is very low. Far more likely are changes in oil availability resulting from political events. We conclude that the ability to achieve political agreement between producers and consumers of oil will have more impact on supply adequacy than will discoveries or scientific advance.

NOTES

1. Ultimately producible oil is defined as the total amount of oil that will have been produced "when it is all over."
2. The curves of Figure 9-3 are formally the first derivative of the logistic curves of Figure 9-2. The formal expression for the logistic curve is $Q(t) = Q(1 + e^{-ct})^{-1}$, where t is the time and c is a constant.
3. This includes forecasts made during the past seven years.
4. This assumes a ten-year time lag between discoveries and production.

REFERENCES

Adelman, M. A., 1972, *The World Petroleum Market*. Baltimore: Johns Hopkins University Press.

Atkins, James, 1978, "U.S. Energy and Middle Eastern Oil and Politics," in *Natural Resource Policies Proceedings*, Vol. 1, pp. 345–58. Madison: University of Wisconsin Press.

Al-Janabi, Adnan, 1979, "Optimum Production and Pricing Policies," paper presented to the first Arab Energy Conference, Abu-Dhabi, March 8, 1979.

Al-Wattari, Aziz, 1980, "The Case of Petroleum," in *Resources and Development* (P. Dorner and El-Shafie, eds.), pp. 205–51. Madison: University of Wisconsin Press.

American Petroleum Institute, 1980, *Two Energy Futures: A National Choice for the 80's*. Washington, D.C.: American Petroleum Institute.

American Petroleum Institute, American Gas Associa-

tion, and the Canadian Petroleum Association, 1979, *Reserves of Crude Oil, Natural Gas Liquids, and Natural Gas in the United States and Canada as of December 31, 1978*, Vol. 33.

American Petroleum Institute, American Gas Association, and the Canadian Petroleum Association, 1976, *Reserves of Crude Oil, Natural Gas Liquids, and Natural Gas in the United States and Canada as of December 31, 1975*, Vol. 30.

American Petroleum Institute, 1975, seminar on reserves and productive capacity, Washington, D.C.

Arps, J. J., M. Mortada, and A. E. Smith, 1971, "Relationship Between Proved Reserves and Exploratory Effort," *Journal of Petroleum Technology* **23** (6), 671.

Atlantic Richfield Company, 1980, "Higher Prices to Spark Offshore EOR," *Oil and Gas Journal*, April 21, 1980, p. 42.

Averitt, Paul, 1960, "Coal Resources of the United States—A Progress Report," United States Geological Survey Bulletin, Number 1275. Washington, D.C.: Government Printing Office.

Barraclough, Geoffrey, 1978, "Waiting for the New Order," *New York Review of Books* **25** (16), 45.

Blair, John M., 1976, *The Control of Oil*. New York: Vintage Books.

Bohi, D. R., and U. Russell, 1978, *Limiting Oil Imports*. Baltimore: Johns Hopkins Press.

Brown, Michael, 1979, *Laying Waste—The Poisoning of America by Toxic Chemicals*. New York: Pantheon.

Chamberlain, Edward, 1950, *Theory of Monopolistic Competition*. Cambridge, Mass.: Harvard University Press.

Chase Manhattan Bank, 1980, "Oil Prospects in the CPE's," *The Petroleum Situation* **4** (12).

Cram, Ira H., 1971, "Introduction" in *Future Petroleum Provinces of the United States—Their Geology and Potential*, pp. 1–33. American Association of Petroleum Geologists, Memoir 15.

Day, David T., 1909, "The Petroleum Resources of the United States," United States Geological Survey Bulletin, Number 394. Washington, D.C.: Government Printing Office.

Department of Energy, U.S., 1980, *1980 Annual Report to Congress*, Vol. 3, Forecasts: Energy Information Admin., DOE/EIA-0173(80)13. Washington, D.C.: Government Printing Office.

Department of Energy, U.S., 1979, "A Comparative Assessment of Five Long-Run Energy Projections," DOE/EIA/CR-0016-02, prepared by A. S. Kydes and J. D. Pearson. Washington, D.C.: Government Printing Office.

Department of the Interior, U.S., 1956, Report of the Panel on the Impact of the Peaceful Uses of Atomic Energy to Congressional Joint Committee on Atomic Energy. Washington, D.C.: Government Printing Office.

Duncan, D. C., and V. E. McKelvey, 1963, "U.S. Resources of Fossil Fuels, Radioactive Minerals and

Geothermal Energy,'' Appendix A, pp. 43–45. Federal Council for Science and Technology Research and Development on Natural Resources.

Elliott, Martin A., and Henry R. Linden, 1968, ''A New Analysis of U.S. Natural Gas Supplies,'' *Journal of Petroleum Technology* 20 (2), 135.

Exxon Company, 1978, *Energy Outlook: 1978–1990.*

Gibson, Paul, 1980, ''We're Betting Five Inches Against the World,'' *Forbes* 125 (3), 44.

Halbouty, M., 1970, *Geology of Giant Petroleum Fields.* Tulsa, Okla.: American Association of Petroleum Geologists.

Hendricks, T. A., 1965, *Resources of Oil, Gas, and Natural Gas Liquids in the United States,* United States Geological Survey Circular, Number 522. Washington, D.C.: Government Printing Office.

Hewett, Donald Foster, 1929, ''Cycles in Metal Production,'' AIME Technical Publication, Number 183.

Hopkins, George R., 1950, ''A Projection of Oil Discovery, 1949–1965,'' *Journal of Petroleum Technology* 2 (6), pp. 6–9 of Sec. 1, p. 10 of Sec. 2.

Hubbert, M. King, 1974, ''U.S. Energy Resources, A Review as of 1972,'' U.S. Senate Committee on Interior and Insular Affairs, *A National Fuels and Energy Policy Study,* Washington, D.C.: Government Printing Office.

Hubbert, M. King, 1973, ''Survey of World Energy Resources,'' *The Canadian Mining and Metallurgical Bulletin* 66 (735), 37.

Hubbert, M. King, 1969, ''Energy Resources'' in *Resources and Man,* pp. 157–242. San Francisco: W. H. Freeman and Company.

Hubbert, M. King, 1967, ''Degree of Advancement of Petroleum Exploration in the United States,'' American Association of Petroleum Geologists Bulletin, Number 51, pp. 2207–27.

Hubbert, M. King, 1962, ''Energy Resources, A Report to the Committee on Natural Resources.'' National Academy of Sciences, National Research Council, Publication 1000-D.

Hubbert, M. King, 1956, ''Nuclear Energy and the Fossil Fuels'' in *Drilling Production and Practice,* pp. 7–25. New York: American Petroleum Institute.

Jamali, Usameh, 1978, ''The Opportunity Cost of Producing and Selling Oil as Crude'' in *Natural Resource Policies Proceedings,* Vol. 2, pp. 47–58. University of Wisconsin, Madison.

Krasner, Stephen, 1977, ''The Quest for Stability: Structuring International Markets'' in *International Resource Plows* (G. Garvey, L. A. Garvey, and D. C. Heath, eds.), pp. 39–58. Lexington, Mass.: Lexington Books.

Lewin and Associates, Inc., 1976, ''Research and Development in Enhanced Oil Recovery,'' Final Report, ERDA-77-20/1,2,3.

Link, Walter K., 1966, ''An Expert Gives A Geological Approach to Reserve Ultimate of the U.S.,'' *Oil and Gas Journal* 64 (34), 150.

Matheny, Shannon L., 1980, ''EOR Methods Help Ultimate Recovery,'' *Oil and Gas Journal* 78 (13), 79.

McCaslin, John C., 1981, ''Torrid Drilling to Surge Past 70,000 Wells,'' *Oil and Gas Journal* 79 (4), 145.

McCaslin, John C., 1980, ''1980: Biggest Drilling Year Since 1957,'' *Oil and Gas Journal* 78 (4), 126.

McKelvey, V. E., 1973, ''Mineral Resource Estimates and Public Policy,'' U.S. Geological Survey, Professional Paper 820, pp. 9–19.

Menard, H. W., 1976, ''Exploration History and Random Drilling Models'' in *Proceedings of the Annual Meeting,* pp. 1–12. Association of Indonesian Petroleum Geologists, Djakarta.

Menard, H. W., and G. Sharman, 1975, ''Scientific Uses of Random Drilling Models,'' *Science* 190 (4212), 337.

Miller, Betty M., et al., 1975, ''Geological Estimates of Undiscovered Recoverable Oil and Gas Resources in the United States,'' United States Geological Survey Circular, Number 725. Washington, D.C.: Government Printing Office.

Moody, J. D., J. W. Mooney, and J. D. Spivak, 1970, ''Giant Oil Fields of North America'' in *Geology of Giant Oil Fields,* pp. 8–17. American Association of Petroleum Geology, Memoir 14.

Murphree, E. V., 1952, ''Where Will Tomorrow's Oil Come From?'' *Oil and Gas Journal,* November 3, p. 119.

Myrdal, Gunner, 1971, *The Challenge of World Poverty: A World Poverty Program in Outline.* New York: Pantheon.

National Academy of Sciences, 1975, *Mineral Resources and the Environment,* Report prepared by the Committee on Mineral Resources and the Environment (COMRATE), Washington, D.C.: Government Printing Office.

National Petroleum Council, 1976, *Enhanced Oil Recovery,* December.

Nutting, Anthony, 1964, *The Arabs.* New York: New American Library (Mentor).

Office of Technology Assessment, 1980, *Projections of Non-Communist World Oil Supplies,* October. Washington, D.C.: Government Printing Office.

Office of Technology Assessment, 1978, *Enhanced Oil Recovery Potential in the United States.* Washington D.C.: Government Printing Office.

Oil and Gas Journal, 1981, ''Six New Fields Make List of U.S. Oil Giants,'' 79 (4), 153.

Oil and Gas Journal, 1979, ''U.S. Petroleum Industry Will Face Monumental Task in Next Decade,'' 77 (46), 176.

Ortiz, Rene, 1980, ''Facing Up to an Energy Crisis: An OPEC View,'' *OPEC Bulletin* 11 (18), 7.

Penrose, Edith, 1968, *The International Firm in Developing Countries: The International Petroleum Industry.* Cambridge, Mass.: MIT Press.

Prats, Michael, 1978, ''A Current Appraisal of Thermal Recovery,'' *Journal of Petroleum Technology,* August, p. 1129.

Pratt, Wallace E., 1956, ''The Impact of Peaceful Uses

of Atomic Energy on the Petroleum Industry'' in
Joint Committee on Atomic Energy, U.S. Congress, *Peaceful Uses of Atomic Energy,* background material for the report of the panel on the
impact of peaceful uses of atomic energy to the
Joint Committee on Atomic Energy, Vol. 2, pp.
89–105.

Pratt, Wallace E., 1950, ''The Earth's Petroleum Resources'' in *Our Oil Resources* (L. M. Fanning,
ed.), 2nd edn., pp. 137–52.

Renfro, J. J., 1979, ''Colorado's Sheep Mountain CO_2
Project Moves Forward,'' *Oil and Gas Journal* **77**
(51), 51.

Sayigh, Y. A., 1975, ''Oil in Arab Developmental and
Political Strategy: An Arab View'' in *The Middle
East: Oil, Politics, and Development* (J. D. Anthony, ed.). Washington, D.C.: American Enterprise Institute.

Schultz, Paul R., 1952, ''What Is the Future of Petroleum Discovery?'' *Oil and Gas Journal* **51** (12),
258, 295.

Smith, David N., 1977, ''Information Sharing and Bargaining: Institutional Problems and Implications''
in *International Resource Plows* (G. Garvey, L. A.
Garvey, and D. C. Heath, eds.), pp. 85–106. Lexington, Mass.: Lexington Books.

Solberg, Carl, 1976, *Oil Power.* New York: Mentor
Books.

Stalkup, F. I., 1978, ''Carbon Dioxide Miscible Flooding,'' *Journal of Petroleum Technology,* August,
p. 1102.

Steinhart, Carol, and John Steinhart, 1974, *Energy:
Sources, Use and Role in Human Affairs.* North
Scituate, Mass.: Duxbury.

Steinhart, John, Bobbi McKellar, et al., 1980, *Energy
Trends: A Set of Visuals on Energy Issues,* Madison: University of Wisconsin Press.

Theobald, P. K., S. P. Schweinfurth, and D. C. Duncan, 1972, ''Energy Resources of the United
States,'' United States Geological Survey, Circular
650. Washington, D.C.: Government Printing Office.

Tinbergen, Jan, ed., 1976, *RID—Reshaping the International Order: A Report to the Club of Rome.* New
York: Dutton.

Tomeh, George, 1980, ''Interdependence: A View from
the Third World'' in *Resources and Development*
(P. Dorner and El Shafie, eds.), pp. 359–84. Madison: University of Wisconsin Press.

Torrey, P. D., 1960, ''Can We Salvage Another 44 Billion Barrels?'' *Oil and Gas Journal* **58** (24), 97.

Ul Haq, Mahbub, 1976, ''The Third World and the New
International Economic Order,'' Development Paper No. 22. Washington, D.C.: Overseas Development Council.

United States President's Materials Policy Commission,
1952, *Resources for Freedom.* Washington, D.C.:
Government Printing Office.

Uri, N. D., 1978, ''Re-Examination of Undiscovered
Oil Resources in the United States,'' Department
of Energy, DOE/EIA—0103/9.

Weeks, Lewis G., 1960, ''The Next Hundred Years Energy Demand and Sources of Supply,'' *Geotimes* **5**
(1), 18, 51.

Weeks, Lewis G., 1948, ''Highlights on 1947 Development in Foreign Petroleum Fields,'' *American
Association of Petroleum Geologists Bulletin* **32,**
1093.

Zapp, A. D., 1962, ''Future Petroleum Producing Capacity of the U.S.,'' United States Geological Survey Bulletin, Number 1142-H. Washington, D.C.:
Government Printing Office.

Coal—Bridge to the Future: A Summary of the World Coal Study (WOCOL) 10

CARROLL L. WILSON

INTRODUCTION

The world has reached the end of an era in its energy history. Increased supplies of oil, which have been the basis for economic growth in the past few decades, are not expected to be available in the future. The development of a new energy basis for continued economic growth has therefore become an urgent necessity. Building a bridge to the energy sources and supply systems of the next century—whatever they turn out to be—is of crucial importance. We believe that coal can be such a bridge and that it will also continue to serve a vital role into the longer-term future.

A tripling of coal use and a 10–15-fold increase in world steam coal trade would allow the energy problems of the next two decades to be faced with confidence. With such increases in coal use, coupled with a mobilization of other energy sources and vigorously promoted conservation, it becomes possible to see how to meet the energy requirements of moderate economic growth throughout the world. But without such a coal expansion the outlook is bleak. This is the central message of our report.

In the industrialized countries coal can become the principal fuel for economic growth and the major replacement for oil in many uses. Coal may also provide the only way for many of the less developed countries to obtain the fuel needed for electric power and industrial development, and to reduce their dependence upon oil imports.

Increased use of coal will require large investments, but no greater than those required for other fuels, such as oil, gas, and nuclear energy. Countries now heavily dependent upon oil must build coal-using facilities on a large scale before they can use the coal they

Dr. Carroll L. Wilson is Mitsui Emeritus Professor of Problems of Contemporary Technology in the School of Engineering, Massachusetts Institute of Technology, Cambridge, Massachusetts 02139. Since becoming a professor in the Sloan School of Management at MIT in 1960, he has organized and directed several global studies, including a study of critical environment problems (*Man's Impact on the Global Environment, Assessment and Recommendations for Action*, SCEP Report, 1970), a study of man's impact on climate (*Inadvertent Climate Modification*, SMIC Report, 1971), a world energy assessment by the Workshop on Alternative Energy Strategies (*Energy: Global Prospects 1985–2000*, WAES Report, 1977), and the Report of the World Coal Study (*Coal—Bridge to the Future*, WOCOL, 1980).

will need. Power stations, coal ports, railways, and handling facilities take a long time to plan and build. So do mines and export terminals. Unless decisions to build them are made soon, these facilities will not be ready by the time when this study indicates that they will be acutely needed. It is necessary for governments and industry to act cooperatively so that the required investment decisions are made promptly.

Unlike oil, the reserve base for coal is sufficiently great to support large increases in production for a long time into the future. Moreover, the technology for its safe and environmentally acceptable production, transport, and use is proved and already widely applied in most areas.

We have examined environmental questions with great care and considered the measures that would have to be taken within each WOCOL country to comply with present and anticipated environmental regulations. The technology exists by which exacting standards of environmental protection can be met, and much work is being done to improve it and lower its costs. We are convinced that coal can be mined, moved, and used at most locations in ways that meet high standards of environmental protection at costs that still leave coal competitive with oil at mid-1979 prices.

The present knowledge of the effects on climate of carbon dioxide (CO_2) from fossil-fuel combustion does not warrant a global reduction in fossil-fuel use or a delay in the expansion of coal. However, support for research on the possible effects of increased atmospheric CO_2, including that of global warming, is essential so that future policies may reflect an improved understanding of such matters.

Coal is not in competition with conservation, nuclear or solar energy, or other sources as the sole solution to the world's energy problems. All these will be required if the energy needed is to be supplied. The world, however, needs an incremental energy source as nearly like oil as possible, but with the vital difference that it will be obtainable in

increasing amounts until well into the next century. Ideally, and if it is to fill the role played by oil over the past decades, it should be versatile in application, easily transported and stored, and reasonably priced. The technology for using it should be mature and generally available so that it can be brought into use rapidly, widely, and safely. It should be capable of satisfying strict environmental standards with presently available technology and at a cost competitive with other fuels. It should be obtainable in large quantities and for long enough to justify the investments required to bring it into widespread use. Only coal comes close to meeting these specifications.

This need for coal has been recognized many times, perhaps most clearly and conclusively in the resolution adopted at the Seven-Nation Economic Summit at Tokyo in June 1979:

We pledge our countries to increase as far as possible, coal use, production, and trade, without damage to the environment. We will endeavor to substitute coal for oil in the industrial and electrical sectors, encourage the improvement of coal transport, maintain positive attitudes toward investment for coal projects, pledge not to interrupt coal trade under long-term contracts unless required to do so by a national emergency, and maintain, by measures which do not obstruct coal imports, those levels of domestic coal production which are desirable for reasons of energy, regional and social policy. [1]

Declarations of intent are, however, not self-executing. They require action. We believe that a deeper public understanding of the importance of coal to the world's future is necessary if the numerous steps by local, regional, and national governments, by international agencies, and by public and private investors are to be taken in time to allow expanded use of coal to provide a major part of the energy required for the world's continued economic growth.

PROSPECTS FOR COAL

The dwindling prospects for any substantial increase in the supply of oil at acceptable prices constitute the main reason for the increased importance of coal. Even with the most optimistic forecasts for the expansion of nuclear power and the aggressive development of all other energy sources, as well as vigorous conservation, it is clear that coal has a vitally important part to play in the world's energy future. World coal production in 1977 was about 3400 million metric tons of raw coal. This was a contribution of about 2500 million metric tons of coal equivalent (mtce) or 33 mbdoe,[2] already greater than any energy source except oil.

Projecting Coal Requirements

Our analysis of future world coal requirements is built upon detailed country studies conducted by WOCOL teams as well as special studies for regions not directly represented in this study. The teams developed two reference cases to project their expected range of future coal demand. Case A considers a moderate increase in coal demand to year 2000, whereas Case B assumes a high increase in coal demand, one which would increase world coal supply, trade, and use to what now appear to be close to the feasible upper limits.

Each country team formulated its case estimates of moderate and high increases in coal use within the general economic, technological, and political environment expected within its country. No attempt was made to develop, for these cases, a specific consensus about the general global economic framework over the next 20 years. The WOCOL country studies were based instead on readily available and detailed national energy studies modified to reflect the specific WOCOL focus on projecting a range of future coal requirements. There is considerable richness and diversity in these WOCOL country studies,[3] which describe the coal expansion countries now expect may be required to meet their future energy needs.

Projections of total coal use were developed by market sector in each country. Teams also provided estimates of indigenous coal production and the coal imports which might be required. The feasibility of satisfying coal import requirements was assessed by matching them against the estimates of coal export potential made by the major coal-producing countries, most of whom are represented in WOCOL.

The average growth in total energy consumption was lower in both cases than that used in some other recent energy studies, for example, that of the IEA Steam Coal Study (December 1978). The average growth in energy use for member countries of the Organization for Economic Cooperation and Development (OECD) in WOCOL Case A was 1.75 per cent per annum for 1977 to 2000; that of Case B was 2.5 per cent.[4] These compare with 2.7 per cent per annum for the same period in the IEA study.

Coal Use by Market Sector—OECD

The estimates of total coal use in Cases A and B were developed through a detailed analysis by market sector (electrical, metallurgical, industrial, residential/commercial, and synthetic fuel) within each country.

The major coal use in year 2000, as today, is projected to be in electricity generation, which consumes more than 60 per cent of the total coal. Estimates of coal requirements in the electric market are strongly influenced by assumptions about rates of growth of electricity demand and the expansion of nuclear power, as well as the replacement rate of existing oil-fired capacity. Our analysis indicates that, even under the moderate electricity growth assumptions of Case A (3 per cent/year), coal-fired electric capacity in the OECD will need to more than double from 350 GWe[5] in 1977 to 825 GWe by year 2000. A 1 per cent per year higher electricity growth rate (Case B) leads to a further 500-GWe increase in total electric capacity requirements, about half of which is projected in the WOCOL team reports to be coal-fired. This yields a coal-fired capacity in the OECD of 1090 GWe in year 2000 for Case B. The corresponding projections for coal requirements for the OECD electric market are an increase from 600 mtce in 1977 to year 2000 estimates of 1325 mtce in Case A and 1850 mtce in Case B.

The estimates reflect an increase in the percentage of OECD coal-fired electric capacity, from 32 per cent in 1977 to about 40 per cent in 2000. Significant regional variations continue in year 2000 with coal penetration in the electricity sector being about 85 per cent in Australia, 50 per cent in North America, 35 per cent in Europe, and 15 per cent in Japan.

Coal for metallurgical purposes is the second-largest market for coal today in the OECD, accounting for about 250 mtce/year, or 25 per cent of total coal use. Its growth since 1960, when the total was 190 mtce, has been slow. The WOCOL projections show a moderate increase to 330–375 mtce/year by the end of the century. However the metallurgical coal share of the total coal market is projected to fall to about 15 per cent of total coal use by year 2000.

The use of coal in industry has declined rapidly over the past two decades and now accounts for only about 90 mtce, or 9 per cent of total OECD coal use. This trend is expected to be reversed. In our projections, industrial coal use in the OECD countries expands by two to four times by year 2000. The major expansion is expected to occur after 1985, when the market is projected to grow at 5 to 7 per cent a year. Industries where coal use is expected to increase significantly include cement, chemicals (as fuel and feedstocks), petroleum refining, and paper.

A substantial new market for coal as feedstock for synthetic oil and gas plants could develop in the 1990s. This appears to be particularly likely in the United States. The total OECD market for the coal required for synthetic fuels by year 2000 is estimated to range from a low of 75 mtce in Case A to a high of 335 mtce in Case B. The higher amount of coal would supply 65 large synfuel plants each producing about 50,000 bdoe of synthetic oil or gas.

Coal has nearly disappeared as a fuel for homes and commercial facilities in the OECD countries, with the use in 1977 being less than 50 mtce. This compares with five times as much, or nearly 250 mtce, as recently as 1960. Although WOCOL country studies suggest that the total residential/commercial use of coal may remain insignificant during this century, there are in some countries indications of a revival of interest in coal for use in homes and office buildings. Projections for the United Kingdom, for example, indicate that such coal use could grow from 10 mtce in 1985 to 15–21 mtce by year 2000. Substantial growth in the use of coal for fuel in district heating plants is also projected in some countries, for example, in Denmark.

Total Coal Use—OECD

The projected total coal use in the OECD, by country, as summed from the WOCOL market sector projections, is given in Table 10-1. The projected growth in OECD coal requirements from 1977 to 2000 is substantial, about 1000 mtce/year, or a doubling, in Case A; and about 2000 mtce/year, or a tripling, in Case B. In a number of countries, for example, Australia, Canada, Japan, and the United States, the increase is particularly large. The OECD demand for coal is projected to accelerate sharply after 1985 in the Case B projections.

The Effect on OECD Coal Needs of Limitations on Oil Imports and Nuclear Delays

Our review of the projections in Cases A and B for total OECD oil imports and nuclear power expansion, summed from the individual country estimates, led to the formulation and examination of two additional cases.

Projected oil imports in the OECD increased in both Case A and Case B by about 3 mbd by year 2000, which implied that oil available for OECD countries to import would need to increase from 26 mbd in 1977 to 29 mbd in 2000. This exceeded our most optimistic estimate of the amount of oil that

Table 10–1. Total Coal Requirements in OECD Countries.

		Total[a] Coal (mtce)			
		1985		*2000*	
Country/Region	*1977*	*Case A*	*Case B*	*Case A*	*Case B*
Canada	25	44	41	82	121
United States	509	655	725	1075	1700
Denmark	4.6	10.7	11.1	9.4	20.9
Finland	4.3	4.4	4.4	9	13
France	45	35	59	48	125
Federal Republic of Germany	102	119	126	150	175
Italy	13.5	22.3	22.9	31.5	60.5
Netherlands	4.5	10.4	10.4	23	38
Sweden	2.1	5.1	5.4	17	26
United Kingdom	109	107	115	133	179
Other Western Europe	51	61	83	135	175
Japan	79	97	102	150	224
Australia	38	65	65	141	166
Total OECD[b]	990	1235	1370	2000	3025

[a] Total includes steam plus metallurgical coal.
[b] Totals are rounded.

would be available. Case A-1 was therefore developed to be compatible with the assumption that OPEC oil production does not rise above today's levels, and that about 22 mbd would be available for import to OECD countries in year 2000 (see Figure 10-1). This required a 20 per cent reduction in the total use of oil projected within the OECD countries in year 2000 in Case A.

A 7–10-fold increase in nuclear power capacity in the OECD was also estimated in the reference projections: from about 80 GWe in 1977 to 550 GWe in year 2000 in Case A, and to 775 GWe in Case B. Case A-2 was established to investigate the effects of possible delays in the expansion of nuclear power. The figure of a 30 per cent reduction in the total OECD nuclear capacity projected in year 2000 in Case A was selected merely to illustrate the magnitude of the effects of possible nuclear delays. The Case A-2 nuclear capacity projected in year 2000 was thus about 400 GWe and compares with the 260 GWe now operating or under construction in the OECD countries.

The effects of possible oil limitations and nuclear delays are reported for Case A only, since the coal projections in Case B were already considered near the upper limits of what is considered plausible. Figure 10-2 shows the effects of the two variations of Case A on OECD coal consumption. Case A-1 superimposes on Case A an incremental OECD coal demand of 500 mtce in year 2000, calculated by the WOCOL teams as part of their responses to the oil limitation. In Case A-2 there is a further increase of 300 mtce/year, calculated under the simplifying assumption that coal is substituted directly for the 30 per cent nuclear reduction, which brings the total OECD coal use in Case A-2 to 2800 mtce in year 2000.

Our assessment is that coal will be required to meet half or more of the energy increase in the OECD over the next two decades. Figure 10-3 makes the point vividly. In the past two decades 67 per cent of the total increase in OECD energy use was met by oil. Coal contributed practically nothing to the energy growth. The progressively increasing share of coal in meeting future energy growth in the Cases A and A-2 anal-

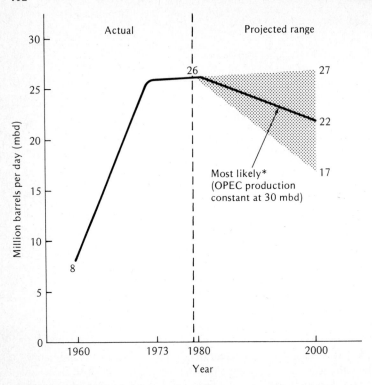

Figure 10-1. Range of net oil imports available to OECD countries (1960–2000). ''Most likely'' corresponds to assumptions of (1) constant OPEC oil production of 30 mbd, (2) an increase in OPEC internal oil consumption from 2.2 mbd in 1978 to 5 mbd in 2000, (3) a decrease in net CPE oil exports from 1.0 mbd to 0 mbd from 1978 to 2000, and (4) constant net oil imports of 3.2 mbd from 1978 to 2000 for other countries outside OECD. Other combinations of assumptions could yield similar results on net oil imports available to OECD countries.

yses is shown in the other diagrams, with the maximum being 67 per cent in Case A-2— exactly the same proportion of the increase taken up by oil in the past.

World Coal Import Projections

The teams from countries likely to import coal each calculated their projected import requirements. This was done by estimating the total indigenous production capacity and subtracting it from the total coal demand. A similar exercise was carried out for the regions not directly represented in WOCOL, to yield total world coal import requirements.

Although the world's present consumption

of coal is large, most coal is consumed in the countries where it is mined. International trade in coal is only about 200 mtce/year (3 mbdoe), or 8 per cent of world coal use. Most coal traded is metallurgical coal for the iron and steel industry. Steam coal represents only about 30 per cent of the international trade (less than 1 mbdoe), and most of this is transported over short distances, such as from Poland to the U.S.S.R. and Western Europe, and from the United States to Canada.

Figure 10-4 shows our projections of world coal import requirements for metallurgical and steam coal in Case A and Case B to the end of the century. World coal trade is projected to increase by three to five times to

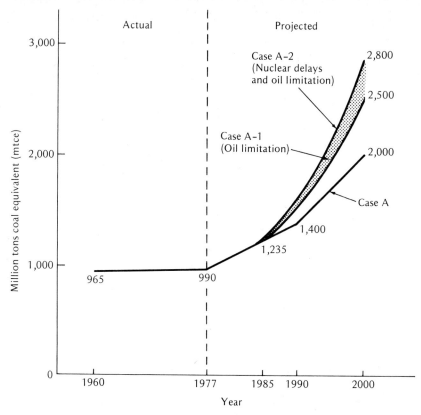

Figure 10-2. The effects of oil limitations and nuclear delays on OECD coal requirements (1960–2000).

560–980 mtce in year 2000. The higher level is equivalent to 13 mbdoe, or nearly half the oil exported in 1979 from OPEC countries.

Demand for steam coal imports, which are shown by country and region in Table 10-2, are projected to increase even more rapidly by about five times in Case A, and about 12 times in Case B. For many nations, the increases are large. Japan, for example, which now imports only 2 mtce/year of steam coal, is estimated to require 25 to 50 times as much by the end of the century, and would become the world's largest steam coal importer. Other large coal importers are projected to include France, Italy, and other Western European countries, for example, the Federal Republic of Germany and the Netherlands, as well as several newly industrializing countries, such as South Korea, Taiwan, and the Philippines in the East and Other Asia region.

World Coal Export Potentials

The feasibility of coal as an international energy source depends on the ability and willingness of coal-producing countries to produce, transport, and export large quantities of coal in time to meet the rapidly growing coal import requirements we project. Coal export potentials were estimated for each of the world's major coal-producing countries, most of which are represented in WOCOL— Australia, Canada, the Federal Republic of Germany, India, Poland, the People's Re-

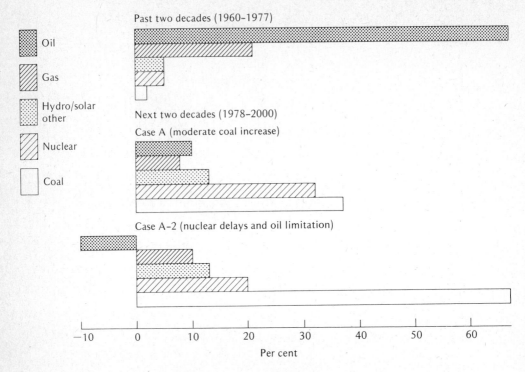

Figure 10-3. Coal's share of the increase in OECD energy needs.

Table 10–2. World Steam Coal Imports by Country and Region (mtce).

Country/Region	1977	1985 Case A	1985 Case B	2000 Case A	2000 Case B
Denmark	4.6	10.7	11.1	9.4	20.9
Finland	4.1	3.4	3.4	7.7	12.4
France	14	11	34	26	100
Federal Republic of Germany	3	9	11	20	40
Italy	2.0	10.3	10.9	16.5	45.5
Netherlands	1.5	7.0	7.0	19.9	34.2
Sweden	0.3	2.9	3.2	14.3	23.1
United Kingdom	1	—	—	—	15
Other Western Europe	7	13	13	32	42
OECD Europe	37	67	94	146	333
Canada	6	6	5	8	4
Japan	2	6	7	53	121
Total OECD[a]	45	80	105	210	460
East and Other Asia	—	5	24	60	179
Africa and Latin America	1	3	3	6	10
Centrally planned economies	17	20	20	30	30
Total world[a]	60	105	150	300	680

[a] Totals are rounded.

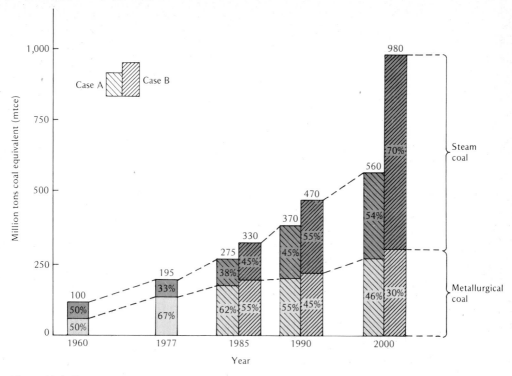

Figure 10-4. World coal import requirements (1960–2000).

public of China, the United Kingdom, and the United States. Current information for Republic of South Africa export potentials by year 2000 was provided to WOCOL by that government's Department of Environmental Planning and Energy. Our estimates of coal export potentials of the Soviet Union, Colombia, and other possible exporters not participating in the study were based primarily on data published by the World Energy Conference, the World Bank, and elsewhere.

Table 10-3 shows 1977 coal exports and projections of coal export potentials for the year 2000 taken from the WOCOL country and regional studies. The estimates shown under "Current Expectations" were taken from WOCOL Cases A and B and represent what coal exporters now expect to be a plau-

sible range of requirements in year 2000. The last column, "Estimated Maximum Potential," represents export levels greater than what coal exporters now expect to supply, but which were indicated as feasible for year 2000 if the demand develops soon enough.

Only Australia (160–200 mtce/year) and the United States (125–350 mtce/year) appear to be capable of exporting significantly above 100 mtce by year 2000. The other significant exporters are projected to be the Republic of South Africa, Canada, Poland, the Soviet Union, and the People's Republic of China.

Some exports are also expected from developing countries such as Colombia, as shown in the estimates for Latin America, Africa, and others. However, these are unlikely to be very large by year 2000. The dif-

Table 10–3. Estimated Coal Export Potentials for Principal Exporters (mtce/year).

| Country/Region | 1977 | Year 2000 | |
		Current Expectations	Estimated Maximum Potential
United States	49	125–200	350
Australia	38	160	200
Republic of South Africa	12	55–75	100
Canada	12	27–47	67
Poland	39	50	50
Soviet Union	25	50	*
People's Republic of China	3	30	*
Federal Republic of Germany	14	23–25	*
India and Indonesia	1	5	*
Latin America, Africa, others	7	25–50	*
World total	200	550–700	930

*Estimates of maximum export potential were not developed.

ficulty of rapid expansion of coal exports from the developing countries, while meeting growing domestic use, is illustrated by the WOCOL estimates for India and Indonesia of less than 5 mtce total coal exports in year 2000. This is despite major expansions projected in coal production for domestic use, from 72 mtce in 1977 to 285 mtce in 2000 for India, and from negligible levels in 1977 to 20 mtce/year by 2000 in the case of Indonesia.

COAL MARKETS AND PRICES

Because of low demand relative to production capacity, a widely distributed resource base, and high transport costs relative to price at the mine, steam coal markets have tended to be geographically limited. Metallurgical coal markets, on the other hand, have been more worldwide than steam coal markets, because metallurgical coal has a smaller, less widely distributed resource base and almost no substitutes in use, and has traditionally received higher prices than steam coal.

Even before the doubling of oil prices in 1979, however, steam coal was competitive in many national and regional energy markets. Since then it has become increasingly

more economically attractive in international markets, especially in the electric power market. This is despite the large maritime transport costs from suppliers such as Australia, Canada, the United States, and South Africa to consumers in Western Europe and Asia, as well as the high costs of environmental control in some of the consuming countries. Further increases in the price of oil should continue to improve the competitive position of steam coal and improve its market penetration in both geographical and end-use markets.

In contrast to an increasingly constrained world oil supply, coal has a very large resource base and considerable scope for expanded production. Given sufficient lead time, it should be possible to continue to increase the supply of coal to meet additional demand at costs that remain competitive. Moreover, the growing number of international coal suppliers, each with different interests, from all world regions including OECD, CPE, and the developing countries, makes the formation of an international coal cartel unlikely.

The growing future demand for steam coal projected by WOCOL will be accompanied by a large expansion in world coal trade. This worldwide market in steam coal will involve many suppliers and will include joint ventures between consumers and producers. Steam coal will be purchased both under long-term contracts and on spot markets. Other arrangements will include captive mines and contract mining. This international steam coal market may tend to acquire some of the characteristics of an international commodity market.

If steam coal is to make the necessary contribution to world energy needs in the coming decades, it will have to remain competitive with other sources of energy, including unconventional oil and gas. In the steam coal market, this competitiveness will be determined by coal's price per unit of obtainable heat, after adjusting for handling and utilization, and other quality characteristics.

Over the next two decades coal will primarily go into the heat and steam-raising markets. It is here that interfuel competition is strongest, although substitution is constrained by the user facilities and infrastructure in existence as a result of previous energy investment decisions. The outcome of the ongoing nuclear debate and the success, or otherwise, of the energy conservation programs now being proposed and implemented will also affect the demand for coal and the prices it can command in the energy market.

Over the long term, the real cost of coal is likely to rise for a number of reasons. The development of new and more costly mining facilities to replace or expand existing capacity; meeting new health and safety requirements; increasing expenditures on environmental protection and land reclamation; and meeting the rising costs of labor will all contribute to future cost increases. On the other hand, increasing productivity, greater mechanization, larger ships, and other economies of scale should help to moderate such increases.

The short-term behavior of prices can, of course, differ substantially from long-term trends. If production capacity exceeds demand, as it did in 1979, prices will tend to move toward the level of the variable cost of marginal mines rather than being determined by the price of alternatives. On the other hand, if the quantity of coal demanded at a particular price exceeds the supply, because of short-term supply and demand inelasticities, spot prices will tend to rise as producers and transporters attempt to capture the increased short-term revenue. However, such price increases would also provide economic signals to some coal suppliers to increase coal production, and thus set forces in motion to eliminate continuation of the short-term advantage. Moreover, a large proportion of the future trade in coal will not be priced on a spot basis. It will be the subject of long-term contracts with greater price stability, and with prices more reflective of the cost of new as well as existing production and transport facilities.

These long-term production and transport costs will set the lower boundary to future prices. The upper level will be determined by the price of the best available alternative. However, in the long run, because of its abundance, and provided a free and competitive international market is maintained, there is little reason to expect that steam coal prices will be directly coupled to world oil prices. This will be particularly so as oil is increasingly removed from the heating market for use in transport and as a specialized petrochemical feedstock.

ENVIRONMENT, HEALTH, AND SAFETY

General Issues

Public acceptance of a large expansion in coal use will be influenced by evidence that coal can be mined, moved, and used in ways that permit adherence to high standards of environmental quality. In this study we have examined the environmental standards currently in effect or under consideration in WOCOL countries; we have considered the environmental control technologies that are available to meet these standards, and we have estimated their costs. Our findings are summarized here.

The principal areas of public concern include land reclamation after surface mining; subsidence from underground mining; acid drainage from the refuse from coal mines and coal preparation plants; emissions from combustion such as SO_2, NO_x, and particulates; safe disposal of ashes; and the possible effects of CO_2 on climate. The applicability and priority of these concerns vary from country to country depending on a number of circumstances. Except for the CO_2 question, however, technology is available to meet these concerns and to comply with the most stringent of the current environmental standards in each WOCOL country at costs that leave coal competitive with oil at mid-1979 prices in most areas. There is no practical method of controlling CO_2 emission from the combustion of fossil fuels and from other sources, and further research is needed on the possible effects of increased CO_2 emissions on global climate. Control of long-range transport of gaseous and particulate emissions may also require new forms of international cooperation.

Difficult choices are required to achieve an appropriate balance between the degree of control of emissions and the resource costs necessary to achieve that control; these, in turn, must be related to the benefits arising from the use of coal. But each increment of further control increases the cost, and as total control is approached the costs become very large. Each country makes its own judgments based upon its own circumstances; we have assumed that national environmental standards will continue to reflect these judgments.

Beyond this, however, it is clear that further research is needed to lower the costs of installing and operating pollution control devices and to improve their effectiveness. Moreover, additional research is required to identify those effluents causing the greatest risks to assure that adequate, but not excessive, control is exercised. Nevertheless, within what can be reasonably anticipated to be applied as regulation, coal can be expected to remain competitive with other fuels in most locations.

Mining Issues

It is now possible with current technology and at a reasonable cost to restore most surface-mined lands to a condition equal to or better than their original condition. Reclamation is generally easier for flat areas. On steep mountain slopes, in arid regions, or in certain ecological regions, reclamation is more difficult. Mining may not be permitted in some such areas under the new laws in effect in several major coal-producing countries. Land areas critical for other purposes, such as agriculture, may also be excluded from mining. Australia, Canada, the Federal Republic of Germany, the United States, and the United Kingdom all have comprehensive legislation for the control and reclamation of surface-mined lands. Even with the possible limitation of mining in some areas, because of the cost of reclamation or because of regulations limiting their accessibility, there are still more than sufficient coal resources in mineable terrain to meet the coal demands projected by this study.

Underground mining can cause land subsidence, i.e., sinking into the area that has been mined. Where the room-and-pillar method of mining has been used, the problem of subsidence is usually not great. However, the room-and-pillar method leaves as much as half the coal in the ground. Another approach is to allow the surface of the land to subside after the removal of the underground coal, and to provide for quickly and fairly carrying out repairs or giving compensation for any damage that may occur. This approach can be used only under certain geological conditions and usually when the longwall method of mining is used. It is currently employed in the United Kingdom and other European countries.

Large quantities of solid waste are produced from surface and underground mining, as well as from coal preparation plants.

This must be disposed of in surface piles, which can be landscaped; as landfill; by returning to the mine; or by use as a construction material. In areas where the waste material contains contaminants, such as high levels of sulfur, prevention of leaching requires careful control of water flows near the storage area, or may even require special ponding arrangements.

Occupational health and safety are important concerns in coal mining. The major occupational health problems associated with underground mining has been the lung (pneumoconiosis, or black lung) disease caused by the breathing of coal dust. The greatest safety hazards in underground mining have been from gas explosion and flooding. Reduction of dust and gas levels by much-improved ventilation and filtration systems; dust suppression by water-spraying or the laying of powdered limestone; continuous monitoring of air quality; and the application of strict work rules and practices have done much to reduce the risk of lung disease and to improve safety by reducing the risk of explosion. Moreover, increasing mechanization and better equipment have reduced the exposure of the work force per unit of coal produced. In mines where the best current practices are observed, the accident and illness rates are now comparable with construction work and many sectors of heavy industry. As coal production expands, extending the application of these health and safety practices and continuing worker training should ensure acceptable occupational health and safety conditions.

Surface and underground coal mining can both be managed with acceptable environmental, health, and safety consequences at reasonable costs. This does not, however, mean that mining procedures in some countries may not need to change. In the United States, for example, producers are now allowed little flexibility in meeting current standards. As experience is accumulated it may be possible to meet the required standards at lower costs through different tech-

niques, and such flexibility should be encouraged.

Coal Transportation and Storage Issues

Coal can be transported by conveyor or truck for short distances, or by train, barge, ship, or pipeline for long distances. The principal environmental disturbances are generally dust, noise, congestion, and runoff of contaminated water. In contrast with oil, the risk of environmental effects from spills are very small, but there are risks from accidents with trains or trucks.

Dust can be controlled by spraying, compacting, or covering. Water runoff can be controlled by careful design of coal storage facilities and treatment of the water where the coal contains soluble or leachable contaminants. Other environmental effects can be controlled by the appropriate design of storage facilities.

Coal Use Issues

Virtually all coal use involves combustion of at least part of the coal with a release of a number of solid and gaseous substances. Most countries now have air quality standards and regulations that limit the rate of allowable emissions from sources such as coal-fired electric power plants. The major emissions that are regulated include sulfur dioxide (SO_2), particulate matter (total suspended particulates, or TSP), and nitrogen oxides (NO_x). The problem of reduced visibility in cities such as London has been greatly diminished by the use of smokeless fuels or the change to appliances using gas, oil, or electricity by residential and commercial consumers.

Technologies for controlling the solid and gaseous emissions from coal combustion already exist, and improved technologies, either to lower costs or to reduce emissions even further, are being developed. Greatly reducing such emissions results in high costs for emission control. Because there are substantial areas of disagreement among experts as to the effects of these emissions, it is not surprising that national policies differ widely on emission control goals and strategies. Cleaning up some of the emissions, especially sulfur, may create new waste disposal problems, such as limestone sludge from flue gas desulfurization.

The extent of the environmental impact from coal-based synthetic fuels production is not clear. Synthetic fuel processes are complex. There are currently very few data available on the characteristics of emissions from the various possible conversion processes because no commercial-scale plants have been built except in South Africa. The major environmental problems will be control of the production and release of potential carcinogens (primarily, complex organic compounds) during the coal conversion and possible toxic materials in the waste. Because the basic costs of coal conversion will almost certainly be high, the additional costs of controlling emissions will probably be an acceptable fraction of the total cost, given the market values and clean nature of liquid and gaseous fuels.

There are some issues on which joint action by nations appears necessary. For example, acid rain resulting from the long-range transport of emissions, including those from coal burning, is acute in some regions and may require early actions by nations in such a region, for example, Western Europe. New mechanisms for international cooperation may be needed. OECD studies of transnational pollution are leading to some agreement on sources of sulfur in the atmosphere (domestic or imported) and on procedures for consultation.

COAL RESOURCES, RESERVES, AND PRODUCTION

World Coal Resources and Reserves

There are vast resources of coal in the world, far in excess of those of any other fossil fuel. This resource base is sufficient to support greatly expanded worldwide use of coal well into the future.

Coal resources vary substantially in quality, reflecting the complex chemical structure of coal. The energy content, or calorific value, of the coal and the sulfur, ash, and moisture content are some of the key elements of quality. Systems of coal resources classification have been developed to deal with these complexities; however, the use of these systems varies from country to country.

The most commonly used classification standards, are those established by the World Energy Conference (WEC), which has defined *geological resources* of coal as a measure of the amount of coal in place and *technically and economically recoverable reserves* as a measure of the quantities that can be economically mined with current mining technology and at current energy prices.[6] The WEC also subdivided coal into two major calorific categories: *hard coal,* defined as any coal with a heating value above 5700 kilocalories per kilogram (equivalent to 10,250 Btu per pound) on a moisture- and ash-free basis, and *brown coal,* any coal with a lower heating value. Bituminous and anthracite coal fall into the category of hard coal, whereas lignite and subbituminous coal are considered by the WEC as brown coals. However, some countries, for example, the United States and Canada, use the term *brown coal* generally to refer only to low-calorific lignite and not to the extensive resources of subbituminous coal that they possess.

Table 10-4 illustrates the quantities of geological resources and technically and economically recoverable reserves of hard coal and brown coal by region on a tons-of-coal-equivalent (tce) basis. The estimates are based on the 1978 WEC report[7] as updated by some WOCOL country teams, such as Australia.

Coal resources and reserves are geographically widely distributed, with more than 80 countries reporting coal deposits. However, ten countries account for about 98 per cent of the currently estimated world resources and

90 per cent of the reserves. Moreover, four countries, the Soviet Union, the United States, the People's Republic of China, and Australia, possess almost 90 per cent of the total world coal resources and 60 per cent of the reserves.

The magnitude of the world's coal resources and reserves is difficult to comprehend. Technically and economically recoverable coal reserves currently amount to about 660 billion tce, or approximately 250 times the 1977 world production. Moreover, estimates of reserves increase as exploration reveals further quantities of extractable coal and as economics and technology change. As a result of increased exploration, stimulated by the oil price increase since 1973, estimates of the world's technically and economically recoverable reserves have increased by about 185 billion tce. This increase is equivalent to about 70 years' production at the 1977 rate of production.

World Coal Production

World coal production in 1977 was about 2500 mtce—equivalent on an energy basis to 33 mbdoe, or more than the total OPEC oil production. Table 10-4 shows 1977 world coal production as well as projections of cumulative production for WOCOL Case B to year 2000. The WEC, the World Bank, and other sources were consulted for the coal production estimates for countries and regions outside WOCOL.

Three countries—the United States, the Soviet Union, and the People's Republic of China—were responsible for nearly 60 per cent of the total coal produced in the world in 1977. The next six largest producers, Poland, the Federal Republic of Germany, the United Kingdom, Australia, the Republic of South Africa, and India, accounted for a further 25 per cent. In 1977 the United States accounted for 50 per cent of the total production outside the centrally planned economy countries. Production in the developing countries is similarly concentrated, with India being by far the largest coal producer and consumer.

Even with the significant expansion in world coal production projected by WOCOL, the cumulative production during 1977–2000 would use up just 16 per cent of the world's present technically and economically recoverable coal reserves.

Table 10–4. World Coal Resources, Reserves, and Production.

	Resources (billion tce)	Reserves (billion tce)	Percentage of World Reserves	1977 Production (million tce)	Cumulative Production 1977–2000 (billion tce)	Cumulative Production as % of Reserves
U.S.S.R.	4860	110	16.6	510	18.0	16
United States	2570	167	25.2	560	25.0	15
China	1438	99	14.9	373	20.0	20
Australia	600	33	5.0	76	4.2	13
Canada	323	4	0.6	23	1.8	—a
Federal Republic of Germany	247	34	5.1	120	3.1	9
United Kingdom	190	45	6.8	108	3.0	7
Poland	140	60	9.0	167	6.7	11
India	81	12	1.8	72	3.9	31
South Africa	72	43	6.5	73	3.3	8
Others	229	56	8.4	368	14.0	25
	10,750	663	100.0	2450	103.0	16

Figures are rounded throughout this article, billion = thousand million.
a Canadian reserves not comparable.

Coal's Role in Developing Countries

Because oil was inexpensive, convenient to transport and use, and readily available until the early 1970s, exploration for coal in the developing countries has been much less widespread and less intensive than exploration for oil and natural gas. Much of the present resources of coal are located in the northern temperate zone. Although the southern hemisphere is less favorable for coal deposits from a geological viewpoint (i.e., less extensive large sedimentary basins), there are large coal resources in Australia and the Republic of South Africa, and there is some optimism that expanded exploration in the southern hemisphere and in the less developed regions of the northern hemisphere will result in the discovery of significant new coal reserves. For example, recent exploration in the southern part of Africa, particularly Botswana and Tanzania, as well as in Indonesia, is yielding favorable results. The world's coal resources and reserves could be significantly larger and more widely distributed geographically than was previously thought.

The World Bank, in its recent publication *Coal Development Potential and Prospects*, notes that coal production in the developing countries accounted for only about 5 per cent of the total 1977 world coal production. About 50 developing countries have known coal resources, and about 30 of these are currently producing coal. A large expansion in coal production and use was projected in this report for the developing countries. In addition to meeting an increasing share of their domestic energy needs, some of these countries, notably Colombia but also Indonesia and Botswana, for example, may have significant potential for exporting coal in the future.

Rapid coal development in the developing countries will require both increased production and increased domestic use of coal. Many of the developing countries have neither the financial resources nor the technical and managerial know-how to launch major coal development programs on their own.

Thus international, regional, and bilateral agencies, as well as private mining companies, are likely to be required to play a major role in supporting the developing countries in an analysis of their coal potential, in an assessment of the role that coal could play in their total energy supply balances, and in providing, where appropriate, financial and technical support for the implementation of coal projects.

Coal-Mining Methods

The choice of whether coal will be mined by surface or underground methods is determined almost entirely by the geological characteristics of the deposit. In general, only deposits that are at shallow depths can be surface mined, but sometimes when the coal occurs in multiple seams, it becomes possible to surface mine at considerable depths.

Surface mining is usually less costly than underground mining and permits the recovery of a higher proportion of the coal in place. Surface mining is also less labor intensive than is underground mining. The majority of the world's coal deposits are, however, accessible only by underground mining. Even though surface mining will expand significantly over the coming decades, particularly in some countries, for example, the United States, it is probable that by the end of the century and thereafter the greater part of the world's coal production will come from underground mines.

Underground mining is usually carried out by either the room-and-pillar method or the longwall method. In the room-and-pillar method, which is the more common in the United States, the coal is extracted leaving behind pillars of coal that support the roof and that considerably lessen, if not eliminate, the possibility of ground subsidence above the mine. However, this method leaves about 40 to 50 per cent of the coal in the mine.

The longwall method extracts the coal across the whole cutting face of a seam. Powered supports are used to hold up the roof temporarily while the coal is cut and removed. The supports are then removed and the roof is allowed to collapse. The longwall method allows a higher proportion of the coal to be removed than does the room-and-pillar method, and, because it tends to be more heavily automated, it has a generally higher productivity. It is the more commonly used method in European mines.

Future Prospects

Coal reserves are sufficient to support a large expansion of coal use well into the next century and beyond. The technology for coal mining is highly developed and steadily improving. Although a large fraction of currently known coal reserves are located in relatively few countries, coal resources exist in many countries throughout the world. There are good prospects that these coal resources can be developed to support both domestic use and additional exports in many more countries in the future.

Transport Systems

Perhaps the most important aspect of all in looking at future transport of coal is the need to consider transport within the context of a coal chain (Figure 10-5). Coal transport systems—inland transportation, export ports, maritime transport, and local distribution—are the essential links between producers and consumers, although frequently independent of both. Unless transport development is integrated with the planning for mines, power stations, and other coal facilities, delays will occur and coal trade will fail to expand at the rate required.

TECHNOLOGIES FOR USING COAL

Established Technologies

Coal is already used extensively, and the technology for expanding its use is well tried and readily available. This applies particularly to the direct combustion of coal to produce heat and steam in electric power generation and in industry. Engineers have been designing and using coal-fired boilers for the past 200 years, and the technology has reached an advanced state. Much progress has also been made over the past few decades in improving the efficiency of electricity generation. The best modern coal power stations can now achieve an efficiency of up to 40 per cent, close to the maximum possible with conventional materials.

Substantial developmental activity is also in progress to further improve control of particulate and gaseous emissions from combustion, in order to meet increasingly strict environmental regulations. Considerable progress has been made in this area in recent years. Reductions of 99.8 per cent in particulate emissions are, for example, achievable using either electrostatic precipitators or bag houses. Most of the emphasis at present is on reduction of sulfur oxide emissions. Wet scrubbers in which the sulfur is absorbed are in commercial use, and regenerative scrubbing systems, which avoid the sludge disposal problems of the wet systems, are at an advanced stage of development.

There have also been advances in the technology for smaller-scale uses of coal in residential/commercial markets. The technology is now available for burning bituminous coal in clean air zones in some countries. Such methods, including automatic underfeed stokers, are suitable for district heating plants, apartment houses, hospitals, and other small- or medium-scale applications. Systems for further improving the convenience of coal use in the domestic sector are also under development.

Developing Technologies for Direct Use of Coal

Many countries are active in developing new technologies, or improving existing ones, to widen the markets for coal combustion. These efforts are directed primarily to increasing the efficiency and flexibility of coal combustion, extending the range of coals that may be used in individual furnaces, and decreasing the

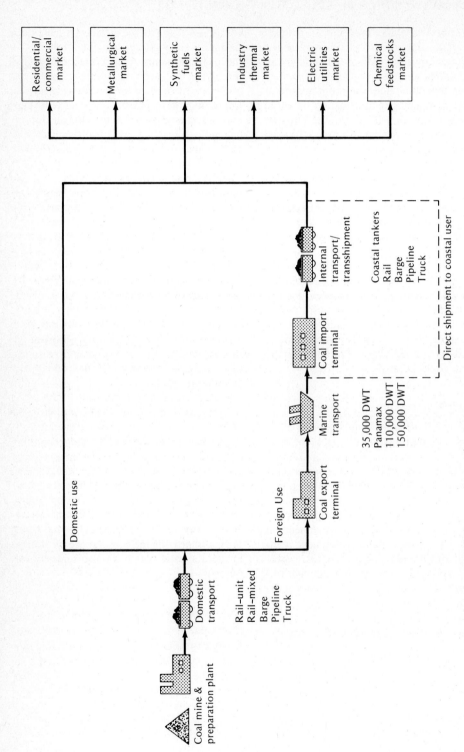

Figure 10-5. Generalized domestic and international coal chains and markets.

environmental impacts associated with coal combustion.

Fluidized bed combustion is one of the most promising of the new technologies for coal combustion. Fluidized bed combustion at atmospheric pressure has been developed to a stage where manufacturers are prepared to give performance guarantees on industrial-scale boilers. It is also anticipated that this technology will be developed so that it can be utilized for centralized power generating stations (in multiples of 200–300 MWe units) within the next ten years. Pressurized fluid bed combustion units are at a substantially earlier stage of development than the atmospheric pressure units.

The fluidized bed combustion method offers a number of important advantages over conventional boilers. The absorption of sulfur oxides within the bed minimizes the problem of removing these from the flue gas. Furthermore, the low combustion temperature reduces the formation of nitrogen oxides substantially. It is also possible to use lower-quality coals which yield very high quantities of ash. Finally, because fluidized bed units are more efficient than conventional boilers, they can be smaller than conventional units of the same capacity.

Progress is also being made in the development of technology for the dispersal of pulverized coal in oil, called coal-oil mixtures (COM), to the extent of 20 to 50 per cent coal, and for the use of COM as a fuel in furnaces and boilers designed for oil firing. There is considerable interest, particularly in the United States and Japan, in the prospects for using COM fuels as a way of accelerating the substitution of coal for oil in existing power plants. For new boiler capacity, the most economic approach will continue to be designing directly for coal combustion.

Developing Technologies for Conversion of Coal to Gaseous and Liquid Fuels

The development of technologically reliable and economically viable methods for the large-scale production of gaseous and liquid fuels from coal could greatly widen the scope of markets for the use of coal in the 1990s and beyond.[8] Interest in technologies to convert coal to oil and gas has greatly intensified recently as prospects for the future availability and price of conventional petroleum sources, especially oil, have worsened.

Because the ratio of hydrogen to carbon in coals is usually less than in petroleum-based fuel, the conversion of coals to gaseous or liquid fuels requires complex chemical changes and the introduction of additional hydrogen, as well as the parallel elimination of oxygen, sulfur, and nitrogen.

Gasification of coal involves the reaction of the coal with steam and requires a heat source, usually provided by having oxygen present. Several grades of gas, in terms of their calorific or heating value, are produced from coal. There is also substantial interest in upgrading such gas to produce methane, i.e., substitute natural gas (SNG), which offers the possibility of supplementing natural gas supplies in existing pipeline systems and for existing appliances.

There are many coal gasification processes under development. The variety reflects the great heterogeneity and variability of coal as a raw material and the complex chemistry involved in conversion. The Lurgi process is probably the best known. It was developed to the stage of large-scale commercial operation using subbituminous coals and lignites before World War II, and further developed in the early 1950s to permit use of noncaking bituminous coals. It is now used in the Republic of South Africa to produce medium-calorific-value gas for use as a fuel as well as an intermediate synthesis gas for the manufacture of motor fuels and various chemicals using Fischer–Tropsch synthesis. Several Lurgi plants are also in operation in Eastern European countries.

The Winkler fluidized bed and the Koppers–Totzek (KT) gasification processes are also in industrial use at present to produce an intermediate synthesis gas in fertilizer and ammonia plants, respectively. A number of new gasification processes are also under development in several countries, with the common objective of allowing gasification to proceed at elevated pressures.

There are three general approaches to the production of liquid fuels from coal: (1) pyrolysis, (2) hydrogenation, and (3) gasification followed by conversion of the synthesis gas to liquids using Fischer–Tropsch or other technologies. All coal liquefaction processes require some gasification of coal as well to produce the additional hydrogen required. Liquid fuels were produced from coal and coal tar in Germany during the Second World War using hydrogenation and the Fischer–Tropsch method. Liquefaction plants were also commissioned in other countries. However, the increasing availability of cheap oil and natural gas after World War II removed much of the incentive for any further work on liquefaction and brought about the shutdown of existing coal liquefaction plants.

The only coal liquefaction plant in commercial use today is located in the Republic of South Africa and has been operating since 1955. This plant (SASOL) which uses the gasification/Fischer–Tropsch approach, has an annual production capacity of 240,000 tons/year (4800 bdoe) of liquid fuels. A second plant is now being commissioned that will increase production significantly, and a third plant has been announced.

Several other coal liquefaction processes are under active development at present. These aim to produce a range of products from clean liquid power station fuels through to a synthetic crude oil capable of being refined to a variety of liquid fuels. There is now significant government support for process development in several countries, as reflected in the WOCOL projections.

Lead Times and Opportunities

It is important to distinguish between the time needed to build a plant using commercially demonstrated technology and that needed for the development of a new technology to the point at which it can be considered ready for commercial use. Experience shows that a period of about 30 years is necessary for the full-scale development of any major new processing technology from the initial concept. It is therefore impossible to make precise estimates of the timing of future commercial use for new technologies still under development and which still face technical problems that must be resolved.

Even conventional power stations, which are now usually built in units of 400 to 750 MWe in the industrialized countries, require a period of six to seven years for completion, about four years of which is needed for construction. In some cases, the total timespan from project initiation to commissioning of the plant can be up to ten years, the extra time arising from administrative procedures, including the need for approvals from regulatory bodies that often involve public hearings and delays. For the established gasification technologies and the Fischer–Tropsch liquefaction process, construction times are comparable with those for power stations.

This question of lead times is crucial. Substantial expansion in coal use within this century must be based largely on currently available technology, that is, conventional power stations, industrial boilers and furnaces, and established domestic and commercial uses. If coal-derived liquids and gases are to make any significant energy contribution before year 2000, the aggressive development programs currently planned for pilot plants and prototype large-scale facilities must be followed up quickly by large commitments for commercial facilities.

The technology upon which to base a rapid expansion of coal use is well established and steadily improving. Moreover, the continuing technological advances in the newer combustion technologies, such as fluidized bed combustion and coal-oil mixtures, offer an opportunity for a widening of the application of coal combustion or improvement of the environmental characteristics of its use. New gasification and liquefaction processes are undoubtedly going to be essential elements in the future energy strategies of many nations in the 1990s and beyond. The long time required for the commercialization and widespread application of new technologies demonstrates clearly the need for a sustained research and development commitment, to ensure that advanced coal technologies are available to support coal's continued role in the energy systems of the next century.

Capital Estimates for Coal Supply

We have found the concept of "coal chains" a useful device for visualizing the links in the system from mines to users and for illustrating the capital costs for each link as well as lead times for planning and construction of facilities. The data below have been assembled for the average capital costs in U.S. 1978 dollars per annual ton of new capacity.

Mines	$53 per annual tce
Inland transport	$23 per annual tce
Ports (loading and receiving)	$23 per annual tce
Ships	$59 per annual tce

Overall, our analysis of capital markets and needs indicates that:

The aggregate amount of capital required to finance the coal expansion projected is well within the capacities of capital markets, representing only a small fraction (about 3 per cent) of the aggregate capital formation in the WOCOL countries. Individual projects may in some cases have some difficulties securing capital.

Most of the capital for coal expansion is needed to build coal-using facilities.

Lead times are long, and early decisions are required by consumers if the necessary coal supplies are to be available when needed.

Provided that an adequate and stable framework is available, we are confident that the capital and resources needed by the coal market as envisaged in this study will be forthcoming. Governments can positively assist in the energy and coal market adjustment process by developing clearly defined energy and environmental objectives, and by adopting a consistent, efficient set of energy policies to achieve those objectives. This would include removal or modification of obstacles to coal production, international trade, and use that are not clearly in the long-term national interest.

NECESSARY ACTIONS

The lead time for individual coal projects is long, usually five to ten years. Even more significant, the large-scale industrial developments necessary to realize the coal expansion projected by WOCOL consist of many interweaving international coal chains that will take sustained effort over all of the next two decades to complete. They must be started now.

The problem is that decisions have to be taken well before the coal is actually needed, in the face of a range of uncertainties. These uncertainties cannot yet be resolved, and, therefore, the prudent approach is to put the emphasis on insurance, with each country developing, as far as economically practicable, the sources of energy it has available. In this way the security both of that country and of the world as a whole is increased.

To this end, Europe can obtain greater security of energy supply by displacing oil imports both by building up its own production of coal where this is appropriate and by increasing its imports of coal. These actions need not be in conflict; both are necessary. A complementary development of indigenous coal production and coal imports will lead to the maximum displacement of oil and increased long-term energy security.

In producing countries, major government decisions will be required to allow production and exports to reach the levels we have projected in this study. For example, in the United States many approvals by local, state, and federal agencies are required for the siting of new coal ports. In addition, there will be a need for an acceleration in the leasing of western coal land, as well as the construction of slurry pipelines to supplement railway expansion and carry part of the large volume of western coal to its markets.

Nations have different ways of achieving similar objectives because of the different roles played by governments in the coal and other energy industries. Thus the balance between legislative action and economic incentive required to facilitate the necessary expansion in the production and use of coal will vary between countries. International agencies such as the IEA and the World Bank can also assist. Many actors are involved and each has a part to play.

We believe that a broadly based national and international program of action is required to:

Ensure that full and clear information on the urgency of the world's energy problem and on the essential role that coal must play is made available to governments, decision makers, planners, and the public everywhere.

Strengthen international cooperation in the development of expensive new technologies.

Expand research and development into methods of environmental control of all aspects of coal production and use, and ensure that the results are applied to improve public acceptance of the expanded use of coal.

In particular, we believe that governments and public authorities have a major part to play and should give serious consideration to the following actions:

Define clearly their policies with regard to production and consumption of coal.

Facilitate and/or establish, as far as they are involved, the necessary infrastructure for the increased production, transport, and use of coal.

Avoid unnecessary delays in planning and licensing procedures.

Facilitate the expansion of free international trade in coal.

Promote, whenever they think it appropriate, a rapid substitution of oil by coal, both domestic and imported, especially in regions that are, or will be, heavily dependent on oil imports.

Coal can provide the principal part of the additional energy needs of the next two decades. But the public and private enterprises concerned must act cooperatively and promptly if this is to be achieved.

NOTES

1. Abstract from Tokyo Communiqué of Seven-Nation Economic Summit, June 1979. Countries included Canada, France, the Federal Republic of Germany, Italy, Japan, the United Kingdom, and the United States.
2. One mbdoe is equivalent to 76 mtce/year.
3. Details of the methods and assumptions used, as well as the results of the projections made by the WOCOL country teams, are described fully in Chapter 2 of *Coal—Bridge to the Future*. The full WOCOL country reports are contained in Volume 2, *Future Coal Prospects: Country and Regional Assessments*.
4. OECD growth rates resulting from WOCOL analyses refer to overall averages for the OECD as derived from individual country projections in the WOCOL studies.
5. Each GWe of electric capacity requires 2 mtce/year of coal if the plant is operated at a 65 per cent capacity factor.
6. World Energy Conference, *Coal Resources*, IPC Science and Technology Press, June 1978, provides a complete description of the WEC standards, which are based on specific definitions of both the depth and the thickness of coal seams, and describes in more detail other aspects of coal reserves and production.
7. Ibid.
8. A more detailed treatment of gasification and liquefaction technologies is provided in Chapter 7 of *Coal—Bridge to the Future*.

SUGGESTED READINGS

Edward D. Griffith and Alan W. Clarke, "World Coal Production," *Scientific American* **240** (1), 38 (1979).

S. C. Morris et al., "Coal Conversion Technologies: Some Health and Environmental Effects," *Science* **206** (4419), 654 (1979).

Harry Perry, "Coal Conversion Technology" in L. C. Ruedisili and M. W. Firebaugh, eds., *Perspectives on Energy,* 2nd edn., p. 181. New York: Oxford University Press, 1978.

W. Peters and H.-D. Schilling, "An Appraisal of World Coal Resources and Their Future Availability," Executive Summary, Bergbau-Forschung GmbH, Central Research and Development Institute of the Hard Coal Mining Industry, Essen, Federal Republic of Germany, 1977.

H.-D. Schilling, "Coal Resource Assessment for the World Energy Conference, 1977" in M. Grenon, ed., *Future Coal Supply for the World Energy Balance,* Third IIASA Conference on Energy Resources. New York: Pergamon Press, 1979.

Joseph F. Wilkinson, "The 1980 Forecast: Mixed Sunshine and Clouds," *Coal Age* **185** (2), 58 (1980).

Carroll L. Wilson, ed., *Energy: Global Prospects 1985–2000,* Report of the Workshop on Alternative Energy Strategies (WAES). New York: McGraw-Hill, 1977.

Carroll L. Wilson, director, *Coal—Bridge to the Future.* New York: Ballinger, 1980.

World Energy Conference, *World Energy: Looking Ahead to 2020,* World Energy Conference Conservation Commission, 1978.

Gas Supply: The Future Looks Bright   11

HENRY R. LINDEN

There has been extensive debate since 1973 on how best to solve the energy problem facing the United States. As is generally the case in very complex issues which cut across many conflicting interests, it has been difficult to reach a consensus. During this period, the Congress and executive branch were each basically split into two opposing factions: those giving priority to conservation and the environment and those giving priority to supply. This division was reflected in contradictions within the policies of the Carter Administration and in the laws passed by Congress.

One basic shortcoming in proposals to ease the oil import problem has been the low priority given to expanded usage of gas to replace the use of oil in stationary heat applications. The potential to reduce the nation's reliance on imported oil through gas substitution is very large and must be examined and given fair consideration as one of the key elements of a long-range solution to the energy problem. The examination should include not only policy issues but also the issue of adequate Research and Development (R&D) to facilitate realization of this potential.

This article discusses the potential for increased gas supply and use to reduce oil imports and then recommends policy guidelines that need to be given serious consideration to implement the "gas option" as an integral part of the solution.

POTENTIAL SUPPLIES OF GAS FOR THE UNITED STATES

In order to develop a perspective on the potential of expanded gas supplies, it is necessary to dispel some widely held beliefs: that since 1973 the gas industry has had a chronic gas shortage; that the gas industry does not have a credible case in saying now that it has supplies through at least the year 2000 far in excess of residential, commercial, and high-priority industrial and agricultural requirements; and that there is a high probability that we shall again find ourselves suddenly short of natural gas for priority uses during a winter as severe as 1976–77 if there is a major relaxation of gas market restrictions.

In 1973, we found ourselves in a natural gas supply situation where we were running out of natural gas that could be produced at a very low cost—we were indeed running out

Original article by Henry R. Linden. Mr. Linden is President of the Gas Research Institute, 10 West 35th Street, Chicago, Illinois 60616. This article is adapted from a talk presented at the Second Annual Gas Industry Bankers Conference in Washington, D.C., December 2–4, 1980.

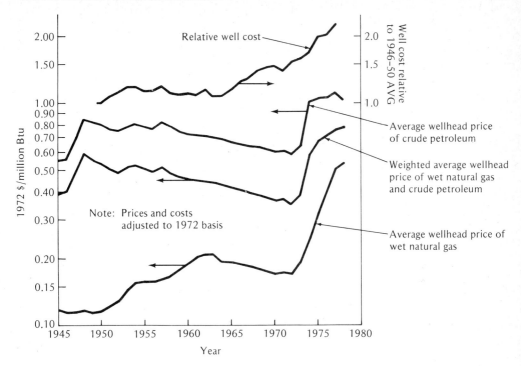

Figure 11-1. Historic trends of wellhead prices of crude petroleum and natural gas and well cost.

of 20–30 cents per Mcf average wellhead price gas. Moreover, the deflated average wellhead price of gas had actually been *dropping* since the early 1960s while deflated well costs were *increasing* (Figure 11-1).

Subsequent events have proved that, as many of us said at the time, our *resources* of natural gas were enormous; the problem was our *reserves* of conventional natural gas.

Let us underline the definitions: the *resource* of a natural commodity is the portion of the total amount of that commodity believed to exist in nature that is thought to be recoverable by the reasonable extension of existing technology; the *reserve* is the portion of the amount known to exist that can be economically produced at today's prices and technology. While proven reserves of conventional natural gas at year-end 1979 were 195 Tcf, estimates of the remaining recover-

able resources range typically from 700 to 1200 Tcf (Table 11-1).

The increase in natural gas wellhead prices from 20–30 cents per Mcf to the current and expected prices under the Natural Gas Policy Act has converted *resources* into *reserves* at an accelerated pace. We have not yet caught up with production rates, and so proven reserves are still declining, but we believe we can reduce this rate of decline to the point where we still can produce 12–14 Tcf of conventional natural gas from the lower 48 states in the year 2000 (Table 11-2). To these projections of conventional natural gas supplies for the lower 48 states must be added the Alaskan supply potential, which is far greater than indicated by the proven reserves of 32 Tcf. Also, none of the huge unconventional natural gas resources are included in the proven reserves, or even in the highest of the remaining recoverable resource esti-

Table 11–1.   Estimates of Remaining Recoverable U.S. Conventional Natural Gas Resources at Year-End.

Year-End	Source	Potential Supply (Tcf)		Proven Reserves[a]	Total
		Old Fields	New Fields		
1979	American Gas Association	—	—	194.9	—
1978	Potential Gas Committee	199	820	200.3	1219
1977	Shell	—	315 (150–500)	208.9	—
	U.S. Department of Energy	—	—	207.4	—
1976	Potential Gas Committee	215	733 (708–758)	216.0	1164
	U.S. Department of Energy	—	—	213.3	—
1974	Exxon	111 (56–321)	287 (127–657)	237.1	635 (420–1215)
	U.S. Geological Survey (Miller et al.)	201.6	322–655		761–1094
	National Academy of Sciences	118.6 (calc.)	530		886
	Average, some major oil cos. (Garrett)	100	500		837
1973	Moody	52	485 (320–700)	250.0	787 (622–1002)
	Mobil Oil (Moody and Geiger)	52	443	250.0	745

[a] American Gas Association.

mates: the 1219-Tcf year-end 1978 estimate of the Potential Gas Committee. As consumer needs dictate, we can anticipate that the future combination of prices, markets, and technology will convert these huge resources into producible and deliverable gas reserves.

It is evident that uneconomical resources can become economically producible reserves as a consequence of several factors. As the price goes up, reserves automatically increase. New technology that lowers production costs can have the same effect. The availability of adequate markets is, of course, a prerequisite because there is no incentive to convert resources into reserves if you cannot sell the product.

Because of the huge potential, let us consider domestic unconventional natural gas resources in more detail.

The Gas Research Institute (GRI) has per-

formed an analysis to forecast the production potential of unconventional natural gas resources under various sets of assumptions concerning wellhead price and new technology development. The relatively near-term (i.e., available in significant quantities in the 1990–2000 time frame) resources include:

Western tight gas sands.

Eastern Devonian gas shales.

Methane from coal seams.

In addition, GRI also analyzed the potential of methane production from the geopressured zones underlying the Gulf Coast area.

This analysis has shown that a large production potential exists (Figure 11-2). By 1990, unconventional sources, with existing technology, could supply annually an addi-

Table 11–2. Year-End 1979 Natural Gas Reserves and Reserve Additions Required To Support Lower 48 States Supply Projections Through 2000.

	Lower 48 States	Total U.S.
Cumulative production	—	577.7 Tcf
Proven reserves	162.98 Tcf	194.92 Tcf
Reserve additions in 1979	13.73 Tcf	14.29 Tcf
Production in 1979	19.69 Tcf	19.91 Tcf
Reserve: production	8.3 years	9.8 years
Required Additions		
Cumulative production, 1980–2000 (assumes decline from 19.7 Tcf in 1980 to 17 Tcf in 1985 to 13 Tcf in 2000)		335.1 Tcf
+Reserves required at year-end 2000 (assumes reserve: production = 7.5 years)		+97.5 Tcf
Total gas required		432.6 Tcf
−Proven reserves at year-end 1979		− 163.0 Tcf
New reserve additions required		269.6 Tcf
Average annual lower 48 reserve additions required, 1980–2000		12.8 Tcf/year

tional 1 Tcf (above current production of 0.8 Tcf of western tight sands gas and 0.1 Tcf of eastern Devonian shale gas) at wellhead prices in the $3–6 per Mcf range (1979 dollars). Gas production could be significantly increased by the year 2000 to a level of from 2.6 to 4.5 Tcf annually with existing technology. The successful development of advanced technology has a major impact and could result in annual production levels ranging from 5.1 Tcf to 8.9 Tcf priced at $3 and $6 per Mcf, respecitvely, (1979 dollars) by the year 2000—the equivalent of 2.5 to 4.4 million barrels of oil per day. Actually, in a recent study by the AGA Gas Supply Committee, the expected range of additional western tight sands and eastern Devonian shale gas production in 2000 was downrated to 1.5–5 Tcf because of a variety of expected institutional barriers and resource and technology uncertainties.

The most significant finding of the GRI analysis was that the maximum quantity of gas that could be produced is at least as technology dependent as price dependent. Doubling the wellhead price from $3 per Mcf to $6 per Mcf using existing technology increased total potential production from the three near-term resources by about three fourths, whereas, at each of the price levels, advanced technology has the potential to double gas production. The potential impact of higher wellhead prices and the development of new technology on the size of the recoverable resource base of each of the three near-term unconventional gas resources were similar, i.e., the size was quite responsive to price increases but also doubled or tripled with new technology.

GRI's analysis was compared with other recent studies. One study was performed by Lewin and Associates, and results were published in 1978. The other study was performed by the National Petroleum Council (NPC) and was released in part in 1980. These studies were reduced to constant 1979

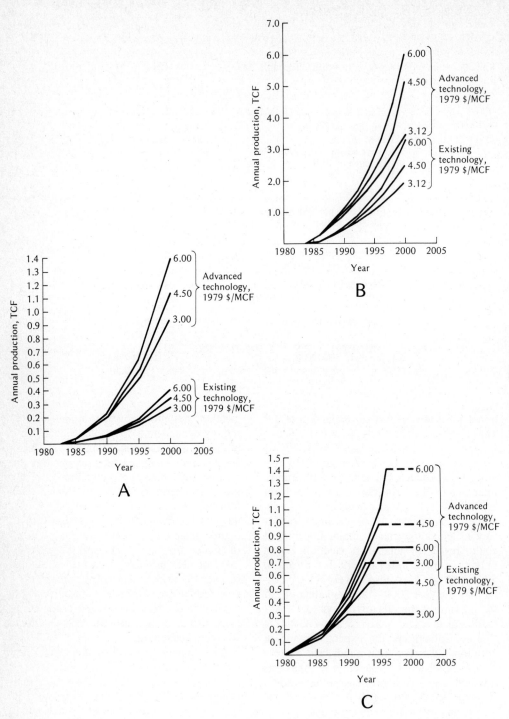

Figure 11-2. Annual production estimates from various sources: (a) coalbed methane; (b) western tight gas sands; (c) Devonian shale.

dollars, and gas price levels were interpolated to permit a consistent comparison. In general, GRI estimates of the recoverable resource bases and production were consistent with the Lewin and NPC studies, as shown in Tables 11-3 and 11-4. All of these studies indicate the tremendous potential for increased gas supplies from unconventional resources with the development of advanced technology.

If one considers the total range of gas resources available, then the potential to expand gas supply and utilization is very large. One new factor is the substantial expansion of underground storage, which will make recurrence of the problems of the winter of 1976–77 highly unlikely. Considering the aggregate supplies available from temporary increases in lower-48 production and increases in Canadian imports, in conjunction with greater use of lower-48 storage and the excess delivery capability of the existing transmission and distribution system, relatively rapid displacement of 1.0–1.5 million barrels per day of fuel oil with 2–3 Tcf of natural gas per year seems feasible. Even greater substitution of gas for oil is feasible with an allowance for the substantial lead times required to (1) develop additional lower-48 production capability, including production from unconventional natural gas and coal gasification sources, (2) secure increased Canadian and Mexican imports and establish acceptable levels and sources of imports of liquefied natural gas (LNG), (3) complete and expand the Alaskan pipeline system, (4)

Table 11–3. Estimates of Unconventional Natural Gas Resources.

	Price (1979 $)	Volume (Tcf)		
		GRI	Lewin	NPC[a]
Western Tight Gas Sands				
Existing technology	3.12	30	90	140
	4.50	45	104	160
	6.00	60	—	175
Advanced technology	3.12	100	170	185
	4.50	120	185	215
	6.00	150	—	230
Eastern Devonian Shale				
Existing technology	3.00	10	6.1	9.4
	4.50	15	9.6	16.6
	6.00	25	—	21.0
Advanced technology	3.00	10	12.1	15
	4.50	30	19.5	24
	6.00	45	—	29
Coal Seams				
Existing technology	3.00	10	—	6.0
	4.50	15	—	16.8
	6.00	30	—	25.3
Advanced technology	3.00	30	1.3–19.2	—
	4.50	40	1.6–24.6	—
	6.00	60	—	—
Geopressured Zones				
Advanced technology	—	—	3.6	0.24–0.57[b]

[a] NPC estimates for western tight sands are preliminary and incomplete.
[b] Projection is based upon "best prospects" known to date.

Table 11–4. Estimates of Unconventional Natural Gas Production Potential.

| | Price (1979 $) | Volume (Tcf) | | | |
| | | 1990 | 2000 | | |
		GRI	GRI	Lewin	NPC[a]
Western Tight Gas Sands					
Existing technology	3.12	0.48	1.99	4.1	—
	4.50	0.53	2.51	—	—
	6.00	0.57	3.31	—	—
Advanced technology	3.12	0.93	3.44	6.8	—
	4.50	1.02	5.21	—	—
	6.00	1.12	6.04	—	—
Eastern Devonian Shale					
Existing technology	3.00	0.31	0.31	0.19	0.42
	4.50	0.36	0.55	0.33	0.78
	6.00	0.40	0.82	—	0.93
Advanced technology	3.00	0.38	0.70	0.40	0.64
	4.50	0.46	1.00	0.73	1.06
	6.00	0.51	1.50	—	1.25
Coal Seams					
Existing technology	3.00	0.06	0.29	—	—
	4.50	0.07	0.35	—	1.5[b]
	6.00	0.07	0.42	—	—
Advanced technology	3.00	0.22	0.95	—	—
	4.50	0.23	1.2	—	—
	6.00	0.24	1.4	—	—
Geopressured Zones					
Advanced technology	—	—	0–1	—	0.01–0.03[c]

[a] NPC estimates for western tight sands are incomplete.
[b] Production is by vertical well technology; price sensitivity projections are unavailable.
[c] Projection is based upon "best prospects" known to date.

reinforce the interstate gas transmission system, and (5) further expand underground storage and extend the distribution system to offer reliable service to the potential new gas customers.

In Tables 11-5 to 11-7, three recent gas supply projections are compared:

The relatively pessimistic National Energy Plan II (NEP II) projection by the Department of Energy.

The very authoritative and extensively documented projections on a scenario basis by the AGA Gas Supply Committee.

The current GRI projection.

There is an industry consensus endorsing the conventional natural gas supply projections for the lower- 48 states contained in NEP II—i.e., a gradual decline from about 20 Tcf in 1980 to 12–14 Tcf in the year 2000. In NEP II, this decline is expected to be balanced by increases in Alaskan gas and unconventional sources. Coal gasification is also expected to make a significant contribution by 2000. Thus, NEP II expects total supplies in the year 2000 to be 18–24 Tcf, roughly maintaining the current level.

Each of the four scenarios considered by the AGA Gas Supply Committee represents different national policy regarding the sources of gas supply. The self-sufficiency scenario

Table 11–5.   Gas Supply Projection by NEP II.

	Quads/year		
	1977	1985	2000
Natural gas			
Conventional lower-48	19.5	16–18	12–14
Alaska	0.1	0.8–1.0	1–2
Unconventional	—	0.3–0.8	1–5
Synthetics			
From coal	—	0–0.1	1–2
From petroleum	0.3	0.2–0.5	0.2–0.5
Total domestic	19.8	18–20	16–22
Imports	1.0	1.8–2.2	1.5–2.0
Net withdrawals from storage	−0.6	—	—
Total supply	20.3	20–22	18–24

From U.S. Department of Energy.

is based on a strong policy of minimizing energy imports. Alaskan gas, coal gasification, and tight formations are sources whose production would be maximized; Canadian, Mexican, and LNG imports would be minimized. The North American focus scenario assumes that the United States plans to minimize the effect of any interruption in the international flow of oil. Lower-48, Alaskan, Canadian, and Mexican sources would be emphasized along with coal gasification and unconventional natural gas from tight formations. The moderate world imports scenario is based on a national policy of maintaining a broad mix of fuels from diverse sources. Additional imports would be encouraged so long as this would lead to a further diversification of supply and contribute

Table 11–6.   Supply Projections for the Year 2000 Based on AGA Gas Supply Committee Scenarios.

	Volume (Tcf)			
	Self-Sufficiency	North American Focus	Moderate World Imports	World Conventional Gas Emphasis
Lower- 48	12–14	12–14	12–14	12–14
SNG from liquid hydrocarbons	0.3	0.3	0.3	0.1
Alaskan	3.0	3.0	1.5	3.0
Canadian	1.0	2.0	2.0	2.0
Mexican	0.1	2.0	2.0	2.0
LNG imports	0.7	0.7	2.5	4.0
Coal gas	3.5	3.5	1.5–2.5	1.5–2.5
Tight formations	1.5–5.0	1.5–4.0	1.5–3.0	1.5–3.0
Misc. new technologies	1.0–2.5	1.0–2.5	1.0–2.5	1.0–2.5
Total	23.1–30.1	26.0–32.0	24.3–30.3	27.1–33.1

Table 11–7. GRI Ranges of Potential Gas
Supplies.

	Volume (Tcf)	
	1985	2000
Natural gas		
Conventional	16–18	12–14
Alaskan	0.1–0.8	1.5–2.0
Unconventional	0.2–0.6	2.6–9.9
Synthetic gas		
From petroleum	0.1–0.5	0.1–0.5
From fossil fuels		
(coal, peat, oil shale)	0–0.1	1.5–4.5
From biomass and wastes	—	0.1–1.0
Pipeline imports		
Canadian	1.0–1.5	1.0–1.5
Mexican	0.1–0.5	0.5–1.0
LNG (incl. Alaskan)	0.5–0.8	0.5–1.5
Total	18.0–22.8	19.8–35.9
Probable range	18–22	25–30

to supply stability. The scenario emphasizing world conventional gas is based on the
previously mentioned fact that the depletion
of the world's huge remaining conventional
natural gas resources is progressing at only
one half the rate of the depletion of the
world's crude oil resources. It also recognizes the growing worldwide need for fuel
and accepts the idea that worldwide trading
of natural gas by pipeline and LNG will increase significantly by 2000.

It should be noted that the AGA Gas Supply Committee category "tight formations"
includes only western tight sands and eastern
Devonian shale gas. Coal seam methane is
included under "new technologies."

Available gas supply for the year 2000 in
all four scenarios exceeds 23 Tcf. The upper
limits of the gas supply estimates vary between 30 and 33 Tcf in 2000. In no case was
gas supply so severely limited that maintenance of the gas share of 25 per cent of total
U.S. primary energy requirements appeared
unlikely.

In contrast with the AGA estimates, GRI's
estimates of unconventional gas production
are based on the sum of the individual poten-

tials of these resources. However, actual total volumes that will be produced will depend on the competitiveness of individual
sources with other gas supply options; availability of capital, drilling equipment, and
supplies; competitiveness of gas with other
energy forms; U.S. policies concerning imports; and other nontechnical factors. Therefore, the range of 2.6 to 9.9 Tcf in 2000 (including 0 to 1 Tcf of geopressured zone
methane) is probably too high and too wide.
The range of other supply options is similar
to that assumed by the AGA Gas Supply
Committee, although somewhat more conservative in regard to imports. In aggregate,
GRI agrees that there is a supply capability
of 25 to 30 Tcf in the year 2000, which should
be cost-competitive with other energy supply
options and with conservation in the year
2000 at a projected total energy demand on
the order of 100 to 120 quads.

The conventional wisdom in 1973 and in
some circles still today is that we are running
out of natural gas. The reality is that in 1973
we were rapidly running out of cheap gas,
but at today's energy prices potential gas
supplies are extensive.

PROSPECTS FOR ALASKAN GAS AND VARIOUS FOREIGN AND UNCONVENTIONAL GAS SOURCES

A key factor to consider in regard to lower-
48 gas supplies is that huge conventional
natural gas resource bases exist in Alaska,
Canada, and Mexico—far beyond what is
currently counted in the reported proved reserves (Table 11-8). Cumulative gas discoveries in Alaska through 1979 total 33.8 Tcf.
About 7.8 Tcf of this was discovered in
southern Alaska or the Naval Petroleum Reserve near Point Barrow and is primarily in
nonassociated gas fields. The Prudhoe Bay
field contains about 26 Tcf of associated-
dissolved gas reserves. Major commercial
production of Alaskan gas began in the 1960s
and has largely been exported to Japan as liquified natural gas (LNG). Besides the current proven reserves of about 32 Tcf in

Alaska, there is a potential of 189 Tcf. Total remaining recoverable resources are 221 Tcf.

The Canadian Petroleum Association estimated Canadian reserves for year-end 1979 at about 89 Tcf. The Canadian Geological Survey estimate of total remaining recoverable resources is about 310 Tcf.

Mexico has tremendous reserves of crude oil and natural gas. As of March 1980, proven reserves were estimated to be over 50 billion barrels of oil and gas in oil equivalents. Probable and potential reserves are estimated to be at least 240 billion barrels. The Mexican government estimates that 29 per cent of proven reserves are gas reserves. This means that 84 Tcf of gas is proved and, if the same applies to probable and potential reserve estimates, an additional 400 Tcf is potentially available, for total remaining recoverable resources of 484 Tcf. These estimates compare to U.S. proven reserve estimates of 195 Tcf and additional potential resources of at least 500 Tcf.

In regard to deliverable supplies from Alaska, Canada, and Mexico (Table 11-9), the AGA Gas Supply Committee made the following assumptions:

Alaskan supply estimates assume that both North Slope gas and southern Alaskan gas will be marketed in the lower-48 states. The

Table 11-9. Alaskan, Canadian, and Mexican Gas Supplies Estimated by AGA Gas Supply Committee.

	Volume (Tcf)		
	1980	1990	2000
Alaska	—	0.8–1.4	1.5–3.0
Canada	1.0	1.0–1.7	1.0–2.0
Mexico	0.1	0.1–1.0	0.1–2.0

low estimate is consistent with only a slight upgrading of currently approved projects. The high estimate requires that a second major transportation system become operational in the 1990s.

Canadian supply estimates are based on a continuation of current licenses and deliveries. The low estimate assumes no new programs over the next 20 years other than to replace expired licenses. The higher estimate assumes the Canadians' development of their frontier resources in addition to the current export programs. Frontier development in Canada is probably dependent upon being able to sell additional gas to the United States in this time period.

The high Mexican supply estimate assumes that Mexican oil exports continue to increase and that the economic advantages of selling the associated gas to the United States outweigh other factors in Mexican national policy formulation. The low estimate assumes that the present amount of gas made available to the United States will not be increased. Little doubt exists as to Mexico's capability to export the high volumes if it so chooses.

LNG trade occurs on a worldwide basis from North Africa to Europe and the United States, from the Middle East to Japan, from Oceania to Japan, and from Alaska to Japan. Proven world natural gas reserves of about 2130 Tcf could be tapped for the LNG market (Table 11-10). As noted before, the annual depletion rate for world conventional gas resources is only half that for oil. This, along

Table 11-8. Alaskan, Canadian, and Mexican Conventional Natural Gas Resource Bases.

	Volume (Tcf)	
	Proven Reserves	Total Remaining Recoverable
Alaska	32[a]	221[b]
Canada	89[a]	310[c]
Mexico	84[d]	484[d]

[a] Reserves of crude oil, natural gas liquids, and natural gas in the United States and Canada as of December 31, 1979.
[b] Potential Gas Committee, as of December 31, 1978.
[c] Canadian Geological Survey, as of December 31, 1976 plus January 1980 revision of western Canadian reserves.
[d] Minister Florencio Acosta, *The Role of Oil in the Mexican Development Plans,* Embajada de Mexico, June 12, 1979.

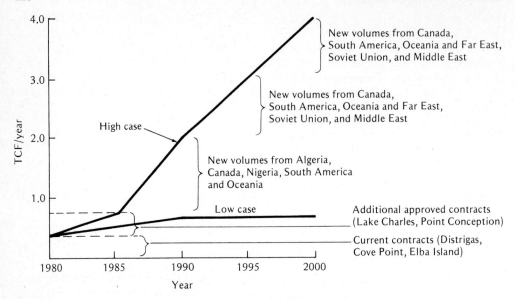

Figure 11-3. LNG imports projected by AGA gas supply committee.

with the lower form value of gas, should theoretically lead to greater price stability for international gas trade than for oil trade.

The AGA Gas Supply Committee's projection of LNG imports reflects this substantial availability of gas resources from overseas. The high case of Figure 11-3 also reflects optimism that U.S. regulatory policy will support foreign resource development. Countries or regions included as possible exporters have all expressed interest in developing LNG projects for export to the United States. The low case of Figure 11-3 assumes no new projects are approved other than supplemental volumes to achieve full capacity use of existing or currently approved facilities. The low case is the more likely case and may even be optimistic because gas exporters are increasingly insistent in their demands for crude oil parity despite the obvious flaws in the economic justification of such demands. This recently resulted in the interruption of supplies from the El Paso-I Algerian liquefied natural gas export project.

As noted earlier, coal dominates U.S. fossil-fuel resources. On the basis of Btu's, coal represents about 80 per cent of the total U.S. fossil-fuel resource base. Recoverable U.S. coal resources on an oil equivalent basis are twice as great as the total recoverable world crude oil supply (1750 billion barrels). A single high-Btu commercial-scale coal gasification facility (250 million cubic feet

Table 11-10. Proven Natural Gas Reserves Potentially Available for LNG Market at Year-End 1979.

	Volume (Tcf)
Algeria	132.0
Canada	85.5
Nigeria	41.4
South America	77.5
Oceania	37.0
Far East	116.2
U.S.S.R.	900.0
Middle East	740.3
Total	2129.9

From *Oil and Gas Journal* **77** (53), 70 (1979).

per day of production) is estimated to consume some 3 quads, or 0.2 billion tons, of subbituminous coal or lignite during 20 years of operation. About 216 billion tons, or 4500–5000 quads, of coal are currently recoverable in the United States. This is sufficient for 20 years of operation of at least 1500 such plants, or of an equal number of 40,000–45,000-barrel/day coal liquefaction plants or of 2000–3000-megawatt power stations. Obviously, this exceeds by far the aggregate need for such coal conversion facilities in the foreseeable future. The quantities of coal available east and west of the Mississippi are roughly equivalent, although they differ in rank and in the proportions accessible to surface mining (Figure 11-4). Although not all coals are suitable for any given coal gasification technology, alternative conversion technologies can accept different types of coal. Research and development efforts have expanded and are expected to continue to ex-

pand the range of coal feedstock that specific conversion methods can accept.

While high-Btu synthetic gas is pipeline quality, medium-Btu gas is suitable only for on-site combustion or moderate-distance transportation. Medium-Btu gas could be provided to industrial customers to replace conventional natural gas, whose use may be prohibited by regulatory constraints. The AGA Gas Supply Committee estimated that by the year 2000, 1.2 to 2.8 Tcf could be produced in 13 to 31 high-Btu plants, and that the high-Btu gas equivalent of about 0.5 to 1 Tcf of medium-Btu gas could be produced (Table 11-11). However, these potential quantities are not additive because they compete, in part, for the same markets and are constrained by several other factors. Thus, the Committee estimated by year 2000, total production from coal gasification could range between 1.5 and 3.5 Tcf.

Although it is true that many uncertainties

Figure 11-4. Demonstrated U.S. coal reserve base by rank and potential method of mining.

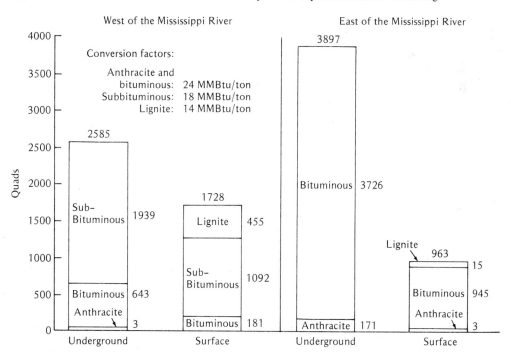

Table 11–11. Production from Coal Gasification in the Year 2000 Estimated by AGA Gas Supply Committee.

	Tcf of High-Btu Equivalent	
	Low	High
High-Btu	1.2 (13 plants)	2.8 (31 plants)
Medium-Btu	0.5	1.0
Trade-off	−0.2	−0.3
Total	1.5	3.5

overhang the supply potential and cost of Alaskan gas and of the various foreign and unconventional sources of gas, in the aggregate, even when severely discounted, these sources constitute a very large and cost-effective supply of energy for the United States relative to many of the nongas options which are being pursued.

One large potential source of gas that has been given wide publicity is methane recovered from geopressured brines. The definition of the geopressured methane resource most used in the literature is that it consists of gas in solution in the brines. However, free (mobile) gas and residual (immobile) gas may also be produced in conjunction with solution gas in the Gulf Coast area. Estimates of the gas in place and ultimately recoverable vary widely. A summary of recent estimates is shown in Table 11-12. These estimates pertain to solution gas only. Quantification of solution gas plus mobile gas and solution gas plus residual gas is a difficult if not impossible task. At this time, key parameters that must be determined in the field are the magnitude of the resources and the potential for economic production. There have even been questions raised that as a result of extensive growth faulting, specific reservoirs may be too small to sustain production long enough for a suitable economic return. The basic data and, more specifically, the field experience are far too limited to make accurate projections of the resource size.

Geopressured-zone methane and other relatively poorly defined miscellaneous sources of gas as categorized by the AGA Gas Supply Committee are listed in Table 11-13. Since each of these miscellaneous sources will respond differently to policy and R&D that affect their development, the extremes of the estimates for these sources are not additive. Therefore, the Committee estimated a total supply range of only 1.0 to 2.5 Tcf in the year 2000.

Table 11–12. Summary of Potential Gas Resources from Geopressured Zones.

		Volume (Tcf)	
Date	Source	Gas in Place	Ultimately Recoverable Gas
1975	USGS, Bull. No. 726	24,000	110–770
1976	Brown, Hudson Institute	60,000–100,000	5000
1976	Jones, Louisiana State University	50,000–100,000	260–1200
1976	Hise, Louisiana State University	3000	125
1976	Dorfman, University of Texas	5700	340
1977	DOE (MOPPS)	150–2000	500
1977	Lewin & Associates	860	42
1978	Wallace (USGS)	3200	—
1978	USGS, Bull. No. 790	5700	150–1550
1978	TRW	3000–5000	150–2000
1978	University of Texas	6000	250

Table 11–13. Supply from Miscellaneous New Technologies in Year 2000 Estimated by AGA Gas Supply Committee.

	Volume (Tcf)
Coalbed methane	0.29–1.4
Geopressured-zone natural gas	0.02–1.00
SNG from peat	0.18–0.90
SNG from oil shale	0.10–0.72
GAS from biomass	0.04–0.14
GAS from urban wastes and animal residue	0.23–0.80
In-situ coal gasification	0.05–0.18
Estimated supply range	1.0–2.5

POTENTIAL OF GAS SUBSTITUTION FOR OIL

In 1979, 2.3 million barrels per day of distillate fuel oil, kerosine, and jet fuel and 2.4 million barrels per day of residual fuel oil were used in stationary heat energy applications in the residential, commercial, industrial, and power plant markets (Figure 11-5).

Figure 11-5. Fuel switching to increase transport fuels and reduce oil imports. (Data from EIA June 1980 MER and 1979 Annual Report to Congress)

	Refined petroleum product demand (1979), 10^6 bbl/day
Gasoline	7.07
Jet fuel	1.07
Other transport fuel uses	1.63
Stationary fuel uses	
LPG	0.98
Distillate	2.31
Residual	2.38
Raw material and other uses of refined petroleum products	3.00
Total	18.43

Displace with gas and coal–and nuclear–based electricity to make more transport fuels

Displace with gas and coal

Data Source: EIA June 1980 MER and 1979 Annual Report to Congress

Theoretically, this entire 4.7 million barrels per day of stationary heat energy demand could be replaced by natural gas if it were deliverable to the potential points of use. Substitution could be immediate where dual fuel capability already exists, as in many industrial and power plant uses, but would take longer in the residential and commercial market and in some industrial markets where replacement of oil-burning with gas-burning equipment is required. In practice, of course, substantial direct substitution of coal, especially for residual fuel oil in large industrial and power plant uses, as well as some substitution of electricity, would prove to be more cost-effective.

The American Gas Association estimates that the immediate replacement potential of gas for stationary fuel oil uses is about 1.1 million barrels per day. AGA also estimates that about one third of the 16 million oil-heated residences use gas for nonheating purposes (cooking, drying, and/or water heating) and could therefore be likely targets for conversion to gas heating to replace about 340,000 barrels per day of oil. An indication of readily attainable residential space heating conversion rates to gas can be obtained from the 365,000 conversions reported by AGA for 1979 and the 380,000 conversions projected for 1980, equivalent to an annual saving of about 25,000 barrels per day of fuel oil.

In addition to this 4.7 million barrels per day of nominally gas-substituable oil demand, there is also a demand for 1 million barrels per day of liquefied petroleum gas for stationary heat energy applications in the residential and commercial market that is included in the total 18.4 million barrels per day oil demand statistics for 1979. Inasmuch as this demand is very likely in areas where natural gas is not available, it is not a promising target for gas substitution.

The most desirable targets for substitution with gas are, of course, the stationary heat energy uses of distillate fuel oil. This would translate quickly into increased domestic transport fuel supplies at minimum cost. To convert displaced residual fuel oil from stationary uses to transport fuels would require substantial refinery investments and lead times, although much less than would be required to produce equivalent amounts of synthetic transport fuels.

CONCLUSIONS

An increase rather than a decrease in gas use between now and the year 2000 is not only feasible but would also be the most cost-effective option to the consumer. Expanded gas substitution for oil usage in stationary heat applications will free domestic liquids for the critical transport fuel market and reduce the nation's dependence on imported oil.

To make this expanded use of gas—i.e., the "gas option" —a key element in an orderly transition to renewable or inexhaustible energy sources would require:

Assurance of both a fair return to the producer and the most cost-effective option to the consumer by oil and gas prices being allowed to compete at the burner tip, or, more generally, for all energy prices to compete on the basis of total service costs.

Removal of restrictions on energy use in terms of market allocation and selective price regulation so that each energy source can seek its highest form value.

A balanced federal research and development program that does not arbitrarily predetermine the preferred energy sources and energy service options. Selective public investment in research and development distorts consumer options as much as selective price regulation or market allocation.

A sound, equitable, and mutually beneficial energy policy with respect to Canada and Mexico.

SUGGESTED READING

Henry R. Linden and Joseph D. Parent, *Perspectives on U.S. and World Energy Problems*. Chicago: Gas Research Institute, April 1980.

Synfuels: Oil Shale and Tar Sands   12

JOHN WARD SMITH

INTRODUCTION

Oil shale and tar sands occupy a unique position in the diverse energy sources assembled into the group entitled "Synthetic Fuels" or "Synfuels." First, both oil shale and tar sands are already fossil (not synthetic) fuels. Second, the basic extraction treatments for oil shale and tar sands yield oil. This oil can feed directly into and help support our well-established dependence on petroleum and natural gas, which now contribute a major part of the U.S. total energy supply. Third, costs for producing major amounts of oil from oil shale and tar sand resources appear significantly lower than those associated with coal hydrogenation, the only other synfuels source of significant amounts of petroleum-like materials.

For a variety of reasons, the world learned to prefer petroleum and natural gas as its primary energy source. Energy sources are not interchangeable in their applications even if they can be calculated as energy equivalents. For example, petroleum and natural gas supply an invaluable chemical raw material for

which nothing else except oil from oil shale and tar sands is a satisfactory substitute. The use of other energy sources as substitutes for many petroleum applications requires apparatus which does not exist. Only lengthy and expensive development can supply the necessary hardware. For example, transportation gobbles up a major share of our petroleum, but coal-fired automobiles and trucks are not an immediate prospect. The possible contribution to U.S. transportation energy by the synfuel alcohol is limited by a basic physical fact—alcohol and gasoline do not mix in quantity. Although engines fueled by just alcohol are practical, the prospect of substituting an alcohol engine in every U.S. car and truck is monstrous. In addition, the supply chain to generate and distribute fuel alcohol does not exist. Any such development requires gigantic capital investment and time.

Time may be the more stringent limitation. M. K. Hubbert's (1956) classic petroleum depletion curves are as inexorable as ever, and so we have a limited time to move from petroleum dependence toward a new

Original article by John Ward Smith. Mr. Smith is a consultant serving the synthetic fuel industry from Laramie, Wyoming. Until 1980 he managed the Resource Characterization Division at the Laramie Energy Technology Center, DOE, the primary U.S. government research and development facility working on energy production from oil shale and tar sands. He has published more than 100 papers describing oil shale and tar sands resources and what can be done with them.

energy mix. It took civilization 200 years to move from a wood-fueled to a coal-fueled economy and 60 years to move from dependence on coal to dependence on oil. These moves were driven by cost factors, just as our present move must be. Current projections (Hubbert and Root 1981) indicate that in the next 30 years or so we must manage most of the move from petroleum. A longer-term answer, perhaps an ultimate answer, might be nuclear fusion, apparently more distant than 30 years. Whatever the next energy mix is, oil shale and tar sands may, by supplying oil, provide a little additional time to make the necessary move.

The factors involved in the evaluation of the time extension available from oil shale and tar sands will be summarized. These include the resources available to development, characteristics of the raw materials, production methods and production records, environmental and socioeconomic effects, the political and economic factors influencing production, and the prospects for major production. World resources and world development experience will be included. However, the U.S. resource and experience bear most strongly on the extension of time for resource transition offered by development of oil from oil shale and tar sands. The United States as a major consumer of oil is the first country to feel the pinch of resource exhaustion.

Oil shale and tar sands are very different materials. Most of what follows must separate them completely. They have one additional major factor in common—as resources they are both usually relatively lean ores. This means that huge quantities of the parent rock must be treated to generate significant production of oil or energy.

DEFINITION
Oil Shale

Oil shale is sedimentary rock containing solid, combustible organic matter in a mineral matrix. The organic matter, often called kerogen, is largely insoluble in petroleum solvents, but decomposes to yield oil when heated (Smith and Jensen 1977). Although *oil shale* is used as a lithologic term, it is actually an economic term referring to the rock's ability to yield oil. No real minimum oil yield or content of organic matter can be established to distinguish oil shale from other sedimentary rocks. Additional names given to oil shales include black shale, bituminous shale, carbonaceous shale, coaly shale, cannel shale, cannel coal, lignitic shale, stellarite, torbanite, tasmanite, gas shale, organic shale, kerosine shale, coorongite, maharahu, kukersite, kerogen shale, algal shale, and "the rock that burns."

Tar Sands

Tar sands are petroleum deposits distinctive in that their petroleum is so viscous that primary production is impossible. The term *tar sands* has come to include an assorted group of materials. The rock types include limestone, dolomite, conglomerate, and shale in addition to consolidated sandstone and unconsolidated sand. The petroleum material in tar sands may be solid or semisolid. It is frequently called bitumen. One thing it definitely is not is tar, a product manufactured by destructive distillation. An arbitrary division between tar sands and "heavy oil" has been imposed on the group of petroleum deposits described by the petroleum industry as "dead oil." The "heavy oil" is also highly viscous petroleum which can, however, be produced by inherent reservoir energy. This "heavy oil" primary production is usually not at economical rates. Only an arbitrary viscosity boundary which includes the specification "at reservoir temperature" can divide these two. This article will make no major effort to distinguish between tar sands and heavy oil, because the same production methods are applicable to both materials. Additional names applied to tar sands include oil rock, rock asphalt, bituminous sandstone, impregnated rock, and oil sands, the term now applied in Canada to the huge Alberta deposits.

WORLD RESOURCES

Resource values estimate how much of a material exists. This will be reported here. The term *reserves* means how much of a resource can be produced or recovered under a specific but usually unstated set of production and economic conditions. "Reserves" for oil shale and tar sands are really nonquantifiable, changing continually with technological development and with economic conditions. Consequently no "reserves" estimates for oil shale and tar sands will be included.

Oil Shale

The world's oil shale deposits represent a vast store of fossil energy. Sediments containing organic matter meeting the oil shale definition occur on every continent and range from Cambrian through Tertiary in geologic age. The United Nations study (1967) "Utilization of Oil Shale" plotted organic matter content against the geologic time scale to point out that major maximums of organic matter production and preservation in oil shale occurred during the Devonian, Permian, and Tertiary periods. These maximums appear to correspond to times when major mountain building occurred.

The potential shale oil represented by the world's oil shale deposits was estimated by Duncan and Swanson in 1965. Their evaluations are summarized in Table 12-1. The values given for known resources refer only to evaluated resources. To these values Duncan and Swanson added possible extensions of known resources and geologically based estimates of undiscovered and unappraised resources to obtain their estimate of order-of-magnitude values for the total in-place oil resource in the world's oil shale deposits.

The size of the potential shale oil resource is staggering. Just the richest part of the world's evaluated oil shale resource (9.1×10^{11} barrels, or 1.3×10^{11} metric tons, see Table 12-1) is equivalent to the world's crude oil reserves at the end of 1979 (6.4×10^{11} barrels, or 1×10^{11} metric tons) (Doscher 1981). These petroleum reserves represent only 4 per cent of the projected total resource of rich oil shale. The relative size of the world resource for three grades of shale—rich, medium, and lean—is an excellent indicator of the distribution of richness of oil shale. The rich resource makes up less than 1 per cent of the total resource. The resource estimates in Table 12-1 are separated into three grades according to oil yield, recognizing that the richest deposits are probably more amenable to economic development. Because many factors besides richness affect the economics

Table 12–1. Shale Oil Resources of the World.

	25–100 Gal/Ton Oil Yield		10–25 Gal/Ton Oil Yield		5–10 Gal/Ton Oil Yield	
	Known (10^9 bbl)	Total (10^9 bbl)[a]	Known (10^9 bbl)	Total (10^9 bbl)[a]	Known (10^9 bbl)	Total (10^9 bbl)[a]
Africa	100	4000	Small	80,000	Small	450,000
Asia	90	5500	14	110,000	—[b]	590,000
Australia/New Zealand	Small	1000	1	20,000	—[b]	100,000
Europe	70	1400	6	26,000	—[b]	140,000
North America	600	3000	1600	50,000	2200	260,000
South America	50	2000	750	40,000	—[b]	210,000
Total	910	17,000	2400	325,000	2200	1,750,000

From Duncan and Swanson 1965.
[a] Order-of-magnitude values.
[b] Not estimated.

of development, the grade designations in Table 12-1 have a somewhat limited economic significance.

The units used in Table 12-1 warrant comment. Oil shale resources are evaluated in terms of the oil they potentially can generate. The Fischer assay procedure, now standardized by the American Society for Testing and Materials (ASTM 1980), provides the analytical basis for resource evaluation. While oil yield in weight per cent is also determined by Fischer assay, the United States expresses oil yield in the volume unit of gallons per ton. This unit is converted from weight per cent oil yield by multiplying by a conversion factor, 2.4, and dividing the result by absolute oil specific gravity determined at 15.6 °C. The metric volume unit, liters per metric ton, corresponds to gallons per ton. One gallon per ton is equivalent to 4.17 liters/metric ton. Oil shale resource values are expressed in barrels of oil in the United States. A barrel of oil, the 42-gallon volume unit used in Western petroleum commerce, has no direct equivalent in metric countries. A barrel of oil represents 0.159 kiloliter, or cubic meter. Specification of oil density, a variable, is necessary to convert the volume unit, barrels, into the metric weight trade unit, metric ton. The 1975 World Energy Conference agreed to define a barrel of oil as 0.145 metric ton, and 1 gal/ton × 0.29 as 1 kg/metric ton, approximations that ignore density variations in oil.

Oil shale deposits are relatively common throughout the world. Each deposit is individual, with its own characteristics. Consequently most information on oil shale deposits has been accumulated only locally. Published information tends to appear only in local literature. One major attempt to assemble this mass of local oil shale deposit information into a world report has been published. This Russian compilation, available in English translation (Matveyev 1974), covers the entire world except the U.S.S.R. itself. Although this little book has faults arising primarily from the scope of the problem and from the necessity of assembling the diverse local information (Smith 1977), it presents a broad picture of the world's oil shale resources.

This picture is by no means complete. Most sequences of sedimentary rock contain some materials which could be classed as oil shale. To estimate the order of magnitude of the world's total oil shale resources (Table 12-1), Duncan and Swanson (1965) applied a statistical approach that considered the amount of sedimentary rock in the world and its probable organic matter content. The fossil energy represented by the values in the order-of-magnitude oil shale resource estimates is many times that estimated for the world's coals. Comparison of the values for the known and the estimated oil shale resources in Table 12-1 emphasizes how little is known of the world's oil shale resources. Much less effort has been invested in finding oil shales than in finding oil and gas hydrocarbons. Oil shale deposits became known from outcrops, from unrelated geological investigations, or by accident. Outcropping deposits generally have been detected, but subsurface deposits must be located by drilling. Much oil shale remains to be discovered. For example, Morocco located more than 25 additional oil shale deposits beyond the three that were known in 1975 through comprehensive coring of the country.

Because of intense petroleum exploration, the sedimentary rocks of the United States are rather well described. However, even the U.S. oil shale resource is not totally known. More than 20 per cent of the country's area is underlain by sediments meeting the oil shale definition. Figure 12-1 locates many of the oil shale deposits in the United States. Only the Tertiary Green River Formation in adjoining corners of Colorado, Utah, and Wyoming is reasonably well evaluated. This 16,500-square-mile (43,000-sq-km) area (Figure 12-2) is the world's largest deposit of high-grade oil shale. Smith (1981a) recently concluded that the total Green River Formation resource of all grades of oil shale represented 1,800,000,000,000 barrels (256×10^9 metric tons) of potential shale oil.

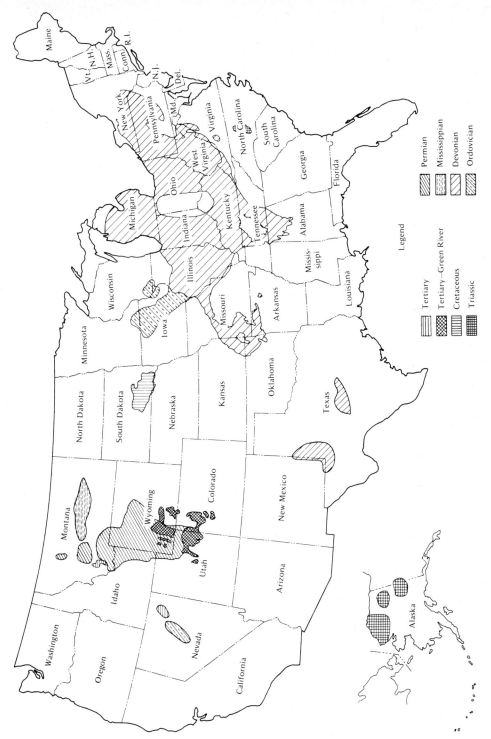

Figure 12-1. U.S. oil shale deposits.

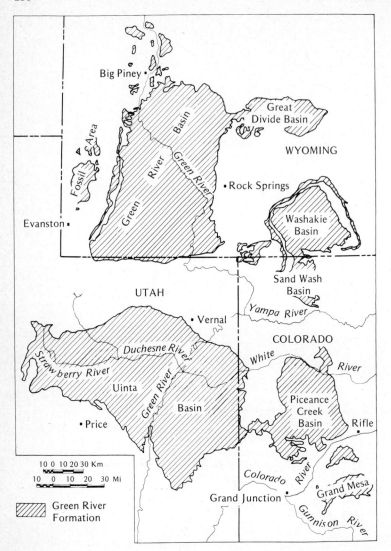

Figure 12-2. Green River Formation.

Total recovery of this resource would represent many years of petroleum supply to the United States. Smith (1981a) points out, however, that stratigraphic and geographic variations in the nature of the Formation's resource preclude complete development by any single method. He also suggests that the United States has perhaps at least another 500 billion barrels (73 billion metric tons) of potential shale oil in lower-grade deposits, but concludes that if the Green River Fromation cannot be developed economically, none of the other deposits can be.

Tar Sands

Tar sand deposits are distributed throughout the world. Estimates of world resources range

Figure 12-3. Alberta tar sand deposits.

posit is actually a group of deposits in Alberta, Canada (Figure 12-3). Together these deposits represent about 1×10^{12} barrels (150×10^9 metric ton) of oil in place (Gray 1981). All of the Alberta deposits in Figure 12-3 are lower Cretaceous in age. The largest deposit, Athabasca, occurs in sands of the McMurray Formation. The Orinoco Heavy Oil Belt in eastern Venezuela is a gigantic petroleum resource not yet well evaluated. Meyer and Dietzman (1979) estimate that the resource in this deposit is as large as the known resources in Alberta. Many smaller deposits are known throughout the world. Of those outside the United States, the largest appear to be the deposits at Bemolanga, Malagasy, and at Olenek, U.S.S.R. These appear to be orders of magnitude smaller than the two giants (Meyer and Dietzman 1979).

Tar sand deposits of the United States are indicated in Figure 12-4. Total resources estimated for the U.S. deposits are summarized in Table 12-2 (Cupps et al. 1979). The bulk of the U.S. tar sand resource occurs in eastern Utah (Figure 12-5). Although Table 12-2 gives a range of potential resources, much of the resources are not adequately defined. For example, the P.R. Spring and Hill Creek deposits are actually one continuous deposit, and recent drilling in the P.R. Spring deposit expanded its known resource by a factor of 80.

from 2.5×10^{12} to 6×10^{12} barrels (4 to 9 $\times$ 10^{11} metric tons) (Gray 1981). Tar sand deposits, like oil shale deposits, have not been actively explored for, and only the obvious ones are known. For example, bitumen has recently been discovered in Paleozoic carbonate rocks beneath the major Canadian deposits in Alberta. Preliminary estimates on this new resource indicate that it is at least as large as that in all the major deposits of Alberta. Even the best known are still being explored. In all probability, more tar sand resources remain to be found.

The world's largest known tar sand de-

RESOURCE CHARACTERISTICS

The total resource of potential oil represented by the world's oil shale and tar sand deposits indicates that massive development might supply sufficient oil to permit our dependence on oil to continue. Complete production of just the Green River Formation oil shale resource or the Athabasca tar sand resource would supply the entire world demand for decades. However, this comforting generalization is a bit more difficult to achieve than it is to state. Characteristics of the two resources help to explain the difficulties involved in such production.

Figure 12-4. U.S. tar sand deposits.

Figure 12-5. Utah tar sand deposits.

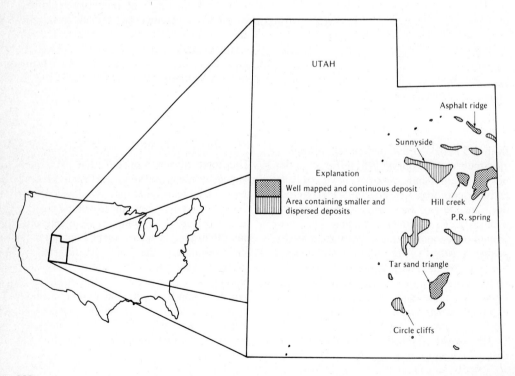

Table 12–2. Tar Sand Resources of the United States.

State and Deposit	Estimated Resources (millions of barrels)	
	Low	High
California		
Edna	141.4	175
Oxnard	565	565
Santa Maria	500	2000
Richfield	40	40
Sisquoc	29	106
South Casmalia	46.4	46.4
North Casmalia	40.0	40.0
Other deposits	16.0	20.2
California total	1407.8	3092.6
Kentucky	33.5	37.3
New Mexico: Santa Rosa	57.2	600
Texas: Uvalde	124.1	3000
Utah		
Tar Sand Triangle	12,504	16,004
P.R. Spring	4000	4500
Sunnyside	3500	4000
Circle Cliffs	1000	1507
Asphalt Ridge	1000	1200
Hill Creek	300	1160
San Rafael Swell	385	470
Asphalt Ridge, Northwest	100	125
Raven Ridge	75	100
Whiterocks	65	125
Wickiup	60	75
Argyle Canyon	50	75
Other deposits	1255	171.3
Utah total	23,164.5	29,512.3
U.S. total	24,787.1	36,242.2

Origin

Oil Shale

Oil shale is very fine grained rock lithified from lacustrine or marine sediments relatively rich in organic matter. Most sedimentary rocks contain small amounts of organic matter, but oil shales usually contain substantially more. Specific geochemical conditions are required to accumulate and preserve organic matter, and these were present in the lakes and oceans whose sediments became oil shale. Garrels and Christ (1965) define these conditions in terms of oxidation-reduction potential (Eh) and acid-base con-

dition (pH) of the water in and around the sediment. Organic matter accumulates under the strongly reducing conditions and neutral or basic pH present in euxinic marine environments and organic-rich saline waters. The organic-rich sediments which became oil shale accumulated slowly in water isolated from the atmosphere. This isolation was achieved by stagnation or stratification of the water body and the accompanying protection of its sediments (Smith and Jensen 1981).

Tar Sands

The geologic setting for accumulation of large tar sand deposits was organic-rich source beds supplying hydrocarbons to porous sediments with a regional cap rock to force the hydrocarbons to flow laterally. The biggest deposits appear to be lodged in deltaic sediments. A homocline with up-dip stratigraphic convergence may have served to aid hydrocarbon accumulation. Mechanisms for turning migrating oil into the heavy residue, still under debate, include bacterial degradation and meteoric water washing. Most of the world's tar sand deposits are in Cretaceous and younger sands. A few, like the Cambrian Siligir deposit in the U.S.S.R. and the Cretaceous Uvalde deposit in the United States, occur in fractured carbonate rocks.

Nature of the Rock

Oil Shale

Because they are very fine-grained rocks, oil shales generally have low porosity and permeability. Many but not all are laminated and fissile as a result of the method of their deposition. The resultant rock tends to be strong and suitable for mining. Quartz, illite, and pyrite (sometimes with marcasite) occur in virtually every oil shale. Clay, smectite, and feldspars are found in many oil shales. Many oil shales contain carbonate minerals, such as calcite, and some, notably the Green River Formation of the United States and the Timahdit deposit of Morocco, contain dolomite as a structural mineral. The oil shale minerals were probably formed by chemical processes related at least in part to the presence of organic matter in the sediment.

Organic matter in oil shale is usually disseminated throughout the rock. The organic matter and the mineral matter in the shales are intimately mixed and rather difficult to separate from each other either physically or chemically. The physical properties of oil shale are strongly influenced by the proportion of organic matter in the rock. The decrease in rock density with increasing organic content illustrates this most graphically. The mineral components have densities of about 2.6 to 2.8 g/cm^3 for silicates and carbonates and 5 g/cm^3 for pyrite, but the density of organic matter is near 1 g/cm^3. Larger fractions of organic matter produce rocks with appreciably lower density. The equation that quantifies this relationship is

$$D_T = \frac{D_A D_B}{A(D_B - D_A) + D_A}$$

Here, D_T = density of the rock, A = weight fraction of organic matter, D_A = average density of organic matter, and D_B = average density of mineral matter (Smith 1969). This relationship is general and applicable to any oil shale.

Behavior of the oil shale rock in mining or processing is affected by the volume of organic matter in it. If oil shale has a continuous phase, it is the organic matter. This fact helps in production of oil from oil shale. On thermal decomposition, much of the organic matter volume disappears from the shale by volatilizing. This permits organic vapor to escape from the rock through the resulting void space. The marked increase in volume fraction of organic matter in oil shale with increasing organic weight fraction can be demonstrated by using the above equation. For example, in Green River Formation oil shale the average density of the organic matter (D_A) is about 1.07 g/cm^3, and the average density of the mineral matter (D_B) is about 2.72 g/cm^3. Thus the density of an oil shale containing 4 weight per cent organic matter

(a lean shale yielding about 2.3 weight per cent oil, or about 6 gallons of oil per ton of shale) is calculated to be 2.56 g/cm^3. The organic matter occupies about 10 volume per cent of this rock. An oil shale containing 15 per cent organic matter (a shale yielding about 10 weight per cent oil, or about 26 gallons per ton) has a density of 2.21 g/cm^3, and the organic matter occupies 31 volume per cent of this rock. An oil shale containing 39 weight per cent organic matter (a relatively rich shale yielding about 20 weight per cent oil, or about 52 gallons per ton) has a density of 1.86 g/cm^3, and the organic matter makes up about 52 volume per cent, or more than half the rock's volume. In the Green River Formation, in oil shales containing 8 weight per cent (about 20 gallons per ton) or more organic matter, the organic material is the largest component by volume, and the physical properties of the organic matter predominate in determining the physical properties of the rock. The organic matter makes this rock tough and resilient and difficult to crush. The richer Green River Formation oil shales tend to deform plastically under load (Smith and Jensen 1981).

When oil shale is heated to about 500 °C, the organic matter in it decomposes to yield condensible oil, water, noncondensible gases, and high-carbon coke. This thermal decomposition is the only economical method now available for producing oil and recovering the energy value of the organic matter in oil shales. This means that the entire oil shale rock must be heated to produce oil, even though a very large fraction of the rock is mineral matter. The thermal load of the mineral matter in oil shale emphatically affects processing economics.

Thermal decomposition behavior of the organic matter in oil shales varies from deposit to deposit. In each deposit, however, the organic matter tends to be the same because the reducing conditions associated with oil shale development digested and homogenized the organic debris. The resulting organic matter (kerogen) is best described as a high-molecular-weight organic mineraloid of

Table 12–3.   Relationship Between Organic-Carbon-to-Organic-Hydrogen Ratio and Conversion of Oil Shale Organic Matter to Oil by Heating.

Deposit Sampled	Carbon-Hydrogen Ratio	Organic Carbon Converted (wt. %)
Pictou County, Nova Scotia	12.8	13
Top Seam, Glen Davis, Australia	11.5	26
New Albany Shale, Kentucky	11.1	33
Ermelo, Transvall, South Africa	9.8	53
Cannel Seam, Glen Davis, Australia	8.4	60
Garfield County, Colorado	7.8	69

indefinite composition. Variations in the hydrogen content of this organic matter are significant because the fraction of organic matter is converted to oil with increasing hydrogen. To illustrate this relationship, Table 12-3 compares the proportion of organic carbon converted and recovered as oil during Fischer assay with the weight ratio of organic carbon to organic hydrogen in several oil shales. For petroleum, the carbon-hydrogen values range from 6.2 to about 7.5; for coal they range upward from 13. The carbon-hydrogen values for organic matter in the world's oil shales range from near petroleum to near coal. Determination of the carbon-to-organic-hydrogen ratio is a difficult analytical process. Organic hydrogen is particularly difficult to determine for this ratio but can be obtained by low-temperature ashing (Smith and Futa 1974). A ^{13}C nuclear magnetic resonance technique for typing oil shales promises to make this evaluation much easier, as illustrated in Figure 12-6 (Miknis and Maciel 1981).

Tar Sands

Tar sands are porous rocks with pores partially or totally filled with bitumen. The nature of the rock matrix containing the heavy petroleum is entirely dependent on the way the rock was formed. Enough of the deposits occur in sand and sandstone that the name *tar sands* is not totally wrong.

Two factors are required to characterize a tar sand resource. The first is the nature of the bitumen. This is extremely heavy oil with absolute densities ranging up to 1.15 g/cm^3. The range in API (American Petroleum Institute) gravity units is from perhaps -10 to 18 as an approximate upper limit. Viscosity of this petroleum in place ranges from high (10,000 centipoise) to ridiculously high (1,000,000+ centipoise), so that the oil cannot move readily through the carrier rock. However, the change of viscosity with temperature is an important characteristic of the bitumen, particularly significant to many production methods. Figure 12-7 illustrates the nature of this change. Most tar sand bitumens become quite fluid at or below 175 °C.

The nature of the porosity in the host rock is a parameter important to oil production from tar sands in place. Stratigraphic distribution of both bitumen and pore saturation defines what is worth working and what can be worked. Horizontal permeability and its directional character with and without oil, functions of pore distribution and saturation, are important to production in place, as is the rarely reported vertical permeability. Methods for determining these properties have not been standardized, and comparing the limited available information is extremely hazardous. One other property of the tar sand rock significant to some surface process methods is the ability (or lack of it) of the bitumen to wet the mineral matter. Like oil shales, tar sand deposits are individuals.

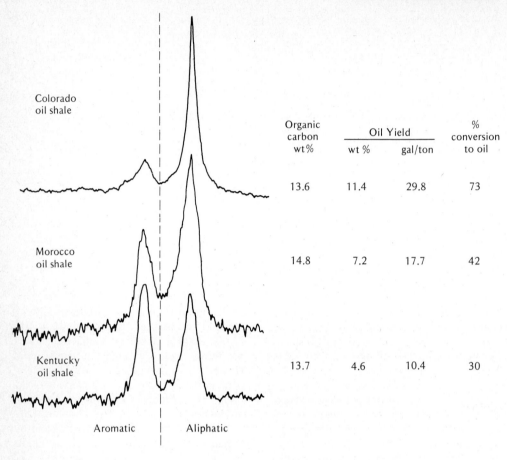

	Organic carbon wt%	Oil Yield		% conversion to oil
		wt %	gal/ton	
Colorado oil shale	13.6	11.4	29.8	73
Morocco oil shale	14.8	7.2	17.7	42
Kentucky oil shale	13.7	4.6	10.4	30

Aromatic   Aliphatic

Figure 12-6. ^{13}C nuclear magnetic resonance response of three shales related to organic-matter conversion to oil.

PRODUCTION METHODS

Oil Shale

The two general approaches to recovering shale oil from oil shales are (1) mining, crushing, and above-ground heating, called conventional processing, and (2) in-place processing. The basic problems facing conventional processing are the handling and heating of huge amounts of low-grade ore and the disposal of huge volumes of spent shale, the residue remaining after oil production. The in-place approach largely avoids the problems of handling and disposal but faces a different basic problem—the impermeability of the oil shale beds.

To make a major contribution to petroleum supply, the conventional processing must operate continuously on a huge scale. Oil shales are lean ores. Mining and heating 1 ton of relatively rich oil shale (25 gallon per ton, 104 liters/metric ton) produces only 0.6 barrel (0.087 metric ton) of oil. To obtain this oil, the ore must be mined, transported, crushed, and heated and the residue disposed of. The residue from heating, called spent shale, is nearly as large in weight and volume as the raw rock after crushing. Min-

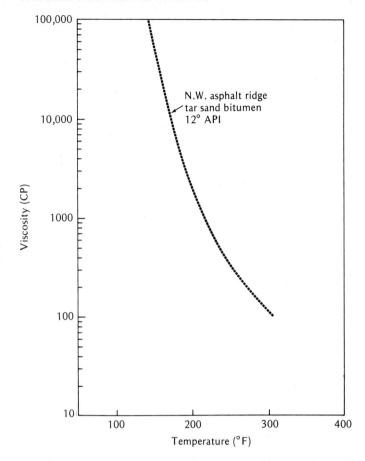

Figure 12-7. Change of viscosity with temperature for a Utah tar sand bitumen.

ing and materials handling on a scale never achieved before is required for oil shale to make a major contribution to the world's oil supply by conventional processing.

In conventional processing, oil shale must be heated to produce oil. This process is called retorting, and the apparatus to accomplish it is called a retort. More than 2500 oil shale retorts have been patented over the years (Klosky 1949, 1958a, 1958b, 1959), and every one will produce oil. In order to reach the throughput in surface retorts necessary to produce oil in major volume, retorting must be done continuously. Two general systems for heating a continuous stream of oil shale are outlined in Figure 12-8. In the internally heated system, the oil shale furnishes its own heat when part of the organic matter is burned inside the retort. Differences in the many tested retorts operating by internal heating are aimed at facilitation of process control and materials handling.

In the externally heated system, fuel burned in an external furnace heats a heat carrier. In one form the heat carrier is the retort itself, operating as a rotary kiln. In other forms the hot heat-carrier material joins the raw shale flow. In directly heated retorting systems, the heat carriers tested include ceramic balls, recycle gas, sand, and spent

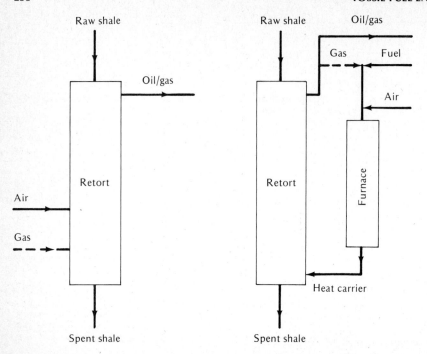

Figure 12-8. Oil shale retorting systems: left, internally heated; right, externally heated.

shale. These carriers circulate either with or counter to the flow of shale. Fuel for heating the exterior furnaces is usually retort off gas, although the carbon of spent shale, part of the product oil, and locally available coal or gas have been considered. Many retorting systems, including both internally and externally heated designs, have been tested on a pilot or semiworks scale. Examples of each are currently operating semicommercially at production rates near 600 metric tons (4000 barrels) per day.

If oil shale could be heated in place, much of the expensive materials handling could be avoided. Research and development efforts toward in-place processing have concentrated on the creation of permeability in the impermeable oil shale. These efforts have taken place primarily in the United States. In-place processing may be accomplished by two means: (1) a borehole technique, in which oil shale is first fractured underground

and heat is applied, and (2) a process in which some rock is first removed by mining, the remaining oil shale is fragmented into the voids created by mining, and heat is applied. These two methods are referred to as in-place (no mining) and modified in-place (some mining) processing. Both of the general heating methods outlined for surface processing (Figure 12-8) have been applied to in-place processing. Although materials handling problems are greatly diminished by in-place methods, process control and product recovery are much more difficult than in surface processing.

Oil produced from oil shale usually requires upgrading to make it easier to transport and refine.

Tar Sands

As with oil shale, tar sand development takes two general forms—conventional development by mining and surface processing, and in-place development. Tar sand mining and surface processing also present a massive materials handling problem. Current conventional development involves strip-mining of the deposit followed by separation of the bitumen from the rock. A variety of separation techniques, including retorting, solvent extraction at normal and elevated temperatures, and detergent stripping, have been proposed and bench-scale tested. However, the only large-scale operations use the Clark hot-water process to separate the tar from the sand. Hot water softens the bitumen and breaks its bonds to the mineral grains. The process is simple, but it requires approximately equal volumes of water and tar sand as feed material. It functions well only if the bitumen saturation exceeds 10 weight per cent (about 25 gallons per ton, or 104 liters/metric ton) and will work only if the native mineral matter is water wet. Before any water can be recycled, finely divided clays must be settled out, a process that requires huge tailings ponds and a very large captive volume of process water. Upgrading of the heavy oil is required to produce a useful product oil.

Most of the world's tar sand resources are located below currently strippable depths. Although hydraulic mining of tar sands is being tested experimentally, subsurface mining has not been seriously tested because of the general weakness of the rock. Consequently, much of the world's tar sand resource may be developed in place only. Various techniques have been tested through pilot-scale operations. Most attempt to lower the oil viscosity by raising its temperature, then encouraging it to flow. This presents several difficult problems. The first is that enough of the essentially immobile oil must be contacted to make the effort worthwhile. Another is that even if the oil becomes fluid, it must be driven from an energyless deposit. Yet another is that subsurface permeability does not follow engineering specifications. Finally, if the liquid oil is driven into cool rock, it resolidifies and plugs its flow path. All of these difficulties vary with individual deposits. Each deposit and perhaps each site will have its own peculiarities of porosity, permeability, oil properties, and oil saturation and distribution. Pilot-scale testing must be carried out on a site-specific basis before any commercial development is practical.

Two general methods for producing oil from tar sands in place are combustion and injection of hot water or steam. Both techniques have their roots in the petroleum industry's secondary and tertiary recovery methods. Figure 12-9 gives a generalized diagram of the steam injection method. Numerous variations of this technique, many of which use additives aimed at lowering and maintaining lowered oil viscosity, have been proposed and tested experimentally. The most straightforward variation, formally titled cyclic steam injection but affectionately known as huff-and-puff, involves the forcing of steam into the formation followed by pressure release. This permits condensed steam to flow back, carrying with it some liquid oil. This procedure's advantage is that it is low in capital cost and can be applied in a single borehole. Its disadvantage is that it contacts only a very limited part of the resource.

General combustion techniques for production of oil from tar sands are shown in Figure 12-10 (Cupps et al. 1976). Forward combustion drives a fire front through the tar sand formation. The oil is heated, partially cracked, and driven toward the production well by the air steam. Coke deposited from cracking fuels the combustion. Creating permeability paths and preventing their being plugged by condensing heavy oil are problems of this procedure. In reverse combustion, a fire front initiated at the production well is designed to travel toward the injection well. Heated and cracked oil must travel through the fire front and the zone of hot rock to reach the production well. This prevents formation plugging. However, maintaining the progress of reverse combustion requires carefully controlled air flow rates (Land et al. 1975, 1977). Variations in formation permeability make this difficult to achieve.

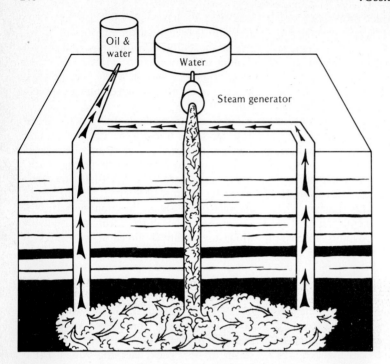

Figure 12-9. Production of oil from tar sands by steam injection.

HISTORICAL AND CURRENT PRODUCTION

Oil Shale

Commercial production of oil from oil shale has a long and checkered history generally controlled by the availability of lower-priced petroleum. For example, a small oil shale industry conducted by more than 50 companies existed in the United States and Canada at the time (1859) Drake discovered petroleum at Titusville, Pennsylvania (Gavin 1922). The oil produced was primarily used for lighting. Cheaper petroleum stopped shale development.

In special situations where other fuels were in short or uncertain supply or where energy transportation was difficult, energy development from oil shale has been carried out commercially. A French operation initiated

in 1838 is probably the earliest commercial energy development recorded, and Scotland, Canada, and Australia produced shale oil commercially before 1870. From 1875 to 1960, energy equivalent to about 250,000,000 barrels (3.1×10^7 metric tons) of oil was produced from Europe's oil shale deposits. Most of this production was derived from deposits in Estonia and Scotland. Low-priced oil from the Near East and improving oil transport systems stopped most oil shale developments.

World War II caused sharp increases in petroleum demand and disrupted both petroleum production and petroleum distribution, reactivating interest in oil shale development. Oil shale production operations during and since World War II have been conducted in Germany, France, Spain, Manchuria (China), Estonia and other areas of the So-

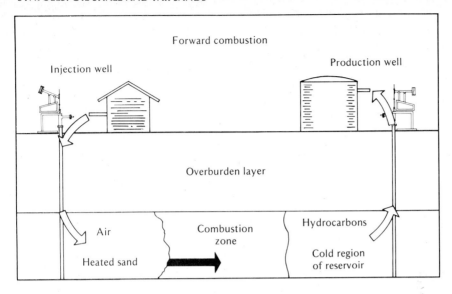

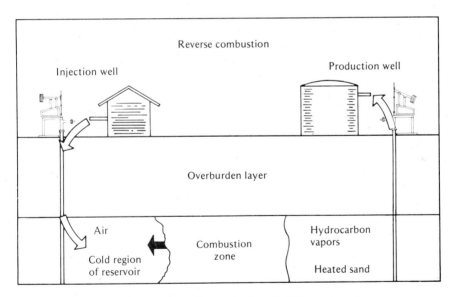

Figure 12-10. Combustion techniques for production of oil from tar sands.

viet Union, Sweden, Scotland, South Africa, Australia, and Brazil.

Australia operated an oil shale plant at Glen Davis, New South Wales, during World War II. Problems associated with the mining of thin seams in competition with natural oil closed this plant about 1950. France pioneered destructive distillation of oil shale at Autun and operated plants on three other deposits after World War II. All had ceased to operate by 1957. Germany operated oil shale plants on the Jurassic deposits in Wurttemberg during World War II, but these developments did not survive postwar economics. Scotland, an oil shale pioneer, continued to produce shale oil by mining and processing the Carboniferous Lothians oil shale deposits until 1963, having exhausted their richer shale. In South Africa, the South African Torbanite Mining and Refining Company, Ltd., began operations on a deposit near Ermelo, Transvaal, in 1935. This grew into a large-scale operation that exhausted the 20 to 30-million-ton oil shale deposit by 1962. This was a truly commercial development, privately financed and paying dividends. An integrated company operating at Puertollano in Spain's Ciudad Real Province produced gasoline, diesel and fuel oils, lubricants, and other by-products on a small scale starting about 1922. An enlarged company created in 1942 by the National Institute of Spain built a new installation which incorporated a low-temperature hydrogenation plant to upgrade shale oil. Production was discontinued about 1960. The Swedish government built a large plant at Kvarntorp in Narke Province during World War II to produce oil from the Alum black shale. This plant was in full production by 1947 but closed in the 1950s. In conjunction with this plant, Sweden tested underground gasification of oil shale by electrical heating. In this procedure, known as the Ljungstrom method, hydroelectric power available during times of low demand was used. Its proponents claimed that the calorific value of the oil and gas vaporized was about three times that of the energy used to produce them. The availability of low-prices oil from the Persian Gulf stopped these developments.

Two countries are currently operating major oil shale developments. These are China and the U.S.S.R. Oil shale operations have been conducted in China for more than 50 years. The initial development was near Fushun in the northeastern section of China, formerly called Manchuria. In this area, a thick (450-ft, 150-m) deposit of low-grade oil shale (15 gallons/ton, 63 liters/metric) overlies a thick seam of Tertiary coal. The oil shale was removed as overburden in order to open-pit mine the coal. An oil shale industry grew up around the processing of the raw oil shale by-product. A vertical combustion kiln called the Fushun retort has been in successful operation since 1930. Two plants at Fushun use this retort to produce oil. A similar plant using the Fushun retort has been established on a shallow, strippable deposit of Eocene oil shale in central China. Although the Chinese have experimented with other retorts and production methods, only the relatively simple Fushun retort has given satisfactory service. Total 1980 production from all three plants was only about 6000 barrels (1000 metric tons) per day. The Chinese feel that the shale oil competes successfully with foreign natural petroleum.

The U.S.S.R. is mining Ordovician oil shale by stripping shallow beds centering in Estonia. Some subsurface mining is also conducted. The rich oil shale occurs in several thin beds with an aggregate thickness of about 10 feet (3.3 meters) or 50-gallon/ton (210-liter/metric ton) shale. About 70 per cent of the 30×10^6 tons per year being mined (1980) from these beds is burned directly to produce electric power. Shale oil is also produced in two different styles of retorts. The Kiviter retort is a directly heated vertical kiln requiring use of shale chunks. The Galoter retort processes fine shale, burns the residual carbon on the spent shale in an exterior furnace, and transfers the heat to the raw shale, using the shale ash as the heat carrier. Oil production from these retorts has been relatively small. The Galoter operates as a production retort, processing 500 metric tons of shale and producing about 100 metric tons (650 barrels) per day. The Kiviter retort has been tested at production about 70 per cent as large.

Brazil currently has the largest oil shale pilot plant. A Brazilian company, Petrobras, has been operating the internally heated Petrosix retort with a production capacity of 50,000 metric tons (350,000 barrels) of oil per year. Raw material for this plant is the Permian Irati shale. In the pilot-plant area, the oil shale, about 10 meters thick, occurs in two layers separated by a limestone bed. The upper shale layer, about 6 meters thick, yields about 6.5 weight per cent (17 gallons/ton) oil. The lower layer, about 4 meters thick, yields about 9.4 weight per cent (25 gallons/ton) oil. In 1980 Brazil announced a decision to begin development of a 50,000-barrel/day (8000 metric tons) shale oil production complex.

The history of U.S. oil shale shows the boom-and-bust patterns generated by oil availability. During the 1920s a promotional boom centered on Green River Formation oil shale. Many of the promotional techniques used then sound valid today—oil shale can provide oil. The development of the East Texas oil field stopped this. Although many major developments of Green River oil shale have been announced over the years, none has reached production. In 1980, plans and investments were again on the table. The Colorado natives, inured to the promotional talk, say they'll believe it when they see it.

Tar Sands

Tar sand development has only a short history. North American Indians 200 years ago used Athabasca tar to seal seams on canoes. The primary historical use of tar sands has been in the building of roads. In the town of Vernal, Utah, every driveway is paved with tar sands from the Asphalt Ridge deposit. The Uvalde deposit in Texas has more than 20 quarries expressly developed to produce road-paving material.

Production of oil from tar sands on a commercial scale was initiated in 1967 on the Athabasca deposit of Alberta. Great Canadian Oil Sands (now Suncor), a subsidiary of Sun Oil, initiated strip mining of the deposit. The mine-run material fed to extraction averages about 10 weight per cent bitumen, 85 per cent sand and clay, and 5 per cent water. Hot-water extraction recovers 90 per cent or more of the bitumen, which is converted to lighter products by delayed coking. Suncor's current production is 45,000 barrels/day (6500 metric tons), but they are in the process of increasing capacity to reach 58,000 barrels/day (8400 metric tons) by 1982. Syncrude Canada, Ltd., constructed their immense project on an adjacent lease. They also are strip mining. They initiated production in mid-1978. By July 1979 they had reached production of 100,000 barrels/day (15,000 metric tons) building toward a permit level of 129,000 barrels/day (19,000 metric tons). With a peak mining rate of 300,000 tons/day the Syncrude project is currently the largest mining operation in the world. Gray (1981) estimates that the strip-mineable area of the Athabasca deposit at depths of 45 meters or less represents 60 billion barrels (9 billion metric tons) of in-place bitumen. He estimates that this resource may ultimately support five or more surface mining plants operating at the same time. In 1979, Alsands, a consortium of nine major oil companies, received preliminary regulatory approval to construct a similar plant by 1986 to produce 140,000 barrels (20,000 metric tons) per day.

The U.S. resource has almost no strippable tar and sand deposits. None of the deposits look large enough to support such large scale operations.

BY-PRODUCT POSSIBILITIES

Both oil shale and tar sand developments currently in progress require the handling of massive amounts of material. Utilization of any of this material would assist the handling problem. However, the mineral material generated is usually useful only for filling holes, or building mountains if you run out of holes. Returning spent oil shale to subsurface mines may permit recovery of a larger fraction of the resource. Some ammonia may be produced from oil shale. Some small operations have managed to use spent shale to make cement. However, no consumption pattern appears capable of absorbing even a minute fraction of the vast amount of waste material. Only the saline mineral section of Colorado's Green River Formation, which offers production of alumina, natural sodium bicarbonate (nahcolite), and soda ash in addition to oil (Smith 1981b), appears to offer any significant product value. Development of oil shale and tar sands on a massive scale must depend on the value of the oil produced.

ENVIRONMENTAL AND SOCIOECONOMIC FACTORS

The environmental effects of oil shale or tar sand processing differ from other energy-producing processes only in scale. Oil shale and tar sand development tend to concentrate and localize the environmental effects because of the huge scale of the production plants. Petroleum wells must be dispersed because of the nature of the production technique. Consequently, their visibility is minimized. The concentration of production from oil shale and tar sands intensified the visibility of environmental effects but eases their management. Landscape, wildlife, and vegetation disturbances; air quality degradation; and water pollution are all negative possibilities. Careful management can minimize or mitigate these effects. The primary environmental area where oil shale and tar sand processing differs from most other fossil-fuel processes is in the amount of spent ore that must be managed. Every ton of shale oil produced by surface methods leaves 10 tons of solid waste. Every ton of tar sand bitumen also leaves about 10 tons of residual solid. Even in-place processing leaves material that must be managed to minimize subsequent leaching of undesirable effluents. Surface disposal of tar sand residual sand requires settling ponds. Surface disposal of the residual material from both oil shale and tar sand processing will usually require a landfill arrangement designed to minimize water flow through the residue, coupled with revegetation. If oil shale is mined subsurfacely, part of the solid waste may be disposed of as mine fill.

Water requirements have frequently been cited as a major deterrent to oil shale and tar sand development. Both major tar sand resources have copious water, but the Utah tar sands of the United States are in arid lands. The supply of water for oil shale development has frequently been viewed as a major limiting factor. However, water requirements can be modified technically. For example, in pilot-plant operations in Colorado, the Tosco process apparently requires three barrels of water per barrel of produced oil; the Paraho retort process requires only 0.5 barrel; and some in-place combustion processes must clean up product water for discard. Water requirements are a technical and economic challenge, not an absolute limiting factor.

The three major synfuel energy resources—the Green River Formation, the Athabasca oil sands, and the Orinoco heavy oil deposits—all occur in areas of limited population. This requires the importation and establishment of an adequate work force. New communities must be constructed and utility corridors developed and established. This will create both work opportunities and substantial dislocations in the local population. The establishment and financing of a comprehensive plan to accommodate new population and to ease local dislocations are the primary socioeconomic requirements of oil shale and tar sand development. Scrimgeour (1980) outlined a plan for accomplishing this for Green River Formation oil shale development. Handling socioeconomic effects represents a significant but not prohibitive cost to development.

Each individual process applied to individual deposits of oil shale or tar sands in the world will generate a specific set of environmental problems. Environmental effects and their risk factors must be evaluated individually for each development in the planning of appropriate mitigation measures.

DEVELOPMENT COSTS

Tar Sands

Development costs and their escalation are available for Alberta tar sands, where a development record exists. Suncor started production in 1967 as pioneers. Their initial installation costs, postulated at $200 to $250 million, escalated dramatically as they progressed through a learning cycle. The final installed cost was substantially larger than the initial estimates. Pioneer operations frequently incur learning expenses, but Suncor is in the process of expanding their production from 45,000 to 58,000 barrels/day (6500 to 8400 metric tons). Syncrude Canada, Ltd., started operation in 1978 on their $2.5 billion strip-mining installation and are nearing production permit level of 129,000 barrels/day (19,000 metric tons). Alsands, a group of nine companies, has applied for permits to develop strip-mine production in another area of the Athabasca deposit. They plan a daily production of 140,000 barrels (20,000 metric tons) by 1986 at a capital cost of about $5 billion. Operating costs for a 45-million-barrel/year (7 million metric tons) plant are estimated at $300 million.

Pilot and experimental programs are in progress to recover bitumen from Alberta deposits which cannot be strip mined. Several of these are producing oil, one to the level of 2 million barrels (300,000 metric tons) per year. The primary effect of these will be to increase the developable resource. Major cost reductions are not expected.

Oil Shale

Estimation of installation costs for production of oil from oil shale has less to go on than the existing tar sand production in Canada. Meaningful costs are difficult to obtain from existing production in China and Estonia. However, that production is continuing. Cost estimates for production of shale oil in Colorado have a remarkable history of escalation. The estimated price of a barrel of shale oil newly produced has scooted along just above world petroleum prices for more than 30 years. Plant cost estimates for production of 50,000 barrels (7200 metric tons) of shale oil per day by mining and surface retorting are about $1.5 billion and climbing.

These costs are comparable to costs for production of oil from tar sands.

Some experimental and pilot programs indicate that significant cost reductions may be possible in specific cases. These projects are techniques aimed at production from the oil shale in place. One mining proposal, initially postulated by Ertl (1948, 1965) and reevaluated by Lewis (1980), also offers a significant possibility of cost decrease and maximum resource recovery. This proposal involves strip mining of the entire 1200-square-mile (31,000-sq-km) Piceance Creek Basin of Colorado. The inability of government agencies to handle such a huge concept makes its application unlikely.

Costs for production of oil from other oil shales will be larger because richness and organic-content conversion factors for oil yield are lower than those for the Green River Formation.

GOVERNMENT AND POLITICAL FACTORS

Any operation as large as oil shale or tar sand production draws politicians and government regulators as efficiently as a dead mule draws flies. Myriads of regulations result, few of which stem from a comprehensive current view of world energy problems or adapt to current production problems. The sheer mass of these forms a deterrent to development, and a few of them create major barriers.

Oil Shale

Because commercial production of oil from Green River Formation oil shales has been considered imminent for 60 years, regulations have clustered around oil shale production. However, the most onerous governmental regulation imposed on this immense fossil-energy deposit is its withdrawal from leasing. In April 1930, President Herbert Hoover signed Executive Order 5327, "temporarily" withdrawing federally owned oil shale from leasing or "other disposal" pending evaluation of the resource. This withdrawal is still in effect! It places on top of perfectly functional land management regulations an unnecessary additional control. Because the U.S. government owns more than 80 per cent of the Green River Formation resource, including the thickest and richest deposits, oil shale is largely not available to development. Experimental development of a new resource requires two land positions—a place to test an idea and a resource for development if experimentation is successful. These have not been available under the withdrawal order. The U.S. government has operated the withdrawal order to direct and control oil shale development efforts. Exceptions to the withdrawal have been made by modification of the original executive order by add-on executive orders. Four tracts for commercial shale oil production were leased by the U.S. government in 1974 by bonus bidding. The current federal attitude is that no more oil shale land will be leased until these prototype tracts are in production. Production on these tracts is still pending. One additional limitation imposed on oil shale development by federal land policy is the 8-square-mile (21-sq-km) lease size. This may curb major installations. Of the privately held oil shale land, the largest block belongs to the Union Pacific Railroad along their rail route in Wyoming. This shale is not optimum for initial development. Seven major oil companies hold most of the rest of the fee land. Other developers are effectively shut out. Although the total resource is large enough to support massive shale oil production, legal availability may be a major deterrent to development.

The environmental and socioeconomic impacts of oil shale operations also complicate oil shale development. It has been estimated that in the United States at least seven years is required between the time a lease is offered and the time mature production is reached. Much of this time is spent preparing detailed environmental impact statements; obtaining required federal, state, and local licenses and permits; and preparing to meet housing and service requirements.

Competition between federal and pro-

vincial or state (for example, Colorado) agencies for the right to regulate shale production provides another source of compounding regulation. This and the other governmental factors described above tend to make potential developers very nervous about the huge investments required.

Tar Sands

In Canada, the Province of Alberta maintains an absolute right to control production quantities and both the Alberta and the Canadian government maintain an absolute right to set product prices. Initially conceived as ways to protect the position of existing producers and consumers, these regulations are stifling tar sand development in Canada. The export of product oil to the United States has been periodically controlled or banned, and controlling ownership of each project is required to be with Canadian companies. In addition, Alberta is in contest with the Canadian central government on rights to regulation. Such political discomfort discourages multi-billion-dollar investments.

In the United States, the largest of the tar sand deposits are federally owned. Although no "tar sand withdrawal" analogous to the oil shale withdrawal exists, the federal government maintained for years that tar sands could be leased separately from petroleum. Because tar sands are petroleum, this created a definition contest which ultimately stopped tar sand leasing for years. This problem was solved in 1981 by Congress who cleverly enacted a law that declared that tar sand was petroleum! But 15 years of experimental opportunity had disappeared. The largest of the U.S. tar sand deposits, the Tar Sand Triangle, lies in a remote area of southeastern Utah. Wilderness area and federal recreation area definitions severely restrict major development of this deposit.

CAN OIL SHALE AND TAR SANDS BUY TIME?

Resources available for oil production from oil shale and tar sands form no limit of themselves to the time extension available from their development. Resource development always proceeds from the easiest, richest, and cheapest toward the more difficult as the more accessible materials are exhausted. The three huge and best-known deposits—the Green River Formation oil shale, the Althabasca tar sand group, and the Orinoco heavy oils—offer an oil resource in place of 4×10^{12} barrels $(6 \times 10^{11}$ metric tons). This is twice the expected ultimate world production of petroleum. Hubbert and Root (1981) forecast that the middle 80 per cent of this ultimate petroleum supply will be consumed in the 60-year period between 1967 and 2027. By then the oil production rate must have fallen to about 11×10^9 barrels $(1.6 \times 10^9$ metric tons) per year from the current 22×10^9 barrels $(3 \times 10^9$ metric tons) per year. Enhanced recovery of petroleum from known oil fields is the best supplementary oil source. However, Doscher (1981) points out the limiting factors on this production. Any petroleum contribution created from oil shale and tar sands can help to make up that production deficit. Production of 10 per cent of these resources over a 40-year period would compensate for the deficit.

The development of new resources is a learning process, which takes time. Hubbert and Root (1981) indicate that world production of petroleum will continue to rise, peaking at 36×10^9 barrels $(5 \times 10^9$ metric tons) per year in 1997, given an orderly evolution. Adequate time for synfuels development is available if the earth were a unified whole. It is not. At the time of this projected peak in world production, the United States will have produced 90 per cent of its producible oil, and costs of additional domestic production will skyrocket. Energy supply has become a political weapon to which the United States will become increasingly vulnerable without supplementary oil production from synfuels. Consequently, the United States has less time than the rest of the world to develop synfuels production as a useful petroleum supply.

The size of the development required to provide significant amounts of oil is startling. Oil shale plants (not yet in operation)

on the Green River Formation are usually planned at 50,000 barrels (7250 metric tons) per day. This is a nice number, selected because it is big and round. Two plants providing this amount of oil each would double Colorado's current petroleum production. But it would take 120 plants this size to supply the 6×10^6 barrels (870,000 metric tons) of oil currently (1981) being imported daily to the United States. However, any contribution is worthwhile. It has been estimated that one job disappears for each $10,000 that leaves the United States in payment for oil. One oil shale plant producing 50,000 barrels (7250 metric tons) of oil per day means that 65,000 jobs per year stay home. Money is said to circulate 18 times per year in the gross national product if it stays in the United States. One oil shale plant might raise the U.S. gross national product $11 billion per year.

Alberta's tar sand developments indicate recognition of the value of production. Current and proposed developments are expected to raise annual production to 350 million barrels $(50 \times 10^6$ metric tons) by the 1990s. This is half of Canada's current consumption and about 16 per cent of current U.S. imports! This demonstrates that continuous progress toward production can generate significant production volumes. It does not all have to be done instantaneously.

Capital costs of developing oil shale and tar sands are also startling. The 50,000-barrel/day (7250 metric tons) oil shale plant is (in 1981) estimated to cost $1.5 billion to start up. These costs are rising insistently with inflation. A Canadian tar sand plant planned to produce 140,000 barrels (20,300 metric tons) of oil per day from sands in place was estimated in 1979 to require $4.7 billion dollars for capital and start-up costs. Reports in 1981 place these costs closer to $11 billion. At $1.5 billion per 50,000-barrel/day (7250 metric tons) oil shale plant, the total current investment required to generate the 120 oil shale plants needed to provide the 6 million barrel (870,000 metric tons) daily imports of the United States is $180 billion. But acceptance of this cost is not a magic answer, because the developments cannot be done simultaneously. It is prevented by the sheer volume of the industrial equipment required.

Nonacceptance of these costs creates a search for other options. One is to plan, organize, and develop a satisfactory substitute energy system. This will also entail capital costs. In fact, it will entail capital costs at a level undreamed of and totally unevaluated by government officials blithely shuttling energy equivalency numbers. Not only are major capital costs necessary to a new energy mix, but two additional capital costs appear that are not part of the extension of the oil supply with oil from oil shale and tar sands. These are distribution costs and consumption costs. Since transportation is a major consumer of oil, the alcohol-fueled car furnishes a good example of these. First, all automobiles and trucks would require replacement at least of their engines. A completely new manufacturing system for engines would be required together with parts, supplies, new repair manuals, and development and testing of engine design. Then an alcohol distribution system must be built capable of providing year-round fuel to engine specifications country or worldwide in all climates and weather conditions. Prevention of water absorption by alcohol and control of its evaporation are additional problems requiring solution. These costs are on top of the capital costs required to develop an alcohol supply large enough to provide the requirements of the population. If you think this is exaggerated, try a similar analysis with coal, electricity, methanol (wood alcohol), or (heaven help us) hydrogen, all of which have been nominated as replacements for petroleum used in transportation energy. In addition, these alterations also must be accomplished in 30 years. This is only one of many petroleum applications, all met only by oil from whatever source. The costs for oil shale and tar sand development should now appear more reasonable.

Costs for petroleum production are relatively low now, particularly in the Near East. But these costs rise with diminishing supply. Hubbert and Root (1981) predict that 90 per cent of the U.S. total petroleum supply will have been produced by 1997. As this happens, production costs escalate and demand exceeds supply. As the costs for production of petroleum rise, oil shale and tar sand production will become economical. In a free market, oil production from Canada's tar sands are economical now. Production costs for oil shale and tar sands have consistently risen with and stayed above world oil prices because oil controls the cost of many of the materials required. As these additional oil sources begin to produce oil, the cost tie with petroleum will weaken.

Most of the objections to development of oil shale and tar sands, particularly spent shale disposal, water requirements, and environmental effects, are technical problems. These yield to planning and experience and are not a major factor in the determination of the value of synfuels development. Governmental controls, however, are severe limits to the necessary development. Elected officials are not normally conditioned to assimilate the larger view required to understand the energy problem the human race faces. Petty political bickering creates an extreme danger because of the short time available for solutions.

If oil shale and tar sands cannot be developed, there are still a couple of options. One is continuous, fervent prayer that an additional 2 trillion barrels (300 billion metric tons) of petroleum be discovered or that practical nuclear fusion be achieved tomorrow. The other is to accept the rising cost of oil by lowering oil consumption and largely discontinuing its use by 2015–20. This will produce declining productivity and, ultimately, decay and regression of civilization.

In his article "Exploring Energy Choices," Doscher (1981) summarized the problem as follows:

The availability and cost of energy and the ultimate limits on its use pose grave questions for the well-being of humankind. The energy crisis is a crisis which far transcends any local and temporary implications. The whole energy question poses a challenge of the most fundamental sort to world social structure and to society's ability to make a new technological and scientific response.

Development of oil production from oil shale and tar sands will not solve civilization's energy problems forever. But it may buy time enough to find a solution.

REFERENCES

American Society for Testing and Materials, 1980. Standard Test Method for Oil from Oil Shale (Resource Evaluation by the USBM Fischer Assay Procedure), Method D 3904-80.

Cupps, C. Q., C. S. Land, and L. C. Marchant, 1976. "Field Experiment of *In Situ* Recovery from a Utah Tar Sand by Reverse Combustion" in J. W. Smith and M. T. Atwood, eds., AIChE Symposium Series 155, Vol. 72, *Oil Shale and Tar Sands*. New York: American Institute of Chemical Engineers.

Cupps, C. Q., L. C. Marchant, and J. J. Stosur, 1979. "Tapping the Vast Tar Sand Reserves of the U.S.," *World Oil* **189** (4), 73.

Doscher, Todd, 1981. "Petroleum Enhanced Recovery" (p. 514), "Petroleum Reserves" (p. 529), and "Exploring Energy Choices" (p. 13) in S. P. Parker, ed., *McGraw-Hill Encyclopedia of Energy*. New York: McGraw-Hill.

Duncan, D. C., and V. E. Swanson, 1965. *Organic Rich Shale of the United States and World Land Areas*. Washington, D.C.: U.S. Geological Survey Circular 523.

Ertl, Tell, 1948. "Mining of Colorado Oil Shale," *Oil and Gas Journal* **47** (24), 116.

Ertl, Tell, 1965. "Mining Colorado Oil Shale," Proceedings of the 2nd Oil Shale Symposium, *Colorado School of Mines Quarterly* **60** (3), 83.

Garrels, R. M., and C. L. Christ, 1965. *Solutions, Minerals and Equilibria*. New York: Harper and Brothers.

Gavin, Martin J., 1922. *Oil Shale, An Historical, Technical, and Economic Study*, U.S. Bureau of Mines Bulletin 210.

Gray, G. R., 1981. "Oil Mining" (p. 488) and "Oil Sand" (p. 492) in S. Parker, ed., *McGraw-Hill Encyclopedia of Energy*, 2nd edn. New York: McGraw-Hill.

Hubbert, M. K., 1956. *Nuclear Energy and the Fossil Fuels: Drilling and Production Practice*. American Petroleum Institute.

Hubbert, M. K., and D. H. Root, 1981. "Outlook for Fuel Reserves" (p. 45) in S. P. Parker, ed., *McGraw-Hill Encyclopedia of Energy*, 2nd edn., New York: McGraw-Hill.

Klosky, S., 1949. *An Index of Oil Shale Patents*, U.S. Bureau of Mines Bulletin 468.

Klosky, S., 1958a. *Index of Oil Shale and Shale Oil Patents, 1946–56, Part I: U.S. Patents*, U.S. Bureau of Mines Bulletin 574.

Klosky, S., 1958b. *Index of Oil Shale and Shale Oil Patents, 1946–56, Part II: United Kingdom Patents*, U.S. Bureau of Mines Bulletin 574.

Klosky, S., 1959. *Index of Oil Shale and Shale Oil Patents, 1946–56, Part III: European Patents and Classification*, U.S. Bureau of Mines Bulletin 574.

Land, C. S., F. M. Carlson, and C. Q. Cupps, 1975. *Laboratory Investigation of Reverse Combustion in Two Utah Tar Sands*. ERDA/LERC/RI-75/2.

Land, C. S., C. Q. Cupps, L. C. Marchant, and F. M. Carlson, 1977. *Field Test of Reverse Combustion Oil Recovery from a Utah Tar Sand*. ERDA/LERC/RI-77/5.

Lewis, Arthur E., 1980. "Oil Shale: A Framework for Development," *13th Oil Shale Symposium Proceedings*, pp. 232–7. Golden, Colo.: Colorado School of Mines Press.

Matveyev, A. K., ed., 1974. *Deposits of Fossil Fuels: Oil Shales Outside the Soviet Union*. Boston: G. K. Hall (English translation).

Meyer, R. F., and W. D. Dietzman, 1979. *World Geography of Heavy Crude Oils*. Report 68, United Nations Institute of Training and Research (UNITAR), 1st International Conference on the Future of Heavy Crude and Tar Sands.

Miknis, F. P., and G. E. Maciel, 1981, "C-13 NMR Studies of Oil Shale Evaluation and Processing," *14th Oil Shale Symposium Proceedings*, pp. 270–80. Golden, Colo.: Colorado School of Mines Press.

Russell, P. L., 1980. *History of Western Oil Shale*. East Brunswick, N.J.: Center for Professional Advancement.

Scrimgeour, Donald P., 1980. "Mitigating the Social and Economic Impacts of Oil Shale Development," *13th Oil Shale Symposium Proceedings*, pp. 238–45. Golden, Colo.: Colorado School of Mines Press.

Smith, John Ward, 1969. *Theoretical Relationship Between Density and Oil Yield for Oil Shales*, U.S. Bureau of Mines Report of Investigations 7248.

Smith, John Ward, 1977. *Sedimentary Geology* **19**, 75.

Smith, John Ward, 1980. *Oil Shale Resources of the United States*, Colorado School of Mines Resource and Energy Resources Series, Vol. 23, No. 6.

Smith, John Ward, 1981. "Alumina from Oil Shale," *Mining Engineering* **33** (6), 693.

Smith, J. W., and K. Futa, 1974. "Direct Determination of Organic Hydrogen in Oil Shales by Low Temperature Ashing," *Chem. Geol.* **14** (1/2), 45.

Smith, J. W., and H. B. Jensen, 1977. "Oil Shale" in Daniel Lapedes, ed., *McGraw-Hill Encyclopedia of Science and Technology*, Vol. 9, 4th ed., p. 352. New York: McGraw-Hill.

Smith, J. W., and H. B. Jensen, 1981. "Oil Shale" in S. P. Parker, ed., *McGraw-Hill Encyclopedia of Energy* (p. 494). New York: McGraw-Hill.

United Nations, 1967. *Utilization of Oil Shale: Progress and Prospects*, U.N. Publication No 67.II.B.20. New York: United Nations.

Atmospheric Carbon Dioxide: What To Do?   13

GREGG MARLAND AND RALPH ROTTY

Industrialized society has reached a scale where it can have a major impact on the regional and global environment. As we begin to understand and to appreciate the character and magnitude of the various impacts, society is forced to make judgments on its priorities and on how to deal with the great variety of assaults on the environment. The way in which these environmental threats are dealt with depends on many factors:

a. How certain are we of the consequences?
b. Are they local, regional, or global in scale?
c. Are the adverse impacts suffered by the same people who reap the benefits (in time and space)?
d. Are they clearly apparent and widely understood within common experience or more obtuse and "mysterious"?
e. Are they cumulative or not, persistent or transient?
f. Are they irreversible or susceptible to correction?
g. To what extent are they linked to activities of major social or economic importance?
h. Can we deal with them conclusively and, if so, at what cost?
i. How serious do we judge the consequences of not addressing them?

There are some issues (e.g., smoking) where the consequences are largely personal and society has chosen to leave many key decisions to the individual. There are others (e.g., trace metals) where the consequences have been judged sufficiently widespread and serious that we have moved promptly to try to eliminate the assault. On other issues (e.g.,

Dr. Gregg Marland is a senior scientist at the Institute for Energy Analysis, Oak Ridge Associated Universities, Oak Ridge, Tennessee 37830. His research work has included geologic aspects of energy resource evaluations, especially fossil fuels and geothermal energy; atmosphere-ocean interactions; and environmental and marine chemistry.

Dr. Ralph M. Rotty is a senior scientist at the Institute for Energy Analysis, Oak Ridge Associated Universities, Oak Ridge, Tennessee 37830. His scientific work has included studies of future energy supplies, energy policy, and especially energy-related impact on regional and global climate.

From Consensus 1980 (2/3), 41 (Fall 1980). Reprinted by permission of the authors and Consensus, published by Studiecentrum Voor Kernenergie, S.C.K./C.E.N., B-2400, Mol, Belgium.

atmospheric SO_2) the effects are poorly defined but largely reversible (that is, the atmosphere would clear itself rather quickly if emissions ceased), and we are adjusting our response as the affected area broadens and the consequences become more acute. One can even cite issues (e.g., fluorocarbons from spray cans) where the environmental consequences were shrouded in uncertainty but we chose to act because the social and economic consequences of action were seen as small compared to the perceived danger (Dotto and Schiff 1978).

As the general pressure on the environment increases, we are being confronted more and more with serious conflicts between economic and environmental goals and between the need for early commitments and the lack of full understanding of the environmental interactions. (We realize, too, that continuing with the status quo is as much a commitment as is any corrective program we might choose to pursue.) Retrospective studies will show that the best decisions have not always been made, but in making decisions we must balance the uncertainties of the impact against the risks associated with making the wrong commitment.

A FUNDAMENTAL CONFLICT

It is hard to envision a more fundamental conflict of priorities than one in which an essential by-product of the world's principal energy production system is seen as a potential threat to the global climate. Currently, almost 85 per cent of global primary energy is generated by the combustion of fossil fuels (coal, oil, and natural gas), and carbon dioxide (CO_2) is discharged at the rate of 19.1×10^9 tons (19.1×10^{12} kg) per year (Rotty 1977). The rising concentration of carbon dioxide in the earth's atmosphere is developing as a test of our ability to confront a potentially important environmental issue in the face of great uncertainty and great social and economic implications. Cooper (1978) has already suggested that, given the short-term economic and social consequences of

prohibiting fossil-fuel use, society will be unable to reach such a decision and will be left to adjust to whatever climate change occurs. Dyson and Marland (1979) had pointed out earlier that the commitment for action would be confounded by the scale of global cooperation required and by the likelihood of unequal distribution of impacts.

Atmospheric carbon dioxide has not yet figured in major policy decisions, although the words are well embedded in the vocabulary of most energy analysts. The issue was raised as an argument against pursuing U.S. President Jimmy Carter's massive program for synthetic fuels development, although in this particular case the argument seems to have been rather quickly defused by recognition that the incremental effect on atmospheric CO_2 is relatively small (L. Carter 1979, Maize 1979).

It is essential that society make a major effort to understand all aspects and consequences of the ''carbon dioxide issue,'' but we must also recognize the uncertainties that remain and avoid arbitrary responses. In this article, we outline the issues surrounding the carbon dioxide buildup in the atmosphere, with emphasis on the related uncertainties. The underlying questions are (1) how long will it take us to determine with reasonable certainty whether or not a disruptive climate change is imminent, and (2) if such a determination is made, will mankind have the will and the capacity to deal with this threat?

19.1×10^9 TONS PER YEAR AND RISING

On one point we are certain—man has caused the release of large amounts of carbon from storage as fossil fuel. When 1 kg of an average bituminous coal is burned, for example, the 0.7 kg of carbon that it contains combines with 1.85 kg of atmospheric oxygen to produce 2.55 kg of carbon dioxide. If we try to account for the quantity of fuel burned annually in the world, make minor adjustments for natural gas flaring, and add CO_2 produced during cement manufacture,

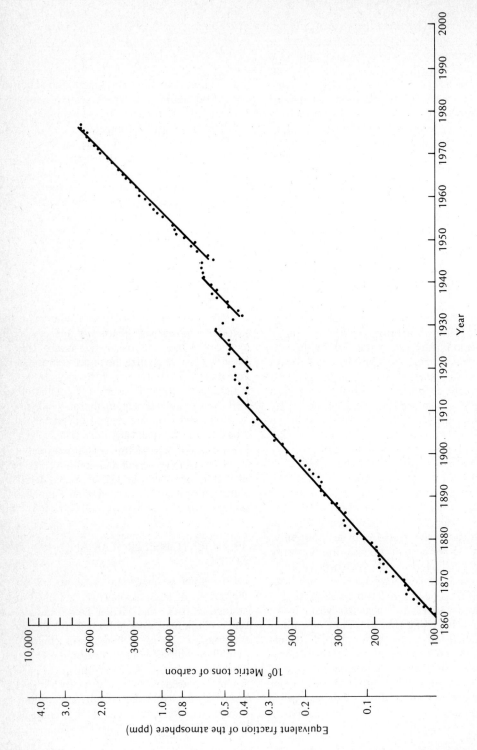

Figure 13-1. The global release of carbon to the atmosphere (as carbon dioxide) from burning fossil fuels and manufacturing cement.

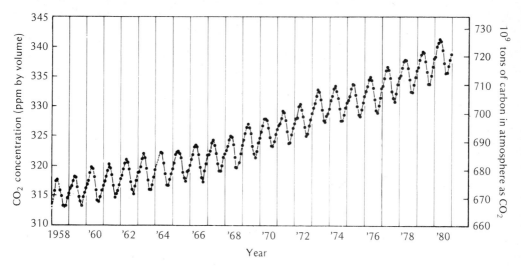

Figure 13-2. The atmospheric carbon dioxide concentration at Mauna Loa Observatory.

we can plot the rate of this release as a function of time. The semilog plot in Figure 13-1 shows that this release has increased exponentially over recent decades with only temporary dislocations at times of global economic trauma—such as World War II and the Great Depression. The carbon dioxide release has grown at a nearly constant 4.3 per cent per year since 1948, and the 1978 value of 19.1×10^9 tons contains 5.2×10^9 tons of carbon.

Another aspect of the carbon dioxide issue on which there is no longer any doubt is that the concentration of CO_2 in the atmosphere is increasing. Callendar (1938) believed that this increase was evident before 1938, but it was not until 1958 that a high-quality monitoring system was installed. Now, with over 20 years of data from the Mauna Loa Observatory (Figure 13-2), and shorter data sets from other sites, there is no doubt that the atmospheric CO_2 concentration is increasing. In fact, the annual increase in the amount of carbon dioxide in the atmosphere has been shown to be equivalent to about half of the amount released by fossil-fuel burning. There are now some 2620×10^9 tons of carbon dioxide (715×10^9 tons of carbon) in the at-

mosphere, giving a concentration of 335 parts per million by volume. The 1978 production of 19.1×10^9 tons from fossil fuel resulted in an increase in the atmosphere of 9.5×10^9 tons of CO_2 (2.6×10^9 tons of C), as represented by an increase of 1.2 ppm in carbon dioxide concentration. (Thus, it is evident that some of the CO_2 from fossil fuels is sequestered elsewhere.) The total increase in the atmospheric CO_2 concentration has amounted to over 6 per cent since 1958 and, based on our best estimates of the preindustrial level, perhaps 12 to 15 per cent since the turn of the century.

While it is clear that man is altering the global carbon cycle through fossil-fuel burning, it is not at all clear whether man's other activities are having a net effect on atmospheric carbon. We can list activities like forest clearing which decrease the amount of carbon in the living biosphere (and probably including carbon in the soil as well), but it is not obvious whether contemporary agricultural practices lead to an increase or a decrease in global soil carbon. Wood harvested for lumber or paper may end up being stored for decades, phosphorous released to the environment probably ties up carbon in organic matter in eutrophic lakes and estuaries, and an increase in atmospheric CO_2 may fertilize plants and lead to increased uptake into the biosphere.

When the accounts are fully drawn, we find cause for great uncertainty. There are those who conclude that man's tinkering with the biosphere results in a carbon flux to the atmosphere as large as the fossil-fuel-related flux (Woodwell et al. 1978), and there are those who as stoutly insist that there is no evidence for a net change in the total biospheric carbon (Broecker et al. 1979). The question is by no means a trivial one. If the only significant anthropogenic source of CO_2 is fossil-fuel combustion, it can be shown that one half of this CO_2 is accumulating in the atmosphere. If, on the other hand, other activities of man are generating an equal amount of CO_2, then only one fourth of the carbon from anthropogenic sources is accounted for in the atmosphere and the other three fourths must be accounted for elsewhere. Ultimately, much of the carbon will find its way into solution in the world's oceans, but mixing of the deep ocean occurs on a time scale of hundreds of years and we do not know accurately how much anthropogenic carbon is being taken up by the oceans now. The accounts must, of course, balance; the fossil-fuel flux of CO_2 must equal the sum of the increase in the atmosphere and the ocean and net changes in the biosphere. The current debate underscores how poorly we understand the global carbon cycle and how difficult it is to project future atmospheric concentrations.

It is, after all, the future about which we are concerned. How fast, and how high, will the concentration of atmospheric CO_2 rise? We shall return shortly to discuss global climate models and what they imply as a consequence of increasing atmospheric CO_2, but one senses intuitively that if man were to substantially increase the atmospheric concentration of any gas that absorbs solar or terrestrial radiant energy, this could not go unnoticed for very long in the climate system. The earth receives radiant energy from the sun, mostly in visible and near-infrared wavelengths, and maintains thermal equilibrium by re-radiating an equal amount of energy to space, mostly in infrared wavelengths. We know that CO_2 (like water) absorbs radiation in bands within the infrared wavelengths at which the earth radiates energy to space. The net result of having CO_2 and H_2O in our atmosphere is to retain a blanket of warm air at the earth's surface and to thus provide a hospitable climate on an earth that, when viewed from space, has a blackbody radiation temperature of about $-18\,°C$. In the extreme, high surface temperatures on Venus are due partly to the high concentration of energy-absorbing gases—mostly CO_2—in its atmosphere.

The 12 to 15 per cent change in atmospheric CO_2 which we have already experienced seems not to have produced any measurable temperature increases which can be unambiguously attributed to this cause (Angell and Korshover 1979). This is not, however, inconsistent with the climate models, and we retain the suspicion that there may be some threshold above which such changes will be observable outside the range of natural climate variability. Efforts to anticipate future atmospheric concentrations are, however, encumbered with great uncertainty. We can speculate on future fossil-fuel use by making estimates of world population and economic activity and, hence, energy consumption. However, the error bands are large, and, when we acknowledge so much uncertainty regarding other CO_2 sources, the future emissions are really educated guesses. In addition, there is the question of where it is all going. Although we suspect that the ocean and its carbonate sediments provide a large sink, we are very uncertain regarding the rate at which this mechanism can operate.

A SCENARIO FOR FUTURE ATMOSPHERIC CO_2

As a framework for our thinking, we have developed a scenario for future fossil-fuel consumption and considered the range of opinion on biospheric changes. While the fuel-use scenario is highly speculative, it does incorporate reasonable expectations for world population and an effort to consider the aspirations and expectations of the developing world. Details of the scenario are available in Rotty and Marland (1980), but suffice it to say here that global energy use in 2025 reaches 850×10^{15} kilojoules (kJ) per year (as compared with 260×10^{15} kilojoules per year in 1975). It is important to note, too, that if our scenario comes close to describing the reality of 2025, the geopolitical distribution of CO_2 sources will be notably different from what it is today (Figure 13-3).

Figure 13-4 approximates the increase in atmospheric CO_2 according to this 850×10^{15} kJ fuel-use scenario and uses a wide uncertainty band to accommodate the debate about current biosphere changes (and, hence, the fraction of the anthropogenic flux that remains in the atmosphere). The lower bound is based on the supposition that the man-induced CO_2 release from the biosphere is presently at a rate of 10^9 tons of carbon per year (hence, the fraction of anthropogenic CO_2 which remains in the atmosphere is 43 per cent). The upper bound assumes that there has been no net man-induced CO_2 release from the biosphere and hence that 53.5 per cent of the total anthropogenic CO_2 flux remains airborne. The lower bound is based on Bolin's (1977) estimate of changes in the biosphere plus the CO_2 production from fossil fuels. Table 13-1 compares the Bolin estimates for biospheric production with that from fossil fuels. That even short-term changes in the mass of the biosphere can have a significant impact on the CO_2 concentration in the atmosphere is amply illustrated by the magnitude of the annual cycle of CO_2 concentration as seen at Mauna Loa (Figure 13-2).

Given this range of assumptions, we see the atmospheric CO_2 concentration reaching 600 ppm by volume somewhere in the 2060 to 2100 interval. Acknowledging the recent

Table 13–1.   Anthropogenic CO_2 Release to the Atmosphere.

	Estimated Present Production (10^9 tons C per year)	Total to Date (10^9 tons C)
Fossil Fuels [a]		
Coal burning	1.683	
Lignite burning	0.265	
Petroleum	2.350	
Marketed gas	0.679	
Flared gas	0.116	
Cement manufacture	0.098	
Fuels total	5.191	150
Forests		
Developed countries	0 ± 0.1	
Developing countries		
Forestation	-0.3 ± 0.1	45 ± 15
Deforestation	0.8 ± 0.4	
Fuel wood	0.3 ± 0.2	
Changes in soil organic matter	0.3 ± 0.2	25 ± 15
Total	6.0 ± 0.6	220 ± 30

From Bolin 1977.
[a] Fossil-fuel data are for 1978 (from Rotty 1980).

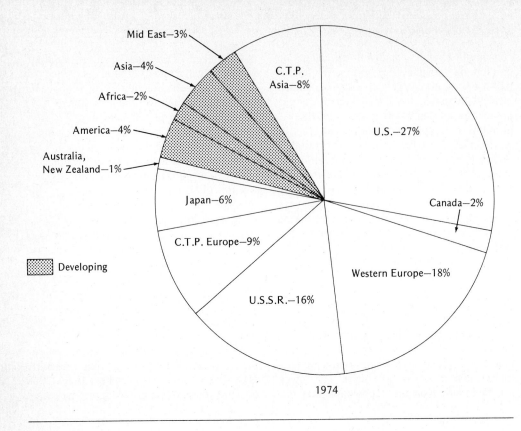

Mid East—3%

Asia—4%

Africa—2%

America—4%

Australia,
New Zealand—1%

C.T.P.
Asia—8%

U.S.—27%

Canada—2%

Japan—6%

Developing

C.T.P. Europe—9%

Western Europe—18%

U.S.S.R.—16%

1974

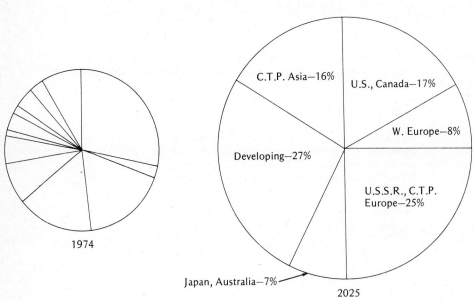

C.T.P. Asia—16%

U.S., Canada—17%

W. Europe—8%

Developing—27%

U.S.S.R., C.T.P.
Europe—25%

1974

Japan, Australia—7%

2025

Figure 13-3. The distribution of fossil-fuel CO_2 sources, by geopolitical region according to historic data and our 850×10^{13} kilojoules per year scenario for 2025. The 1974 diagram is repeated to show the relative magnitude of the total fossil-fuel source in 2025 as compared to 1974. The abbreviation C.T.P. denotes countries with centrally planned economies.

256

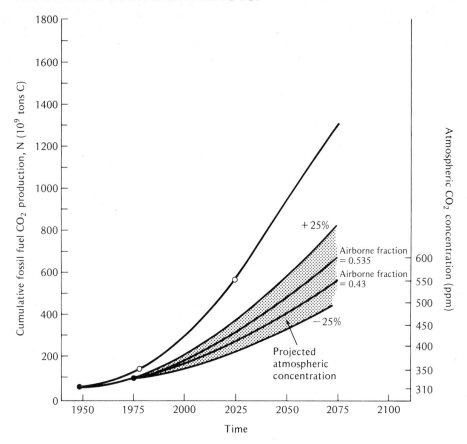

Figure 13-4. Cumulative CO_2 production from fossil fuels and projected atmospheric concentration of CO_2. The upper curve shows the total carbon release, beginning with the 58×10^9 tons of carbon in fuels burned before 1948 and passing through our projected value for 2025. The lower family of curves shows the corresponding rise in atmospheric concentration due to continued fossil-fuel use. These curves assume, respectively, that 53.5 per cent of the released CO_2 is retained in the atmosphere (i.e., fossil-fuel burning has been the only anthropogenic source of CO_2 during the 1958–80 period) and that 43 per cent of the released CO_2 is retained in the atmosphere (i.e., man-induced changes in the biosphere have released 20 per cent as much CO_2 since 1958 as has fossil-fuel burning). Values plus and minus 25 per cent are shown to give an idea of the effect of uncertainties. An appropriate fraction of the future CO_2 release from man-induced changes in the biosphere would have to be added to give the actual atmospheric CO_2 concentration.

debate on synthetic fuels, we note that fuel-use efficiency does play a role. If, for example, a given energy demand is met by burning natural gas, CO_2 is produced at a rate of 0.050 g per kJ, whereas the rate becomes 0.134 g per kJ if the same energy demand is met with synthetic gas produced at 66 per cent thermal efficiency from coal. The dif-ferences are that extra coal must be burned, and extra CO_2 produced, in order to produce gas from coal; plus the fact that burning natural gas (mostly methane, CH_4) derives a large fraction of its energy from oxidizing hydrogen, whereas coal has a much smaller hydrogen content. Nonetheless, for any reasonable synthetic-fuel penetration rate, the net

Table 13–2.   Carbon Content of Fossil Fuel as Compared with Atmospheric
Concentration of Carbon.

Resource Category	Total Contained Carbon ($\times 10^9$ tons)	Contribution to the Atmosphere if 40% Remains Airborne (ppm by volume)
Present atmosphere	715	335
Ultimately recoverable conventional petroleum resources	230[a]	43
Ultimately recoverable conventional natural gas	143[b]	27
Current recoverable coal reserves	440[c]	83
Ultimately recoverable coal resources	3510[d]	660
Total[e]	3883	730

Note: All resource estimates are from the 1977 World Energy Conference (WEC 1978) and are converted
to carbon as prescribed by Keeling (1973).
[a] Corresponds to about 300×10^9 tons of oil (includes deep offshore and polar resources).
[b] Corresponds to 273×10^{12} m^3 of gas.
[c] Corresponds to 636×10^9 tons of coal (adjusted to 29.29×10^6 kJ per ton).
[d] Assumes that 50 per cent of total world coal resources are ultimately recoverable and thus corresponds
to 5063×10^9 tons coal equivalent.
[e] Coal reserves are a subset of coal resources and so are not summed into the total.

effect on atmospheric concentration is on the order of only a few per cent over the next 25 to 50 years.

In an age of concern about the future supply of oil and natural gas, it is appropriate to wonder about our capacity to alter the atmospheric CO_2 level. The atmosphere contains about 715×10^9 tons of carbon as CO_2, and Baes et al. (1976) have estimated that the living and nonliving (including humus and new peat) biosphere contain about 680×10^9 and 1080×10^9 tons of carbon, respectively. Table 13-2 shows the large quantities of carbon estimated to exist in various fossil-fuel resource categories and prompts us to conclude that retention of 40 per cent (or more) of this carbon in the atmosphere could result in an atmospheric concentration more than three times the current value. Inclusion of nonconventional and dilute fossil-fuel resources, such as geopressured methane and shale oil, would suggest even greater potential for high atmospheric CO_2 levels. We note that the largest resources lie in only a few

countries (for example, about 90 per cent of the coal resources of Table 13-2 are in the United States, the U.S.S.R., and China), and this may be of considerable importance if the global society should ever come to grapple with atmospheric CO_2 in a serious way.

CLIMATE MODELS SUGGEST TEMPERATURE WILL INCREASE

Our models of future energy consumption, along with our current, imperfect understanding of the carbon cycle, thus suggest we may be confronted with twice present concentrations of atmospheric CO_2 some 75 to 125 years hence. Even if we could answer all of the questions associated with the carbon cycle, the fundamental question underlying the "CO_2 issue" would remain: What is the effect of increased atmospheric CO_2 concentration on the global climate? To attempt an answer we must again rely on models. Revelle and Suess (1957) have called the CO_2 release a great "geophysical experiment,'' and there are a variety of mathematical attempts to predict its conclusion. Although meteorologists and climatologists have developed a broad understanding of the physical interrelationships that drive our climate system, this same knowledge helps us to understand how much we do not know and how much simplification must be incorporated in order to develop manageable mathematical models. Our confidence in the various models is bolstered somewhat by the fact that most of them agree with one another (and with our intuitions from basic physics) that an increase in CO_2 will result in a global warming—although Watts (1978) points out that quantitative agreement is sometimes fortuitous, given the variety of boundary conditions.

The most frequently cited predictions for CO_2 effects on climate are those obtained from the three-dimensional general circulation model of Manabe and Wetherald (1975). Even in this sophisticated and comprehensive model, assumptions of fixed cloudiness, idealized geography and topography, no seasonal variation, and a swamp [1] ocean were made. The model assumes a stable equilibrium climate and that increasing the CO_2 concentration to 600 ppm is not a large enough perturbation to force the model into a markedly different state. For a "doubling" of atmospheric CO_2, the Manabe–Wetherald model predicts a 2.9 °C increase in average global surface temperature (with much larger increases at high latitude) and a general increase (7 per cent) in the activity of the hydrologic cycle. Although Manabe and Wetherald warn that "it is not advisable to take too seriously the quantitative aspects of the results," most readers do.

Schneider (1975) has summarized the results of various climate models (which predict surface temperature changes of 0.7 to 9.6 °C for a carbon dioxide doubling) and has tried to explain the origin of their differences. Schneider concludes that although the effects of unknown or improperly modeled feedback mechanisms (particularly clouds) could modify the result by "severalfold," a "state-of-the-art, order-of-magnitude estimate" of average global surface temperature change would be 1.5 to 3 °C for a doubling of CO_2.

Watts (1978), in another review, focuses on the importance of feedback mechanisms—particularly cloud amount and albedo (the fraction of solar radiation reflected by a surface)—and tries to place rational bounds on the magnitude of these effects. Some effects should amplify the warming and some may reduce it; some will give an immediate response and some will serve as delay mechanisms. None the less, despite the wide disparity of results from various climate models, there is widespread agreement that a general warming of the earth will occur if atmospheric CO_2 levels rise as predicted.

We recognize that global climate is a constantly variable matter with variations that include periodic fluctuations driven by, among other factors, the earth's orbital parameters and solar cycles. Major Pleistocene ice advances are associated with a mean surface temperature 5 °C lower than present, and there is no reason to believe that the earth has escaped the mode which yielded the alternating glacials and interglacials. Broecker (1975) has suggested that the reason a CO_2 climate effect is not already obvious is that it is currently being countered in a decreasing-temperature portion of the natural cycle. When the natural cycle enters a rising-temperature segment, Broecker suggests augmentation by inflated atmospheric CO_2 may result in a rapid temperature increase.

Through the course of these natural fluctuations, total climate variability has been restrained within relatively narrow bounds by a variety of negative feedbacks that we hope are properly identified and represented in our models. Assumptions on cloud amount, cloud height, and relative humidity, for example, can have a very major impact on model results. The one fluctuation on which we do have data and some ability to test our models is the seasonal variation—and recent models generally try to incorporate this feature as a test of their response (Ramanathan et al. 1979).

In trying to envision real-world consequences of climate change we are also limited somewhat by the impracticality of modeling regional variations. Although the Manabe–Wetherald model shows only latitudinal averages with no geographic or orographic effects, it is important that we see latitudinal migration in factors like the precipitation/evaporation ratio. We need to know much more about the temporal and regional implications of a global climate change.

As easy as it is to locate apparent defects and uncertainties in climate models, we cannot help being impressed with the consensus that an increased atmospheric CO_2 concentration will result in global and regional changes in temperature and other climate factors. A panel recently convened by the U.S. National Academy of Sciences looked for flaws in several climate models and concluded that a doubling of atmospheric CO_2 concentration would produce a $3\,°C \pm 1.5\,°C$ rise in mean global surface temperature (Wade 1979). While the uncertainty may be large, they found no physical effect that they felt could reverse the concluded warming or reduce it to negligible proportions.

WHAT IS TO BE DONE?

We then confront the question of whether these changes are necessarily to the detriment of mankind. Man being traditionally suspicious of change, we are tempted to envision that any change is for the worse, and Budyko and Vinnikov (1977) point out the extent to which the world economic and political system is adapted to the present distribution of climate. Realistically we suspect that a major shifting climate would operate to the physical detriment of some geographic areas while other areas would experience an apparent improvement. However, on a political, interdependent earth it is not clear that there can be any "winners" if such a change of significant magnitude occurs. If, as suggested by Mercer (1978), the changes are sufficient to produce a major short-term decrease in the volume of the Antarctic ice cap and a subsequent rise in sea level, major negative impacts would be virtually universal.

When the CO_2 issue is fully laid out, we end up with the widely held sense that at some poorly defined time in the next century the earth's atmosphere will reach twice its current CO_2 concentration, and that such a change in CO_2 level will probably be accompanied by changes in climate—including a global warming and changes in atmospheric circulation patterns.

Carbon dioxide has another effect outside of its climate interactions. The initial assumption was that, being an essential requirement of the photosynthetic process, excess CO_2 would fertilize plant growth. Slightly elevated CO_2 levels produce favorable responses in most greenhouse plants, but recent suggestions by Strain (1978) regarding effects on plant productivity, water-use efficiency, reproductive potential, phenology, nitrogen fixation, and growth form raise many questions regarding the effect of higher CO_2 concentrations on ecosystem structure and function.

In summary, our present understanding of the CO_2 issue is not sufficiently clear that current policy decisions, which will affect a major global economic activity, are likely to be influenced in fundamental ways—and yet the implications are too disconcerting to simply ignore. Our ultimate choices, if adverse effects are finally demonstrable, can be characterized as "fix," "change," or "cope," and it may be advisable to do some preliminary thinking along these lines. Dyson and Marland (1979) and Baes et al. (1980) have reviewed some possible "fixes" without finding any very attractive,[2] and we really cannot have a very good idea of what "cope" implies without a better understanding of the climate changes which would occur. The predictable trend for response thus becomes one of "change." Although a switch to alternate energy sources has immense economic and sociopolitical implications, such a switch must be contemplated *if* (and/or when) the CO_2 issue is finally demonstrated to be of genuine concern. At the time that CO_2-related concerns become better defined, we shall have to somehow balance the costs of the effects against the impact of restyling the global energy supply system. For now, we need more information, better understanding, better models in many areas, and flexibility—we do not need quickly contrived, short-sighted policy decisions.

NOTES

1. The "swamp" ocean acts as a source of moisture to the atmosphere but has no capacity for heat storage or heat transport.
2. All of the options for collecting and disposing of carbon dioxide from power plants seem to require one third to one half of the plant's power output to run the CO_2 disposal system.

REFERENCES

Angell, J. K., and J. Korshover, 1979, "Global Temperature Variation Surface–100 mb: An Update into 1977," *Monthly Weather Review* **106,** 755; also, "Estimate of Global Temperature Variations in the 100–30 mb Layer Between 1958 and 1977," *Monthly Weather Review* **106,** 1422.

Baes Jr., C. F., H. E. Goeller, J. S. Olson, and R. M. Rotty, 1976, "The Global Carbon Dioxide Problem," ORNL-5194, Oak Ridge National Laboratory, Oak Ridge, Tenn.

Baes Jr., C. F., S. E. Beall, D. W. Lee, and G. Marland, 1980, "The Collection, Disposal, and Storage of Carbon Dioxide," paper presented at Energy/Climate Interactions Workshop, March 3–7, Munster, Federal Republic of Germany, proceedings in press.

Bolin, B., 1977, "Changes of Land Biota and Their Importance for the Carbon Cycle," *Science* **196,** 613.

Broecker, W. S., 1975, "Climatic Change: Are We on the Brink of a Pronounced Global Warming?" *Science* **189,** 460.

Broecker, W. S., T. Takahashi, H. J. Simpson, and T.-H. Peng, 1979, "Fate of Fossil Fuel Carbon Dioxide and the Global Carbon Budget," *Science* **206,** 409.

Budyko, M. I., and K. Y. Vinnikov, 1977, "Global

Warming" in W. Stumm, ed., *Global Chemical Cycles and Their Alterations by Man.* Berlin: Dahlem Konferenzen.

Callendar, G. S., 1938, "The Artificial Production of Carbon Dioxide and Its Influence on Temperature," *Quarterly Journal of the Royal Meteorological Society* **64,** 223.

Carter, L. J., 1979, "A Warning on Synfuels, CO_2, and the Weather," *Science* **205,** 376.

Cooper, C. F., 1978, "What Might Man-Induced Climate Change Mean?" *Foreign Affairs* **56,** 500.

Dotto, L., and H. Schiff, 1978, *The Ozone War.* Garden City, N.Y.: Doubleday.

Dyson, F. J., and G. Marland, 1979, "Technical Fixes for the Climatic Effects of CO_2" in W. P. Elliott and L. Machta, eds., *Proceedings of a Workshop on the Global Effects of Carbon Dioxide from Fossil Fuels,* pp. 111–18. Miami Beach, Florida, March 7–11, 1977, U.S. Department of Energy, CONF-770385.

Keeling, C. D., 1973, "Industrial Production of Carbon Dioxide from Fossil Fuels and Limestone," *Tellus* **25,** 174.

Keeling, C. D., R. B. Bacastow, A. E. Bainbridge, C. A. Ekdahl Jr., P. R. Guenther, L. S. Waterman, and J.F.S. Chin, 1976, "Atmospheric Carbon Dioxide Variations at Mauna Loa Observatory, Hawaii," *Tellus* **28,** 538.

Maize, E., 1979, "CO_2 Threat from Synfuels Unfounded, Report Says," *Energy Daily* **7** (160), 1 (August 21).

Manabe, S., and R. T. Wetherald, 1975, "The Effects of Doubling the CO_2 Concentration on the Climate of a General Circulation Model," *Journal of Atmospheric Science* **32,** 3.

Mercer, J. H., 1978, "West Antarctic Ice Sheet and CO_2 Greenhouse Effect: A Threat of Disaster," *Nature* **271,** 321.

Ramanathan, V., M. S. Lian, and R. D. Cess, 1979, "Increased Atmospheric CO_2: Zonal and Seasonal Estimates of the Effects on the Radiation Energy Balance and Surface Temperature," *Journal of Geophysical Research* **84,** 4949.

Revelle, R., and H. E. Suess, 1957, "Carbon Dioxide Exchange Between Atmosphere and Ocean and the Question of an Increase of Atmospheric CO_2 During the Past Decades," *Tellus* **9,** 18.

Rotty, R. M., 1977, "Global Carbon Dioxide Production from Fossil Fuels and Cement, AD 1950–AD 2000," in N. R. Anderson and A. Malahoff, eds., *The Fate of Fossil Fuel CO_2 in the Oceans,* pp. 167–181. New York: Plenum Press.

Rotty, R. M., 1981, "Data for Global CO_2 Production from Fossil Fuels and Cement," in B. Bolin, ed., Carbon Cycle Modeling, SCOPE 16. New York: John Wiley, pp. 121–5.

Rotty, R. M., and G. Marland, 1980, "Constraints on Fossil Fuel Use," in W. Bach, J. Pankrath, J.

Williams, eds., Interactions of Energy and Climate, Federal Republic of Germany: D. Reidel Pub. Co., pp. 191–212.

Schneider, S. H., 1975, "On the Carbon Dioxide-Climate Confusion," *Journal of Atmospheric Science* **32,** 2060.

Strain, B. R., 1978, "Report of the Workshop on Anticipated Plant Responses to Global Carbon Dioxide Enrichment," Department of Botany, Duke University, Durham, N.C.

Wade, N., 1979, "CO_2 in Climate: Gloomsday Predictions Have No Fault," *Science* **206,** 912.

Watts, R. G., 1978, "Climate Models and the Prediction of CO_2-Induced Climatic Change," ORAU/IEA-78-24(M), Institute for Energy Analysis, Oak Ridge Associated Universities, Oak Ridge, Tenn.

Woodwell, G. M., R. H. Whittaker, W. A. Reiners, G. E. Likens, C. C. Delwiche, and D. B. Botkin, 1978, "The Biota and the World Carbon Budget," *Science* **199,** 141.

World Energy Conference, 1978, World Energy Resources: 1985–2020. Guildford, United Kingdom: IPC Science and Technology Press.

NUCLEAR FISSION AS AN
ENERGY SOURCE

The Dresden nuclear power station near Morris, Illinois, contains three units. Unit 1, in the spherical containment building on the right, was the world's first privately owned and operated nuclear plant. It came on line in 1959 and is currently shut down for chemical cleaning. It is a boiling water reactor (BWR) rated at 215 MWe. Units 2 and 3 are housed in the rectangular building on the left and came on line in 1970 and 1971. They are also BWR's and are each rated at 832 MWe. (Commonwealth Edison Company)

View of the exterior Westinghouse turbine and generator at the San Onofre nuclear generating station near San Clemente, California. The San Onofre station provides 430 megawatts (electric). The sphere houses a pressurized water reactor. (Westinghouse Electric Corporation)

The control room of the San Onofre nuclear generating station. Any one of approximately 60 types of malfunction will shut down the plant automatically. (Southern California Edison Company)

Fueling the San Onofre nuclear generating station. Note the bundle of fuel rods making up this fuel assembly. (U.S. Atomic Energy Commission)

The Zion nuclear power station, completed in 1974, has a total capacity of 2100 MWe—enough to supply a city of one million people. These two pressurized water reactors (PWR's) were supplied by Westinghouse Electric Corporation. The Zion station was built at a cost of $583 million and is located on the shore of Lake Michigan halfway between Chicago and Milwaukee. (Commonwealth Edison Company)

INTRODUCTION

Nuclear power has emerged as perhaps the most controversial issue to characterize the ongoing debate over the direction of U.S. energy policy. Unfortunately, this debate has often suffered from an excess of dogmatism and emotion and a deficit of information and analysis. Certainly, no major energy resource can pose the questions surrounding its utilization in quite as stark and dramatic a fashion as did the Three Mile Island accident. As a result of TMI, however, many of these issues have been analyzed in depth by both public and private studies. In this section, we present basic information on nuclear fission itself, analyses of the most serious problems involved in the harnessing of nuclear energy, and a description of the TMI accident and its meaning.

Proponents of nuclear energy point to the relatively pollution-free power to be obtained (compared with coal plants); the relative economic advantage of fission over fossil energy; the "perfect" safety record of the commercial nuclear industry even after the worst accident in the history of nuclear power; and the large potential fuel resource, particularly with the breeder program. Opponents warn of the environmental risks of radiation emission in the nuclear fuel cycle, uncertainties inherent in the analysis of catastrophic reactor accidents, the lack of fully tested and implemented radioactive waste storage procedures, and the increased risks of nuclear proliferation and terrorism implicit in a "plutonium economy." Certainly all who have studied nuclear energy recognize the imperatives of keeping nuclear wastes out of the biosphere and nuclear materials out of the hands of terrorists. Differing perspectives on nuclear power arise from differing evaluations of the feasibility of satisfying these imperatives.

Since nuclear fission does pose such severe questions for society, it is essential that its risks and benefits be widely understood. Public policy is determined as much by perceptions of the truth as by the truth itself. Hence, we present both factual information on the characteristics of nuclear fission energy and a range of interpretations on the implications of nuclear energy for the social system.

To begin the discussion of nuclear energy we present *A Summary of the Committee on Nuclear and Alternative Energy Systems Report (CONAES)* by Harvey Brooks. This committee was struck by the National Academy of Sciences–National Research Council to resolve some of the issues surrounding nuclear energy. Several members of the CONAES Committee have contributed articles to the present or former editions of this book, and as frequently happens when strong-minded people with different perspectives meet, it was impossible to arrive at consensus on a number of important issues. Since one's opinion on the desirability of nuclear power is part of a broader outlook involving judgments on the feasibility of solar energy, conservation savings, and expanded coal production, these issues are presented as well. The format used is to present the polar positions on issues on which there was divergent opinion and then to indicate how the committee split. This format highlights the areas requiring further study and is very helpful in delineating issues dealt with in considerable detail elsewhere in the book.

James J. Duderstadt and William Kerr present a comprehensive survey of fission reactor technology, ranging from the basic fission process itself through advanced reactor design concepts, in *Nuclear Power Generation*. The wealth of material includes a description of reactor operations, the nuclear fuel cycle, and safety and environmental aspects and then outlines procedures on the licensing and regulation of nuclear power plants. The article concludes with a concise introduction of the issues involved in nuclear energy, with a perceptive analysis of the social and political factors underlying the nuclear debate.

Nuclear fission would certainly be an ideal energy source if it were not for two serious associated risks: (a) the radioactive by-products of the nuclear fuel cycle, which are dangerous to living things, and (b) the basic fuel in the proposed plutonium fuel cycle, which is bomb-grade material. The first of these problems is analyzed in considerable depth by Bernard Cohen in *Impact of the Nuclear Energy Industry on Human Health and Safety*. The data on routine emissions, risks of major accidents, problems of transportation and waste disposal, and plutonium toxicity are considered together with the known biological effects of radiation to calculate the detrimental effects on human health. Although the author in no way suggests ignoring the potential hazards posed by each of these aspects of nuclear power generation, he does suggest that a realistic perspective on these risks involves their comparison with the risks involved in other human activities.

In *An Analysis of Nuclear Wastes,* the League of Women Voters Education Fund presents a very detailed and complete review of this troublesome aspect of nuclear power. First the physical origin of wastes and the biological effects of radiation are discussed, with a good explanation of the controversy relating to the "linear hypothesis" of radiation damage. Categories of waste arising from various stages of the fuel cycle are described along with proposals for the permanent disposal of radioactive wastes. An interesting aspect of the problem is that uranium mine tailings have been the worse offenders to date in causing actual exposure of human populations. The federal-state controversy over disposal site selection is seen in terms of the natural tendency of people to want the benefits of nuclear power without paying the costs of assuming responsibility for the management of the radioactive by-products. This analysis implies, as have several previous scientific studies, that a technical solution of radioactive waste management is the easiest part; "the most critical issues yet to be resolved are social, political, and institutional."

At the heart of much of the debate over nuclear power are differing perceptions as to the risk of catastrophic accidents. And no discussion of nuclear risks is complete without a discussion of the most serious commercial reactor accident to date. In *Three Mile Island: The Accident That Should Not Have Happened,* Ellis Rubenstein presents a concise narrative of the sequence of events during the accident, with a minimum of personal interpretation or judgment. The almost incredible sequence of operator errors, any one of which, had it been avoided, would have saved the reactor, is well documented along with some indication of what design or equipment failure led to each error. A detailed understanding of the accident scenario, such as that published by the Nuclear Safety Analysis Center, was the first step in a sweeping reevaluation of nuclear safety by those responsible for nuclear operations. One of the most prominent, influential, and critical assessments of the TMI accident was that conducted by the President's (Carter) Commission on the Accident at Three Mile Island, a summary of whose findings we present next.

In the summary of *The Need for Change: The Legacy of TMI,* the Kemeny Commission identifies the underlying causes of the accident as "people-related problems and not equipment problems." In enumerating serious weaknesses in the operation, management, and licensing procedures, the members of the commission call for changes which they feel are absolutely necessary (but may not prove sufficient) for avoiding future TMI-like accidents. Weaknesses include lack of adequate operator training, lack of adequate emergency planning, poor communications, and almost total confusion in managing the accident and dealing with the public. The Nuclear Regulatory Commission is strongly criticized, both for its management policies prior to TMI and its performance during the accident. The mountain of NRC regulations "tends to focus industry attention narrowly on the meeting of regulations rather than a systematic concern for safety." The Kemeny Commission's main thrust was to stress the need for change in the attitudes of the utility and NRC personnel responsible for reactor safety. There is evidence from the widespread changes in licensing procedures, operator training and examination programs, and reorganized NRC and utility structure that this advice has been well received.

We conclude our discussion of nuclear fission as an energy source by considering *Environmental Liabilities of Nuclear Power* by John P. Holdren, one of the most vocal and influential critics of nuclear energy. He begins the discussion by asking three fundamental questions of energy philosophy, the answers to which will largely determine one's stand on nuclear energy: How safe is nuclear? How safe is safe enough? Who should decide? Stressing the non-technical nature of such issues, he then lists the liabilities of nuclear power in order of decreasing risk. At the top of the list is the problem of nuclear weapons proliferation, and recent events in the Middle East support this evaluation. Next in order of risk he ranks theft of weapons-grade materials, sabotage and accidents, and, finally, at the bottom of the list, routine radioactive emissions. In emphasizing the "human factors" involved in any evaluation of these risks, he makes the point that it is impossible to quantify any estimate of the associated risk.

The material presented here will greatly assist the reader in interpreting conflicting claims in the nuclear debate. We have also found it helpful in interpreting the nuclear controversy to appreciate the symbolic aspects of nuclear power. Many of the really emotion-stirring social concerns of former years have been resolved (e.g., the Vietnam war, legal rights for minorities, and the most obvious pollution hazards). This leaves nuclear power as the lightening rod that attracts much of the reformist sentiment as well as the alienation and discontent present in a free society. Nuclear power symbolizes bigness, impersonal corporate and governmental bureaucracy, and complex technology, which many who long for the simple life see as the root of all evil. On the other hand, to those who see science and technology as servants of humankind, nuclear power represents a powerful source for human welfare and the symbol of humankind's desire to "beat their swords into plowshares." Both of these symbols provide strong motivation for the intensity of the nuclear debate.

A Summary of the Committee on Nuclear and Alternative Energy Systems Report (CONAES)   14

Today's energy problems are primarily rooted in two interrelated facts: the increasing geographical separation between the most readily producible oil and natural gas resources and the locations where they are consumed, and the increasing geological inaccessibility of new oil and gas resources in the face of growing demand.

These two facts have led to increased politicization of the international oil market, as well as to a time lag between changes in the world price of oil and gas and the adjustment of investment patterns and energy-using infrastructures to these prices. This is because energy demand is very inelastic to price in the short run. Changes in the ratio of energy consumption to gross national product (E/GNP ratio) can occur, for the most part, only slowly, as the average efficiency of energy-using equipment changes when existing equipment is replaced or new equipment purchased.

Two decisions must be crystallized in the near future: the selection of a mix of sustainable energy sources that can be developed and relied upon for the long term and a viable plan for the marshaling of resources for a transitional period leading from present technology and primary fuels to those of the future (past 2010).

The National Academy Committee on Nuclear and Alternative Energy Systems (CONAES)[1] sees this transitional period as lasting from 30 to 60 years, depending upon several factors not completely foreseeable now, and in some instances subject to policy control. These factors are:

Future rates of economic growth in the world, and especially in the highly industrialized countries.

Success in reducing the growth of demand for oil and gas, through improved technical efficiency in energy end use and through the timely deployment of nonfossil energy technology.

Dr. Harvey Brooks is Benjamin Peirce Professor of Technology and Public Policy at Harvard University, Cambridge, Massachusetts 02138. He has served the U.S. government in many advisory capacities. He is the founder and editor-in-chief of *Physics and Chemistry of Solids,* author of *The Government of Science* (1968) and co-chairman of the NAS-NRC Committee on Nuclear and Alternative Energy Systems (CONAES).

From *Bulletin of the Atomic Scientists* 36 (2), 23 (1980). Reprinted by permission of the author and *Bulletin of the Atomic Scientists.* Copyright © 1980 by the Educational Foundation for Nuclear Science, Inc.

The seriousness in planning for the possible climatic changes resulting from worldwide fossil-fuel combustion.

For the transitional phase, coal (including synthetic fuels derived from coal), certain unconventional and expensive forms of oil and gas, and currently developed and deployed forms of nuclear power are available. The duration of the transition will depend on the extent to which they can be extracted and used in an environmentally acceptable manner.

The only long-sustainable energy alternatives available with some degree of confidence on a scale sufficient to make a meaningful contribution to U.S. and world energy requirements are several forms of solar energy and nuclear fission in the form of breeder reactors.

Thermonuclear fusion offers sufficient potential advantages over fission, and possibly solar energy, to warrant continued vigorous research, taking full advantage of the high level of international collaboration that presently exists, but it is too early to say whether fusion can be prudently included among the candidate technologies for long-term sustainable energy sources.

Geothermal energy offers some promise of a supplementary transitional energy source, but as a candidate technology for a long-term sustainable source it offers a much smaller potential resource base than the solar energy and fission breeder options.

In the course of its deliberations, CONAES debated a great many energy policy issues. The committee reached a consensus on some, a partial consensus on many others, and agreed to disagree on several. I list below what I consider to be some of the most significant issues in the current energy debate, and for each issue suggest an intermediate position between strongly stated alternative views. In most of these judgments I have presented what I interpret as the views of CONAES committee members, but I must emphasize that this is a personal interpretation, which may be biased by my own opinions.

CONSERVATION POTENTIAL

Here there were two distinguishable positions:

A. Over a time period on the order of 20 to 30 years, the potential for curtailing growth of energy consumption without adverse effects on employment and consumption is very high, owing to the high long-term price elasticity for energy and to the relatively small fraction of GNP represented by energy production. The E/GNP ratio, which is a measure of the energy intensity of the economy, could be reduced to from one third to one half of its 1973 value in the United States, with only very small effects on aggregate personal income. Econometric estimates based on long-term price elasticities deduced from historical and cross-national comparisons are consistent with engineering estimates of the technical potential for more energy-efficient capital and consumer durables.

B. The substitution elasticity for energy with respect to other factors of production is likely, in practice, to be much more modest than indicated by econometric or engineering estimates. These are necessarily based on many unverified and often unverifiable assumptions regarding individual behavior and future political climate and the detailed workings of the economic system. Hence efforts to curb energy growth are likely to have adverse effects on economic growth and employment, and to be particularly disadvantageous to the poorer segments of the population, especially in view of the high prices required to achieve the largest values of projected savings. Estimates of future conservation potential are too uncertain to be relied upon for prudent planning and cannot safely be used to justify any relaxation of effort toward development of all economically and technically promising energy supply alternatives. This is especially true of those which are already well advanced toward commercialization.

These two positions have widely differing policy consequences. If **A** is correct, then the highest priority in national policy should attach to the raising of prices, subsidization of conservation investments, and establishment of mandatory performance standards for equipment. Under hypothesis **A,** a Btu saved costs less than a Btu of additional supply over a factor of 2 or 3 in E/GNP.

If **B** is correct, then the greatest emphasis should be on expansion of supplies, especially through new sources and conversion technologies, by government subsidy if necessary. The emphasis should be on synthetic fuels from coal and oil shale and on unconventional oil and gas sources to replace declining domestic production of conventional oil and gas and oil imports.

Existing evidence, including the response of the U.S. economy to the 1973 price jump, is somewhat, but not conclusively, in favor of hypothesis **A.** The next few years should greatly enhance our ability to distinguish between the hypotheses of high and low long-term substitution elasticity.

CONAES is in agreement that conservation through improved end-use efficiency of equipment should have the highest priority in national energy policy, but is somewhat divided on the relative emphasis to be given to enhanced supply, and the degree to which supply should be subsidized by the federal government.

PRICING POLICY

With respect to pricing policy, one can distinguish three positions:

A. Prices should continue to be controlled in the interest of equity and of cushioning the economy against the disruptive effect of rapid changes in the world price of petroleum. Transition to a new supply/demand equilibrium should be gradual. In order to achieve this, two subalternatives are open. One would emphasize supply and would provide heavy federal subsidies to develop and deploy new supply alternatives, keeping prices at least temporarily well below those required by production costs. The second subalternative would emphasize demand and would provide federal tax benefits and subsidies to industrial and individual consumers to invest in more energy efficient buildings and equipment and, where applicable, renewable energy supply alternatives. In addition, mandated energy performance standards of progressively increasing stringency would be enacted. Because of the impact of U.S. demand on world oil prices, the per capita collective benefit of conservation or alternate domestic supply investments would be expected to exceed substantially the private benefit to consumers, computed on the basis of a given set of price assumptions.

B. Domestic energy prices in the United States should be permitted to come into market equilibrium with world prices as rapidly as possible, in order to induce conservation and to encourage the development of more secure alternative domestic energy sources, which will become economical in the future at higher prices. In general, because the incremental cost of new energy supplies will generally exceed the average cost of energy, and hence may not be fully reflected in market prices, it will be necessary to provide temporary federal subsidies or price guarantees in order to bring about the commercialization of new energy technologies and supply sources before they are economically competitive.

C. The price of energy based on nonrenewable domestic resources should be raised above world prices according to a predictable schedule in order to further encourage conservation and the commercialization of alternate sources. Taxation on energy is necessary to ensure that its price reflects the true marginal or replacement cost of incremental supplies. Two approaches are possible: one would be to levy excise taxes on all nonrenewable energy sources, designed to reflect as far as possible their marginal supply costs as well as their social costs; the other would be to impose import surcharges or market quotas on oil imports, allowing domestic prices to rise in accordance with domestic market forces. This policy does not exclude modest tax benefits or subsidies for conservation investments, such as those now on the books, but it relies primarily on the price mechanism to induce both conservation and new supply. Higher prices are almost certain to induce more conservation than could be achieved by conservation subsidies and mandatory standards with prices controlled at a lower level.

It was difficult to ascertain the consensus in CONAES among these alternatives. It would probably best be described as converging on policy **B**, but with considerable sentiment in favor of **C**. Some members were prepared to see higher subsidies for conservation with less emphasis on supply, but none favored price controls or mandatory performance standards beyond those that would be economically rational at projected prices. Those who favored policy **C** tended to favor the alternative of tariffs or import quotas because it would focus on the most serious supply problem, oil imports.

RAPIDITY AND EXTENT OF EXPANSION OF COAL USE

With respect to coal, one can distinguish two views:

A. Environmental and health problems (sulfates, water requirements, strip mining, carbon dioxide production) associated with the expansion of coal will make it difficult, if not impossible, to rely on coal beyond two to three times current production (30 to 45 quads per year), and therefore priority should be given to early and rapid expansion of nonfossil alternatives (breeders, solar, with consequent emphasis on electrification).

B. Technological advances in methods for clean utilization of coal, foreseeable in the next 30 years, will greatly reduce the barriers to coal use and will make it a good transitional solution to energy needs until renew-

able energy sources are available for deployment on a large scale.

On balance, CONAES seemed to lean toward position **A** and would oppose any energy policy that might preclude the possibility of reducing relative dependence on coal in the future, especially after 2000. However, vigorous pursuit of clean coal utilization technology, including synthetics and new combustion methods for utilities, was favored, recognizing that the degree to which these measures can be counted upon to offset the environmental problems in **A** is problematic.

RAPIDITY AND EXTENT OF EXPANSION OF NUCLEAR POWER

Here two positions were in evidence:

A. For many reasons nuclear power should be regarded as a last resort. It should be relied upon to the minimum degree possible in the next 20 years and be phased out as soon thereafter as possible in favor of renewable energy sources. This implies that the breeder should be developed only as a ''back-burner'' option and that synthetic fuels should be selected in preference to more rapid electrification.

B. On balance, nuclear power is the safest and least environmentally damaging among the presently available alternatives, and its expansion should be pushed along with early development and deployment of the breeder and with substitution of electricity for fluid fuels wherever it can be economically justified.

The Committee seemed to be nearly evenly split on positions **A** and **B,** with some spectrum of intermediate positions. However, there was a consensus that the breeder option has to be maintained, although a majority cautioned against any actions that would commit the country to its early commercialization. This caution was dictated mainly by the possible dangers of proliferation.

The cautious position on nuclear power, while conceding that some nuclear expansion will be required in the next 10 or 15 years, centered primarily on the problems of proliferation and diversion and on reactor safety issues. This group particularly emphasized human fallibility and the inadequacy of institutions in carrying out what might otherwise be acceptable technical solutions in principle. From the technical standpoint, waste management was regarded as a less serious problem, but its practical implementation could entail some overriding of local opposition, which several members regarded as undemocratic.

The expansionist position on nuclear power maintained that all the technical problems are manageable, that the economics are likely to be the most favorable among all alternatives in the long run, and that acceptable technical solutions for safety, proliferation, and waste management will not present insurmountable institutional and political obstacles, given honest and responsible political leadership. All the health hazards associated with nuclear power are small compared with those of available alternatives and compared with the effects associated with the geographical variations of background radiation.

RAPIDITY AND EXTENT OF POSSIBLE EXPANSION OF SOLAR ENERGY USE

There are two positions which may be discerned:

A. The environmental and sociopolitical advantages of solar energy justify major governmental subsidies and incentives to encourage its rapid adoption. And, once the political will to move forward exists, the cost disadvantages of the solar options will tend to decrease. Compared with alternatives on the basis of marginal rather than average costs, solar energy looks more competitive than it does in conventional calculations. In view of the probable future volatility of fossil-fuel prices and the uncertain future of nuclear power, it is an option whose economic attractiveness is likely to improve rapidly. An adequate national commitment to deployment of solar energy in all its feasible forms can avoid introduction of the breeder and

permit phasing out of nuclear power well before high-grade uranium resources have been fully committed.

B. Many forms of solar energy (solar heating and cooling, solar process heat, solar photovoltaics, limited biomass) are quite promising and have the potential for contributing a major fraction of U.S. energy needs several decades into the twenty-first century. Largely because of the immaturity of the technologies and today's poor economics, however, the promise of solar energy is primarily long-term, and it promises only a very modest contribution to reducing the growth of demand for traditional energy sources within the CONAES time frame (1975 to 2010). Research should be intensively pursued, and development of a solar industry encouraged through removal of political and institutional obstacles, but not through major subsidies. Solar energy cannot be counted upon as an alternative to expansion of nuclear power and synthetic fuels in the next 20 to 30 years.

The majority CONAES position was closer to **B** than to **A.** While there were varying degrees of optimism about the penetration rate of various solar technologies, there was agreement that it would be small, even with the pricing policies represented by option **C** in the above section on pricing policy. Subsidies much larger than those offered by current legislation, plus politically mandated use of solar energy in certain applications, regardless of cost, would be necessary to achieve major solar penetration by 2010. None of the participants believed that solar, in combination with vigorous conservation, could eliminate the need for continued use of nuclear energy or the expansion of coal use in the period 1975 to 2010. There was a considerable difference in view as to how large government subsidies for solar energy could be justified. The preferred alternatives were solar photovoltaics for electricity and vigorous effort to develop adequate technology for the production of liquid or gaseous fuels from solar energy. Wind, biomass, ocean thermal energy conversion (OTEC), solar towers, and additional hydro were regarded as having limited potential for market penetration, at least in the United States.

BREEDERS VERSUS ADVANCED CONVERTERS

Here, again, there were two positions:

A. Breeders, and specifically the liquid metal fast breeder reactor (LMFBR), represent the long-range nuclear alternative which dominates the widest spectrum of possible future situations with respect to electricity growth rate, uranium resources, and proliferation concerns after the turn of the century. Development of other alternatives, though desirable in principle, is likely to interfere with early commercial availability of the breeder while offering compensating advantages only under rather special combinations of assumptions about the future. Therefore, the United States should push forward with its breeder program, including early operation of an intermediate-scale prototype similar to the Clinch River breeder reactor (CRBR).

B. A reduced rate of projected electricity growth, due to conservation and higher prices, greater abundance of high- and intermediate-grade uranium ores than now estimated, and technical difficulties and cost escalation of the liquid metal fast breeder reactor make it likely that the breeder will not become economically competitive with other energy sources until well into the twenty-first century, if ever. On the other hand, efficient advanced converters, especially ones that could be operated initially on a once-through cycle—such as a slightly enriched version of the Canadian heavy-water reactor (CANDU)—could greatly extend uranium resources with reduced proliferation risks. Possibly, it could also avoid the necessity for introducing breeders at all before solar energy becomes available on a large scale. Advanced converters would be useful, either by themselves or operated symbiotically in conjunction with breeders, and would pro-

vide increased options for dealing with proliferation and diversion.

The members of CONAES were about equally divided between these two positions, but in agreement that the option of eventually deploying breeders should be maintained for the present.

Largely owing to their differing views on the urgency of breeder timing, members were also divided on the importance of early operation of an intermediate-scale prototype. There was also a difference of opinion regarding the United States' ability to take advantage of foreign breeder programs if it proved necessary to catch up on breeder development later.

SERIOUSNESS, TIMING, AND RESPONSE TO THE CARBON DIOXIDE PROBLEM

There were again two positions:

A. Evidence is now sufficiently conclusive that the carbon dioxide buildup from fossil-fuel combustion will cause serious climatic changes, beginning early in the twenty-first century, so that commitment to future expansion of fossil fuel should be cautious, quite apart from possible resource constraints. High priority should be given to development and early deployment of nonfossil alternatives. A concerted international approach to the carbon dioxide problem should be taken by the major energy-consuming countries before it becomes a source of international conflict and mutual recrimination. Production of secondary energy sources from coal, including synthetic fuels and electricity, tends to accelerate the appearance of the problem because of greater carbon dioxide output per unit of useful final delivered energy.

B. Climatic effects of carbon dioxide, while a real possibility, can be postponed by slowing energy growth through conservation. This postponement of the problem leaves plenty of margin for better assessment of its seriousness and for the design of strategies to mitigate it to the extent necessary.

The buildup time scale of decades leaves considerable time for adjustment, and, in any case, the existence of a serious problem is predicated on predictions from complex atmospheric models which are still speculative and leave out important possible countervailing factors.

The members of CONAES leaned toward **A**, but with a warning against overreaction. Although caution in commitment to large-scale expansion of the use of fossil fuels is warranted, this need have little impact on planning for the near term—through 2000. On the other hand, the carbon dioxide problem is one factor in the Committee's consensus that the breeder option should not be foregone for the present and that solar research and development deserve high priority.

AIR POLLUTION REGULATIONS FOR COAL

Two postions were voiced:

A. The regulations resulting from implementation of the 1977 Clean Air Act have little or no scientific basis in research on health effects of air pollutants from coal combustion. They seriously inhibit the expanded use of coal (although they are not the only factor) and should be relaxed. At most, the regulation of coal-burning sources should be based on ambient standards for sulfur dioxide, nitrogen oxides, and total particulates. Additional measures—such as prevention of significant deterioration in clean areas, offsets in nonattainment areas, and the requirement for scrubbers on all coal-burning sources regardless of the sulfur content—should be relaxed as having no valid basis.

B. There is sufficient uncertainty about adverse health effects due to sulfates and respirable particulates at emission levels below those required by existing ambient sulfur dioxide and total suspended particulates (TSP) standards, so that continuation of current emission control measures for new sources is prudent until much better laboratory and epidemiological information is available. While present requirements add significantly to the capital cost of new plants, they do not critically affect either the choice of coal as a boiler fuel or the future average cost of electricity up to the year 2000. But the cost of retrofit, if present ambient standards prove to be insufficiently protective of public health, would be expensive as well as disruptive. This fact fully justifies a very conservative approach to the performance requirements for new sources.

There appeared to be no clear consensus in the Committee between **A** and **B**; the majority seemed to lean toward **B**. There is agreement on the inadequacy of the present scientific basis for emission regulations of coal-fired plants. Differences of opinion relate mainly to where the burden of proof should lie. Expeditious resolution of the current technical uncertainty would have a high payoff in terms of possible capital cost savings in future coal-fired generating plants, if research showed that present regulations were unnecessarily stringent. Thus the benefit-cost ratio for research on the health effects of coal combustion could be very high. Also, because of this uncertainty, a mix of coal and nuclear among new plants is preferred because their potential hazards are both uncertain and different.

WATER AS A CONSTRAINT ON THE EXPANSION OF COAL AND ELECTRIC POWER

Two views were put forward:

A. Studies made for CONAES strongly indicate that a coal utilization level between two and three times the current level, especially if it includes 8 to 12 quads per year of coal use for synthetics, would severely strain water supplies in many areas of the country. This is primarily because of competition with other uses, such as irrigation and municipal water supplies, in coal development areas. This will represent a significant inhibition on the expansion of both coal and electric power generation generally.

B. Coal and electric power generation

make a minor contribution to total water use, and their increased requirements for the future could be easily offset by progress in the development of more water-efficient processes (including dry cooling of power plants), particularly in synthetics production. Moreover, changes from present wasteful irrigation practices, the transport of fresh water between basins, and the tapping of new ground water sources (including brackish water) could make even more of a contribution to offsetting future requirements. Hence, fresh water should not be viewed as a significant constraint on coal or electric power expansion in the foreseeable future, given reasonably prudent planning for water management.

Most participants were closer to **A** than to **B**. Although water efficiency in use may be greatly improved, the political difficulties involved in the allocation of water among competitive uses, and in large-scale, interbasin transfer projects, will in practice place significant inhibitions on coal and power development. Most projections of future water use do not go beyond the year 2000, which is about the time in which water may first become a serious constraint on the expansion of coal and electric power. The problems of water requirements for energy production need much more attention than they have received, especially on a disaggregated regional basis and including the period beyond 2000.

ELECTRIFICATION VERSUS FLUID FUELS

Possibly more than one third of the end uses of liquid and gaseous fuels could be replaced by electricity, which can be generated from a large variety of fuels. There is an issue of whether it is more desirable to replace declining natural oil and gas supplies by synthetics or by electricity. Two positions emerged:

A. There exists a large infrastructure for the distribution and utilization of fluid fuels, and many parts of this system must be fully utilized in order to be economical. In contrast, replacement by electricity requires new utilization investments as well as additional generating and transmission facilities. Thus synthetic fuels as well as enhanced production of natural fuels (especially gas from unconventional sources) can make better use of existing investments than electricity can. Thus a policy which emphasizes an increasing of the supply of fluid fuels, including synthetics, is likely to be economically superior to a policy emphasizing accelerated electrification.

B. When flexibility and controllability in end use, and the possibility of using heat pumps, are taken into account, the efficiency of conversion from primary energy resources to usable energy is about the same for electricity as for synthetic fuels. Furthermore, electricity can be generated today from nonfossil sources; thus it would contribute less to air pollution, and particularly to carbon dioxide production, than synthetics, if it were generated from nuclear or other nonfossil primary sources. Although the electricity option is very capital intensive, this may be largely offset in many applications by its greater flexibility and controllability in end use. Furthermore, the future costs of electricity, especially nuclear-generated electricity, are likely to be more predictable than those of synthetics. If national policy were neutral with respect to the choice between electricity and synthetic fuels, electrification would be likely to expand more rapidly than is presently projected. This would be generally beneficial to the environment and possibly less costly to taxpayers.

It appears that most members of CONAES are closer to **A,** although **B** has significant support from those members who are favorable to the expansion of nuclear power and early deployment of the breeder. The Committee agreed that the economics of electricity and of synthetic fuels are comparable, probably favoring one or the other in different locations and in different specific end-use applications. But the advantages of the electricity option are considerably reduced if present political constraints on the expansion of nuclear power continue.

PROLIFERATION AND NUCLEAR POWER

Two major positions may be distinguished:

A. The problems of proliferation of nuclear weapons or diversion of nuclear material by terrorists or dissident groups are real; thus the worldwide deployment of civilian nuclear power contributes significantly to international risk from this source. Because the risk is greatly enhanced if reprocessing of nuclear fuel becomes widespread and early commercialization of fast breeders is encouraged, the United States can contribute greatly to alleviating this problem through setting a good example to other countries by postponing reprocessing of light-water reactor (LWR) fuel and deferring major commitments to development and deployment of fast breeders. These measures will buy time to negotiate a strengthening of international safeguards for nuclear fuel. Since neither reprocessing nor breeders are likely to be economically competitive in the near future, the costs of deferring them are not likely to be high. Therefore, the Carter Administration's policy canceling the CRBR and delaying reprocessing represents a relatively modest sacrifice to avoid a significant risk.

B. The U.S. example is unlikely to be followed because this country is favored with alternative energy sources and a large conservation potential, to a degree not open to most other nations. Present U.S. policy is more likely to throw doubts on the assurance of future fuel supplies for other nations and to increase competition for OPEC oil. The implication of civilian nuclear power and the fuel cycle in the potential for proliferation appears to be minor; hence the negative aspects of U.S. foreign policy that result from a slowing of the introduction of advanced technologies and more resource-efficient reactors are likely to far outweigh the small risk of proliferation from this source. In addition, U.S. actions are more likely to stimulate foreign acquisition of advanced proliferation-prone nuclear technologies than to discourage them, and the withdrawal of the United States as a major participant in the worldwide nuclear power industry is more likely to eliminate its influence over future developments in the field than otherwise. The risks of war resulting from international rivalry for access to scarce fossil fuels would far outweigh those resulting from proliferation of nuclear weapons derived from a civilian power industry.

The members of CONAES were almost equally split on this issue. The only recommendation which could be agreed upon was that the outcome of present policies is uncertain, and that they ought to be carefully re-evaluated after no more than five years to determine whether they are really achieving the purposes for which they are established. Most members of the Committee were skeptical that any technical measures, including various proliferation-resistant improvements of the nuclear fuel cycle, could have an important impact on the risk of proliferation in the foreseeable future. However, a majority believed that buying time to negotiate better institution arrangements was justified.

ENERGY POLICY AND THE THIRD WORLD

The U.S. demand for oil on the world market is a significant long-term influence on the world price of oil, and thus affects the ability of the non–oil-producing developing countries to purchase oil and finance their economic growth. There is thus the issue of how U.S. energy policy, and that of the other industrialized countries, should be configured to maximize the chances for development in the Third World. Two points of view are cited:

A. Since that oil which is cheap to produce is being exhausted, it is important that this low-cost resource be saved as much as possible to facilitate development in the poorest countries. This places an obligation on the industrialized countries to curtail their demand for imported oil in order to keep it as cheap as possible. They are in the best position to do this since they can more readily afford the necessary investments in improved end-use efficiency and in capital-intensive energy conversion technology based on cheap and abundant raw materials, such as breeders and advanced solar technologies.

B. The non–oil-producing developing countries cannot aspire to the oil consumption levels and material lifestyles now enjoyed by the developed countries, and the latter must in fact begin to retreat from their own energy-intensive modes of life. Thus the developing countries should attempt to leap-frog technology in the industrialized nations by developing non–energy-intensive, decentralized rural economies based on local and largely renewable resources, stressing local self-reliance and self-sufficiency. The policy of the developed countries should be to encourage and assist the developing countries in the adoption of these alternate technologies while discouraging their adoption of large-scale, inappropriate, capital-intensive technologies such as nuclear power.

All members of CONAES agree that a reduction of oil imports by the developed countries would be beneficial to them, and in this respect the Committee supported **A.** However, there is less agreement on how far this policy requires the rapid expansion of capital-intensive energy technologies, especially the breeder. Many members would prefer to emphasize conservation and the rapid deployment of renewable energy sources, with public subsidies where necessary.

Many members are also skeptical of the feasibility or likelihood of the mode of development contemplated in **B.** They believe that the industrialized countries will have to continue to sell industrial equipment, including nuclear technology, to the developing world in order to help finance their own oil imports, reduced though these may be. These members point out that much of the developing world has not been explored for oil and that, given their relatively modest requirements, the Third World countries can develop their own oil and gas resources with the aid of equipment and technology from the industrialized countries. Furthermore, they point out, it is unrealistic for the developed countries to try to influence the development paths of the less advanced countries by denying certain kinds of technology to them. Nuclear power, for example, is likely to make an increasing contribution to the energy needs of the developing world, especially in the rapidly growing middle-income countries; the political costs of discriminatory policies, which smack of colonialism in the eyes of these countries, far outweigh the risks of inappropriate modes of development. In the view of most members of CONAES, the interdependence of developed and developing countries represents a sounder route to world development than does a policy of local self-sufficiency.

Note

1. Committee on Nuclear and Alternative Energy Systems (CONAES), 1979, *Energy in Transition, 1985–2010*. San Francisco: Freeman. *Co-Chairmen:* Harvey Brooks and Edward L. Ginzton. *Executive Director:* Jack M. Hollander. *Members:* Kenneth E. Boulding, Robert H. Cannon, Jr., Richard R. Doell, Edward J. Gornowski, John P. Holdren, Hendrik S. Houthakker, Henry I. Kohn, Stanley J. Lewand, Ludwig Lischer, John C. Neess, David J. Rose, David Sive, and Bernard I. Spinrad.

Nuclear Power Generation   15

JAMES J. DUDERSTADT AND WILLIAM KERR

For the past three decades, an extensive international effort has been directed toward the harnessing of the enormous energy within the atomic nucleus for the peaceful generation of electric power. Nuclear reactors have evolved from research tools into mammoth units that drive hundreds of central-station electric generating plants throughout the world today. Nuclear power technology has been developed to the point where it stands ready to meet a significant fraction of the world's requirements for electric energy. Nuclear plants currently supply about 11 per cent of the electricity generated in the United States (and a considerably larger percentage in many European nations). Some projections have suggested that nuclear power could supply well over half of our electric energy needs in the next century. It is fortuitous that this technology has matured at just that point

in history when our society is rapidly exhausting its reserves of conventional fuels.

Yet never before has there been so much controversy over the role that nuclear power is to play in our future. Just as the fruits of decades of research on nuclear power generation are to be harvested, there has arisen a very real concern over whether this source of energy should be implemented at all. Are the potential benefits of nuclear power generation worth the risks posed by this technology? Certainly, events such as the accident that occurred at the Three Mile Island nuclear power plant in 1979 cause serious doubts as to the safety of this energy source. Moreover, such plants must have a significant impact on their environment. What about the dangers and the environmental impact of the mining of uranium ore and its conversion into nuclear fuel? What are the dangers of

Dr. James J. Duderstadt is a professor of nuclear engineering and Dean of the College of Engineering at the University of Michigan, Ann Arbor, Michigan 48109. He is author of numerous publications and six textbooks in the areas of nuclear reactor analysis, radiation transport, kinetic theory, and controlled thermonuclear fusion.

Dr. William Kerr is a professor of nuclear engineering and Director of the Michigan Memorial Phoenix Laboratory, University of Michigan, Ann Arbor. He is author of numerous publications in the area of nuclear reactor analysis and safety and has served as member and past chairman of the Advisory Committee on Reactor Safeguards of the Nuclear Regulatory Commission.

This original article by Professors Duderstadt and Kerr is an updated version of the original article by the same title prepared by Professor Duderstadt for the second edition of this book.

the by-products of nuclear power generation, such as plutonium and radioactive wastes? Will the worldwide development of nuclear power technology contribute to the spread of nuclear weapons? These, as well as a host of related questions, swirl about the debate over the future role of nuclear power in our society.

In this article, we shall concern ourselves with a discussion of such questions. We shall begin with an introduction to the fundamental concepts involved in nuclear fission chain reactions and consider the principal types of nuclear power systems utilized throughout the world today, including advanced reactor types such as the fast breeder reactor. We shall summarize several operational aspects of nuclear power plants, including their safety and environmental impact and their associated nuclear fuel cycle.

BASIC CONCEPTS OF NUCLEAR POWER GENERATION

A nuclear reactor is a device in which a controlled nuclear fission chain reaction can be maintained. In such devices, neutrons are used to induce nuclear fission reactions in the atomic nuclei of heavy metals such as uranium or plutonium. These nuclei fission into lighter nuclei called fission products. The fission process is accompanied by the release of energy plus several neutrons. These neutrons produced by fission can then be used to induce further fission reactions, thereby propagating a chain of fission events. In a sense, a nuclear reactor is simply a sufficiently large mass of fissile material, such as the isotopes uranium-235 and plutonium-239, in which a controlled fission chain reaction can be sustained. A sphere of U-235 metal slightly over 8 centimeters in radius can support such a chain reaction and therefore would be classified as a nuclear reactor.

But a modern power reactor is considerably more complex than a simple sphere of metal. It contains not only a lattice of carefully refined and fabricated nuclear fuel, but also the means for cooling this fuel as fission

energy is released while maintaining the fuel in a precise geometric arrangement with appropriate structural materials. A mechanism is also provided to control the chain reaction. The surroundings of the reactor must be shielded from the intense nuclear radiation generated during the fission reactions. Fuel-handling equipment is necessary for loading and for replacing nuclear fuel assemblies after the fission chain reaction depletes the concentration of fissile nuclei. The reactor must also be designed so that it will operate economically, reliably, and safely. These engineering requirements make the configuration of a nuclear power reactor quite complex indeed.

Fission Chain Reactions and Nuclear Criticality

To understand the principal concepts underlying nuclear reactor operation, we need to look at the fission chain reaction process. To maintain a stable fission chain reaction, a nuclear reactor must be designed so that, on the average, exactly one neutron from each fission will induce yet another fission reaction. That is, the production of neutrons from fission must be balanced against their loss by either leakage from the reactor or absorption in nuclear reactions.

For example, suppose that in a particular nuclear system, more neutrons are lost by leakage and absorption than are produced in fission. A self-sustaining chain reaction cannot be achieved, and we say the system is subcritical. One way to alter the system so that there is a more favorable balance between production and loss is simply to make it bigger. Then the probability that a neutron will leak out before being absorbed by a nucleus is decreased, since the average distance a neutron has to travel to leak out increases in such a way that the neutron will undergo more collisions on the way. An alternative approach would be to increase the relative concentration of fissile nuclei. By adjusting the fuel concentration and the size of the reactor, we can balance neutron production against loss and achieve a self-sustaining chain reaction, a critical system.

Actually, it is appropriate to dismiss neutron leakage and reactor size from further consideration since most modern power reactors are so large that few neutrons leak out, usually less than a few per cent. In fact, the size of a power reactor is determined not by the desire to minimize neutron leakage but rather by the need to provide enough space for coolant flow to remove adequately the heat produced by the fission reactions. In reactor design, one first determines how large the reactor must be to accommodate adequate cooling for a desired power output. Then one determines the fuel concentration that will yield a critical reactor of this size.

Suitable Fuels for Fission Chain Reactors

The necessary fuel concentration will depend sensitively on the fuel type. Nuclear engineers characterize the suitability of a material for sustaining a fission chain reaction by a parameter denoted by the Greek letter eta:

η = average number of neutrons produced by fission per neutron absorbed by fuel nucleus

This parameter characterizes not only the relative propensity of a fuel nucleus to fission, but also its ability to shed further neutrons in the fission process, neutrons that can be used to sustain the chain reaction. Evidently, if the fission chain reaction is to proceed in a self-sustained manner, we must utilize fuels that are characterized by values of η greater than 1 since some neutrons will always leak out or will be absorbed in nonfuel materials in the reactor.

There are few heavy nuclei that have values of η sufficiently greater than 1 to be of interest as candidates for nuclear reactor fuel. These fissile materials include the isotopes of uranium, U-233 and U-235, and of plutonium, Pu-239 and Pu-241. Unfortunately, only the isotope U-235 occurs in nature, and then

only as a small percentage (0.711 per cent) of naturally occurring uranium, which is primarily composed of U-238. To utilize uranium as a reactor fuel effectively, it is usually necessary to increase the concentration of U-235, that is, to "enrich" natural uranium in this isotope through the use of elaborate and expensive isotope separation methods.

The other fissile isotopes can be produced artificially by bombarding certain materials with neutrons. For example, U-238 and thorium-232 can be transmuted into the fissile isotopes Pu-239 and U-233, respectively, by exposing them to neutrons in a reactor. In fact, there is usually a sufficient number of excess neutrons produced in a fission chain reaction (since η is greater than 1) so that fertile materials such as U-238 can be transmuted into fissile isotopes such as Pu-239 as the reactor operates. Actually, this conversion of fertile into fissile materials occurs in all modern power reactors since they contain substantial amounts of U-238, which will be transmuted into plutonium during normal operations. For example, a light-water reactor will contain a fuel mixture of roughly 3 per cent U-235 and 97 per cent U-238 in freshly loaded fuel assemblies. After a standard operation cycle of three years, this fuel will contain roughly 1 per cent U-235 and 1 per cent plutonium, which can then be separated out of the spent fuel and refabricated into fresh fuel elements for reloading. This process is referred to as plutonium recycle.

These considerations suggest that it is possible to fuel a reactor with Pu-239 and U-238 and then produce directly the fuel (Pu-239) needed for future operation. Indeed, it is even possible to produce more Pu-239 than is burned, that is, to "breed" new fuel, if η is large enough. To be more precise, one can introduce the conversion ratio:

$$CR = \frac{\text{average rate of fissile nucleus production}}{\text{average rate of fissile nucleus consumption}}$$

From this definition it is apparent that consuming N atoms of fuel during reactor operation will yield $CR \times N$ atoms of new fissile isotopes. Most modern light-water reactors are characterized by a conversion ratio of about 0.6. In contrast, gas-cooled reactors have a somewhat higher conversion ratio of 0.8 and are sometimes referred to as advanced converter reactors. For breeding to occur, the conversion ratio must be greater than 1. For this to happen, η must be greater than 2, since slightly more than one fission neutron is needed to maintain the chain reaction (if we account for neutron leakage or nonfission capture), and one neutron will be needed to replace the consumed fissile nucleus by conversion of a fertile into a fissile nucleus.

To achieve this, we take advantage of the fact that the parameter η depends not only on fuel type but on the average speed of kinetic energy of the neutrons sustaining the chain reaction as well. In general, η becomes larger as the neutron energy increases. Since the average energy or speed of fission neutrons is quite large (fast neutrons), one can maximize the value of η and hence the breeding ratio by simply designing the reactor so that these neutrons do not slow down during the chain reaction. Reactors that optimize the breeding of new fuel by utilizing fast neutrons are known, naturally enough, as fast breeder reactors. A more careful comparison of η for various fissile isotopes makes it apparent that the optimum breeding cycle for fast reactors uses U-238 as the fertile material and plutonium as the fissile fuel. Careful breeder reactor designs can achieve conversion ratios of 1.3 to 1.5 based on this cycle.

But there is a significant drawback to such reactor designs. The probability of a neutron inducing a fission reaction decreases with increasing energy, so that the minimum amount and concentration of fissile material is required for a chain reaction sustained with low-energy, or slow, neutrons. For this reason, most power reactors are designed so that the fast fission neutrons are slowed down or moderated to enhance the probability that they will induce fission reactions. In such reactors, materials of low mass number, such as water or graphite, are interspersed among the fuel elements. Then, as the fission neutrons collide with the nuclei of these "moderator" materials, they rapidly slow down to energies comparable to the thermal energies of the nuclei in the reactor core—which explains the term *thermal reactors* used to describe such designs. Most nuclear power plants use thermal reactors since these systems require the minimum amount of fissile material for fueling and are the simplest and least expensive reactor types to build and operate, even though they are incapable of achieving a net breeding gain.

It is possible, at least in theory, to achieve breeding even in thermal reactors, if a Th-232/U-233 fuel cycle is used. Unfortunately, the maximum breeding ratio one can achieve in practice with this cycle appears to be only slightly greater than 1 (CR is about 1.02). Although this marginal conversion ratio would permit thorium to be used as reactor fuel—in contrast to conventional thermal reactor types such as light-water reactors, which use uranium—it would not produce enough additional fissile material (U-233) to fuel new reactors. A light-water breeder reactor based on the Th-232/U-233 fuel cycle went into operation at Shippingport, Pennsylvania, in 1977.

Nuclear Reactor Operation

The final parameter of interest in fission reactor design is the fuel concentration, which can be chosen to balance neutron production (fission) and loss (leakage and absorption). The fuel concentration can be altered by adjusting the enrichment of the fuel (that is, the percentage of fissile material in the fuel), the density and geometry of the fuel, and the quantity of nonfuel materials in the core. One refers to the amount of fissile material required to achieve a critical fission chain reaction as the critical mass of the fuel. This amount depends sensitively on the composition and geometry of the fuel. For example, the critical mass of a sphere of pure U-235 metal surrounded by a natural uranium reflector is only 17 kilograms. In contrast, the fuel-loading required for a modern light-water reactor is roughly 100 tons of 3 per cent enriched uranium.

It should be noted, however, that at the beginning of its operating cycle a nuclear power reactor is always loaded with much more fuel than is required merely to achieve criticality. The extra fuel is included to compensate for those fuel nuclei destroyed in fission reactions during power production. Since most modern reactors are run roughly one year between refuelings, a sizable amount of excess fuel is needed initially to compensate for fuel burnup as well as to facilitate changes in the reactor power level.

To compensate for this excess fuel, that is, to readjust the balance between neutron production and loss, one introduces materials into the reactor that absorb neutrons from the chain reaction, thereby canceling out the effect of the excess fuel. These absorbing materials can then be withdrawn as the fuel burns up. They can also be used to adjust the criticality of the nuclear reactor in such a way as to control the chain reaction. This might be accomplished in a variety of ways. For example, the neutron absorber might be fabricated into rods that can be inserted into or withdrawn from the reactor at will to regulate the power level of the reactor. That is, one can withdraw the absorber, or control, rods to make the reactor slightly supercritical so that the chain reaction builds up. Then when power has reached the desired level, the rods can be reinserted to achieve a critical, or steady-state, chain reaction. Finally, the rods can be inserted still further into the reactor to shut the chain reaction down.

The longer-term changes in fuel concentration due to burnup are usually compensated for by neutron absorbers fabricated directly into the fuel or dissolved in the coolant. These absorbers, or poisons, will burn up as the fuel burns up, thereby balancing out the excess core fuel. Eventually the fuel will be

so depleted that the reactor can no longer be made critical, even by withdrawal of all the control rods, and the reactor must then be shut down and reloaded with fresh fuel.

An important facet of reactor operation is the stability of the reactor. The probabilities of various neutron interactions (fission or absorption) depend quite sensitively on the temperature and therefore on the power level of the reactor. Nuclear reactors can be designed with negative feedback so that increasing power, and therefore temperature, "brakes" the chain reaction, thereby decreasing power again. The most significant feedback mechanisms involve a decrease in moderator density, and therefore a decrease in moderation, with increasing temperature in thermal reactors and an enhanced tendency of nonfissile material to absorb neutrons with increasing temperature in all reactor types. These mechanisms can be made so effective that reactors tend to operate in a stable fashion. In fact, power reactors are designed to be quite incapable of operating significantly above their designed power level. That is, if somehow the reactor is inadvertently made supercritical—a control rod is accidently withdrawn while the reactor is operating at full power, for example—the power level of the reactor and hence the reactor temperature will increase only slightly before negative feedback returns the reactor to a critical or subcritical state by slowing down the chain reaction. Such inherent feedback mechanisms, coupled with the dilute nature of the reactor fuel, can eliminate any possibility of a runaway chain reaction and remove the concern about possible nuclear accidents involving the chain reaction itself.

A nuclear reactor generates enormous quantities of hazardous radioactive material (primarily fission products). If somehow the reactor were to crack open and release this material into the environment, it could pose a significant danger to public health. Therefore a number of physical barriers are designed into the reactor to contain radioactive materials and prevent their release to the environment under any conceivable condition.

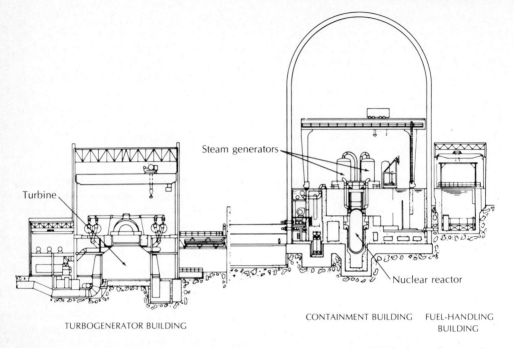

Steam generators

Turbine

Nuclear reactor

TURBOGENERATOR BUILDING

CONTAINMENT BUILDING   FUEL-HANDLING
                                                                  BUILDING

Figure 15-1. A schematic diagram of a nuclear power plant. (G. Masche, Westinghouse Systems Summary, 1971)

Precautionary measures include cladding of the fuel pellets in metal tubes, encasement of the reactor core in a steel pressure vessel, and the surrounding of it all with concrete shielding and steel-lined, reinforced concrete walls.

To guard against even low-probability accidents, reactor designers consider very serious events such as a core meltdown accident, in which the reactor core suddenly loses cooling, because of a break in a large coolant pipe, for example. Although the chain reaction would shut down immediately because of the loss of moderation, the heat produced by the radioactive decay of fission products would be sufficient to melt the reactor fuel elements if auxiliary cooling were not provided. The molten fuel could then slump to the bottom of the reactor vessel, possibly melting through it and the concrete floor of the reactor building and releasing fission products to the environment. We shall return

in a later section to discuss how engineered safety systems are incorporated into reactor design to guard against such an occurrence.

NUCLEAR POWER GENERATION

Nuclear power reactors produce heat that can be used to generate electric energy, usually by way of a steam thermal cycle. In this sense, the primary function of a reactor is that of an exotic heat source for turning water into steam. Aside from the nuclear reactor and its associated coolant system, nuclear power plants are remarkably similar to large fossil-fuel-fired power plants (Figure 15-1). Only the source of the heat energy differs— nuclear fission versus chemical combustion. Many components of large central-station power plants are common to both nuclear and fossil-fuel units.

The current generation of power plants operates on a steam cycle in which the heat

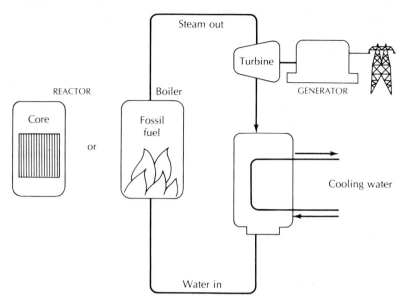

Figure 15-2. Components of electrical generating plants.

generated by combustion or fission is used to produce high-temperature steam. This steam is then allowed to expand against the blades of a turbine (Figure 15-1). In this way, the thermal energy of the steam is converted into the mechanical work of turning the turbine shaft. This shaft is connected to a large electric generator that converts the mechanical turbine energy into electric energy that can be distributed over an electric power grid. The low-pressure steam leaving the turbine is condensed into liquid in a steam condenser so that it can be pumped back to the steam supply system to complete the cycle. The condenser requires large quantities of cooling water at ambient temperature, usually obtained from artificial cooling ponds or cooling towers.

Nuclear Steam Supply Systems

At the heart of a nuclear power plant is the nuclear steam supply, which produces the steam used to drive the turbine generator. The system generally consists of three major components: (1) the nuclear reactor that supplies the fission heat energy, (2) several primary coolant loops and pumps that circulate a coolant through the nuclear reactor to extract the fission heat, and (3) heat exchangers, or steam generators, that use the heated primary coolant to turn water into steam. The steam supply system in a modern nuclear power plant is completely enclosed within a containment structure designed to prevent the release of radioactivity to the environment in the event of a gross failure of the primary coolant system. This nuclear island within the plant is the analog to the boilers in a fossil-fuel unit (Figure 15-2).

The primary component of the steam supply system is the nuclear reactor. A modern power reactor is a very complicated system designed to operate under severe conditions of temperature, pressure, and radiation. The energy released by nuclear fission reactions appears primarily as kinetic energy of the various fission fragment nuclei. The bulk of this fission product energy is rapidly deposited as heat in the fuel. This heat is extracted

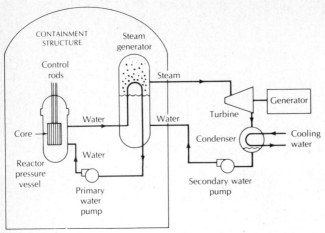

(a) Pressurized-water reactor

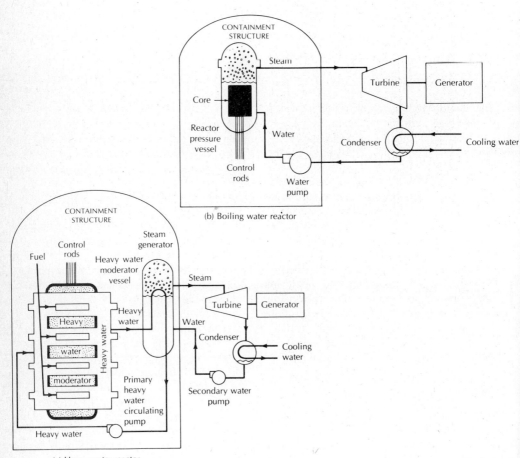

(b) Boiling water reactor

(c) Heavy-water reactor

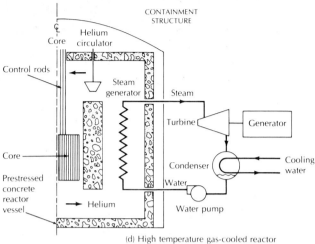

(d) High temperature gas-cooled reactor

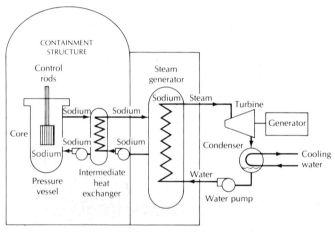

(e) Liquid metal fast breeder reactor

Figure 15-3. Principal types of nuclear steam supply systems. Schematic diagrams of (a) pressurized-water reactor, (b) boiling-water reactor, (c) heavy-water reactor, (d) high-temperature gas-cooled reactor, (e) liquid metal fast breeder reactor.

by the primary coolant flowing between the fuel elements and transported by this coolant to the steam generators.

A variety of coolants can be used in the primary loops of the steam system (Figure 15-3). In fact, nuclear reactor types are usually characterized by the type of coolant they use, such as "light-water" reactors or "gas-cooled" reactors. There are also various possible steam system configurations. For example, one may actually produce the steam in the reactor itself. Or one may use a liquid or gas primary coolant such as high-pressure water or helium to transfer the fission heat energy to a steam generator. In the liquid metal cooled fast breeder reactor, an intermediate coolant loop is used to isolate the water in the steam loop from the chemically

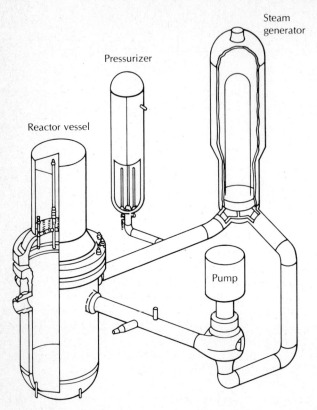

Figure 15-4. A nuclear steam supply system (NSSS) including a nuclear reactor, a coolant loop, and a steam generator. (Systems Summary of a Westinghouse PWR Nuclear Power Plant, G. Masche, 1971)

reactive and highly radioactive sodium in the primary loop.

The most common coolant used in power reactors today is ordinary water, which serves as both a coolant and a neutron-moderating material. There are two major types of light-water reactors (LWRs): pressurized-water reactors (PWRs) (Figure 15-3a) and boiling-water reactors (BWRs) (Figure 15-3b). In a pressurized-water reactor (Figure 15-4), the primary coolant is water, maintained under high pressure (150 atmospheres) to allow high coolant temperatures (300°C) without steam formation within the reactor. The heat transported out of the reactor core by the primary coolant is transferred to a secondary loop containing the working fluid (the steam sys-

tem) by a steam generator within which the inlet water is converted into steam. Such systems typically contain from two to four primary coolant loops and associated steam generators. In addition, a surge chamber, or pressurizer, is connected to the primary coolant loop to regulate the high primary system pressure as it accommodates coolant volume changes in the primary loop. The primary loop also contains coolant pumps, as well as auxiliary systems that control coolant purity and inject new makeup water or control absorbers into the coolant.

In a boiling-water reactor, the primary coolant water serves not only as moderator and coolant but also as working fluid since the system pressure is kept sufficiently low

(70 atmospheres) that boiling and steam formation can occur within the reactor. In this sense, the reactor itself serves as the steam generator, thereby eliminating the need for a secondary loop and heat exchanger. Since there is an appreciable steam volume in the primary loop, a pressurizer is not required to accommodate pressure surges.

The coolant water rising to the top of the BWR core is a wet mixture of liquid and vapor. Moisture or steam separators must be used to separate off the steam, which is then piped outside the reactor pressure vessel to the turbine and then through the condenser before it is pumped back into the core as liquid condensate. The saturated liquid that is separated off by the moisture separators flows downward around the core and mixes with the return condensate. This recirculation is assisted by pumps.

In both PWRs and BWRs, the nuclear reactor itself and the primary coolant are contained in a large steel pressure vessel designed to accommodate the high coolant pressures and temperatures. In a PWR, this pressure vessel must be fabricated with thick steel walls to contain the high primary coolant pressures. The BWR pressure vessel need not be so thick-walled, but it must be larger to contain both the nuclear reactor core, which has a lower power density than that of the PWR, and the steam moisture separating equipment.

The direct cycle of the BWR does present one major disadvantage. Since the working fluid passes through the reactor core before passing out of the containment structure and through the turbine, one must design the system to deal with the resulting radiation hazards. The primary coolant water must be carefully treated to remove any impurities that might become radioactive when exposed to the neutrons in the reactor. Even with this purification, the primary coolant will exhibit significant induced radioactivity, and therefore the turbine must be heavily shielded.

A closely related class of reactors uses heavy water, D_2O, as moderator and either D_2O or H_2O as primary coolant (Figure 15-3c). The most popular heavy-water reactor is the CANDU-PHW reactor (Canadian deuterium uranium pressurized heavy-water reactor). This reactor uses a pressure tube design in which each coolant channel in the reactor accommodates the primary system pressure individually, thus eliminating the need for a pressure vessel. As with a PWR, the primary coolant thermal energy is transferred by means of a steam generator to a secondary loop containing light water as the working fluid. One major advantage of heavy-water reactors is their ability to use natural uranium (with only 0.711 per cent U-235) as fuel as a result of the superior neutron-moderating properties of deuterium. More recently, heavy-water pressure tube reactors have been designed that produce H_2O steam directly in the core in a manner similar to a BWR (the CANDU boiling light-water reactor and the steam-generating heavy-water reactor).

Gas-cooled nuclear reactors have been used for central-station power generation for many years. The earliest such power plants were the Magnox reactors developed by the United Kingdom, which used CO_2 as the coolant for a natural uranium-fueled, graphite-moderated core. More recently, interest has shifted toward the high-temperature gas-cooled reactor (HTGR) that uses high-pressure helium to cool an enriched uranium/thorium core moderated with graphite (Figure 15-3d). To date, all such reactors have been operated with a two-loop steam thermal cycle similar to that of a PWR, in which the primary helium coolant loop transfers its thermal energy through steam generators to a secondary loop containing water as the working fluid. The HTGR is capable of operating at relatively high temperature, thereby producing high-temperature (400°C), high-pressure (60 atmospheres) steam with an attendant increase in thermodynamic efficiency. Moreover, HTGRs have the potential for running in a direct cycle configuration using high-temperature helium to drive a gas turbine (with thermal efficiencies approaching 50 per cent).

The HTGR exhibits several other advantages. The use of helium as a coolant not only allows higher operating temperatures at moderate pressures but also provides flexibility in the selection of optimum coolant temperature, pressure, and flow rate conditions. It also mitigates the effects of any loss-of-coolant accident. Since the coolant always remains in the gaseous phase, the worst that can happen in the event of a rupture of the primary coolant loop is a loss of pressure. And in an HTGR, only a modest circulation of helium at atmospheric pressure is needed to remove the radioactive decay heat given off by the core following shutdown. The higher operating temperatures in the HTGR give an edge to a U-235/Th-232/U-233 fuel cycle over the enriched uranium/plutonium fuels of light-water or heavy-water reactors (although these latter reactor types could also be fueled with a uranium/thorium mixture).

The gas coolant does lead to low power densities and therefore to large reactor sizes. Furthermore, since the fissile material in such reactors is highly enriched U-235 (which is then mixed with thorium), the HTGR presents a problem from the viewpoint of proliferation of nuclear weapons materials. Nevertheless, these reactors have been under development in both the United States and West Germany.

Gas coolants have also been proposed for use in fast breeder reactors (the gas-cooled fast reactor, GCFR). Because of the very high power density required by such reactors, extremely high coolant flow rates would be required. Nevertheless, the large conversion ratios (CR of 1.5) achievable in the GCFR make it a promising alternative to other fast reactor designs that use liquid metals such as sodium.

Although sodium could be used in thermal reactors if alternative moderation was provided, its primary advantages occur in fast breeder reactors, which require a primary coolant with low moderating properties and excellent heat transfer properties. We have noted that the nuclear steam supply system for the liquid metal cooled fast breeder re-

actor (LMFBR) is actually a three-loop system since an intermediate sodium loop is used to separate the highly radioactive sodium in the primary loop from the steam generators (Figure 15-3e).

Nuclear Reactor Components

To introduce the components and system that make up a nuclear power reactor, we shall consider a pressurized water reactor. The reactor proper consists of a core containing the fuel, coolant channels, structural components, control elements, and instrumentation systems (Figure 15-5). The core is a cylinder-shaped lattice roughly 350 centimeters in height and consisting of long fuel assemblies, or bundles. Each fuel assembly is composed of several hundred long metal tubes, the fuel elements, or fuel pins, that contain small ceramic pellets of uranium dioxide (Figure 15-6). Most modern power reactors use such ceramic fuels as either an oxide, carbide, or nitride to facilitate high-temperature operation. The fuel element tube, or cladding, is either stainless steel or a zirconium alloy, which not only provides structural support for the fuel but also retains radioactivity produced in the fuel during operation. The primary coolant flows up through the fuel assemblies between the fuel elements, extracting fission heat. Fuel is loaded into a reactor core one fuel assembly at a time. A typical power reactor core will contain hundreds of such fuel assemblies.

The reactor core itself, the structures that support the core fuel assemblies, the control assemblies, and the coolant circulation channels are all contained in a pressure vessel that is designed to withstand the high pressures of the coolant and damage from the radiation generated in the core. The pressure vessel has inlet and outlet nozzles for each steam loop. The cap, or head, of the vessel can be removed for refueling and maintenance. The reactor pressure vessel, along with the rest of the nuclear steam supply system, is enclosed in a reactor containment structure that prevents the release of radioactivity to the

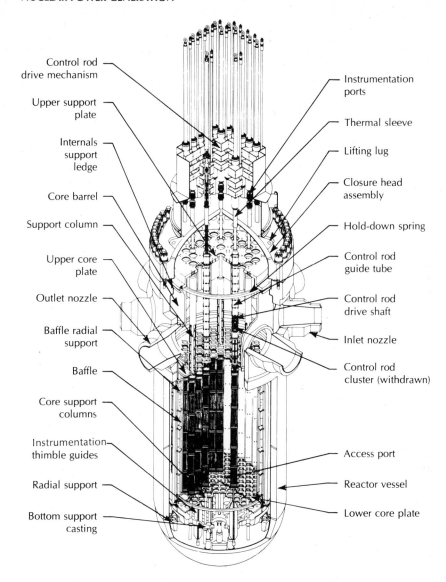

Control rod
drive mechanism

Upper support
plate

Internals
support
ledge

Core barrel

Support column

Upper core
plate

Outlet nozzle

Baffle radial
support

Baffle

Core support
columns

Instrumentation
thimble guides

Radial support

Bottom support
casting

Instrumentation
ports

Thermal sleeve

Lifting lug

Closure head
assembly

Hold-down spring

Control rod
guide tube

Control rod
drive shaft

Inlet nozzle

Control rod
cluster (withdrawn)

Access port

Reactor vessel

Lower core plate

Figure 15-5. A pressurized water reactor (Westinghouse).

environment in the event of a gross failure of
the reactor coolant system. This nuclear island within the plant is usually a steel-lined
concrete structure and contains not only the
reactor itself, but the coolant recirculation
pumps, steam piping, and other auxiliary
systems.

Nuclear Reactor Safety

A recent report of a study of the risks posed
by nuclear reactors contains the following
statement:

Risks have been frequently categorized according
to several dichotomous factors such as whether

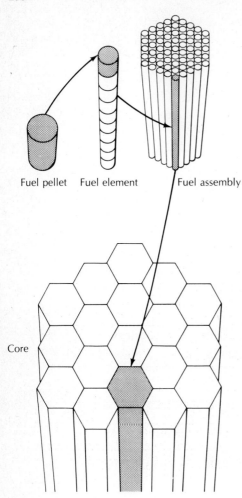

Fuel pellet   Fuel element   Fuel assembly

Core

Figure 15-6. Fuel arrangement in a reactor core.

the exposure to risk is voluntary, new, common, catastrophic, dreaded, lethal, or man-originated. Nuclear power is unique in that it is in a category by itself on these perceptual scales. It is perceived as new, uncommon, dreaded, most likely lethal, involuntary, and potentially catastrophic.

It has been suggested by some observers that much of the opposition to nuclear power is due not to concerns about risks from nuclear power itself but rather to concerns about society and the direction of its future development. However, it is likely that some pub-

lic concern has been generated by the large amount of effort that has been expended to reduce the risk of operating nuclear power plants. No other industry has been subjected to so searching an investigation of risks.

The principal risk from nuclear power plants comes from the large quantity of radioactive fission products generated during operation. In normal operation almost all of these are contained inside the reactor vessel, where they cause no harm. Only if there occurs an accident that causes the release and transport of large amounts of radioactive material from the reactor to populated areas will people be injured. Light-water reactors of the kind being built today simply cannot explode like a bomb. An explosive device requires material containing a high concentration of plutonium or U-235. Light-water reactor fuel typically contains uranium with only about 3 per cent U-235, a concentration much too low to support a nuclear explosion.

Nuclear reactors are designed, constructed, and operated under rigorous standards for design, carefully selected materials, detailed inspection procedures during construction and operation, and highly trained operating staffs. These efforts are intended to make the probability of an accident that will release fission products from the reactor vessel extremely low. But beyond efforts at prevention, which are aimed to ensure that serious accidents will have a very low probability, features are added to mitigate the consequences of even the low-probability accident. This approach to risk reduction is often described as "defense in depth." After having used unusual care in design, construction, and provisions for operating the plant, those responsible for safety go beyond these efforts and include structures, systems, and emergency procedures that reduce risk even further.

In normal operation and even for those abnormal events that might be expected to occur during the life of a plant, a number of barriers are designed to contain the fission products. The first of these is the ceramic fuel pellet itself, which entrains most of the non-gaseous fission products and greatly inhibits the diffusion of gaseous fission products out of the fuel. The fuel pellets are contained in metallic tubes, or cladding, which are designed to retain even the gaseous fission products that build up in the gap between the fuel pellet surface and the cladding tube. The fuel elements are contained within a steel pressure vessel, 20 centimeters thick, which serves as yet a third barrier. The primary coolant loop piping is some 8 to 10 centimeters thick, and the coolant water is circulated through filters to remove any radioactive material. The pressure vessel is surrounded by 2 to 3-meter-thick concrete shielding. Only a very severe accident will lead to release of radiation or radioactive material beyond these barriers.

However, to cope with even a low-probability event, an additional barrier is added in the form of a containment structure. A typical containment structure has a 1-meter-thick reinforced concrete wall lined with a leak-tight steel shell, typically about 0.5 to 1.5 centimeters thick, which is designed to prevent the release of radioactive material even in the event of a major rupture of the primary coolant system. The plant itself is located within an exclusion area to which the operating utility controls access and which separates the plant from the public.

In addition to the usual systems for controlling power production, nuclear plants are equipped with highly reliable safety systems designed to take over and restore the plant to a safe condition in the event of a control system malfunction or an operator error. Furthermore, the plants are equipped with added systems which protect against very unlikely but potentially severe accidents, such as a complete rupture of the largest coolant pipe connecting the reactor vessel to the steam generator. (These emergency systems remove decay heat even after such an accident, thus preventing major fuel damage.) The containment structures are designed to withstand aircraft crashes as well as severe natural phenomena, such as tornados. All the systems important to safety are designed to

survive the largest anticipated earthquake. In addition, an emergency plan that can provide, for example, evacuation of the surrounding population is required by federal regulation.

In the more than four hundred reactor years of commercial reactor operating experience in the United States, there has not been a reactor accident in which a member of the public has been injured or in which damage to off-site property has occurred. Nevertheless, after the accident at the Three Mile Island nuclear power plant in 1979, additional efforts at risk reduction have been undertaken. These include additional refinements of control systems and of systems designed to remove fission product decay heat from the reactor in the event of an accident. Further emphasis is also being given to the training of operators and to the providing of technical staff support for dealing with accidents.

Environmental Impact of Nuclear Power Plants

A major source of the environmental impact of any large power plant is the material released into the air and the water in the neighborhood of the plant. Large fossil-fuel plants, even with the use of the latest systems for removing combustion products and sulfur from the exhaust, discharge large quantities of waste in solid and in liquid form that go directly into the air or into nearby lakes and streams. Because nuclear plants do not produce any combustion products, they are free from this environmental insult. However, they do release small quantities of radioactive materials into the environment during normal operation (Table 15-1).

All large electric generating plants which use a thermal cycle produce significant quantities of waste heat. A modern generating plant will release to the environment 50 to 70 per cent of the thermal energy generated by combustion or by fission. A nuclear plant that generates 1000 MWe will discharge about 2000 MW of heat into a nearby lake or river or into the atmoshere. The same size fossil-

Table 15–1.   Waste Material from a 1000-MWe Power Plant.

	Coal-Fired	Nuclear
Thermal efficiency (%)	39	32
Thermal wastes (MW)	1570	2120
Solid wastes		
Fly ash or slag (tons/year)	330,000	0
(cubic feet/year)	7,350,000	0
(railroad cars/year)	3300	0
Radioactive wastes		
Spent fuel (assemblies/year)	0	160
(railroad cars/year)	0	5
Gaseous and liquid wastes (tons/year)		
Flyash (particulates)	2000	0
Sulfur dioxide	24,000	0
Carbon dioxide	6,000,000	0
Carbon monoxide	700	0
Nitrogen oxides (as NO_2)	20,000	0
Mercury	5	0
Arsenic	5	0
Lead	0.2	0
Radioactive gases or liquids		
(maximum whole-body dose at		
plant boundary in mrem/year)	1.9	1.8

fuel plant, because of a somewhat higher thermal efficiency, will discharge about 1500 MW as heat. Hence, although both plants produce large quantities of waste heat, the type of light-water reactor being operated in the United States today discharges the larger amount. Because of the large quantities of waste heat produced, most modern large power plants now being built, whether fossil-fuel or nuclear, use cooling towers which, much in the manner of an automobile radiator, discharge the heat directly to the atmosphere.

Although nuclear plants release some radioactive materials to the environment during operation, they must be designed and operated so that the radiation exposure produced by such release is far below that judged to have any significant public health effects. In practice, the amount of radioactive materials released from nuclear plants is kept so low that it produces an increased radiation exposure to an individual near the plant of not more than about 1 per cent of that which would be received from such natural sources as cosmic rays or naturally occurring radia-

tion in the environment. Those living some distance away receive an even smaller exposure (Table 15-2).

One important difference in environmental effects that could have important long-range consequences is the absence of carbon dioxide (CO_2) production by nuclear power plants. A recent report by the Council on Environmental Quality calls attention to the increase in atmospheric CO_2 concentration being observed with our increasing use of fossil fuels. The report concludes that although the probability of a severely damaging worldwide temperature rise caused by the greenhouse effect associated with CO_2 buildup cannot be predicted with certainty, it is sufficiently great that it represents an unprecedented risk.

The Licensing and Regulation of Nuclear Power Plants

The Nuclear Regulatory Commission (NRC), a federal agency established by the U.S. Congress, has been given the responsibility

Table 15–2. Average Environmental Radiation Doses to the General Population.

Source	Dose (mrem/year)
Natural radiation [a]	
Cosmic radiation	28
Cosmogenic radionuclides	0.7
External terrestrial	26
Radionuclides in the body	24
Rounded total	80
Manmade radiation [b]	
Medical and dental X-rays	50–100
Fallout (average from 1951 to 1976)	7
Luminous watches	
Radium	3
Tritium	0.5
Airplane travel	0.3 mrem/hour
Color television	Negligible
All consumer products	<1
Nuclear power (400-gigawatt) [c]	0.2
Rounded total	60–110
Total background dose	140–190

[a] National Council on Radiation Protection, Report No. 45 (1975), p. 108.
[b] Report of United Nations Scientific Committee on the Effects of Atomic Radiation (1977).
[c] B. L. Cohen, *American Scientist* **64,** 550 (1976).

for licensing and regulating the construction and operation of nuclear power plants. The NRC must determine that plants are operated without producing undue risks to public health and safety, that the environmental impact of construction and operation is evaluated according to provisions of the National Environmental Policy Act, and that operation does not violate federal antitrust laws.

An application for a license to construct a nuclear power plant must be accompanied by a detailed description of the proposed plant site, a voluminous report which describes the plant in sufficient detail that a safety evaluation of plant operation can be made, an environmental report which permits the environmental costs to be compared with the benefits of operating the plants, and financial and legal information from which can be determined the licensee's capability to pay for construction, operation, and decommission-

ing, as well as conformity to antitrust regulations.

Over a period of perhaps two years, the NRC technical and legal staffs, with the aid of consultants, examine this documentation, typically assisted by several meetings with the applicant (Table 15-3). It is not unusual for the applicant to be asked for additional information. Frequently there will be negotiations that may result in changes in the proposed structures, equipment, or method of operation. If the NRC staff is finally able to approve the application, public hearings are then held by an Atomic Safety and Licensing Board (ASLB) appointed for this purpose by the NRC. In the course of the public hearings, intervenors may participate by presenting evidence and by cross-examining members of the staff of the NRC and of the applicant. If after hearing and considering the available information and testimony the board concludes that a license should be granted, it makes this recommendation to the NRC, which may then grant the license.

When a licensee is within about two years of completing construction, the process is repeated, with current information added to that contained in the original construction permit application. This time the application is for an operating license.

In addition to the detailed review of the application and the granting of the operating license for the plant, the NRC also licenses those who operate the plant. Several years of study followed by operation of elaborate and detailed plant simulators and by experience in operating plants are required before an operator can be licensed.

The NRC staff also inspects the plant during construction to ensure that the required high standards of materials quality and of construction practice are being followed, and to determine if the finished product is acceptable. Once the plant goes into operation, NRC teams make periodic inspections of the plant and its operators. At some plants inspectors are assigned to the plant full-time. Current plans call for all plants eventually to have an on-site NRC inspector.

Table 15–3. Steps in the Licensing of Nuclear Power Plants.

1. Applicant applies for Construction Permit (CP) by submitting Preliminary Safety Analysis Report (PSAR) to Nuclear Regulatory Commission (NRC). Environmental Report submitted concurrently. Reports cover safety, environmental impact, financial capability, and antitrust implications.
2. NRC staff reviews, and on basis of review, which includes correspondence and meetings with applicant, prepares Safety Evaluation Report (SER) and Draft Environmental Statement (DES).
3. Advisory Committee on Reactor Safeguards (ACRS) reviews safety-related part of application, making use of PSAR and SER and holding meetings with NRC staff and applicant. ACRS prepares report to NRC.
4. Atomic Safety and Licensing Board (ASLB), using PSAR, SER, DES, and ACRS report, holds public hearings. The Board may receive comments and testimony either on the documents noted above or on other issues raised by interested parties. It may also ask for any additional information deemed relevant. It recommends for or against granting a Construction Permit. Recommendation goes to NRC, which can reverse ASLB. Any party in the proceeding can appeal to ASLB or NRC.
5. If application is approved, a Construction Permit is granted, permitting the applicant to build plant.
6. Near the end of construction, applicant prepares and submits Final Safety Analysis Report (FSAR) along with updated Environmental Report. The above review process is repeated, with the exception that if the application is not contested, the ASLB is not required to hold public hearings.
7. If safety, environmental impact, and financial–antitrust reviews are approved, the NRC grants Operating License (OL).

The process of getting a plant licensed, built, and in operation may require 10 to 12 years. During this period, new information or experience may dictate changes in plant design or operation to decrease risk. And even after the plant goes into operation, changes can be required if the NRC concludes that a significant increase in safety can be achieved thereby.

THE NUCLEAR FUEL CYCLE

The safety and environmental impact of nuclear power plants have become the subjects of increased public concern during the past decade. But the generating plants are only one aspect of nuclear power. More recently, public attention has encompassed the entire nuclear fuel cycle—the operations involved in the preparation, utilization, reprocessing, and disposing of nuclear fuels.

Nuclear fuels are totally different from fossil fuels in several respects. Nuclear fuel material, such as uranium, must undergo a number of sophisticated and expensive processing operations before it is inserted into the reactor core. It is then "burned" in the reactor for several years before being removed. Even after several years of use, the fuel possesses a significant concentration of fissile material. Therefore, after the spent fuel is removed from the reactor core, it can be chemically reprocessed to extract the unused fissile material, which can then be refabricated into new fuel elements. The by-product waste from the reprocessed fuel includes highly radioactive fission products, and its disposal requires great care.

The safety and environmental impact of each nuclear fuel cycle activity must be examined carefully in any consideration of the future role of nuclear power. Moreover, since the primary economic advantages exhibited by nuclear plants are a consequence of their extremely low fuel costs, the economics of nuclear fuel preparation, reprocessing, and disposal must also be considered.

The various stages of the nuclear fuel cycle are illustrated in Figure 15-7. These include mining, milling, enrichment, fuel fab-

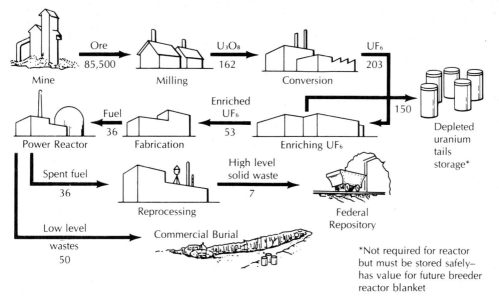

Figure 15-7. Annual quantities of fuel material required for routine (equilibrium) operation of 1000-MW light-water reactor. (The Nuclear Industry, USAEC Report WASH-1174-73, 1973)

rication, fuel burning, spent fuel storage and decay, spent fuel reprocessing, and radioactive waste disposal.

Mining

Most of the uranium ore mined in the United States comes from the sandstone deposits in the Colorado Plateau and the Wyoming Basin. Both underground and open-pit techniques are used in a manner similar to that used in the mining of other low-grade ore. There is considerable uncertainty about the extent of our domestic uranium ore resources. Present U.S. Department of Energy estimates place proven plus probable plus possible resources of high-grade uranium ore at $890,000 + 1,395,000 + 2,080,000 = 4,365,000$ tons. These estimates suggest that there should be no difficulty in fulfilling lifetime ore commitments for light-water reactors through the year 2000 for installed capacities ranging from 300 to 500 GW. Beyond this point, further expansion of light-

water reactor capacity becomes dependent on additional sources of uranium (Table 15-4).

Milling

Milling is necessary to extract and concentrate uranium from the raw ore. The ore is first pulverized, and then a solvent extraction process is used to produce yellowcake, a crude oxide containing some 70 to 90 per cent U_3O_8. The mill tailings produced in this process are mildly radioactive and represent a hazard if not handled properly.

Enrichment

Essentially all power reactors (with the exception of the Canadian heavy-water reactors or the early British gas-cooled graphite-moderated reactors) utilize enriched uranium, that is, uranium with higher than the natural 0.711 per cent concentration of U-235. The enrichment of uranium is a difficult and expensive process since it involves sep-

Table 15–4.   Lifetime Uranium Requirements for Various Reactor Types and Fuel Cycles.[a]

Reactor Types	Fuel Cycle	Uranium Requirements (in short tons)
Light-water reactor	U, no recycling	6410
	U, U recycling	5280
	U and Pu recycling	4340
	Th + U, U recycling	3650
Heavy-water reactor	Natural U, no recycling	5263
	Natural U, Pu recycling	2861
	Pu-Th, U recycling	2210
High-temperature gas-cooled reactor	U-235 + Th, U recycling	2970
Liquid metal cooled fast breeder reactor	U + Pu, recycling	60

[a] Based on a 1,000 megawatt-electric plant operating for thirty years at 80 per cent capacity factor. From T. H. Pigford, *IEC Fund* **16,** 61 (1977).

aration of two isotopes, U-235 and U-238, that have little mass difference and essentially no chemical difference. A variety of techniques have been used or proposed, including electromagnetic separation, gaseous diffusion, gas centrifuge or nozzle separation, and laser isotope separation. At the present time, gaseous diffusion continues to be the most common method of uranium enrichment, although both gas centrifuge and nozzle separation plants are under construction and laser methods hold promise for the future. Enrichment likely will remain an expensive process; it presently accounts for some 30 per cent of nuclear fuel costs and some 75 per cent of the energy investment in nuclear power generation.

Fuel Fabrication

Enriched uranium is chemically converted into a ceramic powder such as UO_2 or UC and compacted into small pellets. These pellets are loaded and sealed into metal tubes to produce reactor fuel elements, which are then fastened together into bundles known as fuel assemblies.

Fuel Burnup in the Reactor Core

The fuel assemblies are loaded into the reactor core for power production and are typ-

ically irradiated for several years. The fuel lifetime may be limited by criticality, that is, the fissile concentration may drop too low to sustain the chain reaction, or by radiation damage sustained by the fuel elements during operation.

Spent Fuel Storage and Decay

After being irradiated in the reactor core, the fuel is intensely radioactive with fission products and other radioactive nuclei as a result of neutron absorption. The spent fuel is removed from the core and stored in water pools in the plant for several months to allow the short-lived radioactive nuclei to decay away. It is then loaded into heavily shielded and cooled casks for shipment to reprocessing or storage facilities by either truck or rail. The shipping containers are carefully designed, and prototypes are tested to ensure their integrity in the event of any conceivable shipping accident.

Spent Fuel Reprocessing

The spent fuel discharged from a nuclear power reactor contains a significant quantity of fissile material. For example, each kilogram of spent light-water reactor fuel contains roughly 8 grams of unused U-235 and

6 grams of fissile plutonium. This material can be extracted and reloaded into fresh fuel elements. The principal scheme for commercial recovery of uranium and plutonium from low-enrichment light-water reactor fuels is the Purex process. This method has been used on a commercial basis throughout the world for almost two decades. However, the plutonium separated from spent reactor fuel is a possible source of material for nuclear weapons. Therefore the U.S. government has deferred indefinitely the reprocessing of spent commercial power reactor fuel. A number of other nations are either operating or constructing spent fuel reprocessing facilities and have plans to proceed with plutonium recycling in light-water reactors.

Radioactive Waste Disposal

Considerable public attention concerning the nuclear fuel cycle has been directed toward the disposal of high-level radioactive waste produced by nuclear power reactors. Much of the radioactivity produced in power reactors will decay away quite rapidly following reactor shutdown and removal of spent fuel elements from the reactor core. However, a significant fraction of the high-level radioactivity induced in the fuel is due to fission products and actinides that will remain toxic for hundreds or even thousands of years. To date, there has been no commitment to a specific scheme for permanent disposal of these wastes. In the event that the present moratorium on spent fuel reprocessing continues (a "stowaway" fuel cycle), the radioactive spent fuel assemblies will be placed indefinitely in a retrievable storage facility. The most attractive proposal for the handling of high-level waste involves its conversion to a stable form, such as glass or cement, followed by encapsulation in stainless steel canisters. The waste would then be shipped to a federal waste repository for permanent disposal. This would involve deep burial beneath the earth in rock formations characterized by exceptional geologic stability.

This sequence of operations characterizes the uranium-plutonium fuel cycle used for light-water reactors. Even this cycle has a number of possible options: (1) a "throwaway" cycle in which the spent fuel is treated as waste for permanent disposal, (2) a "stowaway" fuel cycle in which the spent fuel is stored in a manner that does not preclude the later recovery and utilization of its fissile energy content, (3) reprocessing to recover uranium only, and (4) reprocessing to recover both uranium and plutonium. The choice among these various options is not dictated by ore resources and economics alone, but by political considerations as well because of the potential impact of the nuclear fuel cycle on nuclear weapons proliferation.

A variety of other fuel cycles can be employed in different reactor types. For example, the thorium/U-233 fuel cycles can be used in light-water, heavy-water, or high-temperature gas-cooled reactors, although in all cases fissile material such as U-235 or plutonium must be added to thorium to achieve criticality. The greatest potential for resource utilization is provided by breeder reactors such as the liquid metal cooled fast breeder reactor and the gas-cooled fast breeder reactor. In these reactor types, some 30 to 50 per cent more fissile material is produced by neutron transmutation of uranium-238 into plutonium than is consumed during power production. In this way the breeder reactor can utilize some 60 to 70 per cent of the energy available in natural uranium compared with the roughly 1 to 2 per cent used by light-water reactors. The more efficient utilization of available uranium fuel resources is the primary justification for such breeder reactors. These reactors can be fueled either with natural uranium (or thorium) or with depleted uranium from the tails produced by enrichment plants.

THE FAST BREEDER REACTOR

We have noted that the present generation of light-water fission reactors can effectively utilize only a small fraction of our uranium ore resources. Hence the resource base available to this type of reactor is rather limited, being of the same magnitude as the domestic reserves of petroleum and natural gas. Although these reserves should prove sufficient to fuel several hundred of these power plants and for their operating lifetimes, it is apparent that this type of nuclear reactor must be regarded as only an interim, relatively short-range energy source.

A far more attractive reactor concept is the breeder reactor, which can utilize as much as 70 to 80 per cent of the available uranium (or thorium) resources by converting U-238 (or Th-232) into plutonium (or U-233), which could then be reprocessed into fuel for either the original or other breeder reactors. The reserves of high-grade uranium and thorium ores are sufficient to support a fast breeder reactor economy for thousands of years. In fact, we have already stockpiled over 200,000 tons of U-238 as the tails of the uranium-enrichment plants used in the light-water reactor fuel cycle. Furthermore, sufficient plutonium has been produced in light-water reactors to date to provide the initial fuel charge for several breeder reactors.

The fast breeder reactor occupies a rather unique position among our long-range energy options, since its scientific and technical feasibility and, to some extent, its commercial viability have already been established (in sharp contrast to other long-term options such as solar, geothermal, or nuclear fusion power). Fast breeder reactors have been built in this country and abroad for several decades. Indeed, France, the United Kingdom, and the Soviet Union have been successfully operating demonstration breeder reactor power plants for several years. Furthermore, the French are bringing on line a commercial prototype breeder reactor (Super Phenix) rated at 1250 MWe and expect to market commercial fast breeder reactors by the mid-1980s.

But there are major hurdles that must be overcome if we are to realize even a small fraction of the enormous potential of this energy source. Unfortunately, the most serious barriers to successful deployment of the breeder reactor are of a social and political rather than a technical or economic nature. Foremost among these barriers is the reluc-tance of the public to support breeder reactor development because they perceive such reactors as inherently more dangerous than light-water reactors. In fact, the safety and environmental impact of a breeder reactor power plant is quite similar to that of more conventional light-water reactor power plants.

An equally serious problem involves the fuel cycle of the breeder reactor that utilizes larger amounts of plutonium as fuel than does the light-water reactor. This poses a problem from the viewpoint of the proliferation of strategic nuclear weapons material—a problem that is similar in nature but considerably more serious than that posed by plutonium recycling in light-water reactors.

THE FUTURE OF NUCLEAR POWER

So what is the future of nuclear power? What role should it play in meeting the energy needs of our society, in helping to decrease our dependence on foreign oil and to slow our rush toward the exhaustion of conventional energy resources? We believe that nuclear power is a necessary component in our future if we are to meet our energy needs and those of developing nations in the face of declining reserves of conventional fossil fuels. There appears to be no viable alternative. Coal production and conservation measures alone cannot be expected to fill the gap created by declining petroleum and natural gas reserves. New energy sources, such as solar electric power or nuclear fusion, simply will not be available for massive implementation until well after the turn of the century—if then.

A considerable degree of caution and conservatism must be exercised in speculating about the future of nuclear power, however. The massive implementation of this energy source poses enormous difficulties, not the least of which is an adequate public acceptance of this still rather largely misunderstood technology. Equally serious are the enormous capital requirements of nuclear plants and the bewildering complexity of regulations that threaten to delay and choke off further nuclear plant construction. The climate of uncertainty created by confusing government policy changes and inaction on such critical issues as radioactive waste disposal and spent fuel reprocessing have inhibited expansion of nuclear power generation. Furthermore, the perceived connection between nuclear power development and nuclear weapons proliferation has given rise to a bitter international debate and stimulated a variety of political attempts to restrict the transfer of nuclear technology, equipment, and materials.

Even if these difficulties can be overcome rapidly, the present generation of nuclear power reactors represents a relatively short-term source of energy, roughly comparable to our fluid fossil fuel resources. Without the introduction of advanced converter or breeder reactors, we shall rapidly exhaust our uranium and thorium reserves, and nuclear fission power will cease to be a viable technology shortly after the turn of the century.

If nuclear power is to be more than a temporary energy source, we must look beyond the present generation of nuclear fission reactors to advanced nuclear technologies, such as the breeder reactor. This technology could supply all of humankind's energy requirements for thousands of years to come. It can be regarded, in effect, as an infinite energy source much like nuclear fusion and solar power. But in contrast to these alternatives, the breeder reactor is both scientifically feasible and commercially viable. Demonstration fast breeder reactor plants have been operating successfully in Europe for several years. Commercial-scale prototype breeder reactors are under construction, and the first commercial sales of fast breeder reactors are anticipated later this decade.

But a massive implementation of the breeder reactor faces enormous barriers of a nontechnical nature. Foremost among these are the political responses to fears that breeder reactor technology may accelerate the proliferation of nuclear weapons. This has already led the United States to postpone fast breeder technology and its plutonium fuel cycle in favor of alternative nuclear technologies. Then, too, there is significant public opposition to breeder reactor technology. The potential of these factors for blocking breeder reactor implementation should not be underestimated.

Nuclear power is a highly controversial subject, and opinions concerning its future role differ greatly. It is our belief that nuclear power is essential if our society is to bring into balance its very real needs for energy with its sources of energy supply. But we also acknowledge that there are genuine problems and concerns about nuclear power that must be addressed if this technology is to play a significant role in our future. As nuclear engineers, we feel that these problems can be solved and that nuclear power can be made a safe and suitable source of energy.

There seems little doubt that nuclear power will remain costly. It will continue to concentrate a sophisticated technology in the hands of a few. Certainly, too, a nuclear power station will have a significant impact on its natural environment. Events such as the Three Mile Island accident remind us that nuclear power will never be absolutely safe. It will always present some risk to our society, but then the same can be said for all of our future energy options, whether they be nuclear power, or petroleum, or coal, or even conservation or solar power. One cannot judge nuclear power in a vacuum; rather one must compare its advantages and disadvantages against those of alternatives. Certainly nuclear power presents us with a trade-off, a balance between benefits and risks. But that is a trade-off that has always been a part of the progress of the human race.

SUGGESTED READINGS

General Review

H. Bethe, "The Necessity of Fission Power," *Scientific American* **234** (1), 5 (1975).

J. Duderstadt and C. Kikuchi, *Nuclear Power: Technology on Trial*. Ann Arbor: University of Michigan Press, 1979.

S. M. Keeny et al., *Report of the Nuclear Energy Policy Study Group: Nuclear Power Issues and Choices*. Cambridge, Mass.: Ballinger, 1977.

National Research Council, *Report of the Committee on Nuclear and Alternative Energy Systems*

(*CONAES*). Washington, D.C.: National Academy of Sciences, 1979.

K. F. Weaver, "The Promise and Peril of Nuclear Energy," *National Geographic* **155**, 459 (1979).

Technical Aspects

Harold M. Agnew, "Gas-Cooled Nuclear Power Reactors," *Scientific American* **244** (6), 55 (1981).

T. M. Connolly, *Foundations of Nuclear Engineering.* New York: Wiley, 1978.

J. J. Duderstadt, *Nuclear Power.* New York: Marcel Dekker, 1979.

S. Glasstone and A. Sesonske, *Nuclear Reactor Engineering,* 2nd edn. Princeton, N.J.: Van Nostrand, 1981.

R. L. Murray, *Nuclear Energy,* 2nd edn. New York: Pergamon, 1980.

Status, Economics, and Reliability

American Nuclear Society, "World List of Nuclear Power Plants," *Nuclear News* **24** (2), 59 (1981).

Atomic Industrial Forum, "Economic Status of Nuclear Power," *AIF INFO*, March 1981.

D. Rossin and T. A. Rieck, "Nuclear Power Economics," *Science* **201**, 582 (1978).

Nuclear Power, Radiation, and the Environment

"Comparing Energy Technology Alternatives from an Environmental Perspective." Washington, D.C.: U.S. Department of Energy, Report DOE/EV-0109, February 1981.

B. L. Cohen, "Impacts of the Nuclear Energy Industry on Human Health and Safety," *American Scientist* **64**, 550 (1976).

G. G. Eichholz, *Environmental Aspects of Nuclear Power.* Ann Arbor, Mich.: Ann Arbor Science, 1976.

National Research Council, "Report of the Committee on Biological Effects of Ionizing Radiation (BEIR)." Washington, D.C.: National Academy of Sciences, 1979.

Risk and Nuclear Reactor Safety

American Physical Society, "Report to the American Physical Society by the Study Group on Light-Water Reactor Safety," *Review of Modern Physics* **14**, 546 (1976).

M. W. Firebaugh, ed., *Acceptable Nuclear Future: The Second Era.* ORAU/IEA—80-11(P), August 1980. Available from Documents Librarian, Institute for Energy Analysis-ORAU, Box 117, Oak Ridge, Tenn. 37830.

M. W. Firebaugh and M. J. Ohanian, eds., *An Accept-*

able Future Nuclear Energy System, ORAU/IEA—80-3(P), March 1980. Available from Documents Librarian, Institute for Energy Analysis-ORAU, Box 117, Oak Ridge, Tenn. 37830.

Fred Hoyle and Geoffrey Hoyle, *Commonsense in Nuclear Energy.* London: Heineman Educational Books, 1980.

H. Inhaber, "Risk with Energy from Conventional and Nonconventional Sources," *Science* **203**, 718 (1979).

"Reactor Safety Study." Washington, D.C.: U.S. Nuclear Regulatory Commission, Report 1400, 1975.

Fred H. Schmidt and David Bodansky, *The Energy Controversy: The Fight over Nuclear Power.* San Francisco: Albion, 1976.

Alvin Weinberg, "The Future of Nuclear Energy," *Physics Today,* March 1981, p. 48.

The Nuclear Fuel Cycle

American Physical Society, "Report to the American Physical Society by the Study Group on Nuclear Fuel Cycles and Waste Management," *Review of Modern Physics* **50**, S1 (1978).

"Final Generic Environmental Statement on the Use of Recycle Plutonium in Mixed Oxide Fuel in Light Water Cooled Reactors (GESMO)." Washington, D.C.: U.S. Nuclear Regulatory Commission, Report NUREG-0002, 1976.

Radioactive Waste

American Physical Society, "Report to the American Physical Society by the Study Group on Nuclear Fuel Cycles and Waste Management," *Review of Modern Physics* **50**, S1 (1978).

B. L. Cohen, "The Disposal of Radioactive Wastes from Fission Reactors," *Scientific American* **236**, 21 (1977).

International Aspects of Nuclear Power

W. Epstein, "The Proliferation of Nuclear Weapons," *Scientific American* **233** (4), 18 (1975).

"Nuclear Power and Nuclear Weapons Proliferation," Report of the Atlantic Council's Nuclear Fuels Policy Working Group, Washington, D.C., 1978.

D. J. Rose and R. K. Lester, "Nuclear Power, Nuclear Weapons, and International Stability," *Scientific American* **238** (4), 45 (1978).

Breeder Reactors

T. Alexander, "Why the Breeder Reactor Is Inevitable," *Fortune,* September 1977, p. 123.

Georges A. Vendryes, "Superphenix: A Full-Scale Breeder Reactor," *Scientific American* **236** (3), 26 (1977).

Impacts of the Nuclear Energy Industry on Human Health and Safety   16

BERNARD L. COHEN

Although many people believe that the effects of radiation on human health are poorly understood, at least the upper limits of these effects are well known from incidents in which people have received large doses of radiation. The data have been analyzed in reports of three prestigious groups of radiation biomedical experts—the National Academy of Sciences–National Research Council Committee on Biological Effects of Ionizing Radiation (BEIR),[1] the United Nations Scientific Committee on Effects of Atomic Radiation (UNSCEAR),[2] and the International Commission on Radiological Protection.[3] Their conclusions will be used here to estimate the effects on human health of radioactivity released into the environment by various aspects of the nuclear energy industry.

The principal health effects of radiation are acute radiation sickness, cancer, and genetic defects. Acute radiation sickness, which can be fatal in a matter of days, results from exposures in excess of 100 rem (1 roentgen-equivalent-man = energy deposit per unit mass of tissue, in units of 100 ergs per gram, times relative biological effectiveness). Gamma-ray exposures of about 500 rem (as it enters the body) without medical treatment and of 700 to 1400 rem with various levels of treatment[4] have a 50 per cent probability of causing death. If death does not result, the patient recovers after several weeks. There have been seven fatalities (none since 1961) from acute radiation sickness in the United States—all of them workers on nuclear projects where something went wrong.

Cancer induction by radiation is a much broader threat, and there is a rather substantial body of data on human victims. The largest single source is the Japanese atomic bomb survivors. About 24,000 people were

Dr. Bernard L. Cohen, an experimental nuclear physicist, is a professor of physics and the Director of the Nuclear Physics Laboratory at the University of Pittsburgh, Pittsburgh, Pennsylvania 15260. For the past five years, he has been active in research on the environmental impacts of nuclear power, especially regarding plutonium and waste problems. He is past chairman of the American Physical Society Division of Nuclear Physics and a member of the National Council of American Association of Physics Teachers, the Health Physics Society, and the American Nuclear Society. Also, he is current chairman of the Division of Environmental Sciences of the American Nuclear Society.

The original article was from American Scientist 64 (5), 550 (1976). Reprinted by permission of the author and Sigma Xi, the Scientific Research Society of North America. Copyright © 1976 by Sigma Xi. This article was updated by the author for this third edition of Perspectives on Energy in January 1981.

exposed to an average of about 130 rem, and more than 100 excess cancer deaths resulted up to 1972. Almost 15,000 people in the United Kingdom who were treated with heavy doses of X-rays for ankylosing spondylitis, an arthritis of the spine, received an average whole-body exposure of almost 400 rem, which resulted in more than 100 excess deaths. Among 4000 uranium miners who received average doses to the lung approaching 5000 rem from radon inhalation, deaths have also exceeded that toll. Several situations have caused about 50 excess cancer deaths, including those involving 775 American women employed in painting radium numerals on watch dials between 1915 and 1935, almost 1000 German victims of anky-losing spondylitis treated with ^{224}Ra, and fluorspar and metal miners exposed to radon gas. Several other situations that led to ten or so excess deaths have also been studied.

In assessing these data, it is usually assumed that the cancer risk is proportional to the total exposure in rem. This "linear–no threshold" hypothesis involves a very large extrapolation; it assumes, for example, that the probability of cancer induction by 1 mil-lirem (mrem), which is typical of most ex-posures of interest, is 10^{-5} times the proba-bility of induction by 100 rem, which is the region for which most data are available. For a number of reasons, however, this assump-tion seems more likely to overestimate than to underestimate the effects of low dosage. That there are mechanisms in the body for repairing radiation damage is well estab-lished[5]: chromosomes broken by radiation have been observed under a microscope to reunite, and 1000-rem doses that are lethal to mice in a single exposure have little effect when distributed over several weeks. Rap-idly multiplying cells are more susceptible to radiation injury than normal cells because there is less time for repair between cell di-visions—this is the basis for cancer radiation therapy. In addition, cancer induction by ra-diation is known to be a multievent process: if cancer is caused by a single hit on a single cell, the risk would be proportional to dose,

but if it is caused by hits by separate particles of radiation, risk would be proportional to the square of the dose. Good evidence exists that the latent period before cancers develop (10 to 50 years for large exposures) increases with decreased exposure, and for small ex-posures it may well exceed life expectancy. For these reasons, all of the monitoring groups mentioned above have acknowledged that the linear–no threshold hypothesis is conservative, more likely to overestimate than to underestimate effects of low doses.

The linear–no threshold hypothesis greatly simplifies calculations, making total effects proportional to the population dose in "man-rem," the sum of the exposures in rem to all those exposed. I shall use it here with the understanding that it yields upper limits for radiation effects. In particular, I shall use the BEIR estimate of 120×10^{-6} cancer deaths per man-rem, which means that for every rem of radiation a person receives to his whole body, his probability of ultimately dying of cancer is increased by 1.2 parts in 10,000 above the normal probability (16.8 per cent for the average American). On this basis, 1 rem of whole-body radiation reduces life ex-pectancy by a little less than one day.[6] For perspective, it may be noted that the life ex-pectancy reduction from smoking a single cigarette is equivalent to that of 8 mrem.

If radioactive material enters the body, ex-posure of individual organs is more of a problem than whole-body exposure, and the BEIR report gives the risks separately. For example, the number of cancer deaths per million man-rem exposure to bond surfaces is 0.6; to thyroid, 7; to lungs, 27. Studies of the survivors of the Japanese atomic bomb-ings show no evidence that the incidence of diseases other than cancer is affected by ra-diation.[7]

Genetic defects, which normally occur in about 3 per cent of all live births,[2] number about 100,000 per year in the United States; they are generally caused by spontaneous mutations in the sex cells. Because there is no evidence from human data of genetic de-fects in offspring from radiation received by parents, all estimates of such damage are based on animal data. Averaging the BEIR and UNSCEAR estimates gives 150×10^{-6} eventual genetic defects per man-rem exposure of the entire population.[1,2] The defects range from color blindness or an extra finger or toe (usually removed by simple surgery shortly after birth) to serious deformities that make life very difficult and include diseases that develop much later in life. Studies of the Japanese survivors of the atomic bombings[8] have yielded no evidence for excessive genetic defects among offspring—a result that virtually ensures that the above estimate is not too small.

ROUTINE EMISSIONS FROM NUCLEAR INSTALLATIONS

A light-water reactor consists of long, thin rods of UO_2 (fuel pins) enriched to about 3 per cent in ^{235}U, submerged in water. In the proper geometry, this arrangement permits a chain reaction in which a neutron striking a ^{235}U nucleus induces a fission reaction, which releases neutrons, some of which induce other fission reactions. Each fission reaction releases energy (about 200 MeV), which is rapidly converted to heat, warming the surrounding water. The reactor therefore serves as a gigantic water heater; as water is pumped through at a rate of thousands of gallons per second, it is heated to 600°F. The hot water may then be converted to steam either in the reactor (boiling-water reactor, BWR) or in an external heat exchanger (pressurized-water reactor, PWR); the steam then drives a turbine which turns a generator to produce electric power.

The fuel is in the form of UO_2 ceramic pellets about 1.5 centimeters long by 1 centimeter in diameter. About 200 of these pellets are lined up end to end inside a zirconium alloy tube (cladding), which is then sealed by welding. The reactor fuel consists of about 40,000 of these fuel pins, or a total of about 8 million pellets.

When a ^{235}U nucleus is struck by a neutron and undergoes fission, the two pieces into which it splits are ordinarily radioactive—this is the principal source of radioactivity in the nuclear energy industry. The pieces fly apart with considerable energy (about 80 per cent of the 200 million electron volts of energy released is in the kinetic energy of these pieces), but they are stopped after traveling only about 0.02 millimeter, and so nearly all the radioactivity remains very close to the site of the original uranium nucleus inside the ceramic fuel pellet. The same is true for the second most important source of radioactivity, the neutrons captured by the uranium to make still heavier radioactive nuclei—neptunium, plutonium, americium and curium—called actinides.

Although nearly all the radioactive nuclei remain sealed inside the ceramic pellets, a few of the fission products have a degree of mobility, and some small fraction of them eventually diffuse out of the pellets but remain contained inside the cladding. During the operation of the reactor, about one or two per thousand of the fuel pins develop tiny leaks in the cladding, releasing the radioactive material that has diffused out of the pellets into the water. Chemical cleanup facilities remove the radioactive material from the water, but they do not remove the gaseous fission products, which include krypton and xenon isotopes and iodine. In addition, ^{14}C, formed when neutrons strike ^{14}N impurity in the fuel and induce (n,p) reactions, is in gaseous form as CO_2. The iodine is so volatile that, despite elaborate equipment for trapping it out, a fraction of 1 per cent of it comes off with the gases. These gases, including also small fractions of 1 per cent of a few other volatile fission products, are held for some time within the power plant to allow the short half-life activities to decay. Eventually they are released into the environment. This is one way in which the public is exposed to radiation by routine operation of a nuclear plant.

Nuclear Regulatory Commission regulations require that no member of the public, including those living closest to the plant, receive a radiation dose from these emissions larger than 5 mrem per year to the whole body, or 15 mrem per year to the thyroid. It is estimated that if all the electric power now used in the United States (approximately 400 million kilowatts) were produced by light-water reactors, the average American would receive an average annual exposure from the carbon-14, krypton, and xenon of about 0.02 mrem per year.[9] The iodine released yields exposures to human thyroids of the same order of magnitude,[10] mostly due to the concentration in cow's milk of material settling on grass, but this does considerably less harm because it is less effective in inducing cancer.

Another source of routine emissions of radioactivity is tritium (^{3}H). In about one fission reaction out of 500, the uranium nucleus splits into three parts (ternary fission) rather than two, and in about 5 per cent of these cases (once in 10,000 fissions) one of the three pieces is ^{3}H. Other sources of ^{3}H in reactors are neutron reactions in boron and ^{2}H. The difficulty with ^{3}H is that it mixes with the ordinary hydrogen in the water and cannot be separated from it; thus, when water is released from the plant, it releases some ^{3}H into a nearby river, lake, or ocean. The Nuclear Regulatory Commission maximum of 5 mrem per year of whole-body radiation from this water is calculated on the assumption that a person derives all his drinking water and fish from the plant discharge canal and swims in it for an hour a day. It is estimated that, if all our power were nuclear, the average American would receive less than 0.01 mrem per year from this source.[9]

When as much of the fuel in a reactor has been burned up as consistent with proper operation, it must be removed from the reactor and replaced with fresh fuel (typically one third of the fuel is replaced in one such operation per year). The spent fuel is currently being stored indefinitely in the power plant to allow short half-life radioactivity to decay away. When the nuclear fuel cycle becomes operational, the spent fuel will be shipped to a fuel reprocessing plant. There the fuel pins are cut into pieces of manageable size and

dissolved in acid, and the solution is chemically processed to remove the uranium, which is useful for making fresh fuel, and the plutonium, which may be used for making future fuels. Everything else is classified as "waste"—including all the fission products that contain the vast majority of the radioactivity, the uranium and plutonium that escaped removal in the chemical processing (typically 0.5 per cent with current technology), and the other actinides produced. The eventual disposal of this waste is an important topic that will be discussed below.

When the fuel pins are dissolved, the gases once again present a problem. Since xenon has no long half-life isotopes, no radioactive xenon is present, but krypton-85 has a 10-year half-life, and in current technology all of it is released to the atmosphere from a tall stack. The tritium is released as water vapor in the same way, as is also the ^{14}C as CO_2. It is estimated that if all our present electric power were nuclear, the average American would be exposed to 0.03 mrem per year from the krypton-85, 0.03 mrem per year from the tritium,[11] and 0.11 mrem per year from carbon-14. EPA regulations to go into effect in 1984 require that nearly all of the krypton-85 and much of the carbon-14 be recovered.

When these exposures are added to the 0.02 mrem per year from the power plants, we see that the exposure to the average American from routine emissions if all our power were nuclear would be about 0.19 mrem per year. If we compare this average exposure with other radiation we experience,[12] we find that it is less than 1/500 of our exposure from natural radiation (U.S. average = 100 mrem per year) and about 1/300 of our exposure from medical and dental X-rays. Natural radiation varies widely, from an average of 200 mrem per year in Colorado and Wyoming to 60 mrem per year in Florida. In parts of India and Brazil, where average exposures are 1500 mrem per year from monazite sands rich in thorium and uranium, studies of the population have revealed no unusual effects. Even within a single city, exposures may vary considerably:

the radiation level is 10 mrem per year higher[13] in Manhattan (built on granite) than in Brooklyn (built on sand). Building materials such as brick and stone also contain significant quantities of radioactive material. Living in a typical brick or stone house (rather than a wooden one) adds about 20 mrem per year and in nine months causes more exposure than a lifetime of all-nuclear power. The 0.19 mrem per year from routine emissions if all our power were nuclear is substantially exceeded by even such minor sources of radiation[12] as luminous-dial watches (1 mrem per year) and airplane flights (0.7 mrem per coast to coast flight at 30,000 feet) and is comparable to the average child's dosage from watching television (0.3 mrem per year from X-rays).

The consequences of 0.19 mrem per year of whole-body radiation to the average American may be readily calculated using the risk estimates given in the previous section: about six additional cancer deaths per year and seven additional genetic defects. Uranium and its daughters are released from various other elements of the nuclear fuel cycle besides the power plant and fuel reprocessing facilities, but because these involve alpha-particle emitters, which do their damage to a few specific organs following inhalation, the effects cannot be expressed in terms of whole-body radiation. The numbers of deaths caused by these agents if all our power were nuclear are[14]: uranium ore mills, 0.16 per year; conversion facilities, 0.09 per year; enrichment facilities, 0.01 per year; fuel fabrication plants, 0.03 per year; transportation, 0.08 per year. These represent a negligible addition to the six fatalities per year estimated above, which is thus the first entry in Table 16-1.

When uranium is separated from the mined ore in the ore processing mill, the residue containing radium and its precursor, thorium-230, is accumulated in large piles, typically 1 square kilometer in area by 5 meters deep. The radon gas emitted from the tailings piles resulting from one year of all nuclear power (2.5 piles of the above dimensions) is estimated[15] to cause 0.8 lung cancer

Table 16–1. Annual Cancer (plus Acute Radiation Sickness) Deaths due to Radiation from Aspects of a U.S. Nuclear Energy Industry Generating 400 Million Kilowatts of Electricity.[a]

Source	Deaths per Year
Routine emissions	6.0
Reactor accidents	10.0 (960)[b]
Transportation accidents	0.01
Waste disposal	0.4
Plutonium (routine lease)	0.1
Total	16–20 (1000)[b]

[a] Does not include effects of long half-life radioactivities, sabotage, or terrorism.
[b] Numbers in parentheses represent worst claim by critics of the nuclear energy industry.

per year in the United States (my estimate[16] is an order of magnitude smaller)—but this situation will continue for tens of thousands of years if it is not remedied. Mill tailings have caused local problems, for example, in Grand Junction, Colorado, where they were used for building construction,[17] and in Salt Lake City, where there is a large pile within the city limits. The Department of Energy has an active program for solving these problems in areas where mills have been abandoned, and NRC now requires careful attention to them in mills currently operating.

For perspective, it might be pointed out that if nothing were done about the piles, the tailings from one year of all nuclear power would increase the natural radon background by only 1 part in 6000 and would contribute less radon than is currently being released by phosphate mining. The whole problem would, of course, be grossly reduced with breeder reactors, and one might question the validity of equating health effects tens of thousands of years in the future with those of the present day.[18]

POWER PLANT ACCIDENTS

In routine operations in the nuclear industry, nearly all the radioactivity produced in reactors ends up as waste at the fuel reprocessing plant, and this waste, as we shall see, is not difficult to dispose of safely. However, the danger exists that, as a result of some sort of accident in the reactor, an appreciable fraction of this radioactivity will be released into the environment at the power plant. If this should happen, the potential for damage is very great.

Because essentially all the radioactivity produced is sealed inside the UO_2 ceramic fuel pellets, the only way for it to be released is for these pellets to be melted. Thus any reactor accident of large consequences must involve a "melt-down." Since the melting temperature of UO_2 is over 5000°F, whereas normal operating temperatures are near and below 1000°F, a melt-down cannot result from small abnormalities. One possible cause of a melt-down might seem to be a large reactivity excursion produced by withdrawal of control rods too far and too fast. Several mitigating effects would follow, however. Initially, the power level would escalate rapidly, but as a result the reactor would heat up. As the temperature increases, uranium-238 captures more neutrons in the "resonance region" (due to Doppler broadening of resonances), leaving fewer neutrons to be slowed down to thermal energies where they induce fission in uranium-235; moreover, as the temperature increases, water becomes a poorer moderator (in a BWR, more water boils into steam, which is essentially an elimination of the moderator). Hence, these reactors have a large "negative temperature coefficient of reactivity," which works powerfully against large reactivity excursions. In addition, the emergency insertion of control rods (called "scram") is such a simple operation that it would rarely fail. Because it depends only on gravity (PWR) or on stored fluid pressure (BWR), it does not require electric power.

These safety features are challenged rather frequently. About once a year on the average, a generator is suddenly taken off-line because of some abnormal electrical occurrence, and when this happens, the turbine that drives it can no longer accept steam. The re-

sulting back pressure causes the steam bubbles to collapse in a BWR, thereby suddenly increasing the reactivity, which greatly increases the reactor power. If the emergency control rods should fail to insert, the pressure would build up dangerously. This accident is called ATWS (anticipated transient without scram), and in many reactors the scram system is the only defense against this once-a-year challenge. A backup "poison-insertion system," which would inject boron solution (a strong neutron absorber), would be another defense, but there has been strong resistance to this because, if it were to be activated unnecessarily, it would keep the reactor shut down for many hours.

Safety experts agree that the most likely cause of a reactor fuel melt-down is not a reactivity excursion but a loss of coolant accident (LOCA) resulting from a large leak in the cooling water system. The water temperature in these reactors is about 600°F (high temperatures produce high efficiencies), and to prevent or control boiling, the pressure must be very high, about 1000 pounds per square inch in the BWR and 2200 pounds per square inch in the PWR. If there were a rupture in the high-pressure system, the water would flash into steam and escape at a tremendous rate (this is called a "blow-down"), leaving the reactor core without coolant. Loss of the water moderator would immediately shut down the chain reaction, but it would not, of course, halt the radioactivity decay process.

The power generated by the radioactivity in the fuel pellets immediately after the chain reaction stops is very substantial—about 6 per cent of the full power level of the reactor— and it is easily enough to eventually melt the fuel. In the PWR, if the fuel were left without coolant for 45 seconds (possibly even for 30 seconds), the temperature would get so high that bringing in water might do more harm than good; at high temperatures, water reacts chemically with the zirconium fuel cladding in an exothermal reaction, which would raise the temperature still more. Thus, if cooling is not restored within about 45 seconds, the reactor may be doomed to melt-down (in a BWR, this critical time is about 3 to 5 minutes), releasing the radioactive material. Complete melt-down of the fuel would take about 30 minutes, and after an hour or so the molten fuel would melt through the reactor vessel. An appreciable fraction of the radioactivity would at this point come spewing out in the form of a radioactive dust or gas.

A great deal of engineering effort has been expended to minimize the probability of a LOCA, to reduce the chance that a LOCA would lead to a melt-down, and to mitigate the consequences of a melt-down if it should occur. Quality standards on materials and fabrication methods match or exceed those in any other industry, and rarely is expense an issue in this regard. A very elaborate program of inspections is maintained during the fabrication stage, including X-raying of all welds, magnetic particle inspections, and a very elaborate series of ultrasonic tests aimed at detecting imperfections that might lead to failures of materials. Once the reactor comes into operation, periodic shutdowns are scheduled for extensive ultrasonic and visual inspections. (It was in these visual inspections that the hairline pipe cracks in a few BWR's were discovered in early 1975).

Since large leaks develop from small leaks, the next line of defense is in systems for detecting small leaks if they should occur. Since the water is at high temperature and under high pressure, any leak would release steam, which would increase the humidity. Two types of systems, based on different physical principles, are used to detect increases in humidity around the high-pressure system. Because radioactivity in the water would emerge with the steam, two particularly sensitive types of systems capable of detecting increases of air-borne radioactivity are also used.

If, in spite of these precautions, a LOCA should occur, the next line of defense is the emergency core cooling system (ECCS), an elaborate arrangement for injecting water back into the system to reflood the reactor core (because all pipes enter the reactor vessel above the core, reflooding is possible unless the rupture is in the lower part of the reactor vessel itself; the latter is of very thick, high-quality steel, and its rupture is orders of magnitude less probable than breaks in piping or in external components). The ECCS is a highly redundant system, and simple failures of pumps or valves would not prevent its operation. All estimates indicate that it is about 99 per cent certain of delivering water in the event of a LOCA.

Nevertheless, there has been extensive controversy over whether the ECCS will prevent a melt-down in the event of a large LOCA. By the time the water from the ECCS fills the reactor vessel up to the bottom of the fuel pins, the latter are quite hot, and the water flashes into steam. The steam exerts a back pressure that retards flooding to a rate not much more than 2.5 centimeters per second, and there is a period during which the heat transfer is principally by water droplets entrained in steam. This type of heat transfer is not well understood, and there is considerable uncertainty about cooling by the water-steam mixture during the initial blow-down. In order to assess these problems, engineers conducted experiments with full-length, electrically heated fuel-pin mock-ups and developed empirical "theories" to explain their observations. They then used these empirical theories in computer codes to calculate the operation of the ECCS. This procedure is highly inaccurate, and the engineering approach to such a situation is to apply a factor of safety.

In 1971, the Union of Concerned Scientists (UCS) headed by Henry Kendall studied these matters and declared that they did not consider the factor of safety adequate. In the controversy that followed, the Atomic Energy Commission (AEC) organized hearings that lasted more than one year to consider the question, and several AEC safety experts came forward to support Kendall's contention. As a result of the hearings, the AEC increased the factor of safety and some reactors were forced to reduce their power levels until fuel pins with smaller diameters could be installed. The AEC safety experts who had supported Kendall said (according to an AEC statement) that they were satisfied that the precautions were adequate. In 1978, the LOFT (loss of fluid test) reactor was completed, allowing large-scale testing of the computer codes, and they were found to be highly conservative. Presumably, then, this episode is settled.

If a LOCA should occur and the ECCS should fail to perform its function, there would be a melt-down. In order to mitigate the consequences, the entire system is enclosed in a "containment," constructed of very thick, heavily reinforced concrete lined with steel plate (the most common type is tested to withstand an internal pressure of 5 atmospheres). The containment is strong enough to repel a wide variety of external threats, including objects a tornado might hurl at it (automobiles, trees, etc.) and conventional explosives and bombs. The function of the containment in a melt-down accident is to contain the radioactive dust for a time. Inside it are systems for pumping the air through filters to remove the dust and sprays for removing iodine. Because the walls are relatively cool, many of the radioactive materials would plate out on them. Thus, if the containment held for at least a few hours, most of the air-borne radioactivity would be removed and the consequences of the accident to the public would not be serious.

The situation could be much more serious if the containment should fail shortly after the molten fuel melts through the reactor vessel. This could happen immediately as a result of a very violent (but highly improbable) explosion as the molten fuel drops into water, or relatively early as a result of high steam pressure if the water sprays designed to condense the steam inside the containment should fail to function. In such situations, the air-borne radioactive dust would be released into the environment, and the results would then depend on weather conditions. Ordinarily, the radioactivity would be widely dispersed and cause little obvious damage,[19] but

if there were a strong temperature inversion the radioactive cloud would be concentrated close to the ground—anyone it passed would be exposed externally and would inhale radioactive dust. There could be thousands of fatalities.

Among the many surveys of the probabilities and consequences of reactor accidents, the most elaborate is the study financed by the AEC and directed by Norman Rasmussen of the Massachusetts Institute of Technology. This project involved 60 man-years of effort and cost $4 million; its results were published in draft form in August 1974 and, after extensive criticism and revision, in final form in November 1975 as the multivolume document WASH-1400. If one is willing to accept the Rasmussen study, answers are immediately available to a wide variety of questions. For example, there would be a melt-down about once in 20,000 reactor-years; if all our power were nuclear (400 reactors), we might expect a melt-down every 50 years on an average. The average annual consequences (1/50 the average consequence per melt-down) would be 0.2 deaths from acute radiation sickness plus 10 eventual cancer deaths (see Table 16-1), and $6 million in damages, mostly in cleanup and evacuation costs. The frequency of accidents of varying severity (as indicated by the number of fatalities) is shown in Table 16-2.

The consequences of the worst possible accident are estimated to be 3500 fatalities from acute radiation sickness plus 45,000 later cancer deaths. An accident of this mag-

Table 16–2. Severity Distribution of Accidents.

| | Frequency (average years between accidents) | | |
Number of Fatalities	Nuclear	Other Man-caused[a]	Natural Disasters
>100	100	1.5	2
>1000	300	25	8
>10,000	10,000	500	50

[a]Includes dam failures, airplane crashes in crowded areas, fires, explosions, releases of poison gas, etc.

nitude is predicted to occur about once in a million years in the United States if all our power was nuclear. This catastrophe would do little property damage, but other accidents, especially those accompanied by widespread heavy rainstorms, would cause few fatalities but would contaminate large areas and require extensive evacuation and cleanup. In the worst accident of this type, expected once in 1 million years, these would cost $14 billion.

Although it is often said that the worst nuclear accident would be much more terrible than any other accident connected with man's production of energy, this is by no means correct. There are hydroelectric dams whose failure could cause over 200,000 fatalities, and such failures are estimated to be much more probable than a very bad nuclear accident.[20] Gas explosions, especially those connected with the transport of liquified natural gas, which have been envisioned, could cause 100,000 fatalities. Moreover, unlike these situations, the great majority of the fatalities from the nuclear catastrophe would be essentially unnoticed, arising from a slight increase in cancer occurrence many years later.

The critics of nuclear power have not accepted the Rasmussen report, estimating that melt-downs should occur ten times more frequently and have consequences ten times more severe[21] than the Rasmussen estimates. They conclude that we may expect 2.4 fatalities per GWe-year, or 960 fatalities per year on an average from 400 reactors.

TRANSPORTATION ACCIDENTS

When spent fuel is shipped from the power plant to the fuel reprocessing plant, there is a possibility of an accident in transit that would release radioactivity to the environment. Several features differentiate such releases from those in power plant accidents. First, only a very small fraction of the fuel in a reactor is involved in any one shipment. Second, the fuel is stored in the plant for about six months before shipment to allow

the short half-life radioactivity to decay away; this reduces the potential danger by two orders of magnitude. Third, and most important, the highest temperatures that would ordinarily be encountered in a transport accident are those of a gasoline or an organic solvent fire—about 1500°F, far below the 5000°F temperature to melt the UO_2 ceramic fuel pellets. Thus, since nearly all the radioactivity remains trapped in the pellets, the major danger is that the cladding tubes will be ruptured, releasing the small fraction of the radioactivity that had migrated out of the pellets to become trapped inside the tubes. This radioactivity is principally krypton-85, which is destined to be released from the fuel reprocessing plant under more controlled conditions. In addition, it is sometimes assumed that, in some unspecified way, a very small fraction of the solid fission products are released.[22,23]

In order to minimize the danger of such releases, spent fuel is shipped in casks costing about $2 million each, designed and prototype-tested to withstand, without release of radioactivity, a 48 km/h crash into a solid and unyielding obstacle, envelopment in a gasoline fire for 30 minutes, submersion in water for 8 hours, and a puncture test. It seems reasonable to expect a high degree of protection against accident damage from such an elaborate effort.

The same precautions are also taken with the waste glass to be shipped from the fuel reprocessing plant to a burial site. Moreover, by the time the glass is shipped, radioactivity has decayed by another factor of 6, and the glass is enclosed in a thick stainless steel container which is far stronger than the fuel pin cladding. Thus, the risk in waste glass transport is considerably less than in spent fuel transport.

Studies of the problem, including estimates of releases and their consequences from all types of waste transport accidents, indicate that if all U.S. power were nuclear there would be an average of less than one fatality per century due to radioactivity releases in these accidents.[24] There would, of course, be orders of magnitude more fatalities from the traffic aspects of these accidents, but coal-fired power requires 100 times more transport and hence would cause 100 times as many transport accident fatalities.

RADIOACTIVE WASTE DISPOSAL

At present, the high-level wastes that accumulate in fuel reprocessing plants are kept in solution and stored in large tanks. There are currently 600,000 gallons of this waste from nuclear power plants, and more than 100 times that much from government operations, principally the production of plutonium for weapons. The plans for waste from civilian power are very different. The Nuclear Regulatory Commission regulations require that it be converted to a suitable solid form (as yet unspecified) within five years and delivered to a government repository within ten years. The first deliveries may be expected in the late 1980s (the little waste that has already been generated is exempted from this schedule by a "grandfather" clause). Plans for the repository are not yet final, but the waste will be buried in some suitably chosen geological formation about 600 meters underground. Probably it will be in the form of glass cylinders 0.3 meters in diameter by 3 meters long (0.2 cubic meters volume) enclosed in a leach-resistant casing, probably of a titanium alloy. A typical power plant would produce about 25 of these per year.

I have evaluated the hazard from these buried wastes previously in both popular-level[25] and detailed papers,[26,27] and so I review the situation only very briefly here. The principal danger once the material is buried is that it might come in contact with ground water, be leached into solution, move through aquifers with the water, and eventually reach the surface, contaminating food and drinking water. The ingestion hazard from the waste generated in one year if all our present electric power were from light-water reactors is shown in Table 16-3.

The material must be isolated from our environment for a few hundred years; before worrying about longer times one should consider some of the other chemical and biological poisons in the earth and those produced by man. For example, a lethal dose of arsenic (As_2O_3) is 3 grams, and we import ten times as much of it into our country each year as we would produce radioactive waste if all our power were nuclear. This arsenic is not buried deep underground; in fact much of it is scattered on the surface as herbicides in regions where food is grown.

The requirement that radioactive waste be isolated for hundreds of years seems alarming to some because few things in our environment last that long. Deep underground, however, the time constants for change are in the range 10^7 to 10^8 years. The following factors provide additional protection against release during the few-hundred-year critical period: (a) The material will be buried in a geological formation that has been free of ground water for tens of millions of years and in which geologists are quite certain there will be no water for some time into the future. (b) If water should get into the formation, the rock would have to be leached or

Table 16–3.   Ingestion Hazard from Waste Generated in One Year if All U.S. Electric Power Were from Light-Water Reactors.

	Years After Reprocessing							
	0	*100*	*200*	*300*	*500*	*1000*	*10^6*	*10^8*
Grams ingested for 50 per cent cancer risk	0.025	0.25	2.5	25	150	400	4000	25,000

dissolved away before water could reach the waste. Even if the rock were salt, dissolution would typically require thousands of years. (c) The waste package will be surrounded by backfill material, which swells when it becomes damp and thus would keep the water away. (d) The leach-resistant casing should keep water out for many thousands of years. (e) Once the water reached the waste glass, the latter would be leached at a very low rate. (f) Ground water flows through aquifers rather slowly, requiring typically 1000 years to reach surface waters from a depth of 600 meters. (g) Most of the radioactive materials would be held up by ion exchange processes, traveling 100 to 10,000 times slower than water.

A quantitative evaluation of the hazard from the waste throughout its existence may be obtained by using a model in which the waste is buried at random locations throughout the United States but always at a depth of 600 meters. It is assumed that an atom of buried waste is no more likely to reach the surface and get into a person than is an atom of average rock submerged in ground water. It can be shown[27] that the latter probability is 10^{-12} per year, and if this probability is applied to the radioactive waste, it turns out that we can expect 0.4 eventual fatalities from the waste generated by one year of all nuclear power (Table 16-1).

By comparison, burning up the uranium to produce this waste would save about 50 fatalities due to radon emissions. Thus, on any long-time scale, nuclear power must be viewed as a method for *cleansing* the earth of radioactivity. It should be noted that this estimate is based on no surveillance, since there is no surveillance of the radium with which it is compared in the model.

Estimates were made of the fatalities that would result from releases as air-borne particulates and as gamma-ray emissions from the surface of the ground, and from releases by natural cataclysms and human intrusion, but all of these predicted fewer fatalities than the ground-water-ingestion pathway considered above.

HAZARDS FROM PLUTONIUM TOXICITY

One of the radioactive materials produced in reactors—plutonium—is too valuable to be buried with the waste, since it can be used as a fuel in future reactors. However, plutonium has received a great deal of bad publicity because of its toxicity and its potential as a material for nuclear bombs. I have considered the problems arising from plutonium toxicity extensively in another paper[28] and shall treat them only very briefly here. Claims of great harm resulting from plutonium toxicity are commonly based on the assumption that all the plutonium under consideration will find its way into human lungs and the acceptance of the "hot particle" theory of alpha-particle carcinogenesis, in which it is assumed that concentration of alpha-particle emitters in a relatively few particles causes a few cells of the victim to be exposed to much more than the average amount of radiation and consequently to a greatly increased risk of cancer. The hot-particle theory has now been studied and rejected by many prestigious official groups, including the NCRP, the NRC, the AEC, the British Medical Research Council, a committee of the National Academy of Sciences, and the United Kingdom Radiological Protection Board. In addition, the ICRP and the U.S. Environmental Protection Agency have inferentially rejected it by not changing their standards on allowable exposure to plutonium. No prestigious or official group has accepted the hot particle theory.

If the usual procedures accepted by these groups are used and if normal meteorological dispersion is assumed, dispersal of reactor plutonium (six times more radioactive than plutonium-239) in a large city would typically result in about 25 eventual cancers per pound dispersed. Seventy per cent of these would result from exposure to the cloud of dust generated by the initial dispersal, and nearly all the rest from resuspension of the dust by winds within the first few months after it first settles on the ground. Less than

3 per cent of the effects are due to exposures during the tens of thousands of years during which the plutonium remains in the soil. About 10,000 pounds of plutonium have been dispersed in bomb tests, whereas it is expected that about 0.01 pound per year would be released from an all-fast-breeder nuclear power industry. About 0.1 fatality per year (Table 16-1) is expected as a consequence of these releases.

THEFT OF PLUTONIUM

Widespread concern has been expressed that plutonium may be stolen from the nuclear energy industry by terrorists for fabrication of nuclear bombs. While this threat cannot be quantified for use in Table 16-1, it cannot be ignored in assessing the environmental impacts of nuclear power.

The principal protection against this threat is to prevent plutonium from becoming available to prospective terrorists. The method for keeping track of it since the 1940s has been to weigh all plutonium entering and leaving a facility, but errors in weighing leave substantial room for undetectable losses. It is not impossible that enough plutonium has already been diverted to make many bombs. On the other hand, there is no evidence that plutonium has ever been stolen in quantities as large as 1 gram.

The AEC has long conducted programs for improving security, but in 1973, T. B. Taylor, a former bomb designer, became disenchanted with the slow progress in these programs and "blew the story open" in a remarkable series of articles in *The New Yorker* magazine and a book[29] in which he gave information on how to make a nuclear bomb. As a result of the publicity he received, plutonium safeguard procedures were greatly tightened and new regulations are constantly being added.

Clearly, the issue of terrorism should have been considered in the 1950s, before the nuclear industry began. It seems almost irresponsible to raise the problem after so much money and effort have been expended on the industry and we need its product so badly. Nevertheless, the threat of terrorism has continued to escalate in importance, and it is now one of the principal points of contention in the nuclear power controversy.

A number of factors should be considered in evaluating the risk of terrorism. First, stealing plutonium would be very difficult and dangerous under present safeguards. Taylor has estimated[30] that a group of thieves would have much less than a 50 per cent chance of escaping with their lives. Fabrication of a bomb from stolen plutonium would also be very difficult, expensive, time-consuming, and dangerous. Estimates vary considerably, but a rough median of the opinions of experts indicates that it would require three people highly skilled in different technical areas a few months and perhaps $50,000 worth of equipment to develop a bomb with a 70 per cent chance of doing extensive damage, and that the people involved would have a 30 per cent chance of being killed in the effort.

Terrorist bombs would be "block-busters," not "city-destroyers." Taylor's principal scenario[29] is that an explosion of this sort could blow up the World Trade Center in New York, killing the 50,000 people that building can contain. Of course, there are many much easier ways to kill as many people (e.g., introducing a poison gas into the ventilation system of the World Trade Center). Terrorists have always had many options for killing thousands of people, but they have almost never killed more than a few dozen. And, of course, plutonium and highly enriched uranium, which would be much more suitable for the making of bombs, could be obtained from sources that have no connection with nuclear electric power.

SOME STATISTICALLY COMPARABLE RISKS

Table 16-1 reveals that estimates based on the acceptance of WASH-1400 predict that an all-nuclear energy economy would result in about 20 deaths per year; critics of nuclear energy claim that the number is about 1000.

I shall now attempt to put these estimates in perspective. Since cancer is delayed by 15 to 45 years after exposure, the average loss of life expectancy per victim is 20 years. The loss of life expectancy for the average American from these 20 deaths per year is then $(20 \times 20$ man-years lost/2×10^8 man-years lived) $= 2 \times 10^{-6}$ of a lifetime $= 1.2$ hours.

Some of us subject ourselves to many other risks that reduce our life expectancy, such as smoking cigarettes. One pack per day $(3.6 \times 10^5$ cigarettes) reduces life expectancy by about 8 years,[31] which, assuming linearity, corresponds to 12 minutes loss of life expectancy per cigarette smoked; thus the risk of nuclear power is equivalent to that of smoking six cigarettes in one's lifetime, or one every ten years.

Traveling in an automobile subjects us to a death risk of 2×10^{-8} per mile, or if 35 years of life are lost in an average traffic fatality, loss of life expectancy is 7×10^{-7} years per mile traveled. The risk of nuclear power is then equal to that of riding in automobiles an extra three miles per year. Riding in a small rather than a large car doubles one's risk[32] of fatal injury, so the 1.2 hours loss of life expectancy from all nuclear power is equivalent to the risk of riding the same amount as at present, but three miles per year of it in a small rather than a large car.

Another risk some of us take is being overweight. If we assume loss of life expectancy to be linear with overweight, the 1.2-hour loss from all nuclear power is equivalent to the risk of being 0.02 ounce overweight.[33]

All of these estimates are based on the government agency projection of 20 deaths per year. If we instead accept the critics' estimate of 50 times as many deaths, the risk of nuclear power is equivalent to the risks of smoking five cigarettes per year, riding in automobiles an extra 150 miles per year, riding in automobiles the same amount as at present but 150 miles of it per year (1.5 per cent) in a small rather than a large car, or being one ounce overweight.

Additional perspective may be gained by comparing the effects of an industry deriving electric power from coal by present technology. The most important environmental impact of coal-fired power is air pollution, which, it is estimated, would cause about 10,000 deaths per year,[34] an order of magnitude more than even the critics estimate would be caused by nuclear power. In addition, this air pollution would cause[34] about 25 million cases per year of chronic respiratory disease, 200 million person-days of aggravated heart-lung disease symptoms, and about $5 billion worth of property damage; there are no comparable problems from nuclear power. Mining of coal to produce this power would cause about 750 deaths[35] per year among coal miners, more than ten times the toll from uranium-mining for nuclear power, and the uranium figure would be reduced about 50-fold with breeder reactors. Transporting coal would cause about 500 deaths per year,[36] two orders of magnitude more than would be caused by transportation for the nuclear industry.

It is important to point out that the numbers in Table 16-1 (and the perspective I have put them in) are based on annual averages. As the critics are constantly reminding us, if their estimates are correct, there might be an accident every ten years with several thousand deaths and every 50 years with tens of thousands of deaths. It is not difficult to make this prospect seem extremely dismal. On the other hand, we should not envision these accidents as producing stacks of dead bodies; the great majority of fatalities predicted are from cancer that would occur 15 to 45 years later. In nearly all cases, the affected individuals would have only about a 0.5 per cent increased chance of getting cancer. The average American's risk of cancer death is now 16.8 per cent; typically it would be increased to 17.3 per cent. For comparison, the average risk varies from 18.4 per cent in New England to 14.7 per cent in the Southeast, but these variations are rarely noticed.

If an area were affected by a nuclear accident and authorities revealed that, as a result, the average citizen's probability of eventually dying of cancer was increased from 16.8 to 17.3 per cent, it would hardly start a panic. We have had some experience with a similar but much more serious situation: when reports first reached the public of the risk in cigarette smoking, tens of millions of Americans were suddenly informed that they had accrued a 10 per cent increased probability of cancer death, 20 times larger than the effects from a nuclear accident. Even that story did not stay in the headlines long, and it produced very little counteraction.

Critics often raise the point that the risks of nuclear power are not shared equally by all who benefit but are disproportionately great for people who live close to a nuclear power plant. This, of course, is true for all technology, but let us put it in perspective. The risk is 10^{-6} per year if we accept the WASH-1400 accident estimates, or equal to the risk of riding in an automobile an extra 50 miles per year or 250 yards per day.

Thus if moving away from a nuclear power plant increases one's commuting distance by more than 125 yards (half a block), it is safer to live next to the power plant. If one prefers the estimates we attribute to the critics (20 times larger), it does not pay to move away if doing so increases commuting distances by more than 1.5 miles per day!

NOTES AND REFERENCES

1. Committee on Biological Effects of Ionizing Radiation (BEIR), *The Effects on Populations of Exposure to Low Levels of Ionizing Radiation*. Washington, D.C.: National Academy of Sciences—National Research Council, 1980.
2. United Nations Scientific Committee on Effects of Atomic Radiation (UNSCEAR), *Sources and Effects of Ionizing Radiation*. New York: United Nations, 1977.
3. International Commission on Radiological Protection (ICRP), Publication 26, *Recommendations of the ICRP*. New York: Pergamon Press, 1977.
4. U.S. Nuclear Regulatory Commission. WASH-1400, *Reactor Safety Study*. Appendix VI. Washington, D.C.: NRC, 1976.
5. NCRP, *Review of the Current State of Radiation Protection Philosophy*. Washington, D.C.: National Council on Radiation Protection and Measurement, 1975.
6. One rem gives 120×10^{-6} probability of cancer death which, on an average, causes 20 years loss of life; thus it reduces life expectancy by 2400×10^{-6} years, which is 0.9 days. Smoking one pack of cigarettes per day (about 4×10^5 cigarettes) reduces life expectancy by about 7.5 years (calculated from Surgeon General's Report on effects of cigarette smoking, 1962) or 2700 days. Assuming linearity, one cigarette then reduces life expectancy by 7×10^{-3} days. This is equal to the effect of $(7 \times 10^{-3}/0.9)$ rem, or about 8 mrem of radiation.
7. Data from Atomic Bomb Casualty Commission are tabulated in B. L. Cohen, *Nuclear Science and Society*, p. 64. New York: Doubleday, 1974.
8. J. V. Neel, H. Kato, and W. J. Shull, "Mortality of Children of Atomic Bomb Survivors and Controls," *Genetics* **76**, 311 (1974).
9. American Physical Society Study Group on Nuclear Fuel Cycles, *Rev. Mod. Phys.* **50**, 51 (1978). This cites several other references which find similar results.
10. Environmental Protection Agency. 1973. Report EPA 520/9-73-003, Part C, pp. 141, 142; the cheapest iodine control systems that reduce site boundary doses to below the required 15 mrem/year give annual population doses to the thyroid of about 7 man-rem for elemental iodine and 13 man-rem for organic iodine; the average exposure is thus of the order of 10 man-rem × 200 sites/2×10^8 people, or about 0.01 mrem/year to the thyroid.
11. Ibid., Part D, p. 10. ^{85}Kr from a plant capable of reprocessing 5 metric tons per day would yield an average dose commitment in the U.S. of 520 man-rem/year. Because such a plant would service about 50 reactors, about 8 plants would be needed, yielding a total of 4200 man-rem. Dividing this by a population of 2×10^8 gives an average dose of 0.02 mrem/year. For tritium, the result is larger, in the ratio of 3700/520. The ^{85}Kr exposure is worldwide, but the tritium exposure is largely limited to the United States.
12. Environmental Protection Agency. Estimates of ionizing radiation doses in the United States, 1960–2000. Report ORP/CSD 72-1.
13. M. Eisenbud, "Standards of Radiation Protection and Their Implications for the Public Health," in H. Forman, ed., *Nuclear Power and the Public*. Minneapolis: University of Minnesota Press, 1970.
14. EPA Report 520/9-73-003, Part B, pp. 39, 88, 109, 128, 146.
15. W. H. Ellett (EPA), personal communication, July 2, 1975.
16. The meteorology in references 14 and 15 is not given in sufficient detail to be followed, but according to reference 14 (and other sources) a standard pile 1 km² in area emits 500 pCi/m²-sec, and 2.5 such piles per year would be accumulated if all our power were nuclear, whereas the 7×10^6 km² of soil in the United States emits an average of about 1 pCi/m²-

sec. Thus, natural radon exceeds the annual increase from tailings piles by a factor of $7 \times 10^6/2.5 \times 500 = 6000$. Reference 15 estimates 4000 deaths per year from natural radon in the United States—which at first seems reasonably consistent with their estimate of effects from tailings piles. However, we give here two alternative calculations of deaths from natural radon which circumvent the rather uncertain meteorology. Reference 2 gives 0.15 rem/year as an average exposure to the tracheobronchial tree which, multiplied by 39×10^{-6} cancer deaths per man-rem from reference 1 and the 2×10^8 population, gives 1200 deaths per year. As another approach, one "working level" (WL) is 150×10^3 pCi/m³ and BEIR gives 0.5 rem to lung per WL-"month" where one working "month" is 170 hours, or 0.02 years; these combine to give $5000/150 \times 10^3 \times 0.02 = 1.6$ mrem/year per pCi/m³. (This is 2.5 times smaller than the value used in reference 14, but more significant than this discrepancy is the fact that the BEIR value was used in assessing lung cancer incidence in miners, which was instrumental in determining the value 39×10^{-6} lung cancers per man-rem). From reference 2, the average ^{222}Rn level in the United States is about 120 pCi/m³, which (when multiplied by the above 1.6) gives an average dose of 190 mrem/year to the lung. Multiplying this by 39×10^{-6} cancers/man-rem and 2×10^8 population gives 1500 deaths per year from natural radon, in agreement with our above estimate of 1200. Since radon from tailings piles is 6000 times smaller, these would give 0.20–0.25 death per year from the latter. However, the latter radon originates in sparsely populated regions and must travel about 1500 miles to reach populous regions; since during this travel more than half of the ^{222}Rn (3.6-day half-life) would decay away, the average toll would be no more than 0.1 death per year from each year of all nuclear power.

17. EPA Office of Radiation Programs, *Summary Report on Phase I: Study of Inactive Uranium Mill Sites and Tailings Piles,* 1974. Phase II, which involves remedial action, began in 1975.
18. B. L. Cohen, "Environmental Impacts of Nuclear Power Due to Radon Emissions," *Bull. Atomic Scientists,* February 1976, p. 61.
19. USAEC, WASH-740, *Theoretical Consequences of Major Accidents in Large Nuclear Power Plants.* Washington, D.C.: U.S. AEC, 1957.
20. P. Ayyaswamy, B. Hauss, T. Hseih, A. Moscati, T. E. Hicks, and D. Okrent, "Estimates of the Risks Associated with Dam Failure." UCLA Report ENG-7423, 1974.
21. Union of Concerned Scientists, *The Risks of Nuclear Power Reactors.* Cambridge, Mass.: USC, 1977.
22. USAEC, WASH-1238, *Environmental Survey of Transportation of Radioactive Materials to and from Nuclear Power Plants,* 1972. See also L. B. Shappert et al., *Nuclear Safety* **14,** 597 (1973).

23. M. Ross has proposed that a rather large fraction of the ^{137}Cs might be released in an accident (*Proc. Int. Symp. on Packaging and Transportation of Radioactive Materials,* Miami Beach, 1974, USAEC doc. CONF-740901, p. 830). However, recent studies by G. Parker and L. B. Shappert (personal communication, letter to U.S. Energy Research and Development Agency, and testimony before Atomic Safety Licensing Board, docket numbers STN50-483, STN50-486) indicate that Ross's estimates were far too large. Far smaller amounts were found to migrate out of the fuel pellets and the Cs was found to be in nonvolatile forms.
24. U.S. NRC document NUREG-0034, *The Transportation of Radioactive Material by Air and Other Modes,* March 1976. Other studies giving much lower results are C. V. Hodge and A. A. Jarrett, EPA Report NSS-8191.1, 1974. Transportation accident risks in the nuclear power industry, 1975–2020; and USAEC, WASH-1535, *Environmental Impact Statement for the Liquid Metal Fast Breeder Reactor,* Sec. 4.5, 1975.
25. B. L. Cohen, "High-Level Wastes from Nuclear Reactors," *Scientific American,* June 1977, p. 21.
26. B. L. Cohen, "High-Level Radioactive Waste from Light-Water Reactors," *Rev. Mod. Phys.* **49,** 1 (1977).
27. B. L. Cohen, "Analysis, Critique, and Reevaluation of High-Level Waste Water Intrusion Scenarios." *Nuclear Technology* **48,** 63 (1980).
28. B. L. Cohen, "Hazards from Plutonium Toxicity," *Health Phys.* **32,** 359 (1977).
29. J. McPhee, *The Curve of Binding Energy.* New York: Farrar, Strauss, and Giroux, 1974.
30. T. B. Taylor, at Hearings of Joint Committee on Atomic Energy, June 1975.
31. U.S. Surgeon General, Report on cigarette smoking, 1962.
32. Insurance Institute for Highway Safety, Vol. 10, No. 12 (July 9, 1975), gives 24.6 fatalities per 100,000 vehicle years registered for small cars and 11.3 for large cars. A similar conclusion is obtained from National Safety Council, *Accident Facts—1965.* Chicago.
33. Being overweight by 25 per cent (about 40 lb for the average adult male) reduces life expectancy by about five years (L. I. Dublin and H. H. Marks, *Mortality Among Insured Overweights in Recent Years.* New York: Metropolitan Life Insurance Co., 1952). Since 1.2 hours is 1.4×10^{-4} years, a linear hypothesis gives the risk to be that of 8 lb/year $\times$ 1.4×10^{-4} year $= 1.2 \times 10^{-3}$ lb $= 0.02$ oz.
34. U.S. Senate Committee on Public Works, 1975, "Air Quality and Stationary Source Emission Control," gives values for an urban plant (p. 631) and a remote plant (p. 626); we use the average of these and multiply by 400 for the number of plants. A larger number of deaths is estimated by R. Wilson, paper presented at Energy Symposium, Boulder, Colo., June 1974.

35. B. L. Cohen, *Nuclear Science and Society*, pp. 139, 140. New York: Doubleday-Anchor, 1974.
36. L. Sagan, *Nature* **250,** 109 (1974).

SUGGESTED READINGS

Jerold O. Barney (ed.), *The Unfinished Agenda: The Citizen's Policy Guide to Environmental Issues*. New York: Thomas Y. Crowell (Harper and Row), Rockefeller Brothers Fund.

Bulletin of the Atomic Scientists **32** (1976). Entire issue on the plutonium economy.

Cyril L. Comar and Leonard A. Sagan in L. C. Ruedisili and M. W. Firebaugh, eds., *Perspectives on Energy*, 2nd edn., pp. 29–45. New York: Oxford University Press, 1978.

Joseph M. Dukert, *Nuclear Power and the Environment*. Washington, D.C.: Energy Research and Development Administration, 1976.

John T. Edsall, in L. C. Ruedisili and M. W. Firebaugh, eds., *Perspectives on Energy*, 2nd edn., pp. 314–29. New York: Oxford University Press, 1978.

E. P. O'Donnell and J. J. Mauro, *Nuclear Safety* **20** (5), 525 (1979).

Leonard Sagan, *EPRI Journal* **4** (7), 6 (1979).

Frank Von Hippel and Robert H. Williams, *Bulletin of the Atomic Scientists* **32,** 18–21, 48–56 (1976).

Alvin M. Weinberg, *Bulletin of the Atomic Scientists* **33,** 54 (1977).

An Analysis of Nuclear Wastes   17

LEAGUE OF WOMEN VOTERS EDUCATION FUND

INTRODUCTION

How serious is the problem of radioactive wastes? One way to gauge the scale of the problem is to tally up how much has already been generated (as of 1979):

Roughly 75 million gallons of highly radioactive liquid wastes.

5900 metric tons of spent (used) nuclear reactor fuel.

66 million cubic feet of lightly radioactive wastes (contaminated work gloves, tools, medical isotopes; etc.).

140 million tons of radioactive tailings left over from uranium mining and processing.

These wastes are the result, mainly, of the successive steps in the production of nuclear weapons or electricity from fission reactors. The types differ significantly in their physical form and in the intensity and nature of the radiation they emit. The most dangerous waste form—the high-level wastes (see below)—comes both from weapons manufacture and from the reprocessing of spent fuel

rods to salvage unused fuel elements. At present, since reprocessing of spent fuel from commercial reactors is not allowed, these rods have to be dealt with as if they, too, were "waste" and stored in a way that gives protection from the high levels of penetrating radiation they emit. The low-level wastes and the tailings require quite different procedures for disposal.

While the search for and the arguments about nuclear waste management systems go on, the wastes continue to accumulate—the most dangerous in "temporary" storage. Decisions about final disposal await consensus, both scientific and political. Finding a permanent solution has now become critical for several reasons.

First, the volume of nuclear wastes will rise rapidly as the number of operating nuclear power plants grows: 76 plants are now in operation and 88 more are under construction. Spent reactor fuel is currently kept in storage pools at each power plant site; however, many of these pools are filling up. Unless utilities can gain state and federal approval to build additional pools or the federal government provides interim storage facili-

This article is an abbreviated version of A Nuclear Waste Primer, 1980, which is available as Pub. 391 from the League of Women Voters of the United States, 1730 M Street, NW, Washington, D.C. 20036. Copyright © 1980 League of Women Voters Education Fund. Reprinted by permission.

ties for spent fuel within the next decade, several power plants could be forced to shut down. In addition, there is a shortage of low-level waste disposal sites to serve nuclear power plants, medical and research facilities, and other industries.

Second, some radioactive wastes have leaked from government holding tanks and burial sites, and abandoned uranium tailings have been dispersed by wind and water. As a result, many public officials and scientists have concluded that existing methods of waste storage and disposal are inadequate and that, if continued, they could cause serious environmental and health problems.

Third, the future of nuclear power is at stake. California, Connecticut, and Maine have passed laws prohibiting further nuclear power plant construction until the U.S. government has demonstrated that the waste can be disposed of safely and permanently. And several states have restricted or prohibited the storage and disposal of radioactive wastes, of whatever radiation level, within their borders.

Most people lump all forms of nuclear waste together and fear it in an undifferentiated way. Critics of nuclear power, recognizing that nuclear waste is the Achilles heel of nuclear power, use the unanswered questions about safe disposal and the evidence of past flaws in waste management as arguments for shutting down all commercial and military nuclear operations.

But it is not just nuclear foes who are calling for nuclear waste management decisions that the nation can, literally, live with. Those who see nuclear-generated electrical power as an inescapable part of our near-term energy mix, those who have confidence in the viability of nuclear energy for the long term, and those who simply want existing nuclear wastes disposed of safely join in calling for reliable solutions.

There are technological questions to be answered—with different answers for different types of wastes. But perhaps the most critical issues yet to be resolved are social, political, and institutional. How much safety

will we require in the transportation, handling, and storage of wastes? How much money are we willing to spend to do the job? Who should pay? How should state and local governments be involved? How can we assure the total isolation of these wastes from the environment for the long periods required to render them harmless? These are questions that not just scientists but citizens and public officials, too, must answer.

Any waste disposal strategy will inevitably involve some risks, just as other technologies do. Society must assess those risks in light of the benefits it receives from the activities that produce nuclear wastes. For example, if citizens decide to shut down nuclear power plants to reduce the accumulation of wastes, they must be prepared to do without the power generated by these plants or be willing to accept the risks associated with other presently available energy sources, such as coal.

While it is true that the responsibility for making the basic decisions about nuclear power in general and nuclear waste in particular is shared, in a sense, by all of us, it is also true that from the start the federal government has had a special role. It is hard to think of any other activity that has been so tightly controlled by ''the feds.'' The federal government, through one agency or another, has developed, promoted, and regulated all U.S. nuclear activities, including the licensing of privately owned nuclear power plants. The federal government has also reserved to itself the responsibility for developing and carrying out a program for the long-term isolation of high-level radioactive wastes. At present, most of these wastes are stored in temporary tanks at federal sites.

Recognizing that it is the federal government that must take the lead in solving the problem, former President Carter in early 1980 announced the nation's first comprehensive waste management program (see the section below on the politics of nuclear waste management). It is a program that tackles the hard decision about disposal of high-level wastes now in storage—opting for disposal in deep underground rock or salt formations. And it acknowledges the part that other levels of government and the public as a whole must play in making the basic policy decisions.

SOURCES

Where do radioactive wastes come from? The dialogue about nuclear waste management would be a lot easier to conduct if we could talk about a single activity and a uniform kind of waste material. But it is not that simple.

It all starts with radioactive elements—radioisotopes—that are energetically unstable: they have too much energy. They become more stable by giving up some of this extra energy—either a particle or a bundle of pure energy (an X-ray, for example).

Some radioactive elements occur in nature, and we put some of these—uranium as a reactor fuel and radium in medicine are examples—to use in what are now routine ways. Some of these naturally radioactive materials eventually find their way into the waste stream. But the bulk of the most hazardous nuclear waste is composed either of the "fission products" that are produced in a nuclear reactor or of materials in or near the reactor that have been made radioactive by bombardment from neutrons and other reactor-produced radiation.

During the last 35 years, military activities have generated the greatest *volume* of nuclear wastes. However, even though the wastes from the commercial sector are now only a fraction of the volume resulting from military development, they are roughly equal in *radiation content*.[1] Furthermore, the volume of commercial wastes will begin to rival that of military wastes by the end of the century and will greatly exceed them in radiation content.

Types of Waste

Radioactive wastes are differentiated by the intensity of their radiation—by the *number* of "rays" or particles emitted per second per unit of volume. They also differ in *physical form,* (liquid, gas, or solid), in *chemical form* (and therefore in their biological activity), and in the *nature* of the particles or "rays" they emit (see Types of Radiation). Since radiation is energy and much of the energy released is trapped and remains in the waste material itself, radioactive waste is physically hot, and how hot a particular kind of waste is influences the manner of its disposal.

The federal government has categorized radioactive wastes on the basis of these differences:

High-level wastes (HLW) are those generated in the reprocessing of spent fuel. They contain virtually all of the fission products and some transuranic elements (see below) not separated during reprocessing.[2] (Usually, transuranics make up no more than a few per cent of the total HLW.) These wastes have high-intensity, penetrating radioactivity. HLW are liquid—unless they have been chemically treated, in which case they may be liquid with settled-out sludge, a damp salt cake, or calcine (a dry granular material). They generate a lot of heat and thus must be handled remotely (i.e., with no human contact).

Transuranic (TRU) wastes come mostly from the reprocessing of spent fuel and the fabrication of plutonium to produce nuclear weapons. The transuranics are the elements that have atomic numbers greater than that of uranium. They are artificially produced by the bombardment of uranium and uranium products within the reactor. Though transuranics are less intensely radioactive and thus generate less heat than fission products, they take far longer to decay. Some have lifetimes (half-lives)[2a] of thousands of years or longer. Plutonium-239, with a half-life of 24,000 years, is considered one of the most hazardous transuranic elements. TRU waste is currently defined as material containing more than ten nanocuries[2b] of transuranic elements per gram of material.

Low-level wastes (LLW) are generated in almost all activities involving radioactive materials. They have low and sometimes potentially hazardous concentrations of radioisotopes. LLW include cleanup, rinsing, and decontamination solutions; contaminated wiping rags, protective clothing, hand tools, and the like; vials, needles, test tubes, and other medical and research materials; and gaseous effluents. Such wastes are classed as "low level" if they contain less than ten nanocuries of transuranic elements per gram of material; some are completely free of transuranic elements.

In addition to these three waste categories, there are two other types of radioactive materials associated with nuclear operations that can be considered as radioactive waste:

Spent fuel is made up of the intact fuel assemblies that are discarded after having served their useful life in a nuclear reactor.

Under present U.S. policy, spent fuel from commercial reactors is not being reprocessed and thus must be stored in large pools on reactor sites until the federal government decides whether to reprocess it or dispose of it permanently. If disposed of, spent fuel would be considered a form of high-level waste, since it contains substantial amounts of fission products and TRU elements.

Uranium mine and mill tailings are residues from uranium mining and milling operations (see Nuclear Fuel Cycle) that contain low concentrations of naturally occurring radioactive materials. It is the very large volume of the tailings (which are in the form of fine sand) and their long lifetimes that make them an object of scrutiny.

Table 17-1 summarizes the existing quantities of each waste type and the sites where these wastes are stored or buried.

Table 17–1.   Quantities of Existing Waste and Spent Fuel (April 1979).

Types of Waste	Volume of Waste (thousand cubic meters)		Major Storage Sites	
	Commercial	Defense	Commercial	Defense
High-level waste (HLW)	2.3	283	West Valley, N.Y.	Hanford, Wash. Idaho Falls, Idaho Savannah River, S.C.
Transuranic waste (TRU)	Volume not available (approx. 125 kilograms)	311	Beatty, Nev. Hanford, Wash. Maxey Flats, Ky. Sheffield, Ill. West Valley, N.Y.	Hanford, Wash. Idaho Falls, Idaho Los Alamos, N.M. Oak Ridge, Tenn. Savannah River, S.C.
Spent fuel	2	0	At reactor sites and reprocessing plants at Morris, Ill., and West Valley, N.Y.	
Low-level waste (LLW)	515 (January 1, 1978)	1470	Barnwell, S.C. Beatty, Nev. Hanford, Wash. Maxey Flats, Ky. Sheffield, Ill. West Valley, N.Y.	Hanford, Wash. Idaho Falls, Idaho Los Alamos, N.M. Oak Ridge, Tenn. Savannah River, S.C. Nevada test site
Uranium mill tailings	125	25	Piles are located in Arizona, Colorado, New Mexico, North Dakota, Oregon, Pennsylvania, South Dakota, Texas, Utah, Washington, Wyoming.	

Nuclear Fuel Cycle

Much of the controversy today stems from disputes over what to do with wastes from commercial nuclear power plants—partly because these plants are big and, unlike many military operations, highly visible; partly because they are the fastest growing element in the nuclear industry; and partly because their regulation is subject to citizen review. But nuclear power plant operation is only one stage in the commercial fuel cycle. Each of the other stages also produces radioactive wastes.

Uranium mining.   Routine ventilation of mines results in the release of radon gas and uranium-bearing dust.

Milling.   Uranium ore is crushed, ground, and chemically processed to produce a compound (U_3O_8) known as "yellowcake." This operation releases small amounts of radon gas and uranium dust. After the refining process, the leftover ore is discharged to a settling pond where the finely ground tailings are suspended in water. The water gradually dissipates through seepage and evaporation, leaving behind a slurry and eventually a relatively dry pile containing radium (which decays into radon) and other radioisotopes.

Conversion.   Yellowcake is converted to uranium hexafluoride (UF_6). Depending on the technique used, the process produces wastes that are mostly solids or a sludge, with a small part discharged as gas. These wastes contain mainly radium and some uranium and thorium.

Enrichment.   With the application of heat, UF_6 becomes a gas that permits the concentration (enrichment) of uranium-235, the uranium isotope required for reactor fuel. In this process, small quantities of radioactive gas are vented directly into the atmosphere and some liquid waste from cleanup operations is diluted and discharged to the environment.

Fuel fabrication.   Enriched UF_6 gas is converted by cooling to solid uranium dioxide (UO_2), which is formed into ceramic pellets that are placed in zircalloy cladding to yield fuel rods. The radioactive wastes resulting from these operations include process off-gases and liquid waste containing very small quantities of uranium and thorium.

Power plant operation.   As the uranium-235 fuel in the reactor fissions and generates heat for electric power production, the fission fragments (products) accumulate and gradually dampen the chain reaction. Spent fuel rods are then removed from the reactor and stored underwater in large pools at the reactor site. Other radioactive wastes generated at a nuclear power plant include fission product gases, such as krypton and xenon; filter media left over from treating contaminated cooling and cleaning water; and miscellaneous solid wastes, such as protective clothing and cleaning paper.

Reprocessing.   Residual uranium and plutonium are separated from the fission products in the spent fuel so they can be used again. Liquid wastes are produced—many fission products (HLW) and transuranic elements (TRU waste). Other wastes include the contaminated fuel cladding that remains after the fuel has been dissolved and radioactive gases, such as krypton-85. In 1976, former President Ford imposed a moratorium, which is still in force, on commercial reprocessing to help preclude the possibility of plutonium's being acquired by terrorists and made into weapons. However, the U.S. defense establishment continues to reprocess its spent fuel, as do several foreign countries.

"Spent" nuclear facilities.   Worn-out, inoperable, or outmoded nuclear power plants and reprocessing plants constitute another source of nuclear wastes. Some parts of an obsolete reactor and its containment building will remain highly radioactive for several hundred years. To decontaminate these facilities would itself generate significant quantities of solid and liquid radioactive wastes.

Table 17–2.   Spent Fuel and LLW Accumulation by 2000.

	Low Growth (148 GWe[a])	High Growth (380 GWe[a])
Spent fuel	71,100 metric tons	97,800 metric tons
LLW	83 million cubic feet	260 million cubic feet

[a] GWe = gigawatt, or one million kilowatts; 148 GWe equals the output of approximately 148 large reactors.

In sum, the front end of the fuel cycle—mining and milling—presently generates the largest quantity of LLW, in the form of uranium tailings; the back end of the cycle—reprocessing—produces virtually all HLW and TRU wastes.

Looking toward the future, it is expected that wastes from military operations will increase by only a small amount but those from commercial activities will continue to grow—how rapidly is dependent on the growth of commercial nuclear power. Table 17-2 shows Department of Energy (DOE) projections for commercial spent fuel and LLW accumulation by the year 2000 under a low- and high-growth scenario for nuclear power.[3]

In addition to U.S. spent fuel, the DOE expects to accept from other nations about 22,000 metric tons of their spent fuel by the year 2000, in support of the U.S. government's nuclear nonproliferation policy—though there is domestic political resistance to the plan. (Under the Nuclear Non-Proliferation Act of 1978, the United States may buy and store spent fuel from foreign countries that use American nuclear fuel or buy American nuclear technology.)

As for uranium mill tailings, industry and government project that the number of uranium mills and consequently the amount of tailings may triple in the next 25 years.

Finally, there are hundreds of government and private buildings and nuclear reactors that will eventually have to be decommissioned, i.e., decontaminated.

During the fission process, neutrons bombard not only the uranium fuel but other parts of a nuclear reactor as well. Some strike the steel structures that support the fuel, the steel reactor vessel that holds both the fuel rods and the coolant, and even the massive concrete containment structure that shields the reactor vessel. Some neutrons are absorbed by atoms of cobalt, iron, nickel, and other elements in the steel, water, and concrete. The resulting isotopes, called "activation products," are rendered unstable—i.e., radioactive—for varying lengths of time.

Because some of the activation products, such as cobalt-60, are dangerous for several decades or more, a nuclear power plant must be closed down in a way that will prevent public access to it or dispersion of radioactive materials. This process is called "decommissioning." There are four generally recognized methods:

Mothballing entails removal of the fuel and radioactive wastes and the posting of a guard. A mothballed plant requires constant security surveillance, periodic radiological surveys, and maintenance.

Entombment consists of the sealing of the reactor with concrete or steel after liquid waste, fuel, and surface contamination have been removed to the greatest extent possible. At present, the Nuclear Regulatory Commission (NRC) requires annual surveillance and periodic maintenance but no security systems to protect against intrusion.

Dismantlement means total removal of all radioactive components from the site to a radioactive waste disposal facility.

Delayed dismantlement involves a combination of two of the above methods. Since immediate dismantlement might expose workers to dangerously high levels of radioactivity, this option is to mothball or entomb a reactor for 30 to 50 years until some of the shorter-lived activation products decay and then dismantle the reactor.

The General Accounting Office (GAO) and other groups have criticized the NRC and its predecessor, the AEC, for providing little guidance for decommissioning of commercial nuclear facilities. As difficult as the question of *how* decommissioning will take place are the two major issues of *who will pay* and *how much*. Existing NRC regulations require those who apply for a license to operate a nuclear power reactor to demonstrate that they can get the money to meet the costs of both operating and decommissioning a plant. However, such bona fides may not mean much 30 or 40 years later, when the time comes to decommission. Will the utility still be solvent? Do decommissioning responsibilities pass on to any successor? Will that successor honor the obligation? If a utility were forced to shut down its reactor prematurely because of an accident, would it find itself in a precarious financial situation and thus be unable to pay for decommissioning? Then what? Would the government—which is to say, all of us who pay taxes—foot the bill?

Aware of these potential problems, the NRC is now reexamining funding options. Some of the alternatives are *prepayment,* whereby funds are set aside before reactor start-up; a *funding reserve* or sinking fund, whereby money is set aside annually to pay decommissioning costs when the time comes; or *insurance.* Some state public utility commissions have already required utilities to adopt one of these arrangements.

A major criterion in evaluating these options is equity: are the costs of decommissioning being paid by those customers who benefit from the facility? While prepayment provides a high degree of assurance that funds

will be available when needed for decommissioning, it would impose greater costs on customers early in the facility's life. Spreading costs of decommissioning over the life of the facility so customers would pay the same amount annually in constant dollars would be more equitable, but if the plant were shut down earlier than expected, sufficient funds for decommissioning might not be available.

Estimates of the cost of decommissioning a large commercial reactor (1000-MW$_e$) range from 4 per cent to 10 per cent of its construction costs (which total about $1 billion), that is, $40 million to $100 million.[4] In general, the more dismantling required, the higher the costs. However, most of these estimates are for the actual technological costs and may not include any contingency funds for unexpected regulatory changes, high interest or inflation rates, and taxes associated with decommissioning. For instance, if a utility were to adopt the delayed dismantling technique, local property taxes paid while the plant is mothballed for 30 to 50 years could add substantially to the costs of decommissioning.

While 60 to 70 small experimental and prototype reactors have been decommissioned in the United States, only one small commercial reactor, the Elk River Plant (58-MW$_e$), has been fully dismantled. Until we have experience with decommissioning large-scale power plants, the costs of the cleanup and the determination of who will really pay for it in the end will remain uncertain.

The nation does not have unlimited time to figure it all out. By the year 2000—less than 20 years away—12 commercial reactors may be candidates for decommissioning. And some have proposed that the Three Mile Island plant be decommissioned immediately.

RADIATION HAZARD

Nuclear wastes are hazardous because they are radioactive; they emit either nuclear particles (alpha or beta particles, see below) or pure energy radiation like X-rays and γ-rays (gamma rays). Because these radiations are energetic, they can cause damage. As they travel through human tissue, for instance, they rip electrons from the molecules and atoms they strike or pass near. This leaves the molecule or atom "ionized," that is, charged electrically. These ionized particles and the ejected electrons can cause death or damage to cells and cell components.

The nature and severity of the damage depend on what is struck, on the amount of radiation—the exposure—and on the sensitivity of the struck cell.

Sources of Radiation

Natural radioactive materials in the earth (primarily uranium, thorium, radium, and potassium) and cosmic rays, filtered through the atmosphere from outer space, immerse us in a constant flux of radiation. In addition to this natural background radiation, people are exposed to several man-made sources of radiation—medical applications, such as X-rays; fallout from nuclear weapons testing; and consumer goods, such as color televisions. Of the approximately 160 millirems [4a] of radiation the "average" person living in the United States is exposed to every year, 53 per cent comes from natural sources and 47 per cent from human activities. Medical applications account for more than 90 per cent of the man-made dose (Figure 17-1).

Some activities, occupations, and areas of the country expose a person to a greater-than-average radiation dose. For example, a person living in Denver, Colorado, at an altitude of 5000 feet receives nearly twice as much radiation from the sun as a person living at sea level.

Nuclear waste properly handled will add miniscule amounts of radiation to the environment. However, if spills or accidents occur, the wastes can present a significant hazard.

Types of Radiation

Radioactive elements found in nature, commercial products, and nuclear wastes give off

Natural sources: 85 mrem

Medical sources: 70 mrem
Fallout: 3 mrem
Misc.: 2 mrem
Occupational: .8 mrem
Nuclear power: .01 mrem

Total: 160.81 mrem (average dose for U.S. citizen)

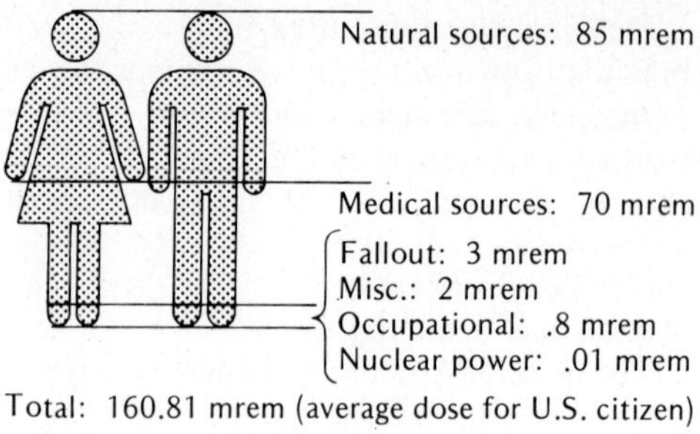

Figure 17-1. Estimated annual whole-body radiation dose in the United States. (Sources: *Nuclear Power: Issues and Choices* and *Energy in Transition 1985–2010*)

three forms of radiation. While all are harmful, they differ in their penetrating power and the manner in which they affect human tissue.

Alpha radiation is the most energetic (densely ionizing) but the least penetrating type of radiation. It can be stopped by a sheet of paper. While alpha particles are unable to penetrate human skin, they may be very harmful if an alpha-emitting isotope enters the body through a cut or by breathing air, eating food, or drinking water. Once inside the body, the radioisotope decays, causing highly concentrated local damage. For example, if an alpha emitter is inhaled, lung tissue could absorb most of the radiation. Long-lived transuranics, such as plutonium, are alpha emitters.

Beta radiation is a more penetrating type of ionizing radiation than alpha. Some beta particles can penetrate skin and damage living cells but, like alpha particles, may be most serious in their effects when beta-emitting isotopes are inhaled or ingested. Most fission products in spent fuel and reprocessed waste (e.g., iodine-131, cesium-137, and strontium-90) are beta emitters. The chemical similarities of some of these radioisotopes to naturally occurring elements in the body make them especially dangerous. For example, the chemical resemblance of strontium-90 to calcium results in its concentration in the bones.

Gamma radiation (high-energy electromagnetic energy waves) has the greatest penetrating power and usually accompanies beta emission. Sunlight and X-rays are similar to gamma rays (they are all electromagnetic radiation) but have different penetrating power. Gamma radiation can penetrate and damage critical organs in the body. The highest levels of gamma radiation usually come from short-lived beta emitters such as krypton-85 (which has a four-hour half-life), a gaseous waste emitted from nuclear reactors and reprocessing plants.

In HLW, beta and gamma radiation dominate for the first 500 to 1000 years; after that, alpha-emitting isotopes in the wastes present the greatest hazard. Since alpha emitters have very long half-lives, some may remain radioactive for as long as a quarter of a million years. TRU waste, TRU-contaminated objects, and uranium mill tailings are major sources of alpha radiation. Low-level wastes emit primarily alpha and beta radiation.

Hazardous for How Long?

How long will nuclear wastes remain hazardous? You have probably heard some experts propose that an isolation time of 300 to 500 years for the HLW is adequate, and others refer to the "million year" waste disposal problem. There are several reasons for the wide discrepancy in estimates. One is the vast difference in half-lives among the different radioisotopes in wastes—about 30 years for the important fission products (strontium-90 and cesium-137), contrasted to 24,000 years for the much less abundant plutonium. Another is the wide latitude available to participants in the nuclear debate in choosing a level of acceptable risk for a given benefit.

Experts who talk of isolating HLW for 300 years base this figure on the average half-life of cesium and strontium and assume that after about ten half-lives ($10 \times$ 30-year average half-life = 300 years), the radioactivity level of the HLW will be low enough (about one thousandth of the original) not to pose a significant hazard. Others who propose isolating wastes up to a million years base their estimate on the longer-lived radioisotopes, such as plutonium. They believe that these radioisotopes, though they are present in much smaller quantities than are fission products, still pose a significant hazard.

One way to put all this in some kind of perspective is to look at Figure 17-2. It compares the relative hazard of reprocessed waste from a commercial reactor if it were buried deep underground with that of naturally occurring toxic ores in the earth's crust. Most of these have half-lives of millions of years and remain hazardous almost forever. Note that if one were to agree to settle for a relative hazard about the same as that of uranium ore, HLW would have to be isolated for something on the order of 1000 years. (Since spent fuel contains substantial amounts of TRU elements, it would take about 250,000 years to reach this hazard level.) To get substantially below that level, however, could take one million years.

Most LLW will decay to the hazard level of uranium ore after 100 years, according to DOE officials. This is because most of the radioisotopes in LLW are shorter lived than those found in HLW (Table 17-3).

Biological Effects

Radiation can kill or damage cells. If enough cells die, the organ they form will die; if crucial organs die, the organism will die. Thus, one consequence of radiation is death.

To be immediately lethal, radiation exposure to the whole body must exceed 1000 rems over a brief period—minutes or hours (such exposures occurred at the Hiroshima

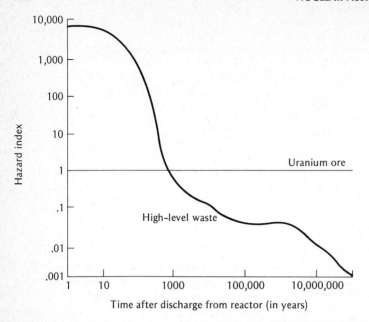

Figure 17-2. Uranium ore versus nuclear waste. (Source: *Radioactive Waste in Perspective*. Westinghouse Advanced Power Systems Division)

and Nagasaki bombings). A dose of 500 rems delivered at one time to the whole body will cause death, on the average, in 50 per cent of the cases.

Table 17–3.   Half-Lives of Typical Radioactive-Waste Components.

Radioisotope	Half-life (years)
Americium-241	460
Americium-242	150
Cesium-135	2,000 000
Cesium-137	30
Curium-242	0.45
Curium-243	32
Curium-244	18
Iodine-129	160,000,000
Neptunium-237	2,100,000
Plutonium-239	24,000
Plutonium-241	13
Radon-226	1600
Strontium-90	28
Technetium-99	200,000
Thorium-230	76,000
Tritium	13

In the range from 500 rems down to 100 rems, radiation sickness occurs and some individuals will die. At lower radiation levels, the consequences are more difficult to predict and detect.

For low radiation doses it is cell damage, not cell death, that is harmful. A dead cell can be replaced. Damage, however, can replicate itself and multiply. The type of damage depends on the nature of the struck cell.

If it is an ordinary cell—bone tissue or flesh, for instance—the damage is confined to the struck organism. This is called "somatic" damage. The most feared type of somatic damage is cancer or leukemia. If however, a reproductive cell is damaged, then a mutation can be caused and the damage transmitted to future generations.

That radiation can cause cancer or genetic mutation is not questioned. What is questioned is the relationship between the dose—particularly doses below one rem—and the number of resulting cancers or mutations. For both types of damage, the latency period—

the time between exposure and the effect—is long. It is 25 or so years for cancer and a generation or more for genetic damage. In both cases there are other possible causes—chemical carcinogens, for example—that can confuse the issue.

Despite uncertainties and difficulties, scientists have developed several models for predicting the effects of low-level radiation. One of the best known models, the linear *no-threshold hypothesis,* is based on two assumptions: first, there is no threshold level below which radiation has *no* carcinogenic effect; second, the incidence of cancer at low doses is directly proportional to the incidence at high dose levels. Using this hypothesis, the known dose-response for high levels of radiation is plotted on a graph and the dose-response for low levels is extrapolated by drawing a straight line from the known data to zero (line *a* in Figure 17-3).

Many scientists believe that the linear no-threshold hypothesis *overestimates* the risks of low-level radiation. They subscribe to the linear *quadratic theory,* which predicts that

cancer incidence is proportionately lower at low doses than high doses, in part because body cells may repair themselves more easily at low doses (line *b* in Figure 17-3).

However, some recent, though highly controversial, studies suggest that the linear no-threshold hypothesis may *underestimate* the risk of cancer from low-level radiation exposure by a factor of ten or more. The low-dose, higher-response curve predicted by these studies might resemble line *c* in Figure 17-3.

In general, risk estimates for low levels of radiation are based on incomplete data and involve a large degree of uncertainty. Federal agencies use the linear no-threshold hypothesis as the theoretical basis for setting current radiation exposure standards because in the absence of firm data, they feel it provides reasonable guidelines for protecting public health.

As for health effects from nuclear wastes specifically, uranium mill tailings pose the most direct health hazard to the general public because they release more radioactivity directly into the atmosphere than any other phase of the nuclear fuel cycle. Specifically, they release radon gas at up to 500 times the natural background rate.

The Environmental Protection Agency (EPA) is the federal agency charged with establishing standards limiting the radiation dose to the general population not only from nuclear power plants but also from other parts of the nuclear fuel cycle. These standards have been revised downward continually since the federal government began setting limits in 1957. Currently, a member of the general population can receive no more than a 25-mrem whole-body dose from nuclear facilities each year. EPA does not claim that this level is a risk-free one, but rather that it is one at which the risk of health effects is balanced against the benefits of nuclear power. Most federal officials believe that this "risk" is well within the range of risks accepted for other methods of generating electricity.

Figure 17-3. Dose response curves. Key: *a* = linear hypothesis; *b* = quadratic theory; *c* = low-dose, higher-response curve. The curves eventually level off and then decrease at high doses of radiation since more cells die than become cancerous. (Source: *Report of the Interagency Task Force on the Health Effects of Ionizing Radiation,* June 1979, U.S. Department of Health, Education and Welfare)

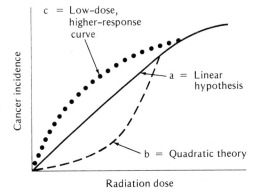

WASTE MANAGEMENT—PAST AND PRESENT

"Benign neglect" characterized the early history of the U.S. nuclear waste management program. While the federal government spent billions to produce nuclear weapons and to commercialize nuclear power during the 1950s and 1960s, it spent only $300 million during that period on researching ways to solve the waste problem. Radioactive wastes were treated, stored, or disposed of with an eye toward convenient, short-term solutions. Even after some nuclear waste storage containers started leaking, efforts to solve the problem were meager.

Only in recent years has the federal government made major efforts to plan for the permanent disposal of nuclear wastes. To understand both the proposals themselves and the public climate in which they are being evaluated, it is necessary to review what government and the nuclear industry have done with nuclear wastes up to now.

High-level Wastes

The atomic weapons program in the early 1940s generated the first HLW, which was stored in large, single-walled, carbon steel underground tanks built at the U.S. Hanford Nuclear Reservation at Richland, Washington. These acidic wastes were neutralized with sodium hydroxide before being pumped into the tanks, in order to impede corrosion. This solution to a problem turned out to be a problem in itself: neutralization increased waste volume and caused a sludge to settle out, making later exhuming and solidification difficult.

Additional tanks built at Idaho Falls, Idaho, and Savannah River, South Carolina, in the early 1950s were of double-wall construction. At Savannah River, the tanks were still carbon steel, and so the HLW was neutralized. At Idaho Falls, the waste was stored untreated because the tanks were stainless steel.

In 1956, the first tank leak at Hanford occurred. Since that time, some 450,000 gallons of HLW have leaked from 20 of the 149 tanks in service there. The surrounding soil has absorbed the wastes, and to date no serious ground-water pollution has occurred from these leaks. In 1960 a tank at Savannah River leaked about 100 gallons of HLW and contaminated some nearby ground water.

In the 1960s, the Atomic Energy Commission (AEC) began to solidify these HLW in order to reduce and stabilize them. The neutralized wastes at Hanford and Savannah River were evaporated down to a damp salt cake. At Idaho Falls, where the HLW had not been neutralized, they could be converted into calcine, a dry granular material that is much easier to store and dispose of than salt cake.

In addition, 20 new tanks with improved leak-detection devices have been built or are under construction at Hanford and 27 at Savannah River. The intent is to eliminate use of carbon steel tanks. But there has turned out to be a serious catch. The still-liquid HLW can be pumped from the old tanks to the new ones, but the sludge (from earlier neutralization) and the salt cake (from later solidification) cannot be redissolved in nitric acid without also dissolving the tank. Furthermore, steelwork protruding into the sludge from the tank floor will interfere with attempts to remove it mechanically. Hydraulic techniques for removing this sludge are now being tested at Savannah River.

Despite problems encountered with the Hanford operation, the AEC approved the same waste storage system—carbon steel tanks and neutralized wastes—for the first commercial reprocessing plant built in West Valley, New York, in the 1960s. During its six years of operation, the plant collected about 600,000 gallons of HLW from the reprocessing of (mostly military) spent fuel. Federal and state authorities are now negotiating over who should be responsible for the long-term monitoring and disposal of these wastes.

Nuclear critics consider the government's

track record for managing HLW dismal. But federal officials counter that, considering the amount of waste stored (about 75 million gallons), the fraction that has leaked is very small—less than 1 per cent. They also emphasize that most of this leakage has occurred at Hanford, where tanks were not as carefully constructed because of wartime pressures and supply problems (e.g., stainless steel was not readily available). Finally, people speaking for the nuclear industry stress that past mistakes in managing HLW have involved mostly military, not commercial, wastes.

Spent Fuel

About 515 tons (10 per cent) of all spent fuel rods from commercial sources are "temporarily" stored in deep-water pools at West Valley, New York, and Morris, Illinois. Why at these two sites? Because they were meant to be reprocessing plants for commercial spent fuel. West Valley did do some reprocessing until it closed in 1972 (that is why it has the HLW that is the subject of so much controversy). The Morris plant never opened because of design problems. And now, since 1976, there is a federal ban on reprocessing.

The Morris site still accepts spent fuel from some utilities, but it is filling up fast, and the state of Illinois has sued DOE, the Nuclear Regulatory Commission (NRC), and General Electric (manager of the site) to prevent any expansion of the facility and to prevent the state's being held liable for the costs of caring for the wastes. The upshot is that most spent fuel rods are being stored in pools on reactor sites around the country.

Since these are also filling up rapidly, many utilities have applied to the NRC for licenses to "re-rack" their fuel rods (pack them closer together) or expand their on-site storage. If the NRC were to license all possible expansions and interpool transfers (that is, shipments of spent fuel from one reactor site to another), U.S. reactors could continue to operate at least until 1990 before their on-site storage pools would be filled.

Some states and communities near reactors, however, are opposing re-racking because of a fear that putting more fuel rods in the same size pool might trigger a chain reaction. This fear seems unwarranted. For such a reaction to occur, the rods would have to be packed more densely than NRC regulations permit. Furthermore, boric acid is flushed into these pools to absorb neutrons that could sustain a reaction. And even if some fissioning occurred, the fuel would grow hot but not explode. Nonetheless, the fear and the opposition exist—and so do technical limits on on-site storage capacity.

Meanwhile, spent fuel rods from nuclear power plants will continue to accumulate as long as the reprocessing ban is in force. Confronted with these facts, DOE is projecting the following *minimum* needs for added away-from-reactor (AFR) storage capacity:

810 metric tons by 1984.

5000 metric tons by 1988.

25,500 metric tons by 1996.[5]

The *actual* needs would depend mostly on the growth of the nuclear power industry.

One other factor could increase these figures. As part of its nuclear nonproliferation strategy, the United States is committed to store spent fuel from reactors in foreign countries that use U.S. technology or nuclear supplies. Honoring this commitment could add 160 metric tons of spent fuel by 1984 and 1465 metric tons by 1996 to the projections. Prime sites for storing these wastes are the existing storage pools located at reprocessing plants in Morris, West Valley, and Barnwell, South Carolina (a plant never completed because of the reprocessing ban). Palmyra Island in the Pacific has also been identified as a potential U.S. AFR site for foreign spent fuel that the United States would accept from nations such as Japan, South Korea, Taiwan, and the Phillipines; however, Australia and New Zealand, countries near the island, are strongly opposed to this proposal.

Congress is now considering legislation that would give DOE the authority to provide AFR storage capacity (licensed by the NRC) for utilities' spent fuel, to collect fees for the service, to take title to the fuel, and to establish a revolving fund to finance the AFR program. This program would be structured to give utilities an "overwhelming" financial incentive to continue to store their spent fuel at reactor sites. For an average-sized nuclear reactor (1000-MW$_e$), the annual costs for storage and disposal at a DOE site (assuming discharge of 30 metric tons of fuel per year) would be about $7 million.

What if commercial reprocessing is sanctioned once again? It is assumed that DOE could return the spent fuel rods to utilities with refund of the charge or keep the rods and provide a credit for the net value of the rods.

Low-Level and TRU Wastes

In the 1940s and 1950s, low-level wastes were buried in shallow pits or packaged in steel drums and dumped at sea. The U.S. government stopped giving out new ocean disposal licenses in 1960, and the last disposal at sea occurred in 1970.

Some land burial sites for LLW were not carefully selected to ensure their isolation from ground water, and several sites are located in areas with heavy rainfall. As a result, some of the radioactive material has migrated from the sites, contaminating soil and nearby streams.

Before 1970, some TRU-contaminated materials were disposed of as LLW and may have also leached from the sites. Since these materials are now defined as TRU wastes, the federal government has exhumed and stored them in a separate facility at one federal LLW site and is considering doing the same at other federal sites.

Of the six commercial sites established to accept LLW, (Figure 17-4), only three are now open: Barnwell, South Carolina; Hanford, Washington; and Beatty, Nevada. The West Valley, New York, site was closed because poor drainage caused the burial trenches to fill with rainwater and overflow. When state authorities in Kentucky discovered that some radioactive material had migrated from a site at Maxey Flats, they put such a stiff surcharge on wastes buried there that the operation soon became uneconomical and the site closed. A Sheffield, Illinois, site closed when it reached its licensed capacity and the state opposed its expansion. In late 1979, the governors of Nevada and Washington closed their sites temporarily because trucks were delivering damaged and leaking nuclear waste containers, most of which were medical wastes.

Each state has always had the option of entering into an agreement with the federal regulatory agency (now the NRC) to regulate commercial LLW disposal sites within its borders and to assume responsibility for perpetual care. All operating sites are in agreement states. On paper, this system looks good; however, after reviewing LLW disposal practices in 1976, the House of Representatives' Committee on Government Operations concluded that "performance of LLW disposal systems has not, to date, been uniformly good" and that "states need to invest more time and money into the perpetual care of these sites." The committee also called for standards to determine unacceptable levels of radioactivity and to signal the need for corrective efforts.

Because Barnwell, South Carolina, is the only open commercial burial site in the eastern half of the country, where the majority of nuclear plants are located, it has been accepting about 85 per cent of all commercial LLW. After the Three Mile Island nuclear power plant accident in March 1979, South Carolina refused to receive any of the voluminous LLW generated by that accident because it felt that South Carolina was already overburdened. In October 1979, South Carolina's Governor Richard Riley announced that over the next two years the state would cut in half the amount of LLW it accepts annually.

Many people assume that most LLW

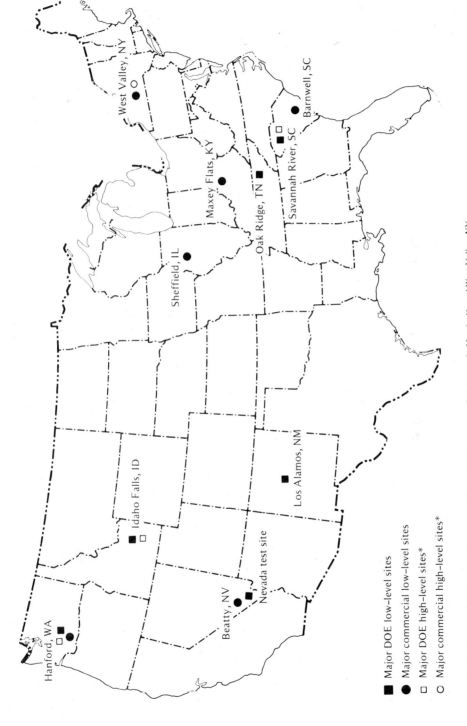

West Valley, NY ○●

Barnwell, SC ●

Savannah River, SC □■

Maxey Flats, KY ●

Oak Ridge, TN ■

Sheffield, IL ●

Los Alamos, NM ■

Idaho Falls, ID ■□

Nevada test site

Beatty, NV ●■

Hanford, WA □●■

■ Major DOE low-level sites

● Major commercial low-level sites

□ Major DOE high-level sites*

○ Major commercial high-level sites*

*Spent fuel is currently being stored on reactor sites and at the reprocessing centers in Morris, IL and West Valley, NY.

Figure 17-4. Major nuclear waste storage/disposal sites.

transported to commercial burial grounds comes from nuclear power plants, but this is not the case. About 43 per cent comes from nuclear plants, 25 per cent from medical and research facilities, 18 per cent from industry, and 14 per cent from government and military operations. All LLW at DOE (federal) sites comes from government research and military activities.

While the burial sites in Washington, Nevada, and South Carolina could accommodate the LLW projected for the 1980s, these states resent being dumping ground for LLW generated in other states. They are pressuring the federal government to help establish a regional network of state-controlled sites. Hospitals and research laboratories are also anxious to have more LLW burial sites established because many of these users, as well as their suppliers (e.g., radio-pharmaceutical firms), have very little space for temporarily storing LLW. When, for example, the Nevada and Washington sites closed simultaneously in October 1979, some hospitals and laboratories were almost forced to discontinue medical treatment and research for lack of a place to store or dispose of the LLW they generated. Although most power plants have ample storage space, their licenses usually limit the quantity of LLW that can be stored on-site.

In response to these pressures, DOE has started to develop a national LLW strategy to present to Congress. Congress is already considering legislation that would permit states to form compacts for investigating, buying, and managing LLW sites.

Uranium Mill Tailings

The most neglected of all radioactive wastes has been uranium mill tailings. Since the tailings did not contain enough radioactive materials to fall under the legal definition of "source material," the AEC, during its 28-year lifetime, insisted that it had no jurisdiction over this part of the nuclear fuel cycle (in spite of the fact that nearly all uranium

produced between 1947 and 1970 was produced for the federal government).

As a result, when a uranium mill closed, the tailings piles were abandoned and left unprotected. These piles, amounting to 27 million tons, are located in Arizona, Colorado, New Mexico, North Dakota, Oregon, Pennsylvania, South Dakota, Texas, Utah, Washington, and Wyoming. A 1976 study done for the Energy Research and Development Administration (ERDA) revealed that radium has leached from each tailings pile anywhere from two to nine feet into the subsoil. While the tailings have not yet contaminated ground water, investigators predict that some piles might do so in the future. Wind has blown the tailings close to buildings and onto land where livestock and wildlife graze. However, the study concluded that hazards from most of these piles are negligible because, although the piles emit radon, they are located in sparsely populated areas. Exceptions include a pile located four miles from downtown Salt Lake City and others near Grand Junction, Colorado, and Durango, Colorado.

Some nuclear critics fault AEC more severely for allowing tailings to be used for building materials. For instance, in the 1960s Grand Junction firms used tailings from a closed mill to make concrete, which was then used for local construction of buildings. For almost two decades, the 30,000 people living in these buildings were exposed to radon levels up to seven times greater than the maximum allowed for uranium miners. (Since the problem was identified, the federal government has provided $12 million to replace the foundations of homes, schools, and churches.) On a smaller scale, tailings have also been used in buildings in Durango, Rifle, and Riverton, Colorado; Lowman, Idaho; Shiprock, New Mexico; and Salt Lake City. Many city streets and foundations of buildings in Denver also contain tailings.

The Uranium Mill Tailings Radiation Control Act requires the federal government to eliminate hazards associated with inactive tailings piles left from past uranium milling operations carried on under AEC contracts. The federal government will pay 90 per cent of the cleanup costs, the state 10 per cent. Final costs will probably be higher than the estimated $140 million because it is likely that a number of piles cannot be stabilized and rendered innocuous in place (e.g., by covering them with loose earth or clay) but will have to be moved and treated elsewhere.

Uranium mill tailings at active mills generally present less risk than abandoned tailings because they are mixed with water and this moisture helps slow the release of radon. However, seepage of radioactive materials from these tailings *has* occurred. EPA has reported some ground water contamination at the Anaconda and the Kerr-McGee mills in the Grants mineral belt in New Mexico, where uranium mining and milling are expected to increase in the future.

One of the worst spills of radioactive wastes in U.S. history occurred in July 1979 at a uranium mine and mill site in Church Rock, New Mexico—also in the Grants region. A muddy mixture of waste material stored behind an earthen dam poured through a 20-foot crack in the dam and gushed into a small stream, allowing 1100 tons of mill tailings to escape during the hour it took workers to seal the crack. Traces of the spill were later found as far away as 75 miles—across the Arizona border. New Mexico health authorities ordered the owner of the mill to recover the waste and clean up any contamination. It is clear that unless tailing management practices improve, water contamination could become a serious problem in this region.

The 1978 Mill Tailings Act clarifies and strengthens the NRC's authority to insist on proper tailings management by uranium mills that it licenses. It requires states that have chosen to license such milling operations themselves to abide by substantive standards at least as stringent as the NRC's. Procedural standards for public hearings and environmental studies also are outlined.

Acceptable long-term management strategies for LLW, though a long way from being fully evolved, show signs of falling into place. What has continued to elude resolution is a comparable strategy for HLW. The nation sorely needs answers to the question of how we can safely dispose of—not just store—HLW (and also spent fuel, if there is not to be reprocessing).

THE SEARCH FOR A PERMANENT SOLUTION

The goal of the federal government's nuclear waste management program, in the words of DOE, which is now the responsible agency, is "to develop the technology and facilities necessary to provide for the permanent isolation of civilian and military wastes from the biosphere so that these wastes pose no significant threat to public health and safety." How best to achieve this goal is a widely debated issue.

Understandably, most of that debate centers on the two "hottest" kinds of radioactive wastes: HLW, in both its liquid and no-longer-liquid states, and spent fuel rods—if they are indeed to end up as waste. As heated as the debate over *interim* storage of HLW and spent fuel has been, it pales beside the intensity of arguments about where to dispose of them *permanently*.

These disputes over permanent disposal are conducted on two planes. On the technical level, experts contend over the best type of disposal method. Even if, as in the case of geologic disposal (see below), consensus begins to emerge, further divisions of opinion crop up. Which kind of geologic formation is best? And, within each, which particular site would be best suited? At this point the political skirmishes begin. Whenever a specific site is identified as suitable for consideration as a permanent repository, the "put it somewhere else" reaction starts another round of controversy.

The array of technical opinion is capsuled here; in the next section, some of the political conflicts and the tools for resolving the conflicts are outlined.

Geologic Disposal

The shorthand term "geologic disposal" stands for permanent disposal in a stable, deep, geologic (rock) formation. Geologic disposal of HLW/spent fuel and TRU waste has at least theoretical acceptance by much of the scientific community. And it has been the focus of federal research ever since 1957, when the National Academy of Sciences (NAS) reported, in a study commissioned by the AEC, that it recommended burial of wastes (HLW and TRU waste primarily) in geologic formations.[6]

The specific kinds of formations that NAS recommended were salt beds and salt domes, and these have been widely regarded, ever since, as the most suitable geologic formations for nuclear waste disposal. Since salt is highly soluble, the very existence of salt formations indicates that they are pretty much free of cracks through which water or brine could travel. And this fact constitutes a major advantage, because the most likely means for wastes to escape from a geologic repository is in water that flows through cracks either to the surface or into an underground aquifer.

Salt formations offer another advantage: because salt flows under heat and pressure, any fractures resulting from an earthquake would heal themselves. In addition, the persistence of salt deposits for 200 million years or more demonstrates their stability.

Within the last few years, however, investigators have discovered some problems with the use of salt formations. Over thousands of years, ground water *has* penetrated and altered salt formations in ways that are difficult to detect from the surface and are not yet fully understood. And very small pockets of brine *have* been found in salt beds and domes. Heat from the radioactive decay of the wastes could burst these brine inclusions, and the brine could then rapidly corrode the waste containers. Finally, if the repository were penetrated by ground water, that water would promptly become saline. And radioisotopes in a saline solution are more likely to migrate

through (instead of being absorbed by) adjacent rock and soil than if the water were salt-free. With these recent findings, public confidence in the suitability of salt formations for long-term waste disposal is not quite what it once was.

There is another drawback to the use of salt formations as a permanent repository: they often exist close to other natural resources such as gas, oil, and gypsum. For example, at the waste isolation pilot plant (WIPP) site near Carlsbad, New Mexico, commercial gas deposits lie below the salt beds and potash deposits above. Some critics would also like to preserve salt domes for use as petroleum storage sites (Strategic Petroleum Reserve).

In 1969, the AEC chose abandoned salt mines near Lyons, Kansas, for a full-scale pilot disposal plant. Kansas state authorities objected on both technical ("It isn't safe enough") and political ("Why pick on Kansas?") grounds. The AEC withdrew its proposal when outside experts demonstrated that previous drilling in the formation and solution mining activities nearby compromised the geologic integrity of the site. The AEC then turned its attention to identifying a more suitable salt formation. In 1975, the AEC's successor, ERDA, announced plans for a waste isolation pilot plant in a salt formation near Carlsbad.

Other geologic formations under consideration for nuclear waste disposal include granite, basalt, shale, alluvium, tuff, and argillite. Unlike salt, all of these rocks are usually fractured to some degree. If the cracks are connected to one another, water can pass through the rock. Heat from radioactive wastes could also increase fracturing. But these rocks, unlike salt, have the ability to chemically absorb most waste elements. Thus, if ground water were to leach wastes from a repository and then move toward aquifers or the surface, absorption by the rock could help prevent the wastes from reaching and contaminating water supplies.

To learn more about the interaction of these rocks with radioactive wastes, DOE is conducting experiments in basalt at the Hanford Reservation and in granite, shale, and other rock formations at the Nevada Test Site.

If nuclear wastes are buried underground in salt or other rock formations, what are the odds that ground water might leach wastes to the environment, causing illness or death? Most studies to date suggest that the odds are very low. For instance, a Swedish study says that not until 200,000 years after burial would the maximum yearly radiation dose in drinking water from a well near a geologic repository rise to one eighth of the average annual radiation from natural sources. An EPA study concludes that the short half-life of some radioisotopes and the absorption of others by the surrounding rock would prevent most from reaching the environment through ground-water pathways during the first 10,000 years.[7]

Of course, there are other sources of risk. Nuclear wastes could be released into the environment by human activities such as exploratory drilling or sabotage, disruptive events such as earthquakes, and engineering-related problems such as shaft seal failures or thermal stress.

One of the most convincing cases for disposing of HLW in stable geologic formations is the "Oklo" phenomenon. Two billion years ago, the intrusion of ground water into a very rich uranium deposit in what is now Gabon, Africa, led to nuclear fission reactions. This "natural" nuclear reactor produced the same types of wastes as a man-made reactor. Studies of the site (near a village called Oklo) show that most of the fission products and virtually all of the transuranic elements, including plutonium, have moved only about six feet from where they were formed.[8]

Other Kinds of Permanent Disposal

While geologic disposal has dominated both science and policy discussions, other answers to the question of "where" have been proposed.

Seabed disposal. It has been suggested that waste canisters could be buried in red clay deposits found in the center of large ocean basins. These deep ocean sites are very stable (undisturbed for as long as 70 million years), virtually bereft of life and isolated from the rest of the planet. Like salt, red clay flows under pressure and its plasticity would tend to seal the waste canisters after emplacement. Although red clay is permeable, water does not naturally flow through these sediments. However, the effects of heat and radiation on ocean sediments are unknown. Also, currents measured near the bottom could conceivably carry across the ocean floor any wastes that leached from the canisters. How far and with what impact would depend on many factors. DOE will not have answers to the engineering problems involved in depositing and, if necessary, retrieving nuclear wastes from ocean basins before 1990.

Antarctic ice sheet disposal. A shallow hole would be drilled in the ice and a waste canister placed in it. The heat given off by the wastes would melt the ice and allow the canister to sink down to bedrock in five to ten years. The melted ice would refreeze as the wastes descended. However, the political problems of using Antarctica for this purpose would undoubtedly be considerable, and the stability of the ice caps over the 100,000 years or more required for the radioactive decay of the wastes is highly uncertain. In 1977, the American Physical Society strongly argued against use of this option.

Transmutation. Some researchers have proposed that HLW be bombarded with neutrons inside a reactor and transmuted into shorter-lived or less harmful substances. Unfortunately, present fission reactors do not do a very good job of altering cesium-137 or strontium-90, two of the most hazardous waste components. Fusion reactors look more promising, but they may not be operating for decades.

Space disposal. In the early 1970s, the National Aeronautics and Space Administration (NASA) concluded that it was technically feasible to rocket nuclear wastes into space. But the risk of a launch accident and the high costs have put this option on the back burner.

Delayed disposal.

Some scientists argue that we can safely delay answering the question of "where" for a century or longer. They recommend keeping spent fuel rods or canisters of solidified HLW in deep-water pools or storing them above ground in air-cooled vaults under continuous human surveillance, while the radioactivity and heat given off by these wastes decreases to more manageable levels. However, others believe that human institutions cannot be depended on to safeguard wastes for this long a time. Also, many people who are critical of the government's past track record on nuclear waste management want a permanent solution now before making further commitments to nuclear power.

Solidification—A Prerequisite for HLW

Almost all strategies for permanent disposal of HLW presuppose that the liquid and sludged HLW will be transformed into a solid, then incorporated into glass or a ceramic, and finally sealed into metal canisters before being inserted into a geologic formation. (Spent fuel would not need to undergo the solidification step, since it is in fact already in ceramic form.) This approach would provide three separate barriers to the release of radioactivity: the chemical form of the waste, its container, and the geologic formation. (Packing material around each canister could add a fourth barrier.)

Most experts who endorse this option do so with a condition attached: that the wastes placed in these formations be retrievable for the first five years or longer, in case problems develop . . . in case better disposal options become available . . . in case the reprocessing ban is lifted and spent fuel can be reused after all.

Just as bedded salt has been considered the most suitable geologic formation for a waste repository for many years, glassification (vitrification) of HLW has been considered the best way to immobilize these wastes beforehand. However, recent studies have cast some doubts on this assumption. They show that while radioisotopes leach from all solid waste forms when they are in contact with water, the leach rate of glass is greatly accelerated by high temperatures and steam over long periods of time (i.e., under the thermal stress conditions imposed by literally hot HLW.)[9] Some scientists believe that this problem could be alleviated by diluting or cooling HLW for several years in order to lower the amount of heat the glass would have to endure.

In 1976, the NRC asked the National Academy of Sciences to prepare a report evaluating alternative technologies for solidifying HLW. The NAS report, released in 1979, concluded that while glass is suitable for use in a first demonstration of a waste solidification and disposal system, it is generally less stable than crystalline materials such as ceramics.[10] (Spent fuel, which is already in ceramic form, has been shown in some studies to have lower leachability than glass.)

At present, the biggest advantage of glass is that the state of the art is more advanced than the ceramics approach. HLW can be glassified in large blocks, whereas ceramic processing is going on only on a small scale in the laboratory. France is already operating a glassification plant at Marcoule, designed to handle all the waste produced in that country's nuclear facilities.

What Other Countries Are Doing

Nuclear waste management is also a very hot political issue in many European countries. Because of environmental opposition and incidents of leaking storage tanks, the governments of Sweden, the Federal Republic of Germany, the Netherlands, and Switzerland have declared that no new nuclear power plants can be built or operated until a safe and permanent waste disposal method can be demonstrated. In 1978, Austrians voted against opening an already completed plant, in part because of their concerns over radioactive wastes. Citizens in France and England have demonstrated against their countries' waste management policies.

To help dispel public opposition to nuclear power, several countries have developed very advanced waste management programs; some are described below.

France is the only European country that now operates large-scale reprocessing and vitrification facilities. Both these facilities are being expanded to handle much larger quantities of spent fuel and HLW. The solid glass blocks of HLW produced will be kept underground in air-cooled storage vaults for 50 to 60 years, then buried at far greater depths in granite (or possibly salt) formations. While France will also reprocess other countries' spent fuel and vitrify the resultant HLW, it will not store these wastes.

The *United Kingdom* now stores its HLW in double-walled steel tanks. It is building a large civilian reprocessing plant at Windscale and will use an HLW vitrification process that its experts say is simpler and less expensive than the French method. The United Kingdom is evaluating permanent HLW disposal in geologic formations, primarily granite and clay.

West Germany has ambitious, though presently postponed, plans to build a huge nuclear complex that would reprocess spent fuel and vitrify and store the resulting HLW. It is also considering direct permanent disposal of spent fuel, after an interim storage period. At present, a salt mine near Asse serves as a pilot repository for disposal of low-level and TRU wastes. These wastes are solidified, packed in steel drums, and stacked by crane in caverns of the salt mine. In the 1980s, the government will experiment with placing spent fuel in this mine, but it has no plans at present to use this site for HLW disposal.

Sweden is considering having its spent fuel reprocessed in another country and then storing the vitrified wastes in a granite forma-

tion. The vitrified wastes would be encased in layers of stainless steel, lead, and titanium before emplacement in granite. (The United States has a cooperative program with Sweden to do field experiments on fluid movement through fractures in granite rock systems.) A mixture of quartz, sand, and bentonite clay, a good absorber of many wastes, would completely surround the canister. Sweden is also considering storage of spent fuel with no reprocessing.

Reprocessing: Is It Essential?

Many nuclear waste disposal plans are predicated on the reprocessing of spent fuel. While it is true that recycling spent fuel does extend fuel *resources,* some experts question the desirability of making reprocessing a basic part of *waste management*. They point out that spent fuel is already a ceramic; hence, one difficult step in dealing with HLW is obviated. They also point out that, while the HLW left over after spent fuel is reprocessed does have a lower content of transuranics than spent fuel has, the reprocessing itself produces TRU wastes. The upshot is that the radioactivity and volume of HLW-TRU wastes and of spent fuel are comparable.[11] In fact, some scientists have pointed out that there is more chance for error and thus greater risk involved in the reprocessing of spent fuel and the solidifying of the resultant liquid HLW than in direct disposal of spent fuel.

Selecting technically sound disposal options is no easy matter. But in order to resolve the nuclear waste issue, the nation has to work out answers that are *fair* as well as *safe,* to the tough question of where and how these wastes are to be finally disposed of.

THE POLITICS OF NUCLEAR WASTE MANAGEMENT

As the AEC's experience in Kansas and the more recent history of WIPP plainly demonstrate, there is more to nuclear waste management than the solving of technical problems. The process by which decisions are arrived at and the degree of trust and mutual regard between levels of government and between citizens and their governments count just as much in determining the outcome of struggles over nuclear waste and, indeed, the future of nuclear power.

There is a natural human tendency for people to want the benefits of nuclear power without suffering the worries or discomforts or risks of coping with the nuclear leftovers. And there is an equally natural tendency on the part of federal officials to want to make decisions without hordes of citizens, or even another set of officials, looking over their shoulders and second-guessing them. But citizens and state and local governments are rightful participants in these decisions. And the principal goal of any political arrangement must be to make it possible for them to play their parts well. It is equally imperative that the net effect of these negotiations and the decisions arising from them be a public perception that risks have been assessed with care and candor and that burdens are being borne equitably.

The State-Federal Standoff

While many states are still receptive to the construction of nuclear power plants, few, if any, are interested in furnishing a site for a permanent HLW (and spent fuel) repository. In fact, more than a dozen states, responding to pressure from citizens, have enacted laws that either flatly prohibit or make difficult the establishment within their borders of disposal facilities for either HLW or LLW. And at least 15 more states, according to NRC, are thinking of following suit.

Why do so many state and local governments want to restrict or prohibit nuclear waste disposal (and even temporary storage)? One major reason is that they believe that the federal government has not made enough progress toward solving the management problems. If these localities are going to have radioactive wastes stored or permanently deposited within their borders, they want assurances that the facilities will be properly managed *now* and *in the future* and will pose no significant risks to citizens.

Adverse experiences with other government projects involving hazardous substances have made states extremely wary of saying yes to nuclear waste facilities. Residents of western states, which have the most favorable conditions for nuclear waste disposal—suitable geology, dry climate, and sparse population—are in an especially mutinous mood. These states have been the sites for many hazardous federally sponsored activities—above-ground atomic bomb tests . . . uranium mining . . . milling and tailings disposal . . . nerve gas production, testing, and storage. As one Westerner put it, "The government has used the wide open spaces as a dumping ground for almost four decades and has inflicted a lot of wounds on us. Well, we've just had enough."

And the impacts of these activities on local populations are just beginning to show up. For instance, recent investigations into the effect of atomic bomb tests conducted in the 1950s revealed that radioactive fallout may have caused an increase in thyroid cancers in southern Utah, Nevada, and Arizona during the 1960s. Although these studies are inconclusive, the widespread publicity they received has made people in western states still more uneasy about the prospect of becoming disposal sites for nuclear wastes.

Where, then, can the U.S. government locate a repository? The General Accounting Office (GAO) reported in 1979 that the federal government could obtain land for a repository within any state without getting that state's consent, a finding that calls into question the legality of state laws on the books. A bill has since been introduced in Congress to give states explicit "veto" power over waste facilities. It would require DOE to:

Notify a state of its intent to explore for a radioactive waste disposal site.

Enable state officials and citizens to review technical, environmental, and safety questions during the planning process.

Grant the state the right to refuse the repository after review, through a public referendum or vote of the state legislature.

Proponents of the legislation believe that it will force a thorough and complete examination of all technical and social issues and will be more likely to lead to a federal decision based on these considerations rather than on what is most expedient.

Some members of Congress believe that states do not need this express veto power because they already have a de facto veto, since there are many different ways a state can block or delay federal activities. They want, instead, to give states incentives, such as money or tax breaks, for accepting repositories. Western states generally oppose special incentives, while the central and eastern states like the idea. All states, however, want the federal government to pay "compensation for the direct and indirect costs of repository siting."[12]

A National Policy

Recognizing the urgent need to resolve nuclear waste issues, former President Carter in 1978 set up the Interagency Review Group (IRG) as a first step toward strengthening and accelerating the federal nuclear waste management program. Its job was to formulate policy recommendations for long-term management of nuclear wastes.

The final IRG report and recommendations, issued in March 1979, formed the basis of the nation's first comprehensive radioactive waste management program, announced by President Carter on February 12, 1980. Key elements, described below, reflect an attempt to give states a voice, but not an overriding voice, in federal siting decisions, to coordinate and speed federal agency actions; and to give citizens, as well as lower levels of government, access points for influencing federal decisions. The timetable for the elements of this program is shown in Table 17-4.

To implement this program, both President Carter and President Reagan have asked for big money. The fiscal 1981 budgets for military and commercial radioactive waste management total nearly $700 million; that is almost $100 million more than is budgeted for all other civilian programs associated with nuclear fission. It is about $150 million over

Table 17–4. Radioactive Waste Management Program.

1980	Drafts of the following standards for the siting and construction of nuclear waste facilities have or will be issued for public review: • EPA standards for the disposal of HLW. • EPA standards for cleanup of inactive uranium mill sites. • NRC technical and procedural standards for HLW repositories. • DOT standards for transport of radioactive materials.
1980 Summer	DOE will release the LLW Program Strategy Document for public review.
1980 Fall	DOE will circulate the draft National Plan for Nuclear Waste Management for public review.
1981	DOE will release the comprehensive National Plan for Nuclear Waste Management.
1981	DOE plans to submit to Congress the final strategy document for LLW.
1985	DOE will announce its selection of a site(s) for a full-scale HLW and TRU wastes geologic repository.
1995	Full-scale HLW/TRU wastes repository will become operational.

the 1980 waste management budget and is vastly greater than budgets of several years ago.

Initial reactions from both environmental groups and industry to the new program have been positive. Industry representatives would have preferred to see repository development on a faster track but can live with the proposed schedule. Environmentalists believe that, overall, the program places priority on the protection of public health and safety.

A Role for Citizens

Long experience suggests that public acceptance of government decisions hinges on a set of conditions that are simple to state but not always easy for government to meet:

Citizens must be involved at every critical stage of the decision-making process.

The process must be reasoned, open, and accessible.

Government must give the public information in understandable language about the technical and institutional aspects of a proposed program.

The process must facilitate discussion in various segments of the public and the government.

If these criteria are important in deciding other controversial issues, they are doubly so in matters relating to nuclear waste—be it choosing a management technique or siting a waste facility.

On these, as on so many other public policy issues, citizens need not be experts in order to make a contribution. The nonexpert can bring perspectives to the dialogue that others may not offer—a longer-range view than most elected officials can afford and "human values" considerations that technical experts may eschew.

Still, even if the process were perfect, achieving meaningful citizen participation in the radioactive waste management program would still be difficult, for several reasons:

1. The technical complexity of the nuclear waste issue will discourage many citizens from participating.
2. Most past decisions on nuclear power and waste issues were made by a small group of technical experts, many of whom continue to feel that citizens do not know enough about the issues to participate— even though citizen participation in standard setting and licensing procedures is now required.
3. Many citizens are aware of past government failings in nuclear management and of the link between radiation exposure and cancer. Laypersons' mistrust of the technical "fix" has also risen, as experts' past "fixes" have unravelled. They may therefore be highly skeptical and suspicious of any government effort to evaluate waste disposal sites and guarantee public health and safety, however sound the scientific data.
4. Management and regulatory responsibilities for nuclear waste are divided. Several federal agencies—notably DOE and NRC —as well as state agencies, control pieces of the regulatory mosaic. Thus, citizens will have to become familiar with each agency's rules for participating in its management or regulatory program.

Citizen Initiatives

Attending workshops and presenting testimony at hearings can be time-consuming and expensive. Thus, citizens should insist that any state participation plan submitted to NRC give financial aid directly to citizen groups or individuals participating in the review process. Citizens can also press NRC or DOE directly to fund a citizen study or report.

Besides financial assistance, citizens will need balanced technical advice on issues. One way to get this advice would be to encourage DOE to sponsor a "science mediation" process. In this process, scientists representing each major position on a scientific question would write a joint paper with the assistance of a mediator, in which the scientists would explain (1) their areas of agreement, (2) their areas of disagreement, (3) each scientist's actual reasons for disagreement on each point, and (4) what information had to be obtained before a sensible decision could be made. The end result would be a single, primary source in which opposing arguments and interpretations are clearly presented and answered point for point by experts in the field. (Sweden used this method in developing its HLW management policy.)

To have this information reach a wide audience, citizens could encourage the Public Broadcasting Service and other networks to present television debates or programs on the nuclear waste issue. Scientists, representatives of public interest groups, and others could also publish articles in wide-circulation magazines, such as *Reader's Digest* and *Popular Science*.

Citizens will need to remember that the wheels of government turn slowly and that many nuclear waste management decisions will take years to fully evolve. But by getting involved now, citizens can help shape the ground rules—the key management plans, strategies, and regulations—and thus help ensure effective and equitable policies in the future.

NOTES AND REFERENCES

1. *Report to the President by the Interagency Review Group on Nuclear Waste Management,* John M. Deutch, Chairman, p. 8. Washington, D.C.: U.S. Department of Energy, March 1979.

2. Transuranic (TRU) elements are elements heavier than uranium (atomic number 92) which are created in the fission process—such as neptunium (93), americium (95), curium (96), and other elements heavier than curium. Transuranic elements make up the bulk of the actinides–actinium (89), thorium (90), protactinium (91), uranium (92), and the other heavier elements—found in HLW.

2a. A *half-life* is the period it takes for any radioactive substance to be reduced by half. The half-life of the radioisotope plutonium-239, for example, is about 24,000 years. Starting with a pound of plutonium-239, in 24,000 years there will be ½ pound of plu-

tonium-239, in another 24,000 years there will be ¼ pound, and so on. (A pound of actual material remains, but it gradually becomes a stable element.)

2b. A *curie* is a measure of the rate of radioactive decay; it is equivalent to the radioactivity of one gram of radium, or 37 billion disintegrations per second. A nanocurie is one billionth of a curie; a picocurie is one trillionth of a curie.

3. *Report of the Task Force for Review of Nuclear Waste Management,* John M. Deutch, Chairman (draft report), pp. 112–13, 141. Washington, D.C.: U.S. Department of Energy, February 1978.

4. R. S. Wood, *Assuring the Availability of Funds for Decommissioning Nuclear Facilities,* p. 4. Washington, D.C.: U.S. Nuclear Regulatory Commission, Office of Reactor Regulation, December 1979.

4a. A *rem* (roentgen equivalent man) measures the amount of damage to human tissue from a dose of ionizing radiation. A *millirem* (mrem) or *millirad* (mrad) is $1/1000$ of a rem or rad. *Man-rem* measures the total radiation dose received by a population. It is the average radiation dose in rems multiplied by the number of people in the population group. A *roentgen* measures the amount of energy lost in air by the passage of gamma rays or X-rays. (When measuring X-rays and gamma rays, one roentgen = one rem = one rad, but this is not true for alpha radiation.) A *rad* measures the amount of energy absorbed per gram of material, such as human tissue, from ionizing radiation.

5. U.S. Department of Energy, *Spent Fuel Storage Requirements—The Need for Away-from-Reactor Storage.* Washington, D.C.: February 1979.

6. National Academy of Sciences—National Research Council, *The Disposal of Radioactive Waste on Land.* Washington, D.C.: 1957.

7. Joint Statement by Dr. James E. Martin and Dr. Daniel J. Egan Jr. of the U.S. Environmental Protection Agency's Waste Environmental Standards Program before the House of Representatives' Subcommittee on Energy and the Environment of the Committee on Interior and Insular Affairs, January 25, 1979, p. 10.

8. George A. Cowan, "A Natural Fission Reactor," *Scientific American* **235** (18), 36 (1976).

9. National Academy of Sciences—National Research Council, *Solidification of High-Level Radioactive Wastes,* pp. 2–3. Washington, D.C.: U.S. Nuclear Regulatory Commission, 1978.

10. Ibid.

11. "Report to the American Physical Society by the Study Group on Nuclear Fuel Cycles and Waste Management," Supplement to *Reviews of Modern Physics* **50** (1), S8 (1978).

12. U.S. Nuclear Regulatory Commission, Office of State Programs, *Means for Improving State Participation in the Siting, Licensing and Development of Federal Nuclear Waste Facilities;* A Report to the Congress, p. 10. Washington, D.C.: March 1979.

SUGGESTED READINGS

International Atomic Energy Agency, *Radioactive Wastes,* December 1978. Georges Delcoigne, Director, Division of Public Information, IAEA, Vienna International Center, Box 100, A-1400, Vienna, Austria.

Richard A. Kerr, "Geologic Disposal of Nuclear Wastes: Salt's Lead Is Challenged," *Science* **204,** 603.

Terry Lash, "Radioactive Waste, Nuclear Energy's Dilemma," *Amicus,* Fall 1979. Natural Resources Defense Council (NRDC), 122 East 42nd St., New York, N.Y. 10017.

League of Women Voters Education Fund, *Anatomy of a Hearing,* 1972, Pub. 108. League of Women Voters of the United States, 1730 M St. NW, Washington, D.C., 20036.

Ronnie D. Lipschutz, *Radioactive Waste: Politics, Technology and Risk.* Cambridge, Mass.: Ballinger, 1980.

U.S. Congress, House Committee on Government Operations, *Low-Level Nuclear Waste Disposal,* House Report No. 94-1320, 94th Congress, 2nd Session, 1976.

U.S. Department of Energy, *Report of the Task Force for Review of Nuclear Waste Management* (draft), February 1978, DOE/ER-004/D.

U.S. Department of Energy, *Spent Fuel Storage Requirements—The Need for Away-from-Reactor Storage,* February 1979, DOE-ET-0075.

U.S. Department of Energy, Draft Environmental Impact Statement, *Management of Commercially Generated Radioactive Waste,* April 1979, DOE/EIS-0046-D. Office of Nuclear Waste Isolation Library, 505 King Avenue, Columbus, Ohio 43201.

U.S. Department of Energy, Public Meetings on Nuclear Waste Management (San Francisco, Denver, Boston), *Consumer Briefing Summary No. 8.* DOE Office of Consumer Affairs, Room 8G082, Washington, D.C. 20585.

U.S. Department of Energy, *Report to the President by the Interagency Review Group on Nuclear Waste Management,* March 1979, TID-29442. IRG-DOE Office of Nuclear Waste Management, Management Support Division, Mail Stop B-107, GTN, Washington, D.C. 20545.

U.S. Nuclear Regulatory Commission, Office of Nuclear Materials Safety and Safeguards, *Regulation of Federal Radioactive Waste Materials.* September 1979, NUREG-0527. U.S.N.R.C., Attention: Publication Sales Manager, Washington, D.C. 20555.

The White House, Office of the Press Secretary, *Fact Sheet: The President's Program on Radioactive Waste Management,* February 12, 1980. Press Release Section, The White House, Washington, D.C. 20500.

Mason Willrich and Richard Lester, *Radioactive Waste: Management and Regulation.* New York: The Free Press, Macmillan, 1977.

Three Mile Island: The Accident That Should Not Have Happened   18

ELLIS RUBINSTEIN

It was not yet 4 A.M. on March 28, 1979. Three Mile Island Reactor Unit 2 was operating at 97 per cent power, and most things seemed normal. Water was being pumped through the reactor core (1) (Figure 18-1), where it was heated under high pressure to prevent it from boiling. Passing out of the core, it proceeded through the steam generator (2), where a heat exchange was taking place. In this heat exchange, water was being transformed by the proximity of the hot core coolant into steam to drive the turbine (3). With heat removed from it, the coolant could return to the core and keep the temperature there within bounds.

The secondary closed-loop water system in a reactor like this runs from the steam generator to the turbine and back through the steam generator. Before 4 A.M., this secondary loop flow seemed normal—that is, water was passing into the steam generator, where it was heated by the hot water passing next to it from the reactor core. The secondary-system water was turning into steam and that steam upon leaving the generator was driving the turbine. It was then being con-

densed back into water by cold water that was circulating out of the cooling tower (4), through the condenser (5), and back into the cooling tower.

TROUBLE AT THE OUTSET

But three things were wrong at Three Mile Island even before 4 A.M. For one thing, there had been a persistent leak of reactor coolant. This leak was known by the operators to be out of the pressurizer (6) through either an abnormally behaving electromatic relief valve (7) or one or both pressurizer safety valves (8). The escaping coolant emptied into the reactor coolant drain tank (9).

The safety valves and relief valve are designed, as their names imply, to relieve abnormally high pressures in the reactor coolant system. The safety valves open automatically on high pressure to prevent rupture of the reactor coolant system. The electromatic relief valve also opens automatically and is to prevent inadvertent and unnecessary opening of the safety valves. And all are designed to channel escaping coolant,

Mr. Ellis Rubinstein is Senior Editor for *IEEE Spectrum*.

From *IEEE Spectrum* **16** (11), 33 (1979). Reprinted by permission of the author and The Institute of Electrical and Electronics Engineers, Inc. Copyright © 1979 by The Institute of Electrical and Electronics Engineers, Inc.

349

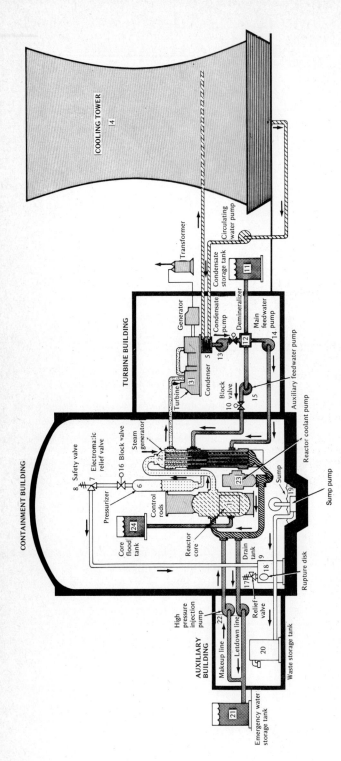

Figure 18-1. Three Mile Island, Unit Two.

which could be radioactive, to a safe holding area—the drain tank.

The problem was that even though coolant was being released by safety valves or the relief valve, the water level in the pressurizer and the pressure in the reactor coolant system were being held at normal levels by the operators. Consequently, they had no reason to be particularly upset by the release. Further, they mistakenly concluded, the Nuclear Regulatory Commission said later, that the leak was within specified limits. In reality it exceeded those limits. Although this does not mean that the core coolant system was experiencing a serious loss of coolant, the leak did play a role in subsequent events in at least one respect: the leakage created ambiguous temperature indications in the drain piping, and these indications later disguised a serious loss of coolant.

The significance of Problem 2 has been debated. Unknown to the operators, two valves (10) were closed, contrary to normal procedures following servicing two days before the accident. These were valves in the auxiliary feedwater system. Ultimately the main feedwater was cut off, triggering the accident. As designed, auxiliary feedwater began to be pumped from the condensate storage tank (11), but the closure of the valves prevented auxiliary feedwater from replacing the main feedwater.

Problem 3 was well known to the operators. They had been working on it for 11 hours prior to the accident. Over that period two shift foremen and various auxiliary operators had been transferring resins from the demineralizer (12) to a resin regeneration tank (not shown in the schematic but connected to the demineralizer). These resins removed minerals from the feedwater system, which must be pure.

Problem 3 came during an apparent resin blockage in a transfer line that forced water back into the condensate pumps' (13) air lines. The details of this process are not important to an understanding of what followed—especially in light of the fact that the problem had happened twice before. What

was important was that the operators, in trying to clear the resin blockage, caused a condensate pump to trip.

It was 4:00:36 A.M. on March 28 when the condensate flow stopped. Within one second the main feedwater pumps (14) stopped, as designed, causing a total loss of feedwater to the generators and an almost stimultaneous—and designed—trip of the main turbine. Thus it was at 4:00:37 that the accident at Three Mile Island began.

AUXILIARY FEEDWATER LOST

Any objective account of events following the loss of feedwater must acknowledge a point of controversy at the outset. Within one second of the feedwater loss and the accompanying tripping of the turbine, all of the three auxiliary feedwater pumps (15) started up as designed. By 14 seconds into the event, they reached full pressure. Their purpose, of course, was to replace the lost main feedwater and prevent the steam generator from going dry. Unfortunately, as previously noted, valves between the auxiliary feedwater valves and the steam generator had been left closed improperly sometime during the previous 48 hours. The result was that there was no flow of auxiliary feedwater. It took the operators 8 minutes to discover this because tags on the control room panel were inadvertently covering the valve position indicators.

But the controversy remains: was the loss of auxiliary feedwater a significant factor in the accident? Babcock & Wilcox, builder of the reactor, says the loss was significant. It suggests that had there been auxiliary feedwater in the system, the temperature of the reactor coolant might have remained relatively stable until the problem of the condensate pumps was corrected and the normal feedwater reinstated.

This view is contested by the utility-sponsored Nuclear Safety Analysis Center, an investigative arm of the Electric Power Research Institute. Its report says:

Further analysis is required to definitely establish whether or not this early unavailability of emergency feedwater had a significant effect on the eventual consequences, other than to add another dimension to the areas of concern within the main control room. However, the current . . . analysis indicates the early unavailability does not significantly affect the core uncovery if the balance of the system thermal hydraulic conditions remain unchanged.

Aside from this controversy, the scenario of the Three Mile Island accident is relatively straightforward: with no water entering the generator and no steam leaving it, no heat was being removed from the reactor coolant system during the first few seconds following the loss of feedwater. The reactor coolant's temperature rose, causing it to expand and creating increased pressure in the overall reactor coolant system. Somewhere between three and six seconds into the incident, the system pressure reached the predesigned point that triggered the opening of the electromatic relief valve (Figure 18-2).

In doing this the system continued to act precisely as it was designed to—that is, the opening of the electromatic relief valve was a control mechanism designed to prevent overpressure in the reactor coolant system. With it open, enough coolant would escape the system loop to return the system pressure to normal.

Before this happened, however, the system pressure continued to increase for a couple of seconds, reaching the automatic reactor trip point at eight seconds into the accident. At the "SCRAM" (trip) signal, the control rods in the reactor core automatically dropped into the reactor core, ending the nuclear reaction and shutting down the reactor within one second. Remaining, though, was the problem of removing the residual heat from the core.

LOSS OF COOLANT BEGINS

Although conditions in the reactor core were still very hot following the reactor tripping, there followed the expected cool-down and attendant reduction of reactor coolant system pressure while coolant was released through the open relief valve. Then one of the most significant events of the entire sequence occurred.

At around 13 seconds, the pressure in the reactor coolant system returned to normal levels, and the electromatic relief valve should have been signaled automatically to close, thereby ending further loss of coolant. In the control room, an indicator noted that this signal had been given. *But the valve never closed.*

It is still not known why this key valve did not close.

Until the valve and its associated control system are actually studied—which may take years until there is sufficient radioactive cleanup for investigators to enter the containment building—no one can be sure why it did not close. But two things are certain: (a) the operators could have closed a block valve (16) manually, thereby mitigating the effect of the stuck-open relief valve and totally preventing subsequent damage to the reactor core, and (b) because the block valve was not closed, a serious loss of coolant occurred for more than two hours, eventually exposing the core and leading to an escape of radioactivity, first into the auxiliary building and finally into the outside environment.

Should the operators have known enough to close the block valve? An indicating light on their control panel indicated that the solenoid designed to close the valve had been activated properly. But there was no indicator to show whether the valve was in fact open or closed.

A second way of determining the position of the valve is by reading the temperature in the pipes leading from the valve to the drain tank. An abnormally high temperature would indicate the presence of escaping reactor water or steam. In fact, such readings were made and high temperatures were noted, but they were thought to be caused by valve leakage, of which the operators were aware before the accident.

A third method of determining if much

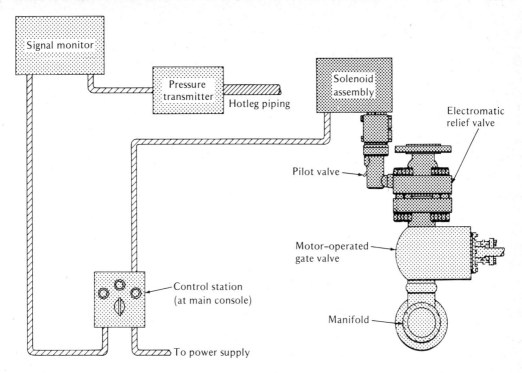

Figure 18-2. The electromatic relief valve can be operated in a manual or automatic mode. In the manual mode the operator can open or close the valve with a switch on the control console. With the switch in the manual open position, the solenoid is energized to open the valve. With the switch in the manual closed position, the solenoid is deenergized to close the valve. With the switch in the auto position, the opening and closing of the valve is controlled by a signal monitor. This monitor responds to a signal from a pressure transmitter in the reactor coolant system. A set of relay contacts in the signal monitor closes when the system pressure reaches the valve opening setpoint, thereby energizing the solenoid. A separate set of contacts in the signal monitor deenergizes the solenoid when the pressure falls to the valve closing setpoint. An amber light on the control console lights up when the operator puts the switch in the auto position. A red light goes on when the solenoid is energized either from the signal monitor or from the manual open contacts.

coolant was escaping through the relief valve was an indicator of pressure in the drain tank. In fact, that pressure was increasing as ever more coolant escaped through the open relief valve until, at about 3½ minutes into the accident, the reactor coolant drain tank's relief valve (17) appears to have lifted. Moreover, as conditions worsened, even the drain tank's relief valve was not sufficient to relieve the increasing pressure of the draining coolant. At about 15 minutes into the accident a disk (18) "blew." It was designed to rupture be-

fore the entire drain tank became endangered. The rupture spewed coolant into the containment building sump (19) and then out into the auxiliary building, where it should have been contained in a series of radioactive waste storage tanks (20). There was no containment because these tanks appear to have overflowed, leading to a general disbursement of coolant. There is some argument as to how soon that coolant became radioactive. Eventually, however, it was transporting radioactivity outside the containment.

All of this could have been forestalled had anyone looked at the drain-tank pressure indicator. However, this indicator was on a panel behind the approximately 7-foot-high primary control-room panels, on which all critical instruments were placed.

Clearly there was ample excuse for the operators in these early minutes of the accident to have missed the fact that leakage was continuing through the relief valve. But there were to be other signs of a serious loss of coolant that would be ignored. For now, suffice it to say that the operators at Three Mile Island did not realize they had a loss of coolant through the relief valve until 142 minutes into the accident. At that point they closed the electromatic relief valve block valve (16), but by then matters had passed beyond the point of no return. By then, the operators had also failed to employ a safety system specifically designed for just such a loss of coolant.

FAILURE TO COMPENSATE FOR COOLANT LOSS

Every nuclear reactor is designed against the potential disaster of a loss of coolant in the reactor core. The reactor at Three Mile Island had both a high- and a low-pressure emergency core coolant system. The low-pressure system will be discussed later; the high-pressure system consisted of a water storage tank (21) and three high-pressure injection pumps (22), which, as their name implies, could inject emergency coolant directly into the reactor coolant system. As the reactor coolant system pressure dropped, because of the open relief valve, it eventually reached a level that automatically initiated the high-pressure injection emergency pumps. In turn, the pumps did their job and began delivering water to the coolant system. The latter's pressure began to rise again. Had these pumps been left to do their designed task, the accident at Three Mile Island could have been averted. But at about 4½ minutes the opera-

tors made their second major misstep: *they began to throttle back one of the emergency pumps and turned off the other completely.*

It was not until 3 hours 40 minutes into the accident that this action was reversed. At that point the pumps came on automatically due to high containment pressure (4 lb/in²). Even then, however, the operators turned the pumps off again, and they continued to fiddle with them until at least 4½ hours into the accident, when they used the high-pressure pumps continuously to inject coolant at a high rate into the reactor core coolant system. By then, as in the case of the block valve closure, the damage had been done.

Why did the operators throttle back an emergency system that was fulfilling the very role for which it was designed? The answer is complicated. When the relief valve opened, the steam in the pressurizer was the first to escape. As might be expected, a surge of liquid coolant from the core coolant system replaced it.

This caused the pressurizer level indication to rise until, at about 6 minutes into the accident, the level went off the scale, indicating that the pressurizer was full of water (Figure 18-3).

Operators call this a "solid pressurizer," and at Three Mile Island they had been trained to avoid this by cutting back on the water added to the reactor coolant system. What the Three Mile Island operators did not realize is that the system was by no means filled with coolant. While the coolant level in the pressurizer was extremely high, the coolant in the reactor coolant system had become a mixture of steam and water with the amount of water rapidly decreasing.

What was happening was that the lack of coolant and the resulting overheating in the early moments of the accident had created voids in the reactor coolant system that were giving the false impression that the system was full of coolant.

It was in part this high level of coolant in the pressurizer that led the operators not to question whether coolant was escaping. Un-

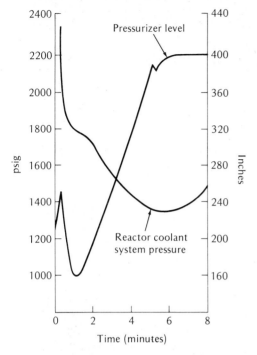

Figure 18-3. "The big leak"—words used by a Babcock & Wilcox expert to describe the small loss-of-coolant accident at Three Mile Island. The problem was that the operators did not realize they were losing coolant, even though the conditions described in the graph (above)—increasing pressurizer level concomitant with decreasing reactor coolant system pressure—signal a loss of coolant. The operators concentrated mistakenly on the pressurizer level alone, even though the indicators were side by side on the control panel.

known to them, the reactor coolant system was becoming, for lack of coolant, a mass of saturated, superheated steam.

THE STRAW THAT BROKE THE CAMEL'S BACK

Among the experts who have reported on Three Mile Island, many seem to feel that the point of no return came somewhere around 100 minutes into the accident. Hindsight would indicate that by the third hour

the core was seriously damaged. By that time the ziracaloy containment (cladding) of the radioactive fuel elements was deteriorating because of oxidation by steam. This exposed the circulating steam and coolant to radioactive fission products. Of about 140 MCi of xenon-133 in the core, about 10 MCi were released into the atmosphere, and of about the same amount of iodine-131, only about 15 Ci were released. The figures are from the Nuclear Regulatory Commission.

Xenon is far less dangerous than iodine. Had as much iodine as xenon been released, the Three Mile Island accident would have been a disaster. However, less iodine than xenon escaped the core. This was because most of the core damage was to the cladding, which primarily yields the noble gases. Iodine is released by damage to the fuel pellets, and this damage was minimal at Three Mile Island. Further, of the iodine that was released, much was absorbed in the escaping water. Additives in the water increased this absorption. And some iodine was separated from released gases by charcoal venting filters. Neither the water nor the filters impeded the release of xenon and the other noble gases, though.

But before that 100-minute mark, there were still four clear possibilities to avoid this. Three have already been noted: (1) the operators could have closed the electromatic relief valve block valve, ending loss of coolant; (2) they should never have throttled down the high-pressure injection pumps in the emergency core cooling system; and (3) they could have restarted those pumps sometime during the first 100 minutes. The fourth choice they had was to let the reactor coolant pumps (23) continue to operate.

At 74 minutes the operators shut off half of the four reactor coolant pumps. At 101 minutes they shut off the remaining two pumps. From all reports, these actions seem to have sealed the reactor's fate.

Babcock & Wilcox experts feel that as long as the pumps were circulating the roiling mixture of water and steam—the remnant of the reactor core coolant—the core, over-heated as it was, would have sustained minimum damage. *But with the pumps turned off, all circulation stopped, and within the next 50 minutes substantial portions of the core became uncovered, damaging the fuel elements and thereby releasing radioactive core fission products and generating the hydrogen that so frightened the public and, briefly, the Nuclear Regulatory Commission directors.*

Why did the operators turn off the reactor coolant pumps? Their reasoning was straightforward: system pressure was low, coolant flow seemed low, and there were high vibrations coming from the pumps. In fact, standing in the control room, they said, they could feel the vibrations.

Unless a nuclear reactor accident involves a loss of coolant, the Electric Power Research Institute states that industry procedure is to protect the pumps from potential damage. But substantial concern has been voiced unofficially by Babcock & Wilcox and several other experts to the effect that such a policy is indicative of utility concern with equipment—or with downtime and potential blackout—rather than with public safety.

At Three Mile Island the operators argue that they had not determined that they were experiencing a loss-of-coolant accident, and so they expected natural circulation of coolant to protect the core in the absence of forced pumping.

Without pumping and with little coolant left, the steam and water in the coolant system separated, stopping all flow, even in the reactor vessel. At about the 2½-hour mark, reactor core temperatures were rising rapidly as the core began to uncover. Between 149 and 750 minutes (12½ hours) into the accident, temperature indications passed the high end of the instrument scale at 620°F. In fact, between four and five hours a digital voltmeter indicated temperatures as high as 2500°F. What existed then was a superheated steam environment with some non-condensible hydrogen in the reactor coolant outlet piping.

The damage had been done. All that remained was trying to recover as rapidly as possible.

REESTABLISHING CONTROL OF THE CORE

The first step in the attempt at core cooldown was taken at 2 hours 18 minutes into the accident, when the operators closed the block valve, terminating further escape of coolant.

For the next 13 hours, they made one attempt after another to remove heat from the core coolant system and to reestablish stable cooling.

The earliest attempts involved methods by which the operators thought they could reestablish forced—or establish natural—circulation of reactor coolant with heat removal through the steam generators.

But, of course, there was too little coolant to circulate naturally. Further, superheated steam and hydrogen had combined in the reactor coolant system, blocking heat removal through the steam generator, even under the by-now-belated application of emergency high-pressure injection of coolant.

A small portion of this hydrogen that was eventually to cause havoc in the recovery attempts was actually intended to be in the system. Its purpose in the reactor design was to combine with oxygen in the coolant to prevent corrosion via oxidation of metal components.

Beginning as early as the one-minute mark of the accident, some of this excess hydrogen began to separate from the increasingly hot coolant and to vent through the open relief valve into the reactor containment building. By itself, this might well have been of no concern, because the volume of the containment building around a pressurized water reactor like that at Three Mile Island is so great that small amounts of leaking hydrogen are not considered harmful.

Never considered, however, was the possibility that operators would override all the emergency systems designed to prevent core uncovery. And it was the high-temperature steam atmosphere created by the core uncovery that triggered a chemical reaction between the circulating steam and the zircaloy metal that clads the fuel elements. It was this situation at Three Mile Island that produced large amounts of excess hydrogen from about the three-hour mark on.

Part of this hydrogen vented into the containment building, part became trapped in the reactor vessel and formed the widely reported "bubble," and part circulated through the coolant system (an illustration of the bubble in the reactor vessel is shown in Figure 18-4). The circulating hydrogen and steam further complicated efforts to reestablish normal circulation and heat exchange.

But what frightened Nuclear Regulatory Commission investigators most in the early hours of the accident were the reactor vessel bubble and the hydrogen in the containment building. Babcock & Wilcox experts have emphasized that there was no cause for fear of an explosion from the hydrogen in the reactor vessel as there was inadequate oxygen present to support combustion.

More serious, however, were the hydrogen pockets forming in the containment building, where there was plenty of oxygen. At about 9½ hours into the accident, the hydrogen in the reactor containment building ignited, as indicated by a 28-lb/in² pressure "spike" (surge). The fact that the building withstood this "burn," as it was designed to, has not made the Nuclear Regulatory Commission sanguine about hydrogen formation in containment buildings.

At Three Mile Island, the operators tried next to depressurize the system. The idea was to mimic a large loss-of-coolant accident— one that might occur with a major break in the core coolant loop. In the event of such an accident, the reactor had core flood tanks (24) that would be activated automatically to flood the reactor. The hope of the operators was to reach that pressure and trigger the opening of the flood tanks. This in turn would initiate a separate heat-removal system that would cool the core coolant by running Susquehanna River water through a heat-exchange

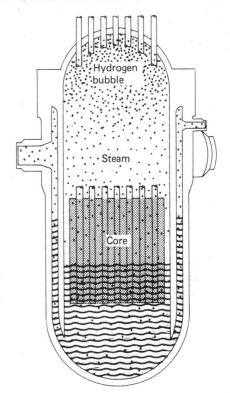

Figure 18-4. The infamous "hydrogen bubble" caused waves of public apprehension following the mistakenly voiced concerns of NRC commissioners that it could rupture the reactor vessel. As shown here, it never was an actual bubble but rather a mixture of hydrogen and steam, with excess hydrogen tending to rise to the top of the reactor vessel. And because there was little oxygen in that vessel, there was no chance of explosion.

device. These plans were abandoned, however, when the operators realized they could not achieve pressures low enough to trigger emergency core flooding. And there is question as to whether the Susquehanna River heat-removal system could have been used inasmuch as it was not designed to handle radioactive liquids and no one knows whether the pump seals would have prevented radioactivity leaking into the auxiliary building.

Five to six hours passed without resolution of the crisis. However, a Nuclear Regulatory Commission report says that "the extended period of low pressure appears to have assisted in the release of hydrogen gas from the reactor coolant system."

Added to the hydrogen that had already escaped through the open relief valve and into the containment building, where it ignited, this new hydrogen release may have been just enough to permit the operators their first success: at about 13½ hours into the accident, they managed to get one of the reactor coolant pumps to start. At 15 hours 50 minutes, based on this preliminary success at reestablishing some forced coolant circulation and heat removal, the operators managed to start a second pump. From then on, it was simply a matter of time before conditions fully stabilized.

The accident at Three Mile Island was over, and the inquiry began.

SUGGESTED READINGS

Harold W. Lewis, "The Safety of Fission Reactors," *Scientific American* **242** (3), (1980).

Thomas G. Lombardo, "TMI plus 2," *IEEE Spectrum* **18** (4), 28 (1981). Special report reviewing the nuclear industry two years after the trauma of Three Mile Island.

Daniel Martin, *Three Mile Island: Prologue or Epilogue?* Cambridge, Mass.: Ballinger, 1980.

Thomas H. Moss, and David L. Sills, eds., "The Three Mile Island Nuclear Accident: Lessons and Implications," *Annals of the New York Academy of Sciences* **365** (1981).

Nuclear Power Plant Safety After Three Mile Island, report of the Subcommittee on Energy Research and Production of the House Committee on Science and Technology, March 1980. Superintendent of Documents, U.S. Government Printing Office, Washington, D.C. 20402.

Ellis Rubinstein, ed., *IEEE Spectrum* **16** (11), 30 (1979). Special issue on Three Mile Island and the future of nuclear power.

The Need for Change: The Legacy of TMI 19

KEMENY COMMISSION

THE CHARGE TO THE COMMISSION

The purpose of the Commission is to conduct a comprehensive study and investigation of the recent accident involving the nuclear power facility on Three Mile Island in Pennsylvania. The Commission's study and investigation shall include:

A technical assessment of the events and their causes; this assessment shall include, but shall not be limited to, an evaluation of the actual and potential impact of the events on the public health and safety and on the health and safety of workers.

An analysis of the role of the managing utility.

An assessment of the emergency preparedness and response of the Nuclear Regulatory Commission and other federal, state, and local authorities.

An evaluation of the Nuclear Regulatory Commission's licensing, inspection, operation, and enforcement procedures as applied to this facility.

An assessment of how the public's right to information concerning the events at TMI was served and of the steps which should be taken during similar emergencies to provide the public with accurate, comprehensible, and timely information.

Appropriate recommendations based upon the Commission's findings.

The President's Commission on the Accident at Three Mile Island was appointed by President Jimmy Carter to investigate the TMI accident. The Commission consisted of John G. Kemeny, Chairman, President of Dartmouth College; Bruce Babbit, Governor of Arizona; Patrick E. Haggerty, Honorary Chairman and General Director, Texas Instruments Inc.; Carolyn Lewis, Associate Professor, Graduate School of Journalism, Columbia University; Paul A. Marks, Vice President for Health Sciences and Frode Jensen Professor, Columbia University; Cora B. Marrett, Professor of Sociology and Afro-American Studies, University of Wisconsin-Madison; Lloyd McBride, President, United Steelworkers of America; Harry C. McPherson, Partner, Verner, Liipfert, Bernhard, and McPherson; Russell W. Peterson, President, National Audubon Society; Thomas H. Pigford, Professor and Chairman, Department of Nuclear Engineering, University of California at Berkeley; Theodore B. Taylor, Visiting Lecturer, Department of Mechanical and Aerospace Engineering, Princeton University; and Anne D. Trunk, Resident, Middletown, Pennsylvania.

This article is excerpted from the *Report of The President's Commission on The Accident at Three Mile Island*. From *Bulletin of the Atomic Scientists* **36** (1), 24 (1980). Reprinted by permission of *Bulletin of the Atomic Scientists*. Copyright © 1980 by the Educational Foundation for Nuclear Sciences, Inc.

OVERALL CONCLUSION

After a six-month investigation of all factors surrounding the accident and contributing to it, the Commission concluded that:

To prevent nuclear accidents as serious as Three Mile Island, fundamental changes will be necessary in the organization, procedures, and practices—and above all—in the attitudes of the Nuclear Regulatory Commission and, to the extent that the institutions we investigated are typical, of the nuclear industry.

This conclusion speaks of *necessary* fundamental changes. We do not claim that our proposed recommendations are sufficient to assure the safety of nuclear power.

Our findings do not, standing alone, require the conclusion that nuclear power is inherently too dangerous to permit it to continue and expand as a form of power generation. Neither do they suggest that the nation should move forward aggressively to develop additional commercial nuclear power. They simply state that if the country wishes, for larger reasons, to confront the risks that are inherently associated with nuclear power, fundamental changes are necessary if those risks are to be kept within tolerable limits.

ATTITUDES AND PRACTICES

Popular discussions of nuclear power plants tend to concentrate on questions of equipment safety. Equipment can and should be improved to add further safety to nuclear power plants, and some of our recommendations deal with this subject. But as the evidence accumulated, it became clear that the fundamental problems are people-related problems and not equipment problems.

When we say that the basic problems are people-related, we do not mean to limit this term to shortcomings of individual human beings—although those do exist. We mean more generally that our investigation has revealed problems with the "system" that manufactures, operates, and regulates nuclear power plants. We are convinced that if the only problems were equipment problems, this Presidential Commission would never have been created. The equipment was sufficiently good that, except for human failures the major accident at Three Mile Island (TMI) would have been a minor incident. But, wherever we looked, we found problems with the human beings who operate the plant, with the management that runs the key organization, and with the agency that is charged with assuring the safety of nuclear power plants.

After many years of operation of nuclear power plants, with no evidence that any member of the general public has been hurt, the belief that nuclear power plants are sufficiently safe grew into a conviction. One must recognize this to understand why many key steps that could have prevented the accident at Three Mile Island were not taken. The Commission is convinced that this attitude must be changed to one that says nuclear power is by its very nature potentially dangerous and, therefore, one must continually question whether the safeguards already in place are sufficient to prevent major accidents. A comprehensive system is required in which equipment and human beings are treated with equal importance.

CAUSES OF THE ACCIDENT

First of all, it is our conclusion that the training of TMI operators was greatly deficient. While training may have been adequate for the operation of a plant under normal circumstances, insufficient attention was paid to possible serious accidents. And the depth of understanding, even of senior reactor operators, left them unprepared to deal with something as confusing as the circumstances in which they found themselves.

Second, we found that the specific operating procedures which were applicable to this accident are at least very confusing and could be read in such a way as to lead the operators to take the incorrect actions they did.

Third, the lessons from previous accidents

did not result in new, clear instructions being passed on to the operators.

In conclusion, while the major factor that turned this incident into a serious accident was inappropriate operator action, many factors contributed to the action of the operators, such as deficiencies in their training, lack of clarity in their operating procedures, failure of organizations to learn the proper lessons from previous incidents, and deficiencies in the design of the control room. These shortcomings are attributable to the utility, to suppliers of equipment, and to the federal commission that regulates nuclear power. Therefore—whether or not operator error "explains" this particular case—given all the above deficiencies, we are convinced that an accident like Three Mile Island was inevitable.

SEVERITY OF THE ACCIDENT

Just how serious was the accident? Based on our investigation of the health effects of the accident, we conclude that in spite of serious damage to the plant, most of the radiation was contained and the actual release will have a negligible effect on the physical health of individuals. The major health effect of the accident was found to be mental stress.

The amount of radiation received by any one individual outside the plant was very low. However, even low levels of radiation may result in the later development of cancer, genetic defects, or birth defects among children who are exposed in the womb. Since there is no direct way of measuring the danger of low-level radiation to health, the degree of danger must be estimated indirectly. Different scientists make different assumptions about how this estimate should be made, and, therefore, estimates vary. Fortunately, in this case the radiation doses were so low that we conclude that the overall health effects will be minimal. There will either be no case of cancer or the number of cases will be so small that it will never be possible to detect them. The same conclusion applies to the other possible health effects.

We conclude that the most serious health effect of the accident was severe mental stress, which was short-lived.

There was very extensive damage to the plant. While the reactor itself has been brought to a "cold shutdown" there are vast amounts of radioactive material trapped within the containment and auxiliary buildings. The utility is therefore faced with a massive cleanup process that carries its own potential dangers to public health. The ongoing cleanup operation at Three Mile Island demonstrates that the plant was inadequately designed to cope with the cleanup of a damaged plant. The direct financial cost of the accident is enormous. Our best estimate puts it in a range of $1 to $2 billion, even if TMI-2 can be put back into operation. (The largest portion of this is for replacement power estimated for the next few years.) And since it may not be possible to put it back into operation, the cost could even be much larger.

The accident raised concerns all over the world and led to a lowering of public confidence in the nuclear industry and in the Nuclear Regulatory Commission.

From the beginning, we felt it important to determine not only how serious the actual impact of the accident was on public health, but whether we came close to a catastrophic accident in which a large number of people would have died. Issues that had to be examined were whether a chemical (hydrogen) or steam explosion could have ruptured the reactor vessel and containment building and whether extremely hot molten fuel could have caused severe damage to the containment. The danger was never—and could *not* have been—that of a *nuclear* explosion (bomb).

We have made a conscientious effort to get an answer to this difficult question. Since the accident was due to a complex combination of minor equipment failures and major inappropriate human actions, we have asked the question "What if one more thing had gone wrong?"

We explored each of several different scenarios representing a change in the sequence of events that actually took place. The greatest concern during the accident was that significant amounts of radioactive material (especially radioactive iodine) trapped within the plant might be released. Therefore, in each case, we asked whether the amount released would have been smaller or greater and whether large amounts could have been released.

Some of these scenarios lead to a more favorable outcome than what actually happened. Several other scenarios lead to increases in the amount of radioactive iodine released, but still at levels that would not have presented a danger to public health. But we have also explored two or three scenarios whose precise consequences are much more difficult to calculate. They lead to more severe damage to the core, with additional melting of fuel in the hottest regions. These consequences are, surprisingly, independent of the age of the fuel.

Because of the uncertain physical condition of the fuel, cladding, and core, we have explored certain special and severe conditions that would, unequivocally, lead to a fuel-melting accident. In this sequence of events, fuel melts, falls to the bottom of the vessel, melts through the steel reactor vessel, and, finally, some fuel reaches the floor of the containment building below the reactor vessel, where there is enough water to cover the molten fuel and remove some of the decay heat. To contain such an accident, it is necessary to continue removing decay heat for a period of many months.

At this stage we approach the limits of our engineering knowledge of the interactions of molten fuel, concrete, steel, and water, and even the best available calculations have a degree of uncertainty associated with them. Our calculations show that even if a meltdown occurred, there is a high probability that the containment building and the hard rock on which the TMI-2 containment building is built would have been able to prevent the escape of a large amount of radioactivity. These results derive from very careful calculations, which hold only insofar as our as-

sumptions are valid. We cannot be absolutely certain of these results.

Some of the limits of this investigation were:

We have not examined possible consequences of operator error during or after the fuel melting process which might compromise the effectiveness of containment.

We have not examined the vulnerability of the various electrical and plumbing penetrations through the walls or the doorways for people and equipment.

The analysis was specific to the TMI-2 design and location (for example, the bedrock under the plant).

We recognize that we have explored only a limited number of alternatives to the question "What if . . . ?" and others may come up with a plausible scenario whose results would have been even more serious.

We strongly urge that research be carried out promptly to identify and analyze the possible consequences of accidents leading to severe core damage. Such knowledge is essential for coping with results of future accidents. It may also indicate weaknesses in present designs, whose correction would be important for the prevention of serious accidents.

These uncertainties have not prevented us from reaching an overwhelming consensus on corrective measures. Our reasoning is as follows: whether in this particular case we came close to a catastrophic accident or not, this accident was too serious. Accidents as serious as Three Mile Island should not be allowed to occur in the future.

The accident got sufficiently out of hand so that those attempting to control it were operating somewhat in the dark. While today the causes are well understood, six months after the accident it is still difficult to know the precise state of the core and what the conditions are inside the reactor building. Once an accident reaches this stage, one that goes beyond well-understood principles, and puts those controlling the accident into an experimental mode (this happened during the first day), the uncertainty of whether an accident could result in major releases of radioactivity is too high.

While throughout this entire document we emphasize that fundamental changes are necessary to prevent accidents as serious as Three Mile Island, we must not assume that an accident of this or greater seriousness cannot happen again, even if the changes we recommend are made. Therefore, in addition to doing everything to prevent such accidents, we must be fully prepared to minimize the potential impact of such an accident on public health and safety, should one occur in the future.

HANDLING OF THE EMERGENCY

We are disturbed both by the highly uneven quality of emergency plans and by the problems created by multiple jurisdictions in the case of a radiation emergency. Most emergency plans rely on prompt action at the local level to initiate a needed evacuation or to take other protective action. We found an almost total lack of detailed plans in the local communities around Three Mile Island. It is one of the many ironies of this event that the most relevant planning by local authorities took place during the accident. In an accident in which prompt defensive steps are necessary within a matter of hours, insufficient advance planning could prove extremely dangerous.

We favor the centralization of emergency planning and response in a single agency at the federal level with close coordination between it and state and local agencies. Such agencies would need expert input from many other organizations, but there should be a single agency that has the responsibility both for assuring that adequate planning takes place and for taking charge of the response to the emergency. This will require organizational changes, since the agencies now best organized to deal with emergencies tend to have most of their experience with such

events as floods and storms, rather than with radiological events. And, insofar as radiological events require steps that go beyond those in a normal emergency, careful additional planning is needed.

A central concept in the current siting policy of the Nuclear Regulatory Commission is that reactors should be located in a "low-population zone" (LPZ), an area around the plant in which appropriate protective action could be taken for the residents in the event of an accident. However, this concept is implemented in a strange, unnatural, and roundabout manner. To determine the size of the LPZ, the utility calculates the amount of radiation released in a *very serious* hypothetical accident. Using geographical and meteorological data, the utility then calculates that area within which an individual would receive 25,000 millirems or more to the whole body, during the entire course of the accident. This area is the LPZ. The 25,000-millirem standard is an extremely large dose, many times more serious than that received by any individual during the entire TMI accident.

The LPZ approach has serious shortcomings. First, because of the extremely large dose by which its size is determined, the LPZ's for many nuclear power plants are relatively small areas, a 2-mile radius in the case of Three Mile Island. Second, if an accident as serious as the one used to calculate the LPZ were actually to occur, it is evident that many people living outside the LPZ would receive smaller, but still massive doses of radiation. Third, the TMI accident shows that the LPZ has little relevance to the protection of the public—the Commission itself was considering evacuation distances as far as 20 miles, even though the accident was far less serious than those postulated during siting. We have therefore concluded that the entire concept is flawed.

We recommend that the LPZ concept be abandoned in siting and in emergency planning. A variety of possible accidents should be considered during siting, particularly "smaller" accidents, which have a higher probability of occurring. For each such accident, one should calculate probable levels of radiation releases at a variety of distances to decide the kinds of protective action that are necessary and feasible. Such protective actions may range from evacuation of an area near the plant, to the distribution of potassium iodide to protect the thyroid gland from radioactive iodine, to a simple instruction to people several miles from the plant to stay indoors for a specified period of time. Only such an analysis can predict the true consequences of a radiological incident and determine whether a particular site is suitable for a nuclear power plant. Similarly, emergency plans should have built into them a variety of responses to a variety of possible kinds of accidents. State and local agencies must be prepared with the appropriate response once information is available on the nature of an accident and its likely levels of releases.

The response to the emergency was dominated by an atmosphere of almost total confusion. There was lack of communication at all levels. Many key recommendations were made by individuals who were not in possession of accurate information, and those who managed the accident were slow to realize the significance and implications of the events that had taken place. While we have attempted to address these shortcomings in our recommendations, it is important to reiterate the fundamental philosophy we stated above: one must do everything possible to prevent accidents of this seriousness, but at the same time assume that such an accident may occur and be prepared for response to the resulting emergency.

PUBLIC AND WORKER HEALTH AND SAFETY

In setting standards for permissible levels of worker exposure to radioactivity, in plant siting decisions, and in other areas related to health, the Nuclear Regulatory Commission is not required to, and does not regularly seek, advice or review of its health-related guidelines and regulations from other federal agencies with radiation-related responsibilities in the area of health, for example, the Department of Health, Education, and Welfare (HEW) or the Environmental Protection Agency (EPA). There is inadequate knowledge of the effects of low levels of ionizing radiation, of strategies to mitigate the health hazards of exposure to radiation, and of other areas relating to regulation setting to protect worker and public health. In preparation for a possible emergency such as the accident at TMI-2, various federal agencies (NRC, Department of Energy, HEW, and EPA) have assigned responsibilities, but planning prior to the accident was so poor that ad hoc arrangements among these federal agencies had to be made to involve them and coordinate their activities.

The Commonwealth of Pennsylvania, its Bureau of Radiation Protection and Department of Health—agencies with responsibilities for public health—did not have adequate resources for dealing with radiation health programs related to the operation of Three Mile Island. The utility was not required to, and did not, keep a record on workers of the total work-related plus non-work-related (for example, medical or dental) radiation exposure.

We make recommendations with respect to improving the coordination and collaboration among federal and state agencies with radiation-related responsibilities in the health area. We believe more emphasis is required on research on the health effects of radiation to provide a sounder basis for guidelines and regulations related to worker and public health and safety. We believe that both the state and the utility have an opportunity and an obligation to establish more rigorous programs for informing workers and the public on radiation health-related issues and procedures to prevent adverse health effects of radiation.

RIGHT TO INFORMATION

We do not find that there was a systematic attempt at a "cover-up" by the sources of information. Some of the official news sources were themselves confused about the facts, and there were major disagreements among officials. On the first day of the accident, there was an attempt by the utility to minimize its significance, in spite of substantial evidence that it was serious. Later that week, the Nuclear Regulatory Commission was the source of exaggerated stories. Due to misinformation, and in one case (the hydrogen bubble) through the commission of scientific errors, official sources would make statements about radiation already released (or about the imminent likelihood of releases of major amounts of radiation) that were not justified by the facts—at least not if the facts had been correctly understood. And the Commission was slow in confirming good news about the hydrogen bubble. On the other hand, the estimated extent of the damage to the core was not fully revealed to the public.

A second set of problems arose from the manner in which the facts were presented to the press. Some of those who briefed the press lacked the technical expertise to explain the events and seemed to be cut off from those who could have provided this expertise. When those who did have the knowledge spoke, their statements were often couched in "jargon" that was very difficult for the press to understand. The press was further disturbed by the fact that, in order to cut down on the amount of confusion, a number of potential sources of information were instructed not to give out information. While this cut down on the amount of confusion, it flew in the face of the long tradition of the press of checking facts with multiple sources.

We therefore conclude that, while the extent of the coverage was justified, a combination of confusion and weakness in the sources of information and lack of understanding on the part of the media resulted in the public being poorly served.

In considering the handling of information during the nuclear accident, it is vitally important to remember the fear with respect to nuclear energy that exists in many human beings. The first application of nuclear energy was to atomic bombs which destroyed two major Japanese cities. The fear of radiation has been with us ever since and is made worse by the fact that, unlike floods or tornadoes, we can neither hear nor see nor smell radiation. Therefore, utilities engaged in the operation of nuclear power plants, and news media that may cover a possible nuclear accident, must make extraordinary preparation for the accurate and sensitive handling of information.

There is a natural conflict between the public's right to know and the need of disaster managers to concentrate on their vital tasks without distractions. There is no simple resolution for this conflict. But significant advance preparation can alleviate the problem. It is our judgment that in this case, neither the utility nor the Nuclear Regulatory Commission nor the media were sufficiently prepared to serve the public well.

THE NUCLEAR REGULATORY COMMISSION

Two of the most important activities of the Nuclear Regulatory Commission are its licensing function and its inspection and enforcement activities. We found serious inadequacies in both.

In the licensing process, applications are *required* to analyze only "single-failure" accidents. They are not required to analyze what happens when two systems fail independently of each other, such as the event that took place at Three Mile Island. There is a sharp delineation between those components in systems that are "safety-related" and those that are not. Strict reviews and requirements apply to the former; the latter are exempt from most requirements—even though they can have an effect on the safety of the plant. We feel that this sharp either/or definition is inappropriate. Instead, there should be a system of priorities as to how significant various components and systems are for the overall safety of the plant. There seems to be a persistent assumption that plants can be

made sufficiently safe to be "people-proof."
Thus, not enough attention is paid to the
training of operating personnel and operator
procedures in the licensing process. And, finally, plants can receive an operating license
with several safety issues still unresolved.
This places such a plant into a regulatory
"limbo" with jurisdiction divided between
two different offices within the Nuclear Regulatory Commission. TMI-2 was in this status at the time of the accident, 13 months
after it received its operating license.

The Nuclear Regulatory Commission's
primary focus is on licensing, and insufficient attention has been paid to the ongoing
process of assuring nuclear safety. An important example of this is the case of "generic problems," that is, problems that apply
to a number of different nuclear power plants.
Once an issue is labeled "generic," the individual plant being licensed is not responsible for resolving the issue prior to licensing. That, in itself, would be acceptable, if
there were a strict procedure within the
Commission to assure the timely resolution
of generic problems, either by its own research staff or by the utility and its suppliers.
However, the evidence indicates that labeling of a problem as "generic" may provide
a convenient way of postponing decision on
a difficult question.

The old AEC attitude is also evident in reluctance to apply new safety standards to
previously licensed plants. While we would
accept a need for reasonable timetables for
"backfitting," we did not find evidence that
the need for improvement of older plants was
systematically considered prior to Three Mile
Island.

The existence of a vast body of regulations
by the Commission tends to focus industry
attention narrowly on the meeting of regulations rather than on a systematic concern for
safety. Furthermore, the nature of some of
the regulations, in combination with the way
rate bases are established for utilities, may in
some instances have served as a deterrent for
utilities or their suppliers to take the initia-

tive in proposing measures for improved
safety.

Previous studies of inspection and enforcement have criticized this branch severely. Inspectors frequently fail to make independent evaluations or inspections. The
manual according to which inspectors are
supposed to operate is so voluminous that
many inspectors do not understand precisely
what they are supposed to do.

Since in many cases the Commission does
not have the first-hand information necessary
to enforce its regulations, it must rely heavily on the industry's own records for its inspection and enforcement activities. It accumulates vast amounts of information on the
operating experience of plants. However,
prior to the accident there was no *systematic*
method of evaluating these experiences and
no systematic attempt to look for patterns that
could serve as a warning of a basic problem.

The Nuclear Regulatory Commission is
vulnerable to the charge that it is heavily
equipment-oriented, rather than people-ori-
ented. Evidence for this exists in the weak
and understaffed branch of the Commission
that monitors operator training, in the fact
that inspectors who investigate accidents
concentrate on what went wrong with the
equipment and not on what operators may
have done incorrectly, in the lack of attention to the quality of procedures provided for
operators, and in an almost total lack of attention to the interaction between human
beings and machines.

In addition to all the other problems with
the Nuclear Regulatory Commission, we are
extremely critical of the role the organization
played in the response to the accident. There
was a serious lack of communication among
the commissioners, those who were attempting to make the decisions about the accident
in Bethesda, the field offices, and those actually on site. We are also skeptical whether
the collegial mode of the five commissioners
makes them a suitable body for the management of an emergency, and of the agency itself.

We found serious managerial problems within the organization. These problems start at the very top. It is not clear to us what the precise role of the five NRC commissioners is, and we have evidence that they themselves are not clear on what their role should be. The huge bureaucracy under the commissioners is highly compartmentalized with insufficient communication among the major offices. We do not see evidence of effective managerial guidance from the top, and we do see evidence of some of the old AEC promotional philosophy in key officers below the top. The management problems have been made much harder by adoption of strict rules that prohibit the commissioners from talking with some of their key staff on issues involved in the licensing process; we believe that these rules have been applied in an unnecessarily severe form within this particular agency. The geographic spread, which places top management in Washington and most of the staff in Bethesda and Silver Spring, Maryland, (and in other parts of the country), also inhibits the easy exchange of ideas.

We therefore conclude that there is no well-thought-out, integrated system for the assurance of nuclear safety within the current Commission.

We have found evidence of repeated in-depth studies and criticisms both from within the agency and from without, but we found very little evidence that these studies have resulted in significant improvement. This fact gives us particular concern for the future of the present Commission.

For all these reasons we recommend a total restructuring of the Nuclear Regulatory Commission. We recommend that it be an independent agency within the executive branch, headed by a single administrator, who is in every sense chief executive officer, to be chosen from outside the Commission. The new administrator must be provided with the freedom to reorganize and to bring new blood into the restructured NRC's staff.

We have also recommended a number of other organizational and procedural changes designed to make the new agency truly effective in assuring the safety of nuclear power plants. Included in these are an oversight committee to monitor the performance of the restructured Nuclear Regulatory Commission and mandatory review by Health, Education, and Welfare of radiation-related health issues.

THE UTILITY

When the decision was made to make nuclear power available for the commercial generation of energy, it was placed into the hands of the existing electric utilities. Nuclear power requires management qualifications and attitudes of a very special character as well as an extensive support system of scientists and engineers. We feel that insufficient attention was paid to this by the General Public Utilities Corporation.

There is a divided system of decision-making within the General Public Utilities Corporation and its subsidiaries. While the utility has legal responsibility for a wide range of fundamental decisions, from plant design to operator training, some utilities have to rely heavily on the expertise of their suppliers and on the Nuclear Regulatory Commission. Our report contains a number of examples where this divided responsibility, in the case of Three Mile Island, may have led to less than optimal design and operating practices. For example, we have received contradictory testimony on how the criteria under which the containment building isolates were selected. Similarly, the design of the control room seems to have been a compromise among the utility, its parent company, the architect-engineer, and the nuclear steam system supplier (with very little attention from the Nuclear Regulatory Commission). But the clearest example of the shortcomings of divided responsibility is the area of operator training.

The legal responsibility for training operators and supervisors for safe operation of nuclear power plants rests with the utility.

However, Met Ed, the General Public Utilities Corporation subsidiary which operates Three Mile Island, did not have sufficient expertise to carry out this training program without outside help. They, therefore, contracted with Babcock & Wilcox, supplier of the nuclear steam system, for various portions of this training program. While Babcock & Wilcox has substantial expertise, they had no responsibility for the quality of the *total* training program, only for carrying out the contracted portion. And coordination between the training programs of the two companies was extremely loose. For example, the Babcock & Wilcox instructors were not aware of the precise operating procedures in effect at the plant.

A key tool in the Babcock & Wilcox training is a "simulator," which is a mock control console that can reproduce realistically events that happen within a power plant. The simulator differs in certain significant ways from the actual control console. Also, the simulator was not programmed, prior to March 28, to reproduce the conditions that confronted the operators during the accident.

We found that at both companies, those most knowledgeable about the workings of the nuclear power plant have little communication with those responsible for operator training, and therefore the content of the instructional program does not lead to sufficient understanding of reactor systems.

It is our conclusion that the role that the Nuclear Regulatory Commission plays in monitoring operator training contributes little and may actually aggravate the problem.

The way that the Nuclear Regulatory Commission evaluates the safety of proposed plants during the licensing process has a most unfortunate impact on the way operators are trained. Since during the licensing process applicants for licenses concentrate on the consequences of single failures, there is no attempt in the training program to prepare operators for accidents in which two systems fail independently of each other.

We agree that the utility that operates a nuclear power plant must be held legally responsible for the fundamental design and procedures that assure nuclear safety. However, the analysis of this particular accident raises the serious question of whether all electric utilities automatically have the necessary technical expertise and managerial capabilities for administering such a dangerous high-technology plant. We therefore recommend the development of higher standards of organization and management that a company must meet before it is granted a license to operate a nuclear power plant.

THE TRANSITION

We recognize that even with the most expeditious process for implementation, recommendations as sweeping as ours will take a significant amount of time to implement. Therefore, the Commission had to face the issue of what should be done in the interim with plants that are currently operating and those that are going through the licensing process.

The Commission unanimously voted:

Because safety measures to afford better protection for the affected population can be drawn from the high standards for plant safety recommended in this report, the NRC or its successor should, on a case-by-case basis, before issuing a new construction permit or operating license: (a) assess the need to introduce new safety improvements recommended in this report, and in NRC and industry studies; (b) review, considering the recommendations set forth in this report, the competency of the prospective operating licensee to manage the plant and the adequacy of its training program for operating personnel; and (c) condition licensing upon review and approval of the state and local emergency plans.

A WARNING

During the time that our Commission conducted its investigation, a number of other reports appeared with recommendations for improved safety in nuclear power plants. While we are generally aware of the nature of these recommendations, we have not attempted a systematic analysis of them. Insofar as other agencies may have reached similar conclusions and proposed similar remedies, several groups arriving at the same conclusion should reinforce the weight of these conclusions.

But we have an overwhelming concern about some of the reports we have seen so far. While many of the proposed ''fixes'' seem totally appropriate, they do not come to grips with what we consider to be the basic problem. We have stated that fundamental changes must occur in organizations, procedures, and, above all, in the attitudes of people. No amount of technical ''fixes'' will cure this underlying problem. There have been many previous recommendations for greater safety for nuclear power plants, which have had limited impact. What we consider crucial is whether the proposed improvements are carried out by the same organizations (unchanged), with the same kinds of practices and the same attitudes that were prevalent prior to the accident. As long as proposed improvements are carried out in a ''business as usual'' atmosphere, the fundamental changes necessitated by the accident at Three Mile Island cannot be realized.

We believe that we have conscientiously carried out the mandate of the President of the United States, within our limits as human beings and within the limitations of the time allowed us. We have not found a magic formula that would guarantee that there will be no serious future nuclear accidents. Nor have we come up with a detailed blueprint for nuclear safety. And our recommendations will require great efforts by others to translate them into effective plans.

Nevertheless, we feel that out findings and recommendations are of vital importance for the future of nuclear power. We are convinced that, unless portions of the industry and its regulatory agency undergo fundamental changes, they will over time totally destroy public confidence and, hence, *they* will be responsible for the elimination of nuclear power as a viable source of energy.

REFERENCE

John G. Kemeny, Chairman, *Report of the President's Commission on the Accident at Three Mile Island.* New York: Pergamon Press, 1980.

SUGGESTED READINGS

''Analysis of Three Mile Island Unit 2 Accident,'' and ''Supplement to Analysis of Three Mile Island Unit 2 Accident,'' Nuclear Safety Analysis Center, EPRI, NSAC-1 and NSAC-1 Supplement, 1979.

''Investigation into the March 28, 1979, Three Mile Island Accident,'' Office of Inspection and Enforcement, U.S. Nuclear Regulatory Commission, NUREG-0600. Washington, D.C., August 1979.

Morris W. Firebaugh, ''Public Attitudes and Information on the Nuclear Option,'' *Nuclear Safety* 22 (2), 147 (1981).

''TMI Action Plan,'' NUREG-0660 and NUREG-0737; and ''Criteria for Emergency Response Facilities,'' NUREG-0696, U.S. Nuclear Regulatory Commission, Washington, D.C., 1981.

Environmental Liabilities of Nuclear Power 20

JOHN P. HOLDREN

There are a variety of questions surrounding nuclear fission that are contested in various ways and by various groups. These include the economics of nuclear power, the adequacy of uranium resources, and the role and timing of the breeder reactor. I confine myself here, however, to the environmental dimension of the nuclear controversy. The essence of this dimension is captured, I believe, in the following hierarchy of questions: How safe is nuclear power? How safe is safe enough? Who should decide?

The first question is to a considerable degree a technical question, but not entirely. The second question is an economic and sociocultural one as well as a technological one, and the third one is primarily a political question. Let me elaborate.

With respect to the first question—How safe is nuclear power?—many specific subquestions can be illuminated by technical insights. For example, How toxic is plutonium? What are the potential consequences of reactor accidents? What is the longevity of radioactive wastes? These are all substantially technical questions. But, in other aspects of the question "How safe is nuclear power?" the human factor looms large. The issues of proliferation, theft of nuclear materials for terrorism or blackmail, and sabotage all have very substantial human components, to which I shall return.

The second question—How safe is safe enough?—is answerable only in the context of the alternatives. That is, one must ask the question "Compared to what?" One must look not only at the risks and benefits of nuclear power compared to each other but also

Dr. John P. Holdren is Professor of Energy and Resources at the University of California, Berkeley, California 94720; Participating Guest in the Energy and Environment Division of the Lawrence Berkeley Laboratory; and Faculty Consultant in the Magnetic Fusion Energy Division of the Lawrence Livermore National Laboratory. Prior to joining the faculty at Berkeley, he was a physicist in the Theory Group of the Magnetic Fusion Energy Division at Livermore and Senior Research Fellow in the Environmental Quality Laboratory and the Division of Humanities and Social Sciences at the California Institute of Technology. Dr. Holdren has published widely in plasma theory, energy technology and policy, regional and global environmental problems, population policy, and technology and policy for development.

From an article in *Social and Ethical Implications of Recent Developments in Science and Technology*, Charles P. Wolff, ed. New York: Plenum, 1981. Reprinted by permission of the author and Plenum Publishing Company. Copyright © 1981 by Plenum Publishing Company.

at alternative courses of action, including other ways to get electricity and including doing without the electricity that otherwise might have been generated by nuclear power. This formulation immediately raises the question of values. That is, as soon as one opens up the issue of alternative kinds of technology versus nuclear power, or of electricity versus not having it, there arises the question of the analyst's relative preference functions for different kinds of costs, different kinds of risks, different kinds of benefits. Herein lies much of the difficulty of environmental assessment for policy formation.

Let me suggest, in fact, that what is needed to answer the question "How safe is safe enough?" is an examination of another, broader, nested set of questions which place the nuclear controversy in context, namely, How much energy do we need? Of the amount that we need, how much should be electricity? Of the amount that we conclude should be electircity, how much should be nuclear? For the amount that is to be nuclear, with what technologies and under what regulations and safeguards should it be provided? And, again, who should decide?

Now, confusion has plagued the nuclear debate in part because people have not been very specific concerning which of these questions they are discussing at any given time. Certainly it is not surprising that an analyst's preferences with respect to the first two of these questions influences his or her conclusions with respect to the others. I will try in this presentation to clarify where my own biases lie in these respects at the outset.

With respect to the first question—How much energy do we need?—my answer is "Less than is generally supposed." What one is really interested in is not energy for its own sake but energy for its contribution to human well-being. The links between energy and human well-being are both two-sided and flexible. That is, getting and using energy can detract from human well-being through environmental and social costs, as well as contribute to well-being through economic benefits. It is certainly possible to suffer from too much energy too soon, as well as from too little too late.[1] The widely presumed link between energy and GNP, moreover, has been shown in much recent work in the past several years to be very flexible indeed. That is, it is possible to reduce considerably the amount of energy needed to create a dollar of GNP and thereby to permit continued increase of well-being (insofar as GNP reflects it) without a parallel growth in energy itself.[2]

With respect to the second question—How much of the energy we need should be electricity?—again my answer is "Less than is commonly supposed." Electricity is thermodynamically the highest-quality form of energy we use. In many of the tasks for which we use it, it is a serious case of overkill, that is, we are using a dearly won and high-quality form of energy for tasks that could be performed with energy much more easily won, of a much lower thermodynamic quality. If one does a better job of matching the end uses of energy and the thermodynamic quality of the energy supplied to carry out those end uses, the fraction of electricity in our economy need not grow as rapidly as many analysts of a few years ago supposed.[3]

The third question—How much should be nuclear?—reveals that we must face up to the fact that one cannot conserve everything. That is, as good as energy conservation is as a safe, practical, and effective energy "source," there will remain a residual demand for energy that must be met, as well as a residual demand for electricity that has to be in that form. In that context, nuclear is one of the alternatives that has to be squarely confronted as to whether it is better or worse than other means available to secure the same benefit. The proper comparison at this time, I believe, is between cleaned-up coal and nuclear. These are the principal electricity generating options at present. However, there has been a tendency to compare dirty-coal systems, as characterized by past performance, with nuclear systems rather highly optimized in terms of their emissions. I think the fair comparison at this point would be to consider what can rather readily be done to clean up coal and then to compare the cleaned-up coal option with the nuclear option from the point of view of environmental effects.

Of course, the strategy of slowing growth, which emerged in my answers to the first two questions, has the benefit of buying some time, reducing the amount of *either* coal *or* nuclear that must be deployed in the short term before the menu of options becomes somewhat broader. That is, there is reason to suppose that there are other ways to generate electricity that will become available over the next 20 to 40 years, and if one is unhappy about the environmental and social costs of both coal and nuclear, which I am, then the slow-growth strategy makes a great deal of sense in terms of minimizing the damage until we have a better set of choices.

The next question—With what technologies and under what regulations and safeguards should nuclear energy be provided?—suggests that it is possible to be a nuclear critic without advocating that all nuclear power be shut down or, in other words, without advocating that we can be absolutely sure at this time that the world can do without this option. We already have some 70 nuclear power plants in this country and a considerable number of others under construction, and I believe it is unrealistic to suppose that the United States will back away overnight from this existing commitment to nuclear power. In this context, the question then arises of how one minimizes the risks with which one ought to be concerned. There are a variety of possibilities. Former President Carter's policy of no reprocessing and no exports of the most sensitive nuclear technologies was one example of a strategy designed to minimize what he perceived to be the adverse impacts most in need of control. A more conservative siting policy with respect to building nuclear power plants would be another such ingredient, and trying to internationalize fuel cycle facilities (that is, if other countries are going to insist on reprocessing, perhaps we can at least persuade them to put those reprocessing facilities under international control) would be yet another.

We come, then, to the last question—Who should decide? The possibilities include technical experts, corporate executives, appointed officials, courts of law, elected representatives, and members of the public. The issues that enter include information (who has it and how good is it?), representation (do those affected have a voice?), accountability (are the decision-makers answerable to those affected?), and efficiencies (is the machinery of the decision-making process capable of producing a result in a reasonable amount of time?). How best to respond to these issues in the case of energy questions—nuclear and otherwise—is obviously a political matter, not a technical one.

Now, from this framework for viewing the controversy, I think there emerge two characteristics that explain the extraordinary prominence of nuclear power as an issue in science and public policy. The first one is the extraordinary diversity of issues that are relevant. The second is the extraordinary number of these in which nontechnical considerations predominate. This notion of the importance of nontechnical considerations is one of the central threads of my argument here.

To illuminate this point further, let me refer to the list of nuclear risks in Table 20-1,

Table 20–1.   A Ranking of Nuclear Risks.

1. Proliferation of nuclear weapons among nations
2. Theft of weapons-usable nuclear material by subnational groups
3. Sabotage and accidents in
   a. Reactors
   b. Fuel processing plants
   c. Spent fuel transportation
   d. Waste repositories
4. Routine emissions and exposures in
   a. Reprocessing plants
   b. Mining and milling operations
   c. Fuel fabrication plants (assuming Pu is recycled)
   d. Reactors

ranked there in my personal order of importance. In what follows, I will survey the main technical and nontechnical issues associated with each of these risks. In so doing I will try to clear up some misconceptions that have been propagated on both sides of the nuclear controversy and to justify my ranking of these risks as a ranking in order of importance.

PROLIFERATION

Until now, the main technical obstacle against the spread of nuclear weapons among nations has been lack of availability of suitable fissile materials, which only the largest industrial nations have had the technical resources to obtain. There have been political and economic obstacles to the spread of weapons, too, of course, but lack of access to fissile materials—which access is spread by the spread of commercial nuclear power—has been the main technical one.

Now, the technical issues relating to the link between nuclear power and weapons proliferation are rather straightforward and not in dispute among well-informed people. These issues include the quantity and form of radioactive materials—bomb-usable materials in particular—throughout the nuclear fuel cycle, the quantity of these materials that would be needed to manufacture bombs, and the costs and technical characteristics of alternative ways not related to nuclear power by which bombs could be obtained.

Now, again, although it is true that the technical facts on these matters are not in dispute among knowledgeable people, some misconceptions have been propagated in some cases by people who ought to know better. One such misconception is the notion that reactor-quality plutonium—the kind of plutonium that comes from commercial nuclear reactors—is not suitable for the production of military-quality nuclear weapons. This is false, and it has been known to be false in the weapons community for many years. People with intimate, official knowledge of the nuclear weapons business (most

recently, Commissioner Vic Gilinski of the Nuclear Regulatory Commission[4]) have stated publicly and unambiguously that, with suitable design, reactor-grade plutonium can be used to produce nuclear bombs of both high and reliable yield. From the point of view of proliferation, this is an extraordinarily important point that, unfortunately, has been widely misunderstood.

There are then a variety of nontechnical or only partly technical issues associated with proliferation that have to be addressed. One is the relation between the number of countries that have nuclear bombs and the probability of war, that is, how much does the probability of nuclear war increase if many countries have bombs as opposed to a few countries having bombs? What are the motivations and incentives that determine whether a given nation acquires a weapons capability or not? What is the deterrence value of the possibility of detection if a country attempts to acquire nuclear weapons? What is the actual as opposed to the theoretical capability to detect violations of safeguards under the Non-Proliferation Treaty, given an inadequate inspectorate in the International Atomic Energy Agency (IAEA) and technically inadequate procedures for determining whether or not a country has diverted materials to make bombs? How does the possibility of coercion and bribery affect the reliability of the safeguard systems that exist to try to detect the diversion of nuclear material?

Some of the arguments that have been put forward with respect to these more qualitative issues are either transparently illogical or wrong. One such argument is that we should not worry about proliferation because the Non-Proliferation Treaty (NPT), and the International Atomic Energy Agency empowered to enforce its safeguards, have taken care of this problem on the international level as best it can be taken care of. I think anyone who has looked carefully at what the NPT says, at what the IAEA's powers are, and at how many people they have to enforce these powers knows this is a joke. In fact, it has been admitted to be such privately by the highest ranking officials in the IAEA with line authority for carrying out that agency's safeguards tasks.[5]

A second assertion often made about proliferation by proponents of nuclear power is that, because there are cheaper and easier ways to get nuclear bombs than through nuclear power, the power-proliferation link can be ignored. That there are easier and cheaper ways than nuclear power to get bombs is true, but largely irrelevant. It is irrelevant because the technically easier ways require a political commitment and a decision to be made by a country long before the bombs actually become available. That is, to take one of these routes, a country must stick its neck out, demonstrating its intentions, in effect, by constructing easily detected facilities explicitly for the purpose of making bombs and thereby subjecting itself to international censure and possibly provoking its potential adversaries into taking similar actions. To be contrasted with this awkwardness, on the other hand, is the quick and relatively easy pathway to bombs that is available once a complete nuclear fuel cycle is in place and operation. In fact, there is a sort of "attractive nuisance" character about commercial nuclear power in this sense. A country may have the best of intentions at the time it deploys nuclear power—that is, it may have no intention at all of developing nuclear weapons from that capability. But when a country has a stockpile of plutonium separated in reprocessing just sitting there, and external circumstances change—e.g., threats materialize which were not there before—a national leader may find the attraction too much to resist and take the steps which can rather quickly convert that pile of plutonium into a pile of bombs.

Finally, there is the argument offered by proponents of nuclear power that every country will *eventually* have nuclear bombs, irrespective of the spread of nuclear power. This argument, too, is true but largely irrelevant. Any country that wants bombs eventually will get them by other means. Fifty years from now, in fact, if I am still alive, I shall doubtless change my ranking in Table 20-1 to put nuclear theft by subnational groups above proliferation, because I believe that 50 years from now either we shall have solved the problem of the spread of nuclear weapons in some fundamental way, or everyone will have them, or we shall all be dead because they have been used. The question in the meantime, however, is the *rate* at which nuclear weapons spread, that is, how many countries get how many weapons how fast. In that respect, it is important to do all that we can, including using the lever of hindering the spread of nuclear power, to buy more time to find ways to make the use of bombs less likely.

Now, why do I rank proliferation above other nuclear risks? I rank it first because risk in the formal sense is the product of the probability of a harmful event times the conseqences of the event if it takes place. Now, we do not know exactly how much the spread of nuclear power contributes to the probability of nuclear war, but it is certainly not inconceivable that the probability of nuclear war is already 1 per cent per year and that the spread of nuclear power could double that figure in the short term. (We do not know this is true—there is no way we can quantitatively defend it—but there also is no way to show that it is grossly wrong.) If nuclear power adds a probability of 1 per cent per year to the chance of nuclear war, and recognizing that the consequences of nuclear war could include the deaths of hundreds of millions to billions of people, then the "expected value" associated with this risk—that is, the probability times the consequences—is very large indeed. Even if the contribution of proliferation to the change of large-scale nuclear war is much smaller than 1 per cent per year, the associated risk is by far the largest of those in Table 20-1.

NUCLEAR THEFT

Consider next the theft of nuclear materials by subnational groups for use in radiological weapons or explosive weapons or simply to make the threat of such use a credible thing for the purposes of extortion, political blackmail, and so on.

The technical issues are, again, relatively straightforward. One wants to know how much material it takes to make a nuclear explosive; one wants to know how toxic plutonium is; one wants to know how good the instruments are that can detect plutonium if someone is trying to sneak it out of your plant.

But again the nontechnical issues dominate. More important than the sensitivity of the instrument watching for plutonium going out the door, for example, is the sensitivity to bribery or coercion of the people who watch the instrument, or the sensitivity of the people who program the computers to get the instrument to sound an alarm automatically. There is simply no way to separate from this problem the human factor—the human frailities—that can intervene in the best laid plans of technologists.

Now, with respect to theft, there have been misconceptions propagated by both sides in the nuclear debate. Some opponents of nuclear power have said that it is easy to make a nuclear bomb from stolen nuclear material. This is not true—it is not easy, it is hard. There are many more incorrect ways to make a nuclear bomb than there are correct ways, and the probability is that people trying to do this would kill themselves before they managed to produce a nuclear explosion.

Nevertheless, it is also not true, as has sometimes been asserted on the other side, that it would necessarily take dozens of people and millions of dollars to manufacture a nuclear bomb. A handful of people, if clever enough and if determined enough, with some small but finite probability could succeed. My view is that it does not have to be easy or safe to be worrisome. One has to ask what probability we are prepared to accept of a nuclear explosion in a major city, where the consequences could be on the order of 50,000 to 100,000 immediate deaths if the explosive were carefully placed (under a large building

or in a football stadium, for example).[6] Obviously, any old place one blows up such a backyard bomb would not create this number of casualties, but anyone clever enough to make the bomb in the first place would be clever enough to use it in a way that could kill on the order of 100,000 people.

Now, the issue of probability times consequences arises again. We do not know what the probability is, but the consequence is very high, not only in terms of the immediate consequences of the explosion but also in terms of the changes in some of the ways our society works that likely would occur if these kinds of threats became credible. Those people who have become concerned with the search-and-seizure procedures that have prevailed in the illicit drug business, for example, will probably be shocked to discover what would be likely to happen if, instead of searching for drugs to seize, the authorities were searching for stolen plutonium after one such device had actually been exploded.

The proponents of nuclear power often say there has been no evidence of nuclear theft, nor has any subnational group exploded a nuclear bomb so far. The same sort of argument might have been made about hijacking airliners say, 15 years ago, before it became a fad. There is also no evidence at this time that special nuclear material has *not* been stolen. Certainly the substantial uncertainty in the inventory of such material—the material unaccounted for (MUF) amounts to 1 to 2 per cent of the total amount of plutonium flowing through the system—permits the possibility that plutonium has already been diverted in quantities sufficient to make a bomb. We do not know that it has, but we also do not know that it has not.

ACCIDENTS AND SABOTAGE

Here, the technical issues are more intricate and more important in terms of the fraction of the problem they represent, but they are still far from the whole story.

Again the main question is one of probability times consequences. The Nuclear Regulatory Commission's Reactor Safety Study (popularly known as the Rasmussen Report)[7] studied both the probability and the consequences and concluded that the worst accident they considered could kill as many as 3000 people outright, produce 50,000 delayed cancer deaths, and produce property damage in the amount $14 billion. The report estimated the uncertainties to amount to multiplicative factors of two to five bigger or smaller on each of these numbers. It calculated the probability of this "worst case" accident to be essentially one in a billion per reactor per year, with an uncertainty range of a multiplicative factor of five bigger or smaller.

Virtually all of the technical reviews of the Rasmussen Report—by the Environmental Protection Agency, by the Nuclear Regulatory Commission itself, by the Union of Concerned Scientists—have concluded that the uncertainties as stated in the Rasmussen Report are understatements (that is, whether or not the "best estimates" stated there are right, the uncertainties are, in fact, bigger than stated there).[8] This is a point to which I shall return. Some of these studies have suggested, moreover, that the "best estimates" in the Rasmussen Report are biased downward by substantial factors, that is, that "best estimates" of the probabilities and consequences of reactor accidents would have been larger if various flaws in the analysis had been corrected. Other reviewers have suggested the opposite. As one might expect, the industry has argued that the Rasmussen Report was too conservative, that things are not as bad as that, whereas environmentalist organizations and environmental bureaucracies like the EPA have argued that the Rasmussen Report overestimated the safety of nuclear power. Limitations of space prevent me from trying to dissect this complex issue in detail here, but I shall make some brief observations on both probability and consequences.

With respect to probability, there are two major uncertainties. One is the probability that the analysts made a mistake. When one confronts a number like one in a billion per reactor per year, in order to arrive at a meaningful personal estimate of what the significance of that number is one has to factor in one's own estimate of the reliability of the analyst, given the complexity of the system being analyzed. This perspective again brings the human factor into what otherwise might have been considered a technical problem. Given the likelihood of oversights and other kinds of errors, it is simply not reasonable to suppose that a system as complicated as a nuclear reactor can be analyzed persuasively at probabilities of this magnitude.

A second issue in probability, which in my view is even more important, is sabotage. If the reactor is entered by a knowledgeable and determined group of saboteurs, all of the calculated failure probabilities become essentially meaningless. The saboteurs can simultaneously disable the various redundant emergency cooling systems, for example, while producing the loss of coolant that would require these systems to function. The Rasmussen Report referred in passing to sabotage by arguing that it could not create an accident bigger than the biggest one considered in the report. It did not point out that sabotage's most important contribution to the issue is on the probability side. In other words, one needs to know how much the "worst case" probability of 10^{-9} per reactor per year would change if we somehow knew how to take account quantitatively of the possibility of sabotage. In this matter, the human factor again looms large and the central questions have no technological answers. How many people will be disgruntled enough to try it? How many of them will be clever enough to succeed? What is an *acceptable* risk of sabotage of a nuclear reactor near a population center?

Now, let me offer some observations about the consequences of accidents or sabotage. Many proponents of nuclear power say that it is meaningful to consider consequences only in the context of the overall risk, that is, whenever we have a high consequence number we must multiply it immediately by its low probability, invariably yielding a low number of expected deaths per year. It is simply not reasonable, in this view, to look at these big potential accidents in isolation from their probability. My view, in contrast, is that we have substantial reason to believe that the probability numbers are garbage, owing to the possibilities of analyst error or deliberate mischief as discussed above. Why, then, should we muddy the waters by taking a number in which we can have some confidence (the estimated consequences if a very large reactor accident takes place) and multiplying it by a probability number in which we can have no confidence at all? The fact is that the maximum potential consequences must themselves be an input to the public-policy decision about nuclear power.

I shall not dwell here on reprocessing plants, spent fuel transport, and radioactive waste repositories. I think they are both less important that reactors in terms of the likelihood that sabotage or accident can cause a really large disaster. In the ranking of risks as a whole, I put *all* of the accident/sabotage hazards behind both nuclear theft and proliferation because I think that in the sense of probability times consequences they are not likely to do as much damage as either of the first two.

ROUTINE EMISSIONS

The most important sources of routine emissions are reprocessing plants and mining and milling operations, the exposures in both cases being both to the public and to workers in these facilities. Next in order are fuel fabrication plants (if plutonium recycle is undertaken). With respect to reactors, I agree that emissions from reactors properly operating are so small that there would be no reason to oppose or even seriously criticize nuclear power on those grounds. That is, if routine reactor emissions were all there were to the environmental and social costs of nuclear power, I would be first in line to be advocating it over most alternatives, and certainly over coal. I do want to add a few details about the radioactive waste problem and

about routine emissions, not because they are important compared with other nuclear risks discussed here—I think they are not—but because they get so much discussion and because a good deal of the discussion they get is wrong.

With respect to radioactive waste, let me state my position at the outset. It is that any of several schemes that have been proposed to deal with radioactive waste would probably be adequate, in the sense that placement of the wastes in any of several kinds of geologic formation would provide secure enough storage that the product of probability of release times the consequences would be considerably smaller than the risks listed higher in Table 20-1. The principal risks of managing nuclear wastes, in my view, occur before the waste has been placed in the ground. This, in fact, is one of the reasons that former President Carter's policy was reasonable in calling for at least a temporary ban on reprocessing. For, while reprocessing makes the waste easier to manage in the long term in certain respects, it increases the amount of above-ground handling and, therefore, it increases the vulnerability precisely where the risk is already greatest.

Of course, the biggest problem with nuclear waste is that we have not decided what to do. The government seems incapable of choosing one thing or another, and it ought to have done so long ago if it expects nuclear power to be an expanding and viable energy source.

At the same time, I cannot resist pointing out that some of the things that are said about nuclear wastes in the public debate make it improperly seem to be an even smaller problem than it is. It has been suggested by nuclear proponents, for example, that the radioactive wastes from nuclear power operations amount to an aspirin tablet's worth of volume per year per person supplied with electricity by nuclear power.[9] In a similar vein, it has been suggested that all of the high-level wastes from running a nuclear reactor for a year would fit under a dining room table. Well, first of all, the principal issue with radioactive waste is not volume but toxicity. Volume gives some measure of how difficult the material is to handle technologically but is not the central issue. And, even confining attention to volume, the aspirin tablet and dining room table examples cited do serious violence to reality.

The degree of this distortion can be understood by reference to Table 20-2, which distinguishes among the various forms in which radioactive wastes find themselves and among the various levels of radioactivity that have to be dealt with. The amount that fits under the dining room table (if you happen to have a table that seats 20, that is) refers only to the category of reprocessed, solid, high-level wastes, that is, putting the fission-product part of the radioactive waste in its most concentrated form in a glass matrix gives you about three cubic meters per large reactor per year. (Such a reactor would meet the electricity needs—residential, commercial, industrial —of about 750,000 people.) At the present time, however, we are not reprocessing. What we have, instead of reprocessed high-

level solids, is spent fuel. That comes out to about 20 cubic meters per reactor per year, and now you need an even bigger dining room table. If one processes the spent fuel into high-level liquid waste, which was done for a small quantity of fuel at the West Valley, New York, plant, one has a volume (depending on the concentration of the liquid) from 25 to 200 cubic meters per year.

In addition, however, there is the category of low- and intermediate-level wastes. Some of these, the so-called transuranic (TRU) wastes, are as toxic or more toxic after several hundred years than the high-level fission products. Clearly, they must also be managed in the very long term, but they are much bigger in volume—on the order of 100 cubic meters per reactor per year. (There are various federal reports on this subject with varying factors of 2 to 3 either way on many of these numbers.) If one recycles plutonium, additional TRU solid wastes from the reprocessing plant could amount to 300 cubic meters per year. Intermediate solids from decommissioning, if one considers those part

Table 20–2. Principal Radioactive Wastes from Generation of One Gigawatt-Year (Electric) in Light Water Reactors.
(1250-MW$_e$ reactor at 80% capacity factor)[10]

	Volume (m^3)	Radioactivity (curies/m^3)[a]
High-level wastes		$4 \times 10^5 - 3 \times 10^7$
If calcined and vitrified, less cannisters	3–4	
If calcined and vitrified, with cannisters	6–10	
As high-level liquids, initial production	230	
As high-level liquids, interim storage	50	
As high-level liquids, most concentrated form	17	
As spent fuel	20	
Cladding hulls and fuel hardware	20	10^4
Reprocessing plant transuranium (TRU) solids	75–110	$0.1 - 10^3$
Low- to medium-level solids at reactor		
Routine operation	300–700	0.1–10
Decommissioning (prorated)	200–600	Not available
Reprocessing plant non-TRU solids	55	1–100
Mixed-oxide fuel fabrication (if Pu recycle) TRU	40–70	500–1000
Plutonium storage (if no Pu recycle)	0.4–7	$5 \times 10^5 - 8 \times 10^6$
Depleted UF$_6$	50	1
Tailings from milling	70,000–180,000	0.005–0.01

[a] 1 curie = 3.7×10^{10} disintegrations per second.

of the radioactive wastes and prorates them over the life of the plant, amount to 100 cubic meters per year. And the big one, mill tailings, contributes 100,000 cubic meters per large reactor per year.

Of course, we have an apples-and-oranges problem here. These are wastes of different characters; they have to be managed in different ways. But, certainly, if one is going to do a comparison with coal on the basis of volume of wastes and count the ash and the carbon dioxide from coal, then it is dishonest not to include these lower-level wastes from radioactive mill tailings and so on. Including them brings wastes from nuclear power roughly to within a factor of 10 in terms of volume to the solid wastes produced by the use of coal.

Let me add a few words about routine emissions. It is often asserted that we know more about the routine effects of radioactivity than about any other pollutant, and this is probably true. That is not the same thing, however, as knowing everything; unpleasant surprises are still possible. For example, although it has long been assumed by most analysts that the "linear hypothesis" about effects of low-dose radiation is conservative (that is, tends to overestimate rather than underestimate the effects), there are some theoretical arguments and the beginnings of actual evidence to suggest that, in some circumstances, low-dose radiation may produce more adverse effects per unit of energy deposited in tissue than does high-dose radiation.[11] It is also far from clear, since the full nuclear fuel cycle is not yet in operation, what the doses to workers in a mature nuclear industry will be. Reprocessing, fuel fabrication if plutonium or uranium-233 is recycled, and decommissioning of worn-out reactors may turn out to be particularly important sources of occupational exposures. Finally, assuming nothing worse than the linear hypothesis, it can be shown that the very long-lived isotopes in uranium mill tailings would be expected to exact a toll of cancers amounting to 90 deaths per reactor-year's worth of mining, these deaths being spread, however, over the next few hundred thousand years.[12] This is a large total number of deaths, but a tiny perturbation in the natural incidence of cancer over so long a period. It is difficult to know how to evaluate the importance of this burden without knowing the effects of alternative energy technologies over comparably long periods. Such information is missing.

Let me now take up some policy issues that emerge from the kinds of considerations I have outlined so far.

COMPARISONS OF RISKS

The most meaningful comparisons are between alternative ways of getting the same benefit, that is, comparing coal versus nuclear power or versus more efficient use of energy to reduce one's electricity needs. Comparisons of the risks of energy with entirely unrelated risks—smoking, automobiles, earthquakes if you live in San Francisco, tornadoes if you live in Minnesota—really tell us very little because the benefits that people get in exchange for accepting those risks are entirely different. For example, those of us who live in the San Francisco Bay area and subject ourselves to a substantial earthquake risk do so in many cases because we perceive ourselves to be getting a set of benefits that can be derived in no other way. That is, if I like the array of environmental characteristics, cultural opportunities, employment opportunities, and so on that exists in the San Francisco Bay area, that is a set of benefits which I cannot derive elsewhere and in exchange for which I am prepared to expose myself to a rather high risk. Similarly, the risks that people expose themselves to every time they get into their cars are accepted in exchange for an enormous degree of personal mobility—one which so far society has not been able to provide in any other way. Nuclear power risks, on the other hand, are accepted in exchange for electricity, which can be provided in other ways. It is with those other ways that the comparisons should be made.

In the same connection, I think one should be very careful in the case of a word like "acceptable." People talk rather casually all the time about what sort of risks are acceptable. We say, for example, that the risks of automobiles are acceptable. Well, they are accepted in exchange for a rather unique set of benefits. We say the risks of earthquakes are acceptable. The risks of earthquakes and tornadoes in general should not be in general called acceptable, they should be called tolerable or endurable[13]; people tolerate them, but because there is almost no place in the country where one is not subjected to some major natural hazard—whether flood, tornado, tsunami, hurricane, or earthquake—these are risks that essentially cannot be escaped. They are tolerated, they are endured, but the active act of acceptance really does not take place.

The second policy-relevant conclusion is the nonuniqueness of nuclear risks. Here, I think, opponents of nuclear power have been guilty of hyperbole and some real misstatements. It is sometimes argued, for example, that nuclear power is the only energy source that imposes burdens on future generations, or that it is the only energy source that creates long-term contaminants, or that it is the only energy source with a substantial link to war. None of these things is true. Burning fossil fuel releases heavy metals whose half-life is forever. This is an important problem and one I am worried about. Importing oil increases the stress on a rather scarce resource in great demand in other parts of the world, and it surely has some connection to the possibility of political conflict and, thereby, the possibility of war. Combustion products of burning fossil fuels are indicted today as likely mutagens; almost certainly we are creating genetic mutations by the combustion of fossil fuels, just as we do by the release of radioactivity, and in so doing we impose burdens on future generations. The question, really, is not whether the risks of nuclear power are unique but whether they are a good trade for the benefits of this particular technology.

Unfortunately, there is no energy source that is free of significant environmental liabilities. Some are assuredly better than others, but the "best" choice can be expected to vary with geography, with cultural and economic setting, and, perhaps, with time. With respect to environmental and social characteristics, then, we should be looking not for the perfect energy source, because there is no such thing, but for the least undesirable *mixture* of energy sources—that is, the mixture that minimizes the environmental and social impacts necessarily associated with getting the benefits of energy.

SOME CONCLUDING OBSERVATIONS

To the extent that the main immediately available energy sources have particularly distressing environmental consequences—and I am as distressed by the consequences of coal as I am by the consequences of nuclear—we should be motivated to reexamine the question of demand as opposed to supply: How much can the growth of demand be slowed down without itself creating serious environmental and social consequences?

In my view, for the short term and the long term, energy conservation (by which I mean not belt-tightening but more efficient use of energy to produce the goods and services we need) is the cheapest, safest, cleanest, fastest, and almost certainly the most job-enhancing energy alternative available to us. It is the one to which we should assign highest priority and with which we can proceed with all deliberate speed.

Among the options beyond conservation (recognizing that we cannot conserve everything), it is possible to draw some distinctions on environmental and social grounds, notwithstanding the fact that nuclear environmental liabilities are not unique in the qualitative sense. Specifically, I regard nuclear power as more troublesome environmentally and socially than its main competitors, for the following reasons.

First, the connection with nuclear war is

more direct in the case of nuclear power than it is with its competitors. Second, the sensitivity of nuclear power to human error is more acute and its vulnerability to intentional disruption is more pronounced than in the case of other energy sources. Finally, and perhaps most important, these problems are less amenable in nuclear power to amelioration by means of technological changes than are the problems of most of the alternatives, that is, what I have called the intractable human factor looms larger for nuclear power than it does for most of the other choices.

Let me add a word about uncertainty. Virtually all absolute statements that are made on these topics have holes in them. I think technologists discredit themselves when they succumb to the policy makers' desire for a one-line answer and understate the uncertainties. This happens all the time when scientists offer technical advice to policy-making bodies. Toxicity of plutonium, for example, will be uncertain by a factor of plus or minus 10 probably for at least another 15 to 20 years for fundamental reasons. The accident risk of nuclear power is uncertain in my view by a factor of, perhaps, 10 on the down side and 1000 on the upside and will remain so for a considerable number of years. Policy makers must ask themselves not whether such-and-such well-defined risks are worth such-and-such well-defined benefits but whether this bag of risks with all the uncertainties that go with it is worth this bag of benefits with the uncertainties that go with that.

Another important issue is public perceptions. Many technologists in this debate seem to behave as if public perceptions can be disentangled from the so-called technical ingredients of the problem. They argue that the public perceives irrationally when it behaves as if it were risk averse. That is, the public does not want to tolerate even a small chance of a very big disaster.

The public is confronted in the comparison of coal and nuclear with a choice between the relatively high routine impact of coal and the arguably small chance of enormous disasters associated with fission. Is the public irrational in weighing so heavily the small chance of big disasters? I am disinclined to say so. I do not think technologists have any business calling the public irrational when it exercises its preferences on dilemmas of this kind. The public perceptions are, irremediably and inextricably, part of the problem.

What, then, is the proper role of technologists in decision making? I am myself a technologist by training, but I have been preoccupied for a number of years with some of the liabilities and shortcomings of technology because I think it important that some of us trained along these lines look at this side of it. I think one of the biggest of the liabilities is to perceive issues as primarily technical where, in fact, the parts of the issue that can be illuminated by technical insights are small. What this amounts to is allocating to technical experts decisions where their technical expertise is of very limited relevance.

This leads me to my last point, which is the question of "trusting the experts." We are asked all the time to trust the experts because these issues are too intricate, too complicated, or require too much education for the public or its elected representatives to grapple with. I think the simple fact is that people do not want to trust the experts, and they do not want to for good reason. Simplicity and comprehensibility of technologies and of the risks associated with them are themselves a virtue. If the nature of a technology is such that an informed and open decision by accountable representatives is impossible in principle, then this is in itself a strong argument against proceeding.[14]

NOTES AND REFERENCES

1. John P. Holdren, "Too Much Energy, Too Soon," *The New York Times,* Op-Ed page, July 23, 1975. See also Paul R. Ehrlich, Anne H. Ehrlich, and John P. Holdren, *Ecoscience.* San Francisco: W. H. Freeman, 1977.
2. Demand and Conservation Panel of the Committee on Nuclear and Alternative Energy Systems, "U.S.

Energy Demand: Some Low Energy Futures," *Science* **200,** 142 (April 14, 1978).

3. Amory Lovins, *Soft Energy Paths: Toward a Durable Peace.* Cambridge, Mass.: Ballinger, 1977.

4. Victor Gilinsky, "Plutonium, Proliferation, and Policy," U.S. Nuclear Regulatory Commission News Release 5-14-76, in *News Releases,* series 2:43, pp. 4–7, week ending November 16, 1976.

5. Slobodan Nakicenovic, former Director of Safeguards Operations, International Atomic Energy Agency, Vienna, personal communication, September 1977.

6. Mason Willrich and Theodore B. Taylor, *Nuclear Theft: Risks and Safeguards.* Cambridge, Mass.: Ballinger, 1974.

7. U.S. Nuclear Regulatory Commission, *Reactor Safety Study: An Assessment of Accident Risks in U.S. Commercial Nuclear Power Plants.* WASH 1400/NUREG-75/014. Springfield, Va.: National Technical Information Service, 1975.

8. U.S. Environmental Protection Agency, *Reactor Safety Study: A Review of the Final Report.* EPA-520/3-76-0009. Washington, D.C.: EPA, 1976. Henry W. Kendall and others, *The Risks of Nuclear Power Reactors: A Review of the NRC Reactor Safety Study.* Cambridge, Mass.: Union of Concerned Scientists, 1977. Frank von Hippel, "Looking Back on the Rasmussen Report," *Bulletin of the Atomic Scientists* **33** (2), 42 (1977).

9. R. Philip Hammond, "Nuclear Power Risks," *American Scientist* **62,** 160 (May 1974). General Electric Corporation, *Nuclear Power: The Best Alternative,* GE Z-6301. 4A. San Jose, Calif.: GE, 1975.

10. U.S. Nuclear Regulatory Commission, *Environmental Survey of the Reprocessing and Waste Management Portions of the LWR Fuel Cycle,* NUREG-0116. Springfield, Va.: NTIS, 1976. U.S. Energy Research and Development Administration, *Alternatives for Managing Wastes from Reactors and Post-Fission Operations in the LWR Fuel Cycle,* ERDA 76-43, Vol. 1. Springfield, Va.: NTIS, 1976.

11. Karl Z. Morgan, "Suggested Reduction of Permissible Exposure to Plutonium and Other Transuranium Elements," *American Industrial Hygiene Association Journal,* August 1975, p. 567. J. Martin Brown, "Linearity vs. Nonlinearity of Dose Response for Radiation Carcinogenesis," *Health Physics* **31,** 231 (1976).

12. Committee on Nuclear and Alternative Energy Systems, *Report of the Risk/Impact Panel.* Washington, D.C.: National Research Council, 1978.

13. The term "endurable" was suggested by Kirk R. Smith.

14. For a more detailed discussion, see John P. Holdren, "The Nuclear Controversy and the Limitations of Decision-Making by Experts," *Bulletin of the Atomic Scientists* **32** (3), 20 (1976).

IV

ALTERNATIVE ENERGY SOURCES

Modules of solar photovoltaic cells are installed at the Lovington Square shopping center, Lovington, New Mexico, in a solar energy research project funded by the U.S. Department of Energy. (American Petroleum Institute)

The Impurity Study Experiment (ISX) device at Oak Ridge National Laboratory. It is being used to study problems and techniques relevant to eventual design of a controlled thermonuclear reactor. Experiments are being performed to study high confinement efficiency (high beta) plasmas; to test hydrogen pellet fueling; to control the influx of impurities; to evaluate ripple injection heating; de demonstrate the feasibility of electron cyclotron resonance heating (ECRH); and to appraise plasma shaping effects. The massive structure on which the letters ISX are mounted is the iron core and the spoke-like structures are the coils used to generate the toroidal magnetic field. The instruments and structures around the periphery of the ISX device include two neutral beam injection heaters and various diagnostics. (Fusion Energy Division of Oak Ridge National Laboratory (ORNL). ORNL is operated by the Union Carbide Nuclear Division for the Department of Energy)

Solar House I, a solar-heated, three-bedroom house, Colorado State University Solar Village, Ft. Collins, Colorado. (Solar Energy Applications Laboratory)

The Sunpak™ solar energy collector developed by Owens-Illinois, Inc. Opposite, top, these Sunpak™ collectors are mounted on a grid system that spans the courtyard and main entrance of the Terraset Elementary School in Reston, Virginia. Nationally recognized for its innovative design, the school won the 1975 Energy Conservation Award from Owens-Corning Fiberglas Corporation. Opposite, bottom, the glass tubular collector uses air or liquid to absorb solar energy. Each collector tube consists of a small feeder tube (shown pulled out here) that distributes fluid throughout the system, an absorber tube in which the fluid is heated, and a clear outer tube, which is evacuated to cut down heat loss. (John E. Hoff, Owens-Illinois, Inc.)

The Geysers Steam Field, Sonoma County, California. This is the largest geothermal power development in the world, generating 502,000 kilowatts (electric) from ten power plant units. (C. C. Newton, Public Information Department, Pacific Gas and Electric Company)

Opposite, MOD-2 wind turbine. The world's first wind energy farm is producing electrical power at Goodnoe Hills, about 13 miles east of Goldendale, Washington, not far from the Columbia River. This unit, the first of three that were designed and built by Boeing Engineering and Construction Company under a program conducted by the U.S. Department of Energy and managed by the National Aeronautics and Space Administration, was completed in December 1980. The rotor for the second unit was lifted into place in February 1981, and all three units began operating in 1981. Bonneville Power Administration buys the output and integrates it into the regional power system.

When its blades are in the vertical position, the wind turbine rises 350 feet above the ground. Its 90-ton blades are 300 feet in diameter. They are mounted to a nacelle, which contains the turbine-generator, gear train, and computer-control equipment. The nacelle rests on a 200-foot-tall tubular steel tower.

INTRODUCTION

Although the fossil fuels and uranium will continue to dominate the energy picture through the end of this century, they both fall into the "finite resource" category and must eventually be replaced by alternative energy sources. In this section, we examine some of the most promising alternatives and try to understand some of the technical, economic, and environmental factors that influence their implementation. Once these factors are appreciated, the development of these energy alternatives can be seen in a more natural historical perspective, and less justification remains for various "conspiracy theories" to explain the present pattern of energy research and development.

With the exception of geothermal energy (whose technical and economic feasibility are both well established) and fusion energy (whose technical feasibility has not been proved), the pattern underlying the discussion of most alternative energy sources is the following: this alternative is now somewhat more expensive than conventional fuels, but as the cost of such fuels escalates and as mass production is applied to the alternative, it will surely become competitive. This argument was proposed in the 1950s and 1960s for nuclear fission and verified in the early 1970s. It was proposed in the 1960s and early 1970s for solar thermal energy, which, as the discussions in this section indicate, is currently economically feasible in certain regions. Although fusion has not yet proved to be scientifically feasible, some fusion optimists are estimating that the cost per installed kilowatt for fusion power plants will be comparable to that of breeder reactors.

This argument has an indisputable logic reinforced by declining domestic oil production which makes it imperative that we press research and development efforts on all energy alternatives that show promise of making a significant contribution to supply. Some critics of fossil-fuel and fission energy oppose new plant construction because "if we can just get along with present facilities

391

for a little while longer, solar or fusion energy will come to our rescue.'' Here, the history of the fission energy program itself provides a useful lesson on the time frame for the introduction of any new, large-scale energy technology. The "proof of principle" for nuclear reactors occurred in 1942, the first commercial reactor came on line in 1957, and the effect of nuclear fission energy was not even measurable until 1970. Even now, over 40 years after its feasibility was proved, it contributes only 10 to 15 per cent of our electrical energy. Because of the huge capital investments involved in research and development, there will simply be no "quick and easy" energy alternative to return us to an era of cheap and plentiful energy.

An important distinction should be made in studying energy alternatives— that between new primary energy source (e.g., fusion and solar energy) and improvements in energy utilization and interconversion techniques (e.g., coal gasification and synfuel production). Both fusion and solar energy promise virtually unlimited supplies of clean energy. The interconversion alternatives promise increased ease of utilization and environmental acceptability of present nonrenewable fuels but generally with a severe loss of overall efficiency.

The energy alternatives we consider all share the characteristic of having an essentially unlimited lifetime. The first five are various forms of solar energy, discussed in order of feasibility and likely time of wide-scale application. Next we discuss the contributions and potential of geothermal energy and close the section with the prospects for fusion, the basic energy source of the sun itself.

We begin the discussion of alternative energy sources with Christopher Flavin's article, *Passive Design: The Solar and Conservation Potential.* The present surge of interest in passive design is shown in historical perspective to be a rediscovery of climate-conscious design recognized and practiced since the days of Socrates. The economic and social forces leading to the abandonment and revival of energy-conscious design are presented in convincing detail. The author describes passive design considerations which simultaneously optimize solar energy collected and minimize heat loss, thereby demonstrating how passive solar design and conservation are integrated concepts. Numerous examples of energy-conscious architecture are cited along with regional variations due to climatic differences throughout the world. Finally, the author considers the problem of existing buildings and the role of public policy in encouraging climate-sensitive design.

The next step in complexity and cost beyond passive solar design, which is primarily architectural, is the "active" collection of solar energy for direct thermal applications, described by John Duffie and William Beckman in their article, *Solar Heating and Cooling.* They describe several experimental solar heating systems that are now operational, the simulation program being used to model the behavior of such systems, and the sensitivity of the economics of home heating systems to such parameters as the collector area. The final economic feasibility of solar heating systems is shown to depend not only on the initial capital costs and on interest rates, but also on public policies such as deregulation of natural gas prices and tax incentives for the installation of solar systems. The present program of deregulation of natural gas and tax incentives for the installation of solar equipment will certainly help reach the goal of 2.5 million solar homes by 1985.

Aspects of one of the truly new forms of solar energy are analyzed by Dennis Costello and Paul Rappaport in *The Technological and Economic Development of Photovoltaics*. This technology for converting sunlight directly into electricity using only semiconductor cells with associated processing electronics has a great appeal because of its beauty and simplicity. With no moving parts, an infinite fuel supply, and no great economy of scale (i.e., small units are just as efficient as large ones), photovoltaic systems appear to be the ideal energy source for a noncentralized, "soft technology" society. Why, then, have solar cells not swept the world? The authors, one of whom helped found the field of photovoltaics, make it clear that the sticking point is economics. With solar photovoltaic systems costing about $10 per peak watt, the electricity produced must sell for $1 to $2 per kilowatt-hour, or about 20 times the price we pay the utilities. However, the authors are optimistic about the future role of photovoltaics and cite market penetration to date, the nature of the "learning curve" in other semiconductor markets, and the private and governmental research programs whose goal is to reduce the cost of photovoltaic systems to a competitive $1 per peak watt by 1986.

Wind power is another historically important form of solar energy which disappeared with the advent of cheap hydro- and fossil-based electricity. David R. Inglis describes, in *Wind Power,* the historical contributions of windmills and the geneology of the modern windmill starting with the Grandpa's Knob experiment in 1939 and culminating with the single-bladed monster, Growian II. The attractive features of windmills include their simplicity, pollution-free operation, and unlimited fuel supply. The unfavorable economics which killed earlier electricity-producing windmills competing with cheap coal is seen to be shifting in favor of windmills as fossil-fuel prices rise and new technology improves the performance of windmills. The author indicates that Boeing can produce the one-hundredth unit of its 2.5-megawatt MOD 2 machine (p. 388) at a cost of $1000/kW, which is cheaper than new nuclear reactors coming on line. However, it would require 750 of these huge, beautiful machines to produce the power of one typical reactor. Dr. Inglis describes the checkered history of federal support for windmill development and urges much more vigorous support to bring this neglected source of energy on line at the earliest possible date.

Perhaps the most promising form of obtaining energy from the ocean is described by Abrahim Lavi and Gay Heit Lavi in *Ocean Thermal Energy Conversion* (OTEC). Although OTEC is generally considered one of the more exotic new forms of solar energy, the concept was first proposed by the French physicist D'Arsonval in 1881. In using the temperature difference between the surface and deeper layers of tropical ocean water (as high as 22 C°), OTEC shares many of the advantages of other forms of solar energy and avoids one of the biggest problems plaguing more direct solar techniques—its intermitency. By using ocean water itself as the storage medium, OTEC eliminates the day-night and cloudy period cycles. The longer-term seasonal cycles match well the seasonal energy needs of the southeastern United States. Problems remain, however, and include low intitial conversion efficiency, corrosion resistance, biofouling, and possible environmental effects. The authors describe two prototype systems, Mini-OTEC and OTEC-1, which are beginning to answer some of the engineering and economic feasibility questions of OTEC.

L. J. Patrick Muffler in his article, *Geothermal Energy,* describes an energy source that has been in continuous operation since 1904. Geothermal energy is the basis for 2.5 GWe of electrical power and over 7.0 GWt of thermal power in 13 countries around the world. In regions with appropriate geological conditions, it can produce electricity more cheaply than any source except installed hydropower. The author describes the geological nature of geothermal energy and the definitions of geothermal resource base and geothermal resource which help us understand why more of the vast amount of the earth's heat is not being exploited. Several examples of existing power stations and interesting thermal applications are cited. Not discussed are the environmental liabilities of geothermal power, which may include land subsidence (sinking), low thermal conversion efficiency (more than five times as much thermal pollution as a coal plant), heavy-metal pollution, and noxious odors (H_2S gas). These problems, apparently, can be controlled or at least lived with, since installed geothermal electrical capacity is growing at a rate of between 7 and 19 per cent annually.

Finally, we turn to *The Prospect for Fusion* by David Rose and Michael Feirtag. In this article, the authors give a qualitative explanation of the basic fusion process, the plasma physics required, and the awesome engineering problems that must be solved before fusion devices produce more energy than they consume. Several techniques for harnessing fusion energy are described, and the Tokamak is shown to be the device most likely to succeed. The authors present a rather somber assessment of the chances of overcoming the extremely difficult engineering problems involved and some good insight on the broader aspects of optimizing the research and development process itself. The prospect of an essentially unlimited energy supply (literally "oceans full of fuel") is too promising to ignore.

In conclusion, we offer the following observations:

1. All the energy alternatives discussed have definite advantages that justify continued research and development.
2. They all have some intrinsic problems of a technological, environmental, or economic nature that have prevented them from contributing significantly to the energy picture to date.
3. A wise energy policy will include a balanced research and development program to assure the optimum deployment of each of these energy alternatives. Any other policy risks the pitfalls of "all the eggs in one basket," which many observers feel represented the U.S. fission reactor program during the last 20 years. With such a balanced program we spread the risks,

and if one approach, for instance, fusion, should prove intractable, the emphasis can easily be shifted to other promising alternatives. Although this approach is more expensive than the policy of ''picking the best alternative and sticking with it,'' it represents the most conservative and safest long-range policy.

Although the energy alternatives presented do not completely exhaust the list of proposals, they certainly include the most promising alternatives and those with the greatest capacity for sizable energy production.

Passive Design: The Solar and Conservation Potential 21

CHRISTOPHER FLAVIN

INTRODUCTION

The energy problem facing the world today has many facets. Transportation, manufacturing, and agriculture are all heavily dependent on expensive fossil fuels. The world's buildings are unfortunately no exception. The heating, cooling, and lighting of buildings now consumes nearly one quarter of the world's annual energy supplies and approximately two thirds of this is derived directly or indirectly from oil and natural gas. The cost of providing fuel and power to this sector has escalated rapidly in recent years, and in many nations buildings have been forced to the front lines in the fight to reduce petroleum use.

The energy problem, then, is in some measure an architectural problem. Modern buildings have until recently been constructed with little heed paid to their energy

efficiency or lifetime fuel costs. The buildings sector, like other parts of the world economy, has been accustomed to low, subsidized fuel prices. The prevailing assumption was that efficiency of energy use was unimportant because technological developments would continue to keep fuel prices low. The price of this shortsightedness is borne by millions of consumers in their monthly fuel bills.

Throughout the 1950s and 1960s, the buildings in many countries were constructed with ever more voracious energy appetites. Lighting levels were increased, air conditioning systems were added, and builders continued to use single-pane windows and minimal insulation. The people who live and work in buildings contributed to the exorbitant energy use as well, setting thermostats too high in winter and too low in summer. The result was predictable: energy use in

Christopher Flavin is a senior researcher with Worldwatch Institute, Washington, D.C. 20036, an independent, nonprofit research organization devoted to the analysis of global energy issues. Mr. Flavin is a graduate of Williams College, and his present research is on energy issues, resource management, agriculture and forestry, and global research and development allocation. He is co-author of the book *Running on Empty: The Future of the Automobile in an Oil-Short World*, New York: W. W. Norton, 1979.

buildings increased more than 5 per cent per year in many countries. The United States, which set the standard for "modern" buildings during the postwar period, is now clearly paying the consequences. Residential and commercial energy use in the United States nearly tripled between 1950 and 1973, and the annual bill for providing fuel and power to those sectors now stands at over $100 billion.

It is not for lack of know-how that the situation is so dismal. Relatively simple changes in design and construction techniques—some of them known to the ancients—could greatly lower the fuel requirements of buildings. And today, well-designed climate-sensitive buildings that use the sun's energy directly to heat the interior and provide light and that use natural breezes for cooling are reducing fuel bills by 75 to 100 per cent, depending on the local climate. Good insulation and windows that face the equator are the basic features of such structures. Many are called "passive solar" buildings since most heating, cooling, and lighting requirements are supplied by sunlight, shading, and natural ventilation. Others rely more on the conservation technique of using very heavy insulation and are known as "low-energy" buildings. In practice, these approaches are often used together in the same structure.

Energy-conscious design, once the domain of backyard inventors and counterculture enthusiasts, is now being practiced by top architectural firms and integrated into suburban communities. Major research efforts under way in more than a dozen countries are refining the state of the art, reducing costs, and starting to apply the new concepts to large commercial structures as well as private residences. Architects and engineers are reassessing the whole range of construction materials and techniques in their attempts to reduce the fuel requirements of buildings. They are also investigating ways to adapt the designs to retrofit existing buildings. The spectacular growth in these fields is rooted in some simple economics: climate-sensitive design is cost-effective at today's fuel prices.

During the 1980s, ignoring energy-saving principles of construction is like throwing money down the drain.

Knowledgeable observers are predicting that the next decade will see some of the most rapid and far-reaching changes in the history of architecture. As one pioneering solar architect noted recently, "Traditionally, architecture has been a response to the times, and energy conservation is the issue of our time." Designers and builders face unprecedented challenges in the years ahead as they attempt to meet the needs of a fundamentally different energy situation. Opportunities for innovation will also be great, however, and it is conceivable that the forces of necessity will breathe new life into the architectural field.

Already, a number of improvements have been made in the way the average building is designed and built. Better insulation and storm windows are among the measures that have dropped the heating requirements of new buildings in many countries by 20 per cent or more in the last few years. The more ambitious design changes—those that would reduce heating and cooling requirements by more than half—have not yet fully penetrated the construction industry. But in a few nations, the day that such designs really take hold is quite close. In the United States, in particular, some large building companies are actively developing passive solar prototypes, and attractive government financial incentives will soon pull them into the mass marketing of energy-saving buildings. Between 10,000 and 20,000 passive solar buildings currently stand in the United States, but there could easily be more than a million by the mid-1980s. A challenging but reasonable goal for most countries would be to cut the energy requirements of *all* new buildings by 75 per cent by 1990.

The need for rapid change is difficult to underestimate. Buildings are long-term investments, and the house that was built with yesterday's $2-a-barrel oil in mind will still be standing when oil hits $50 a barrel in the not-too-distant future. Much of the world is still in the midst of an unprecedented housing boom that will last several more years, and large commercial buildings are being built even more rapidly. Improvements made today in design and construction techniques will have a disproportionately large effect on the comfort and economic attractiveness of the world's building stock at the turn of the century. Equally important are ongoing efforts to retrofit existing buildings with solar and conservation measures. Since 80 per cent of the buildings in use today will still be around in the year 2000, it is essential that they be improved in response to the altered world energy situation. These changes will require substantial economic and institutional adjustments, but in the long run the benefits will unquestionably far outweigh the costs.

ENERGY AND ARCHITECTURE

In this age of standardized buildings and mechanical heating and cooling systems, it is easy to forget that climate-conscious design was once the norm. In some parts of the world, it still is. The pueblos of the American Southwest, the adobe and thatch huts of equatorial Africa, and the cooling towers of Moslem Asia are all testimony to the simplicity and common sense inherent in passive solar design. Built without the aid of architects or engineers, these buildings cleverly use sunlight and natural breezes to heat and cool their interiors.

Over 2000 years ago, Socrates observed that "in houses that look toward the south, the sun penetrates the portico in winter, while in summer the path of the sun is right over our heads and above the roof so there is shade." This basic idea—that the sun describes a lower and more southerly arc in winter than in summer (and a more northerly one in the southern hemisphere)—is applicable everywhere but in the tropics near the equator. It is the central principle in all passive solar design. Two to three times as much sunlight strikes a south-facing wall in winter than in summer, making that the logical side

on which to place windows. The house itself then becomes a solar collector. In the sunny Greek climate, it seems likely that the very simple design of those buildings provided all the heat necessary on two thirds of the winter days. And several Greek cities were planned so that all buildings had exposure to the winter sun.

Passive solar heating was also employed by the Romans, who picked up their knowledge of the basic concepts from the Greeks. The major energy source at the time was firewood and charcoal, but by the fourth century A.D. indigenous wood supplies had largely been stripped from the Italian peninsula. The pressure of fuel scarcity was a strong incentive for solar heating, and Roman architects slowly adapted solar design to the variety of conditions found throughout the Roman Empire. The Roman architect Vitruvius wrote that "it is obvious that designs for homes ought to conform to diversities of climate." Romans developed the first greenhouses as well as solar-heated bathhouses; floor plans indicate residences were organized so that rooms would be comfortable at the time they were most often occupied; and access to the sun was actually made a legal right under the Justinian Code of Law adopted in the sixth century A.D.

Many other cultures developed equally successful climate-sensitive building styles. Homes in ancient China were generally built on the northern side of courtyards, facing the south, and sunlight was admitted through wood lattice windows and rice paper. The Anasazi people of the American Southwest lived in mud or stone buildings constructed against overhanging cliffs that faced south. They were earth-sheltered dwellings, solar-heated in the winter and shaded in the summer, all without benefit of modern building materials or theories. More recently, the early settlers in New England built "salt box" houses that were carefully oriented to face the south. They were two-story dwellings with most of the windows on the front and a long, sloping roof on the north to provide protection from winter winds. It is an astonishingly sensible design for very cold climates and has been undergoing a revival recently.

Other forms of traditional architecture incorporate simple passive cooling techniques effectively. Throughout tropical Asia and South America, simple pole and thatch buildings are open on the sides to allow ample ventilation but are protected from the heat by a thatch roof. Thatch rivals fiberglass as an insulator and is also found atop mud and straw buildings in the sub-Saharan Africa. In Moslem Asia, cooling towers have been in use for a thousand years and help draw air into buildings, providing ventilation and relief from the hot summer climate.

Since the beginning of the Industrial Revolution and the migration to the cities that accompanied it, traditional architectural forms have been abandoned by one culture after another. The climate-sensitive building designs that had been developed were not easy to adapt to cities, and the growth of urban populations encouraged a standardization of architectural styles, particularly in the low-cost housing needed for an expanding labor force. The amount of wood and coal required to heat the buildings of nineteenth-century Europe and North America increased rapidly. Although living conditions improved for many people, they deteriorated for millions of others; heating buildings became a major expense that not everyone could afford.

As modern architecture matured over the last century, energy considerations continued to take a back seat. Architect Richard Stein writes that "during the 1920s many of the most prophetic and influential architects projected the form of the future as being freed from the rigorous demands of climate and orientation." These pioneering designers embraced the notion of using mechanical systems and artificial lighting to provide for needs previously met by sunlight and natural ventilation. The convenience of modern heating and cooling systems and the rela-

tively low cost of new fuels such as natural gas and electricity persuaded consumers as well as builders that worrying about energy efficiency was a waste of time.

In more recent years, the fuel requirements of buildings have continued to grow, more than doubling worldwide between 1950 and 1970. Though a portion of this increase is accounted for by growth in the number of buildings, many structures are simply built to use more energy than those of the past. And because of an unprecedented housing boom, a disproportionately large share of the world's current building stock was constructed during these years. Many of the new building forms were pioneered by American designers and were quickly adopted in Europe and Japan during the postwar reconstruction period. More recently, they have been aggressively marketed throughout the Third World.

There are several common reasons for the unnecessarily large fuel requirements of buildings around the world. Only half the residential buildings in Europe, for instance, have any insulation at all, and storm windows are a rarity there. In the United States, close to one third of the residential housing stock has no insulation, and another 50 per cent is inadequately insulated. In addition, the buildings in many countries, particularly those belonging to the poor, are loosely constructed and "leaky": tiny cracks around windows and in walls and attics can result in considerable heat loss. Another explanation for high fuel use is the usually random siting of buildings, which exposes them to heat-robbing winds in the winter and excessive sunlight in the summer.

This lack of attention to climate-sensitive design or construction techniques, combined with energy-intensive heating, air conditioning, and lighting systems, has proved costly. The results are easily seen in the typical modern office building with its glass facades and mechanical "climate-control" systems in use every day of the year. It is not uncommon to have to turn on a quarter of an acre

Table 21–1.   Energy Use in Residential and Commercial Sectors in Selected
Industrial Countries, 1978.

Country	Residential and Commercial Energy Use (million tons of oil equivalent)	Share of Total National Energy Use (%)	Residential and Commercial Energy Use Per Person (tons of oil equivalent)
United States	442.62	33	2.01
Canada	46.32	33	1.95
Sweden	12.95	38	1.56
Netherlands	21.04	39	1.50
West Germany	78.89	39	1.29
France	50.76	35	0.95
United Kingdom	45.01	31	0.81
Italy	31.80	30	0.56
Japan	56.59	21	0.49

Source: Organization for Economic Cooperation and Development.

of lights to work at a few square feet of desk in these buildings. In private houses and apartments, the rapid spread of air conditioning has been the most important factor in increasing energy use in recent years. Though in many ways the architectural changes of the last 35 years have been a success, from a long-term energy perspective they have been a failure.

There are two main categories of buildings—residential and commercial—and both are big energy users. Together, the residential and commercial sectors account for between 20 and 40 per cent of national energy use in most industrial countries (Table 21-1). Of this, approximately four fifths is used to heat, cool, and light buildings and the rest runs appliances and water heaters. Residential energy consumption exceeds commercial energy use in most countries by 20 to 100 per cent, though the rapid growth of the service sector has meant commercial buildings take up an ever-increasing share.

North Americans warmed to the new energy-intensive buildings earlier and more vigorously than Europeans did. Fuel use per person in the residential and commercial sectors is nearly twice as high in the United States and Canada as in most of Europe.

There are a number of explanations for this large disparity, including the more compact layout of European cities and the tendency of people there to be satisfied with warmer buildings in summer and cooler ones in winter. The industrial country with the best record is Japan, where the per capita fuel use in the buildings sector is only one quarter of the U.S. level, due to the compactness of Japanese buildings and the nearly complete lack of central heating. It is also striking that in Sweden the fuel requirements of buildings are 25 per cent lower than in North America, despite a harsher winter climate. Traditionally higher fuel prices and lower per capita incomes have meant that the energy used in buildings is treated less nonchalantly in Europe and Japan than in North America.

Energy use in the buildings sector has become a major problem in virtually all industrial countries, however, and none can afford to be complacent. In much of Europe, the proportion of national energy resources devoted to buildings is as high as or higher than that in the United States. The U.S. transportation sector claims a large share of available fuel, whereas elsewhere in the industrial world, industry and buildings account for the preponderance of fuel use. Furthermore,

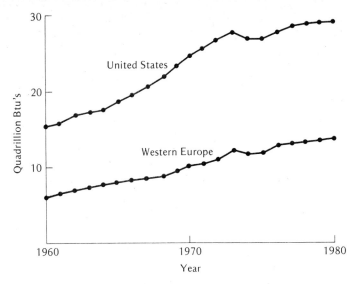

Figure 21-1. U.S. and European energy use in residential and commercial sectors, 1960–80. (Source: U.S. Department of Energy and Resources for the Future)

much of the energy used in buildings in these countries is supplied by petroleum, which makes their positions even more precarious. Nearly all the heating fuel in Japan and 80 per cent of that used in France and West Germany is petroleum, compared with 44 per cent in the United States, where natural gas plays a much larger role. For Europe and Japan, therefore, buildings play a crucial role in any efforts to reduce burdensome OPEC oil bills.

In many nations there are signs that the energy performance of buildings has begun to improve since 1973, albeit slowly. The increase in energy use in the U.S. residential and commercial sectors dropped from a 5.1 per cent annual growth rate in the 1960s to approximately 2 per cent in the late 1970s (Figure 21-1). Most of this improvement came from thermostats being set lower, adjustments of furnaces and air conditioners, and the addition of insulation to existing buildings. Even these simple measures, in combination with slowed economic growth, have been enough to stabilize the energy requirements of most buildings. The remaining

annual increment comes from an increase in the number of buildings. Savings have been more impressive in Europe. Energy use in buildings is increasing at a 1 per cent annual rate in West Germany, has leveled off in Great Britain, and is slowly falling in Sweden. This compares with growth rates that approached or exceeded 5 per cent during the 1960s.

The improvements made so far have been predominantly a response to higher energy prices. The costs of electricity, heating oil, and natural gas have all risen at record rates since the mid-1970s, in marked contrast with the steady and declining prices of the 1960s. In Japan, the real price of heating oil rose 75 per cent between 1973 and 1976 and in Italy it more than doubled. In the United States, the real cost of heating many homes soared, with particularly severe increases for houses heated by oil (Table 21-2). Lower-income groups have been hit particularly hard. In its 1979 annual report, the Tennessee Valley Authority poignantly called attention to one of its customers who paid her electric bill with a Social Security check ''and walked out to

Table 21–2. Average Annual Residential
Heating Bill in the United States, By Fuel,
1960–80.

Year	Oil	Natural Gas (1980 dollars)	Electricity
1960	390	310	960
1970	360	250	630
1975	560	280	710
1980 (est.)	990	420	840

Source: Office of Technology Assessment and U.S. Department of Energy.

face the month of February with less than
$30.'' Though natural gas is still a relative
bargain in the United States, recent estimates
indicate that price decontrol combined with
supply constraints will make it nearly as expensive as heating oil by 1990.

As recently as 1970, economists and planners in many nations were predicting that
residential and commercial energy use would
more than double by the end of the century.
Their predictions were an extrapolation of the
trends from 1960 to 1970, which in 1980
seem like ancient history. The formerly un-
broken curve of spiraling fuel use in buildings has now been interrupted nearly everywhere. Continuing price increases and the
gradual response of consumers and the
building industry are likely to cause a leveling off of the energy requirements of this
sector during the next decade. The major unanswered question is whether architects,
builders, and consumers can actually reduce
the amount of fuel used in buildings by the
end of the century. A growing body of evidence indicates they can.

CLIMATE-SENSITIVE DESIGN

A decade ago, builders and home renovators
interested in cost-effective, energy-conscious
building design would have been in a real
bind. Basic data describing patterns of heat
gain and loss in buildings were hard to come
by, only a handful of architects had come up
with innovative building plans, and financial

information comparing various options was
unavailable. Today the situation is changing
rapidly. Technical journals have appeared and
conferences are scheduled in many countries, national and private laboratories are investing millions of dollars in the development of energy-saving materials and
assemblies, and architectural plans are now
available for everything from a low-energy
office tower to a passive-solar mobile home.

The common goal behind these diverse efforts is to reduce the costly reliance of buildings on fossil fuels. The basic principles being
applied are quite simple: by using insulating
materials to slow the rate at which indoor
temperatures adjust to extreme outdoor temperatures, by admitting as much sunlight as
possible during the winter months, and by
providing various means of shading during
the summer, heating and cooling systems
based on fossil fuels can be greatly reduced
or even eliminated. Different designs are of
course needed for different types of buildings, climates, and economic levels, but in
most cases the underlying principles are similar. In recent years, the international exchange of data and reports has begun to assist the research efforts in many nations.

Glass (and more recently plastic) is the
basic material that makes modern solar heating possible. Grecian solar houses must have
lost heat almost as rapidly as it was collected, a process that is reduced with glass,
which readily transmits sunlight but impedes
thermal radiation, in effect trapping heat in
the building. This phenomenon, known as the
greenhouse effect, is familiar to anyone who
has left a car in the sun on a cool day and
returned to fine it overheated. In its simplest
form, passive solar heating consists of having most of a building's windows on its
southern side (or, in the southern hemisphere, on its northern side). In this way, a
large portion of the heating needs are supplied by the sun. Windows on the east, west,
and north are minimized because they tend
to lose more heat than they gain and because
they can cause overheating problems in the
summer. Many architects now design build-

ings to be elongated on an east-west axis in order to increase the area available for "solar gain" on the south. The proper siting of a solar building is almost as important as the design. Access to the winter sun and protection from cold winds can be facilitated by the correct positioning of the structure.

The first modern solar house was built in Chicago in the 1930s. It differed from a conventional house mainly in that it was carefully sited to take advantage of the sun and had a large window area on the southern side. Similar buildings were constructed over the next two decades in Cambridge, Massachusetts, in Wallasey, England, and elsewhere. They attracted a great deal of attention and convinced some observers that a new physical principle had been discovered. *Business Week* suggested in 1940 that the Chicago house was the "newest threat to domestic fuels." Some of the early solar houses were successful and others were not, but over the years a few persistent designers continued to improve the state of the art.

As research proceeded it became clear that the retardation of heat loss was as essential as the admission of sunlight. The walls, roofs, and windows of conventional houses lose a great deal of heat during cold weather because of radiation and convection. Such a house, when heated only by the sun, cools rapidly after dark. Solar houses developed more recently in Europe and North America have included more than twice as much wall and attic insulation as conventional dwellings have. Most windows are double- or triple-glazed, and vestibules are often used to prevent a sudden loss of warm air when someone enters or leaves the building. In addition, strong emphasis is placed on tightness of construction so that the building has as few air gaps as possible. As much as half the heat loss in a conventional building occurs through direct infiltration of cold air.

Also integral to the success of a passive solar building is a method of heat storage. Through the use of construction materials with substantial capacity to hold heat, a building's ability to store the sun's energy is increased. When the sun sets or when the furnace is turned off, the thermal mass slowly radiates heat, keeping the building warm. Several traditional building materials—including brick, concrete, adobe, and stone—perform this task well and help reduce temperature fluctuation in both winter and summer. In some structures these materials are used in walls that are then insulated on the outside so that the thermal mass is in the interior of the building. Additionally, thermal storage materials can be incorporated in large fireplaces, in secondary walls built mainly to store heat, or in floors. Rock or gravel drainage underneath a dwelling can serve a dual role, acting as a large heat sink. Water is one of the best materials for storing warmth and, though somewhat difficult to use in a building, can be incorporated in a "water wall" or used in a fish pond. Researchers at the New Alchemy Institute in Massachusetts maintain that aquaculture tanks located inside a greenhouse can pay back their cost in heat-storing capacity alone.

In addition to providing heat during the winter months, a climate-sensitive building should be cool and well ventilated in the summer. Fortunately many passive heating features, such as good insulation and thermal storage, help with passive cooling in the summer. Other features that are frequently employed include shades that protect south-facing windows from the high summer sun and ventilation systems that take advantage of natural breezes and thermal convection to keep air moving continuously through a building. Vegetation can be one of the best aids to passive cooling. Deciduous trees protect a house from the summer sun only and often provide a microclimate that is several degrees cooler than in surrounding areas. Pioneering work in this field was done in Japan and the U.S. Southwest. More advanced forms of passive cooling recently developed will play an important role in cooling large commercial buildings.

The beauty of passive solar design is that, though the basic principles are simple, there are a great number of ways to harness the

sun's energy effectively. Solar houses do not
have to be identical, nor need they be dull.
Bruce Anderson, chairman of Total Environmental Action and a solar design expert, notes
that "the variety of climates and personal
tastes will dictate a veritable plethora of
unique solar home designs." Hundreds of
different types of passive solar buildings have
appeared in the last decade as architects attempt to translate the simple principles of
energy-efficient housing into livable, reasonably priced, attractive buildings. Many of the
most efficient new buildings look surprisingly conventional and would easily blend
into any residential or commercial development.

One of the more ingenious designs for solar heating involves the use of a thermal-
storage wall placed several centimeters inside a large expanse of glass on a building's
southern side. The wall, constructed of ma-
sonry or filled with water, is painted a dark
color to absorb heat from the sun; heat collected during daylight hours is radiated to the
rest of the house for many hours after sun-
down. This is known as a Trombe wall,
named after Felix Trombe, who with Jacques
Michel designed buildings using this technique at the French Solar Energy Laboratory
in the late 1950s and early 1960s.

Today, the Trombe wall is a well-
established concept and variations on it can
be found in many parts of the world, using
different wall materials and thicknesses, various techniques of air circulation, and different amounts of insulation and thermal storage. In general, the Trombe wall is an
extremely effective collector of solar energy.
Its chief drawback is that the heat-storing wall
is quite close to a window, causing considerable heat loss through radiation and requiring special thermal shades that are closed at
night. An interesting variation on the Trombe
wall was developed by Steve Baer in New
Mexico in 1972. He mounted 138 barrels of
water behind a south-facing window wall,
providing effective, cheap thermal storage.
Another house in that area uses 2000 old wine
bottles for storage.

A related but distinct method of passive
solar heating is the use of a greenhouse or
"sunspace" on the southern side of a building. An attached greenhouse serves as a natural solar collector that can be easily closed
off from the rest of the building at night. It
can be an interesting addition to a house and
can extend the fresh vegetable season as well
as provide heat. As with other passive solar
systems, the importance of double or triple
glass, tight construction, thermal mass, summer shading, and ventilation is clear. If well-
designed and properly sited, a greenhouse can
supply more than half a building's heating
requirements in sunny climates. The chief
advantage of the solar greenhouse is that it is
an easy addition to conventional designs and
is an attractive option for people who do not
want to alter the basic form of their buildings.

Other solar designers are catching the sun
by moving underground, a seeming contradiction that in fact makes a great deal of
sense. Earth-sheltered buildings, whether
they simply employ a sod roof or are constructed completely beneath the surface, use
the earth as a natural insulator that damps out
temperature fluctuations in both winter and
summer. By exposing such buildings to the
vicissitudes of the weather only on the sunny
side, the sun's heat can be effectively collected and stored. Earth-topped roofs have
the additional advantage of providing natural
evaporative cooling in the summer. In some
climates, watering the roof might be the only
air conditioning necessary. Projects in the
United States and Israel have done much to
advance these designs. What is not yet clear
is whether earth-sheltered buildings can be
built cheaply and made attractive to large
numbers of people. Right now, building underground is an expensive proposition, but
some builders are convinced that the cost can
be brought down substantially.

Other types of climate-sensitive buildings
have been developed in recent years and are
being further refined by research teams and
entrepreneurs. A house developed by Harold
Hay in California uses an enclosed pool of

water on the roof for heat collection as well as for radiative cooling. Another interesting concept, developed by Lee Porter Butler in California and by a Norwegian team, is called the "double-envelope house." It incorporates a greenhouse on the southern side and a continuous air space in the roof, northern wall, and basement to supply heated air to various parts of the building. Both the roof pond design and the double-envelope house have been successful in some custom-built homes, but their large-scale economic attractiveness remains to be determined.

These bold, innovative research efforts are typical of the energy-efficient architecture field in 1980. New design ideas appear regularly, and older designs are being perfected and made less costly. The United States has provided much of the early leadership in climate-sensitive building, but other nations are gaining and may soon be on the cutting edge in particular areas of research. That different national areas of specialization have begun to develop in a field so young should not be surprising: climate is the key factor. Some designs are appropriate in northern, cloudy regions while others are better suited to areas with more sunlight year-round. Darian Daichok of the U.S. Solar Energy Research Institute recently conducted an international passive architectural survey and notes, "Passive research is taking on a distinctly regional flavor. Individual countries are now making major strides in developing buildings that are economical in their climates."

The different paths being pursued by U.S. and southern European designers as opposed to those in Canada and northern Europe illustrate this emerging regional focus. In the former areas, emphasis is placed on obtaining as much winter sunlight as possible. Using Trombe walls and various techniques of direct solar gain through south-facing windows, buildings in these regions can obtain most of their heating needs from sunlight. Designers focus on maximizing solar access as well as providing summer shading and ventilation.

Researchers in Canada and northern Europe are taking quite a different tack. Since the sun is present for only a few hours a day at midwinter and for much of that time is hidden by clouds, relying on sunlight for a large portion of heating needs would be hopeless. Instead, architects are designing super-insulated, very tightly constructed houses with relatively few windows. These buildings are wrapped in polyethylene plastic and special care is taken to seal all trouble spots, resulting in air infiltration that is one tenth that of a conventional house. These buildings are called "low-energy" or "zero-energy" houses, denoting the fact that they require very little heating, even from the sun. In fact, some of them rely as much on the body heat of the occupants and the waste heat from appliances as on heat from the sun, and auxiliary heating is needed only on the coldest of winter days. Typically, a low-energy house is fitted with an air-to-air heat exchanger—a small unit resembling an air conditioner that ventilates the building but prevents heat loss. Without this device the air could become unpleasant or even unhealthy as pollutants such as the radon found in concrete slowly accumulate.

Energy-conscious design is on yet another path in regions that must cope with a hotter climate. Passive cooling research in most countries, including the United States, is ten years behind that on heating. Australia, however, has conducted a major research effort under the auspices of the Commonwealth Scientific and Industrial Research Organization and is now doing extensive work on evaporative and radiative cooling. The results of this research should assist other countries with hot climates. In the case of the Third World, this could be a significant contribution at a time when planners are contemplating the need to expand housing quickly in cities. According to some experts, it should soon be possible to construct all but the largest buildings without air conditioning in most parts of the world. Passive cooling and natural ventilation will keep such buildings comfortable.

The bottom line in determining the success of various climate-sensitive designs will be their cost-effectiveness. This issue has received quite a bit of attention from economic analysts in recent years. Their calculations rest on a comparison of the costs of the solar and conservation features with the value of the fuel saved during the time the building will be occupied. Included in the equations are assumptions about interest rates, the rate of inflation, and the rate of fuel price escalation. Using these criteria, economists can determine a payback period for the money invested in energy-saving features as well as a life-cycle cost for the building. The basic conclusion of these economic analyses has been consistent: a well-designed solar or low-energy building almost always has a lower life-cycle cost than a conventional building does.

Climate-sensitive buildings usually come out ahead in such calculations because they can often reduce fuel bills by 50 per cent at little or no extra cost. For instance, a south-facing window costs no more than one that faces north, and a concrete floor that can store heat is cost-competitive with a wooden one. Options such as using two-by-six wall studs rather than two-by-fours to allow space for extra insulation or employing triple-glazed windows or night shades add only marginally to building costs. Other design possibilities—extensive glazing, a Trombe wall, or a large amount of thermal storage material—can be quite expensive, but if properly employed and suited to the climate of the area they can still be a financially sound investment.

The economic attractiveness of a climate-sensitive building also stems from the fact that expensive central heating and air conditioning systems are not needed in any but the largest buildings or in the most severe climates. Heating needs can be reduced so drastically that localized gas or electric heaters or wood stoves will often suffice. Residential building costs are often lowered by $1000 or more by doing without air conditioners, furnaces, and heat distribution systems.

Flexibility is now the watchword for designers interested in cost-effective solar buildings. Even within a particular region, exclusive reliance on just one design principle is unlikely to consistently yield the "right" answer from a financial viewpoint. Douglas Balcomb of the Los Alamos Scientific Laboratory, a leading expert in the performance of passive solar systems, has found that a mix of passive solar and conservation methods is usually the best bet for the consumer. Using data from a house in Kansas, his analysis suggests that home owners should initially invest in insulation and storm windows, but that beyond $800, investment in passive solar features should be pursued simultaneously. A nearly equal investment in conservation and passive solar measures, Balcomb found, would yield the lowest total cost over the life of the building. Putting the same money into either strategy alone would have resulted in considerably smaller savings.

An important aspect of this flexible approach to design is that passive systems need not be exclusively passive. In most cases, some form of auxiliary heating system makes sense, with the size of the system depending on climate and the local availability and cost of fuels. In regions where electricity or oil rather than natural gas is used, a higher fraction of solar heating is appropriate. And in many cases, it is a good idea to add nonpassive features, such as a fan that moves heated or cooled air to other parts of a building. Fans and even active solar collector systems if properly integrated can improve the financial attractiveness of a climate-sensitive building. By most estimates, however, proper design is the best place to start when planning a building from the standpoint of its energy use. Pump-driven solar collectors, though highly effective for heating hot water, are usually an expensive method of heating a building's interior.

The economics of the energy systems of

buildings is still a relatively young science. Much work remains to be done so that planners, builders, and consumers can have detailed information on the long-run financial picture of a building they want to construct or purchase. The most important results are already in, however. The world's buildings consume billions of dollars worth of fuel unnecessarily each year and fairly simple design changes could have eliminated much of that waste for a relatively small price. From now on, the economic evaluation of a building should include its probable life-cycle cost rather than just its initial expense so that the investments make economic sense in the long run. During the 1980s, to continue to construct houses and commercial buildings such as those that dot the landscape today would be costly, both for individuals and for society as a whole.

BREAKING GROUND

The research and development efforts of numerous innovative architects, builders, and engineers over the last decade have laid the foundation for the transition to climate-sensitive, fuel-conserving buildings. The principles are simple, the necessary materials are readily available, and the buildings are cost-effective at today's prices. The transition may be a gradual as well as a complicated one, however. Institutional barriers, such as the complexity of the building industry and its necessarily acute attention to short-term financial considerations, will make rapid change difficult. Alan Hirshberg of the consulting firm Booz, Allen, and Hamilton notes, "The building industry is very disaggregated and fragmented. You really need to go beyond the traditional research and development. You have to deal with that wide variety of players in the building industry process."

With passive solar architecture, implementation will clearly be more difficult than the basic research. Among the challenges ahead are the education of a whole generation of architects, engineers, and builders; the integration of solar and conservation designs into subdivisions and housing developments; and the transformation of a commercial building industry that so far has been lagging behind. There are a number of signs that some of these changes are beginning to occur. If they continue, climate-conscious homes and offices could take off in a big way in the 1980s.

One of the most encouraging developments is the growing interest of architects in solar design. Heating, cooling, and lighting needs have in recent decades been thought of as unrelated to the design of buildings, and they were often left to the engineers to resolve once the plans were complete. Today a growing number of architects are insisting that energy considerations play a major part in the design process. The early pioneers in climate-sensitive architecture, who often designed homes for their own use, have been joined by hundreds of others. In France, Sweden, and the United States, solar architects can be found in most regions, and plans for a custom-designed solar or low-energy building can be found nearly as easily as those for a conventional one. A recent U.S. government listing of solar design professionals that covered engineers as well as architects included over 1000 firms and individuals. Many of the companies are small, designing fewer than ten buildings a year. However, many solar designers are working to expand their business, and some are drawing up plans for houses that can be marketed to builders and developers.

A more fundamental shift is also beginning to occur as energy considerations enter into the designs of many architects who are not specifically attempting to create climate-sensitive buildings. Here, the architectural schools are playing an important role. Many are for the first time requiring courses in the thermal performance of buildings, and a few are actually teaching passive solar concepts. Often the students have led the faculty in insisting that energy issues be included in the curriculum.

Architects are of course only the tip of the iceberg in the building industry. In the United States, several hundred thousand developers, builders, subcontractors, and suppliers erect more than one million single-family homes, apartments, and commercial buildings each year. Only 10 per cent of these are custom-designed by architects. Building firms must assess the market for each particular project and then choose a design, building materials, and construction methods that appear to satisfy the wishes of their customers and that will provide a decent profit. Unfortunately, passive solar buildings are often thought of as unconventional and costly, a major deterrent to professional builders and developers. In today's real estate market, both single-family homes and commercial buildings are often built on a speculative basis, and the builder must be cautious not to come up with something that will not sell. Features that add cost to a speculative building are naturally shied away from since the selling price is the bottom line and builders do not have to worry about the home's heating bills.

Despite these obstacles, some people within the building industry are beginning to take an interest in energy-conscious construction, particularly in the United States. In the last few years several firms that concentrate exclusively on passive solar construction have been set up. Many have been highly successful. Comunico, a company in New Mexico, has built a large number of custom-designed solar houses that use Trombe walls, attached greenhouses, and other techniques. Comunico is achieving 70 to 90 per cent solar heating in the buildings it has constructed so far, and the group is now developing standardized solar homes that will be suitable for subdivisions. Another successful solar building firm is Green Mountain Homes of Vermont. Established in 1976, the company has built more than 200 solar houses and is now completing a home a week. Most of the components are prefabricated, and it is possible for prospective homeown-

ers to buy the building as a kit and erect it themselves.

Several other U.S. and Canadian firms are building solar houses in large numbers, and hundreds of smaller builders have entered the market. The day when passive solar architecture was mainly the domain of well-to-do owners of custom-designed homes is now ending. As solar builders adopt mass-production techniques and the preassembly of components, dramatic cost reductions are being realized. And one company is now offering a passive-solar mobile home for sale, opening up an entirely new market. Robert Naumann, an engineer with considerable experience with solar construction, echoes the conclusions of many solar entrepreneurs: "The profit margin can be very attractive when the builder keeps cost at a reasonable level. . . . In a progressive community with a good market these types of homes should always be successful." In no other part of the world has solar building taken hold quite so extensively as in North America, though in Denmark and Sweden the low-energy house may soon play an important role in the co-op–dominated building markets in those countries.

Involving the entire building industry in energy-conserving construction will obviously take time, and a continuing flow of information will be essential. It is encouraging that trade associations are introducing educational programs and that industry publications now regularly carry articles emphasizing the need to save energy. Most informed observers seem to believe that government incentives such as low-interest financing and tax credits will also be essential, especially to open up the lower-income housing market. Such programs can help shift the momentum of a rather entrenched industry and can encourage the slightly more costly attention to detail that is crucial if good designs are to result in good houses.

The next important step in the spread of climate-sensitive buildings is the use of these design concepts in housing developments and subdivisions. So far, there are few examples of such large-scale solar building projects, but indications are that many will get off the ground in the next few years. Developers are realizing that solar design and the promise of reduced fuel bills can be a strong selling point, and they are approaching design firms for solar blueprints. The few solar developments that have been constructed so far have been quite successful and are serving as models for other builders.

Village Homes in Davis, California, is a 240-unit solar subdivision developed in the late 1970s. While most of the houses appear at first glance to be quite conventional, they are all passively designed and are sited and landscaped so as to have full access to the winter sun. The members of the community are actively involved in the continuing planning of the Village Homes development, and an architectural review board ensures that one family's buildings and trees do not intrude on another's access to sunlight. The heating and cooling requirements of these buildings are 50 per cent below those of a conventional development. Solar water heaters, extensive bicycle paths, and community gardens further enhance the energy efficiency and livability of this model community. Village Homes has been successful from both a comfort and a financial viewpoint, and it has drawn nationwide attention to the possibilities of solar developments.

Equally as important as the designing of energy-efficient subdivisions is the task of incorporating passive solar features into townhouses, apartments, office buildings, and other high-density urban developments. Siting and orientation are more difficult in these cases, though the advantages of shared walls and less floor space per person help reduce energy requirements. Unfortunately, only modest research efforts have been undertaken so far on climate-sensitive design for such buildings, and it will be some time before the field is as advanced as it is with detached single-family residences.

According to U.S. Department of Energy estimates, even conventional residences in high-density urban neighborhoods, require on average 25 to 50 per cent less heating energy than suburban-style houses do. The inclusion of passive solar features and conservation measures in townhouses can make them an extraordinarily energy-efficient form of housing. Architect Peter Land of Illinois calls for a new emphasis on low-rise, high-density developments that he believes could lower energy and building costs simultaneously. Some of the townhouses he has designed would be open to the south and protected by other buildings on the east and west, and would include an interior patio to admit sunlight. Presumably, solar townhouses could also profit from the heat-conserving techniques developed for low-energy houses in Canada and Scandinavia; since solar access is sometimes limited in cities, a high level of thermal insulation makes sense.

High-rise apartment and office buildings also present unusual design problems that have yet to be fully addressed. There is less latitude in shaping such structures and providing shading. In addition, lighting and cooling in larger buildings usually requires more energy than heating does, so providing daylight and passive cooling for the occupants becomes a major consideration. Research on passive lighting is now under way, and engineer Douglas Bulleit notes that for large buildings "it very strongly appears that daylight is becoming the champion of passive design techniques." Experts in the field are confident that designs that incorporate the use of daylight can make climate-sensitive architecture as cost-effective for large commercial buildings as it is for single-family dwellings.

One of the pioneering architects in the commercial building field is Harrison Fraker, who has developed a number of striking energy-efficient innovations. Among other projects, he has designed a fairly low-cost medical office complex in Princeton, New Jersey, that is expected to use 30 percent less energy than a conventional building. It will incorporate an atrium to provide light and heat, a natural ventilation system, and a roof topped with a rock bed and sprinklers to provide evaporative cooling at night. Another leader in energy-efficient design for large buildings is Gunnar Birkerts of Detroit, who developed a 14-story office building for IBM that was completed in 1979. The building incorporates much less glass than most modern buildings do, yet through an ingenious window design most of the office workers are provided with natural light. Furthermore, the IBM office building includes a unique two-toned facade that allows solar gain on one side but retards heat loss on the other. The world's largest passive solar office tower will soon be standing in Singapore. Designed by a U.S. firm, it will be 40 stories tall and is expected to have energy bills 38 per cent below those of a comparable conventional building.

Modern high-rise buildings are quite complex, and improvement of their energy characteristics is now one of the major research frontiers in climate-sensitive design. In large structures, designs that solve one energy problem can easily aggravate others, and the integration of passive heating, cooling, and lighting into the same building requires complicated models, computer programs, and numerous trade-offs. In addition, most large buildings have mechanical systems to control the building's internal environment, and, while these can be made smaller in passive solar buildings, they cannot be eliminated. Energy-efficient design must therefore be integrated with the mechanical systems. One solar architect notes that "we design to provide comfort and lighting in a passive way for at least 50 per cent [of the energy load] and then use the mechanical systems to handle only the extremes." Recently developed microelectronics-based control systems allow artificial lighting to be adjusted according to the availability of natural light and enable heating to be turned down when people leave a room. Large savings could also be realized by the providing of localized light and heat, making it unnecessary to switch on hundreds of square feet of lighting every time someone enters a room. And giving occupants some control over their own heating,

cooling, and lighting could magnify these savings. The sealed window is one feature of modern office buildings that is ripe for elimination: windows that open would provide cooling breezes on many days when air conditioning is now required.

Considerable research into many of these questions is still needed and is the focus of government support in some countries. In the United States, the Department of Energy is sponsoring a program in which a panel of experts is going through the design process with 30 different design teams, critiquing their plans and offering suggestions. The U.S. government is also providing leadership by example: the Tennessee Valley Authority and the U.S. Solar Energy Research Institute will both soon be housed in passive solar complexes. Such projects will develop a reservoir of experience among private architects and should hasten the day when climate-sensitive design is a standard feature of commercial buildings. This is encouraging, since recent estimates indicate that fuel savings of more than 50 per cent can be achieved on cost-competitive commercial structures, a level of performance that few designers have achieved yet. Though the savings possible in commercial buildings are not quite as great as those in store for residential structures, they are enough to have a substantial impact on world energy needs in the decades ahead.

THE PROBLEM OF EXISTING BUILDINGS

One of the most difficult challenges ahead is improvement of the energy efficiency of existing structures. The majority of the buildings that will be in use in the year 2000 are already standing, and unless their gluttonous fuel appetites are curtailed they will continue to be an unhealthy drain on many countries' fuel supplies well into the next century. For many people, particularly low-income individuals without the financial resources to buy new homes, this is a vital issue. Unfortunately, the need to retrofit existing buildings with solar and conservation measures has received relatively little attention until recently.

Only 1 per cent of most nations' buildings are torn down in a given year, and annual construction accounts on average for 2 to 3 per cent of the building stock. So even if all homes and commercial structures built between now and the year 2000 were solar buildings, they would constitute less than one third of the total at the turn of the century. Relying on this process for a complete transformation of the building stock would take several additional decades. One possible strategy would be to accelerate the replacement of old, inefficient buildings, but the cost would be staggering. It might also be counterproductive. Construction is an energy-intensive process, and it would be difficult to recoup the energy resources squandered in tearing down usable buildings.

Making existing buildings more energy-efficient is thus the logical alternative. Though it is a more difficult, costly, and institutionally complex process than starting from scratch at the design stage, potential fuel savings from such a program in most countries are greater than for the most ambitious new construction programs. This is particularly true in parts of Europe where very few new buildings will be started over the next couple of decades.

Conservation measures that have been taken in millions of residences in Europe and North America since 1973 have commonly resulted in a reduction of fuel bills by 10 per cent or more. These programs usually include weather stripping, storm windows, insulation of the walls if there is a suitable space, and additional insulation of the attic. Savings can be much greater, however. Researchers in West Germany believe that 30 to 50 per cent savings could easily be realized in their residences using conventional conservation measures; in Switzerland, where existing buildings are already quite efficient, 30 per cent savings are expected. In practice, only a few homeowners have been able to reduce fuel needs that much. The success of such programs varies considerably and depends on the skill with which the job is done, as well as on the type and condition of the building.

Potential savings, however, are even greater than those realized in the most thorough conventional program. An interdisciplinary group of scientists at the Princeton Center for Energy and Environmental Studies conducted extensive experiments during the late 1970s using 30 similar townhouses in Twin Rivers, New Jersey. They found that fuel bills could be reduced by a full two thirds through the use of a more comprehensive but still cost-effective program. Storm doors were added, new and more efficient storm windows were developed, and some windows were fitted with insulating shutters to be closed at night. In addition, a tracer test and infrared scan were used to locate remaining sources of air infiltration, which were then caulked and weather-stripped.

According to the Princeton investigators, the key to the success of their program was the thorough evaluation they conducted beforehand of each building's performance. Remedying the thermal deficiencies of an existing house is a complex process in which all of the various problems must be addressed at once. The Princeton scientists found, for instance, that in many houses leaks in the attic were causing warm air to bypass the insulation, which helps explain why conventional insulation retrofits sometimes result in little improvement. Effective retrofit programs will require trained personnel who can evaluate a building's performance and then specify a range of cost-effective measures.

Some builders and homeowners have taken the additional step of converting conventional buildings into passive solar ones. Such a process is actually quite feasible with certain types of buildings that happen to be properly sited and landscaped. The most frequent sort of passive solar retrofit is a solar greenhouse that can be attached to the southern side of a building without the replacing of existing walls. It often makes sense to place vents in the walls and add a fan to circulate the captured heat. In the United States, solar greenhouses have become one of the most popular forms of home improvement; thousands were constructed during the late 1970s, and there is every sign that the interest of homeowners will continue. A number of firms now market prefabricated solar greenhouses, making it possible to "solarize" a house for $2000 to $3000.

In the United States, the problem of dilapidated, inefficient residences of the poor has reached crisis proportions in recent years. Many families in the cold Northeast now spend half their income on heating oil. A national weatherization program for low-income households was included in the 1976 Energy Conservation and Production Act, and the Department of Energy provides approximately $200 million each year to state governments for individual weatherization programs. These efforts have been only partially successful, however. Many of the state programs started quite slowly, and finding sufficient workers and training them has been a problem. In addition, there are serious doubts about whether the funds allocated are sufficient, considering the magnitude of the job ahead. By the end of 1979, only 650,000 low-income houses had been weatherized—less than 5 per cent of the 14 million homes that were eligible. It is estimated that no more than 3 per cent of eligible houses can be retrofitted annually, so that it will take over 30 years to weatherize the entire low-income housing stock. And even those that have been reached are receiving only partial retrofits.

Another serious gap that is slowing the retrofit process is an informational one. Even wealthy homeowners and commercial operators have great difficulty finding someone to provide a comprehensive assessment of the energy performance of their buildings and to recommend improvements. To fill this chasm, Princeton physicist and energy expert Robert Williams suggests a new profession of "house doctor" be created—"one who can quickly identify the important thermal attributes of a building and who is thoroughly familiar with effective retrofits on

most types of housing in the region." The procedure would be intelligence-intensive rather than materials- or labor-intensive. Strict educational and certification standards would of course be necessary to ensure the success of such a program, partly because consumer confidence would be essential. Williams suggests that the house doctor be equipped to make minor on-the-spot retrofits that could save 15 to 20 per cent on space heating and that would serve as additional incentive to have the energy audit done. In most cases, the simple retrofits would involve minor furnace adjustments and the use of a little tape or insulation to plug air leaks. More extensive improvements, such as adding insulating shutters or even a solar greenhouse, would be left to the consumer's discretion after detailed information on cost and the expected payback period had been supplied.

The new Residential Conservation Service (RCS), to be run by the U.S. Department of Energy, will direct the states to develop programs under which utility companies will conduct energy audits and arrange the financing and installation of conservation and solar retrofit measures. Comprehensive training manuals are being prepared, and the energy auditors will have to meet state qualification standards. The Residential Conservation Service has the potential to develop into one of the nation's most important new energy initiatives. Its planners anticipate that the audits could approach the effectiveness of the house doctor concept and that as many as 20 million residences will be audited over the next five years.

BUILDING TO SAVE ENERGY

The introduction of energy standards for new buildings is one of the most common ways national and local governments have tried to influence energy use since 1973. The new laws usually specify certain construction practices, such as insulation levels, and are added to existing building codes. Since people buying a home lack the expertise and equipment to assess the energy performance of a building, many governments believe standards are necessary to assure home buyers they will not later be confronted with bankrupting fuel bills. Among the countries with new building energy standards are France, which adopted them in 1974, and Sweden and West Germany, which followed in 1977. Typical of the new codes is the German ordinance, which includes requirements for insulation and also specifies an acceptable level of heat loss through leaks in the building. In the United States, building codes have traditionally been the responsibility of state governments, and nearly all state codes now include energy-related criteria.

The new standards have generally been quite effective in the United States, the average insulation level in the ceilings of new houses rose 20 per cent between 1974 and 1976. During that same period, the proportion of new houses in the cold northeastern states with double- or triple-glazed windows rose from 72 per cent to 94 per cent. Such measures have helped slow the rate of growth of energy use in buildings from 5 per cent down to 1 or 2 per cent in many countries. In some ways the building codes are like the fuel-efficiency standards for automobiles enacted by the U.S. Congress in 1975. The new regulations are encouraging builders to gear up for a change in construction practices just as consumers are beginning to press for it themselves. By requiring that certain procedures be made standard practice, building energy codes motivate the various sectors of the industry to work together in the development of appropriate materials and components and in their adaptation for particular buildings.

The most rigorous standards on the books so far are those in Sweden and California; each results in houses with heating requirements that are more than one third below those of the average American home (Table 21-3). But even the most demanding of these energy standards does not come close to challenging the efficiency that can be achieved in many passive solar and low-

Table 21–3. Annual Home Heating Costs According to Different Building Standards.

Structure or Standard	Annual Cost[a] (dollars)
U.S. average house, 1978	680
French building code, 1974	500
U.S. building standards, 1978	360
Swedish building code, 1977	230
California building code, 1979	220
Townhouse with retrofit, Twin Rivers, New Jersey	95
Saskatchewan Conservation House	20
Village House I, passive solar	15

[a] Assumes similarly sized houses using oil heat in a similar climate.
Source: A. H. Rosenfeld et al., Building Energy Use Compilation and Analysis and an International Compilation and Critical Review. Berkeley, Calif.: Lawrence Berkeley Laboratories, July 1979.

energy buildings. The Saskatchewan Conservation House and Village House I, a passive solar house in New Mexico, exemplify the innovative designs that can shrink heating and cooling requirements to near zero. In fact, many architects and builders consider even the most stringent current standards to be merely a primitive base level that can be exceeded by 50 per cent or more.

Financial incentives are probably the measures most likely to promote rapid progress in energy-efficient building. As construction costs and interest rates have risen in recent years, the pressure to cut building costs to bare-bones levels has been overwhelming. Many of the possible design innovations and technical fixes that can be applied to buildings require a slightly higher initial investment than for a conventional building. Though the changes may be extremely cost-effective, with a payback period of only a few years, many builders are reluctant to do anything that will raise the base price. And both builders and owners sometimes have trouble getting a large enough loan to pay the extra cost. They are not helped by bankers and lending agencies who are unfamiliar with climate-sensitive design and are worried about the soundness of lending money for such a project.

A major effort to educate the financial community about the common sense and cost-effectiveness of energy-saving buildings would be a significant contribution. In calculating a homeowner's ability to meet mortgage payments, loan officers should be aware of the negligible fuel bills of a passive solar building, which leaves more income available to repay a loan. Though very few banks have begun to think seriously about implementing such criteria, there are now a few exceptions. The San Diego Savings and Loan Association in California is one of several U.S. banks that offer slightly reduced interest rates on passive solar houses. This program brings monthly payments below what they would be for a conventional home, adding to the homeowner's savings from reduced fuel costs.

Tax breaks are also becoming available for energy-saving homes and for retrofitting. More than 20 states in the United States now offer income tax credits for some solar improvements, and property tax exemptions are available in more than 30. So far, most of these tax incentives are for solar collectors only and do not include passive measures that serve multiple functions and are more difficult to define on tax forms. Conservation improvements such as storm windows and weather stripping are also eligible for tax credits in most states and at the federal level, but, again, passive solar is generally excluded. This is a serious gap, which places some of the most financially sound solar and conservation measures at a distinct disadvantage.

Educational programs for consumers, builders, real estate agents, and others can also speed up the transition to climate-conscious buildings. The level of awareness concerning energy use in buildings has risen rapidly, but many people still lack the specific information needed to make wise decisions. Trade associations, community groups, and local governments could sponsor seminars and prepare leaflets with the necessary information. Serious consideration might also be given to requiring the labeling of the fuel requirements of buildings at the time they are sold. Expected fuel use and price could be noted as well as the likely life-cycle cost of the building. This would give buyers a chance to compare the efficiency of buildings, an important factor since some of the climate-sensitive structures are hard to distinguish from conventional ones and the buyer can have a hard time judging competing claims. The fuel bills of solar homes are already an important real estate document in some parts of the United States, as they provide evidence of the energy efficiency of houses.

Because the rapid introduction of energy-efficient design will hinge on cooperation between government and industry, the support of builders is crucial. The building industry has traditionally been skeptical of government programs, but bureaucrats are finding that financial incentives and even technical support and demonstration activities are surprisingly popular. If over the next several years such programs can persuade the majority of builders to at least consider seriously the possibility of building climate-sensitive structures, a major milestone will have been passed. The attractiveness of these buildings should ensure that they will sell themselves after that. It may be that the entire package of government programs—including financial incentives and demonstration projects—can be phased out after only a decade, with the job completed.

The politics behind reducing the energy requirements of buildings are convoluted, yet they are the key to giving the world's buildings a needed face-lift. Builders are in favor of receiving passive solar tax credits but are against new-building energy standards. Consumer groups approve of standards but are opposed to government decontrol of fuel prices. Slowly some areas of consensus are being found, but a lot of careful politicking—and the development of programs with carrots as well as sticks—will be essential if the process is to continue.

These important political efforts could be usefully complemented by slight changes in the life-styles of millions of individuals who live and work in buildings and who regularly make decisions on energy use that range from whether to add a greenhouse to their home to where to set the thermostat. Recent studies have shown that the fuel requirements of a building can vary by as much as 75 per cent depending on the practices of the occupants. A warm sweater is at least as good as investment as storm windows are, and the savings possible through good design can be multiplied by taking sensible steps that do not affect comfort. Simple, unexciting measures such as opening a window on a mildly hot summer day instead of turning on the air conditioner, or switching lights on only when necessary, can yield attractive financial returns. Heightened awareness of seasonal changes and the behavioral shifts that should accompany them could be an important part of the transition to an energy-efficient society.

As support for climate-sensitive design continues to build, it is becoming clear that the new era of more-efficient buildings could have an enormous and beneficial impact on the world energy situation. In the United States, where climate-sensitive buildings will soon be a major factor in the real estate market, programs sponsored by the Department of Energy alone are projected to result in more than 50,000 new passive solar buildings in 1981. This is more than double the number of solar buildings in the country in 1980. The department's goal is to have a half million climate-sensitive buildings standing in the United States by 1986, and there is good reason to think that this is a conservative estimate. With the new Solar Energy and Conservation Bank, the promise of passive solar tax credits, and the recent surge in interest by builders and consumers, the United States may reach that target a year or two ahead of schedule.

A not unreasonable goal for the United States would be to have five million climate-sensitive buildings in place by 1990 and to have such structures dominate the building market during the 1990s. The result, in the year 2000, would be that 10 per cent less energy would be used to heat, cool, and light buildings than if current trends continue. Meanwhile, the fuel requirements of existing buildings could easily be reduced by one third through successful retrofit programs. These combined efforts would yield the equivalent of five million barrels of oil a day by the end of the century, an amount that comfortably exceeds the quantity of fuel to be produced by the national synthetic fuels program under the most optimistic scenario. Moreover, using buildings as an energy source would "produce" fuel at a lower cost and create more than twice as many jobs.

The potential clearly exists to cure most of the energy ills of the world's buildings by the end of the century. Economically and ethically, the decision to begin moving down that path should be one of the easiest energy choices we have to make. Virtually everyone will benefit from a new era of more rational design and construction. And buildings themselves would in a sense be better off. Climate-sensitive structures work with nature rather than against it. As one solar designer recently observed, "Our buildings would be more beautiful if they responded to energy concerns and had a more natural configuration." A more varied and more humane environment could be the result.

SUGGESTED READINGS

Bruce Anderson, *The Solar Home Book: Heating, Cooling, and Designing with the Sun.* Andover, Mass.: Brick House Publishing, 1976.

Bruce Anderson, *Solar Energy: Fundamentals in Building Design.* New York: McGraw-Hill, 1977.

David Bainbridge, Judy Corbett, and John Hofacre, *Village Homes' Solar House Designs.* Emmaus, Pa.: Rodale Press, 1979.

J. Douglas Balcomb, *Energy Conservation and Passive Solar: Working Together.* Los Alamos, N.M.: Los Alamos Scientific Laboratory, 1978.

Robert W. Besant, "Operational Saskatchewan Solar-Conservation House Yields Further Data on Energy Efficient Building Designs," *Soft Energy Notes,* May 1979.

Robert W. Besant, Robert S. Dumont, and Greg Schoenau, "Saskatchewan House: 100 Percent Solar in a Severe Climate," *Solar Age,* May 1979.

Ken Butti and John Perlin, *A Golden Thread: 2000 Years of Solar Architecture and Technology.* New York: Van Nostrand Reinhold, 1980.

"California Orders Its Utilities To 'Unsell' Energy," *Business Week,* May 26, 1980.

California Public Utilities Commission, *Financing the*

Solar Transition, A Report to the California Legislature. Sacramento: CPUC, January 1980.

Peter F. Chapman, "The Economics of UK Solar Energy Schemes," *Energy Policy,* December 1977.

Committee on Nuclear and Alternative Energy Systems, *Alternative Energy Demand Futures to the Year 2000.* Washington, D.C.: National Academy of Sciences, 1979.

Barry Commoner, "The Solar Transition," *The New Yorker,* April 23 and April 30, 1979.

Darian Diachok, and Dianne Shanks, *International Passive Architectural Survey.* Springfield, Va.: National Technical Information Service.

Harrison Fraker Jr., and William L. Glennie, "A Computer Simulated Performance and Capital Cost Comparison of Active vs. Passive Solar Heating Systems," *Proceedings of the Passive Solar Heating and Cooling Conference.* National Science Centre, Box 52, Parkville, Victoria 3052, Australia.

General Public Utilities Corporation, *Conservation and Load Management Master Plan.* Philadelphia: GPUC, March 1980.

Harold R. Hay, "Roof Mass and Comfort" and "Skytherm Natural Air Conditioning for a Texas Factory," *Proceedings of the 2nd National Passive Solar Conference.* National Science Centre, Box 52, Parkville, Victoria 3052, Australia.

Dennis Hayes, "The Solar Prospect," *Worldwatch,* Paper 11, March 1977.

Dennis Hayes, "The Solar Time Table," *Worldwatch,* Paper 19, April 1978.

Eric Hirst, and Bruce Hannon, "Effects of Energy Conservation in Residential and Commercial Buildings," *Science* **205,** (4407), 656 (1979).

Eric Hirst and Jerry Jackson, "Historical Patterns of Residential and Commercial Energy Uses," *Energy* **2,** 131 (1977).

William W. Hogan, "Dimensions of Energy Demand," in H. H. Landsberg, ed., *Selected Studies on Energy, Background Papers for Energy: The Next Twenty Years.* Cambridge, Mass.: Ballinger, 1980.

Henry Kendall and Steven J. Nadis, eds., *Energy Strategies: Towards a Solar Future.* Cambridge, Mass.: Ballinger, 1980.

Los Alamos Scientific Laboratory, *Passive Solar Buildings.* Springfield, Va.: National Technical Information Service, July 1979.

Edward Mazria, *The Passive Solar Energy Book.* Emmaus, Pa.: Rodale Press, 1979.

Robert C. Naumann, "A Builder's Experience with Two Passive Solar Houses in Boulder, Colorado," *Proceedings of the 4th National Passive Solar Conference.* National Science Centre, Box 52, Parkville, Victoria 3052, Australia.

OECD, *Energy Balances of OECD Countries.*

Victor Olgyay, *Design with Climate: A Bioclimatic Approach to Architectural Regionalism.* Princeton, N.J.: Princeton University Press, 1963.

Amos Rapaport, *House Form and Culture.* Englewood Cliffs, N.J.: Prentice-Hall, 1969.

Fred Roberts, "The Scope for Energy Conservation in the EEC," *Energy Policy,* June 1979.

"Roof Ponds Can Work Anywhere," *Solar Energy Intelligence Report,* June 9, 1980.

A. H. Rosenfeld et al., *Building Energy Use Compilation and Analysis (BECA) and an International Compilation and Critical Review.* Berkeley, Calif.: Lawrence Berkeley Laboratories.

Rosalie T. Ruegg et al., *Life-Cycle Costing: A Guide for Selecting Energy Conservation Projects for Public Buildings.* Washington, D.C.: GPO, 1978.

Larry Sherwood, "Passive Solar Systems . . . The Economic Advantages," *Proceedings of the Solar Energy Symposia.* Denver: International Solar Energy Society-American Section, 1978.

William A. Shurcliff, *Superinsulated Houses and Double-Envelope Houses.* Cambridge, Mass.: William A. Shurcliff, 1980.

Robert H. Socolow, ed., *Saving Energy in the Home: Princeton's Experiments at Twin Rivers.* Cambridge, Mass.: Ballinger, 1978.

Solar Energy Research Institute, "Energy Use To Plummet over 20 Years, Solar To Provide 20–33%," *Solar Energy Intelligence Report,* September 15, 1980.

James W. Taul Jr., Carol F. Moncrief, and Marcia L. Bohannon, "The Economic Feasibility of Passive Solar Space Heating Systems," *Proceedings of the Third National Passive Solar Conference.* National Science Centre, Box 52, Parkville, Victoria 3052, Australia.

Tennessee Valley Authority, *Program Summary, Division of Energy Conservation and Rates.* Knoxville, Tenn.: TVA, April 1980.

Mark A. Thayer and Scott A. Noll, "Solar Economic Analysis: An Alternative Approach," *Proceedings of the Third National Passive Solar Conference.* National Science Centre, Box 52, Parkville, Victoria 3052, Australia.

Underground Space Center, University of Minnesota, *Earth Sheltered Housing.* New York: Van Nostrand Reinhold, 1979.

U.S. Department of Energy, *Annual Report to Congress 1979.* Washington, D.C.: GPO, 1979.

U.S. Department of Labor, *Consumer Prices: Energy.* Washington, D.C.: GPO, 1980.

"Utility and Power Picture: Northwestern States Develop Energy Conservation Programs," *Building Energy Progress,* January/February 1980.

Robert J. Wagner, "A Builder's Guide to Solar Financing," *Solar Age,* January 1979.

Robert L. Wagner, "Interest Reduced on Solar Loans," *Solar Law Reporter,* May/June 1980.

Robert H. Williams and Marc H. Ross, "Drilling for Oil and Gas in Our Houses," *Technology Review,* March/April 1980.

Solar Heating and Cooling   22

JOHN A. DUFFIE AND WILLIAM A. BECKMAN

INTRODUCTION

The years since the preparation of this article in 1976 have seen many developments, such as the enactment of tax incentive legislation at the federal and state levels, the development of new manufacturing enterprises, and the growth of a solar heating industry to a gross level of about $0.5 billion per year in the United States. The solar cooling field has not been commercialized and is still very much in an R&D stage. Most of the important ideas in the 1976 article hold today as they did then, and so the article is kept intact. We have added to it this introduction and a few of the more important recent general references in the solar heating and cooling field.

With the development of the "active" solar energy systems noted in this article, there has been in the past five years a strong increase in interest in "passive" solar heating and cooling. Passive systems have collectors and storage units that are integral parts of the building structure and may require less (or no) mechanical energy to operate them. These developments are sufficiently important that a separate article (Article 21) deals with them.

Thermal energy for buildings, supplied at temperatures near or below 100°C, constitutes an imporant segment of the U.S. energy economy and accounts for about one quarter of the nation's energy use. Energy at these temperatures can readily be delivered from flat-plate solar energy collectors, and the solar energy incident on most buildings is more than adequate to meet these energy needs. Flat-plate collectors are manufactured and sold on a small but growing scale in the United States; they have been in use for more than a decade in heating water for buildings in Australia, Israel, and Japan. We expect that solar heating and cooling for buildings, with energy collected by flat-plate collectors, will be the first large-scale application of solar energy.

Dr. John A. Duffie is Professor of Chemical Engineering and Dr. William A. Beckman is Professor of Mechanical Engineering at the University of Wisconsin, Madison, Wisconsin 53706. They are Directors of the Solar Energy Laboratory at the University of Wisconsin-Madison.

From *Science* **191** (4223), 143 (1976). Reprinted by permission of the authors and the American Association for the Advancement of Science. Copyright © 1976 by the American Association for the Advancement of Science.

The basic problem with solar heating and cooling has been that the energy could not, except in special cases, be delivered at costs competitive with costs of energy from other sources. This situation is rapidly changing, and interest in solar energy is increasing almost daily as fuel costs rise. In areas where new natural gas connections are no longer available, where oil is not distributed, and where electrical resistance heating is the only alternative among conventional sources, solar heating is economically attractive.

In addition to technical and economic considerations, several social factors will influence the course and pace of developments. Two examples are worth examining. First, architectural constraints are imposed by the need for collectors to be oriented within rather narrow limits. This will make it difficult to fit solar heating systems to many existing buildings; thus new residential construction will be the easiest starting place for conversion in solar heating. Solar cooling may first be installed in existing low-rise, flat-roof buildings, such as schools and shopping centers, where cooling is usually more important than heating. Second, tax policy is important. Today the installation of solar heating or cooling systems brings an increase in property valuation in most states, and a corresponding modest increase in real estate taxes. Government encouragement to invest in solar energy systems in the form of tax write-offs or other inducements (as are provided for investments by other energy producers) could very rapidly change the competitive position of solar energy in relation to conventional energy sources.

In all buildings, intelligent practices for energy conservation are worth following. The basic advantages of reducing energy needs by good thermal design apply whether buildings are supplied with solar or conventional energy. If solar energy costs the same as an alternative energy source, the value of energy conservation techniques, such as extra glazing on windows and doors or added insulation, is the same whether solar energy or the alternative is being used.

SOLAR RADIATION

The solar constant, that is, the intensity of solar radiation outside of the earth's atmosphere at the mean distance between the earth and the sun, has been determined by measurements from satellites and high-altitude aircraft to be 1.353 kilowatts per square meter.[1] This extraterrestrial radiation, which corresponds closely to that of a black body at 5762 K, is 7 per cent in the ultraviolet range (wavelength less than 0.38 μm) and 47 per cent in the visible range (wavelengths from 0.38 to 0.78 μm), with the balance in the near infrared (largely with wavelengths of less than 3 μm).

Solar radiation is depleted as it passes through the atmosphere by a combination of scattering and absorption; the radiation that reaches the ground—the raw material of this energy source—can vary from almost none under heavy cloud cover to 85 to 90 per cent of the solar constant under very clear skies. Energy rates on surfaces normal to the radiation during good weather are not very high and are typically about 1 kilowatt per square meter (a little more than 1 horsepower per square yard). Solar radiation on the ground consists of a diffuse component that has been scattered by molecules and particulate matter in the atmosphere and, when the atmosphere is sufficiently clear, a beam component that is unchanged in its direction of propagation from the sun. Its spectral distribution is altered in a manner dependent on atmospheric composition, with the major changes due to absorption of ultraviolet radiation by ozone and infrared radiation by water vapor.

There are several sources of solar radiation data. The National Oceanic and Atmospheric Administration weather service measures total (beam diffuse) radiation on a horizontal plane at more than 100 stations. Some stations report daily values, and some report hourly values. These data are available from the National Climatic Data Center.[2] Monthly averages of daily radiation on horizontal surfaces are available for many locations.[3] Daily integrated energy quantities at particular locations vary widely during the year. In Madison, Wisconsin, on a clear January day, energy on a horizontal surface is typically 3 kilowatt-hours per square meter, and July clear-day energy is typically 9 kilowatt-hours per square meter; the corresponding monthly averages of daily radiation on the horizontal surface are 1.8 kilowatt-hours per square meter and 6.2 kilowatt-hours per square meter. Flat-plate collectors sloped toward the south in Madison, with a slope equal to the latitude, will have incident on them an average daily radiation of 3.4 kilowatt-hours per square meter in January and 5.6 kilowatt-hours per square meter in July. These data illustrate the gains to be obtained by orientation of a collector in a favorable manner.

Although solar energy intensity is low, integrated energy quantities may be large. For example, in Madison the annual average solar energy incident per day on an acre of ground is the equivalent of about ten barrels of oil, and on a 200-square-meter house is equivalent to about 25 gallons, which is far more than enough to meet the needs of the building for thermal energy.

CURRENT STATUS

Two major reasons may be cited for the failure of solar energy to be a serious competitor in the energy market in past years. First, the costs of delivering solar energy have been substantially higher than those of other energy sources. Solar energy has not been able to be a competitor to inexpensive natural gas or petroleum. Second, there was no constituency pressuring for solar energy development in a manner similar, for example, to that of the nuclear industry that existed at the close of World War II and gave a substantial impetus to the development of peacetime uses of nuclear energy. The environmental movement of the last five years, the realization that the United States is dependent to an undesirable degree on foreign energy sources, and increasing fuel costs have served to establish

a broad base of interest in the development of solar energy.

The contrasts between the development of nuclear energy and solar energy are striking. After the destruction wrought by the atomic bombs in World War II, a large, concerned constituency, backed by the nation at large, pushed for development of peaceful uses of atomic energy. The result was a program supported by billions of dollars of federal funds over the course of three decades. Solar energy, in contrast, had no such support, and it was only the persistence of a few individuals that kept interest in solar energy alive. Outstanding in this group was Farrington Daniels, who, through his publications[4] as well as through his support of the struggling International Solar Energy Society, served as an elder statesman for solar energy.

During the 1960s, support for solar energy research for applications in the United States was essentially nonexistent. However, one program resulted in economic studies that have become part of the current interest in solar energy. Tybout and Löf, with support from Resources for the Future, developed a series of cost analyses of solar energy for heating and cooling.[5] They indicated that solar heating could be competitive with conventional energy sources in high-energy-cost areas in 1968. They also showed that the combination of solar heating and cooling, which results in higher use factors on the solar energy equipment, is, in most places, more economical than heating or cooling alone. Their two studies were based on optimistic projections of the cost of solar energy equipment ($20 and $40 per square meter of flat-plate collector), but also on 1968 and 1970 energy costs. Later and more detailed studies of cost and thermal performance, based on more realistic collector and energy costs, bear out the same general conclusion that solar heating can now compete with expensive fuels.

By 1972, several dozen solar-heated residences or small laboratory buildings had been constructed and operated. A few of these have been studied, evaluated, and reported.[6] The few air conditioning experiments were confined largely to experimental operation of 3-ton lithium bromide–water absorption machines or analytical studies of system performance.[7] In contrast, the manufacture and sale of solar water heaters to provide hot water for residences and some institutional buildings (hotels, dormitories, and the like) has been a commercial enterprise in Australia, Japan, and Israel for more than a decade. Perhaps a million solar water heaters are in use in these countries.

During the past three years, the availability of funds for experimental programs from the NSF Research Applied to National Needs (RANN) program and the Energy Research and Development Administration (ERDA now DOE), coupled with private and industrial investment, has led to many new experiments and applications of solar heating and cooling in the United States. Public buildings, schools, and a variety of residential buildings with heating or combined heating and air conditioning capacity are being planned and built. Quantitative information is now beginning to come in from these new experiments.[8] The Solar Heating and Cooling Demonstration Act of 1974 should lead to many new solar buildings. In addition to research and development activities, there are now commercial sales of solar heating systems that are installed as operating heat delivery systems rather than as experiments.

SOLAR BUILDING ARCHITECTURE

Several approaches to solar building architecture are evident. A solar-heated house and school are illustrated at the beginning of this section. The basic problem faced by architects and engineers is to integrate collectors into or onto the building in such a way that thermal performance is adequate, while obtaining an esthetically satisfactory structure. In this context, the major variable is the area of collector that must be integrated into the building. Collector area is central to the fraction of heating loads to be carried by solar energy and, ultimately, to cost.

The solutions are mixed. Some collectors have been mounted above flat-roof buildings. To obtain structural or esthetic advantages, other collectors have been built into vertical walls in higher latitudes or placed flat on horizontal roofs in lower latitudes. In addition, collectors have been integrated into the envelopes of buildings at orientations that are near optimum for the best thermal performance.

SOLAR ENERGY SYSTEMS

Systems for producing service hot water, space heating, and cooling are based on the concept of the flat-plate collector. This unique heat exchanger uses a black absorber plate to absorb solar energy. Ducts or tubes carry air or liquid that removes energy from the plate. Layers of air provide transparent insulation between plates and their covers (usually made of glass) and thus reduce upward heat loss. Conventional insulation is provided on the backs and edges of the plates. The collectors are mounted in a fixed position according to the desired use of the energy. Figure 22-1 shows cross sections of air and water heaters.

Figure 22-1. Cross sections of a solar air heater and a solar water heater.

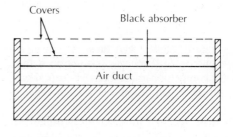

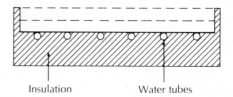

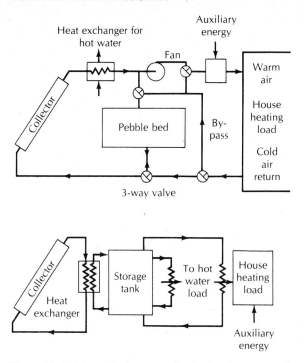

Figure 22-2. Schematic diagrams of solar heating systems based on air and liquid heat transfer media.

The other major component in the system is the energy storage unit, which is designed to accumulate solar energy when it is obtainable and make it available to meet energy needs at other times. Liquid systems usually use insulated water tanks for storage, and air systems usually use pebble beds. A third method of storage takes advantage of the latent heat of a phase transition and has been the object of considerable study.[9] Early work on house-heating applications concentrated on hydrated sodium sulfate ($NA_2SO_4 \cdot 10$ H_2O), which undergoes a phase transition when heated to 32°C. Because phase separation of this hydrate occurs on cycling, other chemical systems that can undergo thousands of cycles without loss of storage capacity are being sought.

Schematic diagrams of liquid and air solar heating systems are shown in Figure 22-2. Both show an auxiliary energy source, which is included in most solar energy systems. In climates in which a high degree of reliability is required of a heating system, the auxiliary source must be capable of carrying the full heating load of the building. If auxiliary energy is added in parallel with solar energy, then the maximum amount of energy can be obtained from the solar system and the balance from auxiliary. Other methods are possible.

For the liquid system, the heat exchanger between collector and storage tank allows the use of an antifreeze solution in the collector loop, which is one of the methods to avoid freezing and reduce boiling problems. Figure 22-2 shows an additional heat exchanger to transfer heat to the building and another to provide service hot water. The technology of solar liquid heaters is very well established, and most of the systems built recently have used liquids for heat transfer.

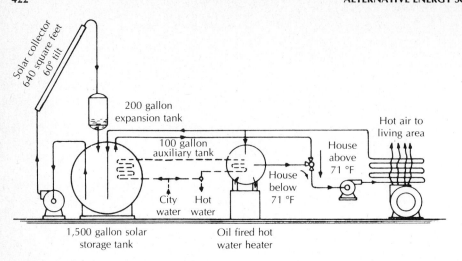

Figure 22-3. Schematic diagram of the heating system in MIT house IV. (Adapted from Engebretson, Reference 6)

Air systems avoid boiling and freezing problems in the collector. In most air systems energy is stored as sensible heat in a pebble bed. A well-designed pebble bed has good heat transfer between air and pebbles, a low loss rate, and a high degree of stratification. Mechanical energy for pumping air can be a significant item of cost, and care is required to design for minimum pressure drops. The design of air systems and the balancing of good heat transfer characteristics against pressure drop are problems that are now receiving adequate attention.

Many of the scores of solar-heated buildings that have been constructed so far have provided reduced fuel bills as well as satisfaction to their owners. The performance of a few of these has been carefully measured and provides a firm base of data on long-term thermal performance. Experiments up to 1961 were very well reported in papers presented at the U.N. Conference on New Sources of Energy in 1961 and summarized by Löf.[6]

Massachusetts Institute of Technology (MIT) house IV, built in 1959, was the last in a series of experiments carried out by H. C. Hottel and his colleagues and represented a cooperative effort of architects and engineers to develop a functional, energy-conserving home with a major part of the energy for space heating and water heating to be supplied from the flat-plate collector. Figure 22-3 from Engebretson[6] is a schematic diagram of the heating and hot water system. The collector had an area of 60 square meters for the 135-square-meter floor area, two glass covers, and a flat black paint, energy-absorbing surface. To avoid freezing, collectors were designed to drain into an expansion tank. The main storage tank capacity was 5700 kilograms. Means were provided for adding auxiliary energy, extracting hot water for household needs, and transferring heat to air that was circulated to the rooms. This solar heating system was operated for three seasons, during which its performance was carefully measured. Data for the first two years are summarized in Table 22-1, which shows how energy requirements for space heating and water heating were met by solar or auxiliary energy. During the first two heating seasons, solar energy supplied 52 per cent of the energy for hot water and heating.

The Denver solar house, built by Löf[6] in 1958, uses air as the heat transfer medium and a pebble-bed storage unit. The ratio of

SOLAR HEATING AND COOLING

Table 22–1. Performance Data for MIT House
IV for Two Heating Seasons (Summarized from
Engebretson, Reference 6).

Item	1959 to 1960 (gigajoules)	1960 to 1961 (gigajoules)
Space heating		
Demand	72.5	70.7
From solar energy	33.6	40.2
Water energy		
Demand	14.7	17.6
From solar energy	8.4	9.7
Total heating		
Demand	87.1	88.3
From solar energy	41.9	49.9
Per cent from solar energy	48.1	56.6

collector area to house area is about 1 to 5, a
proportion much smaller than that of MIT
house IV. This house has served as the Löf
family residence since its construction, and
the equipment has been routinely operated
with only nominal maintenance. The system
performance was measured in 1959 to 1960,
and again in 1974 to 1975. For the period
from December to April, 22 per cent of the
heating and hot water loads were carried by
solar energy during the earlier season and 20
per cent during the later season.

Solar air conditioning technology is not as

advanced as the heating process, since an additional thermodynamic process is needed for
cooling. Several current experiments use absorption cooling cycles that are operated by
heat from flat-plate collectors. These coolers
are the analogs of the gas-fired refrigerators
used in campers, but because of the lower
temperature of fluid from the collectors
(compared to a gas flame), water cooling is
required rather than air cooling. Figure 22-4
shows a diagram of a solar-operated absorption cooling system. The same collector and
storage units that provide winter heating thus
can provide summer cooling.

Colorado State University (CSU) house I
(which serves as an office building) uses a
heating system that differs in some details
from that of the MIT house and also includes
an absorption air conditioner. A glycol-water
solution is used in the collector to avoid
freezing problems and permit collector operation at higher temperatures. A heat exchanger is used to transfer solar heat into the
water storage tank, and additional heat exchangers serve for heat transfer from the tank
to hot water and the building. Thus, the collector supplies energy for three purposes:
space heating, water heating, and air conditioning. A gas furnace provides auxiliary energy for both heating and cooling. The experiments started in August 1974,[10] and for
the first six months of the heating season 86

Figure 22-4. Schematic diagram of a solar-operated absorption air conditioner. AX, auxiliary energy
source. The cooler components are as follows: G, generator; C, condenser; E, evaporator; A, absorber;
HE, heat exchanger to recover sensible heat. (From Duffie and Beckman[18])

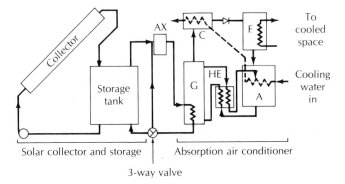

per cent of the space heating loads and 68 per cent of the hot water loads were met from solar energy. Integrated performance statistics of a summer's air conditioning operation are not yet available.

A mobile laboratory [11] developed by Honeywell under NSF-ERDA sponsorship includes a heating and absorption air conditioning system similar to that of CSU house I. In addition, solar heat can be used to vaporize a fluorohydrocarbon, which then expands through a turbine to drive a mechanical air conditioner. Thus, solar energy is converted to mechanical energy, which is used to provide cooling by conventional means. The mobile laboratory is being operated in several locations to gather data and provide a public demonstration.

In addition to closed-cycle absorption cooling, open cycles are of potential interest for solar technology. For example, desiccants can be used to absorb water vapor from room air, which can then be evaporatively cooled; the desiccant is regenerated and recycled. Löf suggested the use of triethylene glycol as the desiccant, with solar-heated air for regeneration [12]; this system is now being evaluated for use on the Citicorp building in New York. Dunkle has designed a cycle with rotary beds of silica gel and rotary heat exchangers. [13] In the Munters (M.E.C.) system lithium chloride is used as the desiccant; the system is being adapted for solar operation. [14]

It is also possible to use a collector as an energy dissipater by designing it to lose heat by convection and by radiation to clear night sky. To accomplish this, the collector must have properties opposite those needed for efficient collection; thus compromises are necessary or movable insulation must be used. Hay designed such a system for a clear, mild California climate. He achieves combined collector-radiator-storage capabilities in the horizontal roof of the building with movable insulation and thereby provides heating in the winter and cooling in the summer. This system was evaluated for one year [15] and kept conditions inside the house within acceptable ranges throughout the period.

Another class of systems combines solar collectors and heat pumps. The heat pump can serve as an independent (auxiliary) source of heating energy, or the collector-storage system can serve as the energy source for the evaporator of the heat pump. The latter system has the apparent advantages of lowering mean collector temperature and raising the mean evaporator temperature of the heat pump (thus improving the performance of each). Systems of this type have been studied experimentally. [16] A simulation study by Freeman compares these methods in one climate and indicates little choice between them, [17] but there remain many unanswered questions on how these combined systems should best be constructed and operated.

PERFORMANCE CALCULATIONS

The general approach to calculating the thermal performance of solar energy systems is to write the equations that describe the performance of each of the components in a system (including collector, storage, controls, pumps, and the like, as well as the building itself) and simultaneously solve the equations, usually hour by hour. Meteorological data for the location in question, which affect both collector and building heating and cooling loads, are used as forcing functions. The solutions are time-dependent temperatures and energy rates. The energy rates can be integrated to give energy quantities over the period of the stimulation. The amount of energy a system is expected to deliver over a year can then be the basis for an economic analysis. These procedures are outlined by Duffie and Beckman. [18]

The most critical and unique component is the collector. Thanks to pioneering studies of collectors by Hottel and his colleagues, begun almost 40 years ago and carried on by others since, [19] methods of predicting collector performance are well established. Based on a detailed analysis, the useful gain of most collectors can be written

$$Q_U = A_c F_R [S - U_L (T_{f,in} - T_a)]^+$$
$$= \dot{m} C_p (T_{f,out} - T_{f,in})$$

where F_R is equal to the ratio of actual energy gain to gain if the whole plate were at the fluid inlet temperature, $T_{f,in}$, and accounts for the material properties and configuration of the plate. This collector heat removal factor takes into account fluid flow rate and temperature gradients along and across the plate and enables the calculation to be made on the basis of $T_{f,in}$ (a very convenient variable). Also, A_c is the collector area, a major design parameter; S equals the absorbed radiation per unit area of collector. It is the product of incident radiation on the plane of the collector, the transmittance of a cover system, and the absorptance of the plate for solar radiation. It is a function of the orientation of the collector, the number of covers, and the properties of the covers and plate for solar radiation. The thermal loss coefficient, U_L, is a function of the number of covers, cover and plate properties for longwave (thermal) radiation, wind speed, and temperatures. Correlations and charts are available to determine this coefficient.[18,19,20] Finally, $T_{f,out}$ is the outlet fluid temperature; T_a is ambient air temperature; and $\dot{m}C_p$ is the product of fluid mass flow rate and heat capacity. The plus sign on the bracket indicates that only positive values are taken. This simulates a controller that turns on the pump or blower whenever useful energy is to be gained from the collector; that is, when the fluid outlet temperature is higher than the inlet temperature.

Included in the equation are a wide range of design parameters and materials properties. For example, the effects of selectivity of the energy-absorbing surface, that is, the absorptance of the surface for solar radiation and emittance for longwave radiation, are implicit in S and U_L.

The equation also illustrates an important determining factor in solar energy system performance. As the collector temperature ($T_{f,in}$) rises, the thermal loss term approaches the absorbed radiation term and collector output diminishes. For most practical designs today, zero output collector temperatures are typically 150 to 175 C° above ambient, and normal operating temperature

ranges are less than 75° C above ambient. So, collectors are uniquely sensitive to temperature and must be designed to operate at minimum temperatures above the levels required.

New collector developments are aimed at increasing absorbed radiation S and reducing thermal losses. Extensive efforts have gone into development of selective surfaces with low longwave emittances to reduce U_L.[21] The practical problem has been to maintain desirable combinations of properties over very extended periods (20 years or more) in oxidizing atmospheres. Many of the surfaces studied are metal substrates with semiconductor coatings, for example, chrome oxide on a bright nickel base. Another approach to control of thermal losses is to evacuate the space between the absorbing surface and the cover, thus reducing or eliminating convection and conduction across the gap. This is done by enclosing the absorbing surface in tubes[22]; elimination of convection and conduction coupled with selective surfaces of low emittances results in very low loss coefficients and allows energy delivery from collectors at substantially higher temperatures than from conventional designs.

Equations based on standard energy and materials balances, rate equations, and equilibrium relationships are available for energy storage, heat exchangers, heating and cooling loads, controls, and other components of solar energy systems. Although the models of some components can be based on physical principles, it may be necessary to fall back on empirical models of coolers, heat pumps, and other complex equipment. The combination of the models of each component provides the basis for system performance calculations.

SIMULATIONS

Physical experiments on solar heating and cooling are indispensable. However, numerical experiments, such as simulations, can yield much of the same kind of information quickly and inexpensively. The effects on long-term system performance of changes in system configuration, materials properties, and component design can readily be assessed in a way that is not practical in experiments. Simulations are also useful in understanding the dynamics of systems (which never operate at steady state) and in selecting and planning experiments. We have developed a modular solar process simulation program, TRNSYS, in our laboratory, and other simulation programs have been described.[23]

The two most obvious design variables of a solar heating system for a particular building are collector area and storage size. To see the effects of these parameters on energy delivered to a building, let us consider the following example. A house in Madison, Wisconsin, is to be provided with solar heating and hot water from a system similar to that of the CSU experiment. The house is a typical moderate-size house with a heat loss

Figure 22-5. Month-by-month performance of heating systems of two collector areas on a Wisconsin house with a floor area of 180 square meters. Incident radiation on the collector is shown by the heavy broken lines. Total heating and hot water load is indicated by the bars; the shaded portion represents the load met by solar, and the unshaded portion, the load met by auxiliary; GJ, gigajoule.

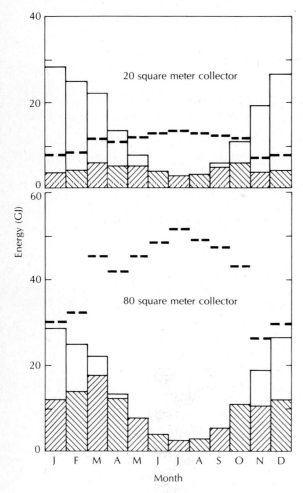

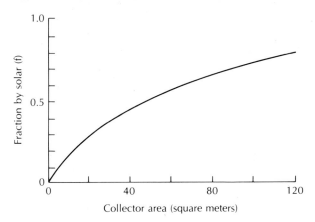

Figure 22-6. Variation of the fraction of the annual total load carried by solar energy with collector area for the Wisconsin example.

rate corresponding to a floor area of 180 square meters and with conventional insulation. A liquid heating collector has two glass covers, a high-absorptance, flat black paint for absorption of solar energy, and is sloped toward the south with a slope equal to the latitude.[24] The storage tank is to be located within the building so that losses from the tank are uncontrolled gains to the building. Hot water demands are typical for a family of four or five.

The results of simulations of this system, with forcing functions of hourly weather data for an average Madison year and a fixed ratio of storage mass to collector area of 75 kilograms per square meter, are shown in Figure 22-5. Incident radiation, total loads, and the load carried by solar energy for two collector areas are shown by month. Monthly collector efficiencies are the ratio of solar energy delivered to the building to incident solar energy on the collector; these are high when heating loads are high relative to the size of the collector and low when loads are low. Thus, these systems tend to be overdesigned for part of the year and underdesigned for part of the year.

Annual performances, expressed as the fraction of loads supplied by solar energy, are shown in Figure 22-6. Since the total loads are nearly independent of collector size, these data also indicate the total amount of solar energy delivered. These numbers are useful in deciding how much collector area should be used on the house and indicate that very large collector areas (relative to the heating loads on the house) are needed to approach 100 per cent solar heating. In other words, the larger the solar energy system, the larger the fraction of the year that it is overdesigned.

What should the storage capacity be? Figure 22-7 shows the effects of storage capacity on annual performance of this system. Below about 50 kilograms per square meter, system capacity drops off rather sharply as tank size decreases. Above 100 kilograms per square meter, there is a slow increase in annual performance as tank size increases. Cost studies by Tybout and Löf,[5] and others, which take into account the cost of tanks as a function of their size, indicate that a slight cost penalty is incurred on going beyond about 100 kilograms per square meter.

There remains a question of seasonal storage from summer to winter. If very large storage systems were used (probably with a volume of roughly the same size as that of the heated space if heat-capacity storage is used), smaller collectors might be possible.

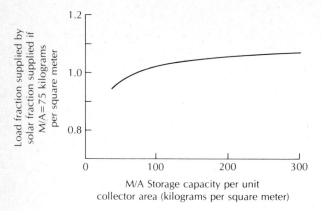

Figure 22-7. Variation of the fraction of the annual total load carried by solar energy with storage capacity for water storage tanks.

Speyer,[25] in 1959, concluded that this is un-economical; reexamination of this possibility with simulation methods would be of interest.

ECONOMICS

Solar energy processes are generally capital intensive; large investments are made in equipment to save operating costs (that is, fuel purchases). The essential economic problem is balancing annual cost of the extra investment (interest and principal, based on reasonable estimate of lifetime) against annual fuel savings. Thermal performance predictions, with estimated equipment and fuel costs, show the effects of major design decisions on annual costs.

An example of annual savings as a function of collector area, on the basis of the performance calculations noted in the previous section, is shown in Figure 22-8. We assume two collector costs, $60 and $100 per square meter, and two conventional energy costs, $5 and $15 per gigajoule. Delivered energy costs in the United States today range from less than $2 per gigajoule for natural gas in the Southwest to more than $15 per gigajoule for demand electric resistance heating in some parts of the Northeast. The collector cost is the major investment and is proportional to

collector area. Storage cost depends only slightly on collector area, and there are other equipment costs that are essentially independent of collector area. Here we have used $500 for the storage and other equipment costs and an annual charge on investment of 12 per cent, corresponding to 10 per cent interest over 20 years.

The curves show distinctly different behavior, with the maximum "savings" at small collector areas for the expensive collector and cheap fuel, and at a collector area of 100 square meters for the $60 per square meter collector and expensive fuel. The savings for the collector cost of $60 per square meter and the fuel cost of $15 per gigajoule are positive over a range of collector areas from 10 to 50 square meters. Significant deviations from the optimum values do not greatly affect savings; thus the selection of a precise value for collector area is not very critical.

There are many assumptions inherent in these curves. Costs of taxes, maintenance, and insurance have not been included. Conventional energy costs were assumed to be fixed over the lifetime of the system. The nature of the equipment for supplying auxiliary energy (as indicated in Figure 22-2) is assumed to be independent of the amount of auxiliary required during a year, whereas in

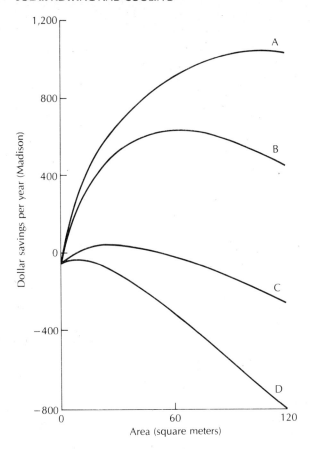

Figure 22-8. Annual savings as a function of collector area for the Wisconsin example. Two collector costs and two conventional energy costs are plotted. C_F, cost of fuel; C_C, cost of collector. Curve A: $C_F = \$15$ per gigajoule; $C_C = \$60$ per square meter. Curve B: $C_F = \$15$ per gigajoule; $C_C = \$100$ per square meter. Curve C: $C_F = \$5$ per gigajoule; $C_C = \$60$ per square meter. Curve D: $C_F = \$5$ per gigajoule; $C_C = \$100$ per square meter.

fact it may change substantially. Costs associated with the time dependence of auxiliary energy needs are ignored; this implies that the auxiliary energy source is stored on site, since utilities could be subjected to unacceptable peak loads by large numbers of solar buildings that draw on them simultaneously only during periods of bad weather.

Nevertheless, some generalizations can be drawn from these analyses. As fuel costs rise and as supplies of low-cost natural gas become increasingly more difficult to obtain,

solar energy will become more competitive and optimum fractions of annual loads to be carried by solar energy will increase. As collector and other solar energy system costs decrease as a result of mass production, by improved technology, or by users "doing it themselves," similar improvements in the relative economics of solar energy will occur.

Finally, political decisions may be made that will affect the extent to which solar energy can be competitive. Deregulation of natural gas prices or further increases in the cost of imported oil will increase the cost of these fuels to consumers and make solar energy more competitive. Tax incentives, such as write-off of investments in solar energy-producing equipment, could make an incremental improvement in solar energy economics.

SUMMARY

We have adequate theory and engineering capability to design, install, and use equipment for solar space and water heating. Energy can be delivered at costs that are competitive now with such high-cost energy sources as much fuel-generated, electrical resistance heating. The technology of heating is being improved through collector developments, improved materials, and studies of new ways to carry out the heating processes.

Solar cooling is still in the experimental stage. Relatively few experiments have yielded information on solar operation of absorption coolers, on use of night-sky radiation in locations with clear skies, on the combination of a solar-operated Rankine engine and a compression cooler, and on open-cycle humidification-dehumidification systems. Many more possibilities for exploration exist. Solar cooling may benefit from collector developments that permit energy delivery at higher temperatures and thus solar operation of additional kinds of cycles. Improved solar cooling capability can open up new applications of solar energy, particularly for larger buildings, and can result in markets for the retrofitting of existing buildings.

Solar energy for buildings can, in the next decade, make a significant contribution to the national energy economy and to the pocketbook of many individual users. Very large aggregate enterprises in manufacture, sale, and installation of solar energy equipment can result, which can involve a spectrum of large and small businesses. In our view, the technology is here or will soon be at hand; thus

the basic decisions as to whether the United States uses this resource will be political in nature.

NOTES AND REFERENCES

1. See, for example, M. P. Thekaekara, *Solar Energy* **14,** 109 (1973).
2. Radiation and related weather data are available from the National Climatic Data Center, Asheville, N.C.
3. B.Y.H. Liu and R. C. Jordan in *Low Temperature Engineering Applications of Solar Energy.* New York: American Society of Heating, Refrigerating, and Air Conditioning Engineers, 1967. G.O.G. Löf, J. A. Duffie and C. O. Smith, *University of Wisconsin Engineering Experimental Station Report No. 21* (1966).
4. The best example is F. Daniels, *Direct Use of the Sun's Energy.* New Haven, Conn.: Yale University Press, 1964.
5. R. A. Tybout and G.O.G. Löf, *Nat. Resour. J.* **10,** 268 (1970); G.O.G. Löf and R. A. Tybout, *Solar Energy* **14,** 253 (1973); *ibid.* **16,** 9 (1974).
6. C. D. Engebretson in *Proceedings of the U.N. Conference on New Sources of Energy,* Vol. 5, p. 159. New York: United Nations, 1964. G.O.G. Löf, M. M. El-Wakil, and J. P. Chiou, *ibid.,* p. 185. R. W. Bliss, *ibid.,* p. 148. G.O.G. Löf, *ibid.,* p. 114.
7. R. Chung, J. A. Duffie, and G.O.G. Löf, *Mech. Eng.* **85,** 31 (1963). J. A. Duffie and N. R. Sheridan, *Mech. Chem. Eng. Trans.* **MC-1,** 79 (1965). The NH$_3$-H$_2$O cycles have also been studied; six papers at the International Solar Energy Society (ISES) meetings concerned NH$_3$-H$_2$O systems.
8. The July 1975 ISES meetings also included 12 papers on performance of solar heating and cooling systems.
9. M. Telkes in F. Daniels and J. A. Duffie, eds., *Solar Energy Research.* Madison: University of Wisconsin Press, 1955. Final summary report to NSF-RANN, *Conservation and Better Utilization of Electric Power by Means of Thermal Energy Storage and Solar Heating,* National Center for Energy Management and Power. Philadelphia: University of Pennsylvania Press, 1973.
10. D. S. Ward and G.O.G. Löf, annual report to NSF-RANN, *Design, Construction and Testing of a Residential Solar Heating and Cooling System.* Fort Collins: Colorado State University, 1975.
11. Honeywell report to NSF-RANN, *Design and Test Report for Transportable Solar Laboratory Program.* Minneapolis: Honeywell, 1974.
12. G.O.G. Löf in F. Daniels and J. A. Duffie, eds., *Solar Energy Research.* Madison: University of Wisconsin Press, 1955. The triethylene glycol system is used in commercial fuel-fired air conditioning equipment.
13. R. V. Dunkle, *Mech. Chem. Eng. Trans.* **MC-1,** 73 (1965).

14. W. F. Rush, J. Wurm, L. Wright, and R. Ashworth, paper presented at International Solar Energy Society meeting, Los Angeles, August 1, 1975.

15. P. W. Niles, *ibid.*

16. F. H. Bridgers, D. D. Paxton, and R. W. Haines, *Heat. Piping Air Cond.* **29,** 165 (1957).

17. T. L. Freeman, thesis, University of Wisconsin-Madison, 1975.

18. J. A. Duffie and W. A. Beckman, *Solar Energy Thermal Processes,* New York: Wiley, 1974.

19. H. C. Hottel and B. B. Woertz, *Trans. Am. Soc. Mech. Eng.* **64,** 91 (1942). H. C. Hottel and A. Whillier in *Transactions of the Conference on the Use of Solar Energy,* Vol. 2, part 1. Tucson: Univ. of Arizona Press, 1958.

20. S. A. Klein, *Solar Energy* **17,** 79 (1975).

21. For example, see H. Tabor in *Low Temperature Engineering Applications of Solar Energy.* New York: American Society of Heating, Refrigerating, and Air Conditioning Engineers, 1967.

22. E. Speyer, *J. Eng. Power* **86,** 270 (1965). D. C. Beekley and G. R. Mather, paper presented at International Solar Energy Society meeting, July 28, 1975. U. Ortabassi and W. Buehl, *ibid.*

23. S. A. Klein et al., *University of Wisconsin Engineering Experimental Station Report No. 38* (1975). See also S. A. Klein, P. I. Cooper, T. L. Freeman,

D. M. Beekman, W. A. Beckman, and J. A. Duffie, *Solar Energy* **17,** 29 (1975).

24. As a rule, a solar collector should be sloped toward the equator with a slope of latitude $+10°$ for winter use, a slope of latitude $-10°$ for summer use, and a slope equal to latitude for year-round use. Deviations of 5° or 10° usually make little difference in annual performance.

25. E. Speyer, *Solar Energy* **3** (4), 24 (1959).

SUGGESTED READINGS

W. A. Beckman, S. A. Klein, and J. A. Duffie, *Solar Heating Design.* New York: Wiley-Interscience, 1977.

J. A. Duffie and W. A. Beckman, *Solar Engineering of Thermal Processes.* New York: Wiley-Interscience, 1980.

Solar Heating and Cooling of Residential Buildings—Design of Systems, S/N 003-011-00089-5. Washington, D.C.: GPO, 1977.

Solar Heating and Cooling of Residential Buildings—Sizing, Installation and Operation of Systems, S/N 003-011-00088-7. Washington, D.C.: GPO, 1977.

Sunset Homeowners Guide to Solar Heating. Menlo Park, Calif.: Lane Publishing, 1978.

The Technological and Economic Development of Photovoltaics   23

DENNIS COSTELLO AND PAUL RAPPAPORT

INTRODUCTION

Technology has often played a key role in
both solving and creating major problems for
society. The depletion of fossil energy resources is currently one of our most pressing
problems. Several technologies are again in
the forefront of the solutions being considered. Solar energy technologies, and photovoltaic technologies in particular, represent one of the more socially attractive of the
technical solutions to the energy problem.
This article reviews the present status of
photovoltaic technologies and investigates
their potential role as an energy supply. In
order to understand the future role of photovoltaics, it is necessary to disassemble the
elements that will determine that role and examine them individually. The first key element is the physical characteristics of the
technology. The second is the photovoltaic
industry, which translates technological advances into commercial products. The mar-

kets and their behavior in response to these
products constitute the next critical element.
Finally, U.S. and foreign governments will
continue to be involved in photovoltaics and
will influence their development. Neither the
technology, the industry, the markets, nor the
public sector alone will dictate the development of photovoltaics. Rather, it will be the
dynamic interaction of all these elements that
determines the energy contribution of this
technology. The four major sections of this
article deal with each of the key elements individually. The final section explores their
interactions.

Definitions

''Photovoltaic technologies'' refer to a set of
solar energy conversion technologies that
convert light energy (photons) directly into
electricity. The phenomenon, known as the
photovoltaic effect, begins with the excitation of electrons in a solid by the absorption

Mr. Dennis Costello is currently on academic leave from the Solar Energy Research Institute (SERI),
Golden, Colorado 80401. He was Manager of the Buildings Division.

Dr. Paul Rappaport was Director and Distinguished Scientist at SERI until his untimely death on April
21, 1980, shortly after he and Mr. Costello completed this review. Dr. Rappaport was one of the
founders of and fundamental contributors to the science of photovoltaics.

of light. This excitation creates a population of both positive and negative charges. These charges are free to move through the semiconductor material only if it contains appreciable numbers of impurity centers. These allow excited electrons to jump across a band of forbidden energy states into the conduction band. When a semiconductor is specially treated (doped) with these impurities, it is capable of converting incident sunlight into electrical energy with relatively high efficiencies compared with other solar conversion processes, such as photosynthesis.

These specially treated semiconductor materials are called photovoltaic cells or solar cell devices. The problems and opportunities facing photovoltaics are similar to those that have faced other semiconductor devices (transistors, diodes, etc) over the past two decades. The majority of commercially available solar cells are produced from single-crystal silicon. A single-crystal silicon solar cell is typically a 0.25-mm-thick wafer which is 5 to 10 cm in diameter. Materials other than silicon, such as gallium arsenide (GaAs) or copper sulfide/cadmium sulfide (Cu_2S/CdS) can also be used to produce solar cells. The output of a photovoltaic cell is direct-current electricity at a low voltage (approximately 0.5 volt and about 30 mA/cm^2).

In order to produce useful energy, the individual cells have to be wired in series to increase the voltage and in parallel to increase the current and then encapsulated for protection from the environment. These groups of cells, termed modules, must then be supported at an angle to face the sun. Support structures must be sturdy enough to withstand winds and the elements. Circuitry (power conditioning) to control and modify the electrical output in appropriate ways is also needed. In some applications, energy storage (e.g., lead acid batteries) must be added to the system to provide power when the modules are not producing power or to supplement their output during variable conditions. A means of regulating the flow of power between the modules, storage, and the load is often necessary. The combination of modules, support structures, control, and storage (if needed) is referred to as a photovoltaic system. It is convenient to talk about the photovoltaic modules and the balance of the system (BOS) as separate entities, especially when referring to costs. The BOS requirements are very much application dependent.

Advantages and Problems of Photovoltaics

Photovoltaics has a number of technical and social advantages over other energy technologies. The first and most obvious advantage is that it utilizes the ultimate renewable resource, sunlight. Second, it converts that resource directly into electricity, a very high quality and versatile form of energy. Third, photovoltaic systems have no moving parts (and operate silently) so that maintenance costs are minor. The efficiency of a photovoltaic system is relatively independent of its size. There is not a critical size as is the case with other solar energy systems, coal plants, or nuclear plants. Therefore, systems can be built at dispersed sites and in small sizes with few penalties in efficiency or cost. In fact, systems located near the load have the cost advantage of reducing electrical transmission losses and costs. The modular nature of photovoltaics allows the technology to be used in applications ranging from fractions of a watt to megawatts.

Another advantage of photovoltaics is that it is environmentally benign in operation. Environmental impacts do occur in the production, assembly, and decommissioning of the system. However, these impacts are relatively minor when compared to the impacts of many other energy technologies.[1] Finally, photovoltaics can take advantage of many of the technical advances achieved in the semiconductor industry over the last 25 years.[2] The high rates of innovation and cost decline that have been the hallmark of the semiconductor industry raise the expectation that photovoltaics will follow a similar development path.

Photovoltaic technologies face a number of problems that presently limit their contribution to solving crucial energy problems. The most obvious is the cost of photovoltaic modules and systems. Today, photovoltaic modules are sold in large quantities for under $10 per peak watt. A peak watt (Wp) is the output of the device measured at its maximum operating capacity under full sun. The average power output (day and night, year-round) at a good site will be about one fifth of the rating in peak watts. A $10/Wp price translates into approximately $1–2/kWh, or 20 times the competitive range for utility-generated power in the United States. Although prices of modules have fallen drastically from their first introduction, further cost reductions are obviously necessary. Energy produced from photovoltaic systems today is extremely expensive for three reasons. First, sunlight is a diffuse energy source. Second, photovoltaic cells are inefficient compared with other electrical generation technologies. Efficiencies of currently available commercial cells range from 10 to 15 per cent. These relatively low efficiencies lead to high BOS costs resulting from the large array areas required for power output. Third, and most important, the processes for producing single-crystal silicon cells are currently very expensive.

A second problem facing photovoltaic systems is that they have not yet been used in many of the large energy-using applications. Utilities, homeowners, insurers, and financial institutions have no experience with the technology. The technical and economic issues surrounding the interface between electric utilities and dispersed photovoltaic systems are among the most crucial. Utility companies have very little experience with small electric generating units located at the point of use. Such facilities will have ramifications on seasonal and daily demand peaks, the dispatch of existing generating capacity, and plans for capacity expansion. The ownership advantages of photovoltaic systems and the value placed on electricity sold back to the utility are key economic issues that have

not yet been resolved. This lack of knowledge and experience with photovoltaics in utility systems will take time to overcome and can slow the diffusion of the innovation into the market place.[3] One solution to this problem is for the federal government to fund a large (approximately 1-MWp) utility demonstration of photovoltaics in the near future.

Most industrial, environmental, and government experts hold the opinion that photovoltaics has the potential to supply a large percentage of the world's electric energy demand. The advantages are seen as long-lasting and matched by only a few other energy alternatives. The problems are viewed as temporary, as evidenced by the advances in efficiency and cost reduction already experienced. The debate in most circles centers around how quickly research and production advances will overcome the cost problems and how rapidly society will adopt the technology.

STATUS OF THE TECHNOLOGY

The wide variety of material and cell concepts that can produce a photovoltaic effect is one of the most attractive features of the technology. It is this variety that makes the probability of significant cost reduction so high. The materials and concepts currently under research are summarized below and are organized into three major sections: silicon technology, concentrators, and thin films.

Silicon Technology

There are a variety of ways to use silicon in the production of photovoltaic cells. Growing a single crystal of semiconductor-grade silicon, slicing it into wafers, and fabricating the wafers into cells is the only well-established method of producing photovoltaic cells. Single-crystal silicon cells, as they are called, are the most widely commercially produced cells today and are the standard of comparison for new materials and concepts being developed.

The production of a single-crystal silicon

solar cell begins with quartz, a relatively pure form of SiO_2. The quartz is refined into a high-purity silicon in a two-stage process. First, it is refined into metallurgical-grade silicon and then into semiconductor-grade silicon. This material, termed polycrystalline silicon, is used to grow large, cylindrical single crystals. The crystal is grown by melting the polycrystalline silicon in a crucible with small quantities of other elements (e.g., boron, phosphorus) called dopants. Next, the mixture is slowly pulled out of the crucible so that it recrystallizes into a cylindrical single crystal, typically 7 to 10 cm in diameter. Experimentation with larger sizes has yielded crystals as large as 30 cm in diameter.[4] The single-crystal growing process is called the Czochralski process, after its inventor.

The single-crystal cylinders are next sliced into flat wafers. Currently, the slicing is done with donut-shaped diamond-tipped saws (called inside-diameter saws). The inside edge of the donut contains the cutting blades. Only one cut is made at a time, producing wafers about ¼ mm thick. Nearly 50 per cent of the silicon material is lost (kerf loss) in the slicing process.[5] The wafers are etched to remove irregularities in the surface and then made into solar cells by the addition of other impurities (dopants). These impurities are diffused into the wafers by heating them in a "diffusion" furnace. The last step in cell fabrication is to add metal contacts to the front and back of the cells. Coatings are also applied to the front surface to reduce energy loss from reflection of sunlight.

In order for the cell to produce useful energy over a long period, numerous cells have to be wired together and protected from the outside environment, i.e., "encapsulated." Usually, the front surface of the cell is covered with glass or plastic. The cells are wired together, placed in a frame for structural support, and sealed from dirt and moisture. The completed unit is called a photovoltaic module and is usually up to 1 m^2 in area.

The process for producing single-crystal silicon photovoltaic modules is currently very costly. The major cost factors are the cost of the raw polysilicon material and the number of steps needed to convert the material into cells. The slicing of the ingot into wafers is a particularly costly process and may be the chief detriment to the use of Czochralski silicon. Any means that can be used to eliminate some process steps, increase the throughput of any of the steps, or to increase the efficiency of the final product will help reduce costs. As might be expected, the most extensive research and development activities are aimed at the most costly production steps. For example, extensive effort has been invested in development of a lower-cost method of producing solar-cell-grade polysilicon material.[6] The goal of this effort is to reduce the cost of the material from its current \$65/kg to \$14/kg.[7]

A number of alternative techniques are being investigated to replace or simplify the silicon crystal growing and slicing steps of the process. One technique is to cast the silicon into bricks and then slice them into square cells. This process, termed the multigrained, or semicrystalline, silicon approach, results in smaller crystals than Czochralski silicon. Experimental semicrystalline silicon cells have attained efficiencies up to 16 per cent.[8,9] The technique has promise in reducing costs because it requires lower-purity raw materials and eliminates the costly production steps of growing single crystals. However, the costly step of slicing the ingot is not eliminated.

The semicrystalline approach is under development by public institutions in the United States and in Germany and by numerous private companies. In Germany, the Wacker Corporation has installed a small production facility and sells some of its product to the United States. Other commercial production lines will probably be installed in the near future.

Another alternative to growing a single silicon crystal is to grow silicon ribbon. The ribbon process begins by melting polysilicon in a crucible. A long, thin ribbon is then pulled slowly out of the crucible. The ribbon, which can be grown up to a width of

100 mm, is next cut into rectangular wafers, with much less material loss than in slicing an ingot. However, efficiencies of ribbon silicon cells are in the range of 8 to 12 per cent, somewhat lower than single-crystal silicon.[10] To date, the rate of ribbon growth has been too low to make the process commercially feasible. However, numerous ways around this problem are being studied. For example, cells having 10 per cent efficiency have been achieved using three related processes (the Westinghouse web dendritic process, the Motorola ribbon-to-ribbon process, and the Honeywell silicon-on-ceramic process).

Concentrators

The diffuse nature of sunlight striking the earth's surface is one of the fundamental reasons that photovoltaic systems are currently an expensive means of producing electricity. One technique to overcome this problem is to concentrate the sun's energy using mirrors or lenses.

Using concentrators instead of flat-plate collectors changes the nature of the problem of cost in four ways. First, some type of mechanical concentrating subsystem must be added to the total cost. Second, the importance of cell cost in the total system cost decreases because the number of cells per unit of electrical output declines significantly. Third, the efficiency of the cell under concentrated sunlight becomes a more important consideration because of the relatively high cost of getting concentrated sunlight to the cell's surface. Many analysts have concluded that cell efficiencies of 30 per cent are required to allow concentrators to be cost effective. Fourth, under proper conditions the efficiency of most photovoltaic cells tends to increase with highly concentrated sunlight.

The discussion of photovoltaic concentrators can be divided into two parts: the concentrating mechanisms and the photovoltaic cell. There are a wide variety of ways to concentrate sunlight on a relatively small surface area. Concentrating devices include compound parabolic concentrators, which aug-

ment sunlight along a line; parabolic troughs, which focus along a line, circular fresnel lines, which concentrate sunlight on a single point; and parabolic dishes, which also focus on a single point. These devices have to move seasonally and/or daily to track the sun's movement across the sky. They have the capability of concentrating anywhere from 2–10 suns (compound parabolic concentrators) to 100–1000 suns (parabolic dish concentrators and fresnel and other lenses).[11]

Efforts to develop better and lower-cost concentrating devices are important to thermal solar technologies as well as photovoltaics. The major problem is to mass produce long-lived and low-cost devices in a manner similar to the way household appliances or automotive parts are currently produced.

The photovoltaic cell used with a concentrator device must be as efficient as possible at high light intensity. At the same time, the cost of the cell can be higher than in a flat-plate system because fewer cells are needed. Some of the most important concentrating photovoltaic cell approaches are modified single-crystal silicon, gallium arsenide alloy cells, vertical and horizontal multifunction cells, thermophotovoltaic cells, multicolor cells, and dye concentrators. Efficiencies achieved for these cells range from 19 per cent (silicon at 300 suns) to 28.5 per cent gallium aluminum arsenide (GaAlAs) and silicon cells illuminated through a beam splitter at 167 suns.[12,13] The American Physical Society's report on photovoltaics predicts the efficiency of the more advanced concentrator cells will reach 26 per cent to 39 per cent.[14] The best efficiency to date for a single junction device has been 24.7 per cent at 178 suns.[15]

Thin Films

An alternative to concentrating sunlight on relatively efficient but costly cells is to let unconcentrated light strike very inexpensive cells. The most inexpensive way to cover a large amount of exposed surface is to spray or coat the area with some type of film. Applying materials to a surface in a thin film that will produce a photovoltaic effect is the subject of considerable research.

Thin film photovoltaic technologies include a large group of materials and processes that are produced in similar ways. Thin films of specialized materials such as cadmium sulfide/copper sulfide (CdS/Cu_2S), polycrystalline gallium arsenide (GaAs), polycrystalline Si, and amorphous hydrogenerated Si are deposited on a metal (or other material) substrate in layers only a few micrometers thick.[16] Thin films have the potential to yield photovoltaic cells of extremely low production cost. Rather than growing a single crystal, slicing, and arranging the wafers into modules, the final product is deposited onto the module in a single step. As well as eliminating many costly production steps, thin films also require less raw material. Material losses due to production handling (e.g., slicing) are eliminated, and the thickness of the final cell is much less than sliced cells or even cells formed from cast sheets.

The major disadvantages of thin films are that they are not yet commercially produced and the efficiency of the cells has to be improved. Efficiency is often relatively low because of the sacrifices made in crystal perfection associated with thin film production processes. Thin films produced to date have not yet exceeded efficiencies of 10 per cent.[17] Attainment of a 10 per cent efficient thin film cell is an important milestone of the technology in the opinion of many experts. They feel the 10 per cent efficiency level must be achieved in order to prevent the encapsulation, array structures, array supports, wiring, and land costs (i.e., BOS costs) from becoming excessive and making the systems too costly.[18]

The thin film approach that has received the largest amount of investigation is the CdS/Cu_2S cell,[19] which is produced by first coating a substrate (often copper) with zinc. CdS is next deposited on the zinc and covered with a layer of Cu_2S by an electrochemical process. The process is being vigorously pursued by the private sector and pilot production facilities have been constructed.[20] The advantages of this type of cell are that it does not have to be as thick as other thin film cells and it has already achieved efficiencies of 9.1 per cent. Uncertainties about the approach center around the useful life of the cell under normal operating conditions, the toxicity of the cadmium, and the limited world supply of cadmium.

Three other thin film approaches are also being researched diligently: polycrystalline gallium arsenide, polycrystalline Si, and amorphous hydrogenated Si. Gallium arsenide has the potential for either single-crystal, high-efficiency cells for concentrators or thin polycrystalline films for flat-plate applications. The layer of gallium required is only a few micrometers in thickness. Although development is still under way, efficiencies of 6.5 per cent have already been achieved.[21]

Polycrystalline Si must be used in thicker films than does gallium because silicon is a less efficient absorber of the sun's energy. Unlike either single-crystal Si or cast semicrystalline Si, thin film Si cells are still in the research stage. Research questions that have not been answered are similar to those facing the other thin film options, including the efficiency, performance stability, and cost of mass producing the cells.

EVOLUTION AND STATUS OF THE PHOTOVOLTAIC INDUSTRY

The technological capability to produce a photovoltaic cell for commercial use was first demonstrated in 1953. During the 1950s, modules were sold for a variety of highly specialized applications, with the majority of sales made to the space industry. The space program continued to be the primary photovoltaic application throughout the 1960s. In the early 1970s, photovoltaics began to be used more extensively for terrestrial purposes.

Between 1973 and 1975, the cumultive industrial production of terrestrial photovoltaic cells totaled well under 1 peak megawatt of capacity. In 1976, the first organized studies of the sales of photovoltaic modules were made by the BDM Corporation[22] and Intertechnology Solar Corporation.[23] The surveys estimated shipments of photovoltaic modules by U.S. manufacturers during 1976 to be approximately 400 kWp. Although estimates of sales of the two surveys differ, several trends were evident. In 1976, sales for terrestrial applications were almost ten times greater than space applications. However, federal government purchases constituted almost one third of the total market. After 1976, only aggregate estimates of module sales are available. Two sources estimate that 1977 sales were approximately 750 kWp, with shipments estimated by one of the sources at about 500 kWp.[24,25] Terrestrial module sales in 1977 have been estimated in a recent industry assessment by Booz, Allen and Hamilton[26] to generate approximately $15 million in revenues to producing companies. Sales for 1978 have been estimated by Booz, Allen and Hamilton to be over 1 MWp, with shipments on the order of 700 to 800 kWp.[27] No figures are available for 1979 sales but discussions with industry indicate that sales exceeded 1 MWp. Current opinion indicates that a major portion of photovoltaic sales for the next 5 to 10 years will be made in international markets, particularly as further decreases in the price of photovoltaic modules are achieved.[28]

Substantial reductions in the price of photovoltaic modules have been realized since the first arrays designed for terrestrial use were manufactured in the early 1970s. The first large federal purchases of modules for terrestrial use in 1975 were at an average price of approximately $20/Wp.[29] This was a significant reduction in the cost from the $200/Wp paid for modules used in space applications. The results of a 1978 Solar Energy Research Institute (SERI) survey of commercial module prices showed that, in small quantities, modules sold for $15 to $36/Wp, depending on the supplier and the quantity purchased.[30] In large-quantity purchases, module prices ranged from $10 to $15/Wp in 1978. Further price declines were evident in 1979 and are expected in 1980. As mentioned earlier, the cost of a complete photovoltaic system is higher than the cost of the modules alone. The exact cost of the system depends on design considerations (i.e., type of modules, power conditioning devices, batteries, supporting structure, and installation). The costs of systems being installed in 1978–79 typically range from $20 to $40/Wp.[31]

In 1979, at least eight U.S. companies were commercially producing terrestrial photovoltaic modules. All of the companies produced single-crystal silicon, flat-plate modules. Production prototypes of polycrystalline and concentrating modules using single-crystal silicon have been successfully produced. Commercial sales activities in these two areas are imminent. At least ten other U.S. companies are investing significant amounts of private funds in research and development of photovoltaic modules. Numerous other companies are participating in federal research and development activities. Plans for the commercial production of thin film cadmium sulfide and single-crystal and special concentrating gallium arsenide modules are now being formulated. Other technologies, such as thin film silicon, gallium arsenide and indium phosphorous, polycrystalline gallium arsenide, and advanced concentrator materials, will require substantial research progress before commercial production planning can begin.

The photovoltaic industry, including companies currently producing and those which are potential producers, can be classified by their affiliations and financial backing. A few of the companies are small businesses which produce photovoltaic modules or systems as their primary product line (e.g., Solenergy). In these companies, current sales and revenue are important to their continued financial viability. Access to large amounts of capital may be relatively difficult to obtain, especially if that investment is considered a high risk by potential investors. The second group of companies are those associated with moderate or large electronics firms (e.g., Texas Instruments, RCA, Motorola). These firms view photovoltaics as another semiconductor product line. The size of these firms and their diverse product lines make the yearly sales revenue from photovoltaics somewhat less important than in the smaller companies. An electronics firm with a photovoltaic product line may consider it a new venture and not require immediate profits from the operation. On the other hand, many of these companies are currently faced with many new venture opportunities. In order for photovoltaics to retain corporate support, it must offer more potential than competing new ventures in electronics.

The third group of photovoltaic companies are those affiliated with major international energy companies (e.g., Exxon, Mobil, Shell, Atlantic-Richfield). These companies are characterized by their desire to diversify into nonoil energy sources and the availability of capital to support new ventures. Another major feature of these companies is their long-term perspective on some of their investments. They appear to be willing to accept a much longer time for return of their investment than electronics firms. This longer time horizon is partially due to the long payback periods required on some of their other investments, such as oil pipelines or seaport facilities. The variety of companies, backgrounds, and perspectives that make up the photovoltaic industry makes it difficult to summarize their viewpoint with general statements. However, it can be concluded that the variety of the industry may give it the versatility necessary to make photovoltaics a viable energy resource in the future. The variety certainly makes government policy decisions more difficult, even those aimed at directly assisting the industry.

EVOLUTION AND STATUS OF PHOTOVOLTAIC MARKETS

The role of photovoltaic markets in the evolution of the technology is often underestimated by the research community. The existence of markets at different price levels was one of the major factors that contributed to the orderly and rapid development of the semiconductor industry. Each time a research advance led to a reduction in semiconductor prices, the technology became competitive in a new market.[32] Sales to that market were large enough to support capital investments in larger and more advanced production facilities. These markets allowed prices to continue to decrease and production volume to increase. The penetration of semiconductors into mass markets, such as consumer applications, would not have been possible without the existence of earlier markets willing to pay premium prices for the technology.

The most widely held view of photovoltaic development is that the technology will have to penetrate a series of premium markets before penetration of large residential, commercial, and industrial electricity markets can begin. These premium markets are crucial to the continued viability and growth of the industry and to continued photovoltaic price reductions. Current and potential markets for photovoltaics can be categorized by the time at which they will be penetrated. The first category is near-term markets, generally involving remote applications. Commercial sales are currently being made in these premium markets. Remote applications are those without access to an electric utility grid. The second category is intermediate markets, those that could be significantly penetrated by the technology if its price was reduced into the range of about $2/Wp for modules. These markets are intermediate in that they lie farther down the demand curve than the current markets for remote applications, but do not require as low a price as the major markets for grid-connected electric power applications. The final category is long-term markets. These are U.S. grid-connected electric power applications, such as residences and central utility power plants. These markets can be penetrated only if major reductions in the price of photovoltaic systems occur.

The major near-term markets of photovoltaics are estimated to be in remote communications facilities and corrosion protection systems. Most of the communication facilities using photovoltaics are radio and microwave repeaters. Impressed-current protection of pipelines and wells is the major corrosion protection application. The radio and microwave repeater markets that photovoltaics are penetrating typically do not have utility power available, and the cost of powerline extensions is exorbitant. These markets and markets for corrosion protection systems are located in the United States and many other parts of the developed and developing world. Photovoltaics are substituted for batteries, thermal electric generators, or diesel generators, which have high maintenance costs and poor reliability.

Photovoltaics are competitive in the communication and corrosion protection markets at system prices from $11/Wp to $53/Wp.[33] Current sales in these markets are small, predictions of future growth are also limited, and published estimates of potential annual sales vary from 3 MWp to 7 MWp.

In order for photovoltaic production volumes to increase substantially, new markets for the technology must be found and penetrated. Following the pattern of semiconductor products, buyers in these markets should also be willing to accept relatively high prices. The intermediate markets that have been identified to date include water pumping for irrigation and potable water use, general power supplies for remote villages and communities, street and highway lighting, and military uses. If prices of photovoltaic systems reach $5 to $15/Wp, they will be competitive in most of these applications.[34] The total potential size of these markets could range in annual sales volume from 50 to 500 MWp. However, almost no commercial sales have yet been made in these intermediate markets, and very little credible information is available on their size and characteristics. In fact, the uncertainty surrounding these markets is one of the major problems facing both government policy makers trying to stimulate the technology and corporate executives faced with production investment decisions.

Only a few facts about intermediate photovoltaic markets seem to be generally accepted. First, most market experts and industry representatives are confident that these intermediate markets will not develop in the United States. They expect the largest opportunities to be in developing nations, with some sales in developed nations. The Solar Photovoltaic Energy Research, Development and Demonstration Act of 1978 (PL 95-590) recognized the significance of these non-U.S. intermediate markets and mandated the preparation of a U.S. government plan "for demonstrating applications of solar photovoltaic energy systems and facilitating their widespread use in other nations."[35] The primary reason for optimism concerning international photovoltaic markets is the high cost of competing energy systems and the sparsely developed electric utility grid systems in many developing nations.

Among the non-U.S. intermediate markets, it is also widely agreed that water pumping is one of the highest potential applications. Both low-lift (0.1 to 0.2 kW) and medium-lift (about 1.5 kW) pumping applications are prime targets for photovoltaic system sales. These applications are appealing because (a) the power requirements are small enough to be fulfilled with a modest photovoltaic array, (b) the demand for irrigation water is often coincident with insolation, and (c) water is easily stored for non-sun periods. Another feature of some of the smaller photovoltaic pumping systems is that the power requirements are so small that no similarly sized gasoline or diesel generator is currently available. Systems being considered in these applications typically use direct-current pumping motors, require very few controls, and do not require storage. These features significantly reduce the system's cost.

Another intermediate market with high potential is general power systems for rural or remote villages in developing nations.

Considerable disagreement exists concerning the size of this market relative to the water pumping market. Proponents of the general power source application maintain that a system put to numerous uses (e.g., lighting, refrigeration, sewing, clothes washing) will have a greater utilization rate and therefore be of more value to the user. Proponents of single-use water pumping systems often raise the problems of obtaining community cooperation and securing financing for the larger general power systems.

The ultimate market for photovoltaic systems is the large electricity-using sectors of the developed world, especially the United States. These markets are currently served by extensive electric grids. In order to penetrate these markets, photovoltaic power costs must be reduced to prices competitive with electric grid-supplied power. The studies that have investigated these long-term markets usually consider three types of photovoltaic systems. The first is a small (3 to 10 kWp) system located on a single residence. The second is a larger system (about 500 kWp) used in a commercial, industrial, or institutional setting. The final system is large (about 100 MWp), centrally located, and part of a utility's generation capacity. The studies indicate that photovoltaic prices for the small residential system (which use utility power as backup and do not have on-site storage) can be competitive in the United States at costs ranging from $2.60 to $1.00/Wp.[36-40] The exact price depends on the geographic location of the system and the price that the utility is willing (or is forced) to pay for power sold back to the grid from the residence. Centrally located photovoltaic systems may have to cost less than $1/Wp to be economically competitive in the United States.[41,42]

THE PUBLIC SECTOR'S ROLE IN PHOTOVOLTAICS

U.S. Department of Energy

The government of the United States and those of many other nations have taken an active role in the development and diffusion of photovoltaic technologies. It is clear that the public sector will continue to play a role through the 1980s. Although debate surrounds the amount of positive influence the public sector can have, it is clear that ongoing and planned activities of these governments will have some effect on the pace and character of photovoltaic development. The following sections describe programs of the U.S. Department of Energy (DOE), laws passed and bills before the U.S. Congress, and activities of Japan, West Germany, France, Britain, and Italy.

The U.S. government has been involved in photovoltaic technology almost since the technology's invention. When the technology was used primarily for space applications, NASA was the primary market and contributed to development. Now that the technology's potential for terrestrial use is of primary concern, the DOE has the lead responsibility within the U.S. government. The goal of the DOE photovoltaics program is to reduce the costs of photovoltaic systems to a level competitive with major U.S. energy markets (i.e., the long-term markets discussed above). This objective is to be accomplished by extensive support of research, development, and demonstration activities and a limited amount of product support. Product support activities include resolution of institutional, legal, environmental, and social issues that are barriers to adoption and dissemination of information on the technology.

The DOE photovoltaic budget in fiscal year 1979 totaled almost $119 million. Annual federal photovoltaic outlays have been growing steadily since the Energy Research and Development Administration (a DOE predecessor) took over the program in 1974. Future plans of DOE call for expenditures totalling $1.5 billion between 1979 and 1988.[43] These funds are supporting four types of activities: advanced research and development, technology development, systems engineering and standards, and systems tests and applications.

The advanced research and development

program supports basic research on a broad range of photovoltaic technology options. The objective of this program is to establish the technical feasibility of various photovoltaic options by obtaining stable, reproducible characteristics and demonstrating laboratory-scale processes that yield consistent performance. Research is supported on advanced materials and cells, including amorphous silicon, thin film polycrystalline gallium arsenide, advanced concentrators, multijunction cells, and a wide variety of other materials. Financial support is given to this wide variety of materials and concepts in order to maximize the possibility of fundamental improvements in the technology. The program can be visualized as a high-risk but high-return portfolio of basic research projects.

The second program area, technology development, investigates the production processes of photovoltaic technologies that have proved to be technically feasible. The emphasis is on the development of large-scale production processes to allow the achievement of stated cost goals. The majority of the federal development efforts to date have been on single-crystal silicon technology. The activities include the design of lower-cost facilities to produce raw silicon, improved crystal growing techniques, better methods of slicing crystals, and many others.

The systems engineering and standards program is a smaller effort than the first two programs (totaling $150 million from 1979 to 1988, compared to $450 to $500 for each of the previous two). The objective of this program is to reduce the cost of photovoltaic systems by improving the design of the system and the efficiency and cost characteristics of the nonmodule components. The program also helps to establish industry performance criteria and standards.

This final element of the current federal program is system tests and applications. Most expenditures in this area are for demonstrating operation of systems in realistic market settings. Some of the major demonstrations funded under the program include a 250-kWp concentrator system at Mississippi County Community College, Blythesville, Arkansas; a 100-kWp single-crystal silicon, flat-plate system on the Natural Bridges National Monument Visitor Center, Blanding, Utah; a 3.5-kWp remote silicon, flat-plate system at the Shuchuli Indian Village, Papago Indian Reservation, Arizona; and a 25-kWp silicon flat-plate system used for irrigation and other agricultural processes at Mead, Nebraska.[44] Other demonstrations funded by the federal government but not officially part of this program include a 1.8-kWp flat-plate pumping and grain milling system in a remote village in Upper Volta and a 500-kWp concentrating power system outside Riyadh, Saudi Arabia.

U.S. Congress

The national interest in technological solutions to the energy problem has led to great interest in photovoltaics by the U.S. Congress, with the most significant action taken by them being the passage of the Solar Photovoltaic Energy Research, Development and Demonstration Act of 1978 (PL 95–590). The law provides for "an accelerated program of research, development and demonstration of solar photovoltaic energy technologies." In most respects, the law merely reinforces the DOE photovoltaic program plan.[45] It calls for the expenditure of $1.5 billion from 1979 to 1988 and includes very aggressive photovoltaic system costs and sales goals (i.e., $1/Wp average system prices and 4 million kWp of cumulative production by fiscal year 1988).

Other legislation has also included photovoltaics. The Energy Conservation Policy Act of 1978 contains a section entitled the Federal Photovoltaic Utilization Act. That act establishes "a photovoltaic energy commercialization program for the accelerated procurement and installation of photovoltaic systems for electric production in federal facilities." This program has an authorized budget of $98 million. An amendment to the National Energy Act of 1978 made another $12 million available for the purchase of photovoltaic systems on federal facilities. The Nuclear Nonproliferation Act of 1978 and the International Development and Food Assistance Act of 1977 emphasize the use of solar and other renewable energy technologies in meeting the energy needs of developing nations. For a more complete review of related legislation see Costello et al.[46]

Activities of Other Governments

The U.S. government has not been unique in its efforts to stimulate the development of photovoltaic technologies. Nations that are competitive with the United States in the export of technology are also very interested in photovoltaics (e.g., Japan, West Germany, and France).

The Japanese government program in photovoltaics emphasizes research on advanced concepts and the development of cell fabrication techniques that could be used in production. Very little effort is spent on system experiments and demonstrations. It is estimated that approximately $2 million was spent on photovoltaic research and development in Japan in 1978.[47] Most of the Japanese photovoltaic activities are carried out under the broader solar energy program "Project Sunshine," launched in 1974. Some additional research is being carried out at several universities (e.g., Kyoto University and Osaka University) and private companies (e.g., Sanyo Electric Company and Fuji Electric Company).

The major European nations undertaking photovoltaic research and development are France, West Germany, Britain, and Italy. The French spent $21 million for photovoltaics in 1979. They have been building terrestrial photovoltaic systems for 18 years and currently export over 90 per cent of their production. The emphasis of their program is on system design, demonstration, and basic research. Numerous demonstrations of photovoltaic water pumping systems in developing nations have been funded by the French government, including three in Upper Volta, two in Mali, one in Corsica, and one in Cameroon.[48] West Germany's photovoltaic research and development program focuses on polycrystalline silicon technology. The major program is the Wacker-Chemtronic/ AEG-Telefunken Polysilicon project. The project will spend $98 million over eight years, of which $78 million is provided by government and $20 million by corporate funds. One of the goals of the project is to produce modules at $2.40/Wp by 1985. The West German government is also spending a limited amount on research into other photovoltaic materials (e.g., amorphous silicon) and on system demonstrations (in West Germany, Iran, Indonesia, Mexico, and India).[49] Italy is investigating a broad range of photovoltaic issues, including research on advanced materials such as GaAs and CdS/Cu_2S, and silicon material and cell development for flat panels and concentrators. Many of the Italian activities are funded jointly with industry. The photovoltaics effort in Britain is very small, with major emphasis on amorphous silicon. Britain sees very little potential for photovoltaic technology in solving its energy problems.

The Commission of European Communities (CEC) also supports a wide spectrum of photovoltaic research activities in Europe. Between 1975 and 1979, CEC provided over $4.3 million for such research. CEC typically contributes 50 per cent of the funding and the contractor provides the remainder.[50]

SUMMARY AND PERSPECTIVES ON THE FUTURE

One of the most popular recent books on U.S. energy problems states, "One can certainly say that photovoltaics thus far constitutes a success story."[51] The success of photovoltaics is evidenced by the rapid decline in prices with relatively modest increases in production volumes. It is also demonstrated by the terrestrial market niche that the emerging photovoltaic industry has gained in a relatively short period.

The success of photovoltaic energy technologies in the future is a key question for government and the private sector, and there is a general optimism about their future for four basic reasons. First, a wide diversity of technical approaches and materials are showing promise as effective solutions to current photovoltaic cost problems. Second, the knowledge gained by industry during the rapid development of semiconductors in the last two decades could be very useful in accelerating the development of photovoltaics. The analogies between semiconductor development and photovoltaic development are numerous; the most striking is the persistent correlation between production volume and unit prices. This experience curve phenomenon is the basis for a significant portion of the optimism in the field. (For a good introduction to this concept, see the report by the Boston Consulting Group.[52]) Other examples of the similarities between photovoltaic cells and semiconductor devices are that the same production processes and machines are often used, the same materials are being investigated, and many of the same companies are involved in both technologies.

The third reason for optimism about the future of photovoltaics is the diversity of private sector involvement and the amount of private funds being invested. As mentioned earlier, the companies involved range from small innovators to semiconductor/electronics firms to major energy companies. Each type of company brings a different combination of technical expertise, management ability, marketing strength, and financial backing. In addition, these companies are looking at a wide variety of ways around the cost problems. This combination of diverse capabilities and diverse areas of investigation significantly increases the probability of producing low-cost commercial photovoltaic systems. The competition among these firms, the added competition from companies in other nations, and the expectations of large profits from lowering costs create an atmosphere that is conducive to rapid industrial innovation.

A final reason for optimism stems from the interest of the governments of the United States and other nations in supporting the technology's advancement. Large expenditures of public funds for photovoltaic research and development are generally agreed to be both a necessary and an appropriate role of the government. Private sector expenditures for basic research and for high-risk, high-return approaches are usually minimal because of the long period before any return can be expected and the uncertainty of useful results. It is appropriate for the government to take a longer-range and broader view of the value of such research expenditures and heavily supplement private sector funds.

Over the next two decades, many decisions by suppliers, buyers, and governments will be necessary to translate this optimism into a significant energy contribution. Publicly available information on such decisions is understandably scarce. With knowledge of innovations now available to the photovoltaic industry and knowledge of how other innovations have diffused into society, the following scenario of photovoltaic's future can be constructed. (Although existing research gives some support to the scenario, it is primarily based upon personal opinion of the authors.) Through 1980, single-crystal silicon technology continued to be the dominant commercial production approach. Private and public investment in a wide variety of research projects continued. Markets for photovoltaic systems in communications and cathodic protection expanded through 1981.

In 1982, new markets in international settings will probably expand rapidly. Water pumping and general remote power sources will be two of the targets of the sales expansion. Semicrystalline and single-crystal silicon technologies will probably dominate commercial sales. It is also likely that suppliers in Japan, West Germany, and France will take significant steps to capture a larger percentage of the expanding photovoltaic markets. Sales of concentrating collectors could prove to be a major competitor to flat-plate systems in this period.

The photovoltaic industry will undoubtedly be undergoing many changes. In the early 1980s, many companies will make serious reassessments of their R&D progress and their corporate commitment to commercial production. Some firms involved only in single-crystal silicon technology may consider diversifying their product lines. Companies which have invested in more advanced concepts will be deciding whether to expand their R&D commitment or invest in production facilities. Some firms may decide to leave the industry because of the competition in cell prices.

Parallel to these private sector reassessments, the federal government may also reexamine its planned photovoltaic expenditures. Decisions will have to be made concerning the role of tax incentives to suppliers and users of photovoltaic systems. If such incentives become law, there could also be some reduction in advanced research expenditures by the federal government. The time period from 1983 to 1986 will be important to the development of new photovoltaic markets. Sales to current markets (e.g., communication and cathodic protection) will probably peak early in this period at sales rates somewhere below 7 MWp/year. However, the continued expansion of new international markets will more than compensate and allow total industry sales volume to grow rapidly. This expansion could add 50 to 100 MWp/year in sales volume to the industry by the mid to late 1980s.

There may also be a small number of U.S. residences equipped with photovoltaics between 1983 and 1986. Many of those residences will be government demonstrations, but a few innovators may purchase systems without direct federal assistance. Regulations and guidelines such as those associated with the Public Utility Regulatory Policies Act of 1978 (P.L. 95–617) will help these innovators overcome institutional barriers to the use of photovoltaics. Many firms in the photovoltaic industry will require commercial results for their R&D investment during the period from 1983 to 1986. A variety of pilot production lines based on advanced photovoltaic concepts will be installed. New products based on thin films may be introduced. Production advances and cost reductions in single-crystal silicon technologies will have to occur rapidly during this period or the market share of single-crystal silicon will be threatened. As prices for all types of flat-plate technologies decline, the market share of concentrating photovoltaics will be threatened. The lower-bound costs of concentrators is probably higher than those for flat plates. As prices approach that lower bound, concentrators will be used only in more specialized markets where they have advantages other than cost. The federal government will continue to stimulate the commercial development of photovoltaics in the 1983 to 1986 period. However, industry and not government action will determine the commercial fate of the technology.

In the late 1980s and 1990s, the development of U.S. grid-connected residential and commercial markets will be essential to the industry's continued growth. Competition from foreign suppliers will be vigorous. The rapid acceptance of innovation and production advances will be the most important factor determining which of the suppliers will prosper. Sales of large central-station photovoltaic systems to utilities to supplement peak and intermediate generation capacities could also begin during this period.

Each step in this scenario assumes that production and technological advances occur rapidly enough to sustain continual reductions in the price of photovoltaic cells well into the 1990s. It is easy to see from the scenario that a long sequence of events will have to take place in rapid succession to ensure the development of the technology. The future of photovoltaics rests on a plethora of decisions by producers, users, and government policy makers. Given the rapid advances that have been the hallmark of the technology and given its technical and market promise, it is likely that photovoltaics will play an increasing role in helping solve world energy problems.

REFERENCES

1. K. Lawrence, *Proceedings of the U.S. Department of Energy Environmental Control Symposium* **3,** 73 (1979).
2. E. Braun and S. MacDonald, *Revolution in Miniature, The Pace of Progress,* pp. 83–101. Cambridge, Mass.: Harvard University Press, 1979.
3. G. A. Lancaster and M. White, *Eur. J. Mark.* **10** (5), 280 (1976).
4. D. Costello, D. Posner, R. Koontz, P. Heiferling, P. Carpenter, and L. Perelman, *Objectives and Strategies of the International Photovoltaic Program Plan,* SERI/TR–52–250, p. 7. Golden, Colo.: Solar Energy Research Institute and Jet Propulsion Laboratory, 1979.
5. H. Ehrenreich, *Principal Conclusions of the American Physical Society's Study Group on Solar Photovoltaic Energy Conversion,* p. 7. New York: American Physics Society, 1979.
6. W. C. Breneman, E. G. Farrier, and H. Morihara, *Proceedings of the 13th IEEE Photovoltaic Specialists Conference,* p. 339. New York: IEEE, 1978. See Also L. P. Hunt, V. D. Dosaj, J. R. McCormick, and A. W. Ravcholtz, *ibid.,* p. 333.
7. D'Alessandro, *Solar Age,* December 1979, p. 13. See also U.S. Department of Energy, *National Photovoltaic Program, Draft Multi-year Program Plan,* p. 2–12. Washington, D.C.: U.S. Department of Energy, Division of Solar Technology, 1979.
8. H. Fisher and W. Pschunder, *IEEE Transactions on Electrical Devices,* pp. 438–41. New York: IEEE, 1944.
9. J. Lindmayer, *Proc. 13th IEEE Photovoltaic Specialists Conf. 1978,* pp. 1096–1100. New York.
10. H. Ehrenreich, Ref. 5, p. 75.
11. H. Ehrenreich, Ref. 5, p. 86.
12. R. I. Frank and R. Kaplow, *Appl. Phys. Lett.* **34,** 65 (1979).
13. R. L. Moon, L. W. James, H. Vander Plas, T. O. Yep, G. A. Antypas, and Y. Chai, *Proc. 13th IEEE Photovoltaic Specialists Conf. 1978,* New York, p. 859.
14. H. Ehrenreich, Ref. 5, p. 95.
15. R. Sahai, D. D. Edwall, and J. S. Harris Jr., *Appl. Phys. Lett.* **34,** 147 (1979).
16. J. C. C. Fan, *Tech. Rev.* **80** (8), 14 (1978).
17. H. Ehrenreich, Ref. 5, p. 106.
18. H. Ehrenreich, Ref. 5, pp. 120–22.
19. D. C. Reynolds, G. M. Leary, L. L. Anter, and R. E. Marburger, *Phys. Rev.* **96,** 533 (1954).
20. *Electronics,* June 22, 1978, p. 42.
21. S. Chu, T. L. Chu, and H. T. Yand, *Proc. 13th IEEE Photovoltaic Specialists Conf. 1978,* New York, p. 956.
22. O. Merrill, *Photovoltaic Power Systems Market Identification,* BDM Corporation, McLean, Va., for U.S. Energy Research and Development Administration.
23. H. Liers, *Photovoltaic Energy Technology Market Analysis,* Inter-Technology Solar Corporation, Warrentown, Va., for the U.S. Department of Energy.
24. BDM Corporation, *Photovoltaic Incentive Options, Preliminary Report,* (BDM/W–78–184–TR), BDM Corp., McLean, Va.
25. *Assessment of Solar Photovoltaic Industry Markets and Technologies, Draft Report,* Booz, Allen and Hamilton, Bethesda, Md.
26. Ref. 25, p. IV–26.
27. Ref. 25, p. III–2.
28. D. Costello, D. Posner, D. Schiffel, J. Doane, and C. Bishop. *Photovoltaic Venture Analysis,* Vol. 3, App. G, Solar Energy Research Institute, Golden, Colo.
29. U.S. Department of Energy, Division of Solar Technology, *Photovoltaic Program Summary,* DOE/ET–9901/1, p. 10.
30. Ref. 28, Vol. 1, p. 4.
31. S. Flaim, *Economic Feasibility and Market Readiness, Draft Final Report,* p. 328. Solar Energy Research Institute (TR–52–055), Golden, Colo.
32. J. E. Tilton, *International Diffusion of Technology: The Case of Semiconductors,* p. 183. Washington, D.C.: The Brookings Institute.
33. O. Merrill, Ref. 22; and H. Liers, Ref. 23.
34. *ibid.*
35. Solar Photovoltaic Energy Research, Development and Demonstration Act of 1978, Public Law 95–590, Section 11a.
36. General Electric Corporation, *Requirements Assessment of Photovoltaic Electric Power Systems.* Electric Power Research Institute, Palo Alto, Calif.
37. General Electric Corporation, *Regional Conceptual Design and Analysis of Photovoltaic Systems.* U.S. Department of Energy, Sandia Laboratories, Albuquerque, N.M.
38. Westinghouse Electric Corporation, *Conceptual Design and Analysis of Photovoltaic Systems.* U.S. Energy Research and Development Administration, Washington, D.C.
39. Aerospace Corporation, *Mission Analysis of Photovoltaic Solar Energy, Conversion: Major Missions for 1986–2000.* U.S. Energy Research and Development Administration, Washington, D.C.
40. Spectrolab, Inc., *Photovoltaic Systems Concepts Study.* U.S. Energy Research and Development Administration, Washington, D.C.
41. D. Costello and D. Posner, *Solar Cells* **1,** 37 (1979).
42. E. DeMeo and P. Bos, *Perspectives on Utility Central Station Photovoltaic Applications,* p. 47. Electric Power Research Institute, Palo Alto, Calif.
43. Solar Photovoltaic Energy Research, Development and Demonstration Act of 1978. See Ref. 35, Section 2(a), Finding 21.

44. U.S. Department of Energy, *System Tests and Applications Photovoltaic Program* (HCP/T4024–01/15), p. 327. Washington, D.C.: GPO, 1979.
45. U.S. Department of Energy, Ref. 7a.
46. D. Costello, R. Koontz, D. Posner, P. Heiferling, P. Carpenter, S. Foreman, and L. Perelman, *International Photovoltaic Program Plan, Volume I,* (SERI/TR–52–361), pp. 1–19. Solar Energy Research Institute, Golden, Colo., 1979.
47. Science Applications, Inc., *Characterization and Assessment of Potential European and Japanese Competition in Photovoltaics* (SERI/TR–8261–1), pp. 4–1 to 4–19. Solar Energy Research Institute, Golden, Colo., 1979.
48. Ref. 47, pp. 3–37 to 3–40.
49. Ref. 47, pp. 3–40 to 3–44.
50. Ref. 47, pp. 3–44 to 3–45.
51. R. Stobaugh and D. Yergin, *Energy Future,* p. 209. New York: Random House, 1979.
52. Boston Consulting Group, *Perspectives on Experience.* Boston: Boston Consulting Group, 1972.

SUGGESTED READINGS

Marshall E. Alper, "Photovoltaics and Solar Thermal Conversion to Electricity: Status and Prospects," *Journal of Energy* 3 (5), 263 (1979).

T. Kidder, "Future of Photovoltaic Cells," *Atlantic Monthly,* June 1980, p. 67.

P. D. Sutton and G. V. Jones, "Photovoltaic Systems Perspective," *Journal of Energy* 4 (1), 7 (1980).

Wind Power 24

DAVID RITTENHOUSE INGLIS

BACKGROUND

The development of civilization has depended heavily on wind power. From ancient times to the nineteenth century, wind-driven ships carried commerce, culture, and conquest worldwide as well as vital local cargo. From the middle ages, windmills grew in power as they rose to meet the increasing needs of the industrial revolution. Their great sails on wooden shafts turned wooden machinery to pump water, grind grain, and power sawmills and other industrial machines. Wind energy is abundant but widely dispersed, and large numbers of mills were required to meet the growing need for power. This expansion was halted as the mills were superseded by steam power in the nineteenth century and internal combustion engines in this century, both more compact and convenient sources of power, depending on the abundance of fuel. Even on into this age of abundant fuel, there were once some six million small windmills pumping water on American farms. In the 1930s, there were

also hundreds of thousands of the faster, two- or three-bladed windmills generating the small amounts of electricity for individual farms, until electric power lines were extended over the countryside.

Now that fuel is no longer cheap and its abundance is in doubt, we are slowly turning again to wind power as a potentially important source of energy to meet appropriate industrial and residential needs. Small windmills for individual residential use are again growing in number, and if millions of people can be induced to install and tend them, they could contribute importantly to national energy needs. In favorable places, they constitute an appropriate technology generating power where it is needed. However, their number is limited by lack of initiative and by lack of population dispersed in favorable windy sites. There is greater promise that really large amounts of power, commensurate with modern industrial and urban needs, could be generated by large arrays of large windmills commercially serviced in windy regions. Thus it would be appropriate that

Original article by Dr. David Rittenhouse Inglis. Dr. Inglis is Professor Emeritus of Physics at the University of Massachusetts-Amherst and has taught at Ohio State University, the University of Pittsburgh, Princeton University, and Johns Hopkins University. He has served for 20 years as Senior Physicist at Argonne National Laboratory and is author of more than 150 technical papers as well as textbooks on mechanics and on nuclear energy. This article contains excerpts from his most recent book, *Wind Power and Other Energy Options*, Ann Arbor: University of Michigan Press, 1980.

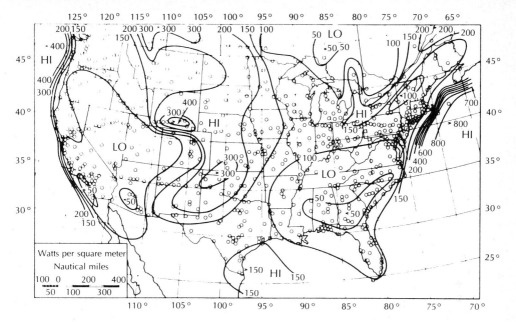

Figure 24-1. Average available wind power. The individual stations (small circles) show fluctuations about the smooth average. The values are in watts per square meter. (From J. W. Reed, "Wind Power Climatology in the U.S.," Sandia Laboratories, SAND 74-0348, June 1975)

the expanding use of large windmills, cut off over a century ago by the advent of cheap fuels, should now in a new age of impending fuel scarcity be resumed with full use of modern materials and technology.

Large windmills close to heavily populated areas may raise aesthetic objections. Some windmills have metal blades that cause distortion of nearby television, for example. However, large windmills will be most economical when located in the windiest regions, and these are mostly remote, unpopulated areas such as the western great plains, broad mountain passes near the West Coast, and the offshore region near the Grand Banks, where wind dynamos could be floated to supply power for the East Coast (Figure 24-1). The hopeful prospect is that great wind farms will be established consisting of thousands of large windmills in arrays perhaps half a mile apart, coupled with appropriate storage facilities and feeding into powerlines to the cities, where it is difficult to find sub-

stitutes for electric power. It may be expected that further innovation and mass production will reduce the costs of large wind dynamos enough that, even with the additional cost of transmission and storage to tide over calm periods, such wind power will be considerably cheaper than newly installed nuclear power.

Isolated pioneers in the innovative development of modern wind power technology have been at work in several countries all through this century, beginning in Denmark, where village electric power was gleaned from the wind even before the turn of the century. Of these efforts in the development of the modern wind dynamo, the largest and most impressive demonstration of all, and one that preceded all but the earliest European efforts, was the wind dynamo built and operated in the early 1940s on a small mountain called Grandpa's Knob, near Rutland, Vermont. As a result of enthusiastic promotion by Palmer C. Putnam, it was built by the S.

Morgan Smith Company, of York, Pennsylvania, a turbine firm interested in developing a new product. It fed 1250 kilowatts, or 1.25 megawatts, into the commercial power grid of the Central Vermont Public Service Company. From the company's first interest and the decision to proceed in 1939, it took only two years to complete the engineering design, build the machine, and start feeding power into the commercial electric grid. Some parts were produced early, in the race with wartime scarcities, and when the design was completed it was found that a part supporting the blades was not strong enough. It could not be replaced in wartime, but it was decided to proceed, despite the risk of breakage. In 1945, when the planned testing period was almost finished, the machine did break, throwing a blade, and the project was abandoned.

As the first big experimental machine, it was not expected to be economical and indeed cost just over $1 million including development. The experience gained was used to design a simpler model of similar size, and it was estimated that building a few of them would entail a capital investment of $191 per installed kilowatt, whereas $125, or 35 per cent less, would be required to be economical for the Central Vermont Public Service Company. The intent was to use wind power as a supplement to the main base of hydroelectric power, saving some of the water power for when the wind was light, but it turned out to be 35 per cent cheaper to use power produced by the cheap coal of that era instead.

In France, in about 1960, the development program of the national utility Electricite de France went directly to large three-bladed wind dynamos. The program developed an 800-kilowatt and a 1-megawatt machine, as well as a few smaller machines. The experiences with these were promising, and plans were instituted to build more large machines with two or four rotors, but the drop of oil prices in 1963 led to termination of the program and neglect of a needed repair on the 1-megawatt machine.

In West Germany, a 100-kilowatt wind dynamo was built near Stuttgart, with special attention to the aerodynamic excellence of its slender fiberglass composite blades. It fed into the electric grid from 1959 to 1968. An ingenious method of hinging the blades to the shaft adopted from this machine contributed to the success of the largest machine (built in 1980) of the U.S. federal wind power program, as we shall see.

Of the three windmills in the megawatt range and several more in the 100-kilowatt range built in the 1940s to 1960s, some had two blades and some had three. It is not clear from that experience which is better. The familiar American farm windmill has a dozen or more blades, almost filling the blade circle, but this does not necessarily extract more power from the passing wind.

Both types of windmill generate about the same amount of power for a given area swept by the blades, the one with two blades usually a bit more than the one with many blades because the former is designed with better aerodynamic efficiency. But they serve different purposes: the one with two blades turns faster to turn an electric generator of reasonable size without too many expensive gears. The slower one with many blades usually pumps water.

The slender fast blades usually move, at their tips, about six times as fast as the wind. After one blade passes, disturbing the air, the wind has not gone far before the next blade comes by, about as soon as it can and still find relatively undisturbed air, so that all the air is used. The wind piles up against the blade circle and makes the pressure greater on the upwind side than on the downwind side. In a way, the pressure difference really supplies the power for either windmill. When a fast blade slices past the wind, it deflects the air that it meets slightly upwind just as an airplane wing deflects downward the air that it meets. After the blade passes, the pressure difference deflects the air back downwind so that it is ready to meet the oncoming next blade at the proper angle to deliver power to it. Thus in the wide gap between the blades the pressure difference is doing work on the air near the blade circle, and that air later passes its energy on to the next blade. The pressure difference is working over the whole area all the time, empty though it may look. (Some air can give energy to other parts of the air by pushing it around!) Things are not always what they seem.

Some fast windmills have three blades rather than two, not to cover more area but mainly to obtain smoother gyroscopic action when the windmill swivels to face into a changing wind. A two-bladed rotor swivels more easily when the blades are in one position than in another and acts as a jerky gyroscope. It may be allowed to swivel only slowly to avoid troublesome vibrations.

RECENT DEVELOPMENTS

More recent development of large wind dynamos continues to be divided between three blades and two. In Denmark three blades have long been favored, and in 1977 the then largest wind dyanmo in the world was completed at Tvind on the North Sea coast by the faculty and students of three schools. The diameter of the circle swept by its three blades is 177 feet, the height of the slender cylindrical concrete tower about the same, and it generates 2 megawatts.

The U.S. federal wind power program has sponsored many designs of small windmills, but in the larger sizes it has been confined to two-bladed machines, following the Grandpa's Knob precedent. After a slow start on windmills in the 100-kilowatt range, the first megawatt-scale machine, known as Mod 1, was completed on Howard's Knob, in North Carolina, in 1979. It was completed seven years after the start of the program, in contrast with the two years or less that it took to design and build the Grandpa's Knob windmill. The General Electric windmill at Howard's Knob has its two blades rigidly fixed at right angles to the shaft. As mentioned above, a two-bladed rotor acts as a jerky gyroscope, making the blades exert troublesome vibra-

tory stresses on the shaft and requiring extra strength. The old Smith-Putnam machine at Grandpa's Knob avoided this by having large hinges to permit a limited flapping motion, as can be seen in the pictures of it. Having three blades rigidly mounted is another way to avoid this difficulty. With two blades, which may be cheaper to build, there is another trick to avoiding this difficulty, known as teetering, a method also used in two-bladed helicopter rotors. The two blades are mounted together like a stiff beam hinged to the shaft at the middle to permit the beam to teeter about the position at right angles to the shaft. The wind pressures on the two blades are almost balanced, and so the stress on the shaft is less than with separate hinges.

This good idea, inherited from the 100-kilowatt West German windmill, is used in the second large machine of the federal program, Mod 2, the first unit of which was completed in Washington state by the Boeing company in 1980 (see photograph, page 388). Two more were built soon after. This tremendous machine, rated at 2.5 megawatts in a moderate wind, has a blade span of 300 feet and a tower height of 200 feet. Most modern machines have blades with adjustable pitch for speed control and to permit their being feathered to minimize forces in a storm. For economy, some have fixed pitch. In Mod 2 it was found cost effective to provide pitch control for only the outer 30 per cent of the blades. With its advanced engineering and clean design, this is the first of the big machines, other than the Grandpa's Knob design, that is considered suitable for quantity production. While the first units built cost a lot more, Boeing estimates that the one-hundredth machine in quantity production would cost $2.5 million in 1980 dollars, or $1000 per kilowatt. Because of unsteadiness of the wind, a wind dynamo in a good location is expected to generate, on the average, only a bit more than one third of its rated power. For nuclear power, with which the cost of wind power should be compared, the corresponding figure, because of servicing time and taking care of various troubles, has in most experiences only been a bit over one half (e.g., in 1980 the plant factor was 51 per cent). These figures mean, incidentally, that it takes about 750 large 2-megawatt wind dynamos to produce the average power generated by one 1000-megawatt nuclear power plant. With this difference in plant factor taken into account, the capital costs for a future nuclear plant that one might decide to build are not much less, if less at all, than the cost of equivalent Mod 2 wind capacity. If one adds the other dollar nuclear costs, including the uncertain costs of fuel and decomissioning, estimated Mod 2 generating costs appear to be less than nuclear.

ECONOMIC AND POLITICAL CONSIDERATIONS

That Mod 2 cost estimate, however, though based realistically on recent experience building a large modern windmill, is not to be taken as the best estimate of future large windmill cost. It should be less. If the market develops in such a way that individual initiatives can be adequately financed, it is likely that competition between suppliers with a variety of designs will reduce costs substantially. Mod 2 is the crowning achievement of a federal wind energy conversion program that seems to have a tendency to inflate development costs. Ever since it was transferred from the National Science Foundation to the Atomic Energy Commission just after its inception in 1972, the wind energy program has been under the wing of the successive administrations of nuclear power, AEC, ERDA, and DOE, with a director recruited from the aerospace industry and with a bias toward awarding contracts to large aerospace companies to the exclusion of independent entrepreneurs who deserved financial assistance.

Until about 1978, the program was hampered by a directive from the nuclear-oriented top management that there should be only one large-scale demonstration project at a time ("sequential development"). The initial development was over-conservatively

kept on the 100-kilowatt scale in contrast with the 1939–41 private effort that went directly to the megawatt scale at Grandpa's Knob. The Department of Energy was organized in 1976 and has, among its many parts, one division charged with research and development and another charged with promoting rapid utilization of new developments. Wind energy conversion was kept confined to the development division, with assistance from the National Aeronautics and Space Administration (NASA). In judging the slow and narrow progress, whether or not this may have unconsciously influenced decisions, it should be appreciated that it was to the personal advantage of the administrators of the wind program to extend the development phase as long as possible.

With all its bureaucratic limitations, the federal program has met with gratifying success in developing one type of large windmill through a succession of models and a process of trial and error. The work involved successive groups of engineers having no previous interest in wind power, although some of them brought relevant aeronautical skills. Even though the trade of millwright is an old one, it is not to be assumed that a greater variety of modern windmills could have been built much more easily under more enthusiastic administration. The proper design of any type of modern windmill is a serious engineering challenge, involving not only efficient and economical construction but, in particular, minimization of vibrations and their influence on materials fatigue and the lifetime of components. Yet one feels that the national interest in the rapid exploitation of large-scale wind power would have been better served if various teams of enthusiastic engineers had been supported in the pursuit of a variety of types of large windmills during the long span of the federal program.

TRENDS IN WINDMILL DEVELOPMENT

On an international scale, some such variety of experience is being acquired, perhaps belatedly, by publicly supported developments in progress in various countries, as well as by a few independent industrial efforts.

Of the national large-scale wind energy programs, that of the West Germans seems to be the most imaginative and innovative, but completion of the very large machines being planned is not expected until the period 1982–85. The first, known as Growian I, is to be quite conventional, rather similar to Mod 2 in size and general characteristics: a two-bladed, downwind rotor and slender tower, but with a different hub structure and taller tower. It is rated at 3 MW in a 26-mile-per-hour wind. The second, Growian II, is to be much larger with only a single blade (!), following an early experimental demonstration by Professor Hütter. The single blade is counterbalanced and teeters, being held almost normal to the axis against the wind by centrifugal force. There is a prototype in nature in a maple seed, which slows its fall and rides on the wind by flying like a helicopter (or autogyro) with a single wing counterbalanced by the seed itself. Growian II will be rated at 5 MW and have a blade circle 475 feet in diameter on a tower 360 feet high. The concept is first to be tested with a one-third scale model.

The third large wind power project is even more imaginative. It is more a matter of direct solar energy than natural wind energy, using solar heat to create an artificial vertical wind to drive a turbine. The solar collector is a very large, circular roof just above the ground connecting at the center to a cylindrical tower up which the air heated under the roof rises to drive a wind turbine—a thermal-upwind power plant. The units being considered are huge, 10 to 100 megawatts. The 100-megawatt one would be 3 kilometers in diameter.

The Swedish megawatt-scale program is similar to the American program, in that it is confined to two-bladed turbines. For the sake of direct comparison of various engineering details, two large windmills with about the same dimensions are being constructed in parallel and due for completion in 1981. The two machines are similar in having about a 250-foot blade span and cylindrical towers 260 feet high, but they purposely differ in many respects: construction by different firms, rated power of 2 or 3 megawatts in winds of 28 or 32 miles per hour, "soft" steel or rigid concrete towers, downwind or upwind rotors with all-glass-epoxy or partly steel blades, and different gear boxes and generators. While well-planned for assessing these design options, the program seems to be progressing no more rapidly than the American program and is indeed about a year behind it in attaining the megawatt scale. The costs to the government for the large demonstrations are about half as great.

The Danish program concentrates on three-bladed rotors. Two wind dynamos of identical dimensions have been built, differing only in control mechanisms and rotor design, again for the sake of testing the viability of variations in these features by direct comparison of the two. Each machine is rated at 630 kilowatts in a 29-mile-per-hour wind, near enough to a megawatt to provide an important variety of experience applicable to megawatt-scale turbines. One of them has a rotor braced with stays, as in the old Gedser mill, but with the stays extending only one third of the radius. Beyond that, the pitch of the blades is somewhat variable, permitting stalling for overspeed control. Small wind dynamos in the range from 10 to 55 kilowatts for connection to utility grids are being demonstrated and manufactured in Denmark by about a dozen different firms. Most of the models are three-bladed, horizontal-axis machines. Most of them are being tested at a government test facility similar to that at Rocky Flats in the United States.

Partly as an offshoot of the Swedish program, the Hamilton-Standard Company of Connecticut, experienced in making windmill blades, teamed up with a Swedish firm to design a 4-megawatt wind dynamo, again a two-bladed machine. It has contracted to construct five units for the Water and Power Resource Service (formerly Bureau of Reclamation) of the U.S. Department of the

Interior for installation at the department's wind power site in windy southern Wyoming in 1981. The sale was, at about $1600 per installed kilowatt, by underbidding Boeing's promotion of Mod 2, a first example of healthy competition in providing megawatt-scale wind power equipment. This marks a beginning of what may become a very large undertaking to supplement hydroelectric power with wind power, letting the wind supply electric power when the wind is strong and saving water in the reservoir to generate power when the wind is weak. This represents the cheapest type of "storage" to back up intermittent wind power. It will doubtless take more than a decade to install enough wind dynamos in wind farms to exhaust the capacity of existing hydroelectric facilities for this purpose. Beyond that, the best method of energy storage with present technology is pumped hydroelectric storage, as already used rather extensively by utility companies to provide peak-load power without building more power plants. In this system, electrical motors pump water from a low reservoir to a high reservoir, and for output the pumps and motors are reversed to serve as turbines and generators as the water flows back down again. One or both reservoirs may be underground, perhaps old mines. In contrast with the wind supplement to existing river-flow hydroelectric capacity, pumped hydroelectric storage requires an extra stage of pumping and is only about 70 per cent efficient, electric back to electric power. In a windy region where the wind is seldom completely calm, only about one quarter of the energy generated may go through storage to experience this inefficiency and the average loss of power thus is small. (See Figure 24-2 for overall efficiencies for various storage techniques.)

The Department of the Interior has long been engaged in building big dams that provide irrigation to be sold to private growers and generate electricity to be sold to utility companies for distribution. Government investment in wind power to enlarge this operation can, if properly administered, provide a competitive market encouraging

independent initiative in the development of a variety of large windmills.

Fortunately, two large utility companies have already in 1980 recognized the importance and prospective economy of large-scale wind power and have budgeted substantial funds establishing the kind of market needed on an appreciable scale. It may be hoped that this market will appear large enough to attract keen competition in economic and reliable design. Southern California Edison's plans call for 320 megawatts of wind-generated electricity, and Hawaii Light and Power has acquired a site on northeastern Oahu, where it has contracted for the installation of 80 megawatts.

One independent private entry into the megawatt field at first seemed to promise a drastic reduction of costs below the Mod 2 level but unfortunately faltered for lack of financing before completion of its first very big machine. The family engineering firm of Charles Schachle and Sons, Moses Lake, Washington, carried out a five-year experimental program with smaller windmills. It involved $2 million of private investment and culminated in the successful construction of an innovative 440-kilowatt prototype windmill intended to be scaled up to much larger sizes. Its three composite-wood blades are supported on a tripod-like tower rotating on a ground-level track to face the wind in place of the usual yaw mechanism aloft. Power is transmitted hydraulically to the electric generator so as to allow the blades to turn at an efficient speed to suit the wind while generating constant-frequency electricity.

The next version proposed was much larger, rated at 3 megawatts in a 40-mile-per-hour wind, and was to have been sold to Southern California Edison. Plans called for installation at a windy site near Palm Springs, for just over $1 million (1978 dollars), or $350 per kilowatt. There were cost overruns for parts, and the firm ran out of funds midway in the project and had to sell out to a large company, Bendix. There were long delays organizing a new division, which imposed corporate overhead costs. Company engineers with no previous windpower ex-

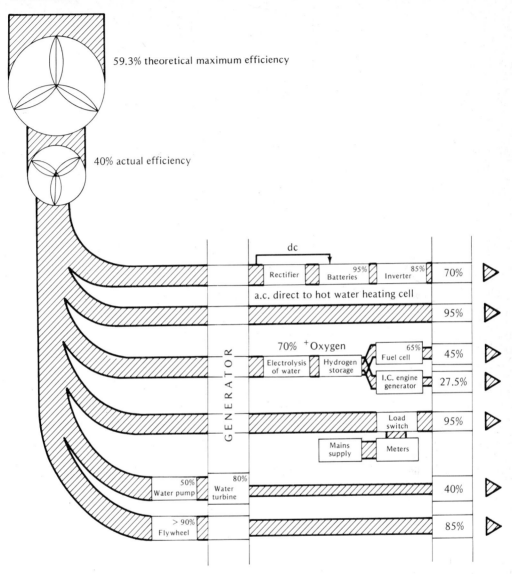

Figure 24-2. Overall efficiency from windmill to end use. (From *Technology Assessment of Wind Energy Conversion Systems*, DOE/EV-0103, UC-58b. Los Alamos Scientific Laboratory, September 1980)

perience disapproved of the hub design and redesigned it, making it weaker and a limitation on output power. When the machine was completed in late 1980, the final cost was about four times that originally anticipated (see photograph, page 3). The moral

of this sad tale seems to be that it is in the national interest to provide financial help for enthusiastic and independent initiatives.

The success with smaller windmills by companies not associated with aerospace indicates that costs much lower than the Mod

2 estimates are possible. Thus, kilowatt-range windmills should be more economical than megawatt-range ones. A small independent firm, U.S. Wind Power of Burlington, Massachusetts, is putting into quantity production a carefully designed, three-bladed, 50-kilowatt wind dynamo and has contracted to erect many of them on a wind farm in California to sell electric power at a competitive 3.5 cents per kilowatt hour. The wind dynamo cost is said to be about $500 per kilowatt. Over 20 of these machines in a group in southern New Hampshire make up the world's first wind farm on such a scale, feeding power into the local grid.

THE ROLE OF GOVERNMENT

One may ask, if wind power is so good, why has it not taken off more rapidly on its own without government help? After all, the automotive industry took off on its own, though it required about 20 years to reach the economy of mass production with the Model T Ford. A big difference is that the automobile had only the horse to compete with, whereas wind power must compete with established power sources that have been kept artificially cheap by government policy through tax incentives, subsidy, and direct federal participation in development, as well as by the bounty of Nature in supplying cheap fuel and the atmosphere as a refuse dump. Without a higher priority of government help than it had through the 1970s, wind power would probably become a major energy source slowly by the next century. With the stability of the national economy being threatened by scarcity and rapidly rising costs of fuel, it is very much in the national interest that wind power should come into prominence more rapidly than that. Technically, and eventually economically, it is entirely possible. It would be appropriate for the government to back wind power as heavily as it has other energy sources, or even more heavily if needed because wind power is more benign.

Energy payback time is another important consideration. It has been estimated that it takes a big wind dynamo about seven months to generate as much energy as was invested in building it and making the material in it. The corresponding time for a nuclear reactor is estimated as two or three years. Also, it takes more than a decade after ordering for a nuclear power plant to start generating power. Even though it may take that long to complete a large wind farm, the first units of it should start generating within a couple of years or so.

Although the Mod-2 and Hamilton-Standard/Swedish developments are constructive steps toward the goal of large wind farms on land, there has been no corresponding progress directed toward floating offshore wind power. Indeed, the DOE has deliberately avoided being so innovative as to investigate this promising prospect. Instead, it paid Westinghouse $500,000 for what was supposed to be a feasibility study but which deliberately undertook to prove that floating offshore wind power would be too expensive. The study ignored good nautical designs that had been suggested and proposed instead to mount existing land-based designs on rafts with many anchors to steady them. The appeal of offshore wind power arises from the fact that the winds are stronger and steadier some tens of miles offshore than on land. Another promising approach that deserves investigation in a broader development program is high-tower wind power, another way to exploit stronger and steadier winds aloft by use of towers some 500 feet high or more on land.

The importance of seeking strong wind regimes for the placement of windmills is accentuated by the fact that the power that a turbine of a given size can extract from the wind is proportional to the third power of the wind velocity, v^3. This is because the kinetic energy brought in by a parcel of air is proportional to the second power, v^2, and the rate at which the wind brings this energy is proportional to another factor, v. This cannot all be extracted because the wind still has energy as it moves out of the way on the downwind side. There is an oversimplified theory that the maximum energy extracted would be 59 per cent of the wind's kinetic energy. In practice, it is usually slightly under 50 per cent.

Congress has displayed impatience with the slow pace of the federal wind energy program and has several times appropriated for it more money than was requested. Legislation in 1975 mandated the offshore feasibility study that was inadequately prepared by DOE. In 1980, legislation recognized the need to establish large wind farms and to back independent initiatives in large-scale wind power development. The Wind Energy Systems Act of 1980 directs the Department of Energy to take new initiatives both to accelerate the construction of large wind turbines and to broaden the scope of the program. It sets a goal of 800 megawatts by 1988 from a total of 387 large (mostly megawatt-scale) windmills, partly or fully financed by the federal government, to be installed at a rate increasing from 10 large turbines a year in 1982 to 162 per year in 1988. The projected cost of the eight-year program is about $1 billion (1980 dollars), the actual appropriation for fiscal 1981 being $100 million, of which $65 million is for large-scale prototypes.

To the previous government role of supporting research, development, and demonstration, this legislation adds the new role called "technological applications," meaning installation of wind energy systems. Besides direct government purchase, both grants and loans covering part of the costs are to be made to outside organizations. The DOE is also to supply funds for large wind systems to the Water and Power Resources Service and the federal power-marketing agencies, such as Bonneville. In a longer perspective, the bill mentions a goal of 1.7 quads per year (or 57 average gigawatts) from wind by the end of the century, but this would seem to justify a more vigorous start now.

By way of encouraging the independent initiatives that might provide some broad experience needed, the DOE is authorized to make grants for demonstrating "a variety of prototypes of advanced wind energy systems under a variety of circumstances and conditions."

The goal of 800 megawatts by 1988 requires a rapid step-up of emphasis on wind power beyond earlier plans, but it is not clear that the provisions in the bill are sufficient to attain this, since the decision to carry out plans is left to a department dominated by interest in other energy sources. The goal is high enough to call for very substantial industrial efforts and aggressive administration, yet it does not represent giving wind power a really high priority. The anticipated $1 billion in eight years is a little over 1 per cent of the $88 billion anticipated for government promotion of synthetic fuels during that period, a competing energy source which requires a correspondingly massive new industrial expansion and which has serious ecological disadvantages. With trends toward budget cutting, it is not even clear that the $1 billion by 1988 will actually be appropriated, making attainment of the goal of the 1980 Wind Power Act still more doubtful.

Even though the goal of achieving 800 megawatts of wind generating power in eight years represents an impressive speed-up of the pace, in the larger perspective of America's enormous industrial capacity this goal seems modest indeed. As an exercise in perspective, the reader might estimate what fraction of the productive capacity of the U.S. automotive industry would be required to produce in a single year enough large wind dynamos to meet this eight-year goal. Assume that about as much industrial capacity is needed to make a million dollar's worth of wind turbines at $1000 per kilowatt as to make a million dollar's worth of autos and trucks averaging $8000 each, noting that there is capacity to produce about 10 million of them a year. Answer—about 1 per cent. In other words, on these order-of-magnitude assumptions, 1 per cent of automotive production capacity could produce roughly the wind power equivalent of one large nuclear or coal-fired generating plant per year.

There is a serious philosophical and economic question concerning whether government should be in business at all, either creating a synthetic fuel corporation or building big hydroelectric dams or separating isotopes to produce nuclear power or rescuing auto companies from bankruptcy or establishing wind farms. Government subsidies and tax incentives, such as the so-called depletion allowances, have helped keep fossil-fuel energy cheap, particularly oil. The government has been most directly involved in the nuclear energy enterprise, not only in preparation of fuel by isotope separation but also in radioactive waste disposal, insurance coverage of calamitous accidents by the Price-Anderson Act, and other props without which the shaky nuclear industry would collapse. Besides, the government gave industry the nuclear technology fully developed in the first place, a spin-off from military developments. There has been political pressure— ironically, largely from proponents of nuclear power—to get the government out of business and let competition in the marketplace make decisions. To be consistent, the government should either get out of the energy business completely, thus phasing out nuclear power and probably deflecting massive capital investment to wind power, or else distribute its support in a way to help desirable new developments like large-scale wind power overcome the head-start that government support has already given to established power sources.

SUMMARY

In light of this help given competing energy sources, it is very much in the national interest that federal funds should be used to accelerate the growth of large-scale wind power. Private enterprise unaided will achieve large goals eventually. We have seen that at least two large utility companies appreciate the potential of wind power already, and this will encourage some competition among suppliers. But the government is in a position to do this in a much bigger way, initially supplementing its hydroelectric facilities. The larger the prospective market, the more capital will be attracted into the competition and the greater the likelihood that new economies will be discovered, further accelerating use of wind power.

The energy of the wind is very diffuse. This means that large numbers of large machines are needed to harness enough of it to supply an important part of our energy needs. This may be seen as a disadvantage, but it is also an advantage, for it means wind power is labor-intensive. Good engineering is needed, but it is not really high technology. Much of the cost is in making big machinery and erecting it in the field, requiring a lot of labor relative to other energy sources. The government is much concerned with unemployment as well as underutilization of industrial capacity. This is one more reason it should be accelerating the use of wind power. Among the promising energy technologies that use no fuel and do not pollute the atmosphere, wind power is the one that is technically ready to go in a big way now. It deserves a very high priority in national planning.

SUGGESTED READINGS

Frank R. Eldridge, *Wind Machines*, National Science Foundation, RANN, Grant No. AER-75-12937. Washington, D.C.: NSF, 1975.

Charles E. Engelke, "A Self-Contained Community Energy," *Bulletin of the Atomic Scientists* **34** (11), 51 (1978); **35** (4), 51 (1979).

Roger Hamilton, "Can We Harness the Wind?" *National Geographic* **148,** 812 (1975).

William E. Heronemous, in L. C. Ruedisili and M. W. Firebaugh, eds., *Perspectives on Energy*, pp. 364–76. New York: Oxford University Press, 1975.

David R. Inglis, *Wind Power and Other Energy Options*. Ann Arbor: The University of Michigan Press, 1980.

David R. Inglis, "Power from the Ocean Winds," *Environment 20*, October 1978.

David R. Inglis, "A Windmill's Theoretical Maximum Extraction of Power from the Wind," *Amer. J. Physics* **47**, 419 (1979).

S. Lieblein, ed., *Large Wind Turbine Design Characteristics*. Washington, D.C.: U.S. Department of Energy, 1979.

M. F. Merriam, "Wind Energy for Human Needs," *Technology Review* **79,** 28 (1977).

Daniel M. Simmons, *Wind Power*. Park Ridge, N.J.: Noyes, 1975.

Bent Sorensen, "Wind Energy," *Bulletin of the Atomic Scientist* **32,** 38 (1976). The ninth article in a series on solar energy.

Ocean Thermal Energy Conversion 25

ABRAHIM LAVI AND GAY HEIT LAVI

An imaginative historian looking at the ways humanity has used the oceans—which bathe three quarters of the globe—could trace the course of the civilized world. First came the drive to explore the shoreline, then the urge to sail beyond the horizon to link continents and to harvest a sea teeming with schools of tuna and pods of whales. Gradually, as science advanced, people discovered that the sea could serve for still more: that the water many fathoms down is rich in nutrients that could—in theory, at least—furnish all of humankind's protein; that the ocean floor is littered with nodules of valuable copper, cobalt, nickel, and manganese; that under the ocean floor lie vast deposits of coal, natural gas, and oil. And today, facing limits to growth on land, we are compelled to explore the ocean frontier for new sources of energy. In the next century, if we are wise in our exploitation of the oceans, their renewable energy may come to provide much of our industrial and domestic power.

Various technologies for gleaning electric power from the sea have been proposed,

though some are further along than others. They include the use of tidal power, ocean currents, wave motion, salinity gradients, and, most promising of all, ocean thermal energy conversion (OTEC). The last is an area of solar-energy research familiar to few outside its still-small circle of specialized engineers and advocates. Of all solar options, it has received the least publicity, and yet its enthusiasts feel it has the greatest worldwide potential.

HARNESSING THE DEEP

Tidal power is the first ocean-energy system to have been commercially developed. An installation in France has been operated successfully for years, demonstrating that the economic uncertainties of tidal power are minimal and that its environmental consequences can be made tolerable. However, an efficient tide-powered turbine requires a site with really powerful tides, and the number of sites that can economically produce power is limited. In North America, engi-

Dr. Abrahim Lavi is a professor in the Department of Electrical Engineering at Carnegie-Mellon University, Pittsburgh, Pennsylvania. He and Gay Heit Lavi are affiliated with Energy Research and Development International, Inc. of Pittsburgh.

neers are considering the Bay of Fundy/ Passamaquoddy, Cook Inlet, Ungana Bay, and Frobisher Bay. But even if every potential site on the globe were harnessed for tidal power, the technology would still contribute only a small fraction of the world's energy needs.

Another energy technology calls for turbines to be turned by the force of ocean currents, much as a windmill's blades are turned by air currents on land. The current-driven turbogenerator, like the windmill, would have to be reoriented continuously to face prevailing currents, since ocean currents, unlike river flow, have variable velocities and at times their course may even completely reverse. The environmental impact of these current-driven systems would probably be benign, but the cost of developing the technology would be quite high. Like tidal-power plants, they would be practical only in a few select spots around the world.

Much more energy could be harvested from the motion of the waves themselves. Interest in wave energy dates back many centuries. For decades, inventive minds have been proposing clever contraptions designed to capture a fraction of what appeared to be a limitless resource. (The U.S. Patent Office issued patents on wave machines as early as 1896.) But commercial development has been slow—so far most applications are aimed at powering remote navigational buoys. Any wave machine, regardless of its operating principle, requires relative motion between two of its elements. Usually, one element is physically constrained or so shaped that its motion relative to a "moving" and "bobbing" element is minimized. A large-scale sea trial of a "pneumatic" wave-energy system has been under way since 1978 in the Sea of Japan, under the auspices of the International Energy Agency. The system consists of two 125-kilowatt turbines, one built by England, the other by Japan.

Ocean-wave conversion has certain inherent limitations as well. As any captain or armchair sailor knows, ocean waves have broad spectra. At times, the power plant would be buffeted by waves too powerful for the machinery to withstand safely, and so its components must be designed to survive storms and hurricanes. During calms, the plant would cease to generate power; to provide energy continuously, it would have to incorporate an energy-storage system.

Still, because the resource is universally available, its magnitude significant, and the technology simple, wave-energy conversion is bound to become commercial within this century, albeit on a limited scale. Growth will depend on the availability of competing energy technologies. It seems realistic to expect that, by the year 2000, wave-energy conversion could supply 10 gigawatts of the world's energy requirements. (A gigawatt is one million kilowatts; the world consumed about 8000 gigawatt-hours—about 240 quads —of power in 1975.)

Yet another ocean-based energy technology relies on salinity gradients. The osmotic pressure difference between salt-free and salty water can be utilized to provide the power necessary to operate a hydroelectric plant. This is possible because the magnitude of the pressure difference between seawater and fresh can range as high as 25 atmospheres. Such a system requires that a selective membrane be stretched between two water masses—at the mouth of a river, for example. Construction of such a huge membrane would be an engineering marvel and might pose serious ecological problems for the river it blocked. Salinity systems, relative to other ocean technologies, remain in the earliest stage of development.

SOLAR-HEATED SEAS

OTEC is a scheme to extract energy from the solar-heated surface of the sea. Of all the ocean-energy systems, it is probably closest to worldwide commercialization. According to its advocates, it promises to contribute most to the world energy budget.

OTEC's elements are little different from those of the conventional steam power plant, and its principles of operation are so simple that it has been called nothing more than "sophisticated plumbing."

In the steam power plant, water circulates between two large containers, a boiler and a condenser. In the boiler, water is heated and the resulting steam then rushes through the blades of a turbine on its way to the condenser. It is the turning of these blades that drives the generator and provides electricity. In the condenser, the steam is returned to its liquid state and recirculated to the boiler, where it can be reheated to repeat the same closed cycle. Conventional steam power plants use coal or oil to boil water and water pumped from a nearby river or lake as the condenser's cooling fluid.

In an OTEC plant, a similar cycle occurs. But, unlike the steam power plant, OTEC requires no unrenewable fuel. Instead, warm water from the ocean surface provides the boiler's heat source and cold water from the ocean depths serves as coolant. Plants would be located in the tropics, where temperature differences between surface and deeper layers of ocean water are greater year round than in more temperate climates. In the tropics, the sun heats a layer about 100 feet deep to temperatures as high as 80 °F. At depths of 1500 to 3000 feet, temperatures drop to as low as 40 °F.

Another difference between OTEC and steam power is that OTEC does not use water to drive the turbine blades. Ammonia, which boils and condenses at the available ocean temperatures, would most likely be used as a working fluid. Because the temperature difference between boiler and condenser is much smaller than that of the conventional steam power plant, the OTEC process is much less efficient and therefore requires large-scale equipment. Enormous amounts of ocean water and working fluid must pass through the cycle to generate enough energy to make the scheme worthwhile.

The pipe used to bring cold water from the ocean depths into the condenser presents one of the system's chief engineering challenges. It is not exactly an off-the-shelf item, and

state-of-the-art technology offers no working model upon which OTEC engineers can pattern their various designs. Imagine a pipe with a diameter of about 50 feet and a length ranging from 1500 to 3000 feet. Remember that such a pipe must be built to withstand tremendous forces and to handle a flow as large as five million gallons of ocean water per minute. At first glance, the prospect of building such a pipe, moving it to its ocean site, and getting it to work seems a monumental and perhaps impossible undertaking. Nevertheless, engineers have developed designs they think will do the job.

The final design will result from trade-offs between cost and efficiency. The longer the pipe, naturally, the deeper it would extend into the ocean. Water pumped from 3000 feet down would be colder than water pumped from 1500 feet. The longer pipe would achieve a lower temperature in the condenser and thus would maximize the temperature differences available to the entire system. But the longer the pipe, of course, the more material is needed to build it, the more it costs to make, and the harder it is to deploy.

OCEAN HIGH-RISE

As an ocean-based system, OTEC requires a plant platform similar to the hull of a ship. Made of steel and (perhaps) prestressed concrete, the platform would have to be equipped with all the safety features of an ocean liner. The deck would have lights, a fog horn, a radar system, and a radio. Viewed from the outside, the plant would very much resemble a supertanker or a large aircraft carrier. The size of the platform would depend on the plant's power capacity. In 1975, TRW designed a 100-megawatt plant calling for a hull 340 feet in diameter and 170 feet high—a floating structure roughly three football fields long and 20 stories high! Recent designs employing more efficient equipment require a somewhat smaller platform.

The platform can be moored in one spot

or given some mobility. If it is to be moored, the line will have to be strong and the anchor may weigh as much as one million tons. Dynamic positioning is a bit more complicated. Electrically driven thrusters, or the trust of warm and cold water discharged from the evaporator and condenser, would be used to create forces that counteract the drag force of ocean currents. A line and anchor might be simpler, but dynamic positioning would permit the plant to "graze," keeping itself on location on top of a seasonally shifting thermal resource.

What distinguishes OTEC from other ocean (and solar) energy systems is its day-night and year-round operation. Alone of all the solar options, OTEC produces baseload power, relying on the ocean for thermal storage. While the plant can, in principle, operate year round, even when the sun fails to shine, its output may vary seasonally as the available ocean temperature difference between surface and deep water changes. Such seasonal variability depends on the distance of the plant site from the equator. For plants located in the Gulf of Mexico and connected to the southeastern United States electric grid, the seasonal variation of plant output conveniently parallels the seasonal demand for electricity, which is high during summer and low during winter.

OTEC could readily supply 30 per cent of the electricity used by the southeastern United States. Economic analyses indicate that the system can be cost competitive with conventional power plants for practically all tropical islands where imported oil is the only available fuel. The annual extractable ocean-thermal resource worldwide is conservatively estimated at 80 quads—roughly one third of the world's total energy consumption in 1975. Engineers concede that, limited by available seawater temperature, OTEC is indeed a less efficient system than standard heat engines fed by coal or oil. But the sun offers OTEC's fuel free of charge, and with the rising cost of fossil fuels, the advantage of OTEC becomes clearer and clearer.

OTEC 1

Though many OTEC proposals have been put forward, problems remain. Much of the innards of the plant must be made of metal that is both a good conductor of heat and a reliable resister of corrosion. Whatever metals are used will necessarily be in constant contact with seawater, and biofouling—the growth of slime—presents a messy engineering problem that will have to be solved.

In an attempt to study these problems, the Department of Energy has heavily weighted its current budget for ocean-energy research and development toward OTEC. Indeed, OTEC will receive more than 90 per cent of the 37 million dollars that DOE requested for ocean energy research and development for 1982. The United States leads the world in OTEC research, followed by the Japanese and the French (the British are more interested at the moment in wave technology).

In 1979, several major energy companies, in cooperation with the state of Hawaii, launched and operated Mini-OTEC, a 50-kilowatt trial system. The test proved beyond any doubt that state-of-the-art machinery can produce electricity from the ocean-thermal gradient. Currently, DOE is conducting an important series of tests of the OTEC concept. A World War II Navy oil tanker, *S.S. Chapatchet,* pulled out of mothballs, has been revamped to carry a 1-megawatt evaporator and condenser. Rechristened the *S.S. Ocean Energy Converter* (OTEC 1, for short), it is floating 18 miles off the northwest coast of the island of Hawaii, supporting several cold-water pipes 2200 feet long and 4 feet in diameter.

OTEC 1 carries no generator because the power-production cycle of the system is considered well understood. Rather, researchers are currently using the prototype to study two major outstanding technical issues. First, there is the problem of biofouling: once a slight slick of slime forms within the system, it grows rapidly. Somehow, the tubes must be kept whistle-clean at all times. One cleaning method employs thousands of sponge-rubber spheres, about the size of pingpong balls, coated with a mild abrasive. The balls are fed in with the water and are carried along by the water itself, squeezing through the narrower pipes and tubes—cleaning as they go. A small amount of chlorine is circulated in the system as well. (The threat of slime must be taken seriously because, as we have seen, OTEC operates on a temperature differential of only about 40 degrees; any extra resistence to heat flow caused by slime within the tubes will lower OTEC's already modest efficiency.)

OTEC 1 will provide other data as well. A research team on board is studying the effects of water discharge and chlorination on the local aquatic environment. The Environmental Protection Agency is monitoring this work closely. Engineers are convinced the system will not damage marine life or pollute the ocean. But bringing deep layers of seawater to the surface steadily and in gigantic quantities might very well have dramatic local ecological effects. There might be unforeseen risks that can be assessed only by on-site testing. Should OTEC's discharge of water disturb the environment too greatly, far-reaching changes in world weather conditions might even occur.

But at least some environmentalists agree with engineers that the plants' environmental effects would be fairly benign. Indeed, Jacques Cousteau, the oceanographer and environmentalist, thinks the uplift of cold, deep water, which is rich in nutrients, might attract tremendous amounts of marine life to the site. He suggests that certain plants could be used to build a mariculture industry—a kind of nursery in the sea.

Whether this turns out to be possible or not, the benefits of OTEC would probably not be confined simply to the production of electric power. Another kind of product has been suggested for the plants, a very important one.

Obviously, energy produced at the plants must somehow be transported to shore. Those plants located just off the coast—off Florida,

say, or Louisiana—would deliver energy directly to the consumer via underwater transmission cables. But plants sited in the open sea near the equator cannot deliver energy directly; rather, they could be used to manufacture certain energy-intensive products that could be shipped ashore.

These energy-intensive products could become a major contribution of the OTEC enterprise. If the cost of conventional fuel continues to rise, which seems inevitable, then cheap electricity produced at sea will become highly cost competitive in the chemical manufacture of such products as ammonia, titanium, and aluminum (7 kilowatt-hours are required to produce 1 pound of aluminum).

OTEC could supply power for mining of deep sea minerals as well. It could liquefy hydrogen gas produced on the site, and the gas could be tanked back to shore. Coal freighted to the site could be combined with hydrogen and then shipped to shore as a synthetic fuel. Desalinization is yet another possibility: warm ocean water would be evaporated and the vapor condensed by cold water from the ocean bottom. All of these options, of course, are predicated on the assumption that OTEC can produce cheap electric power.

ECONOMICS AND THE FUTURE OF OTEC

Probably the first practical commercial market will be larger U.S. islands—Hawaii, Puerto Rico, the Virgin Islands. The fact that they are located in the tropics simplifies the enterprise, and because they are heavily dependent on imported oil—indeed, they often pay twice the national average for their fuel—OTEC is probably cost competitive for them now.

The Secretary of Energy has been authorized by Congress to build a pilot plant in the 40-megawatt range whenever he feels the technology is ready to be implemented. Later, plants would be scaled up to 400 megawatts. Finally, in the 1990s, OTEC may service the entire southern Gulf Coast. Congress has already established a concrete goal for OTEC:

10,000 megawatts of electric power or its equivalent in energy-intensive products by 1999. This is the power equivalent of ten major nuclear plants.

Criticism of the OTEC scheme should not be taken lightly. On the other hand, as we begin to face up to the questions posed by energy shortages, by our oil dependence on unstable foreign suppliers, by objections voiced against fossil-fuel and nuclear alternatives, OTEC—as well as some of the other ocean-energy options—scores high marks. This is a time of ongoing technological revolution. Since World War II, the world has witnessed the birth of an atomic age, a jet age, a computer age, a space age, and soon—possibly—the dawning of a new and spectacular solar age, in which OTEC should play a large part.

SUGGESTED READINGS

Robert Cohen, "Energy from Ocean Thermal Gradients," *Oceanus* **22** (4), 12 (1979/80).

John D. Isaacs and Walter R. Schmitt, "Ocean Energy: Forms and Prospects," *Science* **207** (4428), 265 (1980).

Abrahim Lavi and Gay H. Lavi, "Ocean Thermal Energy Conversion (OTEC): Social and Environmental Issues," *Energy* **4** (5), 833 (1979).

Gregg Marland, "Extracting Energy from Warm Sea Water," *Endeavour* **21** (4), 165 (1978).

Marshal F. Merriam, "Wind, Waves and Tides," *Review of Energy* **3**, 29 (1978).

W. E. Richards and J. R. Vadus, "Ocean Thermal Energy Conversion: Technology Development," *Marine Technology Society Journal* **14** (1), 3 (1980).

Alfred Voss, "Waves, Currents, Tides—Problems and Prospects," *Energy* **4** (5), 823 (1979).

Alfred Voss, "Five New Wave Power Ideas," *Ocean Industry* **14** (2), 68 (1979).

Geothermal Energy   26

L.J.P. MUFFLER

NATURE OF GEOTHERMAL ENERGY

In the broadest sense, geothermal energy is the natural heat of the earth. We know that temperatures increase with depth in the earth, to 200–1000°C at the base of the continental crust and to perhaps 3500–4500°C at the center of the earth. But most of the thermal energy in the earth is far too deeply buried to be tapped by man, even under the most optimistic assumptions of technological development. Although drilling has reached nearly 10 km and may someday reach 15 to 25 km, the depths from which thermal energy might be extracted economically are unlikely to be greater than 10 km.

Even restricting discussion to the outer 10 km of the earth, the amount of geothermal energy is still immense, approximately 2000 times the heat represented by the total coal resources of the world (White 1965). This comparison must be tempered, however, by the fact that most of the geothermal energy is uniformly distributed throughout this 10 km. Most geothermal development to date

has taken place where geothermal energy has been concentrated into restricted volumes in a manner analogous to the concentration of valuable metals into ore deposits or of oil into commercial petroleum reservoirs.

TEMPERATURE DISTRIBUTIONS IN THE EARTH

Perhaps the best set of data to give a perspective on the temperatures and thermal energy in the upper 10 km of the earth was provided by Diment et al. (1975) in the 1975 USGS geothermal resource assessment of the United States. They divided the United States into heat flow provinces on the basis of the component of heat flow that originates from the lower crust and mantle (q^*). They then assigned each heat flow province to one of three types: Sierra Nevada type (low q^*), Eastern type (intermediate q^*), or Basin and Range type (high q^*). They then calculated temperatures at depth on the basis of conventional conductive heat flow theory for various values of radioactive heat production at

Dr. L. J. Patrick Muffler is a geologist for the U.S. Geological Survey, Menlo Park, California 94025. He managed the USGS geothermal resource investigations from 1972 to 1976.

This article is reprinted in modified form from *Energy—For Ourselves and Our Posterity* (W. W. Rubey Volume III), R. L. Perrine and W. G. Ernst eds., Englewood Cliffs, N.J.: Prentice-Hall, in press. Reprinted by permission of the author.

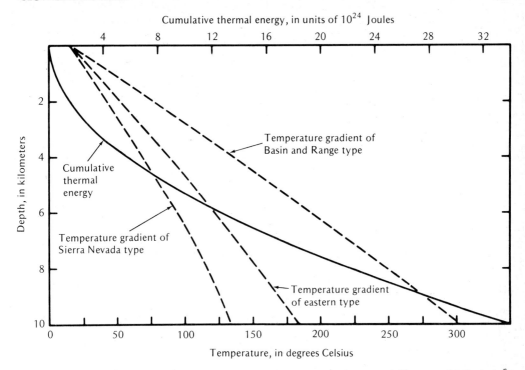

Figure 26-1. Geothermal energy cumulative to 10 km depth in the United States (solid line) and temperatures (dashed lines) in type heat flow provinces of the United States (from Muffler 1979b, as adapted from the "best estimates" of Diment et al. 1975). The temperatures are calculated from the equation for exponential decrease of radioactive heat generation with depth (Diment et al., p. 85), assuming the average radioactive heat production at the surface to be 2.1×10^{-6} J/(m³ s) and the thermal conductivity to be 2.5 J/(m s C°). The Sierra Nevada type has an average gradient of 11 C°/km, the Eastern type an average gradient of 16 C°/km, and the Basin and Range type an average gradient of 28 C°/km.

the earth's surface. From the resultant curves (dashed lines in Figure 26-1), Diment et al. then calculated the cumulative thermal energy in the United States to various depths (the solid curve of Figure 26-1). Note that the thermal energy in the upper 3 km is only about 10 per cent of the thermal energy to a depth of 10 km.

Most geothermal exploration and development to date, however, have been carried out in situations where near-surface temperatures are not controlled by thermal conduction but instead have been increased by the movement of hot water through pores and fractures. The generally accepted model of

such a hydrothermal convection system was developed several decades ago by Donald E. White and is shown in Figure 26-2.

Meteoric water circulates to depths of 3 to 6 km, where it is heated by thermal conduction; the heat source may be an intrusive body that is still hot although not necessarily molten. The water expands upon being heated and moves buoyantly upward in a plume of relatively restricted cross section. The driving force of this large convective circulation system is gravity, effective because of the density difference between cold, downward-moving recharge water and hot, upward-moving geothermal water. An intrusive body

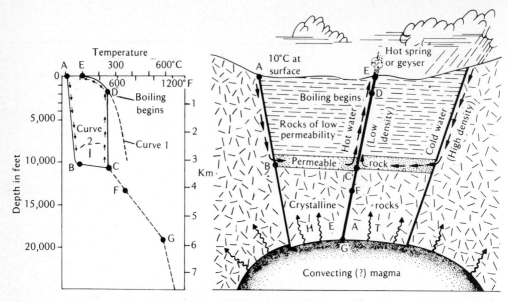

Figure 26-2. Generalized model of a high-temperature hydrothermal convection system with deep convective circulation of meteoric water, heated dominantly by thermal conduction. Curve 1 shows the boiling point of pure water under pressure exerted by a column of liquid water everywhere at boiling, assuming water level at ground surface. Dissolved salts shift the curve to the right; dissolved gases shift the curve to the left. Curve 2 shows the ground-temperature profile of a typical hot-water system. (From White 1967)

is not a prerequisite for a hydrothermal convection system; one can envisage such a system being developed merely by very deep circulation along faults in a region of elevated regional heat flow (Muffler et al. 1980).

At the present time, economical concentrations of geothermal energy for the generation of electricity occur where temperatures of at least 150°C are found in permeable rocks at depths less than 3 km (which is approximately the depth of the deepest commercial geothermal well drilled to date). In such convective geothermal systems, thermal energy is stored both in the solid rock and in the water and steam that fill pores and fractures in the rock. Water and steam serve to transfer heat from the rock to the well bore and thence to the ground surface.

Examples of temperature distributions with depth in hydrothermal convection systems are shown in Figure 26-3. Curves A, B, C, and

D show a range of conductive gradients commonly observed in the upper 2000 m (8 C°/km, 25 C°/km, 50 C°/km, and 75 C°/km). Curve G represents a geothermal system such as that at Wairakei, New Zealand, with hot water at 260°C in a reservoir at depths greater than 600 m. At shallower depths and lower pressures, both water and steam are present and temperatures follow the boiling-point curve for pure water. Curve H represents a geothermal system such as the Salton Sea geothermal system, in which temperatures are above the boiling-point curve for pure water owing to the effect of high salinity. Curve E represents a geothermal system with a high upflow rate; even though the springs at the surface may be boiling, temperatures never get very high at depth. Obviously, the target in conventional geothermal exploration is a system with a temperature distribution like curve G or H.

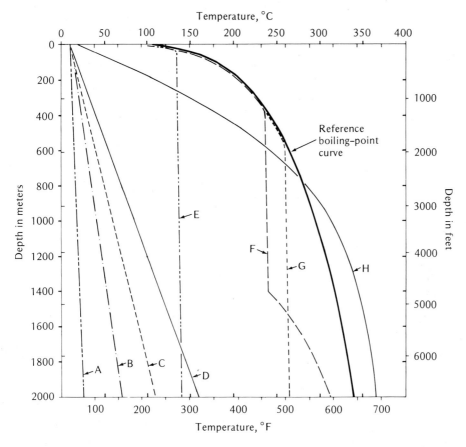

Figure 26-3. Variation of temperature with depth in hydrothermal convection systems. The reference boiling-point curve of pure water is calculated from steam tables assuming boiling conditions throughout the column. (Slightly modified from White 1973)

LIQUID-DOMINATED AND VAPOR-DOMINATED HYDROTHERMAL CONVECTION SYSTEMS

The foregoing considerations apply to a liquid-dominated geothermal system, defined as a geothermal system in which the fluid at depth is a single phase, liquid water. As water at a temperature of (for example) 260°C enters the drill hole, it flashes in part to steam. The steam-water mixture flows up the well to the wellhead, where the steam and water are separated at a pressure of about 500 to 700 kPa. Only the steam can be fed to the turbine to generate electricity; the unwanted water is either discharged to streams or reinjected underground. At a reservoir temperature of 200°C, approximately 80 per cent by weight of the produced fluid is water and about half the produced heat is in the water.

Liquid-dominated hydrothermal convection systems having temperatures less than 150°C are not economical for electrical generation today with direct use of steam in turbines. Instead, it may be possible to employ a binary system that uses geothermal water to heat a low-boiling-point fluid such as freon or isobutane (Milora and Tester 1976, Wahl 1977).

With the exception of a few steam bubbles, the pores of a liquid-dominated geothermal system are filled with water, and the pressure at any point is determined by the weight of overlying water. On the other hand, in a vapor-dominated geothermal system, the fractures and larger pores at depth are filled with steam to the extent that steam is the hydraulically dominant phase (Truesdell and White 1973, White et al. 1971). Pressures, instead of increasing linearly as in a liquid-dominated reservoir, level off in the vapor-dominated reservoir. White et al. (1971) have predicted that there is a body of brine below any vapor-dominated reservoir, within which pressures and temperatures will increase rapidly (curve F, Figure 26-3). Such a deep brine body has not yet been proved or disproved by drilling.

The production of a vapor-dominated geothermal system is far simpler than the production of a hot-water geothermal system. Dry steam is produced at the wellhead with no water and is piped directly into a turbine.

Unfortunately, vapor-dominated geothermal systems are a rarity. Most geothermal systems are of the liquid-dominated type. The known vapor-dominated geothermal systems are Larderello, Travale, and Monte Amiata in Italy, The Geysers and Lassen Volcanic National Park in California, the Mud Volcano area in Yellowstone National Park, Matsukawa in Japan, and Kawah Kamodjang in Indonesia.

EXPLORATION FOR HYDROTHERMAL CONVECTION SYSTEMS

Exploration for hydrothermal convection systems consists of the search for anomalously hot and permeable volumes of rock in the upper crust. In the early days of geothermal development, techniques were borrowed primarily from the more highly developed fields of petroleum and mineral exploration. During the past decade, how-

ever, geothermal exploration has become a specialized discipline with several unusual approaches reflecting the unique character of the resource. Recent comprehensive reviews of geothermal exploration have been prepared by Combs et al. (1980), Diment (1980), and Lumb (1980).

Geologic research in the past 20 years has shown that the outer part of the earth consists of large, rigid plates moving relative to each other at rates of several centimeters per year (Le Pichon et al. 1973). Crust is created at spreading ridges and is consumed at subduction zones (Figure 26-4). Along both types of plate margins, molten rock is generated and moves upward in the crust. The resulting pods of intrusive rock provide the heat that is then transferred by conduction to the overlying convection systems of meteoric water. Producing and explored hydrothermal convection systems (Figure 26-5) lie primarily on or near spreading ridges and subduction zones (Muffler 1976).

Exploration for geothermal energy usually begins with the selection of geologically favorable terrains, usually those with surface hydrothermal manifestations or with young volcanic rocks (see, for example, Luedke and Smith 1978a and 1978b). Favorable tectonic settings include convergent and divergent plate margins, as well as intraplate hot spots (e.g., Hawaii), intraplate rifts (e.g., the Rio Grande rift), and broad zones of extension and normal faulting (e.g., the Basin and Range province of the western United States).

Temperature gradients measured in the upper few hundred meters of the crust give a qualitative depiction of the location of anomalously hot zones, and such temperature surveys have become the primary tool for geothermal exploration in many parts of the world (Combs et al. 1980). However, one cannot predict the temperatures at depth by a simple linear extrapolation of these near-surface temperature gradients. Instead, the gradients usually decrease with depth, owing to the usual decreasing porosity (and thus increasing thermal conductivity) with depth and to the influence of water convection (Combs

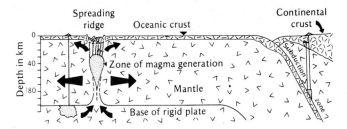

Figure 26-4. Model of development of oceanic crust at spreading ridges and subduction of oceanic crust at consuming plate margins. [Generalized from R. G. Coleman, *Journal of Geophysical Research* **76**: 1212 (1971), Figure 6]

et al. 1980, White 1973). Heat flow, the product of temperature gradient and thermal conductivity, is an independent parameter and thus can be used to outline potentially productive geothermal zones (Combs et al. 1980). However, heat flow data cannot predict the temperature of a hydrothermal reservoir without some independent knowledge of the depth to the reservoir top.

Reservoir temperatures at depth can be predicted, however, by the application of various geothermometers to chemical analyses of waters from thermal springs or shallow wells (Ellis and Mahon 1977, Fournier 1980). One example is the silica geothermometer, based on the increase in solubility of quartz in water as a function of equilibrium temperature (Fournier and Rowe 1966, Mahon 1966). Even if there is near-surface mixing of thermal water with cooler waters, analyses from a variety of warm and cold springs in a given area can be interpreted using various mixing models to give an estimate of the reservoir temperature of the hot component (Fournier 1977 and 1979).

Another geothermometer is the Na-K geothermometer, based primarily on the temperature-dependent partitioning of sodium and potassium between feldspars and solutions (Ellis and Mahon 1977, p. 149). The Na-K-Ca geothermometer (Fournier and Truesdell 1973) incorporates an empirical correction for calcium to give more precise and accurate temperatures, and a similar cor-

rection has recently been developed for magnesium (Fournier and Potter 1979).

Other useful geothermometers are based on the ratio of oxygen-18 between water and dissolved sulfate (McKenzie and Truesdell 1977) and on the ratio of carbon-13 between methane and carbon dioxide in geothermal gases (Truesdell and Hulston 1980).

Ground at depth in geothermal areas is a relatively good conductor of electricity because of high temperature, high porosity, relatively saline water, and alteration minerals of high ion-exchange capacity (such as clays or zeolites). A number of methods, both electrical and electromagnetic, can measure the conductivity of ground at depth and thus delimit geothermal reservoirs. The technique most widely used to date is d.c. resistivity profiling or sounding, using linear Werner, Schlumberger, or dipole-dipole arrays (Banwell and Macdonald 1965, Stanley et al. 1976). The bipole-dipole (or roving dipole) method (e.g., Risk et al. 1970) has also been used extensively, but complex data analysis and ambiguous interpretation detract from its usefulness (Combs et al. 1980, Dey and Morrison 1977). Electromagnetic methods are finding increasing use, despite the complex interpretation necessary. Magnetotelluric techniques have been used for deep investigations of geothermal regions (e.g., Hermance et al. 1976, Ward et al. 1978), and the audio-frequency magnetotelluric method has become a rapid, cost-effective

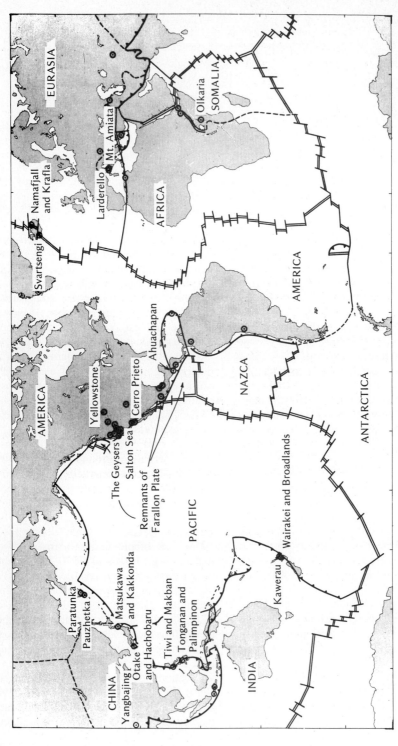

Figure 26-5. Map showing major lithospheric plates (after Le Pichon et al. 1973, Figure 27) and geothermal areas developed or being explored (updated in 1981 from Muffler 1976, Figure 4). Spreading ridges are shown by double lines, subduction zones by barbed lines, and transform faults by single solid lines. Plate boundaries of uncertain nature are shown by dashed lines.

reconnaissance tool for the examination of shallow geothermal conditions (Hoover and Long 1976). Magnetic induction methods have also proved to be useful (e.g., Zablocki 1978). Self-potential (Corwin and Hoover 1979) and induced-potential methods also show some promise.

Passive seismic methods have also found considerable application in the detection of geothermal reservoirs. The incidence of microearthquakes has been shown to be anomalously high in many geothermal areas (e.g., Ward and Bjornsson 1971, Majer and McEvilly 1979), and anomalous amounts of seismic noise may be characteristic of some geothermal areas (Whiteford 1970, Goforth et al. 1972). Pronounced cultural interference affects the latter technique (Iyer and Hitchcock 1976), and its practical use remains to be demonstrated. Recently it has been shown that some large geothermal areas (Yellowstone, The Geysers, Long Valley) are characterized by delays of up to 1 second in the arrival of P waves generated by distant earthquakes (Steeples and Iyer 1976, Iyer et al. 1978), presumably owing to the presence of hot or molten material at depth beneath the geothermal area.

Magnetic and gravity surveys have also proved useful in geothermal exploration, both for regional structural analysis and for the delineation of specific anomalies related to either subsurface igneous rocks (e.g., Griscom and Muffler 1971, Isherwood 1976), the presence of a vapor-filled reservoir (Isherwood 1977), or hydrothermal modification of reservoir rocks (Hochstein and Hunt 1970).

USES OF GEOTHERMAL ENERGY

The most conspicuous use of geothermal energy is for the generation of electricity. Worldwide, the production of electricity from geothermal energy is about 0.1 per cent of electrical generation from all modes; installed capacity in 1980 was approximately 2500 MWe (Table 26-1). Per kilowatt-hour generated, however, geothermal energy is economically competitive with all other generating modes except existing hydropower.

Growth of geothermal electrical capacity with time is shown in Figure 26-6. The nickpoint at about 1977 is due primarily to worldwide reaction to the energy crisis that began in 1973. The dotted extrapolations can be interpreted as upper and lower limits of expected growth in geothermal electrical capacity.

Of equal importance worldwide is the direct use of geothermal energy. Space heating of individual buildings and of whole communities by geothermal energy is becoming common, particularly in Iceland, Hungary, the U.S.S.R., and France. Geothermal energy is also extensively used for agricultural heating in several countries, most notably Hungary and the U.S.S.R, and is used sporadically for product processing (e.g., paper production in New Zealand, drying of diatomite in Iceland). Worldwide use of geothermal energy for space heating and cooling, agriculture, aquaculture, and industrial processes totals over 7000 megawatts thermal (Lienau et al. 1980).

Generation of electricity from geothermal energy in the United States has taken place at The Geysers in northern California since 1960, and the capacity at present is 912 MWe (Figure 26-7). Projected additions should bring the capacity of the field to over 1500 MWe by 1985. Electricity is now being produced from a pilot plant at East Mesa, in the Imperial Valley of California, and additional plants are under construction or in the permitting stages at East Mesa, at Heber (also in the Imperial Valley), in the Jemez Mountains of New Mexico, and at Roosevelt Hot Springs in southwestern Utah.

Geothermal energy has been used directly at scattered localities in the western United States since the last century, but only on a small scale. The most notable direct use has been in individual and community heating systems at Boise, Idaho, and at Klamath Falls, Oregon. Currently, utilization in both cities is being expanded, and geothermal energy is finding increasing uses in agricultural

Table 26–1. World Geothermal Electrical Capacity, 1980.

Country	Field	Electrical Capacity (megawatts)	
		Installed	Under Contruction
United States	The Geysers, Calif.	912	340
	East Mesa, Calif.	10.5	—
	Brawley, Calif.	10	—
	Raft River, Idaho	—	5
	Puna, Hawaii	—	3
Philippines	Tiwi	220	—
	Makban	220	—
	Tongonan	3	112.5
	Palimpinon	3	112.5
Italy	Larderello	403	—
	Travale	15	—
	Monte Amiata	32	—
New Zealand	Wairakei	192	—
	Kawerau	10	—
Japan	Matsukawa	22	—
	Onuma	10	—
	Kakkonda (Takinoue)	50	—
	Onikobe	25	—
	Otake	11	—
	Hachobaru	50	—
	Nigorikawa	—	50
Mexico	Cerro Prieto	150	30
	Los Azufres	—	25
El Salvador	Ahuachapán	90	—
Iceland	Krafla	30	—
	Svartsengi	8	—
	Námafjall	3	—
U.S.S.R.	Pauzhetsk	5	—
China	Yangbajing	1	—
	Fengshun	0.29	—
	Wengtang	0.05	—
	Huailai	0.2	—
	Huitang	0.3	—
	Yingkou	0.1	—
	Ching-shui (Taiwan)	0.4	—
Turkey	Kizildere	0.5	2.5
Indonesia	Kamojang	0.25	2
Kenya	Olkaria	—	30
	Total	2488	713

Data primarily from reports given at the meeting of the Standing Advisory Committee on Geothermal Training (United Nations University) held in Pisa, Italy, in November 1980.

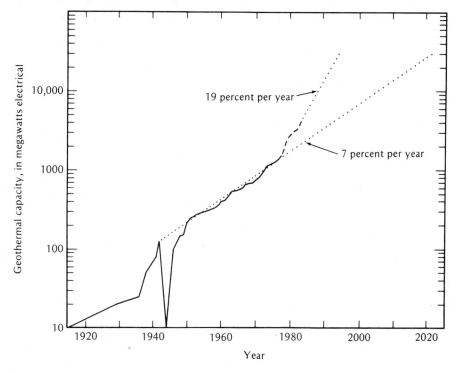

Figure 26-6. Worldwide installed geothermal electrical capacity as a function of time. Dashed line indicates plants under construction or committed up to 1983. Dotted lines are extrapolations of the 1945–77 and 1977–82 segments. (From Muffler and Guffanti 1979, based in great part on data collated by Donald E. White)

endeavors (e.g., heating of greenhouses near Susanville, California) and in industrial processes (e.g., pasturizing of milk in Klamath Falls, Oregon, and dehydration of vegetables at Brady Hot Springs, Nevada).

GEOTHERMAL RESOURCES

Estimates of geothermal resources of the United States vary by as much as six orders of magnitude (Muffler 1973), from a high of 10 billion megawatt-centuries (MWe-c) to several low estimates that cluster around 10,000 MWe-c. To put these figures in context, the electrical generating capacity of the United States in 1979 was approximately 600,000 MWe (U.S. Dept. of Energy, 1979).

Part of the reason for the discrepancy in published geothermal resource estimates lies in a confusion in the terms "resource base," "reserve," and "resource." These terms come from the oil and minerals industries. "Resource base" had been defined by Netschert (1958) and Schurr and Netschert (1960) as "the sum total of a mineral raw material present in the earth's crust within a given geographic area . . . whether its existence is known or unknown and regardless of cost considerations and of technological feasibility of extraction." "Reserve" has been defined by Flawn (1966) as "quantities of minerals . . . that can be reasonably assumed to exist and which are producible with existing technology and under present economic conditions." And finally, "resource" has been defined by Netschert (1958) and by

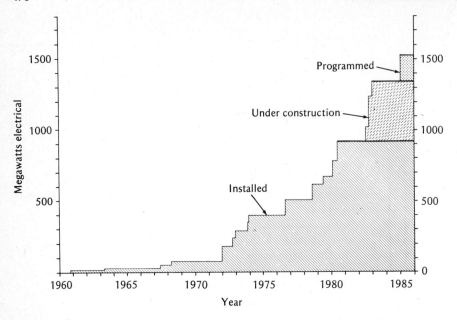

Figure 26-7. Growth of electrical generating capacity of The Geysers with time.

Schurr and Netschert (1960) as "that part of the resource base (including reserves) which seems likely to become available given certain technologic and economic conditions."

The term "recovery factor" is used in the oil and minerals industries to refer to the ratio of the material (or energy) that can be recovered to the material (or energy) in place. Recovery factors can be as high as unity for a few metallic deposits, but for most minerals and energy sources is substantially less than unity. The recovery factor for deep-mined coal is currently about 50 per cent (Schanz 1975). The 1975 USGS assessment of petroleum in the United States (Miller et al. 1975) used a recovery factor of 60 per cent in calculating resources and a recovery factor of 32 per cent in calculating reserves. The recovery factor for natural gas is approximately 80 per cent (Schanz 1975).

With this background, a logical resource terminology specifically for geothermal energy was proposed by Muffler and Cataldi (1978). "Geothermal resource base" was defined as all the heat in the earth's crust beneath a specific area, measured from local

mean annual temperature. "Accessible resource base" was defined as the thermal energy at depths shallow enough to be tapped by drilling in the foreseeable future. The "geothermal resource" was defined as that fraction of the accessible resource base that might be extracted economically and legally at some reasonable future time. The geothermal resource can be divided into economic and subeconomic categories depending on present-day economics.

This logic can be displayed as the vertical axis (degree of economic feasibility) of a McKelvey diagram (Figure 26-8). The degree of geologic assurance is displayed as the horizontal axis and is split into identified and undiscovered components. The geothermal *resource* comprises the upper part of the diagram. The geothermal *reserve* is specified as that part of the resource that is identified and economical today (i.e., the upper left box).

The quantitative relations between accessible resource base, resource, and reserve can be illustrated by the results of a geothermal resource assessment carried out recently in central and southern Tuscany, Italy (Figure

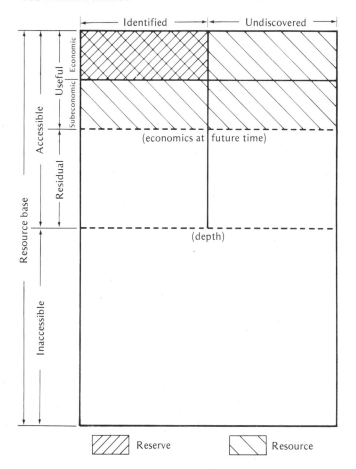

Figure 26-8. McKelvey diagram for geothermal energy, showing derivation of the terms "resource" and "reserve." (From Muffler and Cataldi 1978)

26-9). Note that the resource is only 2.5 per cent of the accessible resource base, whereas the reserve is only 1 per cent of the accessible resource base. Furthermore, from an accessible resource base of $187,000 \times 10^9$ watt-years (thermal), one estimates electricity of only 134×10^9 watt-years (electrical) from the geothermal resource.

USGS GEOTHERMAL RESOURCE ASSESSMENTS OF THE UNITED STATES

In 1975, the U.S. Geological Survey produced the first comprehensive geothermal re-source assessment of the United States (White and Williams 1975). This assessment evaluated geothermal energy in four categories:

1. Regional conduction-dominated regimes.
2. Geopressured-geothermal energy.
3. Igneous-related geothermal systems.
4. Hydrothermal convection systems at temperatures $\geqslant 90°C$.

For each category, the assessment used a two-step process: (1) calculation of thermal energy in the ground and (2) calculation of the recoverable thermal energy (the resource).

In early 1979, the USGS published a refined and updated geothermal resource as-

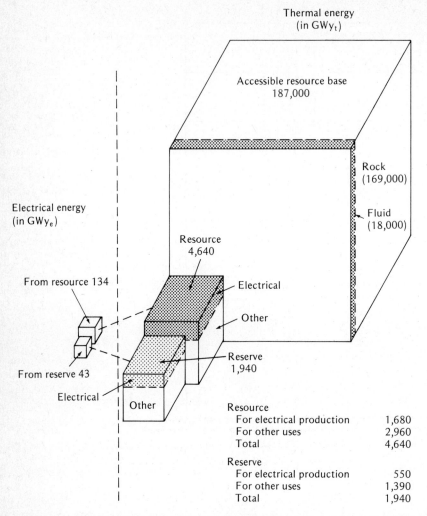

Figure 26-9. Block diagram illustrating the amounts of electrical energy producible from the geothermal resource and from the geothermal reserve of central and southern Tuscany. The volume of the cubes is proportional to the energy content. (From Cataldi et al. 1978)

sessment of the United States (Muffler 1979a) based on data available on July 1, 1978. This assessment in general followed the methodology of White and Williams, and parts of the earlier assessment still valid (e.g., the calculations of thermal energy in regional conductive environments) were not duplicated. Three major items were added to the 1978 assessment: (1) a report describing and

depicting areas favorable for the discovery and development of low-temperature (less than 90°C) geothermal waters from depths less than 1 km (Sammel 1979), (2) a statistical approach to the estimation of geothermal energy in identified hydrothermal convection systems ≥90°C, and (3) three large colored maps depicting a variety of geothermal data.

For the thermal energy in regional conductive environments, the 1978 USGS geothermal resource assessment used the result calculated by Diment et al. (1975) in the 1975 USGS assessment. This number ($33,000,000 \times 10^{18}$ joules) can be viewed as a background value or upper limit for any discussion of geothermal energy in the United States. Included in the number is the thermal energy in geopressured basins, in sedimentary basins at hydrostatic pressure, and in hydrothermal convection systems unrelated to young igneous intrusions.

The geopressured-goethermal resource of the northern Gulf of Mexico Basin to a depth of nearly 7 km was estimated by Wallace et al. (1979) using data from over 3500 wells. Depending on which recovery scheme of Papadopulos et al. (1975) is used, the geopressured-geothermal resource is between 430×10^{18} and 4400×10^{18} joules. This energy consists of nearly equal amounts of thermal energy and energy from methane dissolved in the waters, with a very small amount of mechanical energy from the high pressures. The geopressured-geothermal resource of the northern Gulf of Mexico Basin is equivalent to $75–780 \times 10^9$ barrels of oil, or 8 to 85 years of oil at the present rate of consumption in the United States. Geopressured zones are known in many other sedimentary basins in the United States (Wallace et al., Figure 26), but quantitative estimates were made only for geopressured-geothermal resources of the northern Gulf of Mexico Basin.

Superimposed on the regional conductive environment is geothermal energy related to young igneous intrusions—contained in either magma, solidified igneous rock, hot country rock, or associated hydrothermal convection systems. For such igneous-related geothermal systems, the thermal energy still remaining in silicic intrusions and adjacent country rock was calculated by conductive cooling models using estimates of the size and age of intrusions (Smith and Shaw 1979). This calculation assumes that cooling of the igneous body by hydrothermal convection was offset by the effects of magmatic preheating and additions of magma after the assumed time of emplacement (Smith and Shaw 1975). The total energy in evaluated igneous-related systems was calculated to be $100,000 \times 10^{18}$ joules. Smith and Shaw (1979) further estimate that the energy in systems for which adequate age and volume data are not available is $900,000 \times 10^{18}$ joules, nearly an order of magnitude greater. Rough estimates suggest that 50 per cent of the energy is in magma, 43 per cent in hot dry rock, and 7 per cent in hydrothermal convection systems (Muffler 1979b).

Both the 1975 and 1978 USGS geothermal resource assessments calculated the geothermal energy in identified hydrothermal convection systems, estimated the thermal energy yet to be discovered, and then estimated the fraction of the total energy that might be recovered (Nathenson and Muffler 1975, Renner et al. 1975, Brook et al. 1979). For identified systems, both assessments used a method in which temperature, area, and thickness were estimated for the reservoirs to a depth of 3 km in identified hydrothermal convection systems. Energies were then calculated assuming a constant volumetric specific heat. In 1978, the methodology involved making three estimates for each variable (temperature, area, and thickness) for each system. These estimates are the minimum, most likely, and maximum estimates of a triangular probability density that most nearly approximates the considered estimate of the real probability density of the variable. For temperature, it should be noted that the minimum and maximum values of the triangular probability density are not the minimum and maximum temperatures that might be found in the reservoir but rather the minimum and maximum estimates of the temperature that characterizes the reservoir as a whole.

The means and standard deviations of these triangular probability densities were then calculated and summed analytically to give the mean and standard deviation of the thermal energy in all identified hydrothermal

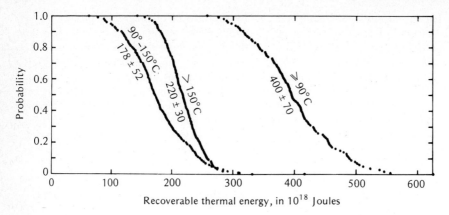

Figure 26-10. Monte Carlo sample distributions for recoverable thermal energy (resource) from identified hydrothermal convection systems. (From Brook et al. 1979, Figure 8)

convection systems $(1650 \pm 140 \times 10^{18}$ joules, excluding national parks). Monte Carlo methods were used to give the resultant probability distributions for the energy in systems of temperature $>150°C$, 90 to 150°C, and the total of all systems of temperature $\geq 90°C$.

A uniform geothermal recovery factor of 25 per cent was then used to calculate the recoverable thermal energy (the resource) in identified hydrothermal convection systems. The probability density of the recovery factor was assumed to be triangular with a minimum of zero, a maximum of 50 per cent, and a most likely value of 25 per cent. Monte Carlo methods were again used to give the probability distribution shown in Figure 26-10.

Electricity was calculated only for those hydrothermal convection systems of reservoir temperature greater than 150°C. First, available work was calculated. The result was then reduced by a utilization factor of 0.4 for liquid-dominated systems and 0.5 for vapor-dominated systems (Brook et al. 1979, p. 26) to give electrical capacity of $23,000 \pm 3400$ MWe for 30 years.

For intermediate temperature hydrothermal convection systems (90–150°C), the 1975 and 1978 USGS assessments tabulated

beneficial heat assuming either a 32 C° temperature drop at 150°C or a 20 C° drop at 100°C; in either case, beneficial heat is 24 per cent of the intermediate-temperature resource.

In addition to this inventory of known hydrothermal convection systems, it was necessary to estimate the geothermal energy yet to be discovered in hydrothermal convection systems. This energy consists of (1) additional energy resulting from upward revisions of the volumes of identified systems, (2) additional thermal energy resulting from upward revisions of temperature estimates of identified systems, and (3) thermal energy in systems not yet identified. In most cases, Brook et al. estimated the undiscovered component by geologic province as a multiple of the identified component. For Island Park, Idaho, and the Aleutian Volcanic chain of Alaska, the undiscovered component was estimated as 1 per cent of the thermal energy contained in the corresponding igneous-related system. The undiscovered component was then broken into high-temperature and intermediate-temperature categories on the basis of extrapolation of number, size, and temperature data of the identified component.

The identified component $(400 \times 10^{18}$ J)

added to the undiscovered component $(2000 \times 10^{18}$ J) gives a total of 2400×10^{18} J, the geothermal resource of all hydrothermal convection systems to a depth of 3 km. This figure is equivalent to 430×10^9 barrels of oil, or 47 years at the present consumption rate of oil in the United States. Depending on the balance between high-temperature and intermediate-temperature fluids in the undiscovered component, this gives electrical energy of 95,000 to 150,000 MWe for 30 years and beneficial heat of 230 to 350×10^{18} J. These ranges reflect the two limiting assumptions for the volume-temperature relation for the undiscovered component. Thus, the higher value for electricity cannot occur with the higher value of beneficial heat.

CONCLUSIONS

Developments in the past few decades have clearly demonstrated that the geothermal energy resources of the world are substantial and that geothermal energy can be an important, if not dominant, contributor to the energy needs of many countries. Although the current usage of geothermal energy in the United States is dwarfed by the traditional fossil fuels, hydropower, and nuclear energy, geothermal resources of several types in the United States have been shown to be large. Geothermal energy thus shows great promise of contributing substantially to the energy needs of the United States, both in the generation of electricity and in direct use.

REFERENCES

Banwell, C. J., and W.J.P. Macdonald, 1965, in *Eighth Commonwealth Mining and Metallurgical Congress, Australia and New Zealand,* New Zealand section, paper No. 213, pp. 1–7.

Brook, C. A., R. H. Mariner, D. R. Mabey, J. R. Swanson, M. Guffanti, and L.J.P. Muffler, 1979, in Muffler 1979a, pp. 18–85.

Cataldi, R., A. Lazzarotto, P. Muffler, P. Squarci, and G. Stefani, 1978, *Geothermics* **7** (2–4), 91.

Combs, J., J. K. Applegate, R. O. Fournier, C. A. Swanberg, and D. Nielson, 1980, in D. N. Anderson and J. W. Lund, eds., *Direct Utilization of Geothermal Energy: A Technical Handbook.*

Geothermal Resources Council Special Rept. No. 7, pp. 2.1–2.16.

Corwin, R. F., and D. B. Hoover, 1979, *Geophysics* **44**, 226.

Day, A., and H. F. Morrison, 1977, *Geothermics* **6**, 47.

Diment, W. H., 1980, in J. Kestin, R. DiPippo, H. E. Khalifa, and D. J. Ryley, eds., *A Sourcebook on the Production of Electricity from Geothermal Energy.* Washington, D.C.: GPO, in press.

Diment, W. H., T. C. Urban, H. H. Sass, B. V. Marshall, R. J. Munroe, and A. H. Lachenbruch, 1975, in White and Williams 1975, pp. 84–103.

Ellis, A. J., and W.A.J. Mahon, 1977, *Chemistry and Geothermal Systems,* pp. 144–59. New York: Academic Press.

Flawn, P. T., 1966, *Mineral Resources,* p. 10. Chicago: Rand McNally.

Fournier, R. O., 1977, *Geothermics* **5**, 41.

Fournier, R. O., 1979, *Journal of Volcanology and Geothermal Research* 5 (1–2), 1.

Fournier, R. O., 1980, in L. Rybach and L.J.P. Muffler, eds., *Geothermal Systems: Principles and Case Histories,* pp. 109–43. Chichester, England: Wiley.

Fournier, R. O., and R. W. Potter III, 1979, *Geochim. et Cosmochim. Acta* **43**, 1543.

Fournier, R. O., and J. J. Rowe, 1966, *Am. J. Sci.* **264**, 685.

Fournier, R. O., and A. H. Truesdell, 1973, *Geochim. et Cosmochim. Acta* **37**, 1255.

Goforth, T. T., E. J. Douze, and G. C. Sorrells, 1972, *Geophysical Prospecting* **20**, 76.

Griscom, A., and L.J.P. Muffler, 1971, Aeromagnetic map and interpretation of the Salton Sea geothermal area, California. USGS Geophysic Inventory Map GP-754.

Hermance, J. F., R. E. Thayer, and A. Bjornsson, 1976, in *Proceedings* 1975, pp. 1037–48.

Hochstein, M. P., and T. M. Hunt, 1970, *Geothermics,* Special Issue 2, Vol. 2, Part 1, pp. 333–46.

Hoover, D. B., and C. L. Long, 1976, in *Proceedings* 1975, pp. 1059–64.

Isherwood, W. F., 1976, in *Proceedings* 1975, pp. 1065–73.

Isherwood, W. F., 1977, *Geothermal Resources Council Transactions* **1**, 149.

Iyer, H. M., and T. Hitchcock, 1976, in *Proceedings* 1975, pp. 1075–83.

Iyer, H. M., D. H. Oppenheimer, and T. Hitchcock, 1978, *Geothermal Resources Council Transactions* **2**, 317.

Le Pichon, X., J. Francheteau, and J. Bonnin, 1973, *Plate Tectonics.* New York: Elsevier.

Lienau, P. J., J. Austin, J. Balhizser, R. G. Campbell, M. Conover, G. G. Culver, L. Donovan, L. A. Fisher, R. L. Miller, R. C. Neiss, and G. Reistad, 1980, in D. N. Anderson and J. W. Lund, eds., *Direct Utilization of Geothermal Energy: A Technical Handbook,* pp. 4.1–4.84. Davis, Calif.: Geothermal Resources Council, Special Report 7.

Luedke, R. G., and R. L. Smith, 1978a, Map showing

distribution, composition, and age of late Cenozoic volcanic centers in Arizona and New Mexico. USGS Miscellaneous Inventory Series I-1091-A.

Luedke, R. G., and R. L. Smith, 1978b, Map showing distribution, composition, and age of late Cenozoic volcanic centers in Colorado, Utah, and southwestern Wyoming. USGS Miscellaneous Inventory Series I-1091-B.

Lumb, J. T., 1980, in L. Rybach and L.J.P. Muffler, eds., *Geothermal Systems: Principles and Case Histories*, pp. 77–108. Chichester, England: Wiley.

Mahon, W.A.J., 1966, *New Zealand Jour. Sci.* **9**, 135.

Majer, E. L., and T. V. McEvilly, 1979, *Geophysics* **44**, 246.

McKenzie, W. F., and A. H. Truesdell, 1977, *Geothermics* **5**, 51.

Miller, B. M., H. L. Thomsen, G. L. Dolton, A. B. Coury, T. A. Hendricks, F. E. Lennartz, R. B. Powers, E. G. Sable, and K. L. Varnes, 1975, *Geological Estimates of Undiscovered Recoverable Oil and Gas Resources in the United States*. USGS Circular 725.

Milora, S. L., and J. W. Tester, 1976, *Geothermal Energy as a Source of Electric Power*. Cambridge, Mass.: MIT Press.

Muffler, L.J.P., 1973, in D. A. Brobst and W. P. Pratt, eds., *United States Mineral Resources*, pp. 251–61. USGS Professional Paper 820.

Muffler, L.J.P., 1976, in *Proceedings* 1975, pp. 499–507.

Muffler, L.J.P., ed., 1979a, *Assessment of Geothermal Resources of the United States—1978*. USGS Circular 790.

Muffler, L.J.P., 1979b, in Muffler 1979a, pp. 156–63.

Muffler, L.J.P., and R. Cataldi, 1978, *Geothermics* **7** (2–4), 53.

Muffler, L.J.P., J. K. Costain, D. Foley, E. A. Sammel, and W. Youngquist, 1980, in D. N. Anderson and J. W. Lund, eds., *Direct Utilization of Geothermal Energy: A Technical Handbook*, pp. 1.1–1.15. Geothermal Resources Council Special Report 7.

Muffler, L.J.P., and M. Guffanti, 1979, in Muffler 1979a, pp. 1–7.

Nathenson, M., and L.J.P. Muffler, 1975, in White and Williams 1975, pp. 104–21.

Netschert, B. C., 1958, *The Future Supply of Oil and Gas*. Baltimore: Johns Hopkins University Press.

Papadopulos, S. S., R. H. Wallace Jr., J. B. Wesselman, and R. E. Taylor, 1975, in White and Williams 1975, pp. 125–46.

Proceedings of the Second United Nations Symposium on the Development and Use of Geothermal Resources, San Francisco, May 1975. Washington, D.C.: GPO, 1975.

Renner, J. L., D. E. White, and D. L. Williams, 1975, in White and Williams 1975, pp. 5–57.

Risk, G. F., W.J.P. Macdonald, and G. B. Dawson, 1970, *Geothermics*, Special Issue 2, Vol. 2, Part 1, pp. 287–94.

Sammel, E. A., 1979, in Muffler 1979a, pp. 86–131.

Schanz, J. J. Jr., 1975, *Resource Terminology: An Examination of Concepts and Terms and Recommendations for Improvement*. Palo Alto, Calif.: Electric Power Research Institute, Research Project 336.

Schurr, S. H., and B. C. Netschert, 1960, *Energy in the American Economy, 1850–1975*, p. 297. Baltimore: Johns Hopkins University Press.

Smith, R. L., and H. R. Shaw, 1975, in White and Williams 1975, pp. 58–83.

Smith, R. L., and H. R. Shaw, 1979, in Muffler 1979a, pp. 12–17.

Stanley, W. D., D. B. Jackson, and A.A.R. Zohdy, 1976, *J. Geophys. Res.* **81** (5), 810.

Steeples, D. W., and H. M. Iyer, 1976, in *Proceedings* 1975, pp. 1199–1206.

Truesdell, A. H., and J. R. Hulston, 1980, in P. Fritz and J.-C. Fontes, eds., *Handbook of Environmental Isotopes*, pp. 179–225. Amsterdam: Elsevier Scientific.

Truesdell, A. H., and D. E. White, 1973, *Geothermics* **2** (3–4), 145.

U.S. Department of Energy, Energy Information Administration, 1979, Annual Report to Congress, p. 139. Department of Energy, DOE/EIA-0173(79)/2, Vol. 2.

Wahl, E. F., 1977, *Geothermal Energy Utilization*, pp. 221–34. New York: Wiley.

Wallace, R. H. Jr., T. F. Kraemer, R. E. Taylor, and J. B. Wesselman, 1979, in Muffler 1979a, pp. 132–55.

Ward, P. L., and S. Bjornsson, 1971, *J. Geophys. Res.* **76** (17), 3953.

Ward, S. H., W. T. Parry, W. P. Nash, W. R. Sill, K. L. Cook, R. B. Smith, D. S. Chapman, F. H. Brown, J. A. Whelan, and J. R. Bowman, 1978, *Geophysics* **43**, 1515.

White, D. E., 1965, *Geothermal Energy*. USGS Circular 519.

White, D. E., 1967, *Am. J. Sci.* **265**, 641.

White, D. E., 1973, in P. Kruger and C. Otte, eds., *Geothermal Energy: Resources, Production, Stimulation*, pp. 69–94. Stanford, Calif., Stanford University Press.

White, D. E., L.J.P. Muffler, and A. H. Truesdell, 1971, *Econ. Geol.* **66** (1), 75.

White, D. E., and D. L. Williams, eds., 1975, *Assessment of Geothermal Resources of the United States—1975*. USGS Circular 726.

Whiteford, P. C., 1970, *Geothermics*, Special Issue 2, Vol. 1, pp. 478–86.

Zablocki, C. J., 1978, *J. Volcanology and Geothermal Res.* **3** (1–2), 155.

The Prospects for Fusion   27

DAVID J. ROSE AND MICHAEL FEIRTAG

Designing a nuclear fusion reactor in the 1980s is a little like planning to reach heaven: theories abound on how to do it, and many people are trying, but no one alive has ever succeeded. Also, the challenge is too important to be ignored—in fusion's case because we are running out of energy. In the short term, for perhaps 25 years, there will be oil. For 50 to 100 years after that, we could fight over the last available drops of oil and burn increasing quantities of coal, probably at great environmental cost. But after that—and preferably before—humankind will have to turn to one or more of the long-term options that nature holds before us. There are only three. One is solar power. A second is nuclear fission—or, more specifically, since rich fissionable material also grows scarce, the development of a so-called breeder reactor that creates more fissionable fuel than it consumes. The third is nuclear fusion. At present, none of these options is sure to work. Each possesses remarkable and perverse difficulties. To bring any one of them to a point at which society can decide whether or not to adopt it as a principal energy resource will cost $10 to $20 billion. To develop alternatives within an option (an alternative breeder, for example) will cost an additional $1 or $2 billion for each new possibility. In short, we are compelled to play a desperate poker game against nature in which each betting chip costs tens of billions of dollars, and thus far we fear to buy the chips. After all, one can imagine the spending in this game of $60 to $100 billion if we are to back all three possibilities, and that is an awesome amount of money—or it seems awesome until it is compared to the capital investment the United States must make by the year 2000 to provide what energy it needs and transform its patterns of use to match new supplies. Even

Dr. David J. Rose is a professor in the Nuclear Engineering Department, Massachusetts Institute of Technology, Cambridge, Massachusetts 02139. He served as a member of the National Academy of Sciences' Committee on Nuclear and Alternate Energy Systems (CONAES), which prepared a study on energy strategies for the Energy Research and Development Administration. He is also a consultant to Congress's Office of Technology Assessment.

Michael Feirtag is a member of the *Technology Review* Board of Editors.

without much growth in energy demand, and even with prodigious energy conservation, that investment will be about $1 trillion, with much more to follow in the twenty-first century. Since the social cost of having *no* energy options in the year 2000 is incalculable, there seems no alternative but to pursue all three long-term options, reassessing them all the while, until we know enough to narrow the choices.

It is in this spirit that we approach controlled nuclear fusion. The subject has a vast scientific literature and a growing technological literature. There are, however, few public explanations; we shall attempt to lessen the imbalance. It is a good time to do so, for during the past few years the pursuit of controlled nuclear fusion has led to a number of so-called reference designs for large fusion reactors; later in this article we shall give some examples. Far from being blueprints for any fusion reactor, the reference designs were started (in 1967) and carried through for almost the opposite purpose—they were meant to ensure that those pursuing controlled nuclear fusion would have to face every problem associated with a given reactor concept. Thus the designs were problem-finders, not problem-solvers, and as such they were spectacular successes. From the tortured and sometimes bizarre schemes to which the researchers were compelled to resort, there emerged long lists of scientific and technological areas needing attention: more radiation-resistant alloys, different systems for cooling the reactor, higher energy-handling ability per unit area of reactor wall, and so forth. Their purpose being accomplished, the current reference designs are now being discarded as their makers turn to the critical problems, and new reference designs in due course should appear. The making of reference designs that uncover problems, and the development of technologies that solve them, are activities that proceed more or less continuously, and if, years from now, the fusion program succeeds, designs will appear for a practical fusion reactor probably unlike anything now envisaged.

But will it succeed? Is civilization mad to persist in a search that seems so complex, so uncertain? The science and some of the technology have progressed to a point where many questions of a decade ago—for example, "Is controlled fusion scientifically feasible?"—now appear obsolete; they have been answered affirmatively as a by-product of the race to develop ever larger experiments. On the other hand, the technological and engineering difficulties are now known to far surpass any original estimates; still another decade or two will be needed to resolve them and decide about fusion, pro or con. In the meantime, fusion's expanding success coupled with its increasingly evident difficulty will remain a hard mixture to manage; it could easily inspire false optimism or false pessimism—and, either way, wrong judgments. Responsibility for keeping a proper sense of perspective falls primarily upon the Department of Energy, the patron of controlled fusion research in the United States— it costs so much that no organization but the federal government can afford to sponsor it. DOE's inclinations, coupled with the prospect of support, could stimulate its clients, the National Laboratories and other research organizations, to adopt a sycophantic, or at least a neutral, stance. When false certainty grows, so critical judgment withers. The true development effort suffers.

How might it happen? The answer is, bit by bit and with the best of intentions. Administrators, overcome by the temptation of so much to administer, may mistake vision for reality. Engineers, meanwhile, may become expert at their task and neglect to maintain constructively critical attitudes. After all, it is human nature to assume that some fault in a grand conception will be solved by future ingenuity, and it is easy to allow a design to become a neat assemblage of what engineers call black boxes. In these ways, an inadequate design becomes the accepted design and finally the official program, attracting to itself all the support and funding that exists. The public is led to believe that technological salvation is imminent. It then develops that the design is un-workable, and the program ends in disaster. An exemplar exists of this sort of calamity. Several years ago, development of the liquid-metal fast breeder reactor was almost halted by administrative rigidity and money—not because that breeder reactor is a bad (or a good) idea, and certainly not because those in charge of the program planned it that way. It just happened. . . . And whether the same thing may now be happening to controlled nuclear fusion is a topic we shall turn to at the end of this article.

ENERGY TECHNOLOGY FOR THE TWENTY-FIRST CENTURY

Strategies to supply future energy are varied: new coal technology, solar energy, perhaps local geothermal energy, accelerated and ad-vanced nuclear fission technology, fusion. All have a long lead time, both for development and for deployment, and most could last for relatively long periods of time if they are de-ployed at all. New technologies that are ex-pected to be extensively deployed early in the next century must be fully developed by the end of this century; thus the main tech-nological lines should be clear and the main difficulties resolved by 1985 or 1990. Ad-vanced fission power systems, including breeder reactors, are being developed in ap-proximate accord with this schedule. Be-cause of fission's cost advantage (relative to fossil-fuel electric power) in most locations, even if uranium prices should rise several-fold, arguments of resource limitation will not apply to nuclear power unless an accept-able breeder reactor cannot be developed by the early twenty-first century. Thus, barring a public decision against it, nuclear power plants could be well developed and installed before the end of the century.

Against this specific situation and on this time scale, fusion must offer some real ad-vantage if it is to supplant advanced fission systems. What then are the benefits? Part of the fuel for a fusion reactor will be tritium—a radioactive isotope of hydrogen. Fortu-nately, tritium is vastly less hazardous than uranium fission products or than plutonium, per unit of energy produced by nuclear re-actions or simply per unit weight. Exactly how much less depends upon the particular route by which the radioactive material is imagined to invade the human body—whether it enters by ingestion, inhalation, and so on. Analysts speak of it being a factor of 1000 or 10,000 less hazardous. Countering that is the fact that such hazards are not solved by merely stating them in simplistic terms. One can imagine a fusion reactor that would work perfectly well, in the technological sense of giving out energy, but that was equivalent in hazard to a fission reactor be-cause its loss rate of fuel, although still mic-roscopically small, was 10,000 times higher than that of a fission reactor. That sort of question can be resolved only on the basis of fairly specific designs.

Some other social advantages of fusion are more certain. First, tritium is no blackmail weapon. Though it can be used in hydrogen bombs, an atomic bomb is needed to set one off, and so a terrorist would have to already own a fission weapon. Second, a fusion re-actor would contain no equivalent of the highly concentrated radioactive fuel used in fission reactors. Thus, a melt-down accident, the most serious calamity possible in a fis-sion reactor, would be impossible in most fusion-reactor designs (the concentrated en-ergy source just would not be there) and ex-ceedingly small in others. To be sure, the structure of a fusion reactor would itself be-come radioactive, to an extent, depending on the materials used to build it. This radioac-tivity would surely complicate operation and maintenance, and the reactor's cold hulk might remain radioactive for many decades after its operating life had ended. Still, cur-rent estimates rate this aspect of radioactivity as much less severe than the difficulties aris-ing from concentrated fuel in fission. Over all, fusion should be "cleaner." If its advan-tages could be coupled with a cost advantage and an engineering study that demonstrates feasibility, an energy-hungry society would have a very good option.

This debate brings us to the timetable for fusion development. To offer a meaningful alternative to fission, fusion should aim for approximately the same time scale—principal engineering problems seen to be resolvable before 1990, say. That is a very short time for a task so large, too short for scientific, technological, and economic feasibilities to be established in conventional and orderly sequence. The various stages must partly overlap and coalesce, so that technology and engineering known to be needed are worked on now. This is the rationale for the reference designs: groups persistently try to design fusion reactors, based on the latest science and technology, then analyze the designs to uncover their shortcomings. From that activity, clues appear about the next set of problems to be studied and also about what decision large experiments should take. If all the problems prove to be tractable, the whole effort will culminate in a well-worked-out fusion concept by the late 1980s, ready for trial on a grand scale.

This leads us to the question "How to do the job?" Development of any difficult modern technological devices, including fusion reactors, tends to suffer from excessive disciplinary reductionism: any single group concerned with only one phase of the work tends to use up all the design flexibility to solve its own problems, thus leaving impossible tasks for everyone else. Each discipline appropriates to itself all the available option space, so to speak. In illustrative exaggeration, we note that the development of fusion concepts has been mainly in the hands of physicists until almost the present time, so controlled fusion has been looked upon mainly as a research problem—specifically, as a matter of confining a plasma (a gas of charged particles). Such physics-oriented research was conducted under the naïve assumption that technology (the development of new materials and processes) and engineering (the application of technology to socially desirable purposes) could do nothing

necessary to make a plasma that precisely suited the plasmologist's fancy. But technology and engineering cannot be conjured up at will, and in fact, they are usually very expensive to extend in new directions. Thus the need for balance from the start. The three major constraints of "plasma dynamics," technology, and engineering must be jointly satisfied if there is to be a happy outcome to the search for fusion.

ENERGY FROM THE NUCLEUS

The idea of controlled nuclear fusion is not new. Ever since the 1930s, when the sun's nuclear cycle was first worked out in detail, people have speculated about its possibility. However, nothing was accomplished until World War II, when the Manhattan Project was set up to build what is commonly called an "atomic bomb," a name that is somewhat misleading. It should be called a nuclear bomb, for "atomic" refers to the whole atom—that is, to the nucleus and the surrounding electrons. An atom may be ionized (it loses one or more of its electrons and thereby becomes positively charged), but the energies involved are modest, on the order of a few electron volts.

Interactions within the nuclei are vastly more energetic; the Manhattan Project developed bombs operating on the principle of breaking up, or fissioning, the nuclei of uranium or plutonium. The energy released by each such fissioning is about 50 million times greater than the energies associated with the outermost electrons in the atom.

Why does fission work at all? In one parlance, it is because the heaviest elements have "excess mass"—more mass than the sum of the atoms into which they fission—so that when their nuclei break up, this mass difference becomes energy, in accordance with Einstein's law, $E = mc^2$. Looking at Figure 27-1, we are not surprised to find that uranium undergoes fission if mass is added to it, such as one more neutron. The facts that ^{238}U can be turned fairly easily into ^{239}Pu, which is useful for fission, and that ^{232}Th can be-

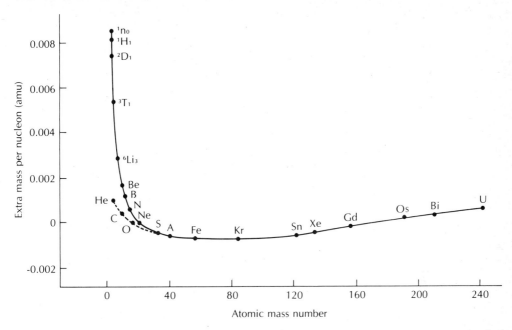

Figure 27-1. The mass of a nucleon—a neutron or a proton—in the nuclei of the various chemical elements. By convention, the ^{16}O nucleus is taken to have a mass of 16 atomic mass units, and other nuclei are then compared with it. It is found that free neutrons ($^{1}n_0$ in the chart) and free protons ($^{1}H_1$ nuclei) have more mass than any neutrons and protons bound within nuclei. This means that when the chemical elements formed, mass disappeared. It became energy, in accordance with $E = mc^2$. The nucleons in the middle-atomic-weight elements—iron (Fe) and krypton (Kr), for example—lost the most mass. Such elements are therefore the most stable: any rearrangement of their nucleons requires energy. On the other hand, breaking up the heaviest nuclei into lighter ones releases energy; this is nuclear fission. So does the combination of the lightest nuclei into heavier ones, and this is nuclear fusion.

come ^{233}U are additional evidence of the general instability of very heavy elements.

Inspection of the curve shows immediately that the middle-weight nuclei are most stable and that a second possibility exists: combining the lightest nuclei and releasing energy in that way—by nuclear fusion. Hydrogen is the lightest element—its nucleus is simply a proton—but the element also has two heavier isotopes: deuterium, in which the proton is joined by one neutron, and tritium, in which it is joined by two. The sun works mainly on a hydrogen fusion cycle, using two protons to make a photon, a positron (a positively charged electron), a deuterium nucleus (called a deuteron), and energy. The

reaction rate is very low, fortunately for us, in that the sun has lasted several billion years, and much of it is still there. But the energy crisis demands reactions that run quicker. The deuterium-deuterium combination is much better, but best of all is the combination of deuterium and tritium. This reaction produces ordinary helium, a neutron, and 17.6 million electron volts. Eighty per cent of the energy, or 14.1 million electron volts, emerges with the velocity of the neutron, and 3.5 million electron volts with the velocity of the helium nucleus (also called an α-particle).

Deuterium is in plentiful supply; it constitutes one part in 7000 of the hydrogen in ordinary water. Since there are some 150×10^6 cubic miles of water in the oceans, the amount is prodigious, enough in fact to last for billions of years, even at the highest rate at which we can use energy without overheating our environment. The problem comes with tritium: it is radioactive, with a half-life of 12.3 years, and so it does not occur in nature. Fortunately, the two isotopes of lithium can be used to make it. High-energy neutrons (above 2.5 million electron volts) can react with ^{7}Li:

$$^7\text{Li} + \text{n[fast]} \rightarrow {}^4\text{He} + {}^3\text{T} + \text{n[slow]}$$

and ^{6}Li absorbs neutrons (the slower they are, the better the absorption) with great appetite:

$$^6\text{Li} + \text{n} \rightarrow {}^4\text{He} + {}^3\text{T} + 4.8 \text{ million eV}$$

Thus we have a generation scheme for tritium. Note the parallel with a fission *breeder*. A fusion reactor is, in fact, a breeder: it uses deuterium and lithium as raw fuels and breeds tritium as an intermediate product that is burned in the reaction. Because only a single neutron appears per fusion event, one might imagine that fusion cannot work: the losses would not let the tritium reproduce itself. Not so, because the two lithium reactions can, in principle, proceed sequentially, using the same initial neutron. On the other hand, other materials in the reactor compete for the neutrons. The result is that fewer than two tritons (tritium nuclei), on average, can be created for each triton destroyed. Still, tritium can be more than replenished.

A second problem now appears. Fission is easy (given some fissionable material) because a neutron is electrically neutral, and so it can enter the positively charged nucleus without encountering any repulsive force. Thus, uranium and neutrons can be at room temperature and yet the uranium will fission, atom by atom, and the fission energy can be turned into heat, say, to boil water. In fact, a water solution of some uranium salts, in which the uranium is suitably enriched with ^{235}U, will go critical—that is, a self-sustaining chain reaction will occur—if the

container is large enough. This is not the case with fusion for a simple reason: the reacting particles are all nuclei, all positively charged. They are therefore all mutually repulsive. Rather than approaching to within a few times 10^{-15} meter of each other so that fusion can occur, a deuteron and a triton are far more likely to repel each other while large distances separate them. Forcing a close approach requires high energy, which requires high temperature—as we shall see, temperature on the order of 10^8 K. All this was well known and applied by nuclear physicists in the late 1940s. The idea of building an "H-bomb" that fuses isotopes of hydrogen occurred to them, as well as a way to attain the high temperature: explode a fission bomb in the middle of a hydrogen-isotope mixture. If the bomb is made large enough or is surrounded by heavy material, then inertia alone will keep it together for the microsecond or so the reactions require.

Clearly that scheme will not do for controlled fusion in a power plant. Gravity will not work either, for only on the scale of the sun (or more precisely on about the scale of Jupiter) would it be strong enough to force the nuclei together. What is left? The reacting particles are fully ionized—they are nuclei with all the electrons stripped from them. Since they are charged, perhaps magnetic forces will confine them.

Thus arose the first epoch in controlled fusion research, starting in the early 1950s. In summary, it was characterized by four realizations. First came measurements of reaction energies and rates between hydrogen isotopes and other light elements, which showed that under proper conditions large energy releases would be possible. Second, the well-known laws of single-particle physics seemed to show how an assembly of high-energy ions and electrons could be confined in magnetic fields long enough to establish the proper conditions. Third, the radioactive isotopes and by-products of fusion were known to be much less hazardous than those associated with nuclear fission: therefore fusion reactors would be simpler and safer than fission reactors. Fourth, deuterium was known to be in plentiful supply. Only the first of these realizations is a prerequisite for making H-bombs. The combination of all four captured the imagination of a sizable and very competent fraction of the physics community. The ensuing search for controlled fusion—the ultimate energy source, the source that powers the stars—has sometimes taken on a moral character, possibly as a reaction to the darker uses to which nuclear energy has been put. Whatever the reason, the efforts exerted by some were like those lavished on Mt. Everest—and a good thing, too, for the 1953 workers had anticipated very few of the complexities in both science and technology that had to be overcome before controlled fusion could reach even today's halfway stage. The whole field of plasma physics—the physics of ionized gases—had yet to be invented at that time; superconducting magnets—absolutely essential for most fusion concepts—were not even contemplated. A controlled fusion reactor seemed simple, but, in fact, it is not. The development of one, assuming that it can be developed, will be scientifically and technologically more difficult than development of a fission reactor in roughly the measure that a fission reactor is more difficult to develop than a coal-burning power plant.

THE DESIGN OF A FUSION REACTOR

It is possible to describe many of the design requirements for a fusion reactor using only simple physical principles and a little common sense. Consider first a deuteron and a triton that are supposed to fuse, and consider also a "target area" associated with one of the particles. The other particle must pass within this area if the two particles are to interact—that is, combine, disintegrate, or even just scatter. The target area will have a different size for each type of reaction, and for a fusion reaction it will grow larger if the incoming particle is more energetic—that is, if it has higher velocity—up to a limit that does not concern us here. The target area is called a reaction cross section and is measured in units of "barns," as in the familiar

taunt, "You couldn't hit the broad side of a barn." One barn is 10^{-24} square centimeter.

Figure 27-2 shows several fusion cross sections for reactions between light nuclei. We see that no reactions involving protons as fuel have usefully large cross sections. Not shown at all is the two-proton reaction by which the sun is burning hydrogen. That reaction has a cross section about 10^{-20} of the deuterium-tritium reaction; accordingly, as we noted, the sun takes billions of years to consume its hydrogen—a time suitable for stars, but not for the fusion reactors we have in mind.

Several lithium reactions are shown; they have smaller cross sections than deuterium, tritium, or ^{3}He reactions because of the higher repulsive force between the particles. Thus, lithium is good for breeding the tritium but not for the fusion reaction itself, and we apparently must consider deuterium, tritium, or helium. In fact, the choice is more limited. It will turn out that the production of more energy than is consumed will be a significant problem in fusion reactors: the constituents must be heated to the reaction temperature, and even then only a small fraction of the hot fuel will react in one pass through the reactor. Power will also be needed for other purposes. This energy-balance difficulty pertains even to the deuterium-tritium reaction, but noting that the deuterium-deuterium and deuterium-^{3}He reactions have cross sections that are smaller by factors of 30 to 100, we conclude that the deuterium-tritium is the fuel of obvious choice, despite the need for tritium breeding and the problems of radioactivity.

Thus, from the illustration, we wish to extract a typical energy and cross section for deuterium-tritium fusion. Although 150,000 electron volts and 5 barns seem ideal, it turns out that the associated temperatures and pressures are too high for many fusion schemes. Let us consider lower energies, say 20,000 to 50,000 electron volts, so that the average cross section for the reacting particles will be about 1 barn—a convenient number to use in our discussions. A gas at 20,000 electron volts has a temperature of

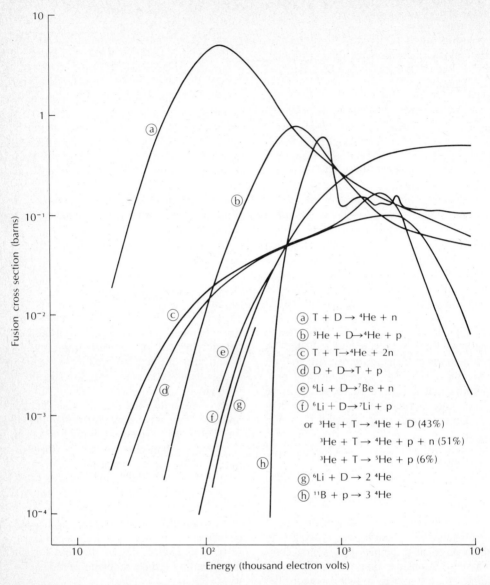

Figure 27-2. Several fusion reactions of interest to an energy-hungry society. The horizontal axis plots particle energy, in electron volts. The vertical axis plots the so-called reaction cross section, a measure of how close two nuclei must come in order to fuse. For the vertical axis, the units are "barns," where each barn is 10^{-24} square centimeter. The fusion of deuterium and tritium is easiest to manage: 20-keV (20,000 electron volts) deuterium and tritium nuclei have, from the chart, a fusion cross section of about 0.1 barn. However, a gas with an average energy per particle of 20,000 electron volts includes a considerable number of more energetic particles, and their fusion cross sections (up to 5 barns) raise the average for the whole group to about 1 barn. It follows that the deuterium and tritium nuclei in a fusion plasma at 20,000 electron volts must approach to within a distance of 10^{-12} centimeter (the radius of a circle 10^{-24} square centimeter in area). One other reaction on the chart—that of boron-11 with a proton—is noteworthy, for it produces no radioactivity and also no neutrons to pass through a reactor wall to damage outside structures. However, a million electron volts is required to start the reaction, making boron-proton fusion remote in humanity's future.

about 2.3×10^8 K, a value far surpassing any temperature ordinarily found on earth. It is exceedingly hot—about ten times the temperature of the sun's core. Will the reacting particles be neutral atoms or will they be ions at this temperature? The question is easily answered by comparing some orders of magnitude. The energy required to ionize a hydrogen atom—that is, to remove its sole electron—is 13.6 electron volts. The energy required to split the two-atom hydrogen molecule into individual atoms is even lower. Thus, an assembly of 20,000 electron volts molecules will decompose into ionized nuclei and free electrons. And at these very high energies, the electrons will not recombine with the ions.

Moreover, these dissociation and ionization processes will occur rapidly. Cross sections for reactions of this sort are typically about the size of the atoms themselves, say 10^{-20} square meter, which is 10^8 times the fusion cross section. Thus, a collection of deuterium and tritium atoms at fusion temperatures will turn into a plasma containing, in this case, deuterium and tritium ions and an equal number of free electrons, in a time very short compared with the time the assemblage would require to undergo appreciable fusion. By this reasoning we can also conclude that a chance neutral atom injected into the fusion plasma will have a very short lifetime. This is not to say that a fusion plasma fully ionizes an arbitrary amount of neutral gas in the vicinity; and task requires energy, and so the plasma cools, and the addition of two much neutral gas will depress the temperature to uninterestingly low values. Also, a truly fearsome atomic-type reaction called charge exchange can take place wherein low-energy atoms (say at room temperature, with corresponding energy of about 0.03 electron volt) enter the plasma and exchange their charge with a high-energy (20,000 electron volts) ion; they become fast atoms that leave the plasma, perhaps to strike and damage the surrounding walls.

Another major question concerns scattering of the ions by each other before they come sufficiently close to fuse—a phenomenon we have mentioned briefly. For 20,000 electron volt ions, the scattering cross section is about 2×10^{-26} square meter, which is 200 times larger than the fusion cross section, and if the energy were less, there would be even more scattering. The implication of this is that the assembly will become randomized; any given initial distribution, even of carefully focused ion beams, will take on the properties of a gas after each ion has been scattered by its neighbors a few times. Accordingly, temperature, pressure, density, and other concepts useful in thinking about ordinary gases will also be useful here. By considering the fusion plasma as a gas, we can derive various combinations of temperature, density, and confinement time that are required if a substantial fraction of the fuel is to undergo fusion. One combination, assuming a plasma temperature of about 10^8 K, is a density of about 10^{14} particles per cubic centimeter and a confinement time of about 1 second. The density is rather low (by comparison, the density of the air around us is about 3×10^{19} particles per cubic centimeter), and so the plasma approximates a good vacuum. But the product of density and temperature is a measure of a gas's pressure, and in this example, the plasma's pressure turns out to be 10 atmospheres. That is not fortuitous. Though large structures can be built to withstand pressures one order of magnitude greater, the extra order of magnitude must be saved to withstand the stresses that arise from high magnetic fields, soon to be described. For now, we note that as far as the physics is concerned, it does not matter whether one achieves high density and short confinement time or low density and long confinement time, at a given temperature; this trade-off can be a matter of technological and engineering convenience. The fractional burn-up of nuclear fuel will be the same.

We have discovered that the reactor contains a deuterium-tritium gas, confined in some way yet to be determined, for some period of time, inside an evacuated chamber—the vacuum being necessary in order to keep the plasma hot and pure. Thus we shall need a vacuum wall. Now recall that 3.5 million electron volts, or one fifth of the energy created by deuterium-tritium fusion, appears in the velocity of the helium ion created in the reactions, and 14.1 million electron volts, or four fifths, appears in the velocity of a neutron. We can imagine the charged particles being slowed down within the reacting plasma itself by electrical forces. On the other hand, the neutrons will not be affected by electromagnetic fields, nor will they be stopped by collisions within any reasonable length of the tenuous gas. They must pass through the vacuum wall into a surrounding "blanket" region of some sort, in which their velocity will be slowed and their energy converted into extractable heat. That was the reason for building the reactor. Moreover, it is in this region that the neutrons must breed tritium as they slow down and are captured in lithium. Our rudimentary fusion reactor now starts to take shape; it is shown in Figure 27-3. The major missing component is a mechanism for plasma confinement, to be discussed shortly.

How thick must the blanket region be? Assume it is made of solid or liquid moderators, absorbers, and so on, all of whose density is typically about 5×10^{28} particles per cubic meter. The cross section for a 14 million electron volt neutron colliding with a nucleus inside such materials is about 2×10^{-28} square meter, or 2 barns, and the corresponding mean penetration distance before the first collision is about 0.1 meter. But the blanket must be much thicker than that in order to stop all but a negligible fraction of the fast neutrons and moderate the energy neutron so that the neutrons can be absorbed in lithium. Calculations show that the blanket must be about ten times as thick, or about 1 meter, and even then further shielding may be required outside.

Almost intuitively, we foresee that the fusion blanket will be a high-technology item and hence expensive. Compounding this cost problem will be the vacuum wall itself, called in the trade the "first wall." Every 14 million electron volt neutron must pass through

it before it regenerates tritium—a circumstance in sharp and important contrast to that in fission reactors, where most of the energy appears with fission fragments, which are stopped in the fuel itself. It takes no great insight to see that material damage via irradiation by 14 million electron volt neutrons will be one of the most severe problems in fusion reactor design. In the first place, some of the neutrons, and all of the X-rays and other radiation from the plasma, will give up energy at the vacuum wall. Calculations of permissible heat transfer show that the total energy incident cannot be allowed to average much more than a few megawatts per square meter, and possibly less. In the second place, one wishes to limit the number of neutrons penetrating all the way through the blanket, so as to minimize damage to outside apparatus, which will doubtless be expensive. (This requirement will certainly be true if we consider putting expensive superconducting magnetic windings outside the blanket, as is contemplated in almost all fusion schemes, for reasons we shall consider shortly.)

All these factors—the need for low values of neutron and energy flux per unit area of vacuum wall, coupled with the high cost of the reactor's technology—together dictate that the reactor must be large. How large? If the vacuum wall surrounds a sphere whose radius is 2 meters, the power created will be about 50 megawatts if the first-wall neutron loading is to be kept at 1 megawatt per square meter. In fact, the power per reactor is liable to be much larger, because of considerations not yet discussed. Some are geometrical ones, such as the circumstance that most fusion concepts work not with a spherical plasma region, but only with a toroidal or donut-shaped one. Thus Figure 27-3 would represent a cut through a torus. If the minor radius to the vacuum wall will be 2 meters, the major radius will be 5 meters or more, the total vacuum-wall area will be 400 square meters or more, and the reactor power will be comparable to that produced by large present-day power plants.

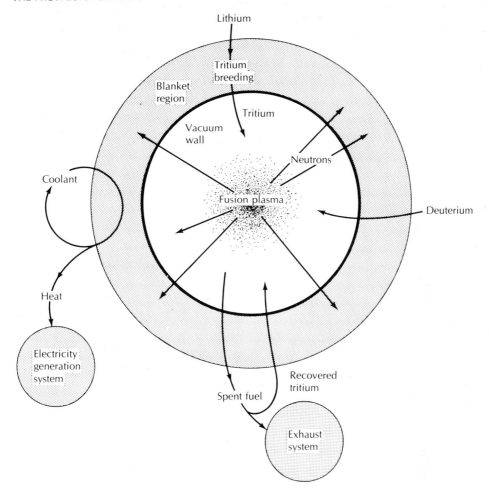

Figure 27-3. Conceptual design of a fusion reactor. At the center of its reaction chamber, gaseous deuterium and tritium are heated to a temperature of 10^8 K, by some mechanism that is not shown—or even suggested—in the drawing. The gas, now a plasma, is confined at that temperature long enough for fusion to occur—again, by a mechanism that the drawing fails to suggest. Neutrons are freed by the fusion reaction. They speed through the so-called vacuum wall and into a blanket region, where their energy is somehow transferred to electricity-generating apparatus—conventional steam-driven turbines, perhaps. Also in the blanket region, tritium is bred from lithium. Meanwhile, spent fuel is somehow removed from the reactor, unburned tritium is somehow recovered, and fresh fuel is somehow injected.

SCHEMES FOR PLASMA CONFINEMENT

Laser Fusion

Last, we consider how to confine the plasma. The only physical confinement principles that exist are:

1. Gravity, which will not work on earth. It can work on the sun only because of that body's mass.

2. Material walls, which are self-defeating: it is not that the plasma would melt the walls, but more likely the plasma's ions and electrons would lose their energy and

cool on their first encounter with their container.

3. A surrounding colder gas, which is not entirely unfeasible but suffers from several problems, among them energy loss by collision with material walls.

4. Static electric fields, which are impossible, not only because oppositely charged particles—nuclei and electrons—must both be confined, but also because the fields required to confine either species would be huge.

5. High-frequency electromagnetic fields, which have been seriously studied; but the prognosis is poor, partly because of the undesirable electrically driven particle motions and partly because the power required for the scheme is very great.

Two principles remain: inertia and magnetic fields. We turn first to the simplest idea of all—inertia, as in the H-bomb itself. Here, nothing confines the plasma; it is heated to a sufficient temperature so rapidly that it fuses before the particles can go anywhere. The idea is to cause fusion to occur in a very small deuterium-tritium pellet, so that the explosion will be very small by nuclear standards and can be managed by a simple mechanical system. The pellet is to be heated by a gigantic laser pulse. Every atom in it is to reach (say) 10,000 electron volts of energy, and a corresponding speed of about 10^6 meters per second. Under these conditions, a pellet 2 centimeters in diameter will disintegrate in about 10^{-8} second. But in that time, and under ideal circumstances, it will fuse about one tenth of its fuel and release more than 10^{11} joules of energy.

This explosion, however, is unacceptably large: it is equivalent to the detonation of many tons of TNT. What to do? Just decreasing the pellet diameter will not work because the pellet will disintegrate more quickly (i.e., each particle will travel a distance equal to the pellet's diameter in a shorter time) and not enough of the fuel will undergo fusion. In principle, there is a way out, based on the fact that the fusion reaction rate increases as the square of the pellet's density. This is the key: smaller pellets will

work if compressed by the laser pulse itself. The pulse is to be very short and very intense; it is to evaporate the outer layers of the pellet so rapidly that the momentum change of the material leaving the surface creates a compressive force in the remaining core. The pellet's density will then rise well above that of the original solid, and fusion can now occur with a smaller total amount of fuel. A 1000-fold compression of the pellet is planned, not only to ensure that the pellet burns well, but also to heat it to the initial fusion temperature. At high density (4×10^{31} particles per cubic meter), the pellet will undergo fusion in about 10^{-11} second.

We now have a fair picture of laser fusion. We imagine a mighty laser, with many beams impinging on the pellet to ensure even illumination, and thus that the pellet will compress. If the illumination were uneven, part of the plasma would spurt out where the intensity of the impinging light is even slightly lower.

Why will this not work? We see at least three reasons. First, the laser necessary to implode the pellet must produce a 1 million joule pencil of light in a pulse lasting a billionth of a second, and it must do so with high efficiency or else it will consume more energy than the fusion reactor can generate. Such a laser does not now exist, and its science remains very unclear. Although there have been remarkable advances in lasers, they have come mainly by extension of well-known scientific principles. Even then, they have come at a very high cost.

Second, laser fusion necessarily implies a pulsed reactor, one that produces nuclear heat in bursts perhaps a fraction of a second apart, but the unfortunate fact is that pulsed reactors must operate under far less internal stress than steady-state reactors, because cyclic stress tends to fatigue materials. Consider the reactor's vacuum wall, and suppose that laser fusion is proceeding in pulses next to it. The vacuum wall is first heated by the fusion-released energy and then cooled by the coolant flowing behind the wall. Thus the temperature throughout the wall cyclically rises and falls to different levels at different depths within it, and because metal expands when it is heated, the result is a cyclic stress in the wall. That stress can be calculated, and both the calculations and the experiments made to date suggest that the laser-fusion schemes thus far proposed will fail from thermal stress alone, unless the walls are placed so far from the pellet that the thermal stress becomes negligible. But in that case, the device becomes too huge and too expensive for the small amount of energy that can be captured—a circumstance that will not please the stockholders.

The third problem may be the worst. Imagine that the fuel starts out as a 1-millimeter pellet, and that the laser light shining upon it creates an inward force on its surface that compresses the pellet by a factor of 1000, to a final diameter of 100 micrometers. Throughout this compression, all the light of the 1 million joule laser must remain focused upon it. Now the light must come from large optical surfaces—mirrors, probably—of most remarkable quality, most carefully placed, for no ordinary searchlight mirror will focus all its energy down on a 100-micrometer spot. There will be perhaps 10 or 20 such mirrors, each 1 meter in diameter, surrounding the little pellet, which compresses and undergoes fusion in 10^{-11} second and releases about 10^8 joules—the energy content of a satchel full of dynamite. But that amount of nuclear heat is worth about 5 cents, even at today's inflated prices, and so the explosion must cause very much less than 5 cents' worth of damage and disalignment to the optical surfaces, considering all the other costs of building and operating the reactor.

Magnetic Mirrors

Evidently the little pellets that can be fried so cavalierly in computor simulations cannot so easily be fried in a fusion reactor. The final possibility is magnetic confinement of a plasma. We therefore turn to an examination of magnetic forces.

Most people have seen iron filings sprinkled on a sheet of paper with a bar magnet placed underneath; it is a ritual of adolescence practiced in high school science classes. A tap of the paper, and the filings arrange themselves into a pattern that suggests a set of curved lines looping from the north pole of the magnet to the south. The lines bunch most closely near the poles and are most widely spaced far from the magnet. These are the so-called magnetic field lines. They are a way to represent a magnetic field which is otherwise an ethereal thing. Its presence cannot be seen until the iron filings line up and reveal it.

For us, the feature of paramount importance about magnetism is that in a straight and uniform magnetic field—that is, in a field that can be represented by evenly spaced, parallel field lines—a charged particle is constrained to follow a circular path around a field line. If an arbitrary constant velocity along the magnetic field is also imparted to the particle, then its trajectory will be a corkscrew with its axis along a field line. In short, the charged particle's trajectory will be tied to the line. Moreover, if the line is slightly curved, the particle will *almost* remain tied. (The importance of the qualification "almost" will appear later, when we describe tokamaks.)

The particle's bondage suggests a scheme for plasma confinement that is most easily understood if we build it up in two stages. The top drawing in Figure 27-4 shows a single magnet coil carrying a current. The resulting magnetic field lines are also shown, and we have no trouble in seeing that the field is strongest in the plane of the coil, where the field lines bunch together, and progressively weaker at increasing distance from the coil, even on or near the axis. Now imagine two such coils, as in the second drawing. Each carries a current in the same direction. If the coils are suitably far apart, some of the diverging field lines from one coil will be "caught" by the other and will pass through it; accordingly, the field's strength will be greatest in the throat of each coil and less in the space between.

This is a simple magnetic mirror. The name describes a charged particle's motion in the field that the two coils create. Consider Fig-

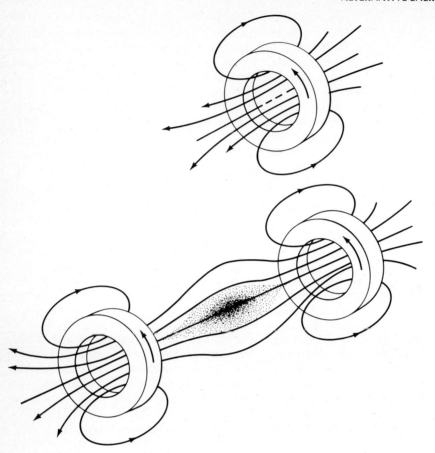

Figure 27-4. The magnetic mirror scheme for confinement of a fusion plasma. In the top drawing, current flows through a single coil. The resulting magnetic field is shown. In the bottom drawing, a second coil is added. Again, the magnetic field is shown. The bottom drawing suggests that plasma can be trapped between the coils of the device. It can, but this simplest mirror configuration is too leaky for a workable fusion reactor.

ure 27-5, which is an abstracted and ideal-ized view of the field lines as they approach and pass through the right-hand coil. Imag-ine that a positive ion is gyrating about the axial field line and that it has some velocity toward the right-hand coil. Now the mag-netic force on the ion is always perpendicular to the field line and to the particle's orbit. Therefore it points toward the axis; this is what makes the particle gyrate around it. But because the field lines converge—because they are not parallel and evenly spaced—a

component of the magnetic force also pushes the ion away from the high-field region. Ac-cordingly, the ion's progress toward the coil is slowed, but because the magnetic field is unchanging, the ion's total energy remains constant—its loss in axial energy is offset by an increase in rotational energy. In the fig-ure, the ion spirals toward the coil until it loses all its axial velocity, and then it spirals back. When it reaches the other magnet coil, it will be reflected there as well. Thus it will remain inside the magnetic cage more or less

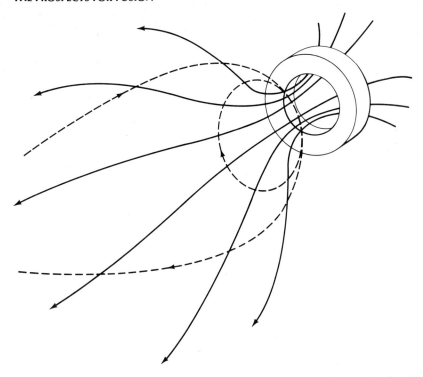

Figure 27-5. The trajectory of a charged particle approaching the right-hand coil of the magnetic mirror device shown in Figure 27-4. At first, the particle spirals toward the mirror region, but its axial velocity progressively decreases. Finally, it is reflected back toward the left. Some particles, however, are not reflected. Those traveling too directly toward the mirror's throat pass through the ends of the device.

indefinitely, unless some other phenomenon intervenes. Seemingly, we need only heat up a fusion plasma and maintain it between two magnetic mirrors and we shall have a fusion reactor.

How can the plasma be heated? Present ideas call for large particle accelerators to produce ion currents totaling thousands of amperes at a voltage of (say) 100,000 electron volts. The ions, as such, would not be able to cross the magnetic field and enter the confinement region; therefore they must be reconverted to neutral atoms before injection. Once inside, the plasma already present will re-ionize them and they will promptly become trapped on magnetic field lines. In sum, the injection process, as now envisaged, proceeds from neutral gas in a storage

tank to energetic ions emerging from the accelerator to energetic neutral particles at the entrance of the fusion chamber and, finally, to energetic ions within the fusion plasma—surely a very roundabout method. Yet it has been tried with good success on a moderate scale, producing, for example, beams of 20 amperes at 50 thousand electron volt energy.

Do we now have a workable fusion reactor? Not yet. Notice that the magnetic field, in its confining effect on charged particles, acts as if it exerted a pressure: where the field strength is high, the pressure pushing against the charged particles is high. On axis in a magnetic-mirror device, the field strength—as well as the pressure—reaches a relative minimum between the two coils, just as one would hope, seeing that a high-energy plasma

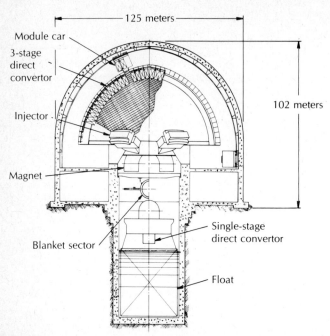

Figure 27-6. A magnetic mirror reference design, as envisaged by the Lawrence Livermore Laboratory. The plasma itself is about 6 meters in diameter. By contrast, the power plant around it is longer than a football field in every dimension. (In most fission power plants, incidentally, the reaction chamber is similarly dwarfed by the reactor superstructure.) The plasma lies within a blanket more than 1 meter thick; heat- and tritium-recovery mechanisms are not shown. As in Figure 27-4, the plasma is confined by two magnet coils. Here, however, both have a complicated shape, and both are superconducting. The lower magnet is underground. It rests on a float so that it can be lowered for maintenance of the blanket. This magnet is more powerful than the upper; thus, particles tend to escape toward the top. Escaping particles—above and below the reaction chamber—make the innate efficiency of the reactor very low. They must be recaptured, and their energy reconverted to electricity, by so-called direct convertors—ion accelerators run backward. The ones at the top are to be more complicated and more efficient, but the design of any direct convertor remains problematic. A "module car" is provided in the design for servicing the upper convertors. Finally, a pair of "injectors" is shown; the design calls for several. They inject high-energy neutral atoms, which are re-ionized by the plasma and thus sustain it.

is to be confined there long enough to undergo fusion. But unfortunately, the field strength also falls off as one moves radially outward, away from the axis. This means that the plasma is unstable: if too much gas is injected between the mirrors, the plasma will pop, like a bicycle tire with too much air pumped into it. The problem can be cured by causing the magnetic field to have a true minimum in the middle of the device. But

there is a cost: the magnetic field and the coils that produce it become very complicated (Figure 27-6).

Having modified the design, do we *now* have a fusion reactor? Still no, for a second problem remains, and it is far more serious: if an ion between the mirror regions happens to be traveling straight along a magnetic field line, or nearly in such a direction, the axial decelerating force acting upon it will be weak.

Thus, ions with velocities directed more or less axially will not be reflected. Instead, they will pass through the high-field region and escape from the ends of the device. We conclude that the magnetic mirror is leaky. But how leaky? The calculations show that each scattering of an ion is fairly likely to send it flying out the end of the magnetic mirror. This means that the ion is unlikely to stay in the device long enough to undergo fusion because, as we found earlier, even a 20,000 electron volt ion is likely to scatter many times before it fuses. For this reason, it appears that the energy released by fusion in a magnetic mirror will be approximately equal to the energy that was required to heat the plasma to fusion temperature in the first place. Now the released energy must be converted, at some loss, from heat to electricity. Moreover, energy is required to remain the steady-state magnetic field that confines the plasma, however poorly, between the mirror regions. In sum, the device seems to be an energy-loser, not an energy-maker.

Immense efforts are now under way to solve the energy-balance problem by developing a way to recover energy from the plasma not only in the form of heat but also directly, as electric power. The idea is to have particles that scatter out the end of the reaction chamber pass immediately into a so-called direct convertor—in essence, an ion accelerator run backward. Here, the ions are made to decelerate. Their kinetic energy, lost in this way, reappears in an external electric circuit as delivered power. Valiant attempts to ameliorate the basic problem have been made—the confinement quality of the magnetic mirror itself—by placing subsidiary mirrors at each end, by injecting ions in artful ways, and so on. Whether these various projects will succeed is still unclear. Yet the magnetic-mirror scheme will be successful in at least some of them if its energy balance is to be made attractive.

We can end this account of magnetic mirrors by noting that nature successfully deploys enormous mirrors above our heads. High-energy ions are produced in space by solar flares or by cosmic rays. They may be trapped by the earth's magnetic field, which looks much like the field of a bar magnet. If trapped, they follow corkscrew trajectories around a field line until they near the north or south magnetic pole, and there, where the field lines close in and the field strength rises, they are reflected. The space around the earth is thus occupied by charged particles shuttling back and forth between mirrors at the polar regions. The heavenly mirrors, however, are no fusion devices. Indeed, they are no better at confining a plasma than their earthly namesakes. Particles almost continuously escape and rain down into the upper atmosphere, where they create the aurora borealis in the Northern Hemisphere, the aurora australus in the Southern Hemisphere. When solar flares modulate the solar wind and rattle the configuration of magnetic field lines far out in space, the aurorae become more spectacular.

Theta-Pinches

A second fusion device employing magnetic plasma confinement is shown in schematic diagram in Figure 27-7. Its principle of operation seems to be fairly simple: electric current is made to flow around a cylinder. In so doing, it creates a magnetic field whose field lines (within the cylinder, at least) are parallel to the cylinder's axis. If cold, dilute plasma initially lies within the cylinder, an immediate increase in the magnetic field strength will compress the plasma initially into a rod-like region at the cylinder's axis; if this increase is sufficiently rapid, the compression can also shock-heat the plasma—sufficiently, perhaps, so that it will start to undergo fusion. The device is called a theta-pinch because the direction in which current flows around the cylinder, by worldwide convention among fusion researchers, is labeled by the Greek letter theta.

The compressing and heating of plasma requires energy. To do it quickly requires that the energy come quickly—in other words, that the power be high. Accordingly, theta-

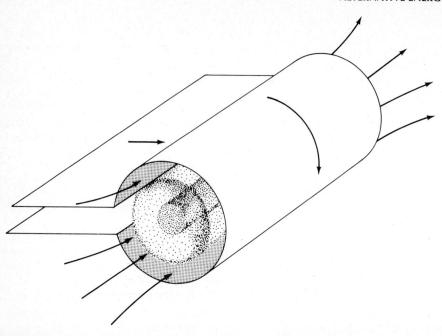

Figure 27-7. The theta-pinch scheme for plasma confinement. Current is made to flow abruptly around a cylindrical surface. This induces a magnetic field that compresses and heats a deuterium-tritium mixture contained within the cylinder. For reasons given in the text, two power supplies and associated switching mechanisms are required, and the larger supply, used for plasma compression, must be superconducting or the reactor will lose energy, not make it. The power supply's design remains to be invented. One major fault of this schematic drawing: if the cylinder in a real theta-pinch reactor were straight, it would have to be several kilometers long or the plasma being compressed would spurt out its ends before fusion could occur. Actual designs call for the cylinder to be bent so that its ends meet, obviating the problem. But other problems persist.

pinches will require enormous power supplies of two sorts. First of all there must be a source of electric current that peaks on the order of microseconds (millionths of a second). It induces a magnetic field that compresses and shock-heats the gas to a temperature of about 10^7 K, or about 1000 electron volts per particle—approximately 100 times the ionization energy. Then a second power supply, cut in by suitable switching, takes over. It furthers heats and compresses the plasma, this time on the order of milliseconds (thousandths of a second). Presumably fusion now occurs. Two stages of heating and compression are necessary because of the differing time scales. To supply energy on

the order of microseconds, capacitors must be used, and capacitors are expensive. To supply energy on the order of milliseconds, cheaper equipment will work. It could be managed, for example, by rotating machinery—a so-called homopolar generator that contains elements that spin at a high rate of speed. When it is slowed, an appreciable fraction of its rotational energy is converted to electrical energy.

Why will *this* confinement scheme, which worked so well at the plasma-physics stage, not work for a commercially interesting reactor? The first problem is that the theta-pinch, like laser fusion, requires a pulsed reactor, with a consequent cycle of thermal stress. Each spurt of fusion in a theta-pinch reactor will last 10^{-2} second, not 10^{-9}, but failure of the vacuum wall through cyclic overstressing will still be a problem. Experiments suggest that only 10^5 to 10^6 cycles can be withstood by a vacuum wall, and thus, with 1 pulse per second, as the proponents of the theta-pinch hope, the reactor will have an operating life of under two weeks. Dismal as this sounds, the true stress problem may be even worse, for in the theta-pinch reactor, the metallic cylinder through which current flows will experience a force that arises because the current sheet lies within the very magnetic field that it creates. This is the familiar force that turns electrical machinery. Here, in the cylindrical geometry of a theta-pinch reactor, it will be equivalent to a pressure coming from within. Because the current comes in pulses, so does the pressure, and the cyclic stress problem is exacerbated.

The second problem is energy balance: in this case, the amount of energy delivered by the power supply as compared with the amount of energy produced by fusion. Consider a 1-meter length of theta-pinch reactor. An input of energy approaching 200 megajoules is required to shock-heat and compress the plasma column contained therein; an output of fusion-released heat approaching 6 megajoules is the result. The reactor, then, is a sure loser, unless the power supply's energy can be recycled with incredible efficiency. But doing so will require superconducting pulsed magnets, and superconducting switches as well. The former are now made in small sizes; however, a design for larger versions is hard to foresee and still harder to develop. The need for these devices suggests that the cost of a theta-pinch will be frightening, if only because the cost of a system often relates fairly closely to the amount of energy it must handle.

Problem number three: suppose that all the other problems have been solved. The fusion reaction proceeds. At the end of each pulse cycle, the magnetic field must drop, in preparation for the succeeding cycle. In consequence, the remaining plasma reexpands to fill the original cylinder volume. Even if fusion no longer proceeds within it, that plasma is still close to fusion temperature. It will damage any material surface it hits, and on the time scale of operation envisaged for a theta-pinch reactor—a fraction of a second per cycle—no way is known by which to protect the vacuum wall from its onslaught. To keep the plasma from hitting the wall, the magnetic field lines would have to change their configuration suddenly at the end of each cycle in such a way that gas nearing the wall would be guided into some receiving area where it could be cooled. The alteration of the magnetic field must be interfaced with the rest of the cycle, and moreover, it must not interfere with the onset of the succeeding cycle. Considering the device's rapid pulsing, its awful energy balance, its precarious technology, and its cost, that feat seems impossible.

One nonmagnetic scheme to save the vacuum wall of a theta-pinch reactor has been explored. The hope was to have thousands of little pipes perforating the vacuum wall. At the end of each cycle, puffs of gas would enter the reaction chamber by way of these pipes and a cooling gas blanket would be created around the plasma. The idea cannot work because the cooling gas obeys the kinetic theory of gases, according to which some atoms in that gas will have higher velocities than others, and they will reach the plasma first. Here, many of them will exchange their charge with ions in the plasma, in the fearsome type of reaction we noted much earlier in this article, and become marauding atoms. Thus, one way or another, high-energy particles will smash into the vacuum wall before the slower particles can enter and erect a gaseous shield (Figure 27-8).

Tokamaks

In searching for another magnetic scheme by which to confine a plasma, we begin again with the basic configuration of a theta-pinch; a long, straight cylinder with current circling through it. Assume that the current travels

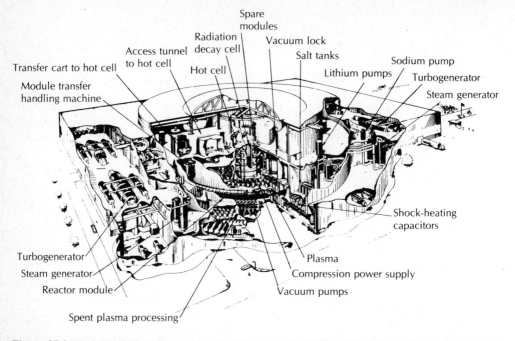

Figure 27-8. A theta-pinch reference design proposed by Los Alamos Scientific Laboratory and Argonne National Laboratory. It suffers from several problems, as described in the text. The reaction chamber is a ring of modules, each working on the principle schematized in Figure 27-7. It is thought that each can be removed for servicing, first to a "radiation decay cell," where, in perhaps one day, it will lose much of its radioactivity, and thence to a "hot cell," where maintenance and repair are to be conducted. Power supplies occupy much of the remaining space around the reaction chamber. Even so, their size may have been underestimated. The reactor is cooled by a series of systems interconnected by heat exchangers. Lithium and lithium salts remove heat from the reactor. Their heat is removed by sodium, which then surrenders it to steam in an exchanger much like the one proposed for the liquid-metal fast breeder reactor. Finally, the steam generates electricity in turbogenerators at either side of the mammoth building.

through windings coiled around the cylinder, as in a solenoid—a helix of electrical conductor. Now the cylinder can be gently bent until it has a circular shape. The cylinder becomes a torus, a donut. Not only do its ends meet, but so do the ends of the field lines within; they are closed loops now.

It might be imagined that the confinement problem is solved—just pump in plasma, and at a suitably impressive magnetic field strength it will remain confined at high temperature and density long enough for fusion to occur. In reality, there are difficulties. The principal one is that the plasma ions cannot be kept from drifting to the sides of the torus because curved field lines in fact yield not the perfect gyration circles we have considered thus far, but almost-circles that gradually drift across the magnetic field. These drifts cause ions and electrons to be lost in opposite directions, since they are oppositely charged, and the separation of positive and negative charges by means of these drifts gives rise to *electric* fields, which cause profound plasma motions. The result is impact of the plasma on the confining vacuum walls in times that are short compared with the desired confinement time. The heart of the

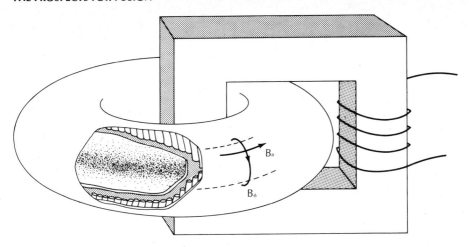

Figure 27-9. The tokamak scheme for plasma confinement. The reaction chamber is toroidal—that is, donut-shaped. Within it, two magnetic fields confine the plasma. The first (B_θ) is induced by current that flows through coils looped around the torus. The second (B_ϕ) is required to correct particle drifts toward the outer torus wall. It is induced by currents in the plasma itself, and those currents are induced by current flowing through the primary coil of a transformer. At present, the tokamak is the main repository of American hopes for fusion.

trouble, expressed in another way, is that a plasma inside the torus experiences a larger magnetic pressure on its inner periphery than on its outer periphery. Thus, it is basically unstable against motion radially outward—motion toward the outer torus wall.

Fortunately, this ill seems curable. The illustration in Figure 27-9 shows the torus in perspective. The windings that wrap around the cylinder are present as before: they induce the magnetic field whose field lines run lengthwise through the donut. But note that the primary coil of a transformer has been added to the device. The secondary coil is the plasma itself. Now a pulse of current sent through the primary coil of any transformer induces a pulse of current in the secondary coil. Here, it induces current in the plasma and also causes the plasma to heat up, just as current sent through a resistor causes the resistor to heat up. The crucial point is that a new magnetic field is created by the plasma's motion. Its field lines circle the plasma in the same way the windings for the first magnetic field circle the torus cylinder. This second

field corrects the separations of positive and negative charges because particles can now flow from the top of the plasma column to the bottom as they gyrate around the new field lines. The result is that the plasma tends to have a uniform electromotive potential. No voltages build up, and in first approximation the plasma is stable. There is a drawback: the unavoidable use of a transformer to generate the second magnetic field means that this device, like a theta-pinch or a laser-fusion reactor, will necessarily be operated in pulses. Here, however, it will be possible to have each pulse last many seconds—perhaps even minutes—and so, while there may still be a thermal-stress problem, it will be much ameliorated. Moreover, the lengthwise magnetic field, which is by far the largest, will not be pulsed at all. It will be steady state, and so it may prove possible, using superconducting coils, to make the energy loss from the main confining system negligible. There will, of course, be losses elsewhere—notably losses from the second, pulsed magnet system—but the overall energy balance may not be the serious problem for this device that it seems to be for the other schemes we have examined. Indeed, it appears at present that the energy balance can be favorable, perhaps by one order of magnitude, which is as good a performance as conventional generating systems can manage today.

The device we are now discussing is the so-called tokamak, first conceived in the Soviet Union and liberally described by the Soviets throughout the mid-1960s, but neglected by the United States until early in 1969, when the late Academician Lev Artismovich gave a series of lectures on the subject at the Massachusetts Institute of Technology. It is currently the principal vehicle by which the United States hopes to realize its fusion ambitions. To some extent, and at the present time, this hope seems justified. For one thing, Alcator, a small device with a tokamak configuration built at MIT's National Magnet Laboratory, has produced the best plasma confinement the world has yet known: the plasma's density was more than 10^{14} particles per cubic centimeter, which is adequate for fusion. The other confinement parameters, however, were not adequate: the plasma's temperature was not 100 million degrees, but only 10 to 15 million, and the confinement time was not 1 second, but only one twentieth of that. Still, these numbers are limited by the size of the device, and Alcator performed roughly as hoped for—well enough to permit new experiments to be conducted on the physics of plasmas. The Magnet Laboratory is building a bigger one.

In tokamaks, the opportunity may exist to solve, or at least bypass, some of the problems unearthed by the other conceptions. Some of the solutions will come at the cost of increased complexity, but at least the researcher has the chance to move on and discover what else lies ahead in fusion development. Consider the problem of plasma hitting the vacuum wall. Although a tokamak may run at almost a steady state for many seconds, the problem persists because plasma confinement inevitably will be imperfect. Even if it magically were perfect, the spent fuel would have to be removed and fresh fuel added. Thus, there is a need for a plasma pump—in practice, a region near the walls of the torus from which the magnetic field lines peel off into a receiving area, carrying plasma with them and keeping it off the vacuum wall. Provision for such a pump is included in the tokamak reference design called UWMAK III, prepared at the University of Wisconsin and shown in Figure 27-10. Here, the simple conceptual design of a tokamak has grown into an immense and immensely complicated structure; the regions at the top and bottom of the plasma are the pump, or graveyard, regions, actually called divertors. Spent plasma arriving there will have a temperature not much below that in the main part of the plasma, and so the pump's surfaces will probably have to be replaced periodically. Still, the damage occurs where it can be planned for, and not at the main vacuum wall.

FUSION'S SHOPPING LIST

Will a tokamak ultimately be the basis for a successful fusion reactor? Notice that our discussion of the device, and the remarkably uncluttered illustration (Figure 27-11), lack all of the following: an injector of fresh fuel; a neutron-moderating blanket; a heat-removal apparatus for the blanket; an arrangement for tritium recovery and tritium breeding; mechanical support for the magnets; thermal isolation of the magnets—they must be kept below 20 K (perhaps less) if they are to be superconducting, while the nearby plasma must be kept at 100 million degrees (perhaps more); any method of heating the plasma to an initial fusion temperature, other than with the transformer (and it can be shown that the resistance heating of the plasma by transformer-induced current is by itself insufficient); and any arrangements for safety, access, or repair. In fairness to tokamaks it should be said that many of these problems are design-independent: they and others will apply in some form to any fusion reactor. Here is our shopping list.

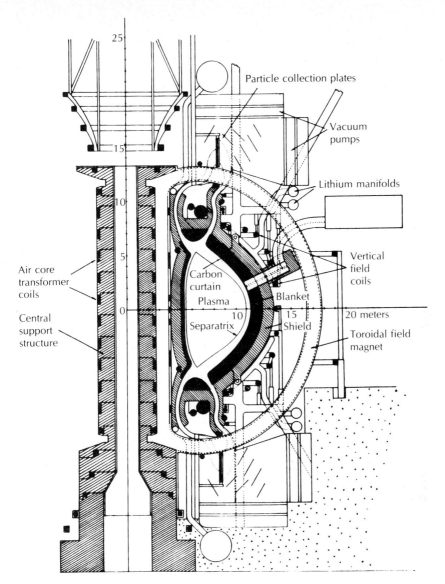

Figure 27-10. Cross section of UWMAK-III, a reference design for a tokamak prepared at the University of Wisconsin. The main magnetic field is induced by a "toroidal field magnet" looping around the reaction chamber. The secondary magnetic field is induced by current in the plasma, and that current is induced by "transformer coils" in the "central support structure." A third field, not mentioned in the text, is produced by "vertical field coils." Like the second field, it corrects particle drifts, and its magnitude is relatively small. The three fields act together to confine a plasma. One field line is shown—the so-called separatrix. Ideally, plasma within that line is confined, while plasma outside it is guided to "particle collecting plates" above and below the reaction chamber. Other details of the UWMAK-III design: The "blanket" contains a lithium-aluminum compound. The lithium is there only to breed tritium, not to cool the reactor. (Cooling is to be accomplished with helium.) The "shield" provides additional protection to outside structures; the blanket alone is deemed insufficient. The vacuum wall has a graphite curtain, intended to protect the wall from radiation and plasma particles. But graphite will trap atmospheric gases during construction and maintenance and then pollute the vacuum during reactor operation.

506

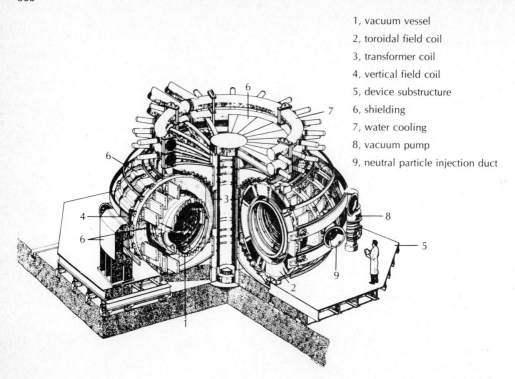

1, vacuum vessel

2, toroidal field coil

3, transformer coil

4, vertical field coil

5, device substructure

6, shielding

7, water cooling

8, vacuum pump

9, neutral particle injection duct

Figure 27-11. The Tokamak Fusion Test Reactor (TFTR), to be constructed at Princeton University's Plasma Physics Laboratory. Over $200 million will be spent to build the device, with its associated facilities, and roughly the same amount will be spent on operating costs through the mid-1980s. The device is not a reactor; it represents an opportunity to create and experiment with a fusion plasma. "Neutral particle injection ducts" will continually introduce hot particles into the torus. Initially, only light hydrogen will be used; toward the end of the experiment's life, a deuterium-tritium mixture may be tried.

Plasma Engineering

Ionized gas is not enough. It must have the right density and the right temperature in the right places. Moreover, it must be pure, in part for reasons we touched upon earlier: impurities cool the plasma and may, after charge-exchange reactions, damage the vacuum wall. An additional reason is that scattering of the plasma's particles by the massive particles typical of impurities leads to loss of energy from the plasma by electromagnetic radiation—for example, by the so-called *Bremmsstrahlung* (literally, braking radiation) created by the deceleration of electrons. Keeping impurities out of a plasma is very difficult. The laws of physics assure us that the boundaries of the plasma will be fuzzy; the best we can hope for is a density that is high in the middle and that tapers at the sides. But then the hydrogen isotopes in the plasma will tend to diffuse outward, while other particles—perhaps heavy niobium or molybdenum atoms freed from the vacuum wall—will tend to diffuse inward. Such phenomena portend future needs for plasmas whose temperature and density gradients are controlled for engineering objectives. Plasma physics will have to become plasma engineering and related to specific fusion systems.

Entrances and Exits

Over $2 billion has been spent around the world on plasma confinement. But how is the plasma to be injected into the vacuum region? And when part of the plasma has fused, how are the helium and the unburnt gas to be removed? In discussing tokamaks, we mentioned divertors to remove the spent plasma; they are a necessary invention for any practical fusion reactor. What about plasma injection? That problem also looks fierce. A neutral beam of deuterium and tritium will not work in a large device; the atoms will be ionized before they get deep enough into the plasma to suit the plasma engineers. Can injection be accomplished with chunks of fuel the size of a grain of rice, or of a pea? The problem resembles in some ways the reentry problem for a space vehicle nearing the earth's atmosphere, but it is harder to solve. Ignoring it is like proposing an internal combustion engine that contains only a combustion chamber and lacks intake valves.

Materials

They must be resistant to an immense flux of radiation, neutrons, and heat and also to strange kinds of corrosion and immense pressures; if the reactor operators had to replace the insides once a week, even if it could be done quickly, the reactor would soon be an economic loser. It will take years to find and develop new materials—longer, probably, than the optimists think and the administrators in Washington proclaim. The vacuum wall, for example, faces an environment that is unimaginably hostile—an energy flux that is higher than anything else ever made, except a nuclear weapon in the process of exploding. There is a serious danger that the vacuum wall will sputter away when high-energy particles hit it, dislodging heavy atoms that will contaminate the plasma. One might conclude that the vacuum wall will have to be outrageously thick if it is to survive many years of operation. On the other hand, thick walls give poor heat transfer and also capture large numbers of neutrons where they should not be captured. Development of a material suitable for the vacuum wall and similar critical locations will be one of the most difficult problems of controlling fusion.

Progress, however, is being made. High-nickel stainless steels developed at Oak Ridge National Laboratory are much more radiation-resistant than any structural metals known a few years ago. Such alloys suggest that the energy flux in the reaction area could safely be increased and that, in consequence, the reactor could be made smaller for the same power output. If it could be made smaller still, perhaps normally conducting coils could be used intead of superconducting ones, with a vast reduction in engineering complexity. But it might then have disastrously poor net power output. The true direction of progress is still uncertain.

Reliability, Repair, and Accessibility

None of the present schemes is credible from this aspect, and in general, the larger and more complex systems are less credible. Again, the most critical item is integrity and possible repair of the vacuum wall. Recall that reactor operation will make it radioactive. The problem cannot be wished away until a fusion demonstrator stage: experience with acceptance of fission reactors shows that the efforts made to develop fusion reactors must include work on these matters, too, or power companies will not be interested.

Plasma Confinement

It appears that simple magnetic mirrors will not work as fusion reactors, nor will fast-pulsed devices such as theta-pinches and systems using lasers. The tokamak configuration remains, but whether it is truly workable as a fusion reactor is not certain either. Fortunately, other ideas are appearing. One, still in its earliest stages of investigation, is the so-called bumpy torus, made by taking a set of coils such as those used in magnetic mirrors and arranging them in a ring, like hoops in a circular croquet game. The plasma confined by this array takes on the shape of a string of sausages. It appears that a bumpy torus may correct the radial instabilities found in a single magnetic mirror, for the stability of a plasma depends on the average history of the particles within it—on the changing strength and orientation of the magnetic field sampled by the particles as they move. Moreover, particles would not be lost out the ends of a bumpy torus, since it has no ends. Finally, a bumpy torus could operate continuously, without pulsing, in the manner of a single magnetic mirror. The scheme is only one of a whole class of steady-state toroidal devices, most of them inadequately explored.

Tritium

The fact that tritium does not occur in nature received little attention until about 1960, because it had been a tacit assumption that, as we reported, one could always make tritium by using fusion's leftover neutron in a reaction with lithium. This means, however, that the fusion reactor must be a complicated breeder reactor from the start. The need for tritium (with its radioactivity) is lamentable, but the cross sections for fusion reactions involving other light nuclei are far smaller than the cross section for the deuterium-tritium reaction. Therefore, far greater values of density, temperature, and confinement time would be required to make the other reactions proceed at an appreciable rate. We noted all this earlier.

Now it may be that only 1 gram of tritium will be reacting in a 50,000-ton fusion reactor at any one time. However, the burn-up of nuclear fuel for each pass through the reactor will be only a few per cent, and so the reactor will likely contain several kilograms of tritium, most of it being separated from the spent plasma and then recycled—all by a process not yet worked out. Perhaps some of the details are yet to be discovered, for controlling tritium is exceedingly difficult. The isotope is a form of hydrogen, and hydrogen diffuses through almost all metals useful for the reactor structure. Those metals that can contain hydrogen include platinum, which is too soft, tungsten, which is almost impossible to machine, and gold.

Lithium

Will there be enough deuterium and tritium to run a fusion-powered economy of the future? Deuterium, at least, is plentiful, as we have seen. To provide an environment in which tritium can be created, every fusion reactor will be surrounded by a lithium-containing blanket in which the lithium-neutron reactions will occur. Therefore we turn to the question of lithium availability.

For the year 2000, it is estimated that an electric generating capability of up to one million megawatts will be needed in the United States. No megawatts in that year will be generated by fusion, but let us imagine similar energy demand at some later time. The potential world reserves of lithium are estimated at 2×10^7 metric tons, enough for 2.8×10^{14} megawatt-hours of electric power production. This is approximately a 30,000-year supply, seemingly enough to last through a long technological age. However, there are complications. First, the inventory of lithium in a fusion reactor must be fairly high if the lithium reactions are to occur before the neutrons are absorbed elsewhere, perhaps by the structural material of the reactor. Second, liquid lithium (or a molten salt that contains lithium) may also be required to remove nuclear heat from the reactor. One might conclude that lithium will be in short supply for fusion reactors. One estimate is that about 9×10^5 metric tons will be required to begin operation of one million megawatts worth of fusion reactors; this is a large drain on lithium resources, but could probably be met by a determined effort.

The lithium reserves quoted above do not

include the lithium in sea water (approximately two parts per million by weight), which is much larger but so far is very expensive to extract. Yet past experience shows that when a serious attempt is made to locate new reserves of previously ignored minerals, more will probably appear. Plainly, the lithium resources could permit vastly more energy generation than petroleum ever did or coal ever will, before heroic measures need be undertaken to exploit dilute lithium deposits. Since complex civilizations have been built on fossil fuel, lithium availability should not be a barrier to fusion development; the supply is short only compared with the essentially endless supply of deuterium.

In view of all the difficulties, we ask again, Are we mad to pursue controlled nuclear fusion? And are all the moribund concepts—laser fusion, simple magnetic mirrors, theta-pinches—a sign of the expensive folly? The answer to the second question is easy, and it is definitely no. But such trials and errors, we have come as far as we have. The experiments have taught us about high-density, high-temperature plasmas, about plasma instabilities, about the damage that high-energy particles and radiation cause in various materials. In any event, the work typifies the way the development of high technology must proceed.

As fusion schemes come and go, it is hard to distinguish between a valuable stepping stone that lets us advance and a cornerstone of the final fusion edifice. How does one tell when the final concept arrives—the one upon which all further efforts ought to be focused? One cannot know in an absolute sense. Judgment must enter in. In our view, the field is still open; the best fusion concept has yet to be recognized. Meanwhile, however, the Energy Research and Development Administration made a long-range plan for fusion and at first gave the impression that it knew pretty well what a fusion reactor will look like. The drafts included trial plans describing in inappropriate detail the development of an "experimental thermonuclear reactor"—an extrapolation, evidently, of current

reference designs. "Start fabrication of magnet coils"; "Complete installation of magnet coils"; "Start test"—this last step in July of 1989! Taken literally, the ERDA exercise suggested to some that the fusion program is fixed in direction through the late 1980s, and this generated vigorous comment in the scientific literature.

Of course, extrapolation of current reference designs would lead to fusion monstrosities: structures 50 meters in diameter, cooled by liquid lithium flowing behind one acre of wall 1 millimeter thick, magically maintained by remote machinery. The superconducting magnets that surround the reaction region, as presently designed, would make it inaccessible for any kind of servicing by an electric utility company. The minimum feasible size of such a reactor might be 10,000 megawatts, too much in one unit for any electric utility. If fusion reactors look like that, officials of the utilities are saying, then we do not want them.

The plans more recently presented by ERDA show a welcome flexibility and no fixed view of what a real fusion reactor will look like. Indeed, the first page points out in effect that such plans are meant to be self-destructing: five years from now we hope to know much more than we do now, just as now we know much more than we did five years ago. So we live in hope.

THE PROSPECT FOR FUSION

The growth of nuclear power fluctuated greatly in the period preceding the early 1970s. Because of hopes for cheap nuclear power, a flurry of orders for fission reactors came in during the mid-1960s, at very low quoted prices—one figure was $130 per kilowatt of installed generating capacity. Reality soon caught up with both the utilities and the reactor manufacturers; it was realized that nuclear power would be more expensive and that all the faults had not been eliminated. Then the orders declined. They picked up again only as fossil-fuel power began to look more expensive, partly because in the early

1970s it came under increasingly strict environmental regulation; virtually every coal-fired electric power plant in the Northeast either closed down or switched to low-sulfur oil. Finally, in 1973, the predictions of fossil-fuel difficulty became known to all, and about 50,000 megawatts of nuclear power-generating capacity were ordered. Continuing fossil-fuel price rises in 1974 reinforced the trend to nuclear power. However, the plants now cost up to $1000 per kilowatt, in 1980s money, for 1980s delivery. The electric utility industry is in real danger of economic collapse; the commitments to new nuclear plants (180 gigawatts, more or less) could break it.

Nuclear fission has had a number of successes and a number of failures. Some of the failures have been organizational in nature—there has been insufficient self-criticism and insufficient internal responsibility. Do not rock the boat, various committees seem to have decided. Give us the money and we shall get around the problems somehow, but let us handle it secretly. This strategy does not work in the long run: criticism will come anyway, and when it does, people will grumble even more, asking, "Why didn't you tell us that before?" Surely it must be possible for a society to face a difficult problem, knowing that an effort must be made though a happy outcome is uncertain. Each citizen seems perfectly able to understand the situation. But when people become members of committees, they dare not say those things that committee members should say to one another.

In efforts to control fusion, some of these problems have already appeared. For example, fusion research was sold for years on the basis that fusion reactors were just around the corner. In the 1950s, engineers were making designs of fusion reactors that employed copper coils and steam pipes instead of the very exotic materials that we now know will have to be developed. At Princeton, researchers planned to build four models of the Stellerator—a fusion device of olden days. They proposed Models A, B, C, and D, each bigger than the one before, and they spent

$30 million on C. Model D was to have been the industrial demonstrator, but it was never built because Model C led to a new plasma science, not to a confirmation of the old. In that way, these expensive experiments showed researchers that an entire field—plasma physics—had yet to be developed. It took about ten years, and great credit accrues to the plasma physicists for managing the feat—in essence, for showing how to contain a Promethian fire. After all, confining a plasma is like taking all the air in a room, forcing it into the center of the room without touching it, and heating it to a temperature of several million degrees. A principal difficulty was expecting and predicting too much too soon. Bit by bit, realism now works its way in. Princeton now plans to build the TFTR—the tokamak fusion test reactor—at a cost of $228 million. It is meant to be operational in the early 1980s, but it will have no engineering for energy recovery. Still, it will (one hopes) confine a plasma well enough for fusion to occur, were it fueled with a deuterium-tritium mixture. (Doing that, however, would cause the TFTR to become radioactive after very few test firings; see Figure 27-11.)*

---

* As of early 1982, the authors' analysis is still essentially correct. However, there have been some recent developments which should soon help verify or disprove their conclusions. For instance, the TFTR is projected to achieve its first plasma in 1983, and the decision was made to use a deuterium-tritium mixture designed to obtain net energy production. The cost of the TFTR is running more than 50 per cent over budget and is expected to reach the $500 million to $1 billion range before the project is completed.

Other nations are also making significant contributions to fusion research. The Joint European Tokamak (JET) is expected to come on-line in 1982. This tokamak will use ordinary hydrogen but should achieve a density-confinement time product which would provide ignition conditions for a deuterium-tritium mixture. The Japanese are building the JT60, which is about the same size as the TFTR but includes many advanced design features. The Russians, originators of the tokamak concept, are working on the T20 tokamak, which is still three to five years away from completion.

All of these machines will provide important experimental information to help clarify and resolve existing plasma problems. One of the major problems remains the parasitic loss of plasma as a result of collisions of the plasma particles. Several theories attempt to explain such losses in terms of turbulent microscopic processes. Experimental data from the new generation of tokamaks will help answer many of these outstanding questions.

The design that currently gets the most money is the tokamak, which has many problems. But at least the difficulties seem to be evenly spread: the confinement time, the divertors that pump the plasma in and out, start-up, access, wall damage, repairability. . . .

Many things must be done, and failing to do any principal one could kill the entire effort. Consider this fable: if you, as the director of an energy utility in the twenty-first century, had a fusion reactor constructed according to 1977 designs, and a pinhole puncture developed in its vacuum wall, you would have to move to Antarctica, and you would be pursued, not necessarily by radiation but surely by outraged investors. Technological problems such as vacuum-wall integrity may yet be the critical ones in controlling fusion.

SUGGESTED READINGS

Steven Bardwell, ed., *Fusion*. Fusion Energy Foundation, Suite 2404, 888 Seventh St., New York, N.Y. 10019.

Lawrence A. Booth, David A. Freiwald, Thurman G. Frank, and Francis T. Finch, "Prospects of Generating Power with Laser-Driven Fusion," *Proceedings of the IEEE* **64,** 1460 (1976).

Steven O. Dean, ed., *Prospects for Fusion*. Elmsford, N.Y.: Pergamon Press, 1981.

Energy Research and Development Administration, *Fusion Power by Magnetic Confinement: Program Plan*. Division of Magnetic Fusion Energy, U.S. ERDA, Washington, D.C. 02545. Five parts, but its "Executive Summary" and "Volume One: Summary" are useful as an overview.

Samuel Glasstone, *Fusion Energy*. Washington, D.C.: U.S. Dept. of Energy, 1980.

A. A. Harms and W. Haefele, "Nuclear Synergism: An Energing Framework for Energy Systems," *American Scientists* **69,** 310 (1981).

Fritz Hirschfeld, "The Future with Fusion Power," *Mechanical Engineering* **99,** 22 (1977).

Gerald Kulcinski, *Fusion Power—One Answer to the U.S. Energy Needs in the 21st Century,* U.W. Fusion Design Report 359, January 1981, available from Fusion Engineering Program, Nuclear Engineering Dept., 1500 Johnson Drive, University of Wisconsin-Madison, Madison, Wis. 53706.

James A. Maniscalco, "Inertial Confinement Fusion," *Annual Review of Energy* **5,** 33 (1980).

William D. Metz, "Fusion Research (11): Detailed Reactor Studies Identify More Problems," *Science* **193,** 38, 76 (1976).

R. F. Post, "Nuclear Fusion," *Annual Review of Energy* **1,** 213 (1976).

V

CONSERVATION, LIFESTYLES, AND ENERGY POLICY FOR THE FUTURE

Aerial view of offshore wells in the Gulf of Mexico. As the demand for petroleum products continues to increase, the emphasis on exploration and production drilling will shift to more offshore drilling. (American Petroleum Institute)

Spoil piles from strip-mining of coal in the 1930s, Montana. With no reclamation, the return of natural vegetation has been slow, even after almost 50 years. (Montana Department of State Lands)

National projections for energy call for the construction of more nuclear power plants similar to this one being constructed some 35 miles south of Miami at Turkey Point, Florida. The two large silo-like buildings house 745-megawatt (electric) power units. (J. R. Lennartson, Westinghouse Electric Corporation)

INTRODUCTION

We have presented the boundary conditions for the energy supply and demand picture, the nature of present fossil-fuel- and fission-based energy sources, and the energy alternatives that are being developed to supplant our diminishing conventional resources. The range of response to those examining the energy situation, and the problems it poses, is extremely broad and reflects the backgrounds, values, and world views of the analysts. In this section, we present a number of articles illustrating the diversity of responses, ranging from straight "supply enhancing" proposals to calls for radical revision of our energy lifestyle.

The analyses presented here are intended to be provocative. Many of the conclusions and cause-effect relationships in one analysis directly contradict those in the next, and some are based on obviously questionable assumptions. Each article, however, also contains perceptive insights into the energy problem. The real future, as it unfolds, will undoubtedly contain elements from all these scenarios, and the game the reader should play is to analyze critically the arguments presented and to consider the alternative energy future that seems most likely and/or desirable. If the most likely future does not coincide with the most desirable future, the political process is available to narrow the gap. This political process is now in full swing on both state and national levels. To understand and contribute intelligently to the political debate, it is necessary to visualize the changes in lifestyle and society that various policy alternatives imply. These articles will assist the reader in this process.

In *Choosing Our Energy Future*, Chauncey Starr states well the case of the technological optimist for an expansionist future made possible by emphasizing the supply side of the energy supply-demand equation. Instead of vindicating the gloom-and-doom predictions of the "limits to growth" school of thought, he feels history shows that technological progress has allowed a "growth of

517

limits.'' By year 2000, there will be 32 per cent more workers in the U.S. job market, requiring an annual growth in the GNP of 2.5 to 3.5 per cent just to maintain our present standard of living. This, in his analysis, will require an increase in electrical production by a factor of 2 to 2.5 by year 2000. After examining all of the alternatives, he sees nuclear energy, ''not as a matter of preference but as a clear necessity.'' This is precisely the ''hard technology'' so abhorrent to those calling for a simpler, decentralized, ''soft'' energy path (see Article 30). However, Dr. Starr suggests that limited expectations due to energy shortages will ''destroy the image of a better future that is so essential for motivating our social institutions.''

Marc Ross and Robert Williams summarized the findings of major conservation studies by the Ford Foundation and the American Physical Society in *The Potential for Fuel Conservation*. The authors make the very useful distinction between quantity and quality of energy in their definitions of ''first-law'' and ''second-law'' thermodynamic efficiencies. Implementation of the technical improvements they suggest would provide no change in lifestyle while reducing the demand for energy by 40 per cent. Such energy thrift would have the effect of moving the clock back 17 years in the upward trend in energy demand. The logic of their scientific arguments is so obvious and the social benefits of their suggestions so attractive that many of their proposals are already a part of national energy policy and practice.

In *Energy Strategy: The Industry, Small Business, and Public Stakes,* Amory Lovins contrasts the energy alternative proposed by Chauncy Starr, which he identifies as ''hard technology,'' with the ''soft technology'' energy option he favors. He believes that hard technology—the continued-growth scenario—is politically, technically, and economically unworkable, primarily because of the enormous capital investments required. In his analysis, new electrical plants, because of their capital-intensive nature, actually destroy jobs rather than create them as the conventional wisdom holds. The concept and properties of soft technology are described in very appealing, humanistic terms and are certainly a useful contribution to energy analysis. Mr. Lovins also identifies very serious problems with the present centralized system, such as the difficulty of utilities in raising capital for new plants. In addition, he strikes a responsive cord in every heart by awakening remembrances of a simpler, pastoral past which probably never was.

One cannot, however, help but wonder why, if soft technology is so attractive, every society with a real choice between the hard and soft paths has chosen overwhelmingly for the hard. The American farmer of the 1930s had a clear choice: keeping the soft path of horses and windmills or opting for the hard path of tractors and the Rural Electrification Administration (REA). The People's Republic of China experimented briefly in the 1960s with the soft path of backyard foundaries and now seeks U.S. technical advice on the building of fission reactors and the world's largest hydroelectric project. What to a soft technologist may appear to be a desirable ''labor-intensive activity'' appeared to the American farmer behind the team of horses as "back-breaking toil," as we know from first-hand experience.

As is apparent from the short editorial comment above, Amory Lovins's writings have an extraordinary capacity to arouse reaction. We present another reaction by A. David Rossin in the article *The Soft Energy Path: Where Does It Really Lead?* His primary objection is that, if society is forced down the soft energy path, the only mechanism for allocating increasingly scarce centralized electricity would be an arbitrary, autocratic government bureaucracy of the very type a soft energy society was designed to avoid. In addition, he notes a basic asymmetry between the present hard energy path and the proposed soft energy path: in the hard energy society, the individual has the choice of opting for a hard or soft energy lifestyle. In a soft energy society, such choice would not exist.

In *A Low-Energy Scenario for the United States: 1975–2050,* John Steinhart and his colleagues cast many of the suggestions of Amory Lovins in concrete terms and, combining these with recent demographic trends and the energy-efficiency improvements suggested by Ross and Williams, project a scenario with a startling 64 per cent reduction in per capita energy consumption by the year 2050. The scenario is proposed not as a plan but rather as a feasibility study for the heuristic purpose of stimulating an examination of the basic values shaping our present attitudes toward energy. The policies recommended are designed specifically to alter lifestyles in gentle ways—designed not to lower the standard of living but rather to provide improved quality of life through a cleaner environment and more rational social patterns. Many of the energy reductions would be achieved through the use of conservation rather than cur-tailment, and this conservation would be "encouraged" by a rather impressive array of proposed federal legislation. The goals of this program are certainly desirable, and the legislation proposed is a plausible route to the goal. The real issue is the desirability of a vast governmental bureaucracy to enforce this new "prohibition" (on energy consumption). Recent political trends seem to be in the opposite direction. In light of the difficulties encountered by the relatively innocuous 55-mph speed limit, the feasibility of much more sweeping and all-inclusive programs seems highly problematic. To the extent that such legisla-tion follows and institutionalizes changing public opinion, however, this scen-ario offers an intriguing and hopeful prospect for the future. In the epilogue, two of the original authors review this scenario in light of recent trends.

Glenn Seaborg presents a valuable synthesis of the more attractive features of both the low-energy scenario of Lovins and Steinhart and the high-technology energy options of Starr in his optimistic perspective, *The Recycle Society of Tomorrow*. He foresees a highly disciplined society of the future in which cooperation, conservation, and ingenuity are dominant values. The "use and discard" ethic will have given way to the "conserve and recycle" ethic, with virgin raw materials used only to replace losses in the recycle process. Energy development will have progressed in all the areas previously discussed, with electronic communications relieving the burdens on energy resources that transportation had imposed. Certainly, this steady-state world, even with its tightly controlled social and physical environment, is vastly preferable to the grim social collapse predicted by the less optimistic world models. The greatest challenge to our society in facing the limitations of a finite world is to resolve the profound ethical questions involved in such social decisions with compassion and wisdom.

Choosing Our Energy Future   28

CHAUNCEY STARR

Written 150 years ago, the words of historian Thomas Macaulay are an apt commentary on today's prevailing mood about energy:

We cannot absolutely prove that those are in error who tell us that we have reached a turning point, that we have seen our best days. But so said all who came before us, and with just as much apparent reason. . . . On what principle is it that when we see nothing but improvement behind us, we are expected to see nothing but deterioration before us?

The attitude that sees "nothing but deterioration before us" is repeated endlessly in much that is now written about the outlook for energy in the United States and throughout the industrialized world. If this perception existed in isolation, it would not be a source of much concern; in that case, the future would take its own course and today's dire predictions would eventually be forgot-

ten. The danger is that in large measure our energy future will be the result of the policies we pursue, and those policies, in turn, are products of the prevailing perception of what lies ahead.

Are we facing a fundamental structural change in the conditions of energy supply with which the world will be forced to live forevermore? I do not think so. I do not believe that the world is imminently running out of its fuel supplies, not even its supplies of mineral fuels. Nor do I think that we are approaching the pollution limits imposed by the carrying capacity of the natural environment.

We are, however, in a period of energy transition away from the transient bonanza provided by low-cost petroleum and natural gas. Because of worldwide inflation in basic energy costs during the past ten years, industrial nations now face two alternatives: ac-

Dr. Chauncey Starr is President of the Electric Power Research Institute (EPRI). Previously he was Dean of the School of Engineering and Applied Science at the University of California at Los Angeles (1967–73), following a 20-year industrial career during which he served as Vice-President of Rockwell International and President of its Atomic International Division. Dr. Starr is a past Vice-President of the National Academy of Engineering, a founder and a past President of the American Nuclear Society, a Director of the Atomic Industrial Forum and the American Association for the Advancement of Science, and a former member of the President's Energy Research and Development Council.

From *EPRI Journal* 5 (7), 7 (1980). Reprinted by permission of the author and *EPRI Journal*. Copyright © 1980 by Electric Power Research Institute, Inc. This article was adapted from Dr. Starr's presentation to the Second Annual Conference on International Energy Issues at Cambridge, England, in June 1980.

commodate to perceived shortages and plan an energy-limited society or exploit new energy sources to ensure their availability, even though they may be more costly than today's sources.

CONSERVATION WITHOUT SHORTAGE

The usual initial response to increased energy costs is to seek a reduction in demand. Conservation is an obvious mechanism, but its mode of implementation depends on whether we face a shortage or an abundant supply at higher unit costs. Survival in an energy-deficient world generates pressures for political intervention, with energy price ceilings and mandatory limits on energy-intensive activities. Some have already been promulgated in the United States, such as requirements on automobile fuel performance, speed limits, room temperature limits, scheduled days for buying gasoline, and restrictions on home and building design. Rationing of fuel and leisure activities is on the horizon. These are the leading edge of a centrally planned lifestyle, the administratively controlled society so abhorred by those of a free spirit.

In contrast, if a successful effort is made to attain energy abundance, albeit at higher cost, conservation will follow naturally as a result of higher prices. Each user will adjust his consumption pattern to match his own resources and needs. He will select his own energy-efficient systems, but he will not face actual shortage. Abundance can result if we stimulate such technologies as shale oil recovery, coal conversion to liquid fuels and gases, nuclear power, and the numerous competitive low-temperature solar alternatives. These are on hand, and every analysis indicates a plentiful mix from them—certainly sufficient to guide our energy transition steps and to avoid serious limitations in the foreseeable future.

We are thus at a key crossroads on energy policy: we plan either to accommodate to enduring shortages or, alternatively, to vigorously promote the development of available energy sources. Higher-cost abundant energy will not have the same result as a truly constraining energy deficiency. Living with selectively affordable energy supplies is different from hand-to-mouth survival on short rations.

The issue goes to the root of our societal perception of the future. Either we foresee a future of shortages arising from limits to growth in a finite earth or we foresee an expansionist future made possible by the "growth of limits" resulting from advancement of technology's frontiers. Historically, as Macaulay's words suggest, the world has always been faced with pending resource limitations, but technology has been able to expand those limits to meet society's needs. This faith in technologic progress is more than wishful dreaming; many of us see the energy paths that can lead us to such an outcome.

The United States provides a case study of these issues. It is, of course, the situation I know best. But there is more to it than that, because what happens here produces worldwide repercussions. Our precipitate movement away from essential energy self-sufficiency, mainly during the past ten years, has been a major factor in producing today's worldwide fuel disorder, and our deliberate movement toward greater self-sufficiency in the future could profoundly encourage new stability in world energy markets.

PROJECTING U.S. ENERGY NEEDS

A key issue for decision makers is the anticipated growth in energy consumption for, say, the next 20 years. In the interconnected net of social expectations, economic growth, and energy demand, a minimal projection of social expectation is that each worker in the 1980–2000 labor force will look forward to the same lifetime pattern of real income for himself as he sees today for other workers of comparable age, skill, and education.

Empirical values for this concept of minimum expectations have been derived by EPRI. The forecast increase in the work force from 1980 to 2000 is about 32 per cent, most of these workers already born. However, the average age of the work force in 2000 will be five years older, and the educational level will be higher: 30 per cent college-educated (versus 16 per cent now) and 60 per cent high-school-educated (versus 50 per cent now). These changes raise the average individual real income by about 25 per cent and the aggregate personal income about 67 per cent, meaning an average annual income growth rate of 2.5 per cent for the intervening 20 years. If the ratio of personal income to gross national product (GNP) remains roughly constant, the U.S. economy must grow at about the same rate to meet this minimal social expectation.

It is doubtful whether the U.S. social fabric can survive with such low expectations of GNP growth in the coming decades. Income must be provided for a growing fraction of retired workers, and the lowest-income sector of our working population is demanding a major improvement in its status; so our minimal social goal is likely to exceed this 2.5 per cent annual rate of increase. How much, of course, is a political uncertainty, as is the maximum rate that we can pragmatically expect to achieve. Happily, GNP growth rates have always managed to provide for more than minimum expectations, except during the depression period of the early 1930s and the 1973 post-embargo recession. For reference, the 1950–60 GNP growth rate was 3.2 per cent annually; the 1962–72 rate was 3.9 per cent.

What does a GNP growth projection of even 2.5 per cent tell us about total energy demand in the United States? EPRI's findings are not startling. Taking into account a 2.5 to 3.5 per cent range of annual economic growth and an expectation of substantial conservation, the total annual energy demand in 2000 will be between about 103 and 138×10^{15} Btu (103 to 138 quads), from the present base of about 80×10^{15} Btu, or roughly from 1.3 to 1.8 times what it is now.

Electricity growth projections are even more uncertain because they involve the ad-ditional factor of interfuel substitutions between electricity and other energy forms. But electricity use has always grown faster than total energy use, and EPRI now foresees electricity use growing about twice as fast. For the year 2000, this means electricity generation at about 2 to 2.5 times today's level.

THREE SUBSTITUTES FOR OIL

Where will the energy come from to meet these needs? Our national problem has usually been presented as one of oil availability, import price, and our resulting negative balance of trade. The understandable public interest in ensuring liquid fuel supplies for transportation has focused political attention and national planning predominantly on that sector. The litany of oil trade numbers and their financial and political implications are constantly reviewed (not to mention disputed), but it is important to recognize that technology can substantially alter the liquid fuel future.

Accelerated near-term electricity production from solid fuels (coal and uranium) can eventually displace a portion of oil use in three areas: under power plant boilers (10 per cent of total oil use), for space and water heating (12 per cent), and in industry (20 per cent). This 42 per cent target is almost as large as all U.S. oil imports. Additionally, electricity can often substitute for gas, which in turn may substitute for oil as a feedstock for the petrochemical industry.

How many electric power plants would all this take? If heat pumps are used to efficiently convert electricity back to heat, each nominal 1000-MWe power plant represents the use of about 30,000 barrels of oil per day. The United States now uses about 17.5 million barrels per day for all purposes, and so 42 per cent of that figure (7.35 million barrels per day) could, in principle, be displaced by 245 coal or nuclear plants. We now have the equivalent of about 580 such plants. An increase of less than half of our present capacity fulfills the need—assuming that com-

Substituting for oil

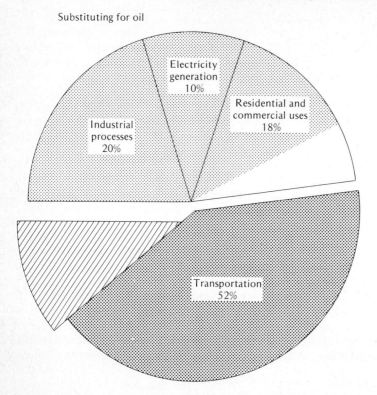

Figure 28-1. How can the pattern of U.S. oil use be changed, and how soon? Coal and uranium can directly replace oil throughout electricity generation and process heat production and indirectly replace oil (by means of electric-powered heat pumps) in many residential and commercial applications. Early substitutions in these three sectors could total 42 per cent of 1979 oil use (nearly as much as the 46 per cent we now import).

By 2000, development of synthetic fuels from coal and shale could yield the equivalent of another 14 per cent of today's use, probably most applicable in the transportation sector.

plete and efficient substitution could actually occur.

Beyond this displacement of natural crude oil alone, however, oil recovery from shale is potentially extremely large and could in time replace all our imported oil. The production process is inherently more expensive than from flowing wells, but the costs are within near-term commercial reach. What are needed now are large-scale demonstrations to settle all the technical uncertainties of environmental issues, equipment reliability, fuel-refining variables, and the like.

There is yet another source of liquid (and

gaseous) fuel. The conversion of coal is a chemical engineering process that has been feasible on a small scale for some time but for which large-scale development remains to be accomplished. This might be done during the coming decade if proposed projects are started soon. Cost projections are promising but still uncertain because full-size engineering subsystems have not yet been demonstrated. The large U.S. coal resources make coal conversion a very attractive long-term objective for liquid fuel supply.

There is no question in my mind that technology is showing us a way to remove the oil squeeze, with room to spare. In industry, in buildings of all kinds, and by utilities, the displacement of oil by coal- and nuclear-powered electricity generation can be significant in the near term. Oil from shale and from coal conversion processes can do even more. We can therefore be optimistic about an eventual assured liquid fuel supply if we take the obvious steps.

But although substitutes for natural petroleum are obviously crucial to our total energy mix and are clearly seen to be technologically attainable, an adequate supply of electricity, in particular, is not thereby assured.

FUELING ELECTRICITY GROWTH

Almost one third of our primary energy equivalent is used to generate electricity today, and by the year 2000 about one half will be so used. Of this electricity, only one third is for residential use; two thirds is for business (commercial and industrial) use. The implications of this deserve attention.

Electric utilities are generally in a good position to supply U.S. needs in the near term. Generating capacity undertaken during the economic growth period of the 1960s and early 1970s has been steadily coming on-line. With the reduced economic growth since 1973, this new capacity provides a reserve that can take care of the more slowly rising needs of the next few years, although probably with some regional shortfalls. But what about the long term? With as many as 12 years passing between decision and availability, deficiencies in electricity supply are not easily rectified.

How can a doubling of electricity output be realized by 2000? Obviously, the bulk must come from coal and uranium. Coal-fired plants now supply about 41 per cent of our electricity needs, and they consume 480 million tons of coal annually in the process—two thirds of our 720-million-ton total production. Considering all the constraints of environmental regulations on end use and the delays and institutional constraints on in-

creasing coal supplies, coal-produced electricity may be limited to slightly more than double that of today—a 4.5 per cent annual growth perhaps yielding a 45 per cent share by the end of the century.

The increasing difficulty of building coal-fired plants is not generally understood. From time of decision to availability now takes 8 to 10 years, of which about half is used for the approval chain. To meet projected environmental standards, environmental controls cost about 60 per cent of the total, and as a result a coal-fired plant now almost approaches a nuclear plant in capital cost. Its fuel costs, of course, are much higher.

Furthermore, the expansion of coal-fired generation may be more constrained if coal is extensively diverted to synthetic liquid fuel production. One ton of coal may yield three barrels of oil. Fully half (360 million tons) our present coal production would be needed to produce about 1 billion (10^9) barrels of oil per year—some 16 per cent of our present use and about one third of our imports.

Hydroelectricity provides about 11 per cent of our generation now, and it can possibly be increased during the next 20 years, but not to the extent of doubling; it may provide about 7 per cent of the output in 2000. Geothermal energy supplies are about 0.2 per cent now and may be about 2 per cent in 2000.

Solar electricity is projected in the February 1979 report to the President, *Domestic Policy Review of Solar Energy*. That interagency forecast for the year 2000 presents estimates of solar-thermal, photovoltaic, and wind generation based on extremely optimistic (in some cases unrealistic) assumptions for technical development. Its projection for equivalent energy displacement by solar resources is 2 to 6 per cent of the electricity demand projected by EPRI.

COMPLETING THE TECHNOLOGY MIX

The total electricity from all these generation sources is 56 to 60 per cent of what is needed

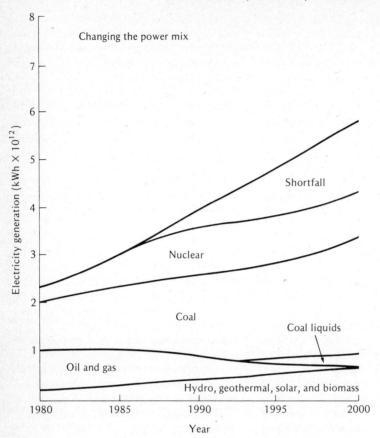

Figure 28-2. As oil and gas are withdrawn from the electricity generation mix, increased use of other utility fuels and technologies may not be enough to meet U.S. requirements by the turn of the century. Nuclear power, in particular, could play a greater role if we reconsider what is wise and possible for that technology.

to meet EPRI's projection of requirements in 20 years. The remainder must come from uranium, coal-derived fuels, oil, and gas. Right now, about one third of our electricity generation comes from oil and gas. Given our need for liquid fuels in transportation and strong federal policy to diminish their use in electricity generation, it is unlikely that synthetics and new oil and gas can be assumed to be available for electricity growth. Oil and gas will probably generate somewhat less electricity than today, perhaps 7 per cent of the year 2000's requirements.

We are left with about one third of our

needs unfilled, even assuming only minimal growth, and nuclear energy is the only source that can fill this gap. Recognizing that these forecasts already include more than a doubling of coal-fired generation, it becomes apparent why utilities see nuclear power not as a matter of preference but as a clear necessity.

The de facto moratorium on nuclear power is thus a prime threat to the secure supply of electricity in the future. Nuclear plants have represented about half of all planned capacity in the United States. Although nuclear power questions have generally been addressed to the utility industry alone, their implications are crucial to the nation's economic future. What is at issue is the future availability of sufficient electric energy to permit the production of industrial goods and services that match realistic projections of the growth of the national economy.

An energy shortfall might take the form of measurable physical shortage, or it might be partially absorbed by gradual adjustment of the economy to a chronic condition of scarce supply in the 1980s and 1990s. Because two thirds of all electricity is ordinarily consumed by commercial and industrial establishments, the impact on our productive sectors would be large, even if shortages were evenly shared by all users. However, past experience indicates that shortages are likely to be allocated—less on voting consumers and more on business establishments. But no matter how shortages are managed, their impact on the economy is bound to be severe.

MOVING PAST THE CROSSROADS

Like the attitudes cited by Macaulay, the prevailing energy outlook for the 1980s is indeed bleak. However, if we can rise to Macaulay's question squarely and objectively, we may be able to *use* the period of the 1980s to implement technical options that will ensure energy abundance in the long run.

The electricity consumption estimates introduced early in this article are based on social expectations of economic growth well below trends of the recent decades. In addition, those estimates assume a very high level of conservation in the years ahead, the result both of higher energy prices and of improved energy-use technologies.

I believe that industrial nations must develop and expand those fuel and electricity sources that technology has brought within the range of economic feasibility. And, of course, they must explore new concepts to ensure energy for the distant future. Energy abundance is a foreseeable goal and at costs that should be acceptable to our economies.

These are our real alternatives: more nuclear power, still more imported oil for elec-

tricity production, or not enough electricity (even with extensive growth of coal-fired plants) to meet minimal social needs. If electricity is not provided, if industry must cut back energy use beyond cost-effective conservation, my concern is that a painful accommodation will be made at the expense of economic output and productive efficiency or by industries' moving to regions where electricity is available. Some industries already foresee regional electricity constraints. Their dilemma generally cannot be avoided by on-site power generation because that alternative either is very costly or requires fossil fuels, usually oil. Local resources, such as wood waste, are possible in only a few regions.

The basic issue remains our dedication to using all the resources of technology to build an energy abundance. The industrial world needs its technical resources for both efficient energy supply and efficient end-use devices. The scale of living in industrial nations does not have to decline. Higher-cost energy does not mean a depressed economy if we permit technology to compensate.

It is a dangerous platitude that we face a future of limited expectations because of energy shortages. Limited expectations lead to policies of retrenchment and cautiously riskless investment. These expectations become self-fulfilling—more important, they destroy the image of a better future that is so essential for motivating our social institutions.

SUGGESTED READINGS

David Bodansky, "Electricity Generation Choices for the Near Future," *Science* 207 (4432), 721 (1980).
C. C. Burwell, W. D. Devine Jr., and D. L. Phung, "Electric Home Heating: Substitution for Oil and Gas," ORAU/IEA 82-3 (M), Oak Ridge, Tenn.: Institute for Energy Analysis, 1982.
W. J. Davis, "Energy: How Dwindling Supplies Will Change Our Lives," *The Futurist* XIII (4), 258 (1979).
Hans H. Landsberg, *Energy Policy—Tasks for the 1980's*. Resources for the Future Reprint 174, pp. 99–130. Originally appeared in Joseph A. Pechman, ed., *Setting National Priorities: Agenda for the 1980's*. Washington, D.C.: The Brooking Institute, 1980.

The Potential for Fuel Conservation 29

MARC H. ROSS AND ROBERT H. WILLIAMS

There is great uncertainty about the capacity of the United States to meet its needs for energy at acceptable social costs in the remaining years of this century. Recent studies suggest that we cannot achieve "energy independence" by 1985 if energy demand continues to increase at historical rates, even if aggressive policies are successful in stimulating new domestic supplies of oil or gas. There is also much concern about the long-term availability of oil and gas. The view of M. King Hubbert, that domestic oil and gas production can be expected to decline indefinitely, is gaining support, although knowledge of the resource base is so poor that determined exploratory efforts may change the outlook. The use of coal and nuclear fission resources presents other problems. Rapidly increasing capital costs for electric generating plants, quality control problems with nuclear power and continuing uneasiness about its risks, and the controversy over the control of sulfur oxide emissions from coal-fired power plants are some of the more pressing problems facing the electric power industry today. With these problems operating in concert, the potential for the expansion of coal and nuclear power, at least for the next decade, is highly uncertain. New resources, such as synthetics from coal and oil from shale, can contribute little to overall supply by 1985, and there may be factors (such as water availability) that absolutely limit their development.

This gloomy scenario has led to increasing attention to fuel conservation, more popularly known as "energy conservation," where supply and demand are brought into balance by emphasizing demand reduction rather than supply increases.

To most people, fuel conservation means such near-term austerity policies as 55-mile-per-hour speed limits and lowered thermostats in winter. But several recent studies ar-

Dr. Marc H. Ross is Professor of Physics at the University of Michigan, Ann Arbor, Michigan 48104.

Dr. Robert H. Williams is a research scientist at the Center for Environmental Studies, Princeton University, Princeton, New Jersey 08540.

From *Technology Review* 79 (4), 49 (1977). Reprinted by permission of the authors and *Technology Review.* Copyright 1977 by the Alumni Association of the Massachusetts Institute of Technology. Much of the information in this article is based on a summer study by the American Physical Society and on research conducted for the Energy Policy Project of the Ford Foundation. Dr. Ross was Director of the A. P. S. study and Dr. Williams was Chief Scientist for the Energy Policy Project.

gue for fuel conservation as a long-term goal—a strategy for simultaneously holding down energy costs, stretching out limited fuel resources, minimizing dependence on foreign energy sources, slowing the introduction of nuclear power to allow time for dealing with unresolved nuclear risks, and protecting the environment. These studies contend that we could reduce our annual growth in energy use from the historical average rate of 3.2 per cent to under 2 per cent. An econometric model by the Ford Foundation's Energy Policy Project suggests that their annual growth rate of 1.8 per cent to the year 2000 could be achieved with little adverse economic effect, and a Conference Board report concludes that historical rates of economic growth could persist over the next decade with energy consumption increasing at perhaps only 1.5 per cent per year.

Such economic studies reflect the growing realization that there is considerable energy "fat" in the U.S. economy today. But these studies do not provide an analytical framework for making quantitative estimates of the potential for fuel conservation.

To assess that potential in quantitative terms, we must answer two questions: what are the real technical possibilities, in both the near and the long term, and what policies are needed to overcome institutional obstacles to the more promising options?

Both of these questions are important. Understanding the technological potential for fuel conservation is essential to establishing realistic goals and priorities. And without effective strategies for overcoming institutional obstacles, conservation goals may never be realized.

In this article, we address primarily the first question, putting forth a framework for estimating the technical potential for fuel conservation. We show that this potential is substantial for both the near and the long term.

The use of energy associated with any product is obtained by multiplying two factors: the demand for the product and the energy required to provide each unit of the product. In this article, we focus our atten-

tion on opportunities for reducing the fuel inputs required to meet existing patterns of consumer demands. Thus, we do not consider here fuel conservation associated with lifestyle changes. (This will be discussed in Article 32.)

WE CONSUME "AVAILABLE WORK," NOT ENERGY

To make quantitative estimates of the potential for saving fuel, a measure of energy performance is needed. Thermodynamics provides a useful framework in which to introduce such a measure.

Energy is never created or destroyed; only its form is changed as processes go on. For example, when fuel is burned chemical energy is converted to thermal energy; the total energy in the system is unchanged. The second law of thermodynamics—which implies that the "disorderliness" of a system always increases—tells us that these changes can occur in only one direction, such that energy loses its "quality," or capacity for performing tasks.

The best overall measure of the capacity for doing any task is *work*—the transfer of the highest-quality energy from one system to another. Physicist Willard Gibbs gave us the concept of *available work*, a measurable quantity that takes into account the quality as well as quantity of energy transformed in any process.

Consider two systems in the environment of the earth. Suppose that a quantity of energy E in one system is transformed so as to do work on the other system. The available work is defined as the theoretical *maximum* amount of work that could be done in this conversion.

If the energy E is of the highest quality, then the available work is

$$A = E \qquad (1)$$

The gravitational energy stored in water behind a dam and electrical energy are examples of energy of the highest quality. Chemical energy (as given by the heat of combustion) is also high-quality energy, for which the available work is approximately E (typically about $0.9E$). In general, for the highest-quality energies we may interchange the terms "energy" and "available work" without substantial error.

However, if the energy E is thermal energy at fixed temperature T, then the available work A is less than E, or specifically

$$A = E\left(1 - \frac{T_0}{T}\right) \tag{2}$$

where T_0 is the temperature of the ambient environment, with both T and T_0 given on the absolute temperature scale (that is, in Celsius units with the zero set at $-273°C$).

The efficiency of an energy conversion system is usually defined as the amount of desired energy or work provided by the system, divided by the energy input to the system. Because it is based on the first law of thermodynamics, which holds that energy is neither created nor destroyed, this concept of efficiency is often called the "first-law efficiency." This efficiency concept enables one to keep track of energy flows and is thus useful in comparing devices of a particular type. However, it is wholly inadequate as an indicator of the potential for fuel savings. Several examples illustrate this point.

Household furnaces are typically said to have an efficiency of about 0.6, meaning that 60 per cent of the heat of combustion of the fuel is delivered as useful heat to the house. This measure suggests that a 100 per cent efficient device would be the best possible. But this is incorrect because a heat pump could do better.

A heat pump is simply an air conditioner operating in reverse. It extracts heat energy from the out-of-doors and transfers it at a higher temperature to the interior space, thereby making available as heat more than 100 per cent of the electrical energy it consumes.

Air conditioners are rated by a coefficient of performance (COP), which is the ratio of the heat extracted to the electric input. A typical air conditioner has a $COP = 2$ (a 200 per cent efficiency), a measure that provides no hint of the maximum possible performance—a COP much greater than 2.

In the most modern fossil-fuel-fired power plants, about 40 per cent of the fuel energy can be converted to electrical energy. In this case the maximum theoretical efficiency is less than 100 per cent, because of the limitations set by the second law of thermodynamics.

In all these examples, the efficiency used is only a partial measure of performance. That is because losses of energy quality, in addition to losses of energy, are inherent to any process. Examples of quality losses are heat flow from higher to lower temperature and the mixing of materials. A more useful measure of efficiency, therefore, would take into account both quantity and quality losses and would show how well a particular energy conversion system performs relative to an ideal one in which there is loss of neither quantity nor quality. The available work concept provides a basis for formulating such an efficiency measure.

The first step in formulating a new efficiency measure is to define the task, such as heating a building, propelling an automobile, or producing steel. The available work consumed in carrying out this task is a direct measure of the expenditure of fuel. In an ideal process this available work would correspond to the absolute minimum expenditure of fuel for the task. But a real process involves losses, so that the actual work consumed, A_{act}, is larger than A_{min}. A suitable measure of efficiency, therefore, is

$$\epsilon \equiv \frac{A_{min}}{A_{act}} \tag{3}$$

This equation for efficiency shows that fuel consumption A_{act} can be reduced either by increasing the efficiency or by modifying the task to be performed (that is, by reducing A_{min}). Because this measure of efficiency shows performance relative to what is possible within the constraints of the second law of thermodynamics, it has been called the "second-law efficiency." In what follows we shall use the term "efficiency" in this sense. This efficiency concept was introduced and applied to a wide range of fuel-consumption activities in a recent American Physical Society study.

The distinction between ϵ and the conventional first-law efficiency can be illustrated with the Carnot engine, an idealized device that makes the fullest use possible of heat E extracted from a reservoir at temperature T. For this engine $A_{act} = A_{min}$ and $\epsilon = 1.0$. In contrast, the first-law efficiency is $A_{min}/E < 1$, suggesting erroneously that this ideal heat engine could be improved upon.

The calculation of A_{min} varies with the task. For a task that involves work W (for example, turning a shaft)

$$A_{min} = W \qquad (4)$$

For the transfer of thermal energy E to a reservoir at temperature $T > T_0$ (for example, heating a room)

$$A_{min} = E\left(1 - \frac{T_0}{T}\right) \qquad (5)$$

It is noteworthy that in this particular example the minimum available work is actually less than the amount of heat delivered, because the ideal process for delivering heat involves use of a heat pump that extracts thermal energy from the ambient environment.

Second-law efficiencies for fuel-consuming activities throughout the economy can be calculated using Equation 3. The results for some important examples are summarized in Table 29-1, which shows that for most activities second-law efficiencies are less than 10 per cent, clear evidence that energy is being used very inefficiently today. (We usually think of efficiency as the ratio of energy or work provided by a particular device to that which was consumed by it. But this conventional measure is wholly inadequate as an indicator of the potential for fuel savings, hence

Table 29–1.   Second-Law Efficiencies for Energy-Consuming Activities.

Energy-Consuming Activities (Current Technology)	Second-Law Efficiency (per cent)
Residential and Commercial	
Space heating:	
Fossil-fuel-fired furnace	5
Electric resistive	2.5
Air conditioning	4.5
Water heating:	
Gas	3
Electric	1.5
Refrigeration	4
Transportation	
Automobile	9
Industrial	
Electric power generation	33
Process-steam production	33
Steel production	23
Aluminum production	13

the authors' emphasis on second-law efficiency, the performance relative to that which is possible for a given task.) For example, the typical household oil-fired furnace has a second-law efficiency of only 5 per cent, compared with its first-law efficiency of 60 per cent. The latter figure, often quoted for household furnaces, gives the misleading impression that only a modest efficiency improvement may be possible, whereas the second-law efficiency correctly indicates a 20-fold maximum potential gain.

How close can we expect efficiencies to approach the ideal limit of $\epsilon = 1$? Examination of experience with high-efficiency systems helps provide insight for making judgments about practical long-term goals. Table 29-1 shows that, contrary to the popular misconception that it is inefficient, electric power generation is one of the more efficient conversion processes in the economy today. (It is only when the power generation system is extended to include especially wasteful uses of electricity that the overall efficiency is often very low; for example, $\epsilon = 0.025$ for electric resistive space heating.) Furthermore, new fossil-fuel-fired plants achieve ϵ of up to 0.40, and combined-cycle systems now being developed (the heat from combustion is used first to drive a gas turbine-powered generator and the gas turbine exhaust is then used to make steam for a conventional steam turbine) are expected to achieve efficiencies approaching 0.55. All these energy conversion processes start with fuel combustion, for which $\epsilon = 0.70$; so, in a sense, combined cycles may achieve 0.55 out of a possible 0.70 in efficiency.

These high efficiencies are associated with highly engineered, costly, and rather inflexible devices, and they may not represent a practical goal for most systems. Nevertheless, study of a variety of high-efficiency devices and processes, some described below, suggests that goals for ϵ in the range of 0.2 to 0.5 are reasonable for ultimate practical systems. Values at the high end of this range are more likely to be characteristic of highly engineered devices designed for specialized tasks; values at the low end are more likely for flexible, less sophisticated devices suitable for wide applications.

Here are some examples of how these concepts of energy and efficiency demonstrate opportunities for substantial fuel savings arising from both efficiency improvements and task modifications.

BETTER MILEAGE FOR MODERATE-SIZE CARS

Consider first the automobile. The task to be performed is the propulsion of a vehicle of given external characteristics (weight, tires, wind resistance, brakes, and so on) under average driving conditions. The efficiency would be the ratio of the theoretical minimum fuel required to perform this task (A_{min}) to the actual amount required by the vehicle in question (A_{act}).

The theoretical A_{min} for today's average 3600-pound automobile is about 1000 Btu per mile. This average automobile actually consumes 9200 Btu per mile (14 miles per gallon), resulting in an efficiency $\epsilon = 0.11$. The

American Physical Society study of energy conservation (see references) identified a series of improvements that could be accomplished with today's technology: a better load-to-engine match could yield an efficiency gain from 0.11 to 0.12, and the use of radial tires, modestly improved streamlining, and a 20 per cent weight reduction would reduce A_{min} to 740 Btu per mile. The combined effect would be to reduce fuel consumption to 6200 Btu per mile (20 miles per gallon). Further improvements possible over the next few years could boost the efficiency to about 0.17 with a more efficient transmission and an improved engine design (Diesel, Rankine, or Stirling), where further improvements in streamlining, development of a better tire/suspension system, and a further 5 per cent weight reduction could bring A_{min} down to 630 Btu per mile. Such improvements could mean that automobiles of the 1990s would differ little in performance from today's cars but would typically travel 30 to 35 miles on a gallon of fuel. The average weight of these cars would be only 25 per cent less than today's average. (Of course, cars smaller than this with comparable fuel economy are being built with present technology.)

WARMING TO THE CONSERVATION TASK

Potential savings in space heating are immense, because there are opportunities for considerably reducing heat losses (i.e., reducing A_{min}) and because present heating system efficiencies are low.

For average U.S. winter weather, the efficiency ϵ of a typical gas furnace system is 0.05, a figure obtained as follows. The task is taken to be the delivery of heat at 86°F into the useful space of a given building (with the level of insulation specified). According to Equation 5, the minimum available work associated with the delivery of heat E, when the outdoor temperature is 40°F is $A_{min} = 0.084E$. Also, we know that, for a furnace that delivers 60 per cent of the energy content of the fuel to the desired space, E

$= 0.6A_{act}$. Using Equation 3 we thus obtain $\epsilon = 0.05$. With a lower outdoor temperature, the efficiency would be higher.

First consider how to reduce the minimum available work (A_{min}) required to heat a typical house. About 60 per cent of the heat losses are from conduction through walls, roof, windows, and floors. The American Physical Society study estimates that with improved insulation and well-designed windows these losses could reasonably be cut to below one third of present values. The remaining 40 per cent of the heat losses are due to heating and humidification of fresh air; studies have shown that typical rates of air exchange in buildings are unnecessarily high, and it is not unreasonable to reduce the ventilation rate some 80 per cent.

These strategies, which could cut total heat losses nearly fourfold, would involve modest innovations in design and development. Such a fourfold reduction would mean that in many buildings no fuel would be needed on average winter days, when a temperature differential of 30 F° is required between outside and inside, to supplement the heat provided by sunlight through windows, by the lights and appliances, and by the body heat of residents.

On colder days a small heat pump or pumps could be used. If outside air is the heat source for an electrically driven heat pump, the heat pump typically delivers two or three times as much heating as is represented in the electricity consumed (that is, $COP = 2$ to 3). If well or lake water (at, say, 55°F) is used as a heat source, a COP of 4 is achievable today, corresponding to $\epsilon = 0.10$.

The net effect of thermal tightening of the building shell and of using such an efficient heat pump would be to reduce the primary fuel consumption for space heating to one eighth of that required for gas furnace heating a house with insulation characteristic of an average house today.

HIGH-RISE COOLING

Consider a typical new office building in New York City with ten stories and one million square feet of office space. Though the efficiency of air conditioning in a typical new office building in New York City is low (likely to be no more than 0.04 on a 90°F summer day), the easiest way to achieve fuel savings is by reducing the air conditioning load—i.e., by reducing A_{min}. In this building, only one sixth of the air conditioning load is due to heat conduction from the outside and solar radiation through windows. Over half of the load is due to heat generated by lighting (about 6 watts per square foot) and about one fifth is due to ventilation (20 cubic feet per minute per person).

It is reasonable to reduce illumination levels to 1.5 watts and ventilation to 5 cubic feet per minute; these efforts to reduce A_{min} would cut the total air conditioning load by more than 50 per cent. (But at least 10 cubic feet per minute of ventilation per person would be required for a building in which smoking is permitted everywhere. In the case of this office building, the extra air conditioning load associated with smoking requires the burning of 20 gallons of fuel oil each hour at the power plant.) Among further practical modifications, the most significant would be the use of heat exchangers to recover "cool" from exhausted air. Pursuing all these measures would lead to about a 70 per cent reduction in the air conditioning load.

An alternative to electric central air conditioning is the heat-driven air conditioner based on desiccation. Such a device, with a $COP = 0.73$, offers the potential of substantial savings if solar energy is used as a partial heat source. The use of sunlight incident on a collector covering the roof of the building as a partial heat source for this device, and the adoption of the load reduction measures considered here, could lead to a system in which the total primary fuel consumption for air conditioning was about one eighth of that in today's large office building with electric-powered air conditioning.

INDUSTRIAL STEAM PRODUCTION

The first-law efficiency for converting fossil-fuel energy to steam in industry is typically an impressive 0.85, but the second-law efficiency is only 0.33 for steam at 400°F. The production of steam through the burning of fuels wastes available work. This is because the high flame temperatures of fuel combustion (up to about 3600°F) represent energy of very high quality, but the temperatures required for industrial process steam are typically 400°F or lower. Substantial fuel savings can be achieved if the high-quality, high-temperature energy available from combustion is first used to make electricity in a heat engine, with the "waste" heat from this device used for low-temperature process steam applications. This "cogeneration" of electricity and process steam is an important application of the more general fuel-saving strategy of "cascading," in which the energy available in combustion is sequentially degraded through a series of uses.

The second-law efficiency for cogeneration is typically between 40 and 45 per cent, compared with about 33 per cent for the separate production of steam and electricity. The resulting savings are actually much more impressive when expressed another way. If only the excess fuel beyond that required for steam generation is allocated to power production, the fuel required to produce a kilowatt-hour of electricity is reduced to about one half of that required in conventional power plants. The national potential is truly great, since process steam is a major energy-consuming activity in the economy, accounting for about 17 per cent of total energy use.

The most promising application of steam-electricity cogeneration appears to be in industrial plants, where electricity could be produced as a by-product whenever steam is needed. A number of cogeneration technologies could be employed. In a steam-turbine system, steam used to drive the power-generating turbine would be exhausted from the turbine at the desired pressure and (instead of being condensed with cooling water, as at a conventional power plant) delivered to the appropriate industrial process. With a gas-turbine system, the hot gases exhausted from a power-generating turbine would be

used to raise steam in a waste heat boiler. The gas-turbine system is the more efficient of the two, typically with $\epsilon = 0.45$ compared with $\epsilon = 0.40$ for a steam-turbine system; in addition, because it produces several times as much electricity for a given steam load, the gas-turbine cogeneration system could yield considerably greater total fuel savings than the steam-turbine system.

Recent studies on the overall potential for cogeneration have been carried out by Dow Chemical Company and by Thermo Electron Corporation. The latter's study shows that by 1985 electricity amounting to more than 40 per cent of today's U.S. consumption (generated with the equivalent of about 135,000 megawatts [electric] of baseload central-station generating capacity) could be produced economically with gas turbines as a by-product of process steam generation at industrial sites. Whereas the gas turbines in most common use today must be fueled with gaseous or liquid fuels, it is likely that over the next decade high-pressure fluidized-bed combustors will be available as an economical method of firing gas turbines directly with coal.

To produce power most economically, an industrial installation that generates electricity as a by-product of process steam production would often produce more electricity than could be consumed on-site. Thus the cogeneration unit should be interconnected with a utility and could substitute for some central-station baseload generating capacity. But such an arrangement is often difficult under existing utility policies. Considerable modification of utilities' transmission, control, and perhaps storage systems may be necessary if interconnected cogeneration capacity is developed on a large scale.

The production of process steam as a by-product of power generation at large central station power plants is an alternative to cogeneration at industrial sites. However, such steam production does not lead to a significant increase in ϵ, since only about 1.5 per cent of the available work originally present in the fuel is discharged in the cooling water, which has an average temperature of about 100°F. If the waste heat is to be useful for industrial processes, power plant operations would have to be modified to produce heat at more useful temperatures (200 to 400°F). But this would reduce the electrical output, and this change could lead to a net loss of available work unless essentially all the heat were put to effective use.

Not only are the potential gains of by-product steam generation small, but there are serious implementation difficulties as well. Because it is uneconomical to transport steam long distances, steam-using industries would have to be near the power plants from which their heat is supplied, and this is a condition difficult to fulfill. There is also a serious mismatch in time: large central-station power plants require six to ten years for construction and are designed for a quarter century or more of service. For these reasons, industrial cogeneration is favored.

THE OVERALL POTENTIAL OF CONSERVATION

Having considered a number of examples, we now turn to an estimate of the overall fuel savings potential in the economy through adoption of measures to increase ϵ and to decrease A_{min}. We take into account technologies that either are commercial now or are likely to be commercial in the near future, and so the estimates we make now are less ambitious than some of the potential savings estimates we have made above. Specifically, we ask what U.S. energy consumption would have been in 1973, had we been a nation of energy thrift.

Our conclusion is that the 1973 living standard of the United States could have been provided with about 40 per cent less energy (see Table 29-2). Nearly 60 per cent of the potential savings lies in four areas: space conditioning, the automobile, industrial cogeneration of steam and electricity, and commercial lighting. Whereas total consumption of electricity is reduced by only 30 per cent in the hypothetical energy budget

for 1973, central-station power generation is reduced by about 60 per cent because of the large amount of power generated by industrial cogeneration and total energy systems.

It is commonly asserted that the automobile offers the greatest opportunities for fuel conservation. Whereas the time scale for substantial improvements is relatively short for the automobile, the present analysis, which is summarized in Table 29-2, points out that there are comparable fuel-saving opportunities in many cases where low-quality, low-temperature heat is required; these areas make up about 40 per cent of U.S. energy use. In typical residential/commercial low-temperature applications (space conditioning, water heating, refrigeration, and so on), efficiencies now in the range of 2 to 10 per cent can be substantially increased. In industrial applications (e.g., process steam) efficiencies are somewhat higher, but opportunities for fuel savings are still substantial. The summary shows that U.S. energy use in 1973 (75×10^{15} Btu) could have been only 44×10^{15} Btu had all the authors' conservation proposals been in place. Note that the potential savings associated with a particular conservation measure in this table sometimes depends on the previously listed measures. For example, the savings associated with a reduced air conditioning load is affected by the previous assumption that all air conditioners are more efficient.

One way to interpret these results is to note that they set the clock back 17 years on energy consumption. That is, growing at the historical rate from the level of this hypothetical energy budget to the actual 1973 level would require 17 years.

We believe that with an aggressive fuel conservation program the fuel-saving technologies of the general type we propose in this table could in fact be brought into use in the United States within two decades. This strongly suggests the possibility of zero energy growth out to the early 1990s without jeopardizing overall economic growth. Through fuel conservation efforts, the growth in aggregate demand for products would be

Table 29–2.   Fuel Conservation by Sector.

	Potential Savings (10^{15} Btu)	Total Energy Demand in 1973 (10^{15} Btu)	Hypothetical Energy Demand with Savings (10^{15} Btu)
Residential Sector			
Replace resistive (electric) heating with heat pump having $COP = 2.5$	0.60		
Increase air conditioning COP to 3.6	0.40		
Increase refrigerator efficiency by 30 per cent	0.27		
Reduce water heating fuel requirements by 50 per cent	1.07		
Reduce heat losses by 50 per cent with better insulation, improved windows, and reduced infiltration	3.30		
Reduced air conditioning load by reducing infiltration	0.42		
Introduce total energy systems into half of U.S. multi-family units (15 per cent of all housing units)	0.31		
Use microwave ovens for one half of cooking	0.25		
Totals	6.62	14.07	7.45
Commercial Sector			
Increase air conditioning COP by 30 per cent	0.37		
Increase refrigeration efficiency by 30 per cent	0.20		
Cut water heating fuel requirements by 50 per cent	0.31		
Reduce building lighting energy by 50 per cent	0.95		
Direct savings 0.82			
Increased heating requirements −0.21			
Reduced air conditioning required 0.34			
Reduce heating requirements by 50 per cent	2.25		
Reduce air conditioning demand through improved insulation (10 per cent)	0.08		
Reduce air conditioning demand 15 per cent by reducing ventilation rate 50 per cent and by using heat recovery	0.10		
Introduce total energy systems into one third of all units	0.64		
Use microwave ovens for one half of cooking	0.06		
Totals	4.96	12.06	7.10
Industrial Sector			
Improve housekeeping measures (better management practices with no changes in capital equipment)	3.85		
Use fossil fuel instead of electric heat in direct heat applications	0.17		
Adopt steam/electric cogeneration for one half of process steam	2.59		
Use heat recuperators or regenerators in one half of direct heat applications	0.74		
Generate electricity from bottoming cycles in one half of direct heat applications	0.49		
Recycle aluminum in urban refuse	0.10		
Recycle iron and steel in urban refuse	0.11		
Use organic wastes in urban refuse for fuel	0.70		
Save from reduced throughput at petroleum refineries	0.87		
Reduce field and transport losses associated with reduced use of natural gas	0.80		
Totals	10.43	29.65	19.22

	Potential Savings (10^{15} Btu)	Total Energy Demand in 1973 (10^{15} Btu)	Hypothetical Energy Demand with Savings (10^{15} Btu)
Transportation Sector			
Improve automobile fuel economy 150 per cent	5.89		
Emphasize fuel savings in other transportation areas (35 per cent savings)	3.20		
Totals	9.09	18.96	9.87
Grand total	31.10	74.74	43.64

compensated for by reductions in the average energy required to provide a unit of product.

Actually, such a focus on 1973 fuel-use patterns tends in two ways to underestimate potential future fuel savings. First, uses such as residential air conditioning and electric resistive space and industrial heating are major growth areas where savings opportunities are substantial; second, fuel-conserving lifestyle changes not considered here are already taking place. The growing shift to small cars is an especially important example of such a change.

Our analysis shows that efficiencies of fuel use are in general higher in the industrial sector (with ϵ typically in the range 0.15 to 0.35) than elsewhere. This suggests that long-term fuel savings opportunities, beyond those indicated in Table 29-2, may be limited in the industrial sector, though this observation must be tested through careful analysis of the major industrial processes. Of course, substantial industrial fuel savings over the long term could be realized by shifting the mix of economic output from energy-intensive products to those that require less energy per dollar of value added. Such possibilities have not been taken into account here, where we have estimated potential savings only for the existing pattern of demand for goods and services.

In contrast, because efficiencies are at present so low in the residential, commercial, and transportation sectors, technologi-

cal innovation in these areas could lead to substantial long-term savings beyond those tabulated, if the appropriate research and development is pursued.

What are the economics of fuel conservation measures? Investments in fuel-saving technology are likely to be costly, but as our supplies of low-cost energy resources diminish these costs may well be less than the costs of increasing our energy supply.

If a particular process requires fuel input at an average rate S_0 (in thermal kilowatts per unit of daily output, say) and costs C_0 (in dollars per unit of daily output), and if the corresponding values for the fuel savings option are S and C, then the capital costs of the conservation option can be expressed as

$$\frac{C - C_0}{S_0 - S} \qquad (6)$$

in dollars per thermal kilowatt saved. This measure of capital costs can be compared with corresponding costs for energy supply options.

We illustrate this with a few examples. Expressing capital costs as 1974 dollars per thermal kilowatt saved, we find that replacing electric resistive heating plus a central air conditioner with a heat pump ($COP = 2.7$) costs \$50 to \$120 per kilowatt saved; retrofitting a house with insulation and storm windows costs \$450 per kilowatt saved; and the installation of waste-heat recovery units on a heat-treating industrial furnace costs $100. The extra capital required to save fuel with a cogeneration system is in fact negative, except for very small plants, simply because the cost of a suitably large combined system is less than the cost of separate electricity and process steam generating facilities.

By comparison, a large coal or nuclear power plant, along with the associated transmission and distribution system, today costs $480 to $650, respectively, per thermal kilowatt of utilized capacity. Only the home retrofitting example is close in capital costs to these investments required for new capacity; it is, in fact, one of the most capital-intensive conservation options. Insulating *new* homes would certainly require less capital.

These numbers are rough, but they illustrate the kinds of calculations that should be performed. A careful comparison is needed of the capital requirements and the life-cycle costs for all major conservation options and the corresponding costs for conventional practices, taking into account the entire system from resource extraction to the point of end use in each case.

RESEARCH OPPORTUNITIES

Research and development opportunities relating to fuel conservation span a wide range, extending from basic physical research through new or improved devices and materials and analysis and instrumentation for existing devices and physical systems. There is also a need for better understanding of systems problems and institutional and policy issues. Selected examples of research opportunities in the first few categories are presented in Table 29-3.

Whereas it is true that enormous improvements in fuel utilization could be made using existing technology, fuel conservation in the longer term could be greatly enhanced by better understanding of a large number of technological questions. Most of the topics on this agenda for new research and development were suggested in a study of energy conservation mounted in 1975 by the American Institute of Physics.

It is clear that much technology for fuel conservation is available today but is not being widely used because of institutional obstacles. Better understanding of these problems and assessments of alternative plans for overcoming them should be given close attention in fuel conservation research programs. Among the more general institutional problems, high priority should be given to methods for providing citizens with life-cycle-cost information on energy-related purchases (e.g., heating and cooling systems and those appliances that consume large amounts of energy); new institutional arrangements for financing fuel conservation measures (examples include home insulating services provided by a utility, with financing charges added to the fuel bill, and regulations that require mortgages to cover additional investments needed to minimize life-cycle costs); the relative merits of reliance on energy price, regulations, taxes, and subsidies for achieving fuel conservation goals; the implications for fuel consumption of alternative courses for economic development; and economic (especially employment) implications of fuel-conservation policies.

One especially attractive feature of research on fuel conservation is that it tends to be less costly than research and development on fuel supply. Moreover, with adequate funding, the time required to generate useful results should often be short relative to that for supply-oriented research.

PUTTING CONSERVATION ON THE NATIONAL AGENDA

In the years since the "energy crisis" dramatized the problems of U.S. energy supply, little has been done to implement serious fuel conservation programs in the United States or even to mount a significant research and development program in this area. This inaction no doubt reflects in part a general lack of understanding of what can be achieved with fuel conservation technology to balance

Table 29–3.   Agenda for New Research and Development.

Building	Transportation	Industry
Basic Research		
Physiological requirements for energy services (lighting, space conditioning) as a basis for revising standards	Studies of the combustion process and development of advanced diagnostics	Heat transfer at interfaces
		Two-phase flow
Aerodynamic studies of buildings	Improvement of combustion efficiency through use of emulsions of fuel with water or methanol	Transport in membranes
		Solid electrolytes
Heat transfer, at a "micro" level, across various boundaries		Charge transfer at electrolyte/electrode interfaces
Improved Devices and Materials		
Thermal diodes (insulation with different heat conductivities in different directions)	Advanced engine cycles (Diesel, Rankine, or Stirling)	Solar-diesel steam generator
		Modulated capacity compressors for heat pumps and refrigeration
Building materials with high specific heats for thermal stability	Battery-electric or hybrid Diesel-electric autos	Materials development for high temperatures
Fuel cells for decentralized power generation	Continuously variable transmission	
	Flywheel storage of braking energy	
New types of windows to control heat loss and insulation	Absorption air conditioning	
Analysis and Instrumentation of Existing Processes		
Diagnostic instrumentation for research and inspection (e.g., local heat-flux meter, air exchange meter, local air velocity meter)	Instrumentation for user feedback such as fuel-flow meter or a specific fuel consumption meter (in gallons of gasoline per ton-mile, say)	Determination of second-law efficiencies of industrial processes
Instrumentation for user feedback (e.g., to signal the consumer when he is consuming electricity at the system peak)	Analysis of energy loss modes (air drag, rolling resistance, braking energy) in different types of trucking	Development of indices measuring energy requirements for various kinds of economic activity
Systems Analysis		
Designs to take advantage of local natural conditions	Strategies for electrification of transportation that would improve electric load factor	Assessment of decentralized versus centralized power generation
Reassessment of practice for designing baseload energy system conditions	Crashworthiness of lighter cars	Materials recycling and reuse of fabricated items in specific industries
Time and zone control of lighting and temperature	Trade-offs in function between tires, wheels, and suspension, aimed at decreasing rolling resistance	Combined electric and heating systems: optimization and development of strategies for integration with utilities

energy supply and demand. It also reflects the existence of institutional constraints, which tend to reinforce this limited vision.

The authors are confident that a better understanding of the potential for fuel conservation will lead to a change in this situation. We have shown that the potential based on existing technology is large, and we suggest that substantial further long-term opportunities could arise if the appropriate research and development is pursued. One of the nation's highest priorities should be to make these opportunities evident to political and industrial decision makers and to citizens generally.

It appears to us that the physical limitations to fuel conservation are considerably less pressing than many of the public risks and capital cost problems posed by substantial increases in energy supply. Furthermore, if our assessment is correct, a shift in emphasis from energy-supply to fuel-conservation technology will involve a net reduction in capital and very likely an increase in productivity of the economy as a whole.

ADDENDUM

In the five years since this article was written, there have been three developments which lead to a somewhat new emphasis:

1. Economic difficulties in the United States have continued to be serious.
2. Energy prices have increased very sharply.
3. Many analyses of the potential for increased energy efficiency have been carried out.

The new emphasis is the cost savings which can be reaped through conservation, or increased energy efficiency. Although most people still perceive the energy problem as calling for new supplies, the message is that efficiency improvement provides a means for major cost savings and thus for enhancing growth and controlling inflation. In contrast, energy supply development is very costly and such activity, whether it be power plant construction, remote oil field development, or synfuels development, implies substantial sacrifices in terms of special tax subsidies and the diversion of scarce industrial capital away from investments which would make U.S. industry more productive.

Another facet of the issue is that new analyses of the potential for conservation show that, since the calculation contained in this article was based primarily on existing technology and energy prices which were low from the perspective of the 1980s, much greater efficiency improvements are justifiable in terms of overall costs. For example, we are beginning to consider automobiles in the 45 to 60 mile per gallon fuel economy range (i.e., 300 to 400 per cent more efficient than the existing average). As another example, we are beginning to consider basic industrial processes with energy requirements 40 to 50 per cent less per unit of product than the present average. (Such process change was largely ignored in the analysis presented in this article.) If such efficiency improvements were realized in the next couple of decades, there would be no fuel and electricity shortages during that period.

Aside from emergency planning, the primary energy issue facing the United States is proper allocation of capital. The nation's economy would benefit enormously if capital were allocated to supply and to efficient use on the basis of even-handed criteria. Under these conditions much of it would flow into efficiency improvement, whereas under present conditions capital is being force-fed into energy supply development, with unsatisfactory results. (M.H.R., February 4, 1981)

REFERENCES

Council on Environmental Quality, "National Energy Conservation Program: The 'Half and Half Plan,'" March 1974.

Energy Policy Project of the Ford Foundation, *A Time to Choose: America's Energy Future*. Cambridge, Mass.: Ballinger, 1974.

E. A. Hudson and D. W. Jorgenson, "Economic Analysis of Alternative Energy Growth Patterns, 1975–2000," a report to the Energy Policy Project, Appendix F in the Ford Foundation Study.

John G. Meyers, "Energy Conservation and Economic Growth: Are They Compatible?" *Conference Board Record,* p. 27, February 1975.

"Efficient Use of Energy, a Physics Perspective," edited by W. Carnahan, K. W. Ford, A. Prosperetti, G. I. Rochlin, A. H. Rosenfeld, M. H. Ross, J. R. Rothberg, G. W. Seidel, and R. H. Socolow, a report of a 1974 summer study held under the auspices of the American Physical Society, in *Efficient Energy Use.* New York: American Institute of Physics, 1975.

J. H. Keenan, *Thermodynamics.* New York: Wiley, 1948.

E. P. Gyftopoulos, L. J. Lazaridis, and T. F. Widmer, *Potential Fuel Effectiveness in Industry,* a report to the Energy Policy Project of the Ford Foundation. Cambridge, Mass.: Ballinger, 1974.

Charles A. Berg, "A Technical Basis for Energy Conservation," *Technology Review,* February 1974, p. 14, and "Conservation in Industry," *Science,* April 19, 1974, p. 264.

R. H. Williams, ed., *The Energy Conservation Papers,* reports prepared for the Ford Foundation's Energy Policy Project. Cambridge, Mass.: Ballinger, 1975.

"Energy Industrial Center Study," report to the National Science Foundation by Dow Chemical Co., the Environmental Research Institute of Michigan, Townsend-Greenspan and Co., and Cravath, Swaine, and Moore, June 1975.

S. E. Nydick et al., "A Study of In-Plant Electric Power Generation in Chemical, Petroleum Refining, and Paper and Pulp Industries," draft report prepared for the Federal Energy Administration by Thermo Electron Corp., May 1976.

SUGGESTED READINGS

Eric Hirst, "Transportation Energy Conservation Policies," *Science* **192,** 15 (1976).

Eric Hirst and John C. Moyers, "Efficiency of Energy Use in the United States," *Science* **179,** 1299 (1973).

G. A. Lincoln, "Energy Conservation," *Science* **180,** 155 (1973).

Marc H. Ross and Robert H. Williams, "Energy Efficiency: Our Most Underrated Energy Resource," *Bulletin of the Atomic Scientists* **32,** 30 (1976).

Marc H. Ross and Robert H. Williams, *Our Energy: Regaining Control.* New York: McGraw-Hill, 1981.

Lee Schipper, "Raising the Productivity of Energy Utilization," *Annual Review of Energy* **1,** 455 (1976).

T. F. Widmer and E. P. Gyftopoulos, "Energy Conservation and a Healthy Economy," *Tech. Review,* June 1977, pp. 31–40.

Energy Strategy: The Industry, Small Business, and Public Stakes 30

AMORY B. LOVINS

Most people and institutions responsible for past U.S. energy policy consider that the future should be the past writ large: that the current energy problem is how to expand supplies (especially domestic supplies) of energy to meet the extrapolated needs of a dynamic economy. Solutions to this problem have been proposed by, among others, ERDA, FEA, the Department of the Interior, Exxon, and the Edison Electric Institute. A composite of the main elements of their proposals might resemble Figure 30-1. This diagram shows a policy of strength through exhaustion—that is, pushing hard on all fuel resources, whether oil, gas, coal, or uranium. Fluid fuels (oil and gas) are obtained from present fields, the Arctic, offshore, and imports. More fluid fuels are synthesized from coal, whose mining is enormously expanded both for that purpose and to supplement vast amounts of nuclear electricity. In essence, more and more remote and fragile

places are to be ransacked, at ever greater risk and cost, for increasingly elusive fuels, which are then to be converted into premium forms—fluids and electricity—in ever more costly, complex, gigantic, and centralized plants. As a result, while population rises by less than one fifth over the next few decades, our use of energy would double and our use of electricity would treble. Not fulfilling such prophecies, it is claimed, would mean massive unemployment, economic depression, and freezing in the dark.

I shall consider in a moment whether such energy growth is necessary. But first let me suggest that it is unwise and unworkable. I do not mean by this that it is *politically* unworkable, though I think that is true: most Americans affected by offshore and Arctic oil operations, coal-stripping, and the plutonium economy have greeted these enterprises with a comprehensive lack of enthusiasm because they correctly perceive the

Amory B. Lovins, a consulting physicist, is British Representative of Friends of the Earth, Limited, London, WIV3DG. His publications include *World Energy Strategies: Facts, Issues and Options* and (with Dr. J. H. Price) *Non-Nuclear Futures: The Case for an Ethical Energy Strategy*, Friends of the Earth/Ballinger, Cambridge, Mass., 1975. His latest books are *Soft Energy Paths: Toward a Durable Peace* (1977) and a supplementary anthology of contextual essays and international soft-path case histories, *Energy in Context* (1977), by the same publisher.

Excerpt from the Senate testimony on alternative long-range energy strategies, as found in the Joint Hearing of the Select Committee on Small Business and the Committee on Interior and Insular Affairs, U.S. Senate, 94th Congress. Reprinted with light editing.

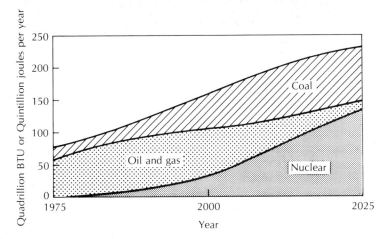

Figure 30-1. An illustrative schematic future for U.S. gross primary energy use.

prohibitive political and environmental costs. Nor do I mean that the policy represented by Figure 30-1 is *technically* unworkable, though there is mounting evidence that we lack the skills, industrial capacity, and managerial ability to sustain such rapid expansion of untried and unforgiving technologies. I mean rather that it is *economically* unworkable.

The first six items in Table 30-1 show why. They state the capital investment typically required to build various kinds of complete energy systems to increase delivered energy supplies by the heat equivalent of one barrel of oil per day. As we move from traditional oil and gas and direct-burning coal systems to frontier oil and gas and to synthetic fuels made from coal, the capital intensity of the systems rises by a factor of about 10. As we move from those systems in turn to electrical systems, the capital intensity rises by about a further factor of 10. Such a hundredfold increase in capital intensity has led many analysts—for example, the Shell Group planners in London—to conclude that no major country outside the Persian Gulf can afford to build these big, complex, high-technology systems on a scale large enough to run a country. The cash flow that electric and synthetic-fuel systems generate is so unsustainable (even for a national treasury) that

they are starting to look like future technologies whose time has passed.

Because Figure 30-1 relies mainly on the most capital-intensive systems, its first ten years—through 1985—entail a total capital investment of over one trillion of today's dollars, three quarters of it for electrification. This sum implies that the energy sector would consume its present quarter of all new private investment in the United States—plus about two thirds of all the rest. We could not even afford to build the things that were supposed to *use* all that energy. Later the burden would grow even heavier.

These astronomical investments would give us so many power stations that, as Professors von Hippel and Williams have shown,[1] we would have to use most of the electricity for very wasteful and economically unjustifiable purposes. Yet because of slow and imperfect substitution of electricity for oil and gas, we would still be seriously short of these indispensable fuels. Even worse, *at least one half* of our total energy growth in the next few decades would be *lost* in conversions from one kind of energy to another before it ever got to us. The efficiency of the fuel chain would plummet. Serious shortages of capital, labor, skills, and materials, as these resources were diverted to

Table 30–1.   Approximate Capital Investment (1976 Dollars) Needed To Build an
Entire Energy System To Deliver Extra Energy to U.S. Consumers at a Rate
Equivalent to One Barrel of Oil per Day (about 67 Kilowatts on an Enthalpic, or Heat-
supplied, Basis).[a]

Traditional direct fossil fuels, 1950s–1960s, or coal, 1970s	$2–3000[b]
North Sea oil, late 1970s	$10,000
Frontier oil and gas, 1980s	$10–25,000+[b]
Synthetic fuels from coal or unconventional hydrocarbons, 1980s	$20–40,000+
Conventional coal-electric with scrubbers, mid-1980s	$150,000[b]
Nuclear-electric (LWR), mid-1980s	$200–300,000[b]
"Technical fixes" to improve end-use efficiency:	
New commercial buildings	–$3000[c]
Common industrial and architectural leak-plugging	$0–5000[c]
Most industrial and architectural heat-recovery systems	$5–15,000[c]
Difficult, extremely thorough building retrofits, worst case	$25,000[c]
Retrofitted 100 per cent solar space heat, mid-1980s, with no backup required, assuming costly traditional flat-plate collectors	$50–70,000[c]
Bioconversion of forestry and agricultural residues, around 1980	$13–20,000
Pyrolysis of municipal wastes, late 1970s	$30,000[d]
Vertical-axis 200-kilowatt-hour (electric), late 1970s	$200,000
Fluidized-bed gas turbine with district heating grid and heat pumps ($COP =$ 2.0), coal-fired, early 1980s	$30,000[c]

[a] The coal-electric, nuclear, and wind systems shown deliver electricity, not heat or fuel. The values given are marginal capital investment to deliver 67 kilowatt-hours regardless of whether it is electric or transient, in accordance with the normal British statistical convention of "heat-supplied basis." As explained in the paper in which these data are derived (A. B. Lovins, "Scale, Centralization, and Electrification in Energy Systems," ORAU symposium "Future Strategies for Energy Development," Oak Ridge, October 1976), these data offer a common basis for further computations concerned with specific end uses. The quality, convenience, reliability, etc., of the form of energy supplied must be taken into account at that time, and this procedure may, for some end uses, significantly reduce the disparity in capital intensity between electric and other systems. Any attempt, however, to generalize such a result would be invalid and would destroy the universal utility of this data base.
[b] These values can be readily calculated from the data base of the Bechtel Energy Supply Planning Model (1975, updated October 1976).
[c] These values include the capital cost of end-use devices to deliver the desired function.
[d] This value excludes a credit for investment saved in equipment to dispose of the wastes normally and to provide the virgin resources substituted for by those recovered from the wastes.

the energy sector, would exacerbate inflation. And every "quad" (quadrillion Btu) of primary energy fed into new power stations would *lose* some 75 000 net jobs, because power stations produce fewer jobs per dollar, directly and indirectly, than virtually any other investment in the economy.

The massive diversion of scarce resources into the energy sector would thus worsen, not correct, the economic problems it was intended to prevent. At the same time it would create serious social and political problems. Reallocating scarce resources to priorities that the market is plainly unwilling to support would require a strong central authority with power to override any objections. Political

and economic power would concentrate in politically unaccountable oligopolies and bureaucracies. A bureaucratized technical elite, politically remote from energy-users, would operate the complex systems and say who could have how much energy at what price. Centralized energy systems would allocate energy and its side effects to different groups of people at opposite ends of the transmission line or pipeline. As the energy went to Los Angeles or New York and the social costs to Appalachia, Navajo country, Montana, or the Brooks Range, inequity and alienation would fuel new tensions, pitting region against region, even as a wider and more divisive form of centrifugal politics pitted central siting and regulatory authority against local autonomy.

Energy supplies would depend increasingly on centralized distribution systems vulnerable to disruption by accident or malice. Electrical grids, in particular, distribute a form of energy that cannot readily be stored in bulk, supplied by hundreds of large and precise machines rotating in exact synchrony across a continent and strung together by a frail network of aerial arteries that can be severed by a rifleman or disconnected by a few strikers. This inherent vulnerability could be reduced only by stringent social controls, similar to those required to protect plutonium from theft, LNG terminals from sabotage, and the whole energy system from dissent. The difficulty, too, of making decisions about compulsory technological hazards that are disputed, unknown, or (often) unknowable would increase the risk that citizens might reject official preferences. Democratic process would thus have to be replaced by elitist technocracy, "we the people" replaced by "we the experts." And over all these structural and political problems would loom the transcendent threat of nuclear violence and coercion in a world dependent on international commerce in atomic bomb materials measured in tens of thousands of bombs' worth per year. The impact of human fallibility and malice on nuclear systems would, I believe, quickly corrode humane values and could destroy humanity itself.

These many side effects of the energy scenario depicted in Figure 30-1 would be even worse in interacting combination than singly. Continued energy waste implies continued U.S. dependence on imported oil, to the detriment of the Third World, Europe, and Japan. We pay for the oil by running down domestic stocks of commodities, which is inflationary; by exporting weapons, which is inflationary, destabilizing, and immoral; and by exporting wheat and soybeans, which inverts midwestern real-estate markets, makes us mine ground water unsustainably in Kansas, and increases our food prices. Exported American wheat diverts Soviet investment from agriculture into defense, making us in-

crease our own inflationary defense budget, which we have to raise anyhow to defend the sea-lanes to bring in the oil and to defend the Israelis from the arms we sold to the Arabs. Pressures increase for energy- and water-intensive agribusiness, creating yet another spiral by impairing free natural life-support systems and so requiring costly, energy-intensive technical fixes (such as desalination and artificial fertilizers) that increase the stress on remaining natural systems while starving social investments. Poverty and inequity, worsened by the excessive substitution of inanimate energy for people, increase alienation and crime. Priorities in crime control and health care are stalled by the heavy capital demands of the energy sector, which itself contributes to the unemployment and illness at which these social investments were aimed. The drift toward a garrison state at home, and failure to address rational development goals abroad, encourage international distrust and domestic dissent, both entailing further suspicion and repression. Energy-related climatic shifts could jeopardize marginal agriculture, even in our own Midwest, endangering an increasingly fragile world peace.

If it were true, as the proponents of the Figure 30-1 scenario insist that there is no alternative to it, then the prospects for a humane and sustainable energy future would be bleak indeed. But I believe there is another way of looking at the energy problem that can lead us to a very different path: one that is quicker, cheaper, more beneficial to the economy, and politically and environmentally far more attractive. Such an energy path, which I shall call a "soft" path, is illustrated in Figure 30-2. It has three components, which I shall discuss in turn:

1. Greatly increased efficiency in the use of energy.
2. Rapid deployment of "soft" technologies (which I define later).
3. The transitional use of fossil fuels.

These three components mesh to make a whole far greater than the sum of its parts: a

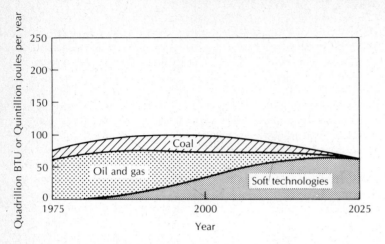

Figure 30-2. An alternate illustrative future for U.S. gross primary energy use.

coherent policy distinguished from the "hard" path of Figure 30-1 not by how much energy we use, but rather by the technical and sociopolitical *structure* of the energy system. This distinction will become clearer shortly.

The hard energy path in Figure 30-1 rests on the belief that the more energy we use, the better off we are. Energy is elevated from a means to an end in itself. In the soft path of Figure 30-2, on the contrary, how much energy we use to accomplish our social goals is considered a measure less of our success than of our failure—just as the amount of traffic we must endure to gain access to places we want to get to is a measure not of well-being but rather of our failure to establish a rational settlement pattern. The cornerstone of Figure 30-2, therefore, is seeking to attain our goals with an elegant frugality of energy and trouble, using our best technologies to wring as much social function as possible from each unit of energy we use.

Many people who have finally learned that energy efficiency does not mean curtailment of functions (that is, that insulating your roof does not mean freezing in the dark) still cling to the bizarre notion that using less energy nevertheless means somehow a loss of prosperity. This idea cannot survive inspection of Table 30-2, which shows how much energy

an average person in Denmark used at various times for heating and cooking (over one half of all end uses). Looking only at the values for 1900 through 1975, one might be tempted to identify increasing energy use with increasing well-being. But if that were true, the statistics for the years 1500 and 1800 would imply that Danes have only just regained the prosperity they enjoyed in the Middle Ages. Deeper analysis shows what is really happening. In 1500 and 1800, Denmark had a wood and peat economy, and most of the heat went up the chimney rather than into the room or cooking pot—just as it does in the Third World today. In 1900,

Table 30-2. Average Per Capita Primary Energy Consumption in Denmark for Heating and Cooking.[a]

Year	Gigacalories per Year
ca. 1500	7–15
1800	7
1900	3
1950	7
1975	17

[a] Source: "Energy in Denmark 1990–2005: A Case Study," Report No. 7, Survey by The Work Group of The International Federation of Institutes for Advanced Study, c/o Sven Bjørnholm, The Niels Bohr Institute, The University of Copenhagen, September 1976.

Denmark ran on coal, burned efficiently in tight cast iron stoves. In 1950, Denmark used mainly oil, incurring refinery losses to run inefficient furnaces. In 1975, the further losses of power stations were added as electrification expanded. This example shows that a facile identification of primary energy use with well-being telescopes several complex relationships that must be kept separate. How much primary energy we use—the fuel we take out of the ground—does not tell us how much energy is delivered at the point of end use, since that depends on how efficient our energy system is. End-use energy, in turn, says nothing about how much function we perform with the energy, for that depends on our end-use efficiency. And how much function we perform says nothing about social welfare, which depends on whether the thing we did was worth doing.

I shall suggest in a moment that a rational energy system can virtually eliminate conversion and distribution losses that rob us of delivered end-use energy. But let me focus first on the efficiency with which that end-use energy can do our tasks for us. There is ample technical evidence that Americans can double this efficiency by about the turn of the century by using only "technical fixes"—that is, measures which (a) are now economical by orthodox criteria, (b) use today's technologies (or, more often, 1920's technologies), and (c) have no significant effect on life styles. Such measures include thermal insulation, more efficient car engines, heat recovery in industrial processes, and cogeneration of electricity as a by-product of process heat.

Over the 50 years shown in Figure 30-2—long enough to turn over much of the stock of buildings and equipment—we can use technical fixes *alone* to treble or quadruple our end-use efficiency, yielding a convex curve of the type shown, despite normally projected growth in population, economic activity, comfort, and equity. People who consider today's values and institutions imperfect are free to obtain the same result by some mix of technical and social changes—perhaps substituting repair and recycling for

the throw-away economy or gradually shifting settlement patterns and concepts of work so as to keep voluntary mobility from being swamped by involuntary traffic. But our end-use efficiency can be improved to and beyond present European levels *entirely* through technical measures if we wish, saving both money and jobs in the process. Table 30-1 shows that conservation is far cheaper than increasing supply. Both technical fixes (such as insulational programs) and shifts toward less energy-intensive consumption patterns produce far more long-term jobs, using existing skills, than alternative supply investments can. Recent input-output studies suggest that conservation programs and shifts of investment from energy-wasting to social programs create anywhere from tens of thousands to nearly a million net jobs per quad saved.

The top curve of Figure 30-2, showing total energy needs of, say, 95 quads in the year 2000, is by no means the lowest that can be realistically considered. Projections are dropping rapidly. Dr. Alvin Weinberg's group at Oak Ridge projects for 2000 a primary energy demand of 101 to 126 quads—the lower value being more likely—even with a modest conservation program. But Dr. Weinberg's end-use energy projections are much *lower* than mine, since he assumes a half-electric economy. Further, a major National Academy of Sciences study is now taking seriously a year-2010 technical-fix projection of around 70 quads of primary energy. One panel is even proposing a value around 40 to 50 quads by assuming some lifestyle changes that could arguably improve the quality of life far more than the lifestyle changes of a high-energy path.

The second major component of the soft path shown in Figure 30-2 is the rapid deployment of soft energy technologies, which I define by five characteristics:

1. They are diverse. Just as the national budget is paid for by many small tax contributions, so the soft-technology component of energy supply is made up of small contributions by many diverse technologies, each

doing what it does best and none trying to be a panacea.

2. They operate on renewable energy flows—sun, wind, forestry or agricultural wastes, and the like—rather than on depletable fuels.

3. They are relatively simple and understandable. Like a pocket calculator, they can be highly sophisticated, but are still technologies we can live with, not mysterious giants that are alien and arcane.

4. They are matched in *scale* to end-use needs.

5. They are matched in *energy quality* to end-use needs.

These last two points are very important, and I must amplify each in turn.

First, scale: we are used to hearing that energy facilities must be enormous to take advantage of the economies of scale. But we are seldom told about the even greater *dis*economies of scale. There are at least five main kinds:

1. Big systems cannot be mass-produced. If we could mass-produce power stations like cars, they would cost less than one tenth as much as they do now, but we cannot.

2. Big systems, being centralized, require very costly distribution systems. In 1972, out of every dollar that U.S. residential and commercial users of electricity paid to private utilities, only about 31 cents went for electricity; the other 69 cents went for having it delivered. That is a major diseconomy of centralization.

3. Associated with the distribution system is a pervasive web of energy losses in distribution.

4. Big energy facilities tend to be much less reliable than small ones, and unreliability is a graver fault in a big than in a small unit, requiring more and costlier reserve capacity. The unreliability of the distribution system compounds this problem and can make a unit of generating capacity some 2.5 times less reliable (from the user's point of view) if sited in a central power station than if locally sited in a small station.

5. Big energy facilities take a long time to build and are therefore especially exposed to interest costs, escalation, mistimed demand forecasts, and wage pressure by unions, which know the high cost of delay.

Other diseconomies of large scale, less expressible in economic terms, are also important. For example, big systems entail the high political costs of centrism and vulnerability that I mentioned earlier. Big systems magnify the cost and likelihood of mistakes. Big systems are also too costly and complex for technologists to play with, and so an important wellspring of inventiveness and ingenuity is dried up. But even ignoring these qualitative effects, one can use orthodox economics, as I have done in the Oak Ridge paper, to reach a basic and perhaps surprising conclusion: that, in general, soft energy technologies are *cheaper* than the big hard energy technologies one would otherwise have to use in the long run to do the same job. Thus, as can be calculated from Table 30-1, a completely solar retrofitted space-heating system, with seasonal heat storage and no backup, has a lower capital cost than a nuclear-electric and heat-pump system to heat the same house; it also has a lower life-cycle cost than a coal-synthetics and furnace system to heat the same house. Making vehicle fuel from forestry and agricultural wastes is generally cheaper than making similar fuels from coal. And so it goes: even using today's rather cumbersome art, the soft technologies compete favorably with their long-run alternatives.

This is not to say that solar heat can always compete with unrealistically cheap gas, but that is irrelevant, for the days of cheap gas are numbered. Congress has already been asked for billions of dollars to subsidize the synthesis of gas from coal at perhaps $25 per barrel equivalent. Of course, the soft technologies would have to be financed at the household, neighborhood, or town scale at which they would be built, but capital transfer schemes already in use can give even householders the same kind of access to capital markets that oil majors and utilities now enjoy.

Soft energy technologies are matched to end-use needs not only in scale but also in energy quality. Table 30-3 classifies our end-use needs by physical type. About 58 per cent is in the form of heat, of which 35 per cent is below and 23 per cent above the boiling point of water. A further 38 per cent is mechanical work—31 per cent to move vehicles, 3 per cent to pump fluids through pipelines, and 4 per cent to drive industrial electric motors. The remaining 4 per cent represents *all* lighting, electronics, telecommunications, smelting, electroplating, arc-welding, electric railways, electric drive for home appliances, and all other uses of electricity other than low-grade heating and cooling. Electricity is a very expensive form of energy: Americans already pay typically from $40 to $120 per barrel equivalent for it. The premium applications in which we can get our money's worth out of this special kind of energy total only about 8 per cent of all our end uses. With improved efficiency, that 8 per cent would shrink to about 5 per cent, which we could cover with our present hydroelectric capacity plus a modest amount of indus-

trial cogeneration. That is, we could advantageously be running this country with no central power stations at all—if we used electricity only for tasks that can use its high quality to advantage, thus justifying its high cost in money and fuels. Those limited premium tasks are already far oversupplied, and so if we make more electricity we can use it only for inappropriate low-grade purposes. That is rather like cutting butter with a chainsaw—which is inelegant, expensive, messy, and dangerous.

This thermodynamic philosophy saves us energy, money, and trouble by supplying energy only in the quality needed for the task at hand: supplying low-temperature heat directly, not as electricity or at a flame temperature of thousands of degrees. Thus, our task is not to find a substitute to produce 1000-megawatt blocks of electricity if we do not build reactors but rather to perform directly the tasks we would have performed with the oil and gas for which the reactors were supposed to substitute in the first place. By thus matching energy quality to end use, we can virtually eliminate conversion losses, just as matching scale virtually eliminates distribution losses. These two kinds of losses to-

Table 30–3.   Per Cent End-Use Energy (Heat-Supplied Basis) Classified by Physical Type, Approximately 1973.

Type	United States (%)		Canada [a] (%)		United Kingdom [a] (%)	
Heat	58		69		65	
Below 100°C		35		39		55
100°C and above		23		30		10
Mechanical work	38		27		30	
Vehicles		31	} 24		} 27	
Pipelines		3				
Industrial electric drive		4		3		3
Other electrical	4		4		5	

[a] Preliminary and approximate data; probably within 5 per cent for Canada and 10 per cent for the United Kingdom.

Source: United States data calculated in A. B. Lovins, "Scale, Centralization, and Electrification in Energy Systems," ORAU symposium "Future Strategies for Energy Development," Oak Ridge, October 1976. Canadian data based on sources cited in A. B. Lovins, "Exploring Energy-Efficient Futures for Canada," *Conserver Society Notes* **1** (4), 5 (1976), Science Council of Canada (150 Kent St., Ottawa). British data estimated by A. B. Lovins from official statistics, including Census of Production data, and consistent with estimates by the Energy Research Group of The Open University (see, e.g., *Energy Policy*, September 1976).

gether make up *more than one half* of the total energy used at the right-hand end of Figure 30-1, yet are all but absent at the same point in Figure 30-2. Delivered end-use energy is not very different in the two diagrams in the year 2025, but it performs several times as much social function in Figure 30-2 as in Figure 30-1, with a corresponding advantage in conventionally defined social welfare. We would literally be doing better with less energy.

Extremely rapid recent progress in developing soft technologies has produced a wide range of technically mature systems—ones that face no significant technical, economic, environmental, social, or ethical obstacles and require only a modicum of sound product engineering. We already have enough mature soft technologies to meet essentially all our energy needs in about 50 years, using only convenient, reliable systems that are already demonstrated and are already economical or very nearly so. This does not assume cheap photovoltaic systems (which will probably soon be, or may already be, available), nor indeed any other solar-electric technologies. Living within our energy income requires only the appropriate use of straightforward solar heat technologies, organic conversion to clean liquid fuels for vehicles, modest amounts of wind collection (mainly nonelectrical), and currently installed hydroelectric capacity.

I believe a *prima facie* case has been made that we already know enough to start planning an orderly transition to essentially complete reliance on soft energy technologies. But this transition, or indeed any other, will take a long time—perhaps 50 years—and so we must build a bridge to our energy-income economy by briefly and sparingly using fossil fuels, including modest amounts of coal, to buy time. We know how to do this more cleanly and efficiently than we are doing now, with technologies flexibly designed so that we can plug into the smaller soft technologies as they come along. Simple, flexibly scaled technologies are also rapidly emerging for cleanly extracting premium liquid and

gaseous fuels from coal, thus filling the real transitional gaps in our fluid-fuel economy with only a temporary (and less than two-fold) expansion of coal mining. We can thus squeeze the "oil and gas" wedge in Figure 30-2 from both sides, making its middle section more slender than that of Figure 30-1 and thereby eliminating much medium-term importing and frontier extraction.

Figure 30-2, in summary, shows a different path along which our energy system can evolve from now on. This path suggests not wiping the slate clean, but rather redirecting our effort at the margin, thus freeing disproportionate resources for other tasks. It does not abolish big technologies, but rather concedes that they have a limited place, which they have long since saturated, and proposes that we can take advantage of the big systems we already have without multiplying them further. Because the long lead time of big systems means that many power stations are now under construction, we can use that backlog, current overcapacity, and the potential for rapid industrial cogeneration to ensure adequate supplies of electricity before improved efficiency starts to bear fruit.

Having these two energy paths before us in outline permits some illuminating comparisons. First, in rates: it appears that the soft and transitional technologies are much quicker to deploy than the big, high technologies of Figure 30-1, because the former are so much smaller, simpler, easier to manage, and less dependent on elaborate infrastructure. For fundamental engineering reasons, their lead times are measured in months, not decades. For similar reasons, as I suggested earlier, they are also likely to be substantially cheaper than hard technologies. They are environmentally far more benign and bypass the risk of major climatic change from the burning of fossil fuel. And they are more certain to work, since the risk of technical failure in the soft path is distributed among a large number of simple, diverse technologies that are in general already known to work and that require an especially forgiving kind of engineering. The hard energy path, in contrast, puts all its eggs in a few brittle baskets—fast breeder reactors and giant coal-gas plants, for example—which are not here and may or may not work.

Hard and soft technologies have very different implications for technologists. Hard technologies are demanding and frustrating. They are not much fun to do and are therefore unlikely to be done well. Although they strain technology to (and beyond) its limits, the scope they offer for innovation is of a rather narrow, routine sort and is buried within huge, anonymous research teams. The systems are beyond the developmental reach of all but a few giant corporations, liberally aided by public subsidies, subventions, and bailouts. The disproportionate talent and money devoted to hard technologies gives their proponents disproportionate influence, reinforcing the trend and discouraging good technologists from devoting their careers to soft technologies—which then cannot absorb funds effectively for lack of good people. And once hard technologies are developed, the enormous investments required to tool up to make them effectively exclude small business from the market, thus sacrificing rapid and sustained returns in money, energy, and jobs for all but a small segment of society.

Soft technologies have a completely different character. They are best developed by innovative small businesses and even individuals, for they offer immense scope for basically new ideas. Their challenge lies not in complexity but in simplicity. They permit but do not require mass production, thus encouraging local manufacture, by capital-saving and labor-intensive methods, of equipment adapted to local needs, materials, and skills. Soft technologies are multipurpose and can be integrated with buildings and with transport and food systems, thus saving on infrastructure. Their diversity matches our own pluralism: there is a soft energy system to match any settlement pattern. Soft technologies do not distort political structures or priorities; they improve the quality of work by emphasizing personal ingenuity, responsibility, and craftsmanship; they are inherently nonviolent and are therefore a livelihood that technologists can have good dreams about.

Soft technologies, unlike hard ones, are compatible with the modern concept of indigenous eco-development in the Third World. They directly meet basic human needs—heating, cooking, lighting, pumping—rather than supplying a costly, high-technology form of energy with imported, capital-intensive, high-technology systems that can enrich only urban elites at the expense of rural villagers. Soft technologies would thus contribute promptly and dramatically to world equity and order. A soft energy path would also do more than a hard path to reduce pressure on world oil and coal markets and to avoid global capital shortages.

An even more important geopolitical side effect of U.S. leadership in a soft energy path is that it promotes a world psychological climate of denuclearization, in which it comes to be viewed as a mark of national immaturity to have or desire either reactors or bombs. Nuclear power programs and their web of knowledge, expectation, and threats are now the main driving force behind proliferation of nuclear weapons. Yet foreign nuclear programs require a domestic political base, both public and private, that does not yet exist and must be borrowed from the United States. Ten years' residence in Europe has led me to the firm political judgment that unilateral U.S. action to encourage soft energy paths at home and abroad, in tandem with new initiatives for nonproliferation and strategic arms reduction, could turn off virtually all foreign nuclear power programs and could thus go very far to put the nuclear genie back into the bottle from which we first coaxed it to emerge. Three decades after we secretly chose a fateful path, which we then tried to force on the world, we can again lead the world by openly choosing, and freely helping others to choose, a path of prudence. But this shift of policy is urgent: we must stop passing the buck before our clients start passing the bombs.

If nuclear power were phased out, here or abroad, but no other policy were changed, the economic, environmental, and sociopolitical costs of centralized electrification would still be intolerable—even with energy conservation. As I suggested earlier, the key distinction between the soft and hard energy paths is in their architecture, the structure of the energy system, rather than in the amount of energy used. The two paths entail difficult social problems of very different kinds: for the hard path, autarchy, centrism, technocracy, and vulnerability; for the soft path, the need to adapt our thinking to use pluralistic consumer choice, participatory local democracy, and resilient design to substitute a myriad small devices and refinements for large, difficult projects under central management. We must choose which of the two kinds of social problems we want. But the problems of the hard path can only be addressed much less pleasantly, less plausibly, and less consistently with traditional values than the more tractable problems of the soft path. The problems of the hard path, too, become steadily harder, whereas those of the soft path become gradually easier.

I believe that to a large extent the soft path would be self-implementing through ordinary market and social processes once we had taken a few important initial steps, including:

1. Correcting institutional barriers to conservation and soft technologies, such as obsolete building codes and mortgage regulations, lack of capital-transfer schemes, restrictive utility practices, and lack of sound information.

2. Removing subsidies to conventional fuel and power industries, now estimated at well over $10 billion per year, and vigorously enforcing antitrust laws.

3. Pricing energy at a level consistent with its long-run replacement cost. I believe this can be done in a way that is both equitable and positively beneficial to the economy, since unrealistically cheap energy is an illusion for which we pay dearly everywhere else in the economy.

I do not pretend that these steps will be easy; only easier than not taking them. Properly handled, though, they can have enormous political appeal, for the soft energy path offers advantages for every constituency: jobs for the unemployed, capital for business people, environmental protection for conservationists, enhanced national security for the military, opportunities for small business to innovate and for big business to recycle itself, exciting technologies for the secular, a rebirth of spiritual values for the religious, world order and equity for globalists, energy independence for isolationists, civil rights for liberals, and states' rights for conservatives. Though a hard energy path is consistent with the interests of a few powerful American institutions, a soft path is consistent with far more strands of convergent social change at the grassroots. It goes with, not across, our political grain.

The choice between the soft and hard paths is urgent. Though each path is only illustrative and embraces an infinite spectrum of variations on a theme, there is a deep structural and conceptual dichotomy between them. They are not *technically* incompatible: in principle, nuclear power stations and solar collectors could coexist. But they are *culturally* incompatible: each path entails a certain evolution of social values and perceptions that makes the other kind of world harder to imagine. The two paths are *institutionally* antagonistic: the policy actions, institutions, and political commitments required for each (especially for the hard path) would seriously inhibit the other. And they are *logistically* competitive: every dollar, every bit of sweat and technical talent, every barrel of irreplaceable oil that we devote to the very demanding high technologies is a resource that we cannot use to pursue the elements of a soft path urgently enough to make them work together properly. In this sense, technologies like nuclear power are not only unnecessary but a positive encumbrance, for their resource commitments foreclose other and more attractive options, delaying soft technologies until the fossil-fuel bridge has been burned. Thus we must, with due deliberate speed,

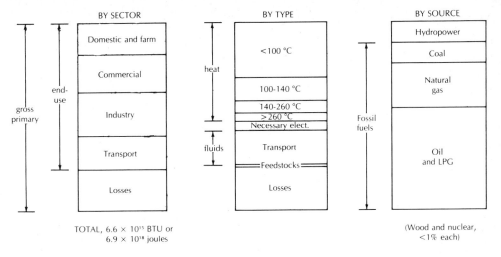

Figure 30-3. Canadian energy use in 1973 (population, 22 million).

choose one path or the other, before one has foreclosed the other or before nuclear proliferation has foreclosed both. We should use fossil fuels—thriftily—to capitalize a transition as nearly as possible straight to our ultimate energy-income sources because we will not have another chance to get there.

To fix these ideas more firmly, I should like now to sketch an example from one of the approximately ten other countries in which soft path studies are under way: Canada, whose energy system is strikingly similar to that of the United States. This example comes from a study[2] I did for the Canadian Ministry of Energy, Mines, and Resources under the auspices of the Science Council of Canada. The data are theirs, the conclusions my own.

Figure 30-3 shows the approximate structure of energy use in Canada. Use by economic sector and by source broadly resembles our own, except that Canada has proportionately more hydroelectricity. The middle bar, showing the thermodynamic structure of end use, reveals that, as in the United States, end-use needs are mostly heat (especially at modest temperatures) and liquids. Indeed, the electrical uses other than low-grade heating and cooling are such a small term that, on an aggregated basis, they

and all high-temperature heat could be supplied by present hydroelectricity.

The Canadian Cabinet has approved technical fixes to improve end-use efficiency. The Ministry calculates that these measures, now being implemented, should hold primary energy use for the commercial and transport sectors and for heating houses roughly constant over the next 15 years despite normally projected growth in population and in sectoral economic activity. However, 15 years is really too short a time for these measures to bear much fruit, since it takes much longer to turn over the stock of buildings and equipment. I therefore looked ahead 50 years—about the end of the lifetime of a power station ordered today. I assumed population and economic growth similar to official projections (with minor and unimportant exceptions), plus technical fixes of the moderate, straightforward type already being implemented. The result over 50 years was a shrinkage of per capita primary energy to one half today's level, or about the present Western European level.

I also applied the same technical fixes not to a standard growth scenario but instead to the per capita activity levels of 1960—which might be a very rough surrogate for a luxurious form of "conserver society." This cal-

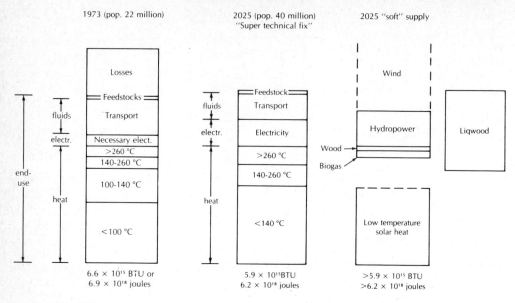

Figure 30-4. Canadian energy use projected to 2025 (estimated population, 40 million).

culation yielded a further factor-of-two shrinkage, to one quarter of today's level, or about the present New Zealand level.

I then returned to my original, higher set of estimates of energy needs for 2025, based on approximately a normal growth projection, and constructed a conservative estimate (exaggerating high-grade needs) of its end-use structure. The result is the middle bar of Figure 30-4, with the current structure on the left for comparison. On the right side of Figure 30-4 I have drawn to the same scale some building blocks of supply.

The block marked "hydropower" is the minimum hydroelectric output already firmly committed for 1985, not counting James Bay. It exceeds the appropriate electrical uses in 2025. Below it is a minor contribution from roughly the present level of rural wood burning and from pyrolysis, anaerobic digestion, burning-energy studies, and so on.

At the lower right in Figure 30-4, I have assumed that 50 years is long enough to meet essentially all low-temperature heat needs with solar technologies. These are already

technically and economically attractive in Canada, and the real question is not viability but deployment rate. If you think 50 years is not long enough to deploy all those devices, you are free to choose a target date later than 2025 and stretch out the transition. The 50 years is merely a first guess.

On the far right side of Figure 30-4 is a box labeled "liqwood." This is the name and number given by the Canadian Forestry Service to the net amount of fuel alcohols that they think they could sustainably produce from forestry. I suspect that closer ecological study will reduce this estimate, but it is still likely to be larger than the liquid fuel requirement for the transport sector (shown in the middle bar), even today (as shown in the left-hand bar).

At the upper right in Figure 30-4 is an open-ended section labeled "wind." It is a very large number, far off scale. The versatility of wind lets it fit anywhere into the end-use structure—pumping heat, for example, or compressing air to drive industrial machines.

Figure 30-4 suggests on its face that the "hydropower" block already committed, plus any two of the other three, will yield *a surplus that matches the structure* of Canadian energy needs in 2025. In practice, one would use a combination of all these sources. The next step in this exercise, which has not yet been done in Canada, is to refine the data, disaggregate by region, and work backwards from 2025 toward the present to see how to get there. Two things make me expect attractive results. First, such a study has already been done in countries less well placed than Canada, such as Denmark and Japan, and the transition looks quicker and cheaper than present policy. Second, if one pairs off the soft energy supply technologies with the hard technologies that one would otherwise have to use in the long run to do the same job, the soft ones are cheaper. This is also true for the conservation and transitional technologies. So when one adds them all up they should still be cheaper.

The value of thinking backwards in this way from where one wants to be in the long run—the right-hand end of Figure 30-2—is that it reveals the potential for radically different paths that would be completely invisible to anyone who merely worked forward in time through incremental *adhocracy*. Such a person could discover only in, say, 2010 that we might have gone in a very different direction if we had thought of it in 1980 before it was too late. To avoid such a trap, we must be unabashedly normative in exploring our goals, then figure out how to achieve them, rather than blindly extrapolating trend into destiny.

Of course, in this short article, I cannot do justice to the richness of the technical background, but I hope I have conveyed the impression that the basic issues in energy strategy, far from being too complex and technical for ordinary people to understand, are on the contrary too simple and political for experts to understand. I believe that as this nation enters its third century we must concentrate on these simple, yet powerful concepts if we are not only to gain a fuller understanding of the consequences of choice but also to appreciate the very wide range of choices available. Only thus can we learn, as Robert Frost did, that taking the road less traveled by can make all the difference.

REFERENCES

1. Frank von Hippel and Robert H. Williams, "Energy Waste and Nuclear Power Growth," *Bulletin of the Atomic Scientists* **32**, 18, 48 (1976).
2. Amory B. Lovins, "Energy Strategy: The Road Not Taken," *Foreign Affairs* **55**, 65 (1976).

SUGGESTED READINGS

Irene Kiefer, ed., "Future Strategies for Energy Development—A Question of Scale," *Proceeding* of a conference at Oak Ridge, October 1976. Oak Ridge, Tenn.: Oak Ridge Associated Universities, 1977.

Amory B. Lovins, *World Energy Strategies: Facts, Issues and Options*. Cambridge, Mass.: Friends of the Earth/Ballinger, 1975.

Amory B. Lovins, "Scale, Centralization and Electrification in Energy Systems," Future Strategies of Energy Development Symposium, Oak Ridge Associated Universities, October 1976.

Amory B. Lovins, *Soft Energy Paths: Toward a Durable Peace*. Cambridge, Mass.: Friends of the Earth/Ballinger, and London: Penguin, 1977.

Amory B. Lovins, *Energy in Context*. Cambridge, Mass.: Friends of the Earth/Ballinger, 1977.

Amory B. Lovins, "Soft Energy Technologies," *Annual Review of Energy* **3**, 477 (1978).

Amory B. Lovins and J. H. Price, *Non-Nuclear Futures: The Case for an Ethical Energy Strategy*. Cambridge, Mass.: Friends of the Earth/Ballinger, 1975.

Michael Stiefel, "Soft and Hard Energy Paths: The Road Not Taken?" *Technology Review* **82** (1), 56 (1979).

A. DAVID ROSSIN

One of the greatest threats posed by our energy problems is that they will lead to greater concentrations of political power and more centralized control over people's lives. Hence the concept of *decentralized* energy is very exciting.

Unfortunately, the proponents of decentralized energy production (the "soft" path) offer a road map that would most likely lead to what they abhor—a society of centralized control. Their road map calls for commitments to phase out conventional methods of supplying electric energy to homes, factories, offices, and farms. The soft path offers instead an array of alternative energy sources, such as solar and wind power, that are said to be benign, environmentally desirable, close to the people, and well suited to local control by individuals or communities. Some of these alternatives may prove viable, but, according to many of the most vocal proponents of the soft path, a crucial part of the strategy is the decreased reliance on transmission lines and power plants, particularly nuclear power plants.

The part of the soft path that is seldom discussed is what might happen if the capability to supply electricity begins to fall short of what people need. If there is not enough to go around, somebody will have to set priorities; and in this event, priorities would be set not by individuals, but by government—big, centralized government. The real question is whether it is desirable to pursue the decentralization of sources of energy as an absolute goal or to permit the diversity we have now, in which anyone who wishes to can use decentralized alternatives. Without sufficient, reliable, centralized electricity, the individual no longer has that choice, the one we take for granted today.

This is not to argue that centralized control is inevitable. But the soft path strategy carries with it the real risk of centralized control of personal decisions. If a free society is to choose to pursue a decentralized energy structure, it should do so only with full knowledge of all the potential consequences.

EXAMINING THE SOFT PATH

The problem is that the soft path just might not lead where its proponents claim. There

A. David Rossin is Director of Research, Commonwealth Edison Company, Chicago, Illinois 60690.

From *The Futurist* **XIV** (3), 16 (1980). Reprinted by permission of the author and *The Futurist*. Copyright © 1980 World Future Society.

is a serious lack of data on the cost and performance of solar, wind, wood-burning, and other soft-path options, especially on a broad enough scale to show how the following of this path might affect the need for more electric power plants. Indeed, decentralized energy systems are generally discussed in the abstract. Alternatives, almost by definition, are said to be better than what works today. Yet we know little about just how decentralized solar or wind systems would work in cities, large towns, or other locations where they might be called upon to serve large numbers of people. We do not know the environmental impacts, the costs, or the resources required.

Yet despite this lack of knowledge about the consequences of following the soft path, the advocates of decentralized energy production are more than willing to do away with the kinds of centralized power sources that supply electricity in the United States today. One of the foremost proponents of decentralized energy options, Amory Lovins, says in his book *Soft Energy Paths: Toward a Durable Peace:* "If nuclear power were clean, safe, economic, assured of ample fuel, and socially benign *per se,* it would be unattractive because of the kind of energy economy it would lock us into." He explains further that it is undesirable because it is centralized, big, and controlled by corporations and requires large amounts of capital. And despite the fact that the soft energy path leads at best through uncharted territory, Lovins is uncompromising. He states that society must choose, apparently once and for all, between the soft path and the kind of centralized systems that supply people's needs today. Moreover, Lovins claims that society must make this choice immediately. As he sees it, additional commitments to centralized systems will make it increasingly difficult to develop soft technologies and have them emerge successfully. He fears that commitments of capital to the conventional systems as we now know them will leave too little for the soft technologies. He proposes that further investment in power plants and transmission lines be prohibited, so that movement along the soft path would be assured. This is the Lovins strategy: block centralized power to force the soft path.

Making the soft path inevitable would not make it inexpensive; despite claims of favorable economics, very few people have chosen to take it. Not only does the energy look expensive compared with what is available, but the capital an individual or community needs to put up is large for what one gets. The soft path would require two to three times as much capital as coal and nuclear would, even if it could be phased in gradually with no surprises. All that capital has to come from somewhere. Furthermore, it is far more difficult for individuals to raise capital than it is for large institutions. The cost of money (interest rates) would be higher, and the cost of the facilities themselves would be higher.

Most people just are not interested, if left to make free choices. Decentralized energy will require subsidies, complete with federal guidelines, inspection, enforcement, and, of course, taxes to raise the money.

If the soft energy path does require substantial subsidies, in a democratic society the representatives of the people would have to vote for them. Taxpayers are also voters, and subsidies would be difficult to sell if more economical and understandable alternatives were available.

In Lovins's words, unless capital, manpower, and expertise are diverted to alternative forms of energy, "the soft path becomes hostage to the hard path." What he means is that decentralized energy will become the way of life only if conventional sources of energy are effectively banned for the long-term future. That is the theory anyway. His failure even to mention the potential risks of the soft energy path, however, raises questions about the credibility of both the concept and its promoters.

The fact of the matter is that alternative energy systems can be built today by those who wish to. Robert Redford has a solar house and Congressman Henry Reuss has a windmill. We have diversity, and if an idea is successful, others can try it. But not everybody wants to!

I would argue, Lovins's opinions notwithstanding, that the energy economy that we might be "locked into" if we further expand nuclear power and other centralized forms of energy production is one of diversity and freedom of choice—including the choice of alternative forms of energy. This kind of energy economy would be characterized, as is today's energy production system, by organizations, both publicly and privately owned, that are *legally obligated* to supply electricity to those who wish to use it for all or part of their needs, whether they have solar panels, windmills, or not.

DECENTRALIZATION AS A PHILOSOPHY

The other part of Lovins's strategy, seldom examined by the public, is revealed in the following quotation from *Soft Energy Paths:*

Many who work on energy policy and in other fields have come to believe that in this time of change, energy—pervasive, symbolic, strategically central to our way of life—offers the best integrating principle for the wider shifts of policy and perception that we are groping toward.

The soft energy path, it seems, involves more than merely changing the way in which energy is produced. It entails, Lovins suggests, "wider shifts of policy and perception"—in short, new concepts not only of energy use but of society, something like a whole new philosophy for society.

If the debate is about the philosophy (indeed, the ideology) of individual self-sufficiency, rather than costs, risks, and benefits, it is appropriate to explore the concepts of decentralization and self-sufficiency themselves. Like other general approaches, decentralization and self-sufficiency are not universally beneficial. How many Americans, for example, would desire to return to a decentralized system for providing their drinking water? What about sewage treatment? Although few Americans remember it,

as recently as 100 years ago some major American cities were without centralized water and sewage systems. The massive effort necessary to create these systems has resulted in incalculable benefits to society, wiping out some of America's most common and deadly diseases. One can hardly conceive of public health authorities, let alone the average person, taking even the slightest chance of a resurgence of typhoid, cholera, and dysentery in order to satisfy a philosophical objective.

Furthermore, the promoters of the soft path admit that it would take time to move from a centralized system for producing energy to a decentralized-energy society. In the transition period, which is generally conceded to be several decades, the centralized system must be kept around. Almost every residence would require central electricity to back up the solar systems or windmills and to supply electricity when the sun is not shining or the wind is not blowing or when the motors, pumps, or seals need repair and the neighborhood fix-it person cannot come. There would still be electric motors, electronic equipment, and electric lights for which no alternate source is available. In New York, Chicago, and other cities where people live in apartments and large commercial buildings are common, the central system would still have to carry most of the load. Moreover, central electric supply would also be essential to supply the tremendous energy needs of the huge industry that would have to be brought into being to manufacture the solar devices and other equipment necessary to make the transition to a soft energy society.

Advocates of decentralization are hard pressed to tell us just how long this transition period will be or just how many more power plants will be necessary to meet people's continuing needs before the transition can be achieved. And how complete must the transition be in order to satisfy its proponents? Power plants and electric grids might just still be needed for industry, commerce, communications, and the offices of the bureaucracy.

The fact is that soft energy systems have little prospect of coming into widespread use unless centralized electric utility supply systems are not only available and reliable, but economical as well. The reason is basic: the homeowner who wishes to enjoy the benefits of a solar energy system will have to pay for it. But at the same time, he will have to continue paying for the conventional electric service that he will not be able to do without. This payment will have to be made either through a utility bill or through taxes.

If, as a result of short-sighted regulation or punitive public policies, the cost of utility electricity is driven steadily up, it will not mean that solar energy will become competitive. What it will mean is that most people will have little money left in the household budget after paying their taxes and utility bills in addition to unavoidable expenditures for food, clothing, and shelter. Few could afford to devote their remaining funds, if any, to another piece of hardware that would supply but a portion of their electrical needs.

When costs go up, everybody has to pay. The impact will fall hardest on the poor and on those with fixed incomes. If the capital comes from government subsidy, its source will be taxation or inflated currency, again hitting everybody, but the poor the hardest.

Experience shows that it is hard to get people to vote for higher gasoline taxes, higher utility rates, or even local school bond issues. Would they do it if they knew that the only rationale for a new tax was largely philosophical and made little sense on economic or environmental grounds?

AN ENVIRONMENTAL IMPACT STATEMENT

Before they undertake major construction projects or other large programs, many industries, including those that build conventional power plants and nuclear reactors, must prepare an environmental impact statement. Such a statement must examine the project itself, its effects on the environment, the alternatives, and even the alternative of not proceeding with the project at all. But no such examination has been offered for the soft path, and the people of the United States need such an examination if they are seriously to consider soft path policies. No one seems to have answered any of the hard questions yet.

What would happen, for example, if a societal decision were made that no more new, large electric generating plants would be built in the United States—either nuclear or coal? Voluntary conservation becomes the national byword, but meanwhile a growing population and the aspirations of people who still hope for a better future for themselves and their children mean growing demands for energy. Whether or not the government has subsidized local solar and wind generators, in a few years there will be gaps in various parts of the nation between the capability of the centralized energy system to supply energy and what is needed to meet the legitimate expectations of individuals, households, communities, and industries. In that event, one traditional political response might well be for the government to take over the generation of electricity and ration it. Then, facing the same realities utilities face now, government would have to build the central generating plants that the various individual utilities were stopped from building because they were "philosophically undesirable."

The cost to every individual would likely be higher because of the delays that forced the crisis. Catch-up building would be necessary instead of prudent and logical planning. Philosophically, the problem is that energy would be supplied by an even bigger and more centralized system, that is, the government, as opposed to local or regional utilities. There might be no time or money for community or neighborhood sources.

But this is just one scenario. Another possibility that requires assessment is the likelihood that the government itself will be unable to build more generating plants. Government's track record in energy policy in recent years is not impressive. Congress and the Administration have hesitated to act, often because of pressure from various special-interest groups. The government might simply be unable to get organized in time, or it could fail to win the political support needed to make firm commitments to build new power plants quickly.

The federal government has failed miserably to implement what the U.S. Congress ordered it to do about nuclear waste management. (In 1970, Congress decreed that only the federal government would be permitted to dispose of high-level wastes. But each year it fails to authorize or appropriate enough money to move ahead.) This is but a small part of the nuclear fuel cycle, and since wastes already exist, the government has no alternative but to build repositories for them. This failure has caused enough public skepticism to put the future of nuclear plants in doubt. If government cannot even authorize the money to store wastes, how can we have confidence that it would build vitally needed power plants? The result could be serious shortages of electric generating capacity.

IMPACT ON THE FUTURE

If energy supply falls short of demand during the promised transition to the soft energy era, on occasion there would be brownouts or blackouts and the emergency curtailment of electric supply to avoid lengthy breakdowns. Short-term emergencies are costly, as any New Yorker knows. But the more serious impacts are long-term—and more subtle. Before blackouts and brownouts become widespread, the fact that possible shortages of electricity are coming would be common knowledge among utility experts, urban planners, and many businessmen. And this knowledge would be crucial to new businesses looking for attractive locations to set up shop.

Businesses do not just appear. Any company considering the construction of a new manufacturing plant that would employ a substantial number of people or a large office building, shopping center, hospital, or housing development asks some basic questions early. They go to the utility company that serves the area in which they are considering

building and ask if they can be assured that electricity will be available at the site both during construction and afterwards to meet their power needs when their plant or shopping center goes into operation. They also want to know if telephone service will be available. They ask the municipality for assurance that water and sewage services will be installed on schedule and at a reasonable cost. They have to know if roads will be built and maintained, and by whom. If they get unsatisfactory answers to one or more of these questions, they will begin looking for a site somewhere else. With it will go the jobs and the benefits that the project was designed to bring.

If they get a lot of negative answers in a lot of different places, they will be receiving a warning that perhaps the market they expect to serve will not be able to buy their products, pay for their services, or get a mortgage for their housing units. Instead of raising more capital and taking more risks, they may decide to stay on the sidelines.

If the time comes when more and more people who would have been the builders and the doers remain on the sidelines because of energy shortages, growth will stagnate and we could enter into what some people remember as a depression. That is a dangerous risk for society to take.

Some studies suggest that if it can be done carefully and gradually, with conservation and introduction of energy-efficient equipment plus solar power and other alternatives, demand growth can be slowed without economic collapse. That is fine if the planners are accurate and society responds perfectly to the signals right on time. But there is no room in their scenarios for any surprises, like the OPEC embargo in 1973–74, the natural gas shortage in 1977, the coal strike of 1978, the winters of 1978 and 1979, or Iran. For when supply and demand do not stay in balance, unattractive things begin to happen.

THE REAL RISKS

There are some people who claim that energy growth and economic growth must be stopped in order to halt ever-increasing environmental problems. The theory is that by halting growth, conservation will become essential, everybody will take only his fair share, and everyone will live happily ever after.

History shows us that things do not work that way. If there is not enough electricity to go around, some decisions will have to be made about who gets priority for the limited amounts that are available. If more than one group wants to build, priorities will have to be established. Who makes these decisions? Decisions could no longer be made on a decentralized, individual basis. They would have to be made centrally. By whom? *By big government.*

Is that really what the proponents of the soft path have in mind? Is that really what people who are impressed by the promises of decentralized energy supply understand as the potential consequence of taking the soft path? Because if necessary commitments to keep up the centralized energy supply system are foreclosed and the extravagant promises of decentralized energy are not achieved, these potential consequences become inevitable.

Here is the basic reality: today we have centralized electric energy supply produced by large utilities. The individual can still make his own decisions about whether or not he wants to buy a new appliance, build a new home that will have electricity in it, or start up his own business and assume that, if he wants to sell something that has to be plugged in, people will be permitted to buy it and use it. He can build his own windmill or his own solar panels or even buy his own diesel generator and generate his own electricity, if he can afford the fuel for it.

Utilities do not decide who gets electricity and who does not. The state gives them a charter that *requires* them to provide reliable service to all customers. If this is to be changed, the decision to do so should be made by the elected representatives of the people (the state legislature), who could order the State Utility Commission to do the allocating. Certainly, utility managers should not be the ones making decisions about which

users are more deserving than others. But if there is not enough capacity to provide electricity to all those who ask for it, someone will have to make exactly that kind of decision.

The need for such far-reaching decisions should not be forced on the public by the actions of a few activists, whether well intentioned or not. This is especially true if these persons have no responsibility to the community, if the citizens have no real say about what they are being coerced into, and if those affected are not informed about what they are likely to face.

The interesting point is that by doing the job their charters of today require, utilities make it possible for *decentralized decision making* to take place at all levels of the society. Here are the real options for the future:

A centralized electricity supply, which provides enough energy for people to use if they choose and thus permits *decentralized decision making,* or

Abandonment or curtailment of centralized energy supply in the hope that decentralized energy will meet expectations, but with the risk of *centralized decision making,* allocation of energy by government, and curtailment of individual freedom.

EQUITY AND RESPONSIBILITY

History suggests that Americans value their personal freedom more than control over how their electricity is produced and delivered. Does everybody really wish to use equal amounts of energy, or is the real issue the freedom of each individual to make his own decisions about uses that he feels are important? Some things that people do with electricity may appear wasteful or capricious to others. Some kitchen appliances are felt to be unnecessary luxuries in the eyes of a young couple who love camping and backpacking. The same things may be necessities to the homemaker with children. What about retired persons who may not be able to do all the physical chores they once could do, and having done them in their younger days remember what life was like before labor-saving electrical devices were available? What kind of society would tell the family who has finally saved up enough money to buy a washing machine and dryer, an air conditioner, or even a color TV that they might not be allowed to use it unless they can get an electricity allocation? And will the next family that hopes to build a new home have to prove a legitimate case for its energy allotment as opposed to competing requests from municipal or industrial applicants?

Another question: what does society do about the individual who does not carry his fair share of the burden? What about the neighbor who fails to provide or maintain his own solar heating system and fails to take adequate care to see that his home and family are warm? The answer is that as a society we will not tolerate children freezing to death in an unheated home. Perhaps a government-administered, tax-supported, energy welfare system will be essential to see that the necessary heat, food, and light are provided to those who cannot, choose not to, or for any reason fail to provide for themselves.

In a compassionate society, electricity prices need to be kept reasonable so that energy use does not become a privilege of the rich but remains available to everyone. If energy supplies can be maintained, however, and prices kept low, people can continue to make their own choices about what switches they wish to turn on or off.

TOWARD ENERGY POLICY

When the concept of energy decentralization is discussed, what is it we should really be concerned about? What are the risks if the promises fail? Certainly, some kind of energy rationing, and the centralized government control that rationing implies, would become real possibilities. Under such a sys-

tem, the threat or reality of acute energy shortages around the country would discourage the individual entrepreneur and curtail individual freedom of choice.

In addition, the restriction of energy supplies could well result in a serious risk of loss of economic growth. Growth has environmental impacts—these are fairly well known—but lack of growth also has impacts on the environment. The risks that accompany *both* need to be studied carefully and discussed openly, along with the risks that a particular energy policy might push things one way or the other. The environmental impact statement for the soft energy path should be published widely. Then the public will be able to rationally evaluate the risks of a failure to have enough power plants.

Which risk is to be most feared: the network of transmission lines and centralized electric power plants that are commonplace today or the possibility of centralized control of how each corporation, group, and person uses electricity? A policy or a philosophy that promises to solve energy problems by curtailing centralized power-generating systems turns out to be bankrupt on close examination.

Proponents of the soft energy path envision not only an energy system, but a way of life in which large power plants—especially nuclear plants—have no place, and perhaps the fact that the soft energy path is as much a philosophy as a platform of policies helps explain the intense polarization that currently marks the nuclear debate. What the advocates of the soft energy path neglect to point out is that without centralized power plants, including nuclear power plants, the threat of energy allocation grows. And the allocation of energy may well mean not decentralization but infringement on individual freedom and the ultimate in centralized control of people's lives.

A Low-Energy Scenario for the United States: 1975–2050   32

JOHN S. STEINHART, MARK E. HANSON, ROBIN W. GATES,
CAREL C. DEWINKEL, KATHLEEN BRIODY LIPP, MARK THORNSJO,
AND STANLEY KABALA

INTRODUCTION

The scenario outlined in this article is an estimate of the lowest-energy-use future the authors consider plausible. The plausibility criterion is applied by eliminating those possible futures that require technological miracles, dramatic changes in human behavior, or complete restructuring of national and/or world political groupings. The results indicate that a 64 per cent reduction in U.S. energy use per capita from 1975 levels can be obtained a few decades into the twenty-first century.

We have attempted to incorporate and extrapolate demographic and settlement relocation trends and to identify some other private trends that might be usefully encouraged. The legislative measures discussed have been proposed at state or federal levels already, although some may be far from enactment now. The changes of lifestyle implicit in this

Dr. John S. Steinhart received degrees in economics and geophysics and is a professor of geology and geophysics and of environmental studies at the University of Wisconsin-Madison.

Mark E. Hanson has degrees in economics, water resources, and land resources. He is an assistant scientist with the Energy Systems and Policy Research Group at the University of Wisconsin-Madison.

Carel C. DeWinkel was awarded degrees in applied physics, civil and environmental engineering, and energy resources. He is currently a research engineer with Wisconsin Power and Light Company.

Robin W. Gates is an energy analyst in the Wisconsin Office of Planning and Energy. He received degrees in environmental planning and resource management and in public administration.

Kathleen Briody Lipp holds degrees in zoology and environmental sciences. She is now an energy forecaster for Wisconsin Power and Light Company.

Mark Thornsjo is a marketing information specialist in energy conservation for Northern States Power Company. He graduated with a degree in urban and regional planning.

Stanley Kabala is director of the Pittsburgh Architects Workshop, which provides architectural and planning assistance for those unable to obtain it elsewhere. He holds a degree in public policy and administration.

This original article is based on a paper titled *A Scenario for Reduced Energy Use and Conservation in the United States: 1975–2050* presented at the American Association for the Advancement of Science Annual Meeting in Denver, Colorado, February 21, 1977.

scenario combine present trends with subjective estimates of how far they might proceed.

We have tried to point out some of the economic and employment implications of the scenario. Quantitative economic analysis is not attempted because such analysis is very sensitive to the projections of future discount rates and future fuel price increases. Plausible (and published) estimates of these two variables differ sufficiently to produce net discount rates for energy costs (rates of deflated energy cost increase minus discount rate) that are either positive or negative. Resolution of these differences requires at least some agreement about the quantitative role of energy price increase in general inflationary pressures. These relationships are disputed.

This scenario is not proposed as a plan. It is offered as one of a growing number of such possibilities in the hope that the arena of public discussion may be enlarged.

Any normative scenario of the future assumes a backdrop of world and domestic conditions as the context in which the scenario's plausibility may be tested. No scenario of modest size can hope to describe even the complexity of the present, let alone the complexities of an inherently unknowable future. The result is that a scenario leaves out more of the future conditions than it describes. Although there is no satisfactory escape from the problem, it may be useful to the reader to draw attention to some views of the future that are in agreement with this scenario and that have contributed to the views expressed.

The resource pressure outlined by Forrester,[1] Meadows et al.,[2] and Mesarovich and Pestel[3] seems to us to be more or less correct. Critics of the several reports of the Club of Rome have raised valid criticisms of some aspects of this work, but no counter models exist and the basic results still stand. The projections for the United States and the world offered by Watt,[4] Heilbronner,[5] Lovins[6] (see Article 30), and Stavrianos[7] and by the Latin American World Model[8] are

very close to our own view of the problems faced. Special attention is invited to the difficulties of managing ever-larger, ever-more complex societal and technological systems. Here, our views correspond most closely with those of Vacca.[9]

The problem is to invent the future. Only the most insistent determinists and some mystical and religious groups demand a single, unavoidable future. In the enormous gap between the unavoidable and the miraculous fall politicians and academics, ordinary people, and fortune-tellers, each in their own way trying to estimate something useful about the future in order to reconcile dreams with possibilities. The general agreement on objectives that characterized the industrial world at the end of World War II has vanished as goals have been met or abandoned. Little was then heard from the three fourths of the world that was poor in the 1940s. Now there are many new voices in the world, and some of them command both money and raw materials. Thus, the future is to be invented in the face of many conflicting objectives—about which there is little agreement—and amidst an uncomfortable frequency of political instability. The process is made even more worrisome by the observable tendency of the industrialized nations to support (or even to bring to power) strong central governments for poor nations, even if (one hopes not because) these governments are highly authoritarian, oppressive, and nonrepresentational.

The energy difficulties of the early 1970s have raised the questions of energy and raw material availability in the developed world, and the joint problems of population growth and food supply in the poor nations of the world have provided new visions of the future, most of them conflict-ridden or downright disastrous. Responses to these visions of the future have been very strange. From the economists, in their role as the leading academic contributors to policy discussion, have come rejections but no comprehensive models that show a less destructive future. Calling for more investment in the poor world, many economists show a faith in technological innovation that is far less unanimous among scientists and engineers.

In a world of growing complexity, there is a greater premium than ever on correct estimation of the future. Nearly every major nation has officially sanctioned bodies to produce forecasts, evaluations, policy impact projections, and other types of assessments. What characterizes these efforts above everything else is their operation within the limits of their sponsors. Under these circumstances, it is not surprising to find that the results seldom challenge any existing institutions. Because these analyses are perceived as value free and "hard-headed," the outcome is often given some special weight. Unfortunately, to attain this attention, the analyses are usually restricted to economic assessment, the more obvious deleterious side effects, and measures to mitigate the problems. Despite inevitable caveats about social effects or political problems, these analyses always purport to offer the best (or the least bad) policy recommendations. But, restricted by the methodology and the nature of the charge, the range of options considered is very limited. Analyses by independent groups or individuals are often far less optimistic than official projections, but in the competition for credibility, they may well sacrifice their strongest arguments in efforts to sound unbiased.

A scenario lies somewhere between a forecast and a fantasy. The objective of this scenario is to explore the possibilities of a particular plausible future. The low-energy scenario was chosen because it provides the least expensive energy options, but upon further examination was found to have some other interesting and possibly desirable properties. For example, all are aware of the environmental effects of intensive energy use: air pollution, water pollution, land degradation from mining and sprawl, inefficient and occasionally excessive production. All these problems are reduced in proportion to our ability to use less energy more effectively.

There are international implications inherent in this scenario. First, a real reduction in energy use by the industrialized nations may offer the only way to avoid direct confrontation of the rich few with the hopes of the destitute many, and the United States is the leading user of energy. Second, because of sheer economic power and considerable domestic resources, the United States could play an exemplary role—as some U.S. politicians believe we now do. Third, the social conflicts and changes in the United States since the mid-1960s make it possible now—in a way that it was not 20 years ago—for different coalitions and ideals to dominate U.S. policy. Indeed, this scenario incorporates several recent social trends in it, on the theory that it is easiest to persuade people to do something that they already intend to do or are resigned to doing. For example, the U.S. Department of Agriculture notes that in 1976 one half of all U.S. households attempted to grow some of their own food. Even though the food system requires 16 per cent of all U.S. energy use,[10] no one seems to have considered the possible implications for this major social shift for energy policy. Finally, the idea of central control appears to be weakening both in the United States and in the rest of the world. There are now new local and regional disputes and power groups in the United States, and in the world the number of independent and independent-minded voices has increased since World War II. This combination of nationalism at the international level and localism within nations might offer new opportunities for a world with less armed conflict (though with more rhetorical shouting).

OVERVIEW

The scenario described in this article results in an energy use per capita in the year 2050 of 13×10^{10} joules, which is 36 per cent of the 1975 U.S. level of 35×10^{10} joules. To accomplish this end, major changes will be required, but these changes will not lead to a

Table 32–1.   Overview of Final Energy Use 1975–2050: Low-Energy Scenario.

	1975 Primary[a] Energy per Capita (10^{10} joules)	2050 Primary[b] Energy per Capita (10^{10} joules)	Primary Energy Savings per Capita (10^{10} joules)	1975 Total[c] Primary Energy (10^{18} joules)	2050 Total[d] Primary Energy (10^{18} joules)
Residential[e]	6.9	2.4	4.5	14.6	6.7
Commercial[e]	5.9	1.8	4.1	12.6	4.9
Transportation	9.2	2.3	6.9	19.6	6.5
Personal	(6.0)	(1.4)	(4.6)	(12.7)	(3.8)
Freight	(3.2)	(1.0)	(2.3)	(6.9)	(2.7)
Industrial[f]	13.0	6.2	6.8	27.6	17
Total[g]	35.0	12	22	74.4	35

[a] Population, 213×10^6.
[b] Population, 277×10^6. Year 2050 primary energy per capita is 36 per cent of that of 1975.
[c] News release, March 14, 1977. "Annual U.S. Energy Use Up in 1976," Office of Assistant Director—Fuels, Bureau of Mines, U.S. Department of the Interior.
[d] Year 2050 primary energy is 47 per cent of 1975 primary energy.
[e] The division of energy into residential and commercial from Bureau of Mines figures follows Lovins "Scale, Centralization and Electrification in Energy Systems" discussion draft.
[f] Miscellaneous and unaccounted for in Bureau of Mines data included in industrial.
[g] May not total 100 due to rounding.

fall in the standard of living. This contrasts with the assertion that further improvement or maintenance of the standard of living is inexorably linked to continued increase in energy use. Dennis Hayes[11] frames the issue this way:

Curtailment means giving up automobiles; conservation means trading in a seven-mile-per-gallon status symbol for a 40-mile-per-gallon commuter vehicle. Curtailment means a cold house; conservation means a well-insulated house with an efficient heating system.

Although this scenario goes considerably further than Hayes does in reducing the use of energy, his distinction is useful.

This analysis takes the traditional sectoral breakdown—residential, commercial, transportation, agricultural, and industrial—and examines each one for possible energy savings. The posited changes in each sector are described, borrowing from the extensive literature in the field, and aggregated into an overall sectoral primary energy use estimates as shown in Table 32-1. Energy sources sufficient to meet this level of use throughout the period are exhibited in Figure 32-5. The depletion of petroleum and natural gas and the termination of nuclear power as well as

the implementation of alternative energy sources during the period are outlined.

To bring about these reductions in energy use, changes will have to occur in social and physical institutions and arrangements. These changes will come not from a master plan but from the evolution of tastes, preferences, and organizations. Thus, new community patterns, already emerging in demographic trends, are outlined in the following section. Shifts in the mix of goods and services and employment are described, including the winding down of the automobile, road building, and food processing industries and the rise of the solar energy, mass transit, telecommunications, housing, construction, and health care industries. Shifts in employment patterns, the distribution of work, leisure, and income, and their meaning are discussed later in this article. These trends anticipate the end of the continued growth of the commodity component of society's output. In any event, endless growth may be neither desirable nor possible.

The final sections treat the policy questions, issues, and measures that are consistent with this scenario as well as the international implications of this scenario.

SETTLEMENT PATTERNS AND CHANGING COMMUNITIES

Many problems of the manmade environment result from sheer size. Getting out of a city of 100,000 is relatively easy even without an automobile; to get beyond the edge of a city of 3 million even with an auto involves a major effort. The air pollution of a city would be less of a problem if it were dispersed over a large area or if cars were restricted—the seriousness of the problem is a function of city size. The magnitude of the problem of solid waste disposal more than doubles with a doubling in the size of a city, not because the volume of waste more than doubles but because all of it must be hauled farther for "disposal." Economies can result from small scale as well as large.

Today's pattern of extremely high and extremely low densities will evolve to a pattern of medium-size cities with moderate densities, separated not by suburban sprawl but by farm and forest land. The satisfactory size for such cities would be small enough to avoid megalopolitan diseconomies of scale yet large enough to carry a vigorous industrial and commercial base and sustain a high level of cultural activity.

Instead of the seemingly endless sprawls we now know, suburbs should be villages, composed of a series of hamlets, closely spaced, that together make up a village and support the services a village can provide. . . . The village may, within its political boundaries, contain, say, 8 square miles, but only 3 square miles should be developed. The average density of the developed area should be 5 dwelling units per acre. The village would have a population, then, of about 35,000. Five square miles should be field and forest, mountain and river. . . . The 5 square miles should, of course, link up with the open spaces of neighboring villages in a linear fashion which leaves undisturbed and undeveloped the major landscape features, including the principal drainage ways, the ridgelines of hills, and outstanding scenic areas. The village should own all this land. Much of it should be put to agricultural or silvicultural use on a lease basis. . . . Many of the prospects, the striking visual landscapes, may once

have been sites for those 1950's-style subdivisions of half-acre, single family houses. They should be razed.[12]

"Impossible to achieve?" asks Charles E. Little after conjuring up this image. Probably so, in terms of the 1950s, 1960s, and 1970s style of development made possible by the private automobile. Probably not, with development to accommodate constrained private transportation.

City planner Patrick Geddes thought that decentralization would be brought about by the development of electricity that would free industry from centralized power facilities. Ralph Borsodi theorized in the 1930s that it could be achieved by the proliferation of power tools that would make possible home-based skilled labor, on the sound but economically heretical assumption that it is more economical to haul machine parts than workers. Frank Lloyd Wright felt the advent of the automobile would bring it about and designed Broadacre City on that premise. None of the technological innovations that could have been the basis for social innovation halted the trend toward centralization and urbanization. The social, cultural, and economic attraction of the city has overwhelmed any competing considerations. Put simply, modern industrial culture has been rich enough to afford the kind of cities it has, even though there have been simpler and more forthright means of providing for human needs.

The costs of Ward's "unintended city," not planned for human purposes but shaped by the hammers of technology, applied power, the overwhelming drive of self-interest, and the single-minded pursuit of economic gain, are now, however, taking their toll in the crime, bankruptcy, blight, and environmental deterioration of cities.[13] The response of modern commerce has been to substitute the blacktop-surrounded, air conditioned, landscaped mall for the central town square or piazza as a place for people to meet and mingle among shops and fountains. This new town center is now far afield, accessible

Table 32–2.   United States Demography, 1970 and 2050.

	1970	2050
Total U.S. population	203.2 million	277 million[a]
Number of cities > 100,000	156	100
Population (% of total)	56.5 million (38)	20 million (7)
Average population	362,000	200,000
Number of cities > 25,000–100,000	760	1500
Population in cities (% of total)	34.6 million (17)	79 million (29)
Average population	45,500	53,000
Number of cities 2500–25,000	5519	8000
Population in cities (% of total)	42.4 million (21)	84 million (30)
Average population	7682	10,000
Rural population (% of total) including urban < 2500 plus unincorporated urban	69.8 million (34)	94 million (34)

[a] Bureau of the Census, Department of Commerce, *Statistical Abstract of the United States, 1977*, U.S. Census, Series II-X projection 1980–2050.

only by car and stripped of all community functions but the sale of commodities. The energy, environmental, and ultimately financial costs of these changes suggest that a metamorphosis to new settlement patterns has become necessary.

These new forms will be characterized by smaller cities, with present large megalopoles multi-nucleated into communities of human scale such as the "urban villages"— Chelsea, Trastevere, Greenwich—in the 50,000 to 100,000 range of population.[13] The density foreseen is along the lines suggested by Little, roughly five dwelling units per acre.[12] The distribution of the population that we envision is indicative of the form of our future settlements (Table 32-2).

The overall effect on America's large cities will be the simultaneous implosion and explosion of population, the implosion the result of the loss of the automobile as the primary means of city transit and the explosion the reestablishment of distinct—as opposed to megalopolitan—urban areas of reduced size. Urban areas will generally contain 200,000 or fewer residents, a population still manageable in terms of the economics of city operation, full pedestrian and bicycle access to all districts, and accessibility to surround-

ing food-producing lands.[14] A fundamental aim in the redesign of the urban areas is maximum accessibility as opposed to maximum mobility. As private transportation becomes more and more costly, and as local grocers, butchers, bakers, druggists, jewelers, booksellers, restauranteurs, and clothiers are found to be able to supply most of local residents' needs, the neighborhood and the town will regain their social and community functions. It may be that the parking lots of today's shopping centers will be the sites of tomorrow's housing.

The path of evolution from the present to this future state has been elusive to planners, government officials, and citizenry who have attempted to face the problems of the present patterns of settlement. We believe, however, that the transformation is possible, and there is considerable evidence, from recent data on the depopulation of the largest urban areas, that migration out of the largest urban complexes is already occurring.

The history of the United States is sometimes written in terms of population migration, and there is no evidence that these migrations have stopped. Figure 32-1 shows the familiar westward and southward migrations. The westward movements continue

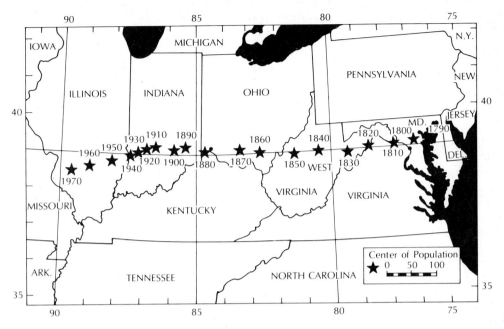

Figure 32-1. Population migration, west and south.

Figure 32-2. Net population migration for the largest megalopoles (Standard Consolidated Statistical Areas), 1970–74.

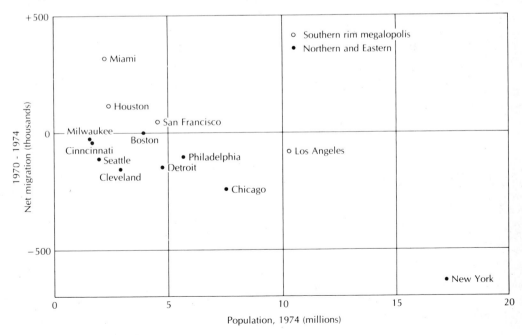

from our earliest days as a nation. The southward migration, recently popular as a theme
for social and political analysts, has continued since 1910 and accelerated since 1940.
Superimposed on these changes has been,
until recent years, migration from rural areas
to cities and from small towns to large metropolitan areas and, within metropolitan
areas, movement of residences from city
center to suburban areas.

In the years 1970 to 1974 an unanticipated
movement began. Ten of the 13 largest megalopolitan areas (the Standard Consolidated
Statistical Areas of the U.S. Census Bureau)
exhibited net population migration out of the
urban areas. Figure 32-2 shows net migration as a function of city size. From the data,
the southern-rim urban complexes show net
gains or small losses, but it does appear that
larger urban areas suffer the most out-
migration.

The changes implied by these trends are
predicated not on comprehensive planning but
on the decision of many people to engage in
activities necessitated by the realities of high-
cost energy. The shifts in settlement patterns
set out in Table 32-2 will result from the encouragement of these trends.

The encouragement we envision includes
legislative action. These acts will facilitate
the abandonment of nonviable populated
areas primarily in the megalopli and the construction of viable communities in small cities and rural settlements.[15]

A major step in the process of repopulat-
ing rural America will be a National Homestead Lease Act. This act will permit individuals to lease, at no cost, tracts of publically
owned land provided the lessee makes his
home on the land. Tracts might be 4 to 40
acres in size, depending upon the location.
Leases will be renewable indefinitely, as long
as the lessee remains living on the tract and
cultivates some portion of it. Incentives
and/or assistance might be offered at the outset. We estimate that 1 to 4 million people
will take advantage of such a program, at a
federal cost less than the current federal job
programs.[16] If 2 million people took advan-

tage of the program, there is enough public
land in unused and abandoned military bases
to support the program.

An indication that the United States has
been reconsidering the depopulation of its
rural areas and the general destruction of rural community life is the restriction of the
corporate industrialization of American agriculture embodied in the Family Farm Act
of 1974.[17] The Act prohibits any corporation
with nonfarm assets of more than 3 million
dollars from owning or operating farms. This
law comes at a time when American agriculture is on the verge of being captured by corporate interests and provides needed protection for the farmer-owner—with no loss of
efficiency. The U.S. Department of Agriculture, usually the ally of large-scale farming,
concluded in a 1968 study that the most efficient organization in American agriculture
is the one- or two-family farm.[18]

A repopulation of rural American will take
some pressure off the largest cities and preserve the desirable aspects of city culture that
have been destroyed by urban growth far beyond the human scale. The substantial migration out of urban areas we have included
as part of this scenario will leave those who
stay behind in possession of cities that could
be made very livable.

As a counterpart to the National Homestead Lease Act, Congress will expand the
National Urban Homestead Act with the
threefold purpose of: (a) providing housing
for those in the cities who cannot fully afford
it, (b) improving the inner cities by putting
to use structurally sound buildings now falling into disrepair whose replacement would
require impossible amounts of both capital
and energy, and (c) accomplishing these
things in a way that would increase the independence and equity of the individual involved. The guiding principle of the law is
the idea elaborated nearly a decade ago by
George Sternlieb, that the basic variable that
accounts for differences in the maintenance
of slum properties is the factor of ownership.[19] The idea for renewal, tried successfully on a small scale by a number of cities

in the 1970s, consists of the selling of publicly owned (abandoned) dwellings to families for a low price in return for the new owner's agreement to bring the building up to city codes within 18 months and to occupy it for at least 5 years.[20] Because few of the prospective owners will be able to obtain loans to finance the rehabilitation of their buildings, public low-interest loans and loan guarantees will be made available on the basis of maintaining future property values and the income arising from the improvements made. The program could be made to serve the goal of energy conservation. For example, in 1976, the Community Services Administration funded an experimental program in which heavy insulation and a solar water-heating system were installed in an abandoned five-story, 11-unit tenement as part of the "sweat equity" renovation of the building by its new owners.

As a necessary complement to the rural repopulation and urban rebuilding in the previous two acts, the National Conurbation Act will provide for the restructuring of present megalopoles. The intent of the Act is to develop a regional pattern of distinct cities and towns intended as no-growth entities, including farms and open space as parts of the whole. As people move out, abandoned unsound structures will be demolished with nothing built in their place.[21] What will take place next—in slum neighborhoods out of necessity, in better-off neighborhoods out of frugality or civic pride—is the conversion of these vacant areas into community gardens and orchards. The cities, straining under the loss of tax base as a result of out-migration, will welcome this use of empty lots. After a time, the trend of building rehabilitation, urban open space accrual, and new building construction will lead to multi-nucleated patterns of compact towns and communities on the sites of present-day megalopoles. This planned development with appropriate zoning ordinances will integrate the energy, agricultural, transportational, and recreational needs of a community while avoiding unnecessary transportation of food and materials from distant regions. Zoning in certain areas will allow a mix of commercial and residential uses on the same block, and in the same building complex, so that minimum travel is needed for shopping and going to work.

Thus, we envision recent trends, with the encouragement and structuring from the legislation, leading to a restructuring of settlements.

RESIDENTIAL ENERGY USE

Space and water heating consumes about 75 per cent of the total U.S. energy used in the residential sector. The rest is for cooling, lighting, air conditioning, and the use of a variety of appliances.

As many citizens have noticed in recent years, significant savings in heating fuels can be obtained by changing the allocation of heat to a home both spatially and temperally (zoned control, day and night thermostat setbacks). An estimated 25 per cent of the heating fuels can be saved and often more.[22] We expect this trend of changing behavior during the heating season to continue, since fuel prices are likely to continue to rise.

Major reductions in both heating and cooling requirements can be obtained by the retrofitting of old buildings and careful design of new buildings. Reducing infiltration, adding insulation, and improving windows of old buildings could result in an average reduction of at least 50 per cent in heating requirements.[23] Renovation of old buildings is consistent with a policy of moderate urban density without autos, since most of the older structures are relatively closely spaced. Up to a 80 per cent reduction in heating needs can be attained for new buildings by the building of smaller homes and use of sound building techniques and correct window placements to gain solar energy.[24,25]

Water heating devices are very inefficient and can take up to 15 per cent of total residential energy demand. Improved insulation, lower temperature settings, heat recovery from other appliances, and reduced hot water use can reduce this energy consumption by 50 per cent or more.[26]

We estimate that the implementation of these measures will decrease the energy consumption per capita for space and water heating to about 30 to 40 per cent of the 1975 level by the year 2050 and in many cases, a smaller percentage. Total residential energy use per capita can therefore be reduced by about 50 to 60 per cent by the year 2050. It is worth noting that space and water heating require only relatively low-temperature heating devices and can therefore be met for a significant part by simple solar (and wind) heating systems.

About 25 per cent of residential energy use is for cooling, refrigeration, air conditioning, and so on. Increasing the efficiency of appliances will have a significant impact. Furthermore, better design of new buildings will make air conditioning unnecessary in a large part of this country by the year 2050. Energy use for home freezers will increase somewhat because of the changing food system. We estimate that energy use for all these devices can be cut by about 50 per cent, or about a 10 per cent reduction in total residential energy use. Electricity will continue to be the major form of energy to be used for these tasks, with air conditioning a possible exception.

Energy-efficient building codes on the national, state, and local levels, and national and state efficiency standards and labeling laws for energy-using equipment, will help bring about these reductions in residential energy use.

The combination of the measures mentioned above will reduce the annual residential energy use per capita of about 6.9×10^{10} joules in 1975 by 65 per cent, to about 2.4×10^{10} joules in the year 2050. Total residential energy use for the year 2050 is estimated at about 6.7×10^{18} joules, of which about 75 per cent, or 5.0×10^{18} joules, is for space and water heating (low-temperature energy needs).

Figure 32-3 shows the space and water heating demand for *both* the residential and

the commercial sector over the period 1975 to 2050 (see Energy Supply).

FOOD AND AGRICULTURE

This scenario includes a regionalization of agriculture as well as changes in the average American diet that will reduce total energy use in food production by 50 per cent over the next 75 years.

Until the 1940s, most regions in the United States produced most of their own food. Specialized, large-scale food production is an invention of the last 25 years. The startling growth in home gardening and local direct-marketing cooperatives in the early 1970s suggest that a trend toward less centralized and specialized food production may already have begun. Large-scale production for export will be with us for quite a while, as developing nations struggle with food production. But only by decreasing the energy subsidy to these crops for export can the United States remain competitive as energy prices rise. It is often overlooked that U.S. food is, at present, expensive by world food cost standards (see Article 5).

The present fossil-fuel subsidy to the American food system in the form of gasoline, machinery, fertilizer, and pesticides is 15 calories for every calorie of food energy consumed.[10] As fossil fuels become scarce, agriculture of this intensity will become impossible, not only in the developed nations upon whose agriculture much of the world has come to depend but also in the less developed countries whose hopes for progress rest on the boosting of food production with machinery and chemical fertilizer.

The American diet and food system will change over the next 75 years. Beef consumption will be cut by half, and the beef that is eaten will be produced from pasture- and range-fed cattle eating grasses that humans cannot digest instead of from feedlot cattle raised on valuable grain. Swine will be fed poor-quality grain and refuse, and land formerly used to grow feed for both cattle and swine will be used to grow grain and

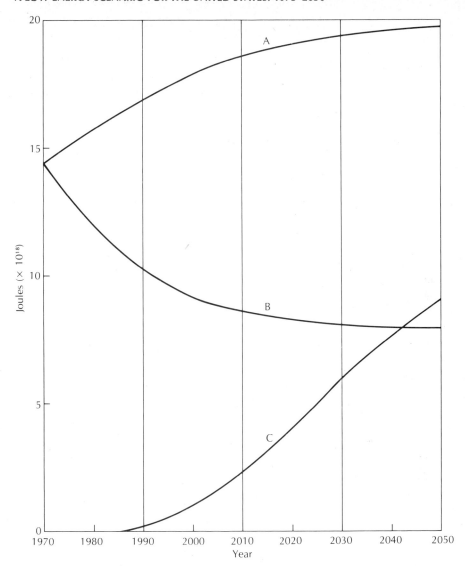

Figure 32-3. Commercial and residential space and water heating projections compared with projections of low-temperature energy supplied from solar and wind sources. Curve A, primary fuel demand for commercial and residential space and water heating at 1973 level of consumption; curve B, low-energy scenario space and water heating demand; curve C, solar and wind low-temperature energy production.

beans for human consumption, much of which will be exported. The reduction in beef protein intake will mean that protein needs can be made up by other sources. Food such as fish, chicken, eggs, soybeans, and dairy products, which have high protein conversion efficiencies and lower energy requirements, will make up the difference.

Fish will be supplied by coastal and local fishing and the extensive use of fish farms. Poultry and egg production will operate on a local basis in a far more energy-efficient manner than the current poultry factories are operated. Consumption of dry beans and fresh potatoes, which has declined in the United States in recent decades, will increase, as will the direct consumption of grains. Consumption of fruits and vegetables—90 per cent of which will be grown and eaten locally—will rise.

Even if direct consumption of grain were to increase to make up for the protein lost due to reduced meat intake, and even if per acre grain production were to drop slightly as a result of the less-energy-intensive use of chemical fertilizers, there would still be substantially more grain available for export.[27] It is not necessarily the point of this reduction in American meat consumption to provide more grain to feed hungry people around the globe. Such grain surpluses will provide reserves for year-to-year fluctuations in production and aid to famine-stricken areas, but we doubt that enough grain could be freed in this way to supply the world, and we think that, in any case, it would be just one more stop-gap measure to alleviate world hunger until population growth exhausts these grains as well. For Americans, the dietary changes will reflect market responses to escalating prices of energy-intensive foods. Maintaining moderate food costs is of considerable interest nationally, even for upper-income families who could pay higher prices, if non-food expenditure levels are not to decline drastically as food prices rise.

In addition to the anticipation that the American diet will include a greater proportion of whole grains, dried beans, potatoes, and vegetables generally, we expect that most produce will be grown by the individual or by local gardening co-ops or farmers for local consumption. It is not difficult to imagine a time when lawns will have very nearly disappeared and houses will be surrounded by gardens and orchards tended by city residents, who will gradually come to refuse to pay the high costs of commercially produced produce. We envision the effect of high food prices on suburbs to be quite startling. Areas that now look like green residential deserts will come to resemble the intensely farmed and carefully tended gardens of England or China.

The most fascinating thing about a widespread resurgence of gardening will be the huge amount of food that will be produced by an activity that seems to us today so much like an avocation. Truly, no one should be surprised by the economic value of home gardens. In fact, the U.S. Department of Agriculture has calculated that a family of four could be well fed without animal products by using only one sixth of an acre. Modern intensive gardening techniques produce still higher produce yields per unit area.

Both the National Conurbation Act and the National Homestead Lease Act support the trends to more local food production and distribution. Energy savings from these measures and agricultural energy savings (see Article 5, for some specific measures) contribute to the reductions in industrial, residential, commercial, and transportation energy use summarized in Table 32-1. The National Container Act will contribute further savings in the food system (see the following section).

COMMERCIAL ENERGY USE

The characteristics of energy consumption in the commercial sector are similar to those of the residential sector. Space and water heating represent about 75 per cent and air conditioning about 10 per cent of the total commercial energy use, all of which require only low-temperature energy sources. Conservation measures and legislation outlined under the residential sector will reduce the commercial energy consumption dramatically. In addition, significant improvements will be made by using "waste" heat generated within buildings. The smaller, nondetached stores and other moderate-size commercial buildings will make extensive use of natural ventilation. These changes result in a reduction of about 70 per cent in commercial energy needs for space heating and cooling and water heating, or about a 60 per cent reduction in the total commercial energy use per capita by the year 2050.

Additional energy savings will occur through the use of natural light and reduced lighting standards.[28] Reduced use of lighting for advertising will have the same effect. Furthermore, changes envisioned in the food system will have a significant impact on commercial energy use: phase-out of fast food services and other "junk" commerce, drastically reduced packaging using standard sizes, and more careful display of goods (e.g., no open freezer cases). The popular fact that McDonald's hamburger chain uses enough energy in a year—largely for packaging—to provide all the electricity needed for the cities of Pittsburgh, Boston, Washington, and San Francisco illustrates the American penchant for throw-away containers.[29] Now, litter and solid waste simply represent resources not being put to best use. As energy becomes more costly, a society that uses packaging and produces garbage on the scale of the United States will find it necessary to reduce the amount of this waste.

Packaging energy use will decline as disposable paper and plastic containers nearly disappear, as commodities that can be shipped, stored, and merchandised in bulk are handled in bulk lots, and as nonessential junk foods, overprocessed foods, and excessive toiletries and cosmetics nearly vanish. The National Container Act will reduce both wastes and total packaging energy. Litter and solid wastes will cease to be a problem when commercial packaging and containers are designed for nearly complete recycling and reuse. About one eighth of local trucking is for garbage and trash disposal. A reduction in the amount of trash produced means less trucking.

We estimate that the combination of these measures will reduce the commercial energy use per capita of about 5.9×10^{10} joules in 1975 by about 70 per cent to 1.8×10^{10} joules

in the year 2050. Total commercial energy use for the year 2050 is estimated at about 4.9×10^{18} joules, of which about 70 per cent, or 3.5×10^{18} joules, is for space and water heating (low-temperature energy needs).[30]

Figure 32-3 displays space and water heating demand (see Energy Supply).

TRANSPORTATION

The transportation energy savings in personal travel and freight in this scenario reduce the per capita direct energy use to 25 per cent of present levels. What is more surprising is that no reductions in social interaction or accessibility will occur. The sources of the energy savings are due to two primary factors. The first is that the redesign of settlements will mean that trips will be far shorter, and the combination of reduced length and provision of alternative modes, including safe walking and bicycle pathways, will markedly reduce the necessity of driving.[31] The second factor is that for the remaining trips that require driving, the vehicles involved will be smaller and far more efficient in terms of energy use per vehicle mile. The same factors apply to freight energy use.

The change foreseen in settlement patterns will reduce not traveling itself but rather the distances spanned and will change the mode of the trip.[32] This will be a reversal of trends in social and physical arrangements that began with the widespread ownership of the automobile and the universal willingness to sacrifice 25 to 50 per cent of urban areas to the operation and storage of the automobile. The pattern of placing numerous competing food and department stores together in shopping centers and malls accessible only by private auto will be replaced by a pattern of decentralized neighborhood stores. Schools and work places will also be decentralized throughout the community. With the emphasis on walking, cycling, and traveling by public transportation, far less land will be given over to roadways and parking, and their associated noise and emissions will be reduced. The net effect will be to make the five dwelling units per acre even more spacious than such units are now.

These changes, partially motivated by the increasing cost of operating automobiles, will reduce average annual automobile vehicle miles per capita from 8000 miles to 3000 miles. After increased use of other motorized forms of transportation is taken into account, a 50 per cent reduction in energy use is anticipated from these measures.

Precedents for this type of transportation future already exist, and others are being developed. In Rotterdam, 43 per cent of all daily trips are by bicycle. The cities of Runcorn, England, (population 70,000) and Port Grimand, France, were planned to function without autos; they have bus, bicycle, and pedestrian access exclusively, except for emergency and special delivery vehicles.

This scenario envisions a heavy emphasis on rapid rail systems for intercity travel, especially for travel under 500 miles, to eliminate the least efficient aspects of air travel. Buses and the automobile will still carry a significant part of the load of intercity travel. The requisite interlinking of these modes, especially at terminals, will be part of the National Transportation Coordination Act. Other acts will include the termination of the Interstate Highway Program (Interstate Highway Termination Act), the termination of air terminal expansion and other airline subsidies (Airline Deregulation Act), and the enforcement of speed limits on auto, air, and truck travel to energy efficient levels.

Complementary legislation will be the Railroad Revitalization Act, which will effect the rebuilding and expansion of most of the existing rail network. This long process will involve some nationalization to provide the requisite capital.

Improvement of personal communications and access will be provided by telecommunications. Many types of interactions and transactions which now require large commitments of time and energy in travel will be met by the rapidly developing telecommunications field.

With automobile passenger and vehicles miles reduced by more than 60 per cent, the next point of focus is the efficiency with which the remaining, still large number of trips are carried out. What is required here is slightly more than has been mandated in the Energy Policy and Conservation Act of 1975: 27.5 miles per gallon for the 1985 new car fleet average.[33] These cars will average 2200 pounds instead of the current 4000 pounds, and they will obtain an average gasoline mileage of at least 25 miles per gallon in town, last 20 years or 200,000 miles, and carry the driver an average of 3000 miles per year instead of the current 8000 miles. These improvements will be mandated in the Fuel Economy Act.

The fuel mix of transportation will shift away from an almost exclusive dependence on petroleum to a mix of electricity, synthetic fuels, hydrogen, and petroleum. Twenty-five per cent of personal travel energy will be electric, mostly for local automobile trips and electrified rail and bus travel. Fifty per cent of freight energy use will be electrified, with electric trains on all main rail lines and electric trucks for local delivery.

These changes in vehicle characteristics result in a reduction of over 50 per cent in energy use per vehicle mile. The combined effects of reduced vehicle mileage and improved efficiency result in a 77 per cent reduction in annual personal travel energy use from 6.0×10^{10} joules per capita in 1975 to 1.4×10^{10} joules per capita in 2050.

Similar savings are seen for freight energy use. Since recreation and farm land will separate communities, a large reduction in ton-miles will be due to changes in the agricultural sector resulting from the decentralization and localization of food production and processing (at present about one half of all trucks haul food and agricultural products). With a trend to more decentralized manufacture in industry and shifts in production noted in the industrial sector, an overall 50 per cent reduction in ton-miles will occur.

Citywide delivery systems will be required as part of the shift to rail transportation and the decline of centralized shopping centers. Intelligently done, the distribution of goods to neighborhood stores should involve little additional time or cost than delivery to centralized shopping centers. This is especially true given a shift away from cross-country truck delivery of freight, which has allowed direct delivery to the central retailer from producers. This does not occur in rail-supplied cities where local distribution from the city railhead to the various neighborhood outlets is by small trucks.

In addition, we estimate that with the elimination of 90 per cent of all air freight, and with 75 per cent of all truck freight carried shifted to rail carriage, total U.S. freight transportation energy use will be reduced 40 per cent.

The combined effects of reduced ton-mile, modal shifts, and increased modal efficiencies, especially for trucks,[34] will result in a 70 per cent per capita freight energy savings from 3.2×10^{10} to 1.0×10^{10} joules per year in 2050. The total transportation energy use, including freight and personal, in the year 2050 will be 2.3×10^{10} joules per capita resulting in a total of 6.5×10^{18} joules.

INDUSTRY

Currently, U.S. industrial production uses 37 per cent of the nation's total energy budget, or 27.6×10^{18} joules annually. This scenario envisions changes in the mix of goods and services produced in the economy. These changes will result in production reductions

in certain energy-intensive industries (Tables 32-3 and 32-4), which will save 3.8×10^{10} joules per capita, or 10.9 per cent of total 1975 energy use.[35] Overall output of goods, though changing in mix, will cease to grow on a per capita basis.

Table 32-3 summarizes the information from *A Time to Choose,*[36] the report of the Energy Policy Project of the Ford Foundation, to show the effect of a change in the output mix and associated cutbacks in production in certain energy-intensive sectors. An example of such a cutback in production is a decline in automobile production due to decreased auto ownership and use and greater auto durability. To project the effects of such changes some hypothetical reductions are considered. Each of the energy-intensive industries in Table 32-3 is cut back 30 per cent with the exception of food and kindred products and primary metal, which are reduced 50 per cent.

The changes in food and kindred products will primarily occur in the food processing parts of the industry as opposed to primary food production. The change in primary metals will be largely due to the changes in automobile production.

It should be kept in mind that these changes result from production changes, not conservation. Conservation measures are separate measures, which can stand by themselves or can be undertaken in addition to the hypothetical cutbacks. In Table 32-4, it is also assumed that overall levels of activity in the rest of the economy would remain as they are

Table 32–3. 1971 Economic Parameters of Energy-Intensive Industry.

Energy-Intensive Sector	Manufacturing Gross Energy Consumption (%)	U.S. Employment (%)	Industrial Production (%)	New Plant and Equipment Investment (%)
Primary metal	26.8	1.6	6.3	3.4
Chemical and allied	17.0	1.1	9.4	4.2
Stone, clay, and glass	7.3	0.8	2.9	1.1
Paper and allied products	7.6	0.9	3.6	1.5
Food and kindred products	6.3	2.1	9.6	3.3
Total	67.8	7.3	31.8	13.6

Table 32–4.   Economic Parameters of Table 32–3 Industries—After Production Change.

Energy-Intensive Sector	Manufacturing Gross Energy Consumption (%)	U.S. Employment (%)	Industrial Production (%)	New Plant and Equipment Investment (%)
Primary metal	19.3	0.8	3.7	1.8
Chemical and allied	16.7	0.8	7.7	3.1
Stone, clay, and glass	7.2	0.6	2.3	0.9
Paper and allied products	7.5	0.6	2.9	1.2
Food and kindred products	4.5	1.1	5.6	1.7
Total	55.2	3.8	22.2	8.7

shown in Table 32-3. Such an assumption cannot be defended except as an heuristic device for interpreting the magnitude of the effects. By the same token, it should be noted that the cutbacks do not take into account secondary effects. With these qualifications, the cutbacks noted will:

1. Save 29 per cent of manufacturing gross energy consumption, or approximately 9.7 per cent of total energy use.
2. Add 3.4 per cent to unemployment.
3. Reduce industrial production 12 per cent.
4. Save or make available 5.5 per cent of new plant and equipment investment capital.

Aside from these changes in industrial production, a significant reduction in industrial energy consumption will be obtained by improved efficiency. Strenuous energy conservation measures will reduce the energy use after the change in production by 3.0×10^{10} joules per capita, an additional 33 per cent. This projection is estimated from data from Ross and Williams[26] (see also Article 29), who based their study largely on the concept of second law of thermodynamic efficiency as outlined by the American Physical Society.[37] Higher efficiencies will be obtained by careful matching of the energy quality or temperature of energy supply and demand. The introduction of relatively small-scale, multipurpose power plants as outlined under Energy Supply will play a major role. We expect that fuel-use taxes will be placed on

industries in the near future to discourage the use of oil and natural gas (Depletable Fuels Tax Act), which will make these multipurpose power plants economically attractive for industry. Table 32-5 shows our estimates of the energy conservation effects in the industrial sector that will be obtained by the year 2050.

The total reduction in industrial energy use due to the change in production and the energy conservation is 6.8×10^{10} joules per capita, which is a 52 per cent reduction from the 1973 to 1975 level of industrial energy use by the year 2050. Total industrial pri-

Table 32–5.   Projected Energy Savings in the Industrial Sector by Conservation Measures, Based on Data from Ross and Williams.[38]

Measure	Reduction (%)
Good housekeeping measures through industry (except for feedstocks)	13
Fuel instead of electric heat in direct heat applications	0.6
Process steam and electric cogeneration	8.7
Heat recuperators or regenerators in 50 per cent of direct heat applications	2.5
Electricity from bottoming cycles in 50 per cent of direct heat applications	1.7
Recycling of iron and steel in urban refuse	0.4
Reduce throughput at oil refineries	3.0
Reduced field and transport losses associated with reduced use of natural gas	2.7
Total reduction	≈ 33

mary energy use in the year 2050 will be 17 $\times 10^{18}$ joules.

ENERGY SUPPLY

Decreasing per capita energy consumption by 64 per cent postpones but does not solve our energy supply problems. Our increasing population combined with drastically decreasing production of oil and natural gas will put considerable pressure on other energy supplies. In this scenario, solar and wind sources will supply a large fraction of our energy needs. Other alternative supplies, such as wood, bio-fuels, geothermal, and steam, will also grow substantially. Less coal, with its associated environmental problems, will be used by the year 2050.

The 2050 electric generating system will be substantially different from that of today. We envision the decline of the nuclear industry. As a result of rising costs and public concern over safety, nuclear power plant construction halts in the mid 1980s, and the breeder reactor fails because of cost, technological problems, and the threat of nuclear proliferation. Solar, wind, water, and geothermal electric generating systems will grow to supply most of the demand for electricity. Marginal cost pricing combined with other load management techniques will be used to tailor the demand curve to follow supply. Daylight peaks may be encouraged to make use of the cheapest solar production stations. As cryogenic storage and electrolysis become less expensive, solar energy may assume a base-load function. Hybrid plants, such as coal-solar or coal-wind, will be built, and wind will be used to generate electricity on a centralized as well as a decentralized basis. The trend in coal plants will be toward small (less than 300 megawatts) multipurpose plants. Industries will generate their own process heat and electricity, selling the excess as district heating expands. The transition to the proposed system will be gradual. As currently operating plants become obsolete, new plants of the types described will take their place. Diverse and innovative electric generation will be the result.

This transition will be facilitated in several ways. We expect to see nuclear construction moratoria and an end to nuclear subsidies and to the federal commercial breeder program. Strong local opposition to nuclear plants is already a frequent occurrence. The cost of solar and wind energy systems, both thermal and electric, will drop as a result of improved techniques and mass production. In some cases, demonstration projects will provide the necessary production demand. A tax and loan guarantee incentive program for the use of renewable (flow) resources on state and federal levels will help this transition. A national strip mining act will set stringent standards for land restoration, so that the continued use of coal will not seriously scar the landscape. Finally, a depletable-fuels tax will be levied on most uses of oil and gas as fuels. It will be a progressive tax, increasing steeply for uses for which there are alternatives. The naturally increasing costs of depletable fuels combined with this tax will improve the economic viability of alternative energy sources.

In our scenario, multipurpose power stations will generally be used in industrial and moderate- and high-density residential and commercial areas. The industrial sector now consumes about one third of the electricity generated in the United States. This electricity is mainly used for motors, lighting, and electrolysis (high-quality energy needs). In addition, about 45 per cent of the total industrial energy consumption is used to generate process steam. Electricity could be produced by the industries in combination with steam generation. Gyftopoulos et al.[39] calculated that with the present requirements for process steam, enough electricity could be produced as a by-product not only to meet industrial electricity needs but to have a surplus to sell. The same type of power plant can also supply both electricity and heat to the residential and commercial sectors, where they can be adapted to the energy-quality needs of the area. We expect that the general increase in fuel prices will make these small but relatively efficient "total energy systems" cost advantageous in the near future. We estimate that coal will be used to fire many of these plants in the industrial sector and some in the residential and commercial sector. Total coal consumption will be slightly lower than the total industrial energy consumption by the year 2050.

Space and water heating by solar energy will be widespread by 2050 and will supply, along with wind energy, most of our low-temperature energy needs. The rapid growth in solar heating has already begun. In the last two years, the number of firms engaged in solar energy production has grown by one order of magnitude. Figure 32-3 illustrates our space and water heating requirements and our estimate for low-temperature solar and wind applications. By the year 2020, nearly all buildings will have solar or wind space and water heating. Approximately 16 per cent of total U.S. energy use is for industrial process heat below 350°F,[40] which can be easily accommodated by solar and wind technologies.

Central station solar thermal electric generation will be available on a commercial scale by 1985. The early solar plants will most likely be coupled with coal-fired plants. These hybrid plants will eventually give way to solar base-load production when cryogenic storage and electrolysis become economical.

One of the few technological breakthroughs we see in our "future" is the declining cost of photovoltaic cells. We forecast that, by 1990, the cost for photovoltaic electric generation will be $1000 (in 1976 dollars) per installed kilowatt of capacity—a little less than the current cost for nuclear electric plants. We base our estimate on the current trend in photovoltaic costs and the expected breakthrough in the edge-defined film growth (EFG) production technique of silicon cells and the refinement of gallium arsenide (GaAs) cells that are capable of high efficiencies (20 per cent) in high-temperature applications.[41] Table 32-6 shows current and

Table 32–6.　Photovoltaic Cost Projections.

Year	Cost per Peak Kilowatt (1976 dollars)	References
1959	200,000	42
1976 (March)	21,000	43
1976 (October)	15,500	43
1979	5000	43
1986	500	42–44

projected photovoltaic costs for panels only. Figure 32-4 is a graph of total solar energy supply projections.

Wind energy will be utilized in central as well as dispersed applications with the capability of supplying a significant part of the electrical capacity in the year 2050. Included in wind systems are generators ranging in size from a few kilowatts for individual homes to installations of 1 to 2 megawatts connected into electrical grid systems to meet part of the electrical demand. The capital costs per installed kilowatt capacity of wind machines and conventional electric generators are now of the same order of magnitude. Wind electric systems with relatively short-term storage systems or an array of wind machines spread over a large area appear to have approximately the same reliability as large conventional generating units.[48] Wind energy conversion systems will also serve as space heating devices for a large area of the United States, since the seasonal availability of the wind corresponds with space heating needs. Present trends of rapid growth in electric space heating[49] could continue, to some extent, with large wind electric systems, if combined with centrally controlled heat storage devices in individual buildings. In addition, individual wind machines for neighborhoods and small towns will be used as "wind furnaces," without the interconnection with the electric grid. Combinations of these wind thermal and electric systems will also be employed. The economics of wind furnaces appears to be as good as that for solar heating devices. No major technical breakthroughs are needed for large-scale applications of

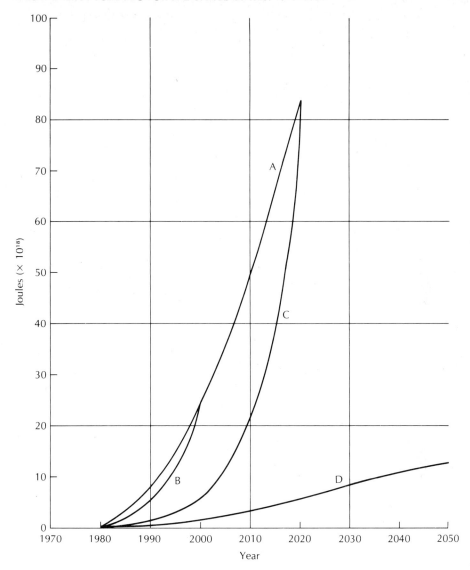

Figure 32-4. Projections of total solar energy supply, 1970–2050. Curve A, data from Morrow[46]; curve B, data from Business Communication Corp.[45]; curve C, data from Solar Energy Panel[47] (includes wind); curve D, low-energy scenario (includes wind).

wind energy conversion systems. Table 32-7 illustrates the role of wind energy conversion systems as a major energy source in the future.

Table 32-7 and Figure 32-5 summarize energy supply in various stages of our scenario. We have separated low- and high-temperature energy supplies to further illustrate the necessity of thermodynamic matching of supply and demand.

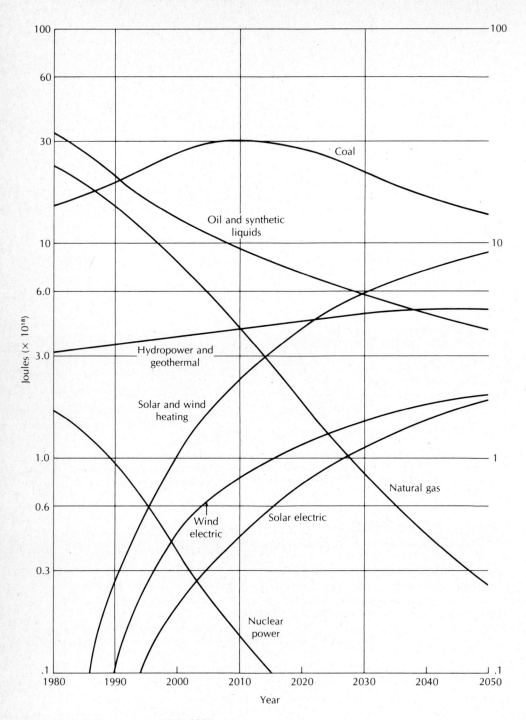

Figure 32-5. Energy supply, 1980–2050.

Table 32–7.　Overview of Energy Supply During the Period 1975 to 2050 (10^{18} joules).

	1975	1980	1990	2000	2010	2020	2030	2040	2050
Solar electric, including photovoltaic	—	—	—	0.20	0.43	0.80	1.1	1.5	1.9
Wind electric	—	—	0.10	0.44	0.80	1.1	1.5	1.8	2.0
Hydro and geothermal	3.3	3.4	3.5	3.8	4.1	4.4	4.7	4.9	5.0
Solar and wind heating	—	—	0.25	1.0	2.3	4.0	6.0	7.6	9.0
Coal	14.1	15	19	26	27	25	21	17	13
Oil, imported[a]	12.9	12	7.0	2.0	—	—	—	—	—
Oil, domestic	21.4	21	14	11	9.5	7.0	5.5	3.4	2.0
Synthetic liquids[b]	—	—	—	0.10	0.12	0.20	0.50	1.2	2.0
Natural gas	21.1	21	15	8.3	4.0	1.8	0.80	0.44	0.25
Nuclear electric generation[c]	1.6	1.6	0.95	0.37	0.15	—	—	—	—
Total[d]	74.4	74	60	53	48	44	41	38	35

[a] No oil for U.S. import available approximately 10 years after expected world peak production.

[b] Synthetic liquids from coal, oil shale, biological waste (bioconversion), etc., will increasingly be used for transportation, combined with domestic oil.

[c] Last nuclear power plant construction in 1985; approximate lifetime, 30 years.

[d] Figures may not add up because of rounding.

EMPLOYMENT, SOCIAL WELFARE, AND LEISURE TIME

We can be sure that any attempt to control energy consumption by restraint of economic output will raise cries of opposition from those concerned with maintaining employment, increasing profits, and getting a piece of the economic pie—labor, industry, and the poor, respectively. A clue to a stable employment policy for a time of economic reorganization in response to reduced supplies of energy may be supplied by the example of Sweden. Sweden has been able to "rationalize" its industry, i.e., carry out structural changes in industries and adapt new techniques of production, and by doing so maintain a highly efficient economy. This has been possible, according to Gunnar Myrdal,[50] because its firm national policy of full employment, in the form of guaranteed jobs, effective retraining programs, and smooth relocation of labor, has effectively banished the threat of technological unemployment. Were the United States to take full employment seriously and guarantee jobs to everyone wanting them, the fears of organized labor of economic reorganization could be

allayed and the transition to a low-energy economy made smoother.

We have set the average work week in 1990 at 30 hours to reflect two changes: a 25 per cent decrease in work time for the average individual, accompanied by a 25 per cent decrease in real income that is offset by an increase in (potentially profitable) leisure time. There would then be a 30-hour work week available for a large bloc of the formerly structurally unemployed, who would fill the work hours left by reduced work time. An average reduction in work hours of 75 per cent would presumably provide employment opportunities for one new worker in every three now employed, assuming no change in total production and labor productivity. The 1976 labor force of 84 million could thus be expanded by some 28 million, just to maintain current output. The Full Employment Act will provide the needed impetus by establishing a 30-hour work week for federal and state jobs by 1985.

How, it will be asked, can we be so sure that people will agree to lose one fourth of their income in exchange for more free time? We cannot be sure, of course, but we may at least interpret, with Dennis Gabor,[51] the increased and endemic employee absenteeism of recent years as a sign that a significant bloc of people have already reached the point where time off does mean more to them than the dollars they forgo in not going to work. The plain fact is that the opportunity has never before existed except for low-paying part-time jobs. One is expected to work the customary full week or not at all. The increased time available may indeed prove more profitable than the equivalent time spent on the job. Some people will want the unpaid time off not just for leisure, but to be able to have time to perform services for themselves. Conceivably this will result in the service being completed more cheaply than if the person remained on the job during that period and paid for the service. Will the commercial and industrial establishment be able to handle the increased numbers of employees? We believe so, since the 1976 auto workers' contract effectively results in nearly one fourth of the year as paid time off. The institutionalization of the option to take one's wealth in the form of leisure time and individually performed service time instead of goods may be the way for a society to smooth the transition to a situation of reduced energy availability. Finally, a society that is at last forced to substitute the serious redistribution of wealth for economic growth as a means of eliminating poverty may find a reduction of the average work week that spreads aggregate labor hours among more individuals to be a quite palatable means of redistributing wealth.

Substantial interregional and interindustrial shifts of labor will take place as part of the adjustment to a low-energy society. The transfer of population and industry from the North and Northeast to the South and Southwest will continue (see Settlement Patterns and Changing Communities), bringing the new industrial states into greater prominence. The factors that encouraged this shift—favorable tax structures, lower wage rates, and a generally unorganized labor force[52]—will produce a less than acceptable situation for workers for as long as it takes

for union strength to develop and state regulatory authority to be applied. The Appropriate Skills Retraining Act, which provides increased federal support for trade and technical schools teaching needed new skills, will ease this transition. The demand for new housing would provide the opportunity to build good, energy-efficient homes for the influx of workers. At the same time, this drain of people from the older metropolitan areas would reduce the population to provide the opportunity to renovate and rehabilitate the older cities, increasing the demand for the labor of those remaining.

Interindustry shifts of labor will be significant. The housing industry, actively trying to fill the need for energy-efficient dwellings to replace wasteful older buildings, will grow still further in response to the need for building reconstruction, rehabilitation, insulation, and solar retrofitting. In a report prepared for the New York State Legislative Commission on Energy Systems, it was noted that conservation measures will lead to increased employment. By saving 1 million kilowatts (1000 megawatts) through energy conservation at $300 per kilowatt, the employment level would be 12,000 people on a 30-year cycle, slightly less than a nuclear or coal plant would employ to generate 1000 megawatts,[53] but significant numbers of people would be employed immediately in the conservation effort. Efforts to conserve energy in existing buildings would employ electricians, carpenters, plasterers, painters, truck drivers, factory workers, engineers, and heating, ventilating, and air-conditioning specialists and other building-trade workers. As the investment per conserved kilowatt increases, so does the employment. It becomes evident that conservation of energy does not reduce employment, as many have claimed, but will instead increase employment.[54]

The rehabilitation of railroads and the construction of mass transit systems will also require an increased labor force. Bezdek and Hannon report the following. Total employment would increase and energy use would decrease if the highway trust fund were reinvested in any of several alternative federal programs; if construction monies were shifted from highways to railroads, the energy required for construction would be reduced by about 62 per cent and employment would increase by 3.2 per cent. Passenger transport by railroad required more labor and less money and energy than transport by automobile in 1963. A similar conclusion was reached in a study substituting buses for automobiles in urban areas.[55] Freight transport by railroad was less expensive in terms of dollars, energy, and labor requirements than was truck transport in 1973. If the monetary savings had been absorbed as a tax and spent on railroad and mass transit construction, about 1.2 million jobs would have been created. Had there been a full shift from intercity car and truck transportation to transportation by railroad, with the savings spent on railroad construction, 2.4 million new jobs could have been created in 1963.[56] This is the trend we envision in railroad and mass transit requirements.

The manufacture and installation of central as well as single-unit solar and wind energy systems will be a burgeoning industry, successfully taking up the slack labor from the declining automobile and fossil-fuel industries, the exception among which will be coal, which will grow in the interim to supply that part of electrical demand not met by renewable resources. Wind generators require no fuel, but the operation and maintenance of a large wind system would require two to four times the labor force on a continuous basis as would an equivalent nuclear or coal power plant.[54] Increasing reliance upon solar energy is likely to have substantial employment benefits. Even though some of the components will undoubtedly require large production assembly, many of the collectors can be produced locally and certainly installed locally with a relatively small capital investment. The installation of solar systems is a labor-intensive process. The FEA's Project Independence Task Force estimates that the solar energy industry could create half a million jobs by the end of this decade.

The need for labor in recycling will increase as this industry expands to handle not only metals and glass but also wood, paper, and cloth containers and to transfer urban sewage sludge to agricultural and silvicultural uses. This new employment will take up the slack created in the throw-away packaging and container industry.

As energy becomes more and more costly, labor will return to agriculture, horticulture, and silviculture, though in new, less physically demanding roles. We noted earlier some of the changes that may take place in an economy working under the new constraints of energy supply, and we suspect that the makeup of organized labor will be changed as well.

POLICY IMPLICATIONS AND IMPLEMENTATION

How are the economic and social changes we forecast to come about? Existing trends and logical thinking will not be enough. Public and private policy changes will be necessary. Although government intervention in our lives has come to be resented, we cannot escape the fact that in the complex society in which we find ourselves, it may be a necessary evil. The specific proposed policies we have discussed in the text are summarized in Table 32-8. Many of the necessary measures have already been considered in some form. None require massive political changes, although many will not see an easy passage. No attempt has been made to be all-inclusive in describing policies that would have an impact on energy use. For example, defense programs can be very energy intensive, but they have not been included.

The direct as well as the indirect impact of these policies must be considered. The effect that they will have on our social, economic, and political environment is a crucial question. How is the striving for material gain to be relaxed, and will people acquiesce to high prices and a reduction in material consumption? Why will individuals cease to be acquisitive? We do not suppose that they will,

Table 32–8. Policy Directions and Timetable.

Policy Area	Description	Implementation Level	Timetable
Transportation	Railroad Revitalization Act—regulatory reform and partial nationalization	National	1978–2000
	Fuel Economy Act—higher fleet mile per gallon requirements, tax penalties for poor economy, and strict speed enforcement	National and state	1976–1985
	Interstate Highway Termination Act—end of federal highway building program and redirection of funds to mass transit and railroads	National	1978–1985
	National Transportation Coordination Act	National	1978–1985
	Airline Deregulation Act	National	1978–1979
Energy and fuel supply	Nuclear power moratoria	National, state, and local	1977–1985
	End of federal nuclear subsidies	National	1977–1985
	Solar and wind incentive programs—tax write-offs, loans, research, product standards, education, and demonstration projects	National and state	1977–1990
	National Strip-Mining and Restoration Act	National	1977–1980
	Depletable Fuels Tax Act	National	1977–1990
	Synthetic Fuels Standards Act—quality and transportation standards	National	1985–1990
Economic reform	Gross National Product redefinition incorporating externalities	National	1977–1985
	Guaranteed Income Act	National	1980–1985
	Corporate Reform Act—effective antitrust regulations, more progressive income tax, and incentives for small businesses	National and state	1978–1985
	World resource price stabilization through "indexing to industrial products and food prices"	International	1980–1990
Energy and resource conservation	Energy-efficient building codes	National, state, and local	1977–1980
	Efficiency standards and labeling laws—all energy-using equipment	National and state	1977–1985
	Marginal cost pricing for electricity	National and state	1977–1985
	National Container Act—standard, recyclable containers; state Container Acts	National and state	1977–1985
Industrial	Multipurpose generation laws—cogeneration and sale of steam, heat, and electricity by industries	National and state	1980–1990
	Utilities cooperation acts—public ownership and energy sales to utilities	National and state	1977–1995
Employment	Full Employment Act—30-hour work week for federal and state jobs	National and state	1980–1985
	Appropriate Skills Retraining Act—increase in federal support for trade and technical schools teaching needed skills	National	1977–1985

Policy Area	Description	Implementation Level	Timetable
Employment	Appropriate Technology Act—funding for appropriate technology research and development	National	1977–1985
Land use	National Homestead Leasing Act	National	1980–1985
	National Urban Homestead Act	National	1980–1985
	National Conurbation Act	National	1990–1995
	Mixed-use zoning and taxation laws	State and local	1980–1990

but we suspect that they will accommodate their wants to their diminished means over time. Economic wants are not unlimited, and the premise of the industrial revolution—that wants expand indefinitely—is simply not so. We would like to think that the "drive beyond consumption" noted by W. W. Rostow is appearing frequently and that the pursuit of Maslow's "higher needs" was the message of the 1960s. All of this would ease our task, but in the end we rely on a conventional mix of economic and legislative measures.

Economic growth has been prescribed as a means of redistributing income and eliminating poverty through the provision of jobs in a bustling economy. Robert Lampmann argued in the late 1960s that economic development had improved the general lot of the poor.[57] But we note that for a significant segment of the population a good part of the time the usual means of securing an income are not effective; there are no jobs. Even in boom times there stubbornly persists a class of "structurally unemployed" living on some form of relief, and usually in slums. Adherence to the idea of aggregate economic growth may have its justifications, but belief in its usefulness in alleviating poverty is among the poorest. The "trickle-down" theory, if it works at all, is not fast enough or equitable enough to meet the needs of all the poeple in the short time available to make the transition to a low-energy economy. In a future of energy and materials austerity, the sort of GNP growth we as a nation have come to expect will be impossible to maintain and will hardly lend itself to the redistribution of income.

We are more optimistic than Lord Keynes, who noted with dread "the readjustments of the habits and instincts of the ordinary man, bred into him for countless generations, which he may be asked to discard within a few decades." We do not think a calamitous transition is inevitable as long as no individuals or no segments of society is grievously displaced or left destitute and as long as the populace generally feels that everyone is bearing his share of the complications, inconveniences, and troubles of the conversion to a low-energy economy. Destitution alone has never been a cause for revolt; history seems to indicate that poverty combined with rising expectations is the formula for revolution. By the same token, a middle class fearful of the loss of jobs and income and led by the proper demagogues to perceive both domestic and foreign "conspiracies" causing and profiting from their hardship is the recipe for an ugly strain of fascism. Individual economic security must be assured. Once it has been, the gradual erosion of the "commodity standard of living" will be accommodated in the same way that fiscal inflationary erosion is accommodated now.

The value of a guaranteed annual income (GAI) in a world of limited growth may be greater than any of its early proponents could have suspected. Proponents of the GAI continually explained how it would increase the opportunities for study, social services, and environmental improvement.[58] One can imagine its prospective value as a damper on economic growth. An income program that adequately supplied what we now refer to as necessities, combined with the option of additional work to pay for luxuries, might generate a class of people who would be willing to live at a lower income level and forgo these luxuries if to get them meant more work. The obvious outcome of this is reduced economic demand and reduced energy demand.

If we are able to assure everyone's economic security, what then? We may take some direction from Abraham Maslow's theory of a hierarchy of needs. He believed that as physical needs are met, higher needs such as love, learning, and self-actualization become our driving forces.[59] We have difficulty talking about transcending needs in the United States, a nation at the pinnacle of wealth, while the cities rot, decent housing is not to be found for many, and adults and children suffer from malnutrition and inadequate basic medical care. Our policies must take the inequities of our present society into account. Only when we are all assured of the basics of life may our higher needs be met. In this process our traditional beliefs, economics, and politics may have to change. In this scenario, there is not a cause and effect relationship, but policy does affect values and values affect policy in a process that has already begun.

INTERNATIONAL CONSIDERATIONS

The most obvious international consequence of this scenario is the reduction in a gradual and orderly fashion of U.S. payments for foreign oil imports. Only the increase of arms sales abroad from about $1 billion to more than $10 billion annually (1970 to 1976) has prevented far more serious balance of payment deficits. How long can we or should we be the world's leading arms supplier? The end of U.S. oil imports will come shortly after 2000 (Table 32-7), about the time world oil production is expected to peak and begin to decline.

Of course, the national security arguments for ending oil imports would be served by this scenario, but we have never been persuaded by these arguments, considering the dependence of the United States on foreign sources for more than 30 critical raw materials.

The nonnuclear future of this scenario offers several international advantages: (a) the effort begun by President Carter to eliminate proliferation of nuclear weapons via nuclear power would be possible without agreements or inspection; (b) the United States, in trying to persuade others to forgo nuclear enrichment technology, would be believable by example; (c) nuclear waste disposal and nuclear accidents would disappear as problems; and (d) the United States could become the leader in solar and wind technologies.

Increasing world oil prices of the early 1970s were an even greater burden to the poor nations of the world than to the United States. With limited foreign exchange and plans for development resting on oil-dependent technology, many poor nations have accumulated growing debts that may threaten the international banking system. At best, plans for development may be delayed by the drain on foreign exchange represented by oil. Are we really helping such nations by urging energy-intensive agriculture on them and offering nuclear power plants and other costly and energy-intensive technologies to them?

Opinions vary widely about the responsibility of the industrialized world to assist in the development of the Third World, but few would defend policies that inhibit attempts by poor nations to effect their own development. If, as poor nations strive to better their lot, we are perceived as outbidding them for scarce resources, conflict is a likely result. How much better to offer developed solar, wind, and geothermal technologies, which may adapt more easily to labor-rich Third World countries, than to offer capital-intensive nuclear plants or oil-dependent machinery. In any case, eliminating U.S. fuel imports will help preserve remaining stocks for developing nations.

We have no illusions about the voluntary redistribution of the wealth of the world, but this scenario offers the chance for a future profitable to both rich and poor. The alternative may mean confrontation between the rich few and the destitute many.

EPILOGUE: THE LOW-ENERGY FUTURE—SIX YEARS LATER

We have been encouraged to update *A Low-Energy Scenario for the United States: 1975–2050*. An update would permit us to include recent trends and the knowledge accumulated since the scenario was originally created, or, as we put it in 1975, "invented." We have chosen not to follow these suggestions. Despite emerging trends and new knowledge which we have monitored with keen interest since 1975, our basic themes and message remain largely unchanged. Furthermore, attempting to "invent the future" differs from traditional scientific inquiry. The scenario gains credibility more by standing the test of time than by including the latest bits of data or marginal advances in knowledge.

Six years ago we began serious efforts to assemble a comprehensive low-energy scenario. During a graduate seminar on energy policy in the fall of 1975, conversation kept returning to energy policy and its purposes. The conventional forecasts of energy needs were extrapolations of the past with minor modifications, suggesting that the future would be like the past—only more so: a most unlikely outcome if the present was any index. Cities needed to change past patterns and improve their decaying cores; minorities and the poor hoped and worked for change; 30 years of farm programs had been drastically reduced—portending changes; a national commitment was under way to reverse a century-long decline in air and water quality, despite continuing conflict about how rapidly and vigorously to pursue these programs; and, seemingly quite apart from programs and policies, demographic changes were under way that were accompanied by obvious changes in lifestyle. In short, what we saw were changes that—whatever else might happen—were not extrapolations of the past.

More specifically, domestic oil and gas production had peaked and seemed hardly likely to go up in the future. Project Independence as policy was blandly oblivious to much of the politics of the real world. (If, after all, we achieved energy independence, what were our foreign policy partners in Europe and Asia to do?) As a last gasp of extrapolation policy, Project Independence was headed for a fast trip to the science-fiction shelf. The scenarios of the Ford Energy Policy Project had hardly been published before the highly optimistic estimates of recoverable oil and gas in the United States that underlay them were revised drastically downward by the U.S. Geological Survey, which had provided the estimates in the first place.

It had long been clear that, as a purely economic matter, conservation of energy was the cheapest incremental way of coping with energy shortages. It was also the only thing that could be done quickly. What was the focus of vigorous dispute in 1975 was how much conservation and how fast. The American Physical Society summer study that year had presented the possibility of dramatically reducing energy use by a better matching of energy expenditure to the specific tasks to be performed. The Energy Policy and Conservation Act of 1975 was under debate, it ultimately passed the Congress, instituting a schedule of automobile fuel efficiency, among other provisions. The estimates of energy conservation, the lengthy reports on energy policy, and even the laws passed, all dealt in part, and implicitly, with changes in the way we live; on the more explicit question of *how things ought to be,* however, they were mostly silent. The noisy public discussion was concerned more with placing blame for problems than with visions of how to solve the problems.

In this setting, we began, with our fellow students in the seminar, to generate a picture of how little energy use might be contemplated without either basic changes in human nature or astonishing technological breakthroughs. This scenario is just one of a great many scenarios that could have been generated with these two strictures. We found that we could not avoid a number of subjective choices for which science provides no guidance. A "value-free" scenario is impossible, and forecasts that purport to be value-free simply choose to hide their assumptions at an earlier and unstated stage of the work.

The major trends we spoke of appear to be continuing: settlement patterns are changing, with emigration from the big cities of the North and Northeast and the continued growth of the southern-rim cities; rural towns are growing (some unhappily, as resource exploitation operations overwhelm them), and renovation is reviving the pedestrian city. Minneapolis is one example of a city with a declining population whose remaining residents continue to rehabilitate old houses and grow gardens on vacant lots. Other anecdotal examples come from Boston, San Francisco, Philadelphia, and Washington, D.C. But in some other areas (for instance, Houston), exurban development continues wildly and major projects for rehabilitation of urban centers are not always successful. There are gradually increasing signs of concern over uncontrolled sprawl, and at the same time the basic extensions of roads, sewers, water, and other services have been subjected to more scrutiny and citizen input almost everywhere.

A surprisingly large number of specific trends and policies anticipated in the scenario have come into existence. The list includes airline deregulation, the end of the interstate highway program, nuclear power moratoria (a number of states have a formal moratorium on new plants, and there has not been a single order for a new plant in the United States in the last two years, which constitutes a de facto moratorium nationwide), energy-efficient building codes adoption in a number of states, a national strip-mining act, national appliance energy labeling, marginal cost pricing for electricity in numerous states, rules governing the selling of electricity into utility grids from industry cogeneration, and, in a number of states, bottle bills. More generally, many of the anticipated conservation trends in the industrial, commercial, agricultural, residential, and transportation sectors have taken place, with industry leading the way.

Because the trends in the scenario and in actual experience have closely coincided since 1975, we could expect that the actual use of energy in 1980 would closely correspond to our scenario for 1980. Indeed, the figures are very close—far closer than most other analyses done at that time using sophisticated forecasting techniques. Table 32-9 compares 1980 actual values to 1975 actual and 1980 scenario values from Table 32-7.

In our scenario, we felt that 1980 was too early to have a noticeable contribution from solar electric, wind electric, solar and wind heating and synthetic liquids, and the 1980 data bear this out. But the growth in wind electric and solar applications to space and water heating is continuing rapidly. Unlike six years ago, some of the largest utilities are testing wind for electric generation. The growth in solar heating of space and water has outstripped government projections. The 1980s will test whether this growth can continue. We are optimistic because opinion polls show, time after time, the overwhelming public desire for solar (conservation finishes second and all else is far behind). Hydro has held its share by growing slightly, and although geothermal has nearly doubled, it still is a very small component.

We expected an increase in domestic coal production and use; the increase occurred, although more rapidly than indicated in the scenario. The production and use of natural gas, imported oil, and domestic oil were expected to decline slightly between 1975 and 1980: natural gas and domestic oil levels are about the same as in 1975, but imported oil grew by a larger amount. Imported oil volume, however, declined 27 per cent between 1977 and 1980, a trend that we expect to continue at a moderated rate. The slight increase in domestic oil production between 1975 and 1980 is attributable to the Alaskan pipeline production, which in 1978 resulted

Table 32–9.   U.S. Energy Supply (10^{18} joules).

	1975	1980 Low-Energy Scenario	1980 Actual[a]
Solar electric, including photovoltaic	—	—	—
Wind electric	—	—	—
Hydro and geothermal	3.3	3.4	3.3
Solar and wind heating	—	—	—
Coal	14.1	15	16.4
Oil, imported	12.9	12	14.1
Oil, domestic	21.4	21	21.6
Synthetic liquids	—	—	—
Natural gas	21.1	21	21.5
Nuclear electric generation	1.6	1.6	2.9
Total	74.4	74	79.8

[a] Energy Information Administration.

in a significant one-year growth in production within an overall context of declining domestic production; despite the Alaskan contribution, 1980 domestic oil production was 10 per cent below 1970 production. Production of natural gas is struggling to maintain a constant level, which in 1980 was 11 per cent below 1973 production. Our scenario for nuclear electric generation in 1980 is considerably below actual data because we did not adequately account for plants under construction; however, our assumption that nuclear power plant construction will end in the mid-1980s as a result of rising costs and public concern over safety (and, we would now add, excess utility reserve margins in much of the United States) appears to be plausible. For two years there have been no orders for new units in the United States, and numerous units have been cancelled.

The conclusion to be drawn from Table 31-9 is that nothing has happened in the last five years to make the low-energy scenario implausible. What we see suggests that much of what is in the scenario is coming about.

One of the strong points of our scenario (we think) is that we tried to approach the solving of energy problems by looking at the secondary and tertiary influences on energy use: do not just make cars more efficient—obviate, as well, the need to use them so much by building residences near work places and by relocating, when necessary, commercial and recreation areas. Our axiom "maximum accessibility, not maximum mobility" remains an imperative. We spoke of the transformation and decline of the gigantic shopping malls that have sprung up around the country and that are generally accessible only to cars. Since we wrote about that problem, the shopping center and mall have come under increasing attack, as formerly bustling downtown areas (primarily in smaller cities) are drained of consumers and hard times beset the businesses that remain.

The Homestead Lease Act and other measures to accommodate "back to the land" folks seem far from enactment, and despite occasional successes some of these efforts appear discouraging. However, population flow into small towns and rural areas continues—in North Carolina, New England, Wisconsin, Arkansas, Colorado, and elsewhere.

We still insist that neighborhood stores are fine and necessary in the pedestrian city, but much more is needed to broaden consumer access to a wide range of goods and services. We note, for instance, that doctors' house calls are making a comeback and phone-order grocery service is again available in most large cities. (In Minneapolis one such service is competitive with the supermarkets in both quality and price, and of course they deliver.) The advent of the moderate-cost home computer could enable the people in our scenario to enjoy the consumer accessibility they have today with little of the commuter inconvenience. Two-way cable TV, a reality now with the QUBE system in Ohio and with View-Data in England, seems finally ready to penetrate the mass markets of the world.

The automobile, it seems, will be with us for a long time to come, but the combination of fleet fuel economy standards and increasing fuel prices is producing major reductions in gasoline use that will continue through the 1980s. The current availability of automobiles capable not only of meeting the 1985 flat standard of 27.5 miles per gallon but, in some cases, of doubling it makes us optimistic about meeting the scenario. Indeed, the energy use envisioned in the scenario may be met with fewer reductions in the use of the private automobile than we originally thought.

Although the automobile will continue to be an integral part of most American lifestyles, its role in the economy will diminish. As we anticipated in the scenario, both the output and the capacity of the automobile and steel industries in the United States have contracted in the last two years. This is partly due to energy concerns and prices, but is also due to the transition in the automobile industry from a growth market to a replacement market.

For renewed rail service, the present prospects are less bright. While there remains ample time for the Railroad Revitalization Act in the scenario, present moves are in the opposite direction. Rail developments in Europe and Japan are very different. In France, for example, passenger service at more than 100 miles per hour now operates routinely and faster service is to come soon.

Amtrak, despite increased patronage (by the summer of 1981 reservations were difficult to get), calls for larger subsidies. No one seems to notice that public tax money provides more than twice the Amtrak subsidy just for the air traffic control system and that state and local taxes provide still more subsidy just for auto parking. But Amtrak may be the wrong way to go. Saddled with 1940s and 1950s heavy equipment (some new equipment began to appear in the last two years), it combines the worst features of public bureaucracy with private operation and maintenance unconstrained by the necessity to make a profit or please customers. In the longer term, we still feel optimistic because the physics is clear. Rails are cheaper and less energy-intensive to install and maintain than roads, and any ground transport will be more efficient on rails than on roads.

Among our second thoughts is a growing feeling that, as others have done, we made too much of legislating the future. We may have outlined too large a role for the government. Grass-roots activity is appearing everywhere. Statistics are hard to come by.

Another common error we share with statisticians and economists is an excessive reliance on averages, large numbers, and statistics generally. The future, like the present, holds diversity, with attendant conflicts over objectives and lifestyles. The real challenge to the government may well be to live with this diversity and conflict—and to derive political choices from it—so that we have a chance to foster any ideas that will work. Our social fabric will be similarly challenged as people with divergent visions of the future try to work and live together.

The early versions of the scenario were presented first to an energy workshop in Europe and then to a meeting of the American Association for the Advancement of Science. Scientists, especially those in the energy area, responded with polite indifference. By contrast, journalists snapped up the available 200 copies and many asked where to write for additional copies. Since then, demand for the scenario has been steady from government agencies, planning groups, and citizens' organizations of various kinds and from abroad, and at length there has been a growing interest (but not necessarily agreement) from scientists as well. The scenario has been translated into three foreign languages that we know of.

The widest exposure of the scenario has come through several feature articles in newspapers and two local television specials in the Midwest. In that sense, we feel that we have been very lucky, for our aim was to bring others into the arena of public discussion of possible futures. These earlier notices, as well as a publication by Friends of the Earth, provide just such an opportunity. If there is a message in all this it is to hold on to your dreams. Expert advice aplenty will be needed, but the shape of our future is a legitimate concern for us all. (J.S.S. and M.E.H., April 1981)

NOTES AND REFERENCES

1. Jay W. Forrester, *World Dynamics*. Cambridge, Mass.: Wright-Allen Press, 1971.
2. Donella H. Meadows, Dennis L. Meadows, Jorgen Randers, and William W. Behrens III, *The Limits to Growth: A Report for the Club of Rome's Project on the Predicament of Mankind*. New York: Universe Books, 1972.
3. Mihajlo Mesarovich and Eduard Pestel, *Mankind at the Turning Point: The Second Report to the Club of Rome*. New York: Dutton/Reader's Digest Press, 1974.
4. Kenneth E. Watt, *The Titanic Effect*. Stamford, Conn.: Sinauer Associates, 1974.
5. Robert Heilbronner, *An Inquiry into the Human Prospect*. New York: Norton, 1974.
6. Amory B. Lovins, "Energy Strategy: The Road Not Taken," *Foreign Affairs* **55**, 65 (1976).
7. L. S. Stavrianos, *The Promise of the Coming Dark Age*. San Francisco: W. H. Freeman, 1976.
8. Amilcar Herrera et al., *Catastrophe or New Society: A Latin American World Model*. Ottawa: International Development Research Centre, 1976.
9. Roberto Vacca, *The Coming Dark Age*. Garden City, N.Y.: Anchor Press, 1974.
10. Booz-Allen Associates, *Energy Consumption in the Food System*, Report No. 13392-007-001. Washington, D.C.: GPO, 1976.
11. Dennis Hayes, *The Case for Conservation*, p. 9. Washington, D.C.: Worldwatch Institute, 1976.
12. Charles E. Little, "The Double Standard of Open Space" in *Environmental Quality and Social Justice in Urban America*, p. 8. Washington, D.C.: The Conservation Foundation, 1975.
13. Barbara Ward, *The Home of Man*. New York: Norton, 1976.
14. Daniel Bell, *The Coming of Post Industrial Society: A Venture in Social Forecasting*. New York: Basic Books, 1973.
15. Although we do not foresee the construction of many complete "new towns" on the order of Columbia, Jonathan, or Soul City, we do anticipate the incremental rebuilding of established towns.
16. It should be noted that the U.S. Census data for recent years indicate that the four highest growth rates in the United States are in rural areas.
17. Walter Goldschmidt, "A Tale of Two Towns" in Catherine Lerza and Michael Jacobson, eds., *Food for People, Not for Profit*, pp. 70–3. New York: Random House, 1975. See also U.S. Senate Small Business Committee, *Small Business and the Community*. Washington, D.C.: GPO, 1946.
18. U.S. Department of Agriculture, *Economies of Scale in Farming*. Washington, D.C.: GPO, 1968.
19. George Sternlieb, "Slum Housing: A Functional Analysis," *Law and Contemporary Problems*, 1966.
20. The abandonment of inner city buildings has been a growing urban problem for some time; one may note that in 1975 some 30,000 buildings were abandoned in New York City alone. Large numbers of these buildings are actually structurally sound but in great need of repair.
21. In the large cities, buildings, and in some cases whole blocks, may be left vacant—much as New York City now has more than 3 square kilometers of vacant office space.
22. D. A. Pilati, *The Energy Conservation Potential of Winter Thermostat Reductions and Night Setback*. Oak Ridge, Tenn.: Oak Ridge National Laboratory, 1975.
23. Note: percentage reductions by behavioral and structural changes are *not* additive.
24. *A Nation of Energy Efficient Buildings by 1990*. Washington, D.C.: American Institute of Architects, 1975.
25. Raymond W. Bliss, "Why Not Just Build the House Right in the First Place," *Bulletin of the Atomic Scientists* **32**, 32 (1976).
26. Marc H. Ross and Robert H. Williams, "Energy Efficiency: Our Most Underrated Energy Resource," *Bulletin of the Atomic Scientists* **32**, 30 (1976).
27. Dwayne Chapman, "An End to Chemical Farming," *Environment*, March 1973, p. 12.
28. A reduction in energy use for lighting will increase the heating needs, but these needs can generally be met without the use of electricity.
29. Bruce Ingersoll, *Chicago Sun Times*, October 30, 1972. This article reports in some detail Professor Bruce Hannon's study of the McDonald's hamburger chain.
30. Bureau of the Census, Department of Commerce, *Statistical Abstract of the United States, 1977*.
31. Work trips have accounted for the largest share of trips, approximately 35 per cent, with business 9 per cent, shipping 15 per cent, and other 13 per cent.

32. Assuming a city of five dwelling units per acre or 10,000 people per square mile with a circular form, the distance from the edge of the city to the center would be 1 mile for a city of 31,000, a 20-minute walk or a 6- to 10-minute bicycle ride. The distance would be 2 miles for a city of 126,000.

33. Fleet averages for recent years have been on the order of 11 to 15 miles per gallon.

34. Cummins Engine Company has estimated that a 25 per cent energy savings is possible with a typical tractor semitrailer combination. With an appropriate engine, transmission, and axle combination, a total of 35 per cent can be saved.

35. This estimate does not include savings from resulting cutbacks in the energy supply industry.

36. Ford Foundation Energy Policy Project, *A Time to Choose: The Nation's Energy Future*, pp. 140–57 Cambridge, Mass.: Ballinger, 1974.

37. American Physical Society, *Efficient Use of Energy: A Physics Perspective. A Report of the Summer Study on Technical Aspects of Efficient Energy Utilization*, in ERDA Authorization-Part I, 1976, and Transition Period Conservation. Hearings before the Subcommittee on Energy Research Development and Demonstration of the Committee on Science and Technology, U.S. House of Representatives, 94th Congress, 1st session, February 18, 1975.

38. Marc H. Ross and Robert H. Williams, *op. cit.* We assume that these reductions in percentage will be approximately the same for the industrial sector with the proposed changes in industrial production.

39. Elias P. Gyftopoulos et al., *Potential Fuel Effectiveness in Industry. A Report to the Energy Policy Project of the Ford Foundation.* Cambridge, Mass: Ballinger, 1974.

40. *Machine Design,* October 7, 1976, p. 4, from ERDA source.

41. Gallium Arsenide Used for Low Cost High Efficiency Solar Cells, *Computer Design,* September 1975, p. 44.

42. Bruce Chalmers, "The Photovoltaic Generation of Electricity," *Scientific American,* October 1976, p. 34.

43. "Cost of Solar Cells Down 26% in Six Months," *Machine Design,* October 7, 1976, p. 10.

44. *Electronics,* April 1, 1976.

45. Business Communications Co., Inc., *Solar Energy: A Realistic Source of Power,* 1975. Includes solar space heating and cooling, solar electric, and solar fuels for transportation and industry.

46. W. E. Morrow Jr., "Solar Energy: Its Time Is Near" in L. C. Ruedisili and M. W. Firebaugh, eds., *Perspectives on Energy,* pp. 336–51. New York: Oxford University Press, 1975. Includes solar heat, solar thermal electric, and electrolysis.

47. Solar Energy Panel, *Solar Energy as a National Energy Source.* College Park, Md.: NSF/NASA, 1972. Solar heating and cooling, solar electric, and wind electric.

48. Bent Sorensen, *On the Fluctuating Power Generation of Large Wind Energy Converters, with and Without Storage.* Copenhagen: Niels Bohr Institute, University of Copenhagen, 1976.

49. Fifty per cent of new dwellings in 1976 have electric heating devices. In certain areas of the Midwest, a 15 per cent annual increase in the number of all electric homes is not uncommon.

50. Gunnar Myrdal, "Is Sweden Richer Than the U.S.?" *Forbes Magazine,* April 1, 1972.

51. Dennis Gabor, *The Mature Society,* pp. 10–11 London: Secker and Warburg, 1972.

52. Barbara Koeppel, "Something Could be Finer Than To Be in Carolina," *The Progressive,* June 1976, pp. 20–23.

53. Included is the labor required for maintenance, material processing, component manufacturing, and construction but not mining. Also note that the numbers of jobs apply to one state only, and conceivably employment levels nation wide would be 50 times as great.

54. James Monroe et al., *Energy and Employment in New York State, Draft Report: A Report to the New York State Legislative Commission on Energy Systems,* Report ES 119, May 3, 1976.

55. Energy Research Group, *Urban Auto-Bus Substitution: The Dollar, Energy and Employment Impacts.* Report to Energy Policy Project, 1973, 1776 Massachusetts Avenue, NW, Washington, D.C., see reference 56.

56. Roger Bezdek and Bruce Hannon, "Energy, Manpower and the Highway Trust Fund," *Science* **185,** 669 (1974).

57. Robert J. Lampman as cited by W. W. Heller, "Coming to Terms with Growth and the Environment" in S. H. Schurr, ed., *Economic Growth and the Environment.* Baltimore: Johns Hopkins University Press, 1972.

58. Warren A. Johnson, "The Guaranteed Income as an Environmental Measure" in W. A. Johnson and J. Hardesty, eds., *Economic Growth vs. the Environment.* Belmont, Calif.: Wadsworth, 1971.

59. See, for example, Marianne Frankenhauser, "Limits of Tolerance and Quality of Life," paper presented at the symposium *Level of Living–Quality of Life,* held at Biskops–Arno, Sweden, in 1974 and published in *Viewpoint,* New York: Swedish Information Service, 1974.

The Recycle Society of Tomorrow 33

These days it is difficult enough to forecast what the world will be like next year, let alone predict what kind of a world it may be in 1994. Although I enjoy future forecasting sessions, I cannot help but feel that many of us are projecting the kind of futures we would like to see, knowing that the world could be that way, hoping it will, rather than trying to base our forecasts on that combination of progress and the obstruction to it due to human foibles and follies that inevitably contribute to future conditions.

In the past we physical scientists have been especially prone to the blue skies approach to the future, tending to see the possibilities of gaining and applying new knowledge, of the "technological fix," and of the value of human cooperation. Those in the social and political sciences and in the business world are apt to be a bit more realistic because they deal more directly with the perversity and irrationality of human nature as well as its admirable features.

In recent years we have also seen the rise

of forecasting by systems analysts, using elaborate computer models and warning us of total collapse based on the projections of current trends. Their studies offer some serious warnings of what could be. But I am inclined to agree with Rene Dubos's statement that "trend is not destiny."

My own thoughts about where we might be, and might be going, in the 1990s are based on what I consider to be a number of imperatives. That is, I think there are things that will have to happen, conditions that will have to prevail, given the physical limitations we face, but also given man's creativity and will to survive. In other words, sooner or later we will stabilize our population, we will minimize our environmental impact and efficiently manage our use of natural resources, and we will achieve a relatively peaceful world with a more equitable distribution of goods, services, and opportunities throughout the world.

The difficulty is not in predicting that we will arrive at these points, or even how we

Dr. Glenn T. Seaborg is currently University Professor of Chemistry, Lawrence Berkeley Laboratory, University of California, Berkeley, California 94550. He previously served as chairman of the U.S. Atomic Energy Commission and chancellor of the University of California. He won the Nobel Prize for chemistry in 1951 as co-discoverer of plutonium and other elements.

From *The Futurist* **VIII** (3), 108 (1974). Reprinted by permission of the author and *The Futurist*. Copyright © 1974 World Future Society.

will arrive, but in predicting when and how much disruption, deprivation, and destruction will take place in the interim. This, in turn, will depend to a large extent on how quickly we grasp and apply certain principles of constructive human behavior, how we balance self-interest with mutual interest, to what degree and how soon we greatly improve cooperation between people and nations. It will also depend—somewhat fortuitously—on the kind of leadership that rises around the world. That is a most important catalyzing agent over which we seem to have little control.

Bearing all this in mind, I shall try to give some approximation of where we might be in 1994—assuming that most things turn out right and that cool heads, kind hearts, and common sense prevail and guide all our other human assets. I will not speculate on what might happen if "the ghost in the machine," as Arthur Koestler refers to man's self-destructive flaws, takes over.

First, let me cover some general conditions that I think will have arrived by 1994, or will be well along in their formation.

TOWARD A "STEADY-STATE" WORLD

Broadly speaking, the 1990s will be a period characterized mainly by the need to stimulate maximum creativity in a tightly controlled social and physical environment. The reason for this is that by the mid-1990s we should be well on our way to making the transition from an "open-ended" world to a "steady-state" one. The United States will be in the forefront of this movement. Others will be following with various degrees of enthusiasm and reluctance, much depending on the sacrifice and cooperation of the advanced nations.

Some of the major characteristics of this transition will be:

1. Movement toward a highly disciplined society with behavior self-modified and modified by social conditions. On the surface, and by the standards of many young people to-day, it will be a "straight society," but a happier, well-adjusted one with a much healthier kind of freedom, as I shall explain later.

2. Organization of a "recycle society" using all resources with maximum efficiency and effectiveness and a minimum of environmental impact.

3. A mixed energy economy, depending on a combination of several energy sources and technologies and highly conservation conscious. During this time we will still be searching for the best ways to phase ourselves out of the fossil-fuel age.

4. Greater progress toward a successful international community spearheaded by the economics of multinational industry, new international trade arrangements that improve the distribution of resources, and a high degree of scientific and technical cooperation.

Let me elaborate a bit on these characteristics, beginning with a few words about social attitude and behavior, as I believe these will be among the biggest determinants of where we are and where we will be going.

A HIGHLY DISCIPLINED SOCIETY IN THE 1990S?

By the 1990s I suspect we will be a society almost 180 degrees different from what we are today, or some think we will be in the future. I see us in 1994 as a highly disciplined society with behavior self-modified by social and physical conditions already being generated today. The permissiveness, violence, self-indulgence, and material extravagance which seem to be some of the earmarks of today will not be characteristic of our 1990 society. In fact, we will have gone through a total reaction to these.

We therefore will have a society that on the whole exercises a quiet, nonneurotic self-control, displays a highly cooperative public spirit, has an almost religious attitude toward environmental quality and resource conservation, exercises great care and ingenuity in managing its personal belongings, and shows an extraordinary degree of reliability in its work. Furthermore, I see such a society as being mentally and physically healthier and enjoying a greater degree of freedom, even though it will be living in a more crowded, complex environment.

All this will not come about by everyone's being made to subscribe and live up to the Boy Scout oath. I think it will come about as an outgrowth of a number of painful shocks—shocks of recognition, not future shocks—we will undergo over the coming years, one of which we are already getting in our current energy situation. The energy crisis is just the forerunner of a number of situations that we will be facing that will change our attitudes, behavior, and institutions—although I do not in any way minimize its importance or its far-reaching effect on all aspects of our lives. We will face a number of critical materials shortages and some failures in our technological systems that will force us into fairly radical changes in the way we are conducting our lives and managing our society.

My reasons for projecting the ''straight society'' I have mentioned for the 1990s spring from the series of reactions that the forthcoming shocks will elicit. The reactions will come in sequence but will also widely overlap. The first period will be one mainly emphasizing conservation and cooperation. Of course, there will be some degree of negativism about the noncompliance with the required changes. And there will be those who, with the usual amount of hindsight, blame others for not being able to anticipate current problems.

But by and large, most people will respond positively as they have in the past in the time of crisis. In fact, after the extended period of comparative affluence and self-indulgence most people have enjoyed in this country, we may witness something of a quiet pride and spartan-like spirit in facing some shortages and exercising both the stoicism and ingenuity to face and overcome them. What is important, though, is that the emphasis will shift from stoicism to ingenuity as we come up with new ideas and technologies to overcome our problems. By the mid-1990s we should be a good way along in this shift, the results of which I shall discuss in a moment. But the results of the changes and transitions we face will have left their effect on our society, for we will have realized that we will never again live in a society where so much is taken for granted—where so many apparently ''knew the price of everything and the value of nothing.'' The environmental movement, the energy crisis, and the problems yet to come will have changed all that well before 1990.

Oddly enough, the kind of general outlook that will prevail in 1994 will be a synthesis of ideas coming out of today's low-technology communes and high-technology industries. We will not see complexity for its own sake. But neither will we be able to maintain the desired quality of life for the number of people present by dependence on something akin to handicraft and cottage industry. High technology, much better planned and managed, and important scientific advances will still be the basis for progress.

But that progress will be guided by many of the new values being expressed by young people today. We will be more of a functional and less of a possessive society, more apt to enjoy a less cluttered life, more inclined to share material things and take pleasure in doing so. This will bring us a different kind of freedom, one more closely related to Hegel's definition when he said, ''Freedom is the recognition of necessity,'' but one also allowing more people to ''do their own thing'' within the framework of a cooperative society.

Let me turn now to some of the physical changes that will be taking place that will accompany these social changes as we move toward and through the 1990s.

HOW THE "RECYCLE SOCIETY" WILL WORK

As I mentioned before, we will be creating a ''recycle society.'' By this I mean not simply one in which beer cans and Coke bottles are all returned to the supermarket, but one in which virtually all materials used are reused indefinitely and virgin resources become primarily the "makeup" materials to account for the amounts lost in use and production and needed to supplement new production to take care of any new growth that would improve the quality of life.

In such a society the present materials situation is literally reversed; all waste and scrap—what are now called "secondary materials"—become our major resources, and our natural, untapped resources become our backup supplies. This must eventually be the industrial philosophy of a stabilized society and the one toward which we must work.

To many who have not thought about it, this idea may sound simple, or a bit confusing. To many who have given it considerable thought, it can be mind boggling and sound physically and economically impossible, given our current state of industry. And to some it may even appear morally objectionable, given the state of development in many parts of the world. To clarify the concept, let me explain some of the things it will and will not involve.

First, it involves a shift in industry to the design and production of consumer goods that are essentially nonobsolescent. This means that products will be built to be more durable; easily repairable with standardized, replaceable parts; accessible; and able to be repaired with very basic tools. (Along these lines, I understand that at a recent international auto show in Frankfurt, Porsche displayed a car designed to have a 20-year life, or 200,000 miles—but then quickly assured the industry it had no intentions of putting the car into production!)

In the recycle society, all products and parts will be labeled in such a way that their use, origin, and material content can be readily identified, and all will have a regulated trade-in value. Many items of furniture, housewares, appliances, and tools, in addition to their low-maintenance qualities, will be multifunctional, modular, and designed for easy assembly and breakdown to be readily moved and set up in a different location when necessary. Their design and construction will also allow for their reassembly and redesign into essentially new products when their owners have different uses for them or seek a change.

When a consumer (it would be more correct to call him a "user") wishes to replace an item or trade up for something better or different, he can return the old item for the standard trade-in price. All stores will have to accept these trade-ins. They thus will become collection centers as well as selling outlets in the recycle society.

Manufacturers in turn will receive and recondition the used products, use their parts as replacement parts in "new" products, or scrap them for recycled material. Since literally everything will be coded and tagged for material content, much of the high cost of technological materials separation will be eliminated. Materials that were mixtures and alloys would be color coded or magnetically or isotopically tagged to facilitate optical or electromagnetic separation.

Recycling and reprocessing will also apply to software—clothing, bedding, carpeting, and all other textile materials, organic and synthetic, and, of course, paper products.

The industrial processing aspects of the recycle society may be the easiest to achieve, as there are already one or two major companies that claim the ability to recycle totally the waste products of selected plants without any economic penalty. By 1994 we should also see extensive recycle of organic material from agriculture and forest industries. Animal waste will find many uses as fertilizer, fuel, and feed. Protein will be grown on petroleum waste and extracted from otherwise inedible plants and agricultural products.

PEOPLE WILL BE BETTER OFF IN THE 1990S

To make the transition to an economic recycle society, much will be required in the way of new legislation, regulation, tax incentives, and other measures that will make the use of secondary materials more economic than of virgin resources. The opposite situation prevails today. In addition to the setting up of new methods of marketing and management required to operate a recycle society, there will be the necessity for a long-term consumer education program. A whole new public outlook will have to be acquired.

An entire society reusing and recycling almost all its possessions, especially after an extended era of conspicuous consumption and waste, will take a great deal of pride in a lifestyle that is extremely creative and varied and based on a new degree of human ingenuity and innovation. The ''recycle society'' of the 1990s will be better off than the affluent society of the 1970s.

Several types of assets will accrue from the movement toward the kind of society I have been describing. One lies in the fact that it will be far less energy intensive. For example, recycled steel requires 75 per cent less energy than steel made from iron ore; 70 per cent less energy is used in recycling paper than in using virgin pulp; 12 times as much energy is needed to produce primary aluminum as to recover aluminum scrap. A society set up to reuse most of its resources systematically and habitually could effect enormous energy savings.

Perhaps even greater would be the reduction in the environmental impact of a recycle society. This would be true for several reasons. An overall one is that such a society would have developed by the 1990s an environmental and conservationist ethic, due to the scarcity of resources as well as the new value placed on land, water, and air. We can see this ethic already in the making today. There is some fear that because of our energy crisis we will severely compromise this ethic, or even abandon it. I do not think this will be the case. Rather I believe we will be making substantial sacrifices in the years ahead to change our lifestyle in order to match our economic and environmental needs.

By the 1990s our industrial and power systems will be much more efficient users of energy; hence, they will not be rejecting as great a percentage of waste heat to the environment. In fact, most systems will probably be planned and designed to make maximum use of waste heat, using it for space heating or possibly agriculture or aquaculture. Waste water and sewage water will also be recycled in industrial and perhaps even municipal water systems. What water is returned to the environment will be as clean as, if not cleaner than, it was when it entered the manmade system.

It would be foolish to believe that by 1994, even with a recycle society as a reality, we will not still be drawing on a substantial amount of new resources. Even with population growth leveling off and economic growth cooling off, we can expect substantial growing demands for new materials resources well into the next century. What this means is that by the 1990s we will have to develop a new level of ingenuity in materials substitution and in what Buckminster Fuller calls ''ephemeralization''—the process of doing more with less. Fuller uses as an example of this, the Telstar satellite, which, while weighing only one tenth of a ton, outperforms 75,000 tons of transatlantic cable.

Communications, of course, have offered the best examples of this ''more with less'' phenomenon, as in electronics we have seen the size of basic devices reduced by a factor of 10 roughly every five years. Today, a single chip of silicon a tenth of an inch square may hold microscopic units that perform the functions of as many as 1000 separate electronic components. Furthermore, silicon is one of the most abundant substances in the earth's crust.

COMMUNICATIONS MAY SUBSTITUTE FOR TRANSPORTATION

By the 1990s substitution not only in materials but in functions—the way we conduct our business and personal affairs—may vastly alter our lives, effecting many savings in energy and time and therefore affecting how we otherwise spend our energy and time. For example, communications as a substitute for transportation can effect such savings to a great extent.

Shopping by telephone and having good home delivery is a very old example of this, one which is largely out of style today. But if one could survey local supermarkets and department stores via videophone and a computer to do some quick comparative shopping and then have the selections delivered, one would have more time, money, and energy (personal and automotive) left for other things. Another aspect of this type of shopping involves a considerable saving in space. A simple warehouse with a small fleet of trucks could service thousands of customers, eliminating the paving over of large parking areas and the operation of an elaborate market.

A society that exercises this option of using communication in place of transportation in many of its activities—whether in shopping, business, or educational activities—can conserve many resources. But it must be one that has learned to substitute other activities for the social and entertainment value that we have come to find in our more random way of life.

In conducting this kind of society, questions that loom larger than those of the technological possibilities are as follows. Assuming a 1994 liberated housewife (if that is not a contradiction in terms) is able to do most of her shopping by video computer and her other chores so efficiently, how will she spend her extra free time? When it is possible to hold national and international conferences via home holography, will we miss the luncheons, banquets, and corridor talk? And what will happen to the Willy Lomans of the world when they can sell their lines long distance in living color, through similar electronic techniques, without covering the territory in person? These are only a sample of the kinds of questions that can be raised when the matter of substituting communication for travel becomes a viable option.

It is possible to speculate that in 1994 we may find a situation in which our working world will be served mainly by communication and public transportation, and the savings from this will allow us to use private transportation in a limited way for recreation and vacations.

WILL URBAN SPRAWL DESTROY THE COUNTRYSIDE?

Much of what I have said to this point will be influenced by, and influence, how we use our space here on earth—how we manage our land, develop our urban areas, place our industry, locate our power system. Do we build out, up, or down? Do we draw people back into the cities, continue to disperse them around cities, or cluster them in new areas, in new cities around new industries? And how much of a planned, concerted effort do we make to do any of these? We are, in fact, just beginning to take a serious look at land management and the control of our populated areas in this country. In the past we have seen our population explode and implode with both good and bad effects, but certainly without much conscious control on our part.

It is difficult to speculate on how far we shall have gone by 1994 in having effected any widespread control or change over today's patterns of growth. Twenty years is not a long time to institute and carry out major changes in land use and population distribution. And yet, unless we do, some of the major effects of the current style of growth could (according to Environmental Protection Agency estimates) lead to some 20 million additional acres being covered by urban sprawl (an area equivalent to New Hampshire, Vermont, Massachusetts, and Rhode Island); more than 3 million acres paved over for highways and airports; and about 5 million acres of agricultural land lost to public facilities, second-home development, and waste control projects. In addition, the approximately 1000 power stations that may be built by the 1990s, together with their cooling facilities, fuel storage and safety exclusion areas, and right-of-way land for power lines, could require another 2 to 3 million acres.

MEASURES TO COUNTER THE DESTRUCTION OF THE COUNTRYSIDE

Much of this is inevitably going to take place before 1994. But I see some of the following as countermeasures and countertrends that may be initiated, or well under way, by the 1990s:

1. A large shift toward clustered, attached, and high-rise housing surrounded by community-owned open lands. This planned housing would rise as "new cities" and as neighborhood communities within larger urban areas. It would help eliminate today's suburban sprawl and preserve more open land for recreation, agriculture, or natural reserves.

2. Increased use of underground space made possible by advances in excavation technology: underground shopping centers, warehouses, recreation and entertainment complexes, rapid mass-transit lines, and power and communications cables.

3. Offshore power plants with extrahigh voltage, and superconducting transmission cables carrying electricity greater distances inland. Cables would be underground and might occupy the same rights-of-way as rapid rail transit systems and cable communication systems. In the 1990s such offshore power-plants would be nuclear electric. In the next century they may include nuclear hydrogen-producing plants and solar-powered electric and hydrogen-producing plants.

4. Integrated industrial complexes planned to concentrate energy sources, materials, and manufacturing in single locations. This would reduce long shipments of fuel and material resources, make more efficient use of waste heat and materials, and confine and control environmental impact.

IMPROVEMENT IN ENERGY SITUATION BY 1994?

Concerning the energy situation, which is uppermost on people's minds today and will certainly have a major bearing on our future, I believe we will see a turning point in our difficulties by 1994. But the intervening years will necessitate bearing some difficulties and hardships because we have not given ourselves the necessary lead time to make an orderly transition to new energy technologies and resources. There is no doubt that we have been shortsighted and complacent about supplies and have overestimated our ability to develop and shake down new technologies and to get them on line economically.

In the 1990s we will still have a very mixed energy economy. By then, oil and gas will be giving way to coal—but grudgingly, as it will take some time to develop and build economical coal gasification and liquefaction systems. Oil shale may be a significant factor by then. And we may even have found a way to retort the oil from the shale via underground heating and explosives (chemical) to avoid stripping and excavation. We will have a growing amount of electricity in some parts of the country supplied by geothermal energy. Solar energy equipment for home heating and cooling will be prevalent on new single-family dwellings and incentives will be introduced to encourage homeowners to retrofit their houses with such equipment if possible. Solar energy may supply a small amount of home use electricity, but the large-scale production of solar electricity may still be a few years off. However, by the 1990s we should see some prototype "solar farms" in the southwestern United States testing out the economical, large-scale conversion of solar energy.

I am confident that by the 1990s we will be well over the difficulties and resistance facing nuclear power today and that more than one third of our electric power will be generated by nuclear plants. The liquid metal fast breeder reactor will have been tested out to everyone's satisfaction by then and coming on line commercially. Other systems, such as the high-temperature gas-cooled reactor, will be adding to our national electric capacity. We may have achieved laboratory success in controlled fusion by 1994 and be building prototype fusion reactors.

We simply must pay the price of pursuing all possibilities in the energy field and at the same time pursue the energy conservation ethic I mentioned before. Any energy bonuses that would come our way through new breakthroughs would not give us energy to squander but would allow a well-planned, equitable increase in the quality of life on a worldwide basis.

A DESIRABLE FUTURE FOR AN INTERDEPENDENT WORLD

This brings me to my concluding thoughts on 1994, which center on global cooperation. In the midst of an energy crisis aggravated by the withholding of oil as a political weapon, this does not seem to be a popular topic. It is quite natural to want to act from a position of strength. With some sacrifices we can.

And yet, in our immediate reaction to strive for energy self-sufficiency, we should not overreact, not delude ourselves into believing that self-sufficiency in energy and other matters is the total solution to national security and well-being. This would lead to a dangerous neo-isolationism at a time when we must move in the other direction—toward greater international cooperation, no matter how difficult and painstaking the process seems at some times. The harsh facts are that we live in a highly interdependent world—one in which there will continue to be some hard bargaining but in which cooperation is growing increasingly important.

By 1994, I see the scales tipping more and more in favor of cooperation over competition. My travels over the world in the past dozen years to more than 60 countries—and most recently to the People's Republic of China—have led me to believe that all nations of the world need each other, that all have something to offer, and all could benefit by a greater exchange of human and material resources, of knowledge and goods.

Over the next 12 years we will have to make enormous strides—together—in controlling population, increasing food production, managing our environment, investigating and controlling the resources of the seas, conducting global research, developing methods to reduce the human impact of natural disasters, and generally uplifting the economic conditions of a large number of the world's peoples. There are no alternatives to these measures—except a tremendous increase in human misery that will ultimately affect all the world's peoples.

Forecasting the world of 1994 involves much more than projecting the trends at hand or even reciting the possibilities ahead. That future will be determined in large part by our considering and choosing values, examining and deciding among alternatives, exercising great will and perseverance, and searching for the leadership that will assemble and catalyze the proper resources to construct a chosen future. To the extent that we can do this we will either be drifting toward the world of 1994 or building it. Most likely, we will be doing a little of each, but I hope we will not be trusting to luck that which we could achieve by a new and concerted human effort.

Finally, . . . 1994 is only 12 years off. We had better get moving!

SUGGESTED READINGS

Dennis Hayes, "Repairs, Reuse, Recycling—First Steps Toward a Sustainable Society," *Worldwatch*, Paper 23, September 1978.

Vaclav Smil, "Renewable Energies: How Much and How Renewable?" *The Bulletin of the Atomic Scientist* **35** (10), 12 (1979).

David C. Wilson, "Energy Conservation Through Recycling," *International Journal of Energy Research* **3** (4), 307 (1979).

Appendix: Energy Definitions and Conversion Factors

The variety of units and the unclear distinctions between energy and power often prove confusing to students of the energy problem. In this appendix we present a brief introduction to these concepts and a table of conversion factors to assist in energy unit conversions. We use the International System (SI) of units to present these concepts and the table at the end of the appendix to express British and other energy units in the SI units.

Energy and Power

As an operational concept, *energy is defined as the ability to do work*. Work, in turn, is defined as a force acting through a distance parallel to the force.

$$W = Fd \qquad (1)$$

For instance, if we push a stalled car a distance of 10 meters (m) with a force of 500 newtons (N), we have performed

$$W = 500 \text{ N} \times 10 \text{ m} = 5000 \text{ N-m}$$

of work. Another name for newton-meter is joule (J). This task required 5000 J of energy.

Energy is available in a variety of forms, falling into the three general categories of (a) *kinetic energy,* (b) *potential energy,* and (c) *rest-mass energy.*

Kinetic energy is the energy of motion and is expressed mathematically as

$$KE = \tfrac{1}{2} \, mv^2 \qquad (2)$$

where m is the mass of the moving body (kilograms) and v is the velocity of the moving body (meters per second). In the example above, for instance, if the car was initially at rest and there was *no* friction involved (an impossible situation), the work performed on the car would result in 5000 J of kinetic energy manifested in the final velocity of the car.

There are several forms of *potential energy*. These include the energy stored in a compressed spring, chemical energy stored in fossil fuels, and gravitational energy stored in matter as it is lifted in the earth's gravitational field. For example, as solar energy evaporates water and lifts it high into the mountains, the water accumulates potential energy, given by

$$PE = mgh \qquad (3)$$

where m is the mass of the water (kilograms), g is the acceleration due to gravity

(9.8 meters per second squared), and h is the height through which the water is raised (meters).

A kilogram of water raised 4000 meters would store 1 kilogram $\times$ 9.8 meters per second squared $\times$ 4000 meters $= 39{,}200$ kg-m²/s² $= 39{,}200$ N-m, or 39,200 J of potential energy. This process is the basis for hydroelectric energy.

The third form of energy is *rest-mass energy*. The conversion of mass into energy is the basic process involved in fission reactors and the experimental fusion devices. The sun, in fact, is just one enormous fusion reactor converting part of the mass of hydrogen into heat and radiation energy. The relation governing this conversion is the famous Einstein equation

$$E = mc^2 \qquad (4)$$

where m is the mass being converted (kilograms), c is the speed of light (3×10^8 meters per second), and E is the energy released (joules).

For example, if 1 kilogram of reactor fuel is converted to energy, we get

$$E = 1 \times (3 \times 10^8)^2 \text{ J}$$
$$= 9 \times 10^{16} \text{ J}$$

of energy released. Thus, nuclear fuels such as deuterium, tritium, uranium-235, and plutonium-239 may be thought of as very concentrated forms of potential energy.

Although energy itself is the fundamental quantity (the quantity we purchase as electricity and gas), a somewhat more intuitive concept is that of *power,* which is defined as *the time rate of doing work.* It is given by

$$P = W / T \qquad (5)$$

where W is work (joules), T is time (seconds), and P is power (watts, where 1 watt $= 1$ joule per second). Thus a typical light bulb may be rated at 100 watts (power) and will therefore consume 100 joules of energy every second it burns. The power of motors, both electrical and gasoline powered, is frequently given in horsepower. The conversion factor is 1 horsepower $= 746$ watts. Therefore a standard U.S. car engine of 300 horsepower would be rated as 223,800 watts, or 223.8 kilowatts.

If we know the power rating of a device and the length of time it operates, we may compute the total energy consumed by rewriting Equation 5 as

$$W = PT \qquad (6)$$

The 100-watt light left on for 10 hours would consume 100 watts $\times$ 10 hours $\times$ 60 minutes per hour $\times$ 60 seconds per minute $= 3.6 \times 10^6$ watt-seconds $= 3.6 \times 10^6$ joules of energy. This amount is often expressed as 1000 watt-hours or 1 kilowatt-hour of energy. The kilowatt-hour is the energy unit used by electric utilities for billing their customers. A kilowatt hour of electrical energy costs in the range of 5 to 10¢, depending on the region of the country.

Other Units and Conversion Factors

For historical reasons or as a matter of convenience, a number of other energy units are used. Below, we define some of the more common units and their conversion factors in terms of the kilowatt-hour.

British thermal unit (Btu) $=$ the amount of heat energy required to raise the temperature of 1 pound of water 1 Fahrenheit degree.

Therm $= 100{,}000$ Btu. Widely used in the sale of natural gas energy.

Quad $= 10^{15}$ Btu. Frequently used to display U.S. energy demand of 70 to 85 quads per year for the near future.

Foot-pound (ft-lb) $=$ the work done by a force of 1 pound acting through a distance of 1 foot.

calorie (also gram-calorie) $=$ the heat energy required to raise the temperature of 1 gram of water 1 Celsius degree.

Calorie (also kilogram-Calorie) = 1000 calories. The average adult consumes between 2000 and 3000 Calories in food energy per day.

Electron volt (eV) = the energy change when an electron falls through an electric potential difference of 1 volt. Atomic processes range in energy from several eV to several KeV (1000 electron volts). Nuclear pro-

cesses typically occur in the MeV range (million electron volt).

Fossil fuel energy units. Many graphs use units of trillion cubic feet (natural gas), million barrels of oil (or the power equivalent of a million barrels/day = mmb/d), and pounds or tons of coal. Although the energy content of each of these fuels varies, average energy equivalents are given in the following table.

Energy Conversion Table.

	Joules	*Kilowatt-hours*	*calories*	*Btu's*
1 Joule (J) =	1	2.778×10^{-7}	0.2389	9.480×10^{-4}
1 Kilowatt-hour (kWh) =	3.600×10^{6}	1	8.600×10^{5}	3413
1 calorie (cal) =	4.186	1.163×10^{-6}	1	3.969×10^{-3}
1 BTU =	1055	2.930×10^{-4}	252.0	1
1 Therm =	1.055×10^{8}	29.30	2.520×10^{7}	1×10^{5}
1 Quad (Q) =	1.055×10^{18}	2.930×10^{11}	1.520×10^{7}	1×10^{15}
1 Foot-pound =	1.356	3.766×10^{-7}	0.3239	1.285×10^{-3}
1 Kilocalorie (Cal) =	4186	1.163×10^{-3}	1000	3.969
1 Electron volt (eV) =	1.602×10^{-19}	4.450×10^{-26}	3.827×10^{-20}	1.519×10^{-22}
1 Barrel of crude oil =	6.12×10^{9}	1700	1.46×10^{9}	5.80×10^{6}
1 Gallon of gasoline =	1.32×10^{8}	36.7	3.16×10^{7}	1.25×10^{5}
1 Ton (2000 lb) of coal =	2.36×10^{10}	6.57×10^{3}	5.65×10^{9}	2.24×10^{7}
1 Cubic foot of natural gas =	1.08×10^{6}	0.299	2.57×10^{5}	1020

From Jackalie Blue and Judith Arehart, *A Pocket Reference of Energy Facts and Figures*. Oak Ridge, Tenn.: Energy Division, Oak Ridge National Laboratory, 1980.